EXCELLENCE IN CONCRETE CONSTRUCTION THROUGH INNOVATION

PROCEEDINGS OF THE INTERNATIONAL CONFERENCE ON CONCRETE CONSTRUCTION, KINGSTON UNIVERSITY, LONDON, UK, 9–10 SEPTEMBER 2008

Excellence in Concrete Construction through Innovation

Editors

Mukesh C. Limbachiya & Hsein Y. Kew
Concrete & Masonry Research Group
Kingston University, London, UK

The software mentioned in this book is now available for download on our Web site at: http://www.crcpress.com/e_products/downloads/default.asp

CRC Press
Taylor & Francis Group
Boca Raton   London   New York   Leiden

CRC Press is an imprint of the
Taylor & Francis Group, an **informa** business

A BALKEMA BOOK

Cover photograph supplied originally by the Concrete Centre – UK

CRC Press/Balkema is an imprint of the Taylor & Francis Group, an informa business

© 2009 Taylor & Francis Group, London, UK

Typeset by Charon Tec Ltd (A Macmillan Company), Chennai, India
Printed and bound in Great Britain by Antony Rowe (A CPI Group Company), Chippenham, Wiltshire

Published by: CRC Press/Balkema
                P.O. Box 447, 2300 AK Leiden, The Netherlands
                e-mail: Pub.NL@taylorandfrancis.com
                www.crcpress.com – www.taylorandfrancis.co.uk – www.balkema.nl

ISBN: 978-0-415-47592-1 (Hardback)
ISBN: 978-0-203-88344-0 (eBook)

Excellence in Concrete Construction through Innovation – Limbachiya & Kew (eds)
© 2009 Taylor & Francis Group, London, ISBN 978-0-415-47592-1

Table of Contents

Theme 3:   Design and construction in extreme conditions

Theme 4:   Protection against deterioration, repair and strengthening

Excellence in Concrete Construction through Innovation – Limbachiya & Kew (eds)
© 2009 Taylor & Francis Group, London, ISBN 978-0-415-47592-1

Preface

The Concrete industry has embraced innovation and ensured high levels of long-term performance and sustainability through creative applications in design and construction. As a construction material, the versatility of concrete and its intrinsic benefits mean that it is well placed to meet challenges of modern construction industry. Indeed, concrete has kept evolving to satisfy ever more demanding design requirements and relentless pressure for change and improvement in performance. This is done through introduction of new constituent materials, technology and construction methods. The current challenges faced by concrete construction may not necessarily be the same as those in the future. However, an ongoing programme of innovation and product development means that concrete should continue to provide cost effective sustainable solution that are able to turn a challenge into an opportunity.

The Concrete and Masonry Research Group (CMRG), part of the Sustainability Technology Research Centre within the Faculty of Engineering at Kingston University, organised this International Conference to discuss how concrete industry has addressed challenges from new materials, technologies, environmental concerns and economic factors to maintain its excellence. This is done by bringing together engineers, designers, researchers and scientists from 27 different countries to celebrate excellence in concrete construction and promote recent innovations in science and engineering. This Conference dealt with key technical, as well as practical achievements under concurrently proceeded five themes; *(i) Innovations and Developments in Concrete Materials and Design, (ii) Composite Materials in Concrete Construction, (iii) Design and Construction in Extreme Conditions, (iv) Protection Against Deterioration, Repair and Strengthening, and (v) Environmental, Social and Economic Sustainability Credentials.* Over 80 papers were presented by authors during this Conference and these are compiled in the CD and a hard bound single volume conference proceedings.

The event was organised with co-sponsorship from the American Concrete Institute and support from Asian Institute of Technology – Thailand, Università degli Studi di Napoli "Parthenope" – Italy, The Hong Kong Polytechnic University – Hong Kong, University of Cape Town – South Africa, Universita' Politecnica delle Marche – Italy, The Concrete Centre – UK, British Cement Association – UK and Wuhan University of Technology, P R China. All there organisations are gratefully acknowledged for their invaluable support. The work of Conference was an immense undertaking and help from all those involved are gratefully acknowledged, in particular, members of the Scientific and Technical Committee for their assistance from start to finish; the Authors and the Chair of Technical Sessions for their invaluable contribution.

The Proceedings have been prepared using camera-ready copy printed from the electronic manuscripts submitted by the authors and editing has been restricted to minor changes where it was considered absolutely necessary.

Mukesh C. Limbachiya
Hsein Y. Kew
Kingston University – London
September 2008

Excellence in Concrete Construction through Innovation – Limbachiya & Kew (eds)
© 2009 Taylor & Francis Group, London, ISBN 978-0-415-47592-1

Organising Committee

Concrete and Masonry Research Group –
Kingston University

Professor M.C. Limbachiya *(Chairman)*
Dr H.Y. Kew
Dr K. Etebar
Dr T. Donchev
Professor S.B. Desai OBE
Dr A. Cheah
Dr J. Omer
Miss D. Petkova

Faculty of Engineering Research Office

Kingston University London

Excellence in Concrete Construction through Innovation – Limbachiya & Kew (eds)
© 2009 Taylor & Francis Group, London, ISBN 978-0-415-47592-1

Scientific and Technical Committee

Professor Dr Ir H.J.H. (Jos) Brouwers
Professor of Sustainable Building, University of Twente, THE NETHERLANDS

Professor Raffaele Cioffi
Vice-Director, University of Parthenope of Naples, ITALY

Professor Peter Claisse
Professor of Construction Materials, Coventry University, UK

Professor Jorge de Brito
Instituto Superior TÃ©cnico/Technical University of Lisbon, PORTUGAL

Professor Satish Desai OBE
Visiting Professor, Kingston University
Principal Structural Engineer – Trenton Consultants, UK

Eur Ing Costas Georgopoulos
Education & Training Manager, The Concrete Centre, UK

Dr Jamal Khatib
Reader in Civil Engineering Materials, University of Wolverhampton, UK

Professor Mukesh Limbachiya (Chairman)
Research Professor, Faculty of Engineering
Director – Sustainable Technology Research Centre, Kingston University, UK

Dr Surendra Manjrekar
Chairman & Managing Director – Sunanda Speciality Coating Pvt Ltd, INDIA

Professor Giacomo Moriconi
Director – Department of Materials and Environment Engineering & Physics – UniversitÃ
Politecnica delle Marche, ITALY

Professor C.S. Poon
Professor – The Hong Kong Polytechnic University, HONG KONG – P.R. CHINA

Professor Rafat Siddique
Professor of Civil Engineering – Thapar University, INDIA

Professor Dr Shui Zhonghe
Professor – Wuhan University of Technology, P.R. CHINA

*Theme 1:   Innovation and development in
concrete materials/design*

Foamed Concrete: Application and specification

R.A. Barnes
The Concrete Society, Camberley, UK

ABSTRACT: The Concrete Society has produced a new guide on foamed concrete, its range of applications and a specification (The Concrete Society, 2008), which this paper aims to summarise. Although foamed concrete was used some 2000 years ago it wasn't until recently that modern foamed concrete began to be developed. In 1987 a full scale trial on the use of foamed concrete for trench reinstatement was conducted in the UK, which led to a further increase in use in this country. Foamed concrete has the following properties: it is lightweight, free flowing and easy to level, it does not require compaction, has good thermal insulation and frost resistance properties, and it is easy to pump, both vertically and horizontally. Its uses include: Trench Reinstatement, Blinding, Filling (basements, pipes, tunnels, subways, mine workings), Building use (under floors and roofing), Soil stabilisation, Reductions in lateral loading, Sports fields and athletics tracks and Sandwich fill for precast units. As well as the composition and production methods, this paper covers some of the practicalities and properties of the material.

1 INTRODUCTION

The term 'Foamed Concrete' may be somewhat misleading in that most do not contain large aggregates (indeed it may be considered to be foamed mortar or foamed grout). It is a lightweight concrete manufactured from cement, sand or fly ash, water and a preformed foam. Its dry density ranges from 300 to 1600 kg/m^3 with 28 day strength normally ranging from 0.2 to 10 N/mm^2 or more.

A widely cited definition of foamed concrete is:

"A cementitious material having a minimum of 20 per cent by volume of mechanically entrained foam in the plastic mortar or grout".

This differentiates it from air entrained concrete which has a far lower volume of entrained air (typically 3–8%), retarded mortar systems (typically 15–22%) and aerated concrete where the bubbles are chemically formed.

In the production of foamed concrete, a surfactant is diluted with water and passed through a foam generator which produces a stable foam. This foam is then blended into a cementitious mortar or grout in a quantity that produces the required density in the foamed concrete.

Surfactants are also used in the manufacture of Low Density Fills (also called Controlled Low Strength Material (CLSM)). In this case, however, they are added directly into a sand rich, low cement content concrete to give 15 to 25% air. Somewhat confusingly, some suppliers of Low Density Fills refer to these materials as foamed concrete, but as the foam is not formed separately to the concrete they are not true foamed concretes.

2 APPLICATIONS

The value of foamed concrete lies in its good void filling ability with a rigid hardened structure which will not deflect under low loading and also the low density where loading on other parts of the structure are critical. Although it will give enhanced thermal and fire rating properties, it is not usually the most cost effective solution for these applications unless access is difficult.

Some examples of the wide variety of applications where foamed concrete has been used include:

2.1 Trench reinstatement

One of the main causes of damage to road pavements are excavations carried out by utilities companies. Settlement of backfill means that the surfacing is damaged and constant patching will be required. Foamed concrete meets the criteria for the ideal backfilling technique in that: it normally requires no compactive effort and does not settle after placing; it does not transmit axle loads directly to the services in the trench; final resurfacing is possible the next day; it is economic; it is readily available; it permits easy re-excavation; it does not require unreasonably complicated equipment or skilled labour.

In order to satisfactorily compact the bituminous surfacing foamed concrete with a compressive strength of approximately $1 \, N/mm^2$ is required. Further guidance on the design of foamed concrete for trench reinstatement can be found in *Foamed concrete for improved trench reinstatement* (Taylor, 1991).

In order to evaluate the suitability of foamed concrete as an alternative to normal granular backfill a trenching trial was undertaken in Wickford in 1988. In 2003 a long term assessment of the results suggested no significant difference in the predicted total life of the carriageway since construction of the reinstatement (Steele et al, 2003).

2.2 *Void filling*

As foamed concrete is flowing, self levelling and self compacting it provides a rapid, effective and competitively priced solution for void filling, and has been used for this application throughout the UK. Its rigidity, range of strengths and densities along with its thermal insulation and controlled water absorption properties make it an ideal choice for a wide range of void filling applications. Old mines and tunnels often lead to ground stability problems and many have been filled with foamed concrete, along with sewers, service trenches and highway structures such as subways and culverts. Collapsed tunnels at the Heathrow Express rail Link were stabilised with $13\,500 \, m^3$ of foamed concrete.

2.3 *Replacement of existing soil*

In areas of weak soil conditions the weight of a foamed concrete layer and the construction on top of this layer (e.g. road or building) can be designed to equal the weight of the excavated soil (balanced foundation). Therefore, the stress in the underlying soil layers is not increased, minimising settlements. Densities in the range $300–600 \, kg/m^3$ tend to be used.

Foamed concrete can be used as a foundation to roads where there are poor underlying ground conditions. It also provides a more stable foundation than light granular material. $27000 \, m^3$ of foamed concrete was placed from a purpose built barge to form a road foundation as part of the London Docklands project (S van Dijik, 1991).

2.4 *Lateral load reduction*

This application has been used in harbour quays (e.g. diaphragm wall or sheet pile retaining wall) where foamed concrete is used as a lightweight backfill material behind the quay. Vertical loads and hence the lateral load are reduced. Settlements are also reduced and maintenance is minimized. Densities in the range $400–600 \, kg/m^3$ tend to be used.

2.5 *Soil stabilization*

To improve the slope stability of embankments part of the soil is replaced with foamed concrete. This reduces the weight which is a major factor in the instability of slopes. Densities in the range $400–600 \, kg/m^3$ tend to be used.

2.6 *Bearing capacity enlargement*

Cast-in-place piles of foamed concrete can be used as skin friction piles in weak soils. Densities in the range $1200 \, kg/m^3$ tend to be used.

2.7 *Raft foundation*

This application has been used in housing. The foamed concrete acts as a lightweight raft foundation and thermal insulating layer. This is protected with a floor screed or a concrete blinding which also acts as a load spreading layer. Densities in the range $500 \, kg/m^3$ tend to be used, average thickness $0.2 \, m$ (R Jones & A Giannakou. 2002). Another application is a raft foundation manufactured with $400–600 \, kg/m^3$ foamed concrete, $0.75 \, m$ thick for dwellings to be built that sit on water in dykes (in Holland mostly). These are also used as floating pontoons in marinas.

2.8 *Roof slopes*

Low density foam concrete has many roofing applications but is particularly suitable for profiling the positive slope to drains on flat concrete roofs. By adding sand to the mix slopes of $16 \, mm/m$ are achievable and the foam concrete surface can be finished with a tolerance of $\pm 10 \, mm$ relative to the required level while maintaining the required slope (L Cox & S van Dijik, 2003).

2.9 *Floor levelling*

Raising the level of an old floor can be expensive when using conventional concrete but placing a foamed concrete sub base on top of the old floor before laying a new concrete floor on top can be more cost efficient. Different densities tend to be used for different layer thickness.

2.10 *Blinding*

Foamed concrete has the advantage of high workability and flexibility of placing over conventional concrete for blinding. Densities of approximately $1200 \, kg/m^3$ are used if thermal insulation is not important and $500 \, kg/m^3$ if it is.

2.11 Sports fields and athletic tracks

To achieve a rapid draining sports field a permeable foamed concrete (density: 600–650 kg/m^3) with a high drainage capacity is used (Darcy permeability 300 mm/hour). The foamed concrete serves as a lightweight foundation and is covered with gravel and/or a synthetic turf for sports fields used for hockey, football and tennis.

2.12 Filling of pipes

Underground fuel tanks, pipelines and sewers which are out of use can cause fire hazards or can collapse. Once filled these structures are supported and blocked by the foamed concrete. Densities in the range 600–1100 kg/m^3 tend to be used.

2.13 Support of tank bottoms

Foamed concrete can be poured under steel storage tanks, which ensures that the whole tank bottom is supported. Densities in the range 500–1000 kg/m^3 tend to be used.

2.14 Shock-absorbing concrete (SACON) (U.S. Army Environmental Center, 1999)

Shock-absorbing concrete (SACON) is a low-density, fibre-reinforced foamed concrete developed in the USA to be used in live fire military training facilities. It was developed to minimise the hazard of ricochets during urban training. As well as reducing ricochets, the shock absorbing properties of this foamed concrete also function to create a medium for capturing small-arms bullets.

3 COMPOSITION

In general, foamed concretes with densities below 600 kg/m^3 consist of cement, foam and water, with the possible addition of fly ash or limestone dust. Higher densities are achieved by adding sand. For heavier foam concrete the base mix is typically between 1:1 to 3:1 filler to Portland Cement (CEM I). At higher densities (above 1500 kg/m^3) there is higher filler loading and a medium concreting sand may be used. As the density is reduced the amount of filler should also be reduced and at densities below about 600 kg/m^3 filler may be completely eliminated. The filler size must also be reduced, first to a fine concreting or mortar sand, and then to limestone dust, pfa or ggbs at densities below about 1100 kg/m^3.

3.1 Cement and combinations

Portland Cement (CEM I) is normally used as the binder but other cements could be used including rapid hardening cement. A wide range of cement and combinations can also be used e.g. CEM I 30%, fly ash 60% and limestone 10%. Cement contents tend to be in the range of 300 to 400 kg/m^3.

3.2 Sand

Sand up to 5 mm maximum particle size may be used but a higher strength is obtained using finer sands up to 2 mm with 60–95% passing a 600 micron sieve.

3.3 Foam

The most commonly used foams are based on hydrolised proteins or synthetic surfactants. Synthetic based foaming agents have longer storage times and are easier to handle and cheaper. They also require less energy to produce foam, however protein based foaming agents have higher strength performance.

The preformed foam can be divided into two categories: wet foam and dry foam.

Wet foam has a large loose bubble structure and although stable, is not recommended for the production of foamed concretes with densities below 1000 kg/m^3. It involves spraying a solution of the agent and water over a fine mesh, leading to a foam with bubbles sized between 2–5 mm.

Dry foam is extremely stable, a characteristic that becomes increasingly important as the density of the foamed concrete reduces. It is produced by forcing a solution of foaming agent and water through restrictions whilst forcing compressed air into the mixing chamber. The resulting bubble size is smaller than wet foam at less than 1 mm in diameter and of an even size.

Foaming admixtures are covered by BS 8443:2005 *Specification for establishing the suitability of special concrete admixtures* (BSI, 2005).

3.4 Other aggregates and materials

Coarse normal weight aggregates cannot be used in foamed concrete as they would sink in the lightweight foam.

3.5 Mix details

The properties of foamed concrete are mostly dependent on the following aspects: volume of foam, cement content, filler and age.

Water/cement ratio has relatively little effect on strength but other factors like filler content and particle size do.

4 PRODUCTION

There are two main methods of producing foamed concrete, namely the inline and pre-foam methods.

4.1 *Inline*

In this case the base mix is put into a unit where it is blended with the foam. The mixing process is more controlled and greater quantities can be more easily produced. It can be split into two processes:

4.2 *Inline system (wet method)*

The base materials are the same as those used in the pre-foam system but are generally wetter. The base material and the foam (dry type – see above) are fed through a series of static inline mixers where the two are mixed together. The foam and the base materials are blended together and checked with a continual on-board density monitor. The output volume is not governed by the size of the ready-mixed concrete truck, but by the density of the foamed concrete – one $8\,m^3$ delivery of base material can produce $35\,m^3$ of a $500\,kg/m^3$ foamed concrete.

4.3 *Inline system (dry method)*

This method is widespread in Europe and is also used in the UK. Dry materials are loaded into on-board silos from where they are batched weighed and mixed on-site as required using on-board mixers. The base mix is then pumped into a mixing chamber where the foam is then added in the same way as the wet inline system. They require large amounts of water at site for mixing. One delivery of cement/fly ash blend can produce up to $130\,m^3$ of foamed concrete.

4.4 *Pre-foam*

In this method the base materials are delivered to site in a ready-mixed concrete truck. The pre-formed foam is then injected directly into the back of the truck whilst the mixer is rotating. This method has the advantage that relatively small quantities can be ordered, for trench fill, for example, however, it does rely on the mixing action of the concrete truck. Densities in the range of $300–1200\,kg/m^3$ can be achieved. These systems are typically foamed air in the range 20 to 60% air. As this normally takes place in a ready-mix truck, the volume of base mortar or concrete mixed in the drum must be reduced to allow for the final volume of foamed concrete. The amount of stable air and hence density is difficult to control precisely so a degree of both under and over yield must be allowed for when estimating deliveries.

Once the foam is formed it is added to the sand cement mortar that normally has a water cement ratio of 0.4 to 0.6. Too wet a mortar leads to an unstable foam, too dry and the pre-foam may not be able to blend with the mortar.

5 PRACTICALITIES

5.1 *General*

On exposed surfaces there will be some shrinkage but this tends to be in the form of micro cracking. Abrasion resistance is not high, especially at the lower densities so a surface coating is usually needed. The air cells are closed and do not immediately fill with water but at lower densities this will progressively occur if there is any pressure head. Foamed concrete is not used in conjunction with steel reinforcement.

5.2 *Formwork*

Formwork needs to be waterproof and able to resist the pressure exerted by the foamed concrete. If cables and pipework are to be incorporated in the foamed concrete, they may need to be loaded or anchored to prevent them from moving and floating. Consideration must be given to the fact that foamed concrete will fill every accessible space and that the surface will be practically horizontal after setting. When casting foamed concrete against the ground it may well be beneficial to use a geomembrane or geotextile.

5.3 *Health and safety*

All the parties involved with the use of foamed concrete should ensure that all works are carried out in accordance with current health and safety regulations. In particular it is important to protect against drowning, which is a risk whilst the foamed concrete remains fluid. Measures to be taken include: the use of warning signs, guarding the construction site and covering the foamed concrete.

5.4 *Pour depths*

In general, the depth of a pour should be limited to a maximum of 1.5 m, thicker pours increasing the risk of segregation and settlement. Where greater depths are required, pours should be carried out in approximately equal layers.

5.5 *Pumping*

Foamed concrete may be placed by pump but pressure involved in pumping reduces the air content and the properties of the foamed concrete should be assessed at the point of placement (i.e. once it has left the pump). For long pump distances the grout and the foaming agent can be pumped separately. The foam can then be formed and pumped up to 100 m then blended with the grout to form the foamed concrete at the point of application.

5.6 *Specifying*

If strength and or density are critical to the application, ensure that they are adequately specified, both maximum and minimum values if necessary. Density measured on site will be wet density but a cube or core will be an air dry density. Air dry density is typically 100 to 150 kg/m^3 lower than wet density. Compressive strength is measured on dry cubes.

Strength and density can be accurately controlled but this will cost more, especially for pre-foam production. In most cases where strength/density are not specified as a critical requirement, control of foamed concrete is quite loose.

6 PROPERTIES

6.1 *Visual appearance*

The foam that is added to the mortar to produce foamed concrete closely resembles shaving foam. Once this is mixed with the mortar the foamed concrete is liquid with a consistency similar to yoghurt or milkshake.

In its hardened state foamed concrete is similar in appearance to aerated autoclaved blocks (or an Aero chocolate bar).

6.2 *Fresh properties*

The foam has a strong plasticising effect and foamed concrete is normally of high workability with slumps ranging from 150 mm to collapse. For most applications of foamed concrete this is an advantage and it can be difficult to make a low slump if this is what is required. Foamed concrete is quite thixotropic and it can be quite difficult to restart the flow once the concrete has been static for several minutes (although this is not always the case).

The high air content of foamed concrete eliminates any tendency to bleed. With its good insulation properties, as the mix temperature increases during setting, the air expands slightly which ensures good filling and contact in confined voids.

If a foamed concrete mix is over sanded or uses an over-coarse sand, segregation or bubble collapse can occur leading to volume loss and/or a weak top surface. Foamed concrete can be pumped but care should be taken to avoid a significant free fall down the last length of pump line as turbulence may destroy the bubble structure.

6.3 *Hardened properties*

As can be seen in Table 1 below the physical properties of hardened foamed concrete relate to the dry density. Thermal conductivity ranges from 0.1 W/mk to

Table 1. Typical properties of foamed concrete.

Dry density kg/m^3	Compressive strength N/mm^2	Tensile strength N/mm^2	Water Absorption* kg/m^2
400	0.5–1.0	.05–0.1	75
600	1.0–1.5	0.2–0.3	33
800	1.5–2.0	0.3–0.4	15
1000	2.5–3.0	0.4–0.6	7
1200	4.5–5.5	0.6–1.1	5
1400	6.0–8.0	0.8–1.2	5
1600	7.5–10.0	1.0–1.6	5

* The guide value indicates the total quantity of water in kg that permeates a 1 m^2 foam concrete surface during 10 years, if this surface is constantly exposed to water with the same pressure as a 1 m water column. The water absorption may vary according to the type of foam used.

0.7 W/mk, whilst drying shrinkage ranges from 0.3% at 400 kg/m^3 to 0.07% at 1600 kg/m^3.

It should be noted that foamed concrete is, in general, not as strong as autoclaved blocks of similar density. If the concrete is saturated at the time of compressive strength testing, a low result will be obtained due to the internal hydraulic pressures set up as the sample deforms under load.

The cellular structure of foamed concrete gives it good resistance to the effects of freeze thaw action. Foamed concrete does not appear to be vulnerable in freeze-thaw situations and specimens of foamed concrete with densities ranging from 400 to 1400 kg/m^3 showed no signs of damage when subjected to a freeze thaw regime with a temperature range of $-18°C$ to $+25°C$.

If a low density foamed concrete has been specified for its lightweight properties then the effect of possible water absorption on the final density should be taken into consideration.

7 QUALITY CONTROL

7.1 *Foam density and stability*

The properties of foamed concrete are highly dependent on the quality of the foam. The wet density of the foam can be simply determined through weighing a known volume of foam e.g. using a glass measuring cylinder or a bucket, and should be done routinely.

The stability of a foam can be assessed by measuring its collapse with time, using a glass measuring cylinder but a wide plastic pipe may be better as it reduces side restraint.

7.2 *Plastic density of the foamed concrete*

The plastic density of the foamed concrete can again be simply determined through weighing a known volume of foamed concrete e.g. using a bucket. The method is outlined in BS EN 12350:Part 6:2000. *Testing fresh concrete: Density* (BSI, 2000e).

7.3 *Consistence and segregation*

As slump is normally high, the slump test is not ideal but can be used to indicate whether the foamed concrete workability is too low.

The consistence of foamed concrete can be quantified by the slump flow test to BS EN 12350-5:2000 *Testing fresh concrete Part 5 Flow table test*, (BSI, 2000d) but without jolting the table.

Segregation of foamed concrete in the fresh state can be detected by foam rising to the surface of the mix, or by the formation of a separate paste at the bottom of the mixer (only noticeable when mixing). Segregation can be quantified through difference in oven dry densities of 25 mm thick slices taken from the top and bottom of a 100 mm diameter core. Another method to quantify segregation is through difference in oven dry densities of horizontal cores taken at different heights.

7.4 *Cube strength*

The foamed concrete should be sampled in accordance with BS EN 12350–1:2000 *Testing fresh concrete, sampling* (BSI, 2000b). Compressive strength can be measured in accordance with BS EN 12350–3:2000 *Testing hardened concrete. Compressive strength of test specimens* (BSI, 2000c).

To manufacture the test specimens 150 mm rather than 100 mm cubes may be required to ensure sufficient accuracy. Disposable polystyrene moulds are often used as the concrete can be left in the mould (with a suitable lid) until testing. The foamed concrete should not be tamped or vibrated into the mould. Having been left covered for at least 3 days the cubes can be demoulded and immediately sealed in plastic bags and cured at $20 \pm 2°C$.

On low density foamed concrete a core is often taken the following day for curing and testing alongside the cubes.

It should be noted that the variability in strength of a foamed concrete is greater than that of a normal concrete.

7.5 *Soundness*

The soundness of the surface of the foamed concrete can be used to assess its strength development. To assess the soundness the BRE screed tester can be used to determine the in situ crushing resistance. For screeds the test is described in BS8204-1:2003 Annex D (BSI, 2003), however, for foamed concrete the penetration of a single drop of the weight should be measured rather than the four successive blows.

8 SPECIFICATION

There is currently no standard specification for foamed concrete in the UK although there are some guidance documents available (Brady et al, 2001). BS EN 206-1:2000 *Concrete – Part1: Specification, performance, production and conformity* (BSI, 2000a) specifically states that it does not apply to foamed concrete.

8.1 *Strength and density requirements*

The specification gives requirements for strength and density limits.

8.2 *Constituent materials*

Constituent materials are covered by the specification, for example:

"Any foaming admixtures shall comply with BS 8443:2005 *Specification for establishing the suitability of special purpose concrete admixtures*. (BSI, 2005)".

8.3 *Production*

As well as specification details covering production, notes are also given, such as:

Note: It is known that there is a correlation between the density and compressive strength of foamed concrete. Comparing the wet density of the foamed concrete at the point of discharge, with the wet density determined during development testing gives an estimate of the likely strength at a given age.

8.4 *General*

General specification clauses are included in this section, and it is noted that for some applications specialist advice should be sought.

8.5 *Safety*

The specification clause on safety is reproduced below:

"Foamed concretes are likely to provide minimal load bearing capacity for several hours after mixing; during this time unguarded concreting works can represent a drowning hazard for site operatives, children and animals."

9 CONCLUSIONS

The Concrete Society has produced a new guide on foamed concrete, which introduces the material, covers its wide range of applications, gives information on its composition and production, includes practicalities of its use, summarises its properties, advises on quality control and includes a specification.

ACKNOWLEDGEMENTS

This paper is based upon The Concrete Society document, *Foamed Concrete: applications and specifications*, the production of which was sponsored by Foam Concrete Ltd, Propump Engineering Ltd, The Highways Agency, London Concrete and the Cement Admixtures Association.

REFERENCES

Brady, K.C., Watts, G.R.A. & Jones, M.R. 2001. *Specification for foamed concrete, Highways agency Application Guide AG39*. TRL, Crowthorne.

British Standards Institute. 2000a *BS EN 206-1:2000 Concrete – Part1: Specification, performance, production and conformity*. BSI, London.

British Standards Institute. 2000b. *BS EN 12350 – 1:2000 Testing fresh concrete, sampling*. BSI, London.

British Standards Institute. 2000c *BS EN 12350 – 3:2000 Testing hardened concrete. Compressive strength of test specimens*. BSI, London.

British Standards Institute. 2000d. *BS EN 12350-5:2000 Testing fresh concrete Part 5 Flow table test*. BSI, London.

British Standards Institute. 2000e. *BS EN 12350:Part 6:2000. Testing fresh concrete: Density*. BSI, London.

British Standards Institute. 2003. *BS8204-1:2003 Annex D (BSI, 2003). Screeds, bases and in situ floorings – Part 1*. BSI, London.

British Standards Institute. 2005. *BS 8443:2005 Specification for establishing the suitability of special concrete admixtures*. BSI, London.

Cox, L. & van Dijik, S. 2003. Foam concrete for roof slopes and floor levelling. *CONCRETE*. February. pp. 37–39.

Jones, R. & Giannakou, A. 2002. Foamed concrete for energy-efficient foundations and ground slabs. *CONCRETE*. March. pp. 14–17.

S van Dijik. 1991 Foam Concrete. *CONCRETE*, July/August.

U.S. Army Environmental Center. 1999. *Shock-Absorbing Concrete (SACON), Bullet Traps for Small Arms Ranges. Cost and Performance Report*. U.S. Army Environmental Center. Report No. SFIM-AEC-ET-TR-99019.

Steel, D.P., McMahon, W. & Burtwell, M.H. 2003. *Long-term performance of reinstated trenches and their adjacent pavements. Part 2: Long term performance of reinstatements in the highway*. TRL Report TRL 573. TRL, Crowthorne.

Taylor, R.W. 1991. *Foamed concrete for improved trench reinstatements*. British Cement Association, Camberley.

The Concrete Society. 2008. *Foamed Concrete: application and specification*. The Concrete Society, Camberley.

Excellence in Concrete Construction through Innovation – Limbachiya & Kew (eds)
© 2009 Taylor & Francis Group, London, ISBN 978-0-415-47592-1

The estimation of concrete quality by power functions

B.K. Nyame
Consultant, London, UK

ABSTRACT: Concrete is the main composite material. The need arises to rationalize computer mix designs for multi-phase concrete. There are only two simple two-phase models for composites – the upper and lower bounds, with quality estimated by the expected values. Otherwise, the models are free two-phase models with varied geometric configurations and whose quality is estimated by power functions. In these 3 papers, two-phase models are (i) used to simply describe the complex distribution of material quantity q_i, and quality Q_i. and (ii) interpolated by the interparticle model – the IM -, using its variable interface vector m. The overall objective is to estimate and control to design for the composite property. Power functions are empirically used to estimate non-linear behavior by curve fitting. This 1st of 3 papers, deals with power function estimations of the concrete property, Q_0,. Firstly, expected values are expressed as power series and analysed for uniqueness. Next, power functions are derived by eliminating the characteristic IM interface vectors. The power mean value is the mean, mode and median of the conventional mean values. The power function, $Q_0 = \prod_{i=1}^{m} Q_i^{qi}$ defines central tendency as the power mean value.

1 INTRODUCTION

Modern structures are designed for a wide range of loading and concrete quality as described in table 1.

A long time before two-phase models, concrete was regarded (Troxell et al. 1968) as a mixture of granular materials, with water added merely for fluidity. The properties of concrete were popular as empirical

Table 1. Quality for concrete construction.

Loading	Structure	Quality
Humidity Temperature – hot & cold.	Pavement Slabs Power Stations, Refractories, LNG tanks	Deformation Cracking & Thermal Resistance
Prestress & Vibrations	Buildings and Bridges	Tension, Deformation, Damping.
Hydraulic	Offshore, pipelines, tanks, tunnels, dams, foundations, marine	Durability & Permeability
Shock waves	Hardened Structures	Impulse Response.
Acoustic waves	Buildings & Studios	Acoustic Impedance
Radiation	Power Stations	Absorption

power functions – or power laws. Powers differentiated the water content as evaporable and non-evaporable at 105°C and then explained the deformation of concrete.

Power functions are popular as empirical equations which describe non-linear behavior of systems.

In 1960, Hansen (1960) introduced the simple two-phase models into concrete technology, for the estimations of elastic modulus by linear elastic analysis. Later, workers such as Hirsch (1962), Counto (1964), Hobbs (1971) and Campbell-Allen (1963) estimated deformation with what was then a new opening to two-phase models.

Interfaces affect strength and fracture mechanisms. Barnes (1976), Alexander et al. (1998), Yeh (1992) and Hughes & Ash (1969), described interface effects by measurements and morphology. Since 1960, two-phase models have only been used as estimation models (Illston et al. 1979) for the concrete deformation.

Illston et al. (1979) outlined some limitations for slow progress of two-phase models as measurement techniques, starting data, moisture contents, the interface effects, the algebraic complications, non-linearity, high strength aggregate have little effect on concrete strengths, or universal non-acceptability of models. Counto (1964) and Hobbs (1971) derived non-linear estimators. For the design of concrete nuclear

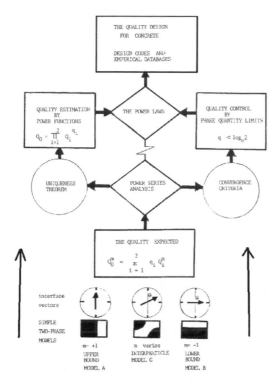

Figure 1. Logical flow diagram for quality estimation, design and control.

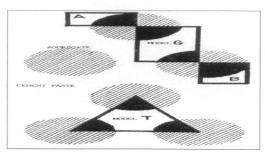

Figure 2(a). The IM (G) interpolates the Upper (A) & Lower (B) Bounds.

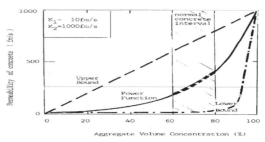

Figure 2(b). The power function for the IM (G) – interpolates the Upper (A) and Lower (B) bound permeability.

heat shields, Browne (1972), point estimated the coefficients of thermal expansion by assuming linearity at normal aggregate volumes concentrations of 60% to 80%.

This paper shows that interface vectors are variables to eliminate from two-phase models on figs 1 & 2 (a). One problem identified was the hope to estimate the non-linear effects using the linear elastic analysis.

Another problem is that upper bounds on phase Q_i^{+1} show direct composite quality increases. However, the lower bounds on phase Q_i^{-1} do not predict direct quality increases as the phase quality is increased.

Firstly, the upper and lower bound expected quality of two-phase models and the 'micro-chip' IM are identified by interface vectors m = +1 or −1 on fig 1 (read from base up) and by configurations on fig 2 (a).

Next, interface vectors are eliminated by applying the uniqueness theorem (Rade & Westergren 1990) to derive power functions for estimation. Using convergence criteria (Rade & Westergren 1990), stable mix compositions are deduced in the 2nd of 3 papers.

Finally, by combining the estimation and the control of stable mix compositions, limit surfaces for design are described. Fig 1 outlines the logical mix design.

2 POWER SERIES ANALYSIS

The expected quality of simple two-phase models by superposition from figure 1 is

$$Q_0^{\ m} = \sum_{i=1}^{2} q_i \, Q_i^{\ m} \tag{1}$$

$$m = \text{cis } 2\theta = \cos 2\theta + j \sin 2\theta = e^{j2\theta} \tag{2}$$

Interface vector, m = + 1 upper, m = −1 lower bound. The expected quality is evaluated by linear flow and linear elastic analyses (Illston et al. 1979). The power series are

$$Q^m = e^{m \ln Q} = \sum_{n=1}^{\infty} (m^n/n!) \, [\ln Q i_1]^n \tag{3}$$

Hence the power series for the expected quality of two-phase models from eqn 1 is

$$\sum_{n=1}^{\infty} (m^n/n!) \, [\ln Q_0]^n = \sum_{i=1}^{2} q_i \sum_{n=1}^{\infty} (m^n/n!) \, [\ln Q_i]^n \tag{4}$$

This vital power series for the expected quality as eqn 4 is investigated for uniqueness (Rade & Westergren 1990), in this paper for quality estimation. It will then be investigated for convergence (Rade & Westergren 1990), in the 2nd paper for quality control. The important equation 4 sums up the linear flow or linear elastic analysis (Illston et al. 1979) of simple two-phase models for the mix design described in the 3rd paper.

By linear elastic analysis, uniqueness of eqn 4 yields the response to loads, which has limits as strengths -the resistance to loads. However, the convergence of vital equation 4 yields the stable response to loads. This links equation 4 to real structural analysis and design (Kong & Evans 1975, Allen 1988), now applied to concrete mix design.

2.1 The power mean value – PMV

Uniqueness (Rade & Westergren 1990), is investigated for equation 4, so as to derive the power function that estimates the power mean value – PMV – which as central tendency, is the mean, mode and median of conventional means, as the rms, arithmetic and harmonic means on fig 4.

Applying the uniqueness theorem, (Rade & Westergren 1990) by comparing coefficients of $(m^n/n!)$ on both sides of equation 4, eliminates the random interface vector m, so that the expected quality is

$$[\ln Q_0]^n = \sum_{i=1}^{2} q_i \Sigma [\ln Q_i]^n \tag{5}$$

From eqn 5, if $n = 1$, then power mean value Q_0 from the power function is

$$Q_0 = Q_1^{q1} \cdot Q_2^{q2} = \prod_{i=1}^{2} Q_i^{qi} \tag{6}$$

Power mean values are powerful 3 in 1 non-linear estimators of the mean, median and mode values as shown on figure 4, but are independent on interface vectors, m and geometric configurations. Except for the upper and lower bounds, free two-phase models, such as the IM, obey the power functions.

2.2 Transcendental power functions

The transcendentals are deduced by comparing eqn 1 to eqn 5, as superposition of phase effects, so that Qi and m of eqn 1 are replaced by ln Qi and n at eqn 5.

By applying the uniqueness theorem (Rade & Westergren 1990) to eqn 5, the 1st transcendental power function for the two-phase composites is

$$\ln Q_0 = (\ln Q_1)^{q1} \cdot (\ln Q_2)^{q2} = \prod_{i=1}^{2} (\ln Q_i)^{qi} \tag{7}$$

In effect, applying the uniqueness theorem[14] to the power series of expected values, the Σ superposition is transformed to Π factors of phase effects in power functions and the 1st transcendental power function.

In lubrication engineering[7] empirical transcendental power functions are used to blend grades of oil. So as a liquid, fresh concrete may obey transcendentals.

Deviations from power functions occur as non-linear effects of by-pass flow around aggregate particles.

In this respect, exclusive powers function, where the exclusion factor h depends on levels of by-pass flow.

Hence $\quad Q_0 = Q_1^{(1-q)} \cdot Q_2^{(1-h)q} \tag{8}$

$h = 1$ for impermeable aggregate on full by-pass flow
$h = 0$ for permeable aggregate with no by-pass flow
If $h = 1$ or no flow occurs through the aggregate, the exclusive power function becomes simply

$$Q_0 = Q_1^{(1-q)}. \tag{9}$$

3 QUALITY ESTIMATIONS – Q_0

Everyone can estimate. Those estimating nothing have estimated zero. Those estimating with the least errors from accepted values give the better estimates. There are 3 types of quality for concrete.

3.1 Steady state – linear flow or no flow

Properties in this group, such as density and porosity are structural and if there are any flows for their measurement, they are linear. Steady state quality of free wc 0.47 mortar samples (Nyame 1985) cured for 28days are described by power functions on figure 3.

3.2 Steady state – non-linear flow

Properties in this group, such as the elastic modulus, permeability and thermal conductivity are structural and depend on the flow of substances or energy, so they have by-pass flow around aggregate particles.

Fig 3 shows that permeability obeys power functions and the 1st transcendental, but with slight deviations.

3.3 Ultimate state – strength

Strength has many ways for estimation. The most important by Weibull's theory (Weilbull 1939) from failure theories and fracture mechanics is that, failure of the weakest links, leads to total failure. This theory explains why high strength aggregate have little effect on concrete strengths (Illston et al. 1979). The 168 year old concept of Feret's law (Feret 1892) that cement concentrations affect concrete strengths remains good estimators. The continuous cement paste, dispersed

13

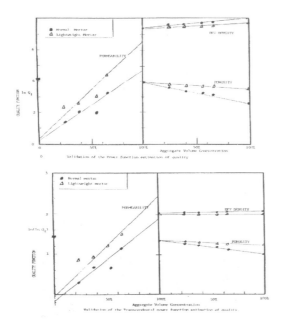

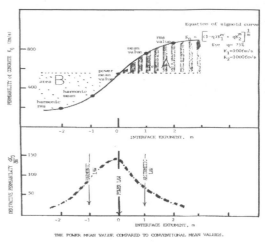

Figure 4. The power mean value is the mean, median and mode of the conventional means i.e. RMS, harmonic and arithmetic means.

Figure 3. Power functions & the 1st transcendental fit density, porosity and permeability data.

with aggregate, is the weakest link in concrete to affect the ultimate composite strength.

The 192 data for 28 day strengths (ACI 1997, Mehta & Aitchin 1990, Jambor 1976, Domone & Soutsos 1995, Parrott 1995, Mangat & Molloy 1995, Egan 1994, Watson & Oyeka 1981, Karihaloo et al. 2001) were analyzed for best fit on figs 5 & 6. The data presented 137 years after the Feret law (Feret 1892) was published and in the 25 years from 1976 to 2001, lacks experimental bias (Illston et al. 1979) in the empirical relations discovered. Volume concentrations based on a unit volume of concrete, cement paste and the free water have been related to the 28 day compressive strength data (ACI 1997, Mehta & Aitchin 1990, Jambor 1976, Domone & Soutsos 1995, Parrott 1995, Mangat & Molloy 1995, Egan 1994, Watson & Oyeka 1981, Karihaloo et al. 2001).

Figure 5 shows non-unique linear relations for strength based on concentrations in a unit volume of concrete. Line (1) for concrete (65%–75% aggregate volume) and cement paste line (3) (0% aggregate) both indicate that concentrations, defined in a unit volume of concrete, is not appropriate for both, but should include additional variables for the aggregate. However, line (2) for fibre and (35%–45% aggregate volume) suggests an extension of line (1) despite the considerable fibre content in the concrete.

Figure 6 shows linear relations for concrete strength, if concentration is defined by volume of either the cement paste, or the free water matrix.

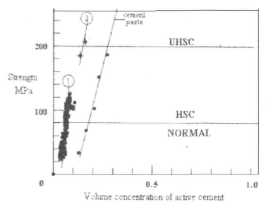

Figure 5. The influence of active cement in concrete on strength.

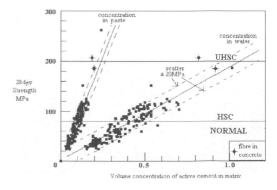

Figure 6. The influence of active cement in matrix on concrete strength.

The linear correlation 92%, based on concentrations per unit volume of the cement paste matrix was:-

$$fcm = 825 \sqrt{(LS)} \left(\frac{vc}{wc + vc} \right)^2 \quad MPa \qquad (10)$$

The linear relation of correlation 93% based on the concentrations per unit volume of water matrix was

$$fcm = 180 \sqrt{(LS)} \frac{vc}{wc} \qquad MPa \qquad (11)$$

Figure 6 shows less scatter for equation 10 than 11. As the free wc tends to zero, equation 10 predicts limited strengths of $825\sqrt{(LS)}$ MPa, but equation 11 predicts unlimited strength, which is unrealistic, so equation 10 was preferred. The most appropriate definitions of the cement concentration are based on a unit volume of cement paste matrix – eqn 10 & 12.

Equation 12 superposes the pozolanic activity of the lime, silica and alumina by $[\sqrt{(LS)}$ & $\sqrt{(LA)}]$ on the 28 day compressive strength of concrete – correlation coefficient of 91% as:

$$fcm = 550 \left[\sqrt{(LS)} + \sqrt{(LA)} \right] \left(\frac{vc}{wc + vc} \right)^2 \quad MPa \quad (12)$$

where L = % lime in cement
    S = % silica of in cement
    A = % alumina in cement
    vc = specific volume of cement
    wc = free water/ cement ratio
    fcm = 28day strength.

and Q_0 = concrete quality
    Q_1 = matrix quality
    Q_2 = particle quality
    q_1 = matrix quantity as volume concentration
    q_2 = particle quantity as volume concentration

Using the power function for density and porosity, and the power functions with the by-pass flows for permeability, the estimators cited help mix designs for concrete and for its several qualities and strength.

4 CONCLUSIONS

1. The power functions and the empirical strength estimators are deduced to design concrete mixes.
2. The power means values – PMV – non-linearly describe all 3 in 1, mean, median and mode values.
3. Power functions estimate concrete quality with the interparticle model – the IM.
4. Power functions non-linearly interpolate simple two-phase models, which estimate on **linear** effects.

REFERENCES

ALEXANDER K, WARDLAW J , GILBERT D Aggregate-cement paste bond and the strength of concrete. Proceedings of an international conference on the structure of concrete. Editors AE Brooks, K Newmann, C&CA, 198, p 59

ALLEN AH Reinforced Concrete Design to BS 8110:E & F N Spon, 1988, p 1, Chpt 1

AMERICAN CONCRETE INSTITUTE Manual of Concrete Practice, 1997, part 1, p 233

BARNES BD Morphology of the paste-aggregate interface. PhD thesis, Vol 1, p 125, Purdue University, Lafayette, Indiana, 1976, JHRP76-13

BROWNE RD Thermal movements of concrete. Cur Practice Sheet 3PC/06/1, Concrete 6, 1972, 51

CAMPBELL-ALLEN D and THORNE CP The thermal conductivity of concrete Magazine of Concrete Research, 15, 43, 1963, pp 39

COUNTO UJ The effect of elastic modulus of the aggregate on the elastic modulus, creep and creep recovery of concrete. Mag of Concrete Research, 16, 1964, 129

DOMONE PL and SOUTSOS MN Properties of high strength concrete mixes containing pfa and ggbs Magazine *of Concrete Research,* 47, 173, Dec 1995, p 355–67, table 1,3

EGAN PJ, Benefits of superplasticising admixtures, *Concrete*, May/June 1994, Vol 28, No 3, p 18–21

FERET R Annales des Ponts et Chausses, Series 7, 4, 5 -164, 1892.

HIRSCH TJ Modulus of elasticity of concrete as affected by elastic moduli of cement paste matrix and aggregate. Proc ACI, 59, 1962, 427

HOBBS JW The dependence of the bulk modulus, Young's modulus, creep, shrinkage, and thermal expansion upon aggregate volume concentration. Materiaux et Construction, 4, 1971, 107

HUGHES BP and ASH JE Water gain and its effects on concrete. Concrete Vol 3, 1969, p 494

ILLSTON JM, DINWOODIE JM, SMITH AA Concrete, Timber and Metals, – Van Nostrand Reinhold Co, 1979, p 280 285

JAMBOR J Influence of water-cement ratio on the structure and strength of hardened cement paste. *Proc of conference on hydraulic cement pastes, Sheffield*, 1976, p175

KARIHALOO BL, ALAEE FJ, BENSON SD A new technique for retrofitting damaged concrete structures, *Concrete Communication Conference 2001*, p 293–304

KONG FK, EVANS RH Reinforced &Prestressed Concrete, Nelson, 1975, p 11, *Magazine of Concrete Research*, 37, 130, 1985, pp 46–48

MANGAT PS and MOLLOY BT, Chloride binding in concrete containing pfa, gbs, silica flume under sea water exposure. *Magazine of Concrete Research,* 47, 173, Jun 1995, p 129

MEHTA PK and AITCHIN PC Microstructural basis of selection of materials and mix proportions for high strength concrete. *Proceedings 2nd International Symposium on high strength concrete*, Detroit, 1990

NYAME BK Permeability of normal and lightweight mortars.

PARROTT LJ, Influence of cement types and curing on the drying and air permeability of cover concrete. *Magazine of Concrete Research,* 47, 171, Jun 1995, p 103–111

POWERS TC, The physical properties of Portland cement pastes. in The Chemistry of Cements, TAYLOR HFW, Editor, Academic Press, Vol 1, Chp 10, pp 392 HANSEN TC Creep and Stress relaxation of concrete Proc of Swedish Cement and Concrete Res Inst, pp 31, 1960

RADE L and WESTERGREN B Beta Mathematics Handbook, 2nd edition, Chartwell-Bratt, 1990, pp 111–112

TROXELL GE, DAVIS HE, KELLY JW Composition and properties of concrete, 2nd edition, McGraw Hill, London 1968, pp 3, 75-, 429-

WATSON AJ and OYEKA CC, Oil permeability of hardened cement paste and concrete. *Magazine of Concrete Research,* 33, 115, Jun 1981, p 85

WEIBULL W A statistical theory of the strength of materials. Proc. of Royal Swedish Institute for Engineering Research, Stockholm, 1939, 151, 5.

YEH JR The effect of interface on the transverse properties of composites. International Journal of Solids and Structures, 29, 20, 1992, p 2493–2502.

Excellence in Concrete Construction through Innovation – Limbachiya & Kew (eds)
© 2009 Taylor & Francis Group, London, ISBN 978-0-415-47592-1

The control of stable concrete quality

B.K. Nyame
Consultant, London, UK

ABSTRACT: It is generally accepted that 'too much' or 'too little' of anything is not good. It means the extremes, like the maximum or minimum quantities control stable quality. His 2nd of 3 papers deals with the control of stable concrete quality. The expected values of simple two-phase models are investigated as power series for convergence to deduce the stable mix compositions. This is the stable quality theory. Concrete has stable quality, if any or all the mix compositions are the basic $1/(e-1)$ to a critical $\ln 2$ aggregate volume concentration i.e. (1) Aggregate volume, AgVol, from 58% to 69%. (2) Water content, W, from $155\,kg/m^3$ to $210\,kg/m^3$, (3) Cement content, C, from $480\,kg/m^3$ to $650\,kg/m^3$ (4) Free w/c ratio, wc, from 0.24 to 0.44. These compositions from the stable quality theory and IM configurations are compared with mix design specifications. The stable mix compositions help to draw *limit surfaces* for mix design in the final of 3 papers.

1 INTRODUCTION

The reduction of variability in quality, particularly the concrete compressive strength, is an important aim in construction quality control (BSI 1992, ACI 1973, BSI 1985). 28 day strength variations (Kong & Evans 1975) by standard deviations, of 4–$5\,N/mm^2$ signify good control, 5–$7\,N/mm^2$ show fair control and 7–$8\,N/mm^2$ indicates poor control. Variability is influenced by properties. Permeability has largest scatter and density, the least variability. Scatter is usually due to inadequate equipment and site production practice. However, the scatters, now deduced, arise from unstable design of fresh mixes.

In general, stability means firm, secure, robust, and 'not fickle', or the *tendency* to maintain or return to steady conditions, after small disturbances or loads. Mechanical stability is a *tendency* to resist collapse. Mathematical convergence is the *tendency* to a point. From the word *tendency* in mechanics, mathematics and everyday use, **stability** also means *convergence*.

In the 1st of 3 papers on estimation the expected quality is transformed to the power mean value by a power series analysis, which tests for *uniqueness*.

This 2nd of 3 papers derives the theoretic conditions to rectify instability in the designed quality but now, power series analysis with *convergence* tests for the series, so as to define the *stable quality* conditions.

The concrete is regarded as a two-phase model of cement paste matrix and aggregate particles. The *stable quality* of concrete is due to the matrix binding the particles to resist failure or collapse and the particles restraining deformation and the flow of fluids, heat and strain energy. These two factors should be optimized to determine *stable quality,* by the control of mix compositions for the min/max and stable matrix or particle volume concentrations.

The simple relation for the volume concentrations of matrix and particles in two-phase models is that they add up to unity or 1. This means that stabilizing the matrix also effectively stabilizes the particles.

2 POWER SERIES ANALYSIS

From the chart, 1st of 3 papers, figure 1 (*read it from base upwards)-* analysis of the expected quality is by the linear flow or linear elastic analyses (Illston et al. 1979) of simple two-phase models, to sum up the **vital** power series.

This *convergence* of the power series of the expected quality is the *stable quality theory*. It is intuitively satisfactory because convergence is tendency to a single point and a stable behaviour.

3 STABLE QUALITY THEORY

The expected quality Q_0 is represented by

$$Q_0{}^m = (1-q)Q_1{}^m + q\,Q_2{}^m \qquad (1)$$

Q_0 for the expected quality has the power series of

$$Q_0{}^m = \sum_{i=1}^{2} (m^n/n!)\,[\,\ln Q_i\,]^n \qquad (2)$$

The non-dimensional form of equation 1 is

$$(Q_{01}{}^m - 1) = q\,(\ Q_{21}{}^m - 1) \tag{3}$$

The power series of this non-dimensional eq 3 is

$$\sum_{n=1}^{\infty} (m^n/n!)\,[\ \ln Q_0\ ]^{\,n} = q \sum_{n=1}^{\infty} (m^n/n!)\,[\ \ln Q_i\ ]^{\,n} \tag{4}$$

This is the **vital** power series, but now, it is tested for *convergence*[5] to **control** stable quality. It sums up the linear elastic analysis[7] or *response* to loads.

3.1 Stable Relative Quality $Q_{21} = Q_2/Q_1$

Using the root test[5] equation 4 converges if

$$\underset{n \to \infty}{\text{Lim}} \quad \frac{(m\,\ln Q_{21})}{(n!)^{\,1/n}} \ \leq \ 1 \tag{5}$$

so that $m \ln Q_{21} \ \leq \ 1$ \tag{6}

since $m \leq 1$ for two-phase models, the stable relative quality of aggregate to paste Q_{21} is in the interval

$$1/e \ \leq \ Q_{21} \ \leq \ e \tag{7}$$

3.2 The Basic Stability

If m of equation 6 is replaced in equation 3, then the stable relative quality of the concrete to paste Q_{01} is

$$Q_{01} = (1 + q\,(e-1))^{\ \ln Q_{21}} \tag{8}$$

The value of Q_{01} converges as a binomial series[5] if

$$q_b \leq 1/(e-1) \tag{9}$$

Aggregate volume concentration q_b, is *basic stability* in a concrete mix. To sum for the basic stability,

$$\mathbf{q_b \ \leq 1/(e-1)}\ \text{ or } 1/e \ \leq \ Q_{21} \ \leq \ e \ \text{ or } \ \tfrac{1}{2} \leq \ Q_{01} \ \leq 2$$

3.3 The Critical Stability

Maximum relative quality of concrete to paste Q_{01} above is 2,

so $Q_{01} = 2^{\ln Q_{21}} = Q_{21}{}^{\ln 2}$ \tag{10}

Therefore the critically stable concrete mix has

$$\mathbf{q_c \leq \ ln\ 2} \quad \text{(over } {}^2/_3 \text{ of volume are particles)} \tag{11}$$

3.4 The extreme min/max Mix Compositions

Any 2 of the 3 main ingredients, cement, water and aggregate define the mix completely. Cement and water is the matrix, and aggregates are the particles.

3.5 The min/max free wc

The fresh paste has min/max concentrations of the cement or water up to ln 2.
For phase 1, the water: $wc/(wc + vc) \leq_{\ln 2}$ For phase 2, the cement $vc/(wc + vc) \leq_{\ln 2}$ If specific volume of the cement is $vc = 0.32$

Then $0.14 \leq wc \leq 0.72$ \tag{12}

is the min/max range of free wc which varies with cement blending that changes its specific volume vc and the water demand of the blended cement mix.

3.6 The Min/max Cement Content

The volume concentration of matrix cement paste is

$$(1-q) = C(wc + vc) \tag{13}$$

For the min/max free wc 0.14 to 0.72, if $vc = 0.32$, the min/max cement contents are

$$0.295 \leq C \leq 0.667 \tag{14}$$

The non-dimensional units are based on unit density of water as in relative density concepts.

3.7 The Min/max Water Contents

The two-phase model is matrix water and particles of the total solids (cement + aggregate).
    Close-pack volume concentration of solids is 0.864

So $\ln 2 \leq (1 - W) \leq 0.864$ \tag{15}

This implies the min/max water contents of

$$0.136 \leq W \ \leq \ 0.307 \tag{16}$$

3.8 The Stable Mix Compositions

The cement content is the power series of

$$C = (1-q)\,[vc(\ 1 + wc/vc)\,]^{-1} \tag{17}$$

The water content is the power series of

$$W = (1-q)\,[(1 + vc/wc)]^{-1} \tag{18}$$

Divide these stable water and cement contents to get the stable free wc range of wc = 0.24 to 0.44.

Table 1.   Stabilizes the Water and Cement Contents.

Condition	Water	Cement
Power Series	Equation 18	Equation 17
Converges if	$W \leq \tfrac{1}{2}(1-q)$	$C \geq \tfrac{1}{2}(1-q)/vc$
Basic q = 1/(e-1)	$W \geq 0.209$	$C \geq 0.653$
Critical q = ln 2	$W \geq 0.154$	$C \geq 0.480$
Optimal – range	0.154 to 0.209	0.480 to 0.653
Optimal – PMV	0.179	0.560

4 STABLE PARTICLE STRUCTURES

The **IM** – interpolated the upper and lower bound two-phase models to derive the power functions for the estimations of quality (figure 2 in paper 1 of 3).

Stable **IM** return to original after small disturbances. Figure 2 describes the **IM** equations and vectors. Figure 3 shows 2 **IM** fault lines that control strength.

(A) If $\theta_1 = 90°$ then fault line FF1 is vertical
   Configuration = Vertical-fault & 2 lower bounds Low compressive strengths fcm
   Particle Volume Concentration, $\mathbf{q_{1V} = 11\%}$
(B) If $\theta_2 = 0°$ then fault line FF2 is horizontal
   Configuration = Horizontal-fault & 2 upper bounds High compressive strengths fcm
   Particle Volume Concentration, $\mathbf{q_{2H} = 56\%}$
(C) If the 2 particles touch each other then
   Configuration = Close-pack & Discontinuous matrix Low compressive strengths fcm
   Highly brittle fracture mechanism
   Particle Volume Concentration, $\mathbf{q_{CP} = 86\%}$

Note that the 50%:50% PMV ≡ Square Root ($\sqrt{}$)

- For basic stable particle volume concentration, $q_b = 1(e-1)$, hence $q_{2H} = 56\% \approx 1/(e-1)$.
- The PMV of $q_{1V} = 11\%$ & $q_{CP} = 86\% \approx (1 - \ln 2)$ i.e. the *matrix* is critical stable or $1 - q = \ln 2$
- Now, the PMV of $q_{2H} = 56\%$ & $q_{CP} = 86\% \approx$ critical stable $\ln 2$.
- Finally, the PMV of $q_{1V} = 11\%$ & $q_{2H} = 56\% \approx 25\%$

This represents the quarter particle structure, as the PMV of the strong horizontal-fault line and the weak vertical-fault line particle structures. The basic stability is definitely the strong double upper bound stable configuration, $q_b \leq 1/(e-1)$, fig 4. The critical stability is the PMV of the strong basic stability and the economy close-pack brittle unstable configuration, $q_c \leq \ln 2$, fig 4.

This validates stable quality theory and particle structures are retained after small load disturbances.

The combinations of the 3 particle configurations are shown on fig 4 and also summarized in table 2.

Table 2. The equivalence to the PMV of 3 particle structures.

		$q_{2H} = 56\%$ $q_b \leq 1/(e-1)$	$q_{CP} = 86\%$
	$q_{1V} = 11\%$		
$q_{1V} = 11\%$	identical	quarters	$q = (1 - \ln 2)$
$q_{2H} = 56\%$ $q_b \leq 1/(e-1)$	quarters	identical	$q_c \leq \ln 2$
$q_{CP} = 86\%$	$q = (1 - \ln 2)$	$q_c \leq \ln 2$	identical

5 DISCUSSION

It is often considered that 'too little' or 'too much' of anything is not 'too good'. This view applies to the design of concrete mixes. The extreme min/max

Figure 1. The Min/Max and Stable Mix Compositions.

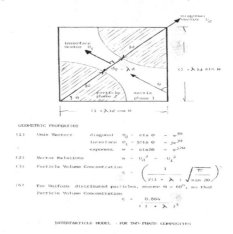

Figure 2. The IM – Configuration

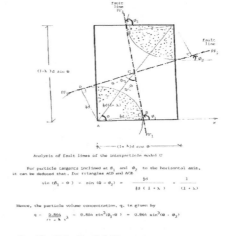

Figure 3. The IM – Particle Tangents.

19

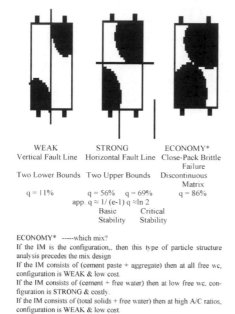

WEAK   STRONG   ECONOMY*
Vertical Fault Line   Horizontal Fault Line   Close-Pack Brittle
                                           Failure
Two Lower Bounds   Two Upper Bounds   Discontinuous
                                           Matrix
$q = 11\%$ $q = 56\%$ $q = 69\%$ $q = 86\%$
         app. $q \approx 1/(e-1)$ $q \approx \ln 2$
                 Basic   Critical
                 Stability   Stability

ECONOMY* -----which mix?
If the IM is the configuration,. then this type of particle structure
analysis precedes the mix design
If the IM consists of (cement paste + aggregate) then at all free wc,
configuration is WEAK & low cost.
If the IM consists of (cement + free water) then at low free wc, configuration is STRONG & costly.
If the IM consists of (total solids + free water) then at high A/C ratios,
configuration is WEAK & low cost.
If the IM consists of (total solids + free water) then at low A/C ratios,
configuration is STRONG & costly.

Figure 4. The IM – Stable Configurations.

and stable mix compositions, on the staircase diagrams of figure 1, indicate the *action limits* when the mix compositions will definitely cause mix problems.

For instance, if the aggregate volume exceeds the critical ln 2 or 69%, then mix problems of cohesion and segregation are likely because of insufficient cement paste matrix to bind the aggregate particles.

Similarly, if the water content is below the minimum of 135 km³, as on figure 1, then the mix is likely to be unworkable, because of the insufficient water, cement paste and any admixtures to lubricate the particles. Such logical 'If'. 'Then' deductions have led to computer algorithms to detect the designed concrete mix problems in the final of these 3 papers.

The min/max cement contents in the old CP110, BS8110 and BS5328 of 250 kg/m³ to 650 kg/m³ all agree with 295 to 665 kg/m³ by stable quality theory.

Water is virtually free. There was the old tendency to increase it on sites to get workable mixes to place. Water is important for quality control as it affects the final concrete porosity, permeability and durability.

Teychenne et al. reported a survey of water contents used for ready mix as 170 kg/m³ to 230 kg/m³. These values agree within 20 kg/m³ or 18% of the figure 1 stable water contents of 155 kg/m³ to 210 kg/m³.

The free wc is more important than the water content for engineering properties of concrete. For complete hydration, minimum free wc is 0.25. This is within the min/max range of 0.14 to 0.72. However, as the

unhydrated cement grains enhance quality, by their restraints, a lower limit than free wc ≥ 0.14 will be

$$\text{free wc} \geq \text{ (min Water/max Cement)} \qquad (19)$$

$$\text{free wc} \geq (135 \text{ kg/m}^3 /665 \text{ kg/m}^3) \geq 0.20. \qquad (20)$$

6 CONCLUSIONS

The stable quality theory, as convergence of the vital power series for the expected quality of two-phase concrete derives the stable mix compositions.

The stable quality theory is validated by an analysis of the particle tangents of the IM. The stable concrete mix compositions are:

(1) Aggregate volume concentrations from the basic $1/(e-1)$ to critical ln 2, which is 58% to 69%.
(2) Water Contents from 155 kg/m³ to 210 kg/m³
(3) Cement Contents from 480 kg/m³ to 650 kg/m³
(4) free wc from 0.24 to 0.44

The min/max concrete mix compositions are:

(1) Aggregate volume concentrations from 0 to 86%
(2) Water Contents from 135 kg/m³ to 305 kg/m³
(3) Cement Contents from 295 kg/m³ to 665 kg/m³
(4) free wc from 0.14 to 0.72

Concrete quality increases with the number of stable and min/max mix compositions to be used for the mix design in the final of these 3 papers.

REFERENCES

ACI Committee, 704, Enchiridion E704-4 Concrete Quality, Detroit, 1973, 26 pp.
BSI BS7850 Total Quality Management Part 1: 1992 Guide to quality improvement methods.
BSI CP110, 1985, Part 1 – Table 50, p103, 20 mm max aggregate sect 6.7.6.1.
ILLSTON JM, DINWOODIE JM, SMITH AA Concrete, Timber and Metals, Van Nostrand Reinhold Co, 1979, p 280–284.
KONG FK and EVANS RH Reinforced & Prestressed Concrete, p 42, table 2.8-1 Thomas Nelson & Sons Ltd, 1975.
RADE L and WESTERGREN B Beta Mathematics Handbook, 2nd edition, Chartwell-Bratt, 1990, pp 111–112.
TEYCHENNE DC, FRANKLIN RE, ENTROY HC, HOBBS DW, NICHOLS JC, Design of normal concrete mixes. BRE Report, Dept of Environment.

Excellence in Concrete Construction through Innovation – Limbachiya & Kew (eds)
© 2009 Taylor & Francis Group, London, ISBN 978-0-415-47592-1

The design of concrete mixes on limit surfaces

B.K. Nyame
Consultant, London, UK

ABSTRACT: Graphical limit surfaces show how parametric factors control the design of systems. Limit surfaces, therefore, facilitate visual control by point estimations in systems design. This final of 3 papers deals with the design of concrete mixes on limit surfaces. The estimation and control of quality of two-phase models are used to generate limit surfaces for mix design. The limit surfaces are drawn on axes of free wc and aggregate volume concentration (Wc, AgVol). There are intersecting networks of water, W, and cement contents, C, so that all mix variables are at particular points on the limit surface. Using stable mix compositions, quality maps are classified for the strengths of concrete. A graphical hypothesis to detect and pre-determine concrete mix problems is advanced.

1 INTRODUCTION

Concrete technologists have abundant accumulated mix designs that proved successful for construction projects. As computer memory can now be increased easily to store substantial concrete mix data, the ready-mixed concrete industry now tends for simple *mix selections* rather than the complex *mix designs*.

The aims of concrete mix design (Kong & Evans 1975) are to deduce the quantities of cement, aggregate, water, admixtures and additives that satisfy the required strength, workability, durability and economy. Similarly, the aims of structural design are to deduce a structure that is *safe* to satisfy its purpose of use and to be economical. *Safety*, from limit state philosophy (Kong & Evans 1975, Allen 1988), is that the structure, with very low probability risks, will not violate limit states of collapse, deflections, cracking, vibrations or fire resistance.

The type of structure must be chosen and structural members designed to be safe, durable and economic. *Structural selections* exist, rather like *mix selections*. The progress is to *design* mixes from their *response* and *resistance to loads*, by eqn 4 – 1st & 2nd papers, just as economic *member sections*, by the structural designs and to select from stored mix design data.

Since Abrams laws (Abrams 1981) of 1918, concrete mixes have been described in quite a few ways, by cement, water and aggregate contents as shown by equation 1, with the parameters Z and n, deduced in table 1.

$$q = [1 - Z(wc + vc)]^n \qquad (1)$$

Table 1. Equation 1 parameters (Z, n) values for estimating the aggregate volume concentrations.

Design codes	Mix specifications	Z	n
Any Code	(C, A, W)	full specification.	
BS8110	(wc, C)	C	+1
CP110	(C, A)	use mix specs.	
CP114	(wc, 1:a:b) by vol.	$-D_2/(a+b)$	-1
CP114	(wc, 1:a:b) by weight	$-D_c/(a+b)$	-1
Road Notes	(wc, Ac)	$-D_2/A_c$	-1
Euro codes	min C	producer decision	

In the 3 papers, aggregate volume concentrations, q, are calculated from equations 2.

$$q = A/D_2 = 1 - C(wc + vc) = (1 - W)(1 + vc/wc) \qquad (2)$$

where
$D_2 = $ density of aggregate (kg/m^3)
$D_c = $ density of cement (kg/m^3)
$C = $ Cement Content (kg/m^3)
$W = $ Water Content (kg/m^3)
$A = $ Aggregate Content (kg/m^3)
$wc = $ free water/cement ratio
1: a: b = 1: fines: coarse ratio for aggregate
$A_c = $ aggregate/cement ratio

2 RESPONSE & RESISTANCE TO LOADS

The *response to loads* is by the power series analysis of two-phase models (eqn 4 in 1st & 2nd papers). By

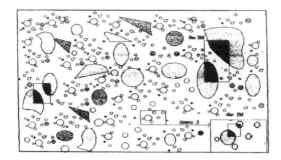

Figure 1. The interparticle model is tessellated from a surface of concrete.

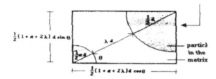

The interparticle model - the IM

The Mix Composition

Engineering Composite size range d > 30mm	Physical Structure 30mm < d < 10 μm	Chemical Structure d < 10 μm
particle volume k	$q = 1 - C_D(wc + vc)$ $C_D vc$	$C_D vc S$
steel	aggregate	blended cement — silica
		$C_D vc A$ alumina
reinforced		
concrete — CONCRETE	CEMENT PASTE — WATER	LIME
matrix volume 1	$1 - q = C_D(wc + vc)$ $C_D wc$	$C_D vc L$

* relative cement content C_D = cement content C kg/m³ / 1000 kg/m³

Figure 2. The volume composition of two phase concrete.

linear elastic analysis (Illston et al. 1979), it shows steady state upper and lower bound responses to loads by deformation or flow/no flow by permeability, porosity or density.

In structural design, loads imposed on all structural elements are estimated by linear elastic or plastic analyses – with beams as horizontal lines, columns as vertical and struts as inclined lines. For mix design, the interparticle model – the **IM** – as on figure 1 is the tessellated structural element which models the concrete mix just as beams, columns and the struts.

The **IM** non-linearly interpolates upper and lower bounds to represent the mix at the engineering, physical and chemical levels, by two-phase models, as shown on fig 2, with particle and matrix volume concentration relations in rows 2 & 4 of figure 2.

The (steel + concrete) model has viable alternatives. The (aggregate + cement paste) and (lime + silica) or (lime + alumina) models are now viable here so that e.g. CEMENT 4933 has 49% lime and 33% silica.

2.1 Mix composition and particle structure

The particle volume concentration q is calculated by the scale factoring principle, that if a volume is q, then

$q^{1/3}$ is the linear scale factor, and $q^{2/3}$ is the area scale factor, so that as shown on figure 1:

$$q^{2/3} = \left(\frac{\text{area of particles in the } \mathbf{IM}}{\text{area of the unit } \mathbf{IM}} \right) \tag{3}$$

Generally, for any particles in the matrix, since both θ and α vary, it follows that

$$q = \left(\frac{\pi(1 + \alpha^2)}{4(1 + \alpha + 2\lambda)^2 \cos\theta \sin\theta} \right)^{3/2} \tag{4}$$

Where   α is ultrafine particle size factor in the **IM**.
λ is interparticle spacing factor in the **IM**.
θ is interparticle centres angle in the **IM**.

Specifically, for uniformly mixed similar particles in the matrix, if θ = 60° and α = 1 then

$$q = \frac{0.864}{(1 + \lambda)^3} \tag{5}$$

Assume that the **IM** for uniform mixes has θ = 60° then $\cos\theta \sin\theta = \sqrt{3}/4$.

The function for the composition and structure by the volume concentration of cement particles in the cement paste matrix and ultrafine size factor α is

$$\left(\frac{vc}{wc + vc} \right)^2 = \left(\frac{\pi}{\sqrt{3}} \frac{(1 + \alpha^2)}{(1 + \alpha + 2\lambda)^2} \right)^3 \tag{6}$$

2.2 Mix composition and concrete quality

Quality has 2 *steady states*: linear and non-linear that are estimated by power functions and the exclusive power functions, which take account of by-pass flows around particles. There is the *ultimate state* compressive strength, fcm, to estimate empirically.

For the mix design, equation 10, (1st of 3 papers) is re-structured to

$$\left(\frac{vc}{wc + vc} \right)^2 = \frac{fcm}{825 \sqrt{(LS)}} \tag{7}$$

equation 12, (1st of 3 papers) is re-structured to

$$\left(\frac{vc}{wc + vc} \right)^2 = \frac{fcm}{550 [\sqrt{(LS)} + \sqrt{(LA)}]} \tag{8}$$

The free wc for any cement with L% lime, S% silica and A% alumina is calculated in mix design from equations 7 & 8 for 28 day required strengths fcm.

Table 2. Min/max and Stable Mix Compositions.

Composition	Min ACT**	Stable range WARNING*	Max ACT**
Water kg/m^3	135	155 to 210	305
Cement kg/m^3	290	480 to 650	665
Aggregate %vol	0%	58% to 69%	86%
free wc	0.14	0.24 to 0.44	0.72

ACT** for ACTION prevents low unpredictable quality
WARNING* prevents uncontrollable, but secures stable quality.

2.3 Concrete quality and particle structure

This is the design of the Particle Structure – the **IM**.

Eqs 6 & 7 for 28 day strength fem are re-structured to

$$\frac{fcm}{825\sqrt{(LS)}} = \left(\frac{\pi}{\sqrt{3}} \frac{(1+\alpha^2)}{(1+\alpha+2\lambda)^2}\right)^3 \quad (9)$$

and re-structured to include the alumina A% as

$$\frac{fcm}{550[\sqrt{(LS)}+\sqrt{(LA)}]} = \left(\frac{\pi}{\sqrt{3}} \frac{(1+\alpha^2)}{(1+\alpha+2\lambda)^2}\right)^3 \quad (10)$$

3 LIMIT SURFACES FOR MIX DESIGN

Figures 4, 5 and 6 are limit surfaces for concretes.

A limit surface is used to **design** by point estimation. It complies with the Action and Warning Limits of mix compositions obtained by power series analysis – or linear elastic analysis summarized on table 2.

It has 2 of 4 mix variables (q, wc, C, W), as x-y axes for the (aggregate, free wc, cement, water) as shown on figure 3. The 3 shown have (wc, q) as x-y axes with intersecting network contours of (C, W) for the Cement and Water Contents of the designed mix.

3.1 Mix descriptions

Figure 4 describes mixes using engineers' jargon.

Workability – Water Contents (w-contours)
It is a **Normal** mix
or **Dry** if W ≤ 160 kg/m^3 and **Wet** if W ≥ 200 kg/m^3
Economy – Cement Contents (c-contours)
It is a **Medium** mix or **Lean** if C ≤ 350 kg/m^3 and **Rich** if C ≥ 450 kg/m^3
Response to Loads -Aggregate Contents (y-axis)
It is an **Optimum** stable mix
or **Basic** if q ≤ 1/(e-1) and **Critical** stable if q ≥ ln 2
Durability- free wc (x-axis).
It is **Durable** if wc≤0.46 or **not Durable** if wc≥ 0.46.

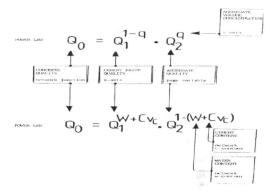

Figure 3. Limit Surface Axes and the Power Functions.

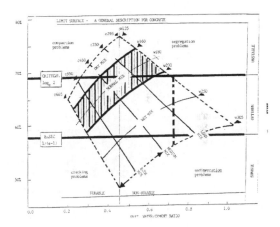

Figure 4. Limit surface for mix descriptions.

If the need is for a Dry, Durable, Normal, Rich Mix a likely point estimate is Mix (0.31, 67%). i.e. OPC C60, 50 mm slump, W = 160, C = 515 kg/m^3.

3.2 Mix strengths

Figure 5 is a 28 day strength limit surface. It shows the strengths at network junctions for the mix water and cement contents. Its boundaries are the min/max mix compositions, indicate the ACTION LIMITS to prevent unpredictably low strengths. The internal boundaries indicate the WARNING LIMITS for mix compositions to prevent uncontrollable strengths but

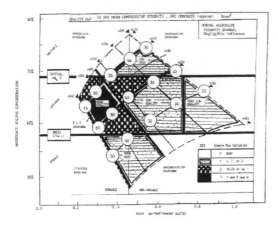

Figure 5. Limit surface for 28 days strengths of concrete.

secure stable quality as on table 2. A limit surface, like a fish, shows high & low strength and high & low cost regions as cement and water contents vary. There is a very low strength region at the fish tail and very high strengths at the fish head. Mix designs lying outside the limit surface or not inside the fish shape will satisfy research & development interests.

3.3 Detection of mix problems

Fig 6 is the limit surface that detects the extent and type of mix problems. The 4 problems occur near the *min/max mix composition* boundary at the 4 corners. All 4 mix problems are:-

– prevented by those mix point estimates within the internal boundaries of *stable mix composition*.
– Segregation, Compaction, Sedimentation, Cracking occurs as the mix position vectors, $\mathbf{R}$cisψ, satisfy the conditions laid down on table 3 below.

The central problem-free mix is (wc = 0.46, q = 65%). A mix positioned in the 10% to 15% target circles violates the *min/max mix compositions* so the extents of mix problems are *major* to *severe*. The mixes in the 10% target circle have *stable compositions* so the extents of problems, if any, are *minor*.

Relative to problem-free (0.45, 65%), mix (wc, q) has polar co-ordinates $\mathbf{R}$cis ψ to detect the problem type, by ψ and its extent, by $\mathbf{R}$,, as defined in table 3.

Complex $\quad \mathbf{R}$ cis $\psi = \mathbf{R}$ cos $\psi + i\,\mathbf{R}$ sin $\psi \quad$ (11)
and $\quad \mathbf{R}^2 = [(wc - 0.45)^2 + (q - 0.65)^2] \quad$ (12)
with ψ = arctan [(q–0.65)/ (wc–0.45)] +180° n (13)

where n = 1, and if wc $\geq$ 0.45, then n = 0.

On figure 6, mix (wc = 0.35, q = 53%) shown by the (+) has *severe cracking* problems – ***severe cracking***? Adjust to (wc = 0.35, q = 65%) causes *major cracking*. It is economic, but its *resistance* to loads is reduced

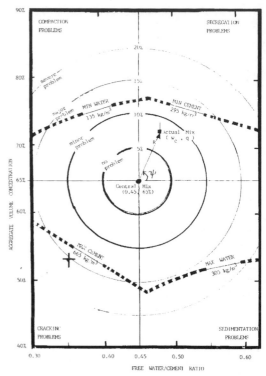

THE PROBLEM-IDENTIFICATION LIMIT SURFACE
FOR CONCRETE MIXES

Figure 6. Limit surface with target circles detects concrete mix problems.

Table 3. The 4 Concrete Mix Problems.

Mix problem	Angle ψ	Extent	R value
Segregation	0° to 90°	None	$\leq 5\%$
Compaction	90° to180°	Minor	$5\% - 10\%$
Sedimentation	0° to $-90°$	Major	$10\% - 15\%$
Cracking	$-90°$ to $-180°$	Severe	$\geq 15\%$

with the higher aggregate volume from 53% to 65%. Using mix position vector, $\mathbf{R}$ cisψ, the limit surface, figure 6, detects mix problems after each mix design.

4 POWER FUNCTIONS FOR MIX DESIGN

The *mix designer* decides the *production control* type as Research, Ready-Mixed, Good, Fair or Poor in order to assign safety factors of 1 to 1.5, for the published strength equations used for programming.

The free wc is calculated from equations 7 & 8 for the required strength fcm. The water content W is calculated by power functions and two-phase **IM** that assume the concrete cone slumps by self-weight.

Admixtures are added to control the Water, Cement or Aggregate contents, as well as the response to elastic loads, brittleness and chemical reliability to indicate durability as 1− (free wc/lime). If wc > lime, then the negatives are chemically unreliable mixes.

The aggregate is proportioned by power functions and two-phase **IM** which use stochastic fractions that fit the BS 882 aggregate grading curves (Kong & Evans 1979).

The range of loads for no cracks is predicted by power functions of stress-strain derived from the two-phase **IM** for a decreased elastic matrix, but an increased cracks/inelastic inclusions at higher loads.

The cements are blended on a risk diagram deduced by linear algebra which warns of unstable cements as strengthening pozolanic reactions occur in two-phase **IM** cements type (L, S) or (L, A). The risk diagram is drawn on x-y axes of Lime vs. (Silica + Alumina)

Limit surfaces with C-W, C-wc, as external x-y axes and internal intersecting contours of the other two mix variables on fig 3, are drawn for other quality and costs. Designs must specify strength, age, slump, aggregate type with density and blend the cements. Curing temperature relations will be included later.

5   CONCLUSIONS

Except for strength and aggregate, two similar mixes conclude the mix design with power functions and the two-phase IM.

MIX COMPOSITION

Designed mix to Research Control.
C 50/150 mm/ 28 day/Cement 4933, vc = 0.34
Lightweight Concrete Batch Vol = 1 m^3

CEMENT 4933	565 R->	475	kg/m^3
free WATER	293 R->	210	
20 mm ssd Lytag		650	
10 mm ssd Lytag		370	
5 mm ssd Sand		205	
FEBFLOW SP3 for 28% wR-> 4.3 litres/m^3			

CEMENT 4933	OPC	csf	pfa	ggbs	metka	+?
kg/m^3	335	50	50	50	0.0	0.0

^{3} RELATIVE DENSITIES ^{3} Cement Coarse Fine ^{3}
^{3} Fresh Mix Fresh Paste^3 4933 Agg   Agg^3
   ^{3} 1.91   1.85 ^{3} 2.95 1.60   2.60^3

MIX QUALITY CONTROL

2.1 fcu = 50 MPa   fcm = 55 to 60 MPa
    Designed mix with 3 stable ingredients
    Mix Adjustments (kg/m^3)

WATER = −85,   CEMENT = −90,   AGGREGATE = +225
Reduced free Wc from 0.52 to 0.44

2.2 Composition: [Wc, AgVol] = [0.44, 62%]
    Collapse SLUMP of 150 mm. Very high. Self-levelling
2.3 Major sedimentation 0.10 cis −11°
    CEMENT 4933 reduces thermal effects by 25%
    FEBFLOW SP3 enhances workability.
2.4 NORMAL, Mix
    Chemical Reliability is 11 % and Brittleness is 3.3
    Linear elastic range [0 MPa to 15 MPa] uncracked
2.5 USE for Plain Concrete, for R.C, for P.C.

Major Sedimentation, 0.10 cis −11°

MIX COMPOSITION

Designed mix to Research Control.
C100/150 mm/ 28 day/Cement 4933, vc = 0.34
Normalweight Concrete Batch Vol = 1 m^3

CEMENT 4933	970 R->	820* kg/m^3
free WATER	363 R->	260*
10 mm ssd Gravel		815
5 mm ssd Sand		390
FEBFLOW SP3 for 28 % wR-> 7.4 litres/m^3		

CEMENT 4933	OPC	csf	pfa	ggbs	metka	+?
kg/m^3	575	80	80	80	0.0	0.0

^{3} RELATIVE DENSITIES ^{3} Cement Coarse Fine ^{3}
^{3} Fresh Mix   Fresh Paste ^{3} 4933   Agg   Agg ^{3}
^{3} 2.28   2.01 ^{3} 2.95   2.60   2.60^3
KEY: * = unstable, R-> = reduction

MIX QUALITY CONTROL

2.1 fcu = 100 MPa fcm = 110 to 115 MPa
    Mix Adjustments (kg/m^3)

    WATER = −105, CEMENT = −150,
    AGGREGATE = +405
    Reduced free Wc from 0.38 to 0.32

2.2 Composition: [Wc, AgVol] = [0.32, 46%]
    Collapse SLUMP of 150 mm. Very high. Self-levelling
2.3 Severe cracking 0.19 cis 263°
    CEMENT 4933 reduces thermal effects by 25%
    FEBFLOW SP3 enhances workability
2.4 WET, Rich Mix
    Chemical Reliability is 36 % and Brittleness is 4.4
    Linear elastic range [0 MPa to 45 MPa] uncracked
    WARNING – CEMENT is over 600 kg/m^3
2.5 USE for Plain Concrete, for R.C, for P.C.
    Severe Cracking, 0.19 cis 263°

REFERENCES

ABRAMS DA Design of Concrete Mixtures, Structural Materials, Res Lab. Lewis, Institute Bulletin, 1, Chicago 1918

ALLEN AH Reinforced Concrete Design to BS 8110: E & F N Spon, 1988, p 1, Chpt 1.

ILLSTON JM, DINWOODIE JM, SMITH AA Concrete, Timber & Metals, Van Nostrand, 1979, pp. 280–285.

KONG FK, EVANS RH Reinforced & Prestressed Concrete, Nelson, 1975, p11 & 19.

Excellence in Concrete Construction through Innovation – Limbachiya & Kew (eds)
© *2009 Taylor & Francis Group, London, ISBN 978-0-415-47592-1*

Improvement in the compressive strength of cement mortar by the use of a microorganism – *Bacillus megaterium*

V. Achal, R. Siddique, M.S. Reddy & A. Mukherjee
Thapar University, Patiala (Punjab), India

ABSTRACT: This study reports the results of compressive strength of cement mortars incorporating microorganism. The effect of addition of microorganism, *Bacillus megaterium*, on the compressive strength of cement mortar cube has been studied. Ordinary Portland cement (OPC) was used to prepare mortar with different cell concentration of microorganism in the mixing water. A significant increase in the compressive strength of cement mortar cube at different ages (3, 7, 14 and 28 days) was achieved with the addition of *B. megaterium*. Increases in compressive strength were observed maximum in case of 10^5 cells/ml of microorganism. This improvement in compressive strength is due to deposition on the microorganism cell surfaces and within the pores of cement–sand matrix.

1 INTRODUCTION

Portland cement concrete has clearly emerged as the material of choice for the construction in the world today. This is mainly due to low cost of materials and construction for concrete structure as well as low cost of maintenance. Therefore, much advancement of Concrete Technology have occurred depending on (i) the speed of construction (ii) the strength of concrete (iii) the durability of concrete and (iv) the environmental friendliness of industrial material like, fly ash, blast furnace slag, silica fume, metakaolin etc. (Mehta 1999).

Compressive strength is one the important parameters in determining the strength of building materials and structures. Compressive strength is relatively easy to measure, and it commonly relates to some other properties, such as tensile strength and absorption of the mortar. The compressive strength of mortar depends largely upon the cement content and the water-cement ratio. The mean compressive strength required at a specified age, usually 28 days, determines the nominal water-cement ratio of the mix. The water-cement ratio determines the porosity of hardened cement paste at any stage of hydration. Thus the water-cement ratio affects the volume of voids in concrete thereby influencing the strength of concrete.

Microbial induced carbonate mineralization has been proposed as a novel and eco-friendly strategy for the protection and remediation of stone and mortar (Adolphe et al. 1990). Microbial mineral precipitation (biodeposition) involves various microbes, pathways and environments. Ureolytic bacteria are found suitable for carbonate (calcite) precipitation. These bacteria are able to influence the precipitation of calcium carbonate by the production of an enzyme, urease (urea amidohydrolase, EC 3.5.1.5). This enzyme catalyzes the hydrolysis of urea to CO_2 and ammonia, resulting in an increase of the pH and carbonate concentration in the bacterial environment (Stocks-Fischer et al. 1999). There are few *Bacillus* species which are considered to produce high amount of urease which helps in calcite precipitation. *Bacillus megaterium* is a rod-shaped, Gram-positive species of bacteria. *B. megaterium* is generally considered a soil microbe (Vary 1994) and has been shown to precipitate carbonate minerals (Cacchio et al. 1993).

Ghosh et al. (2005) reported a significant improvement in compressive strength of cement mortar using *Shewanella* species.

The present work deals with the compressive strength of cement mortars, which is one of the most important parameters influencing the strength of building materials and structures and finally its performance. An attempt also has been made to observe the effect of different concentration of microorganism, *Bacillus megaterium* on the compressive strength of cement mortars.

2 MATERIALS AND METHODS

2.1 *Mortars and concrete specimens*

Ordinary Portland cement (OPC) conforming to IS 12269 (Bureau of Indian Standard, New Delhi, 1987)

Table 1. Chemical compositions of ordinary Portland cement (OPC) used in this study.

Chemical	Constituent (%)
CaO	63.50
SiO_2	19.10
Fe_2O_3	2.90
Al_2O_3	4.00
MgO	2.80
SO_3	2.60
$Na_2O + K_2O$	0.80

Table 2. Physical characteristics of ordinary Portland cement (OPC) used in this study.

Physical property	Value
Consistency of standard cement paste	36%
Soundness of cement	
Initial setting time	3.9 mm
Final setting time	123 min
Fineness by air permeability (cm^2/g)	174 min
	2690 cm^2/g

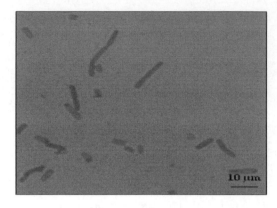

Figure 1. Rod shaped *B. megaterium* as seen under phase contrast microscope.

was used. Chemical and physical properties of cement are presented in Table 1 and Table 2 respectively. Sand from Bhakra canal, Punjab, India having fineness modulus of 2.89 conforming to IS 383-1970 (Bureau of Indian Standard, New Delhi, 1970) was used as fine aggregate. Sterile milliQ water was used through out the experiments to prepare mortars.

2.2 *Microorganism and growth conditions*

Bacillus megaterium MTCC 1684 was used in this study. Morphological study (shape and size) of this bacterium was studied by observing under phase contrast microscopy (Leica DM LS2). The culture was maintained in Nutrient Agar medium (Himedia, Mumbai, India). The final pH of the culture was around 7.4. Liquid culture media consisted of 13 g/L nutrient broth powder (Himedia, Mumbai, India). All the media were sterilized by autoclaving at 121°C for 15 min. Cultures were incubated over-night at 37°C on a shaker at 120 rpm.

2.3 *Preparation of test specimens*

Mortar cubes were prepared with ordinary Portland cement (OPC) of size 70.6 mm as per IS 4031-1988 (Bureau of Indian Standard, New Delhi, 1988). The cement to sand ratio was used as 1:3 (by weight), and the water to cement ratio was fixed at 0.4. Different concentrations of *B. megaterium* MTCC 1684 were used to prepared mortar cubes. The cell concentration was determined from the bacterial growth curve made

by observing optical density at 600 nm and bacterial cells were counted manually. Different concentrations of cells (10^3, 10^5 and 10^7 cells/ml) were obtained by growing culture for different time intervals followed by centrifugation at 8000 rpm for 10 min at 4°C. At the conclusion of centrifuging, the supernatant was removed and cells were washed with milliQ water. In all the cube specimens, cells were mixed along with water. Control mortar cubes were cast without the addition of microbes. All the experiments were performed in triplicates.

2.4 *Compressive strength test*

Mortar cubes were cast and compacted in a vibration machine and after demolding all specimens were cured in air at room temperature until compression testing at 3, 7, 14 and 28 days. Compression testing was performed using automatic compression testing machine, COMPTEST 3000.

3 DISCUSSION OF RESULTS

The main aim of the present work is to study the effect of different concentrations of *B. megaterium* MTCC 1684 on the compressive strength of cement mortar cubes at different ages (3, 7, 14 and 28 days). *B. megaterium* MTCC 1684 is a rod shaped bacterium found to size range between 3–12 μm as seen under phase contrast microscopy (Figure 1), which contains urease activity.

Figure 2 summarizes the compressive strength of cement mortar cubes using microbes and compare with the control. The compressive strength was found to increase by the addition of *B. megaterium* MTCC 1684 as compare to control, where no microorganisms were added. Compressive strength was found to

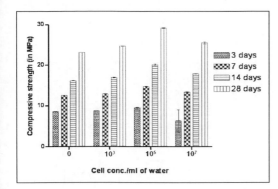

Figure 2. Effect of different concentrations of *B. mega-terium* MTCC 1684 on the compressive strength of cement mortar cubes at different ages (3, 7, 14 and 28 days). Error bars show standard deviation (n = 3).

increase by the addition of microorganism. The greatest improvement in compressive strength occurs at cell concentrations of 10^5 cells/ml for all ages.

This improvement in compressive strength is probably due to deposition on the microorganism cell surfaces and within the pores of cement–sand matrix, which plug the pores within the mortar (Ramachandran et al. 2001, Ramakrishnan et al. 1998, 1999).

4 CONCLUSIONS

Based on the tests performed in this study the following conclusions and suggestions could be drawn:

a) Microorganism like *B. megaterium* could be used to increase the compressive strength of cement mortar thus building materials and structures.
b) According to the results of this study, 10^5 cells/ml of *B. megaterium* MTCC 1684 was proposed as an optimum concentration to achieve maximum compressive strength.
c) This technique could be further exploit to get better results.
d) More researches on the other durability characteristics like permeability, carbonation should be performed in the future.

REFERENCES

Mehta, P.K. 1999. Advancement in concrete technology, *Journal of Concrete International*. 69–75.
Adolphe, J.M., Loubiere, J.F., Paradas, J., Soleilhavoup, F. 1990. Procede de traitement biologique d'une surface artificielle. *European patent* 90400G97.0 (after French patent 8903517, 1989).
Stocks-Fischer, S., Galinat, J.K., Bang, S.S. 1999. Microbiological precipitation of $CaCO_3$. *Soil Biol Biochem.* 31(11):1563–71.
Ghosh, P., Mandal, S., Chattopadhyay, B.D., Pal, S. 2005. Use of microorganism to improve the strength of cement mortar. *Cement and Concrete Research*. 35:1980–1983.
Vary, P.S. 1994. Prime time for *Bacillus megaterium*. *Microbiology*. 140, 1001–1113.
Cacchio, P., Ercole, C., Cappuccio, G., Lepidi, A. 2003. Calcium carbonate precipitation by bacterial strains isolated from a limestone cave and from a loamy soil. *Geomicrobiol. J.* 20, 85–99.
IS 12269: Specification for 53-grade Ordinary Portland cement, *Bureau of Indian Standard*, New Delhi, 1987.
IS 383: Specification for coarse and fine aggregates from natural sources for concrete, Bureau of Indian Standard, New Delhi, 1970.
IS 4031: Determination of compressive strength of hydraulic cement, *Bureau of Indian Standard*, New Delhi, 1988.
Ramachandran, S.K., Ramakrishnan, V., Bang, S.S. 2001. Remediation of concrete using microorganisms, *ACI Materials Journal*. 98 (1):3–9.
Ramakrishnan, V., Bang, S.S., Deo, K.S. 1998. A novel technique for repairing cracks in high performance concrete using bacteria, *Proceeding of the International Conference on High Performance*, High Strength Concrete, Perth, Australia, pp. 597–617.
Ramakrishnan, V., Deo, K.S., Duke, E.F., Bang, S.S. 1999. SEM investigation of microbial calcite precipitation in cement, *Proceeding of the International Conference on Cement Microscopy*, Las Vegas, pp. 406–414.

Excellence in Concrete Construction through Innovation – Limbachiya & Kew (eds)
© 2009 Taylor & Francis Group, London, ISBN 978-0-415-47592-1

Nonlinear analysis of ultra high strength concrete RC structure

H.G. Kwak & C.K. Na
Civil and Environment Engineering in KAST, Daejeon, Korea

S.W. Kim & S.T. Kang
KICT, Seoul, Korea

ABSTRACT: A numerical model that can simulate the nonlinear behavior of ultra high performance steel fiber reinforced concrete (UHPSFRC) structures subjected to monotonic loading is introduced. A criterion to take into account the biaxial behavior of UHPSFRC is designed on the basis of Hussein's experimental result, and the equivalent uniaxial stress-strain relationship is introduced for proper estimation of UHPSFRC structures. The steel is uniformly distributed over the concrete matrix with particular orientation angle. In advance, this paper introduces a numerical model that can simulate the tension-stiffening behavior of tension part of axial member on the basis of the bond-slip relationship. The reaction of steel fiber is considered for the numerical model after cracks of the concrete matrix with steel fibers are formed. Finally, the introduced numerical models are validated by comparison with test results for idealized UHPSFRC beams, and additional parametric studies are followed to review the structural behavior of UHPSFRC beam according to the change in material properties.

1 INTRODUCTION

A rapidly increased construction of long-span bridges and high-rise buildings requires a strength increase of construction materials, and concrete which has become one of the most important construction materials and is widely used in many types of engineering structures will not be exceptional. After introduction of normal strength concrete (NSC) with compression strength ranged from 20 MPa to 40 MPa, its strength has been continuously increased and ultrahigh strength concrete (UHSC) with compression strength of more than 100 MPa has been used experimentally in practice (ACI committee 363 1997, KICT 2005, Wee et al. 1996). However, due to an increase of brittleness in proportional to an increase of compressive strength, very limited application of ultra-high strength concrete has been performed. Accordingly, in attempts to improve the mechanical properties of UHSC, such as strength, stiffness and ductility, ultra-high performance steel fiber reinforced concrete (UHPSFRC) has been developed by adding steel fiber at different volume fractions to the concrete matrix (Kolle et al. 2004).

To be used as a construction material, however, the structural behavior of UHPSFRC members as well as the material properties of UHPSFRC itself needs to be verified, and a lot of relevant experimental studies have been conducted (Kolle et al. 2004, Wee et al. 1996). Within the framework of developing advanced design and analysis methods for UHPSFRC structures, the need for experimental study continues. Experiments provide a firm basis for design equations and also supply the basic information for numerical analysis, such as material properties. In addition, the results of numerical analyses have to be evaluated by comparing them with experiment of full-scale models of structural sub-assemblages or entire structures.

Nevertheless, very little work has been done on the structural behavior of UHPSFRC systems on the basis of finite element analysis, because of the computational effort involved and the insufficient knowledge for the material behavior of UHPSFRC under biaxial stress states. Recognizing that many of the material models for biaxial loading condition have not been fully verified so far, it is one of the intent of this paper to address some of the model selection issues in the numerical analyses of UHPSFRC structures, in particular, with regard to the strength of reinforcing steel and the tension stiffening effect in concrete.

Finally, the introduced numerical models are validated by comparison with test results for idealized UHPSFRC beams.

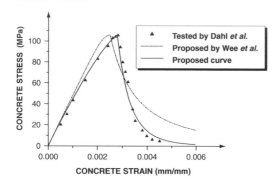

Figure 1. Compressive stress-strain relationship of UHPSFRC.

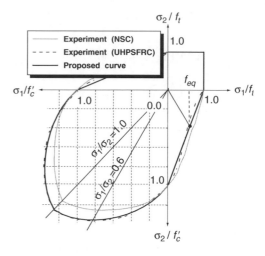

Figure 2. Biaxial strength failure envelope of UHPSFRC.

2 MATERIAL MODEL

2.1 Uniaxial compression of concrete

The uniaxial stress-strain curves of normal strength concrete (NSC) have been proposed on the basis of experimental studies and a lot of idealized curves have also been introduced (Carreira, et al. 1985, Hognestad et al. 1951). Uniformly in the cases of high strength concrete (HSC) and UHPSFRC, the same experimental studies have been conducted to describe the stress-strain relationship of corresponding concrete (Dahl et al. 1992, Kolle et al. 2004). As shown in Figure 1 representing typical compressive stress-strain relationships, and increases of compressive strength accompanies a rapid decrease of ductility in the strain softening region.

In describing the uniaxial compressive stress-strain behavior of UHPSFRC, more attention needs to be given to the definition of strain softening region. However, differently from HSC, it was too much difficult to find any published general equation to predict the complete stress-strain curve of UHPS-FRC in the existing literature. On these backgrounds, this paper introduces a relationship in Eq. (1) with a slight modification of the equation proposed by many researchers (ACI committee 363 1997, Carreira et al. 1985, Hognestad et al. 1951, Wee et al. 1996) on the basis of related experimental data (Dahl et al. 1992), to simulate the nonlinear behavior of structural member composed of UHPSFRC.

$$f = f_c' \left[1 - (1 - \varepsilon / \varepsilon_c)^\alpha \right] \quad \text{for } 0 \leq \varepsilon \leq \varepsilon_c$$

$$f = f_c' \frac{k\beta(\varepsilon / \varepsilon_c)}{k\beta - 1 + (\varepsilon / \varepsilon_c)^\beta} \quad \text{for } \varepsilon \geq \varepsilon_c \quad (1)$$

where $\alpha = \ln(1 - \eta) / \ln(1 - \eta/(\varepsilon_c E_c / f_c'))$; $\eta = f / f_c'$; $\beta = 1/(1 - (f_c' / \varepsilon_c E_c))$; and $k = (50/f_c')^{2.5}$.

2.2 Biaxial stress state of concrete

With an uniaxial stress-strain relationship of concrete, material behavior of concrete under biaxial loading needs to be defined because concrete exhibits strength and stress-strain behavior somewhat different from concrete under uniaxial loading condition due to the effects of Possion's ratio and micro crack confinement (Chen 1982). Fig. 2 shows the biaxial strength failure envelope of UHPSFRC introduced in this paper with that of NSC to compare their differences, and the equation for the failure envelope in the compression-compression region is expressed by

$$f_{2p} = (0.25\zeta^3 - 1.25\zeta^2 + 1.25\zeta + 1) \cdot f_c', \quad f_{1p} = \zeta \cdot f_{2p} \quad (2)$$

where $\zeta = (f_1/f_2) = $ the principal stress ratio; and $f_{ip} = $ the maximum equivalent principal stresses corresponding to the current principal stresses f_i.

After determination of the equivalent concrete compressive strength of f_{1p} and f_{2p} from the biaxial failure surface of UHPSFRC, the equivalent uniaxial stress-strain relationship in the compression-compression region, corresponding to current loading history, is constructed by replacing the compressive strength in Figure 1 with the equivalent compressive strength f_{ip}. In the compression-tension and the tension-tension region, however, the following assumptions are adopted in this paper because the response of typical UHPSFRC members is much more affected by the tensile behavior than the compressive behavior of concrete: (1) failure takes place by cracking when the principal tensile strain exceeds the limit strain and therefore, the tensile behavior of concrete dominates the response; (2) the uniaxial tensile strength of concrete is reduced to the value f_{eq}, as shown in Figure 2, to account for the effect of the compressive stress under biaxial state; and (3) the concrete stress-strain relationship in compression is the same as under uniaxial loading and does not change with increasing principal tensile stress.

The constitutive relationship is intended for simulating the variation of material property corresponding to the stress change. The biaxial stress-strain relationships for concrete material are idealized as incrementally orthotropic, with the axes of orthotropy aligned with the current principal direction of total strain. Thus, the incremental constitutive relationship takes the form (Kwak and Kim 2004, Kwak and Na 2007):

$$
\begin{Bmatrix} d\sigma_1 \\ d\sigma_2 \\ d\tau_{12} \end{Bmatrix} = \frac{1}{1-v^2} \begin{bmatrix} E_1 & v\sqrt{E_1E_2} & 0 \\ v\sqrt{E_1E_2} & E_2 & 0 \\ 0 & 0 & (1-v^2)\cdot G \end{bmatrix} \begin{Bmatrix} d\varepsilon_1 \\ d\varepsilon_2 \\ d\gamma_{12} \end{Bmatrix} \quad (3)
$$

where $(1-v^2)G = 0.25(E_1 + E_2 - 2v\sqrt{E_1E_2})$; $E_i =$ the secant moduli in the direction of the axes of orthotropy; $G =$ the secant shear modulus; and $v =$ Poisson's ratio.

The proposed concrete model accounts for progressive cracking and changes in the crack direction by assuming that the crack is always normal to the total principal strain direction (the rotating crack model); that is the shear strain with normal stress in the cracked stage is ignored.

2.3 Steel and pre-stressing steel

The stress-strain curves for steel are generally assumed to be identical in tension and compression. For simplicity in calculations, it is necessary to idealize the one-dimensional stress-strain curve for the steel element. The normal strength steel is usually assumed to be a linear elastic, linear strain hardening material. However, the normal strength steel embedded in concrete matrix represents different behavior from the bared steel bar because of bond interaction along the steel bar between adjacent two cracks, and it means that the averaged yield stress, which is significantly less than yielding stress, must be used to eliminate an overestimation of the post-yielding behavior of concrete structures in the case of considering the tension-stiffening effect into the stress-strain relationship of concrete. In advance, the pre-stressing tendon is approximated by a series of straight tendon segments maintaining a constant force and sectional area.

3 TENSION STIFFENING MODEL

3.1 Force equilibrium of axial member

A cracked ultra high performance steel fiber reinforced concrete element subjected to axial stress is shown in Figure 4(a), and a part of the element along the

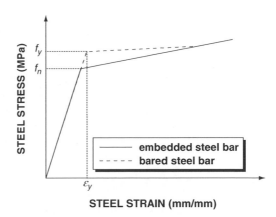

Figure 3. Steel stress-strain relationship.

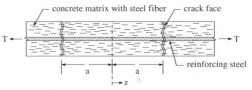

(a) Cracked axial element

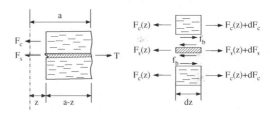

(b) Force equilibrium   (c) Free body diagram

Figure 4. UHPSFRC axial member.

crack faces where the z direction is normal to the crack surface with crack spacing of 2a can be taken as the free body diagram (see Figure 4(b)).

Since the applied principal tensile force of the element T is carried partly by the concrete matrix with steel fiber (F_c) and partly by the reinforcing steel (F_s), the following force equilibrium equation is obtained.

$$
T = F_s + F_c = A_s E_s \varepsilon_s + A_c E_c \varepsilon_c = A_s E_s \frac{du_s}{dz} + A_c E_c \frac{du_c}{dz} \quad (4)
$$

where $E =$ elastic modulus; $\varepsilon =$ strain; $u =$ displacement; $A =$ area; $s =$ steel; and $c =$ concrete.

An infinitesimal element of length dz is taken out from the intact concrete between cracks to obtained

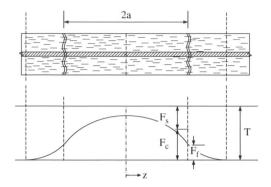

Figure 5. Force distribution of UHPSFRC between cracks.

the equilibrium equations for the concrete and steel (Figure 4(c)). Base on the force equilibrium at the steel-concrete interface, the variation of the resisting force by the reinforcing steel and concrete can be derived in terms of the bond stresses, where $p =$ perimeter, $m =$ the number of steel and $f_b =$ the bond stress with linear ($= E_b\Delta$) or constant value ($= \tau_b$).

$$\frac{dF_s}{dz} = pmf_b, \quad \frac{dF_c}{dz} = -pmf_b \qquad (5)$$

3.2 Linear bond-slip relationship case

Since the bond-slip Δ at the steel-concrete interface is defined by the relative displacement between the reinforcing steel and concrete ($\Delta = u_s - u_c$), the second order differential equation of bond slip with linear bond stress relationship leads to

$$\frac{d^2\Delta}{dz^2} = \frac{d^2 u_s}{dz^2} - \frac{d^2 u_c}{dz^2}, \frac{d^2\Delta}{dz^2} - k^2\Delta = 0, k^2$$

$$= \frac{pmE_b}{A_s E_s}(1 + n\rho) \qquad (6)$$

$\Delta = C_1 \sinh kz + C_2 \cosh kz$ is the general solution, and $C_2 = 0$ is determined because of the slip should be zero at the center. Integration after substituting the obtained general solution leads to the following expressions for the steel and concrete forces F_s and F_s.

$$F_s = \frac{pmE_b C_1}{k}\cosh kz + C_3, \quad F_c = T - F_s \qquad (7)$$

$C_3 = (T - F_f) - (pmE_b C_1/k)\cosh ka$ is obtained from the boundary condition at the crack surface ($F_s = T - F_f$ at $z = a$) as shown in Figure 5.

If it is assumed as the steel fiber force at the crack face (F_f) is proportional to concrete force at the center (F_s), $F_f = \delta F_c(0)$ (Bischoff 2003), the following F_f

can be obtained where δ is coefficient of steel fiber and concrete properties and can be determined by force equilibrium.

$$F_f = \frac{\delta}{1-\delta}\cdot\frac{pmE_b C_1}{k}(\cosh ka - 1) \qquad (8)$$

Moreover, the steel and the concrete forces and displacements are obtained from Eq. (4), Eq. (7) and Eq. (8), and $C_1 = (T/A_s E_s)\cdot(1-\delta)/k(\cosh ka - \delta)$ is determined from the relationship of $\Delta = u_s - u_c$.

$$F_s = T - \frac{pmE_b C_1}{k}\left\{\frac{1}{1-\delta}(\cosh ka - \delta) - \cosh kz\right\}$$
$$F_c = \frac{pmE_b C_1}{k}\left\{\frac{1}{1-\delta}(\cosh ka - \delta) - \cosh kz\right\} \qquad (9)$$

$$u_s = \frac{Tz}{A_s E_s} - \frac{C_1}{1+n\rho}\left\{\frac{kz}{1-\delta}(\cosh ka - \delta) - \sinh kz\right\}$$
$$u_c = n\rho\cdot\frac{C_1}{1+n\rho}\left\{\frac{kz}{1-\delta}(\cosh ka - \delta) - \sinh kz\right\} \qquad (10)$$

Based on the obtained equations, a descending branch of the concrete stress-strain relationship can be determined. The force equilibrium Eq. (4) can be rewritten as Eq. (11) and the average strain in the reinforcing steel and the average stress in the concrete can be obtained as Eq. (12) and Eq. (13).

$$T = F_s + F_c = A_s E_s \varepsilon_{sm} + A_c \sigma_{cm} \qquad (11)$$

$$\varepsilon_{sm} = \frac{u_s(a)}{a} = \frac{T}{A_s E_s}\cdot\frac{1}{1+n\rho}\left\{n\rho + \frac{(1-\delta)\sinh ka}{ka(\cosh ka - \delta)}\right\} \qquad (12)$$

$$\sigma_{cm} = \frac{T}{A_s}\cdot\frac{1}{1+n\rho}\left\{1 - \frac{(1-\delta)\sinh ka}{ka(\cosh ka - \delta)}\right\} \qquad (13)$$

The maximum concrete stress and corresponding strain are obtained from $\sigma_{c,\max} = F_c(0)/A_c$ and $\varepsilon_{crack} = \sigma_{c,\max}/E_c$. Now the non-dimensional form which is point A in Fig. 6 can be determined as follows.

$$\frac{\sigma_{cm}}{\sigma_{c,\max}} = \left(1 - \frac{(1-\delta)\sinh ka}{ka(\cosh ka - \delta)}\right)\bigg/\left(1 - \frac{1-\delta}{\cosh ka - \delta}\right) \quad (14a)$$

$$\frac{\varepsilon_{cm}}{\varepsilon_{crack}} = \left(1 + \frac{(1-\delta)\sinh ka}{n\rho ka(\cosh ka - \delta)}\right)\bigg/\left(1 - \frac{1-\delta}{\cosh ka - \delta}\right) \quad (14b)$$

3.3 Constant bond-slip relationship case

Further deformation leads to the yielding of the reinforcing steel followed by the constant bond stress relationship. The steel and concrete forces and the steel displacement can be obtained as Eq. (15) and Eq. (16)

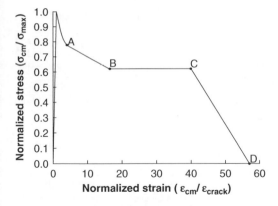

Figure 6. Average tensile stress-average strain.

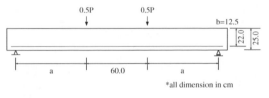

Figure 7. UHPSFRC beam.

by integrating Eq. (5) from the boundary condition at the crack surface ($F_s = T - F_f$ at $z = a$) and the steel fiber force $F_f = \delta F_c(0)$ acting at crack face can be determined.

$$F_s = T - pm\tau_b\left(\frac{a}{1-\delta} - z\right)$$
$$F_c = pm\tau_b\left(\frac{a}{1-\delta} - z\right)$$
(15)

$$u_s = \frac{Tz}{A_s E_s} - \frac{pm\tau_b z}{A_s E_s}\left(\frac{a}{1-\delta} - \frac{1}{2}z\right)$$
$$u_c = n\rho \cdot \frac{pm\tau_b z}{A_s E_s}\left(\frac{a}{1-\delta} - \frac{1}{2}z\right)$$
(16)

Moreover the same procedure of previous case is adapted, the average strain in the reinforcing steel and the average stress in the concrete can be obtained from Eq. (15) to (16), and the maximum concrete stress and corresponding strain are obtained. Now the non-dimensional form which is point B in Figure 6 can be determined as follows.

$$\frac{\sigma_{cm}}{\sigma_{c,\max}} = \frac{1+\delta}{2}$$
(17a)

$$\frac{\varepsilon_{cm}}{\varepsilon_{crack}} = \frac{1-\delta}{n\rho}\left(\frac{T}{pm\tau_b a} - \frac{1}{2}\cdot\frac{1+\delta}{1-\delta}\right)$$
(17b)

The coefficient δ is determined from the force equilibrium Eq. (18) which is the relationship between force equilibrium at the center and that at the crack face. The strain of the reinforcing steel and concrete at the center is identical values because of no slip, therefore, the steel stress and the concrete stress can be expressed by $f_s = nf_t$.

$$T = F_s + F_c = f_y A_s + \alpha f_t A_c = f_s A_s + f_t A_c$$
(18)

Hence, δ can be expressed as Eq. (19) where minus value means that the tensile strength of concrete is not improved by the steel fibers.

$$\delta = 1 - \frac{f_y - nf_t}{f_t}\rho \geq 0$$
(19)

Further deformation leads to the yielding of the steel fiber, the average strain in Eq. (20) which is point C in Figure 6 can be obtained from the steel fiber properties while maintaining the average stress of the concrete $\sigma_{cm}/\sigma_{c,\max} = (1+\delta)/2$.

$$\varepsilon_{sm} = \frac{w_m}{S_m} = \frac{(f_{y,f}/E_f)\cdot l_f}{S_m}$$
(20)

where w_m = average crack width; S_m = average spacing; $f_{y,f}$ = yield stress; E_f = modulus; and l_f = length of steel fiber.

4 NUMERICAL APPLICATION

4.1 Example properties

KICT (2005) tested several types of UHPSFRC beam as shown in Figure 7. The tests were to verify the effect of steel fiber content, steel ratio, shear span ratio and span-height ratio. In Figure 7, two reinforcing steels are located at the bottom of beam specimen to resist the tensile force. Because specimen is subjected to two point loads 30 cm apart from the center, 60 cm length of beam at the center is subjected to bending moment without shear force. The steel properties are yielding stress $f_y = 634.0$ MPa, elastic modulus $E_s = 2 \times 10^5$ MPa, and the concrete properties are compressive strength $f_c' = 146.0$ MPa, tensile strength $f_t = 13.9$ MPa with elastic modulus $E_c = 4.9 \times 10^4$ MPa. In the test, the compressive and tensile strength varies according to steel fiber contents from 1% to 3%, however, 2% of steel fiber content is used to simulate the beam specimen.

5 beam specimens are used for numerical analysis according to shear span ratio and steel ratio. The properties of steel area, length of specimen and shear span ratio are listed in Table 1. The coefficient δ of proposed tension stiffening model is 0.0 to 0.17.

Table 1. Dimensions of example.

Beam	D10L16	D10L20	D10L24	D13L24	D16L24
As	2D10	2D10	2D10	2D13	2D16
a(cm)	39.6	55.0	77.0	77.0	77.0
a/d	1.8	2.5	3.5	3.5	3.5

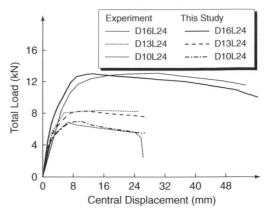

Figure 9. Load-displacement (steel ratio effect).

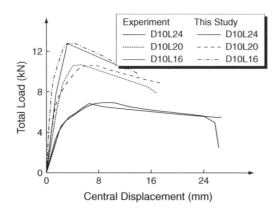

Figure 8. Load-displacement (span ratio effect).

4.2 *Example results*

The analytical responses of beam specimens are compared with the experimental measurement of KICT (2005) in Figure 8 and Figure 9.

The results of various shear span ratio from 1.8(D10L16) to 3.5(D10L24) are plotted in Fig. 8. Failure modes of all beam specimens are bending failure, and the ultimate load capacity decreases and the displacement increases as shear span ratio increases.

The analytical results show that the ultimate loads are similar to experimental results, however, the displacements of analytical results are slightly higher than those of experimental results. This can be explained as follow. The resistant force after cracks formed has increased by steel fibers in beam specimen, and the displacement will be decreased compare to the beam without steel fibers. However, in numerical analysis, this effect cannot be simulated perfectly as it behaves in experiment, and the stress-strain relationship of mold used in numerical analysis could be different from that of specimen.

The results of different steel ratio are shown in Figure 9. The cracking load, the ultimate load and ductility of beam increase as steel ratio increases. The compressive strength increases while the neutral axis is not varied after steel yielding state, the load capacity increases as the steel ratio increases.

The energy capacity of the numerical result is similar to that of experimental result, therefore, the numerical procedure by proposed stress-strain relationship with tension-stiffening effect can simulate the real behavior of beam specimens.

5 CONCLUSIONS

In this research, a numerical method that can simulate the tension-stiffening effect of ultra high performance steel fiber reinforced concrete is proposed. The beam specimen with steel fiber tested by KICT(2005) is selected to verify the validity of proposed model.

Correlation studies between the numerical and experimental results and associated parametric studies led to the following conclusions: (1) tension-stiffening effect plays important role in beam behavior not only plain RC beam but also steel fiber RC beam, (2) steel fibers are important in ultra high strength concrete to improve ductility after crack forms.

ACKNOWLEDGEMENT

The work presented in this paper was funded by Center for Concrete Corea (05-CCT-D11), supported by Korea Institute of Construction and Transportation Technology Evaluation and Planning (KICTTEP) under the Ministry of Construction and Transportation (MOCT).

REFERENCES

ACI Committee 363. 1997. *State-of-the-Art Report on High Strength Concrete*. Detroit: America Concrete Institute

Bischoff, P.H. 2003. Tension Stiffening and Cracking of Steel Fiber-Reinforced Concrete. ASCE *Journal of Materials in Civil Engineering* 15(2): 174–182

Carreira, D.J. & Chu, K.H. 1985. Stress-Strain Relationship for Plain Concrete in Compression. *ACI Journal* 82(6): 797–804

Chen, W.F. 1982. *Plasticity in Reinforced Concrete*. New York: McGraw-Hill

Dahl, K.K.B. 1992. *A Constitutive Model for Normal and High-Strength Concrete*; *ABK Report No. R287*. Denmark: Department of Structural Engineering in Technical University of Denmark

Hognestad, E. 1951. *A Study on Combined Bending and Axial Load in Reinforced Concrete Members: Engineering Experiment Station, Bulletin Series No. 399*. Illinois: University of Illinois

Kolle, B. & Phillips, D.V. & Zhang, B. & Bhatt, B. & Pearce, C.J. 2004. *Experimental Investigation of the Biaxial Properties of High Performance Steel Fibre Reinforced Concrete. Proceeding of 6th EM Symposium on Fibre-Reinforced Concrete*: Italy

Kwak, H.G. & Na, C.K. 2007. Nonlinear Analysis of RC Structures Considering Bond-Slip Effect. *KSCE Journal* 27(3A): 401–412

Kwak, H.G. & Kim, D.Y. 2004. Material Nonlinear Analysis of RC Shear Walls subjected to Monotonic Loading. *Engineering Structures* 26(11): 1517–1533

Wee, T.H. & Chin, M.S. & Mansur, M.A. 1996. Stress-Strain Relationship of High-Strength Concrete in Compression. *ASCE Journal of Materials in Civil Engineering* 8(2): 70–76

Excellence in Concrete Construction through Innovation – Limbachiya & Kew (eds)
© 2009 Taylor & Francis Group, London, ISBN 978-0-415-47592-1

Stiffened deep cement mixing (SDCM) pile: Laboratory investigation

T. Tanchaisawat & P. Suriyavanagul
Kasetsart University Chalermprakiat Sakonnakhon Province Campus, Sakonnakhon, Thailand

P. Jamsawang
School of Engineering and Technology, Asian Institute of Technology, Bangkok, Thailand

ABSTRACT: The low strength and stiffness of Deep Cement Mixing (DCM) pile causes unexpected failure that has been mitigated with the introduction of Stiffened Deep Cement Mixing (SDCM) pile. The SDCM is a new type of DCM pile reinforced by concrete core pile. In this paper, the interface behavior of SDCM pile and its strength have been studied by various laboratory tests. The cement content was varied from 10 to 20% by dry weight of clay and mixed at the water content corresponding to its liquid limit to obtain optimum strengths. The interface friction between the core concrete pile and the cement-admixed clay was studied by means of the direct shear tests and K_o interface shear tests. The 15% cement content yielded optimum interface shear strength. The CIU triaxial compression test of model SDCM pile revealed that the concrete core pile length should be more than 75% of the DCM pile length in order to have significant improvement.

1 INTRODUCTION

Deep cement mixing (DCM) pile has been widely used to improve the engineering properties of soft clay layer. The DCM pile can effectively reduce settlements of full-scale embankments by studies of Bergado et al. (1999) and Lai et al. (2006) whereas the practice research work by Petchgate et al. (2003a, b; 2004) showed that DCM pile has low strength and stiffness and this kind of pile may lead to low bearing capacity and large settlement (Wu et al., 2005). Consequently, DCM pile is not suitable for medium or high design load constructions (Dong et al., 2004). Hence, a new composite structure of DCM with a concrete core has been introduced. This is called stiffened deep cement mixing (SDCM) pile which employs a concrete plug at the center of DCM as shown in Figure 1. The concrete plug or concrete pile having higher strength and stiffness serves to resist compression stress along the pile shaft and takes most of the load and gradually transmits it to the soil-cement through the interface between the concrete core pile and cement-admixed clay. Therefore, the interface shear strength between the concrete pile and cement-admixed clay must be strong enough to develop the necessary load transfer mechanism. A series of pile load tests have been conducted to investigate the behavior of SDCM piles in China by Dong et al. (2004), Wu et al. (2005), and Zheng et al. (2005). Most of the test results were only bearing capacities of SDCM piles.

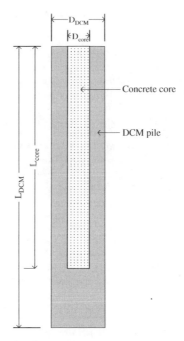

Figure 1. Schematic of SDCM pile.

Kunito & Mashima (1991) and Tungboonterm & Yoottimit (2002) studied bond strength between soil-cement and steel applied for DCM pile with reinforcement of steel beams in order to increase bending and tensile strength and bearing capacity (Kitazume et al., 1996). Bergado et al. (2004) studied shear strength of interface and clay surrounding the DCM pile. However, research work on interface shear strength between concrete pile and soil-cement, shear strength behavior of SDCM pile and consolidation behavior of a composite foundation consisting of a SDCM pile and untreated soil are still limited with insufficient experimental data. This paper aims at studying the behavior SDCM pile in laboratory such as interface, shear strength, and consolidation behavior.

The interface shear strength behavior was studied by the direct shear test and K_o interface shear test. The K_o interface shear test was employed to simulate field condition of the SDCM pile and was successfully modeled in the laboratory to study the interface friction. Furthermore, the effect of the boundary condition was also studied by varying the diameter of the DCM pile but maintaining the same diameter for the inner concrete pile. The influence of the concrete pile length and cement content on shearing behavior of this composite material and the overall strength development was also studied by consolidated undrained (CIU) triaxial compression test by varying the length of core piles and the cement contents.

2 PREPARATION OF CEMENT-ADMIXED CLAY

The base clay utilized in this study was the soft Bangkok clay. The soil samples were taken at the campus of the Asian Institute of Technology, Thailand. Samples were extracted from at 3 to 4 m depth for all tests. The liquid limit is about 103% and the natural water content varies from 76% to 84%. The undrained shear strength obtained from unconfined compression tests ranges from 16 to 17 kPa. Type I Portland cement was used for all tests conducted. The cement content (A_W) was 10%, 15% and 20% considering the current DCM practice in Thailand and the curing time was 28 days. The remolded clay samples at a particular remolding water content were mixed with cement slurry having water cement ratio (W/C) of 0.6. Mixing was done using a portable mechanical mixer until the soil was visually homogenous. All the specimens were prepared with mixing time 3–5 minutes.

The study by Bergado & Lorenzo (2005) confirmed that unconfined compressive strength as well as one dimensional yield stress is greatest when the total clay water content is 105% for Bangkok clay that is near the liquid limit of soft Bangkok clay. Consequently, a total clay water content of 105% was used in this research. The total clay water content (C_w) is defined as follows:

$$C_w = w^* + W/C \, (A_w) \qquad (1)$$

where w^* is the required remolding clay water content.

3 UNCONED COMPRESSION TEST

The frictional resistance between the two interfacing surfaces mortar and DCM has been found to increase with the increase in cement content of DCM. To relate the interface strength with the strength of DCM, a series of these tests has been conducted with different cement contents. The unconfined compression test is a strength index test that can assess the strength improvement characteristics of cement-admixed clay under various conditions of cement content. After thoroughly mixing the clay in the mechanical mixer, cement slurry was added to the clay and mixing was continued until a uniform mixture was achieved. Enough pushing was done into the mold to remove the honey-combed portions at the opposite end. The mold together with the specimen was waxed and kept in the humid room for curing.

After curing, each specific specimen was removed from its mold and made available for the intended tests. Finally, for particular mixing condition, the specimens with smooth surface and with almost the same densities were selected for testing; other specimens that did not qualify were discarded. Specimen dimensions were 50 mm in diameter by 100 mm in height.

4 DIRECT SHEAR BOX TEST

The interface shear strength between the cement mortar and cement-admixed clay was determined by this test. The detail of the apparatus is shown in Fig. 2. The small size of the specimen having diameter (D) of 55 mm and height (H) of 50 mm made possible multiple testing at different parameters. The circular specimens in place of rectangular specimens were chosen to avoid the stress concentrations at sharp corners. The PVC mold with steel base was used to prepare the lower half box with the cement mortar. The preparation of the cement mortar and its casting and curing was carried out. The bottom surface of the prepared cement mortar thus reflects the surface characteristics of the steel. Cement admixed clay was placed over this bottom surface. The upper half part of the mold was then tied to the lower half with metallic belts and the cement admixed clay prepared is pushed into the mold. The molds together with the specimen were waxed and kept in the humid room for curing. The specimens were removed from the molds and checked for their surface smoothness. The specimens with smooth peripheral surface were used for testing. The specimen was set up in the shear box in such a way that the plane of shearing would occur at the interface of the lower cement mortar half and the upper cement-admixed clay half as shown in Figure 2. The rate of shearing applied to all tests was 0.6 mm/min. Normal stresses of 50 kPa, 100 kPa and 200 kPa were utilized in each test condition.

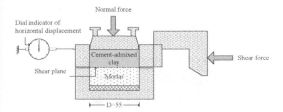

Figure 2. Direct shear box apparatus (unit:mm).

5 INTERFACE SHEAR TEST

The interface shear test in this study refers to the test employed to find the friction between the two shearing surfaces, the outer surface being subjected to the restrained boundary (K_o condition). Obviously, the K_o loading condition of this test brings it closer to the real field condition compared with direct shear test. The purpose of this test is to relate the load transfer to pile movement and it was found that this could be done best by using the continuous loading method. The loading procedure was used to determine the movement of the pile needed to mobilize the resisting forces (load transfer) in the soil. Similar approach has been used but with different confining pressure in the modified triaxial cell by Coyle & Reese (1966) for pile soil interaction. They fixed the loading rate at 1.5 mm/min. This research maintained nearly the same loading rate as the direct shear test for comparison (although not exactly the same due to the different rates available on two machines). The 17 mm diameter core mortar pile was inserted into two DCM sizes diameter 50 mm and 100 mm but with constant height 100 mm as shown in Fig. 3. The ratio of diameter between the core pile and DCM was first maintained 3 based upon the findings for shear strength between cement mortar and DCM and between DCM and clay. For the similar testing condition, the frictional strength between cement mortar and DCM is 2 to 3 times higher between DCM and clay. The corresponding values for the DCM pile and clay were obtained from Tran (2003). The role of the surrounding soil in the real field condition provided restrain to the DCM pile in addition to its frictional support. Hence, the DCM pile can be idealized as a structure subjected to K_o condition loading. The same boundary condition was maintained in this test by enclosing the DCM with the PVC mould.

Further, to study the influence of the boundary condition on the interface strength the diameter of the DCM was increased to 100 mm with the same diameter of concrete core pile. The PVC mold is used to prepare the specimen. Immediately after the pushing of cement-admixed clay into the PVC mold, a cement core pile was inserted at the center of the cement-admixed clay with the help of the machine at a rate of 10 mm/min. The cement core pile was 120 mm long

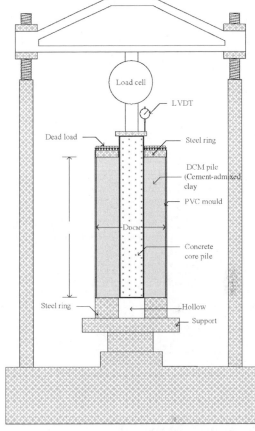

Figure 3. Interface shear test apparatus (unit:mm).

and the whole specimen was 100 mm long. The mold together with the specimen was waxed and kept in the humid room for curing.

After the curing time was achieved, the specimen was taken out from the humid room and wax was removed. The specimen was checked regarding the position of the cement mortar piles. Only those with cement mortar piles with tolerance of 2 mm from the center were used for the test. Both ends of the specimen were trimmed with the wire cutter. The specimen was then placed over the hollow metallic plate as shown in Figure 3. Similar arrangement was made on the top of the specimen in order to apply the vertical load to apply the vertical load over the specimen.

6 CONSOLIDATED-UNDRAINED TRIAXIAL (CIU) TEST

To further investigate the improvements on the strength brought about by the inserted concrete core piles, a series of triaxial compression test under consolidated

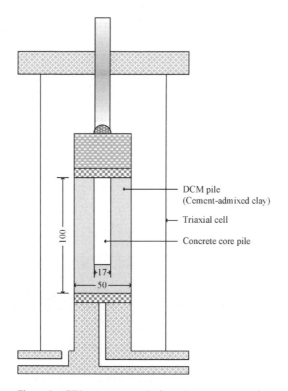

Figure 4. CIU test apparatus (unit:mm).

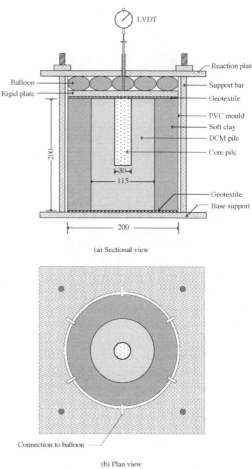

(a) Sectional view

(b) Plan view

Figure 5. SDCM model test apparatus (unit:mm).

undrained conditions has been done by varying the length of core piles. The method of specimen preparation for triaxial test is the same as that of unconfined compression test. The details of the apparatus are shown in Figure 4. After curing the specimen for 28 days, the specimen was prepared for testing. A back pressure of 400 kPa was applied in stages together with the cell pressure to prepare the specimen for saturation. The cell pressure and back pressure were gradually increased with pressure increment of 25 kPa, with the cell pressure being higher by about 10 to 15 kPa for every increment. After full saturation of the specimen, the pre-shear consolidation was applied into the specimen. The back pressure was lowered to 150 kPa while the cell pressure left at 250 kPa thus maintaining the difference between the cell pressure and the back pressure equals to the desired pre-shear consolidation pressure (100 kPa).

After consolidation, the displacement transducer was setup and the loading piston was adjusted to start the shearing. All tests were strain controlled with strain rate of 0.009 mm/min. This rate of strain is the same as that utilized by Lorenzo and Bergado (2004). The length of the concrete core pile has been varied as 60%, 75% and 90% of the length of the DCM in the SDCM test specimens.

7 PHYSICAL MODEL TEST OF SDCM PILE

The particular aspects of any prototypes can be studied using the proper physical modeling in the laboratory scale. Much research has been done on DCM regarding the strength and stiffness but the consolidation behavior of DCM foundation is studied only recently using the physical model in the laboratory (Yin & Fang, 2006). Similar test was performed for SDCM pile. In this test, the influence of the mortar pile length on the settlement behavior of the SDCM pile was studied by making an appropriate physical model test. Due consideration was given on the ratio of the length of concrete core pile to the DCM and the ratio of diameter of concrete core pile to DCM diameter. However, the restriction of the possible largest undisturbed soil sample made it difficult to simulate the boundary conditions as in the field. The details of the laboratory model are shown in Figure 5. The undisturbed soil was collected in a 200 mm diameter PVC pipe. A thin steel tube (shoe cutter) with outer diameter of 115 mm was pushed into the sample by a machine creating a hole at the center of the clay sample. The soil from the tube was collected and mixed with cement to prepare the cement-admixed clay. The cement-admixed clay was returned into the hole compacted by a tamping in four layers, thus, forming compacted DCM at the center of the clay sample.

Afterwards, a concrete pile of 30 mm diameter was inserted into the center of the DCM by a machine at 10 mm/min pushing rate. Wax was applied to the prepared sample to prevent the moisture loss and kept in a humid room for curing for the specified periods. At the end of the curing period, the wax was removed from the sample. On the top and bottom of the sample, a geotextile was placed for the drainage. Then, the sample together with the specimen was placed in the loading frame as shown in Figure 5. The top of the sample was further covered with a steel plate in order to maintain equal strain. The plate was greased on its periphery to reduce friction inside the mould during loading. The air pressure was applied on top of steel plate with the help of balloon. The displacement was recorded with use of a displacement transducer. The load was applied for twelve hours after which the settlement was almost constant. The vertical pressures, σ_v of 80 kPa and 160 kPa were applied. The length of the concrete pile was varied as 45%, 60% and 75% of the DCM pile length. Similar tests were conducted with only clay specimen for the comparison of settlement magnitudes.

8 TEST RESULTS AND DISCUSSIONS

8.1 Unconfined compression test

The purpose of this test is to use it as a reference for the other tests like interface strength tests whenever a comparison is needed. Lorenzo & Bergado (2004; 2006) studied engineering properties of cement admixed-clay at high water content using soft Bangkok clay as the base clay. They found that after-curing void ratio (e_{ot}) and cement content (A_w) are parameters to obtain unique relationship of unconfined compression strength q_u of cement-admixed clay at high water content. The unconfined compressive strength in this study was found to be 527 kPa, 702 kPa and 937 kPa for 10%, 15% and 20% cement content, respectively and also was plotted together with the relationship proposed by Lorenzo and Bergado (2004) as shown in Figure 6.

8.2 Direct shear test

The typical interface shear stress-strain curves at cement content 15% are plotted in Figure 7. When the

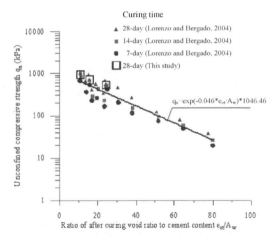

Figure 6. Unconfined compression strength, q_u, versus e_{ot}/A_w ratio.

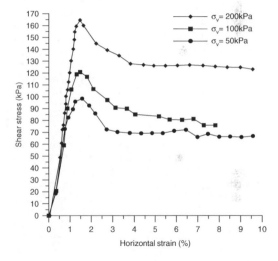

Figure 7. Interface shear stress – horizontal strain curves from direct shear test at $A_w = 15\%$.

cement content on cement admixed clay was 10%, the failure at interface between mortar and clay cement was observed at 3% of horizontal strain. Similarly, the horizontal strain at failure was 1.5% and 1.0% for cement content 15% and 20%, respectively. The peak interface friction angle(δ) varied from 23.3° to 25.6° while the adhesion (c_a) between the interfaces varied from 42.2 kPa to 78.6 kPa over the range of cement contents from 10% to 20% (Figure 8). This indicated that adhesion intercept is very sensitive to cement content compared to friction angle that.

The results from the direct shear test infer that the increased cement content on cement admixed clay induces more brittle failure. Hence, the peak strength

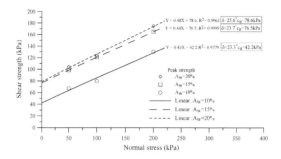

Figure 8. Morh-Coulomb failure.

was mobilized at lower displacements with increasing cement content. The increase in cement content from 10% to 20% reduced the strain needed for peak strength mobilization from 3% to 1%. The increased cement content enhanced the friction angle and adhesion. It is observed that the interface friction angle increased slowly from 23° to 25° while the adhesion between the interfaces increased sharply from 42.2 kPa to 76.5 kPa when the cement content was increased from 10% to 15%. A marginal increase to 78.6 kPa was obtained with further increase in cement content to 20%. The values of friction angle and adhesion obtained over the different cement content can be used for numerical analysis and for design of SDCM piles.

8.3 Interface shear test

The boundary effect has been studied by varying the diameter of the clay cement pile for the constant diameter concrete core pile. The average interface shear strength $\tau_{interface}$ can be calculate from a measured peak load F_{peak} divided by the concrete core surface area A_{core} as follows:

$$\tau_{interface} = \frac{F_{peak}}{A_{core}} = \frac{F_{peak}}{(\pi D_{core}) L_{core}} \tag{2}$$

The concrete core pile surface area, A_{core}, is calculated by multiplying the embedded length, L_s, of the concrete core in contact with the surrounding cement-admixed clay with the perimeter of the concrete core pile, πD_{core}.

The result of the interface shear test with diameter of concrete core pile 17 mm and that of cement-admixed clay 100 mm indicated the maximum interface shear strengths as 86.8 kPa, 144.6 kPa to 174.4 kPa for cement content 10%, 15% and 20%, respectively (Figure 9a). The maximum interface strength was observed for pile displacement around 0.5 mm. Similarity test was conducted with the clay cement pile of diameter 50 mm and core pile of constant diameter 17 mm. The maximum interface strength was found to be 108.8 kPa, 174.5 kPa and 216.5 kPa for cement content 10%, 15% and 20%, respectively (Figure 9b).

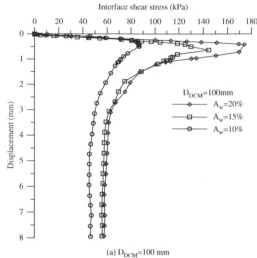

(a) D_{DCM}=100 mm

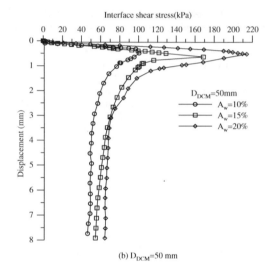

(b) D_{DCM}=50 mm

Figure 9. Interface shear stress versus displacement at various cement content with L_{DCM} = 100 mm and D_{core} = 17 mm.

8.3.1 Effects of cement content on interface shear strength

The results from interface shear tests concluded the increase in interface strength between concrete core pile and clay cement with increasing cement content on clay cement as shown in Figure 10. The compressive strength of the cement-admixed clay also increases with increasing cement content indicating a correlation between them. Consequently, the interface shear strength has been plotted against the unconfined compression strength of the clay cement in Figure 11. The interface strength increased sharply with the increase in cement content from 10% to 15% while it increased at slower rate from 15% to 20%. The interface strength

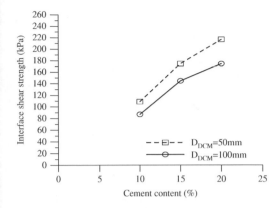

Figure 10. Variation of interface shear strength with cement content from interface shear test.

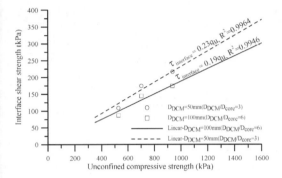

Figure 11. Interface shear strength versus unconfined compressive strength of clay cement.

from direct shear tests are also plotted together with corresponding results from the interface shear tests in Figure 11. The direct shear test yielded lower values of the peak strengths as compared to interface shear test results. Wu et al. (2005) defined the ratio of the interface shear strength τ_u to the unconfined compression strength q_u of the cement-admixed clay cement as adhesive coefficient denoted by α and calculated as follows:

$$\alpha = \frac{\tau_{interface}}{q_u} \qquad (3)$$

The interface shear strength equals to $0.19q_u$ and $0.23q_u$ for $D_{DCM} = 100\,mm$ ($D_{DCM}/D_{core} = 6$) and $50\,mm$($D_{DCM}/D_{core} = 3$), respectively so that adhesive coefficient α is 0.19 and 0.23 for $D_{DCM} = 100\,mm$ and $50\,mm$, respectively. The increased cement content resulted in higher cementation and higher bond strength. Once the displacement exceeded the limit of maximum peak strength mobilization, the bond broke and the strength reduced to similar values irrespective of the cement content. In all interface shear tests, the

maximum interface strength was observed at lesser displacement than in direct shear test. This behavior was similar to the results from pullout test compared with direct shear tests obtained by Chu & Yin (2005).

8.4 CIU triaxial compression test

The typical effects of various L_{core}/L_{DCM} ratios on the deviator stress of the SDCM pile at cement content 15% is shown in Figure 12. The excess pore pressure measurement was used as an indicator for failure on the clay cement at the point where excess pore pressure started to decrease after reaching the peak value was understood as failure. The limited length of the core pile ($L_{core}/L_{DCM} = 0.60$) was unable to bring any changes on the behavior of DCM. Increasing the length of core pile to 75% of DCM height ($L_{core}/L_{DCM} = 0.75$) showed improvement on post-failure behavior of SDCM pile. Though unable to increase the strength, this length of core pile deemed useful on maintaining the post failure strength. The further increase on length of core pile to 90% of DCM ($L_{core}/L_{DCM} = 0.9$) demonstrated the considerable increase in the strength of SDCM pile after the primary failure of DCM. The strength of the SDCM pile may be limited to the strength of the concrete core pile in this case.

8.4.1 Influence of concrete core pile length on failure mode

When the concrete core pile extended only 60% of the total specimen size of DCM, the failure mode was unchanged compared with the case of only DCM. However, the failure became more brittle with the increased cement content from 10% to 20% as expected. When the core pile extended 75% of the total specimen size of DCM, the failure mode was again brittle but the location of the failure plane was affected by presence of concrete core pile. The failure plane did not occur at the full length of concrete core pile. The failure plane was continued within DCM material. The strength at which the DCM material failed was the same. The longer length of concrete core pile, however, helped to maintain the strength of SDCM after the failure of DCM material.

When the concrete core pile extended 90% of the total specimen size of DCM, the failure mode was totally different from the previous cases. There was lateral spreading of the DCM material at the unstiffened end of SDCM. Though the failure on DCM material was observed at the same strength as reflected by the pore pressure development within the specimen, the load carrying capacity of the SDCM increased until the failure was observed in the concrete core pile. Zheng et al. (2005) categorized the failure mode as crush or crack of DCM material when the concrete core pile is shorter. This failure mode was observed in triaxial test when the concrete core pile length extended

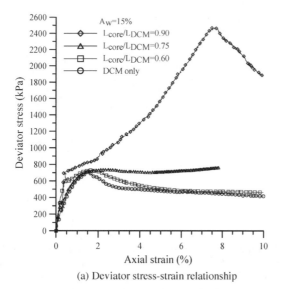

(a) Deviator stress-strain relationship

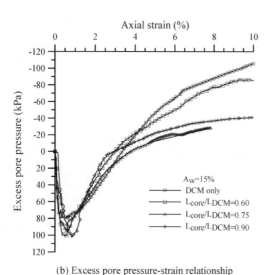

(b) Excess pore pressure-strain relationship

Figure 12. Effect of concrete core pile length on strength of SDCM pile from CU tests at Aw = 15%.

only 60% and 75% of the DCM material. Further, they categorized that the failure on the concrete core pile as a governing factor when its length is sufficient but the cross sectional dimension is comparatively small. This failure mode was observed when the concrete core pile extended 90% of the length of DCM material in the CIU triaxial test.

8.4.2 *Influence of cement content on failure mode*
The CIU triaxial test is used to compare the strength of SDCM for certain core pile length while

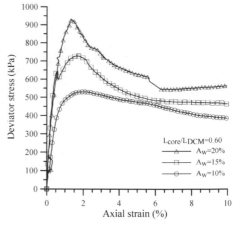

(a) Deviator stress-strain relationship at L_{core}/L_{DCM}=0.60

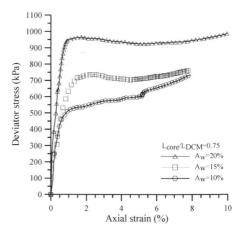

(b) Deviator stress-strain relationship at L_{core}/L_{DCM}=0.75

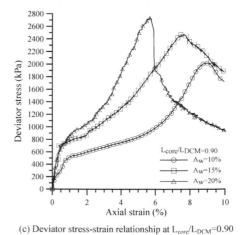

(c) Deviator stress-strain relationship at L_{core}/L_{DCM}=0.90

Figure 13. Effect of cement content on strength of SDCM pile from CU tests with varying L_{core}/L_{DCM}.

varying the cement content on DCM material. The plot of SDCM strength for 60% core pile length ($L_{core}/L_{DCM} = 0.60$) is shown in Figure 13a. The failure mode changed towards more brittle when the cement content increased from 10% to 20%. As explained previously, this length of core pile was practically insignificant to bring the improvement on performance of SDCM pile. The similar plot, Figure 13b, for 75% core pile length ($L_{core}/L_{DCM} = 0.75$) showed that the post failure strength increased when 10% cement content was used. On the other hand, the post failure strength reduced slowly till the 5% shear strain then only it climbed upward for 15% and 20% cement content. This behavior can be explained from Figure 12a. The increased cement content induced brittle failure on DCM. From this study, it is clear that the concrete core pile shared the load only after the failure of DCM material due to the boundary condition for load transfer in triaxial test. The more ductile nature of failure in case of 10% cement content was thus enhanced by the load sharing of core pile to show the effective rise in post failure strength of SDCM. The case was different when 15% and 20% cement content on DCM material. The load taken by the mortar pile was compensated by the loss on strength of DCM material due to brittle failure which resulted into the constant post failure strength of SDCM.

The plot of SDCM strength for 90% core pile length ($L_{core}/L_{DCM} = 0.90$) in Figure 13c shows two important trends in the behavior. The higher cement content tends to mobilize the full strength of the SDCM at small strain. During the test, it was observed that the specimen bulges laterally for the unimproved portion of SDCM which was more prominent for lower cement content with consequent large strain before the full strength of SDCM is mobilized. The strength of SDCM was observed to be higher for higher cement content due to the higher strength of DCM. But the improvement observed in SDCM beyond the differences in the strength of DCM material may have resulted from the possible variation of strength of concrete core pile.

9 CONCLUSIONS

The laboratory investigation of SDCM pile yielded interesting results about the various parameters involved. The friction between the surfaces showed sharp increase in the range of cement content 10% to 15% while it achieved a slower improvement at higher cement contents. Similarly, the failure mode became more brittle and lesser pile displacement was required for peak strength with increased cement content on the cement admixed-clay.

The adhesion intercept is very sensitive to cement content compared to friction angle that were measured. The 15% cement content was shown to yield optimum improvement. The adhesive coefficient α

was found to be 0.19 and 0.23 for D_{DCM}/D_{core} of 6 and 3, respectively.

The CIU triaxial tests of model SDCM pile revealed that the core pile length should be more than 75% of the DCM length to demonstrate any improvement. The concrete core pile extending 90% of the DCM length yielded significant improvement on the strength of SDCM pile, but it should be clear at this point that the failure on DCM was observed nearly at the same strength irrespective of the length of concrete core pile due to the boundary conditions of the triaxial test. However, the mortar pile length influenced the position of failure plane as well as failure mode. The study revealed that the shorter pile length ($L_{core}/L_{DCM} = 0.60$ and 0.75) caused the failure on DCM pile while the longer pile length ($L_{core}/L_{DCM} = 0.90$) but relatively smaller cross sectional area shifted the failure to the concrete core pile.

REFERENCES

Bergado, D.T., Ruenkrairergsa, T., Taesiri, Y. & Balasubramaniam, A.S. 1999. Deep soil mixing to reduce embankment settlement. *Ground Improvement Journal* 3(1): 1–18.

Bergado, D.T., Lorenzo, G.A. & Tran, C.V. 2004. Laboratory shear strength of interface and clay surrounding the DMM pile. *Proceedings of Engineering Practice and Performance of Soft Deposits*, IS-Osaka, Osaka, Japan: 223–228.

Bergado, D.T. & Lorenzo, G.A. 2005. Economical mixing method for cement deep mixing. *Proceedings of ASCE GeoFrontiers 2005 Conference*, GSP-136, Austin, Texas, U.S.A.

Chu, L.M. & Yin, J.H. 2005. Comparison of interface shear strength of soil nail measured by both direct shear box tests and pullout tests. *ASCE Journal of Geotechnical and Geoenvironmental Engineering* 131(9): 1097–1107.

Coyle, H.M. & Reese, L.C. 1966. Load transfer for axially loaded piles in clay. *ASCE Journal of the Soil Mechanics and Foundations Division* 92(SM-2): 1–26.

Dong P., Qin R. & Chen, Z. 2004. Bearing capacity and settlement of concrete-cored DCM pile in soft ground. *Geotechnical and Geological Engineering* 22(1): 105–119.

Kitazume, M., Tabata, T. & Ishikawa., Y. 1996. Model tests on failure pattern of cement treated retaining wall. *Proceedings of IS-TOKYO'96/The Second International Conference on Ground Improvement Geosystems*. Tokyo, Japan: 509–514.

Kunito, T. & Mashima, M. 1991. Effects of soils and mixing conditions on bond strength between soil-cement and steel. *Soil Improvement*: 167–181.

Lorenzo, G.A. & Bergado, D.T. 2004. Fundamental parameters of cement-admixed clay – new approach. *Journal of Geotechnical and Geoenvironmental Engineering. ASCE* 130(10): 1042–1050.

Lai, Y.P., Bergado, D.T., Lorenzo, G.A. & Duangchan, T. 2006. Full-scale reinforced embankment on deep jet mixing improved ground. *Ground Improvement Journal* 10(4): 153–164.

Lorenzo, G.A. & Bergado, D.T. 2006. Fundamental characteristics of cement-admixed clay in deep mixing. *Journal of Materials in Civil Engineering. ASCE* 18(2): 161–174.

Petchgate, K., Jongpradist, P. & Panmanajareonphol, S. 2003a. Field pile load test of soil-cement column in soft clay. *Proceedings of the International Symposium 2003 on Soil/Ground Improvement and Geosynthetics in Waste Containment and Erosion Control Applications*, 2–3 December 2003, Asian Institute of Technology, Thailand: 175–184.

Petchgate, K., Jongpradist, P. & Samanrattanasatien, P. 2003b. Lateral movement behavior of cement retaining wall during construction of a reservoir. *Proceedings of the International Symposium 2003 on Soil/Ground Improvement and Geosynthetics in Waste Containment and Erosion Control Applications*, 2–3 December 2003, Asian Institute of Technology, Thailand: 195–205.

Petchgate, K., Jongpradist, P. & Jamsawang, P. 2004. Field flexural behavior of soil-cement column. *Proceedings of the Fifth Symposium 2004 on Soil/Ground Improvement and Geosynthetics*, King Mongkut's University of Technology Thonburi, Thailand: 85–90.

Tran, C.V. 2003. Composite Interface Strength Between DMM Pile and Surrounding Soil. *M. Eng. Thesis No. GT-02-18*, Asian Institute of Technology, Bangkok, Thailand.

Tungboonterm, P., & Yoottimit, C. 2002. Factors affecting bond strength of steel reinforcement in cement column. *Proceedings International Symposium on Lowland Technology*, September 2002, Saga University, Japan: 193–198.

Wu M., Zhao X. & Dou Y.M. 2005. Application of stiffened deep cement mixed column in ground improvement. *Proceedings International Conference on Deep Mixing Best Practices and Recent Advances*, Stockholm, Sweden: 463–468.

Yin, J.H. & Fang, Z. 2006. Physical modeling of consolidation behavior of a composite foundation consisting of a cement-mixed soil column and untreated soft marine clay. *Geotechnique* 56(1): 63–68.

Zheng, G. & Gu, X.L. 2005. Development and practice of composite DMM column in China. *Proceedings 16th International Conference on Soil Mechanics and Geotechnical Engineering*, Osaka, Japan: 1295–1300.

Excellence in Concrete Construction through Innovation – Limbachiya & Kew (eds)
© 2009 Taylor & Francis Group, London, ISBN 978-0-415-47592-1

Relationship between compressive strength and UPV for high strength concrete containing expanded Perlite aggregate

M.B. Karakoç & R. Demirboğa
Ataturk University, Engineering Faculty, Civil Eng. Dept., Erzurum, Turkey

ABSTRACT: In this study, different expanded perlite aggregate (EPA) percents (7.5%, 15%, 22.5% and 30%) were used in stead of fine aggregate (0–2 mm). The relationship between ultrasonic pulse velocity (UPV) and compressive strength (CS) of high strength concrete (HSC) are evaluated. CS and UPV were determined at the 3, 7, 28 and 90 days of curing period. HSC containing EPA showed good strength development with both wet and dry curing conditions. According to experimental results, both CS and UPV were decreased with increase of EPA ratios for both wet and dry curing conditions. However, with the increase of curing period, both CS and UPV of all the samples are increased. The increment in the CS due to curing period was higher than that of UPV. The relationship between CS and UPV was exponential for percent EPA aggregate both wet and dry curing conditions.

1 INTRODUCTION

Ultrasonic non-destructive testing methods are frequently used to estimate the quality of concrete. These methods are typically based on the measurement of the propagation of velocity, which is closely related to mechanical properties and, more directly, to the modulus of elasticity. Non-destructive testing methods consider concrete as a homogeneous material (Krautkramer & Krautkramer, 1990).

The test is described in ASTM C597, 1991 and BS 1881: Part 203:1986. The principle of the test is that the velocity of sound in a solid material, V, is a function of the square root of the ratio of its modulus of elasticity, E, to its density. The method starts with the determination of the time required for a pulse of vibrations at an ultrasonic frequency to travel through concrete. Once the velocity is determined, an idea about quality, uniformity, condition, and strength of the concrete tested can be attained.

The UPV method, also known as the transit time method, uses a detector to measure the time of flight it takes for an ultrasonic pulse to pass through a known thickness of solid material. The UPV can be written as:

$$V_c(x,t) = x/t \qquad (1)$$

where V_c (x, t) = UPV in concrete; x = propagated path length; and t = transit time.

Nondestructive methods of investigation, such as ultrasonic measurements, are often proposed. The use of ultrasonic methods in cement materials investigations dates more than 40 years ago on both Portland and other special cements (Robson, 1962). The usual parameter measured is the velocity of longitudinal ultrasonic waves in material that, together with the density, enables calculation of Young's modulus of elasticity. The velocities of transversal and surface ultrasonic waves are also measured and relations between thus obtained ultrasonic parameters and bulk elastic modulus of material can be found (Krautkramer & Krautkramer, 1990, Hwa & Lee, 1999, Berthaud, 1991). The influence of aggregate to cement ratio on the CS–UPV dependence had been shown (Krautkramer & Krautkramer, 1990, Neville, 1987).

The ultrasonic method (Filipczynski et al., 1966) is one of the nondestructive testing techniques and is frequently adopted for evaluating the quality of in situ concrete structures. Malhotra (1976) presented a comprehensive literature survey for the nondestructive methods normally used for concrete testing and evaluation. However, a successful nondestructive test is the one that can be applied to concrete structures in the field, and be portable and easily operated with the minimum amount of cost.

The UPV technique is used as a means of quality control of products which are supposed to be made of similar concrete: both lack of compaction and a change in the water/cement (w/c) ratio would be easily detected. The technique cannot, however, be employed for the determination of strength of concretes made of different materials in unknown proportions

(Neville, 1995). It is true that there is a broad tendency for concrete of higher density to have a higher strength (provided the specific gravity of the aggregate is constant) so that a general classification of the quality of concrete on the basis of the pulse velocity is possible (Jone and Gatfield, 1955). Some figures suggested by Whitehurst (1951) for concrete with a density of approximately 2400 kg/m^3 are given as excellent, good, doubtful, poor and very poor for 4500 m/s and above, 3500–4500, 3000–3500, 2000–3000 and 2000 m/s and below UPV values, respectively. According to Jones and Gatfield (1955), however, the lower limit for good quality concrete is between 4100 and 4700 m/s. The measurement of the ultrasonic compressional wave velocity has been used for a long time to evaluate the setting and hardening of cementation systems (Goueygou et al., 2002, Herdnandez et al., 2002, Joeng, 1996, Herdnandez et al., 2000, Tan et al., 1996, Topcu, 1995, Naffa et al., 2002, Koehler et al., 1998, Elvery & Ibrahim, 1976, Sayers & Dahlin, 1993). According to Naik et al., (2004), the pulse velocity for ordinary concrete is typically 3700 to 4200 m/s.

Many investigators have found that the pulse velocity is affected significantly by the type and amount of aggregate (Naik et al., 2004, Bullock & Whitehurst, 1959, Sturrup et al., 1984, Swamy & Al-Hamed, 1984, Anderson & Seals, 1981, Jones, 1954, Popvics et al., 1990). In general, the pulse velocity of cement paste is lower than that of aggregate. Jones (1954) reported that for the same concrete mixture and at the same CS level, concrete with rounded gravel had the lowest pulse velocity, crushed limestone resulted in the highest pulse velocity, and crushed granite gave a velocity that was between these two. On the other hand, type of aggregate had no significant effect on the relationship between the pulse velocity and the modulus of rupture. Additional test results by Jones (1962), Bullock & Whitehurst (1959) and Kaplan (1959) indicate that at the same strength level the concrete having the higher aggregate content gave a higher pulse velocity.

The effect of age of concrete on the pulse velocity is similar to the effect on the strength development of concrete. Jones (1954) reported the relationship between the pulse velocity and age. He showed that velocity increases very rapidly initially but soon flattens. This trend is similar to the strength vs. age curve for a particular type of concrete, but pulse velocity reaches a limiting value sooner than strength. He further concluded that once the pulse velocity curve flattens, experimental errors make it impossible to estimate the strength with accuracy (Jone, 1962).

The pulse velocity for saturated concrete is higher than for air-dry concrete. Moisture generally has less influence on the velocity in HSC than on low-strength concrete because of the difference in the porosity (Jones & Facaoaru, 1969). A 4 to 5% increase in pulse velocity can be expected when dry concrete with high

Table 1. Chemical composition of PC, SF and EPA (%).

Component	PC	SF	EPA
SiO$_2$	19.94	93.7	71–75
Al$_2$O$_3$	5.28	0.3	12–16
Fe$_2$O$_3$	3.45	0.35	–
CaO	62.62	0.8	0.2–0.5
MgO	2.62	0.85	
SO$_3$	2.46	0.34	
C	–	0.52	
Na$_2$O	0.23	–	2.9–4.0
K$_2$O	0.83	–	
Chlor (Cl$^-$)	0.0107	–	
Sulphide (S^{-2})	0.17	0.1–0.3	
Undetermined	0.08		
Free CaO	0.51		

w/c ratio is saturated (Jones, 1962). Kaplan (1958) found that the pulse velocity for laboratory-cured specimens were higher than for site-cured specimens. He also found that pulse velocity in columns cast from the same concrete were lower than in the site-cured and laboratory-cured specimens.

There are many studies related to the UPV; for example, Tharmaratnam & Tan (1990) provided the empirical formula of the combined UPV and ultrasonic pulse amplitude (UPA). Liang & Wu (2002) studied theoretical elucidation of the empirical formula for the UPV and UPA and combined methods. Ye et al., 92004) determined the development of the microstructure in cement-based materials by means of numerical simulation and UPV. Demirboga et al., (2004) also studied the relationship between UPV and CS for concrete containing high volume mineral admixtures.

In this study, the effect of EPA on the UPV and CS of HSC for different curing periods and different curing conditions were investigated. In addition, the relationship between UPV and CS of HSC were derived.

2 EXPERIMENTAL STUDY

ASTM Type I, Portland cement (PC), from Askale-Erzurum, Turkey was used in this study. Silica fume (SF) and EPA were obtained from Antalya Electro Metallurgy Enterprise and Etibank Perlite Expansion Enterprise in Izmir, Turkey, respectively. The chemical composition of the materials used in this study was summarized in Table 1. The physical and mechanical properties of PC were similarly summarized in Table 2. The specific gravity of SF and EPA were 2.18 and 0.28 g/m^3, respectively. Sulphonate naphthalene formaldehyde was used as a superplasticizer, compatible with ASTM C 494 F (high-range water reducer) at a dosage of 2.0 ml/kg of cement.

Table 2. The physical and mechanical properties of PC.

Specific gravity (g/cm^3)		3.13
Specific surface(cm^2/g)		3410
Remainder on 200-mm sieve (%)		0.1
Remainder on 90-mm sieve (%)		3.1
Setting time initial (min)		2.10
Setting time final (min)		3.15
Volume expansion (Le Chatelier, mm)		3
Compressive strength (MPa)	2 days	23.5
	7 days	35.3
	28 days	47.0
Flexural strength (MPa)	2 days	5.0
	7 days	6.2
	28 days	7.7

Table 3. Hardened concrete properties of samples in wet curing.

Samples	Control samples	EPA ratios (%)			
		7.5	15	22.5	30
3-day Compressive strength, (MPa)	53.80	43.03	38.89	34.47	31.80
3-day UPV, (m/s)	4465	4177	4058	3940	3933
7-day Compressive strength, (MPa)	72.74	62.66	53.77	49.75	43.01
7-day UPV, (m/s)	4587	4453	4396	4296	4210
28-day Compressive strength, (MPa)	80.77	70.72	70.52	67.20	53.97
28-day UPV, (m/s)	4681	4599	4507	4436	4410
90-day Compressive strength, (MPa)	84.29	78.30	70.09	67.78	59.39
90-day UPV, (m/s)	4717	4610	4581	4533	4450

The percentages of EPA that replaced fine aggregate in this study were: 0%, 7.5%, 15%, 22.5% and 30%. The work focused on concrete mixes having a fixed water/binder ratio of 0.25 and a constant total binder content of 500 kg/m^3. 7% SF was used in replacement of cement by weight. The maximum size of coarse aggregate was 16 mm. For each mixture, twenty-four samples of 100×200 mm cylinders were cast. The samples were tested at 3, 7, 28 and 90 days for UPV and CS in accordance with ASTM C 597-83 (1998) and ASTM C 39 (1998), respectively.

In the UPV test method (Naik, 2004), an ultrasonic wave pulse through concrete is created at a point on the surface of the test object, and the time of its travel from that point to another is measured. Knowing the distance between the two points, the velocity of the wave pulse can be determined. Portable pulse velocity equipment is available today for concrete testing to determine the arrival time of the first wavefront. For most test configurations, this is the direct compressional wave, as it is the fastest wave. Dry curing is defined as samples were air cured at uncontrolled temperature and relative humidity until the test age and similarly, the wet curing is defined as samples were immersed in $20 \pm 3°$C water until the test age.

Table 4. Hardened concrete properties of samples in dry curing.

Samples	Control samples	EPA ratios (%)			
		7.5	15	22.5	30
3-day Compressive strength, (MPa)	38.50	31.94	27.53	21.49	20.13
3-day UPV, (m/s)	4040	3926	3787	3559	3557
7-day Compressive strength, (MPa)	48.32	43.81	38.22	30.81	29.04
7-day UPV, (m/s)	4257	4177	4097	3883	3876
28-day Compressive strength, (MPa)	56.77	47.62	46.10	39.64	39.62
28-day UPV, (m/s)	4377	4263	4193	4080	4020
90-day Compressive strength, (MPa)	60.00	57.16	50.09	48.44	41.23
90-day UPV, (m/s)	4410	4368	4245	4197	4100

3 RESULTS AND DISCUSSIONS

The results obtained in the tests are shown in Tables 3–4 and Figs. 1–3. They are evaluated and discussed below.

3.1 Dry unit weight

According to the obtained data, it was observed that unit weights of 28 days hardened HSC decreased with increasing EPA in the mixtures, because the specific gravity of EPA is lower than that of traditional fine aggregate. The samples' unit weights changed

between 2388 and 2108 kg/m^3. Demirboga et al., (2001) reported that lower dry unit weights results for concrete containing EPA.

3.2 The effect of EPA on the compressive strength and UPV

The CS and UPV results of the concretes made with EPA were determined at 3, 7, 28 and 90 days. Table 3 shows that EPA reduced CS of the concretes at all levels of replacement at 3, 7, 28 and 90 days in wet curing. Reductions were 20%, 28%, 36% and 41%; 14%, 26%, 32% and 41%; 12%, 13%, 17% and 33%; and 7%, 17%, 20% and 30% due to 7.5, 15, 22.5 and 30 percent EPA replacement of fine aggregate at 3, 7, 28 and 90 days, respectively. The reduction values

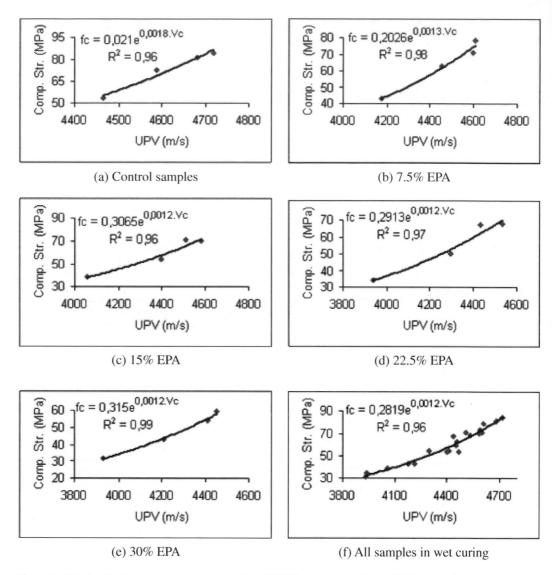

Figure 1.   Relationship between compressive strength and UPV for samples containing EPA in wet curing.

decreased with the increasing curing period. This may be due to the porous structures and weak texture of the EPA (Demirboga et al., 2001).

Table 4 shows that EPA reduced CS of the concretes at all levels of replacement at 3, 7, 28 and 90 days in dry curing. Reductions were 17%, 28%, 44% and 48%; 9%, 21%, 36% and 40%; 16%, 19%, 30% and 30%; and 5%, 17%, 19% and 31% due to 7.5, 15, 22.5 and 30 percent EPA replacement of fine aggregate at 3, 7, 28 and 90 days, respectively.

Goble & Cohen (1999) have concluded that the sand surface area has a significant influence on the mechanical properties of Portland cement mortar. Demirboga et al., (2001) reported results of an extensive laboratory study evaluating the influence of EPA and mineral admixtures on the CS of low density concretes. They concluded that the addition of mineral admixtures increased the CS of concrete produced with lightweight EPA. Ramamurty & Narayanan (2000) concluded that the CS was a function of density of concrete and with an increase of density resulting in higher CS. Gül et al., (1997) reported that the CS decreased because the density decreased with increasing pumice aggregate ratio instead of the normal aggregate. Akman & Tasdemir (1977) concluded that CS decreased because the density decreased with increasing lightweight aggregate ratio instead of the tradition aggregate. Faust (2000) reported that the replacement of natural sand by lightweight fine aggregate reduces the CS.

It can be seen from Tables 3–4 that UPV values decreased with increasing EPA replacement for fine aggregate at the 3, 7, 28 and 90 days curing periods. Maximum reduction occurred at 30% EPA replacement and it was 12%, 8%, 6% and 6% at the 3, 7, 28 and 90 days curing periods, respectively. The UPV values decreased with increasing EPA replacement percentage. However, reduction in UPV values due to EPA replacement was much lower than that of CS. The higher the EPA replacement, the higher the decrease in UPV values, especially at early ages. However, after about the 28th day curing period, the UPV reached a certain value and thereafter increased only slightly.

The UPV values for 3, 7, 28 and 90 days are shown in Tables 3–4. UPV changed between 4465 and 3557, 4587 and 3876, 4681 and 4020, and 4717 and 4100 m/s at the 3, 7, 28 and 90 days curing period, respectively. UPV values decreased with increasing EPA as it occurred for CS. However, the gap between UPV values was much lower than those of CS.

3.3 The effect of curing conditions on the compressive strength and UPV

CS and UPV values of samples cured in wet and dry curing are shown Tables 3–4. CS values of samples in wet curing were higher than those of dry curing throughout the study. For 3, 7, 28 and 90 days curing time, comparing control samples to each other, the reductions of CS of control samples at dry curing were 28%, 34%, 30% and 29%, respectively. The reductions were 26%, 30%, 33% and 27%, 29%, 29%, 35% and 29%, 38%, 38%, 41% and 29%, and 37%, 32%, 27% and 31% due to 7.5, 15, 22.5, and 30 percent EPA for 3, 7, 28 and 90 days curing time, respectively.

The same trend was observed for UPV values. Wet cured samples' UPV values were also higher than those of dry curing. However, reduction percent of the UPV values were less than those of CS. For 3, 7, 28 and 90 days curing periods, comparing control samples and the same percent of the EPA ratios to each other, the reductions of UPV values at dry curing were 10%, 7%, 6% and 7%, 6%, 6%, 7% and 5%, 7%, 7%, 7% and 7%, 10%, 10%, 8% and 7%, and 10%, 8%, 9% and 8%, due to 0, 7.5, 15, 22.5, and 30 percent EPA for 3, 7, 28 and 90 days curing time, respectively.

The effect of age of concrete on the pulse velocity is similar to the effect on the strength development of concrete. Jones (1954) reported the relationship between the pulse velocity and age. He showed that velocity increases very rapidly initially but soon flattens. This trend is similar to the strength vs. age curve for a particular type of concrete, but pulse velocity reaches a limiting value sooner than strength. He further concluded that once the pulse velocity curve flattens, experimental errors make it impossible to estimate the strength with accuracy (Jones, 1962).

Mannan et al., (2002) reported that the loss of moisture in the capillary pores due to evaporation or dissipated hydration may cause reduction in hydration resulting in lower strength. The hydration becomes very slow if water vapour pressure in the capillaries falls below about 0.8 of the saturation pressure (Neville, 1995). In this study, the wet cured samples give higher CS than dry cured samples of concrete. When the samples are directly exposed to air during curing, a large water loss occurs from the surface of the samples. As shown in Tables 3–4, the comparison made between the relationship established for control and HSC containing EPA shows that dry curing resulted in a 32 percent average reduction in CS compared to wet curing CS of HSC containing EPA. The strength reduction can be ascribed to shrinkage cracks. For the samples cured in open air, micro-cracks form on the surface and lead to strength decrease.

Regarding the influence of curing conditions, Ramezanianpour & Malhotra (1995) stated that "if the potential of concrete with regards to strength and durability is to be fully realized, it is most essential that it be cured adequately. They reported that dry curing resulted in 28% drop in CS of silica fume concrete. Meloleepszy & Deja (1992) observed and reported that silica fume mortar containing 5% and 10% silica fume was influenced by curing conditions. Dry air curing resulted in up to 40% reduction in CS of mortar studied. Influence of curing conditions on plain concrete have been extensively studied and results have been established (Neville, 1995, Toutanji & Bayasi, 1999, Mehta, 1986).

3.4 Relationship between compressive strength and UPV values of HSC containing EPA

The general relationship between UPV and CS is pooled together for all results in Fig. 3 for concretes at ages between 3 and 90 days. There was a very good exponential relationship between UPV and CS. Because $R^2 = 0.98$, we can say that 98% of the variation in the values of CS is accounted for by exponential relationship with UPV (see Fig. 3). For all results it was found the following law relating CS (f_c in MPa) to UPV (V_c in m/s):

$$f_c' = 0.2852 e^{0.0012 V_c} \qquad (2)$$

The relationship determined in this study, between f_c' and V_c, fitting the general Eq. (3) was reported by Tharmaratram & Tan (1990) for cement mortar, by

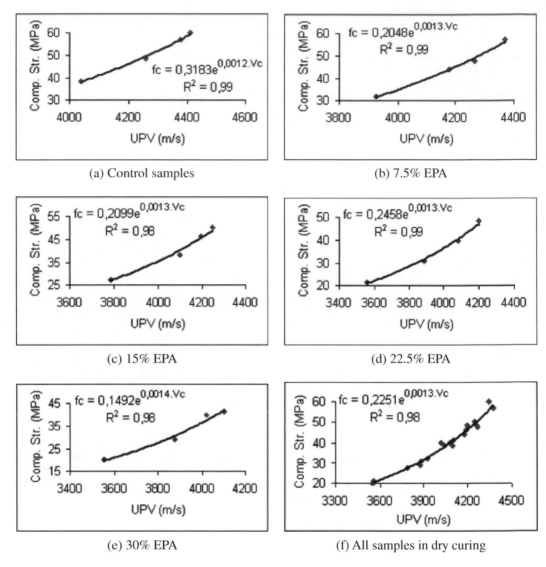

Figure 2.   Relationship between compressive strength and UPV for samples containing EPA in dry curing.

Demirboğa et al., (2004) for mineral admixtured concrete and by Gül et al., (2006) for mineral admixtured mortars.

$$f_c' = ae^{bV_c} \qquad (3)$$

where a and b are parameters dependent upon the material properties.

Turgut (2004) noted that, also, with the formula $f_c = 0.3161e^{1.03V_c}$, obtained with the correlation of earlier researches' findings and this study's findings, approximate value of CS in any point of concrete can be practically found with ignoring the mixture ratio

of concrete through using only longitudinal velocity variable (V_c).

When models of at 0, 7.5, 15, 22.5 and 30 percent EPA in wet curing, separately pooled it was found that the relationships were also exponential (see Fig. 1. a–e). Determination coefficients of 0, 7.5, 15, 22.5 and 30 percent EPA in wet curing were 0.96, 0.98, 0.96, 0.97 and 0.99, respectively. Determination coefficients of all samples in wet curing were 0.96 (see Fig. 1. f). When models of at 0, 7.5, 15, 22.5 and 30 percent EPA in dry curing, separately pooled it was found that the relationships were also exponential (see Fig. 2. a–e). Determination coefficients of 0, 7.5, 15,

54

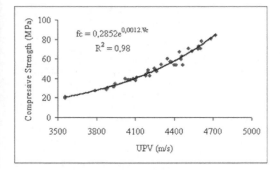

$$fc = 0.2852e^{0.0012.Vc}$$
$$R^2 = 0.98$$

Figure 3. Relationship between compressive strength and UPV for all samples with EPA for wet and dry curing conditions.

22.5 and 30 percent EPA in dry curing were 0.99, 0.99, 0.98, 0.99 and 0.98, respectively. Determination coefficients of all samples in dry curing were 0.98 (see Fig. 2. f).

According to Sturrup et al., (1984), factors other than concrete strength can affect pulse velocity, and changes in pulse velocity may overshadow changes due to strength. Hence, there is no unique relationship between UPV and CS; variations were found between results when wet and dry pastes, mortars and concrete were used. For example, Qasrawi (2000) was reported the relationship between the UPV test result and the crushing cube strength of concrete. The best-fit line representing the relationship is given as:

$$S = 36, 72. V - 129,077$$

where V is the pulse velocity (km/s). The R^2 value was found to be 0.96.

4 CONCLUSIONS

HSC containing EPA replacement induced to reduction in CS at all levels of replacement. The gap in the CS was very high at early age, and also higher than that of UPV values. The effect of curing period of concrete on the pulse velocity is similar to the effect on the strength development of concrete, but UPV reaches a limiting value sooner than strength. The maximum CS and UPV were observed with the control samples. Both CS and UPV were lower for samples containing EPA. However, with the increase of curing period, both CS and UPV of all samples increased. The increment in the CS due to curing period was higher than that of UPV for all level of EPA. Wet cured samples' UPV values were also higher than those of dry curing. However, reduction percent of the UPV values were less than those of CS. A determination coefficient (R^2) of 0.98 indicates a very good exponential relationship between UPV and CS when all results pooled together.

It can be concluded that relationship between CS and UPV is also exponential for HSC containing EPA.

REFERENCES

Akman, MS, Taşdemir, MA, Perlite concrete as a structural material (Taşıyıcı Malzeme Olarak Perlit Betonu), 1st National Perlite Congress. MTA Institute Press, Ankara, Turkey, 1977, pp. 40–48.
Anderson, DA and Seals, RK, Pulse velocity as a predictor of 28 and 90 day strength, ACI J., 78, 116, 1981.
ASTM C 39-96, Standard Test Method for Compressive Strength of Cylindrical Concrete Specimens, Annual Book of ASTM Standards, 1998, Easton, MD, USA.
ASTM C 597-83 (Reapproved 1991), Test for Pulse Velocity Through Concrete, ASTM, U.S.A., 1991.
ASTM C 597-83, Standard Test Method for Pulse Velocity through Concrete, Annual Book of ASTM Standards, 1998, Easton, MD, USA.
Ay, N, Topçu, IB, The influence of silicoferrochromium fume on concrete properties, Cem. Concr. Res. 25 (2) (1995) 387–394.
Berthaud, Y, Damage measurements in concrete via an ultrasonic technique: Part I. Experiment, Cem. Concr. Res. 21 (1991) 73–82.
BS 1881: Part 203: 1986: Measurement of Velocity of Ultrasonic Pulses in Concrete, BSI, U.K., 1986.
Bullock, RE and Whitehurst, EA, Effect of certain variables on pulse velocities through concrete, Highway Res. Board Bull., 206, 37, 1959.
Demirboga, R, Orung, I, Gul, R, Effects of expanded perlite aggregate and mineral admixtures on the compressive strength of low-density concretes, Cement Concrete Research, 31 (11), (2001), 1627–1632.
Demirboga, R, Türkmen İ and Karakoç, MB, Relationship between ultrasonic velocity and compressive strength for high volume mineral admixtured concretes, Cement and Concrete Research 34, (2004), 2329–2336.
Elvery, RH, Ibrahim, LAM, Ultrasonic assessment of concrete strength at early ages, Mag. Concr. Res. 28 (1976) 181–190.
Faust, T, Properties of different matrix and LWAs their influences on the behavior of structural LWAC, Second International Symposium on Structural Lightweight Aggregate Concrete, 18–22 June. Kristiansand, Norway, 2000, pp. 502–511.
Filipczynski, L, Pawlowski, Z, Ultrasonic Methods of Testing Materials, Butterworth, London, 1966.
Goble, C, Cohen, M, Influence of aggregate surface area on mechanical properties of mortar, ACI Mater. J. 96 (6) (1999) 657–662.
Goueygou, M, Piwakowski, B, Ould Naffa, and Buyle-Bodin, F, Assessment of broadband ultrasonic attenuation measurements in inhomogeneous media, Ultrasonics 40 (2002) 77–82.
Gül, R, Demirboğa R and Güvercin, T, Compressive strength and ultrasound pulse velocity of mineral admixtured mortars, Indian Journal of Engineering & Materials Sciences, 2006, Vol. 13, pp.18–24.
Gül, R, Şahin, R, Demirboğa, R, Investigation of compressive strength of lightweight concrete made with Kocapınar's pumice aggregate (Kocapınar pomzası ile üretilen hafif

betonların mukavemetinin araştırılması), Advances in Civil Engineering: III. Technical Congress, Vol. 3, METU, Ankara, Turkey, 1997, pp. 903–912, in Turkish.

Herdnandez, MG, Anaya, JJ, Izquierdo, MAG and Ullate, LG, Application of micromechanics to the characterization of mortar by ultrason, Ultrasonics 40 (2002) 217–221.

Herdnandez, MG, Izquierdo, MAG., Ibanez, A, Anaya, JJ, Ullate, LG, Porosity estimation of concrete by ultrasonic NDE, Ultrasonics 38 (2000) 531–533.

Hwa, LG and Lee, GW, The influence of water on the physical properties of calcium aluminate oxide glasses, Mater. Chem. Phys. 58 (2) (1999) 191–194.

Jeong, H, Hsu, DK, Quantitative estimation of material properties, of porous ceramics by means of composite micromechanics and ultrasonic velocity, NDT E Int. 29(2) (1996) 95–101.

Jones, R and Facaoaru, I, Recommendations for testing concrete by the ultrasonic pulse method, Mater. Struct. Res. Testing (Paris), 2(19), 275, 1969.

Jones, R and Gatfield, EN, Testing concrete by an ultrasonic pulse technique, DSIR Road Research Tech. Paper No. 34 (London, H.M.S.O., 1955).

Jones, R, Non-Destructive Testing of Concrete, Cambridge University Press, London, 1962.

Jones, R, Testing of concrete by an ultrasonic pulse technique, RILEM Int. Symp. on Nondestructive Testing of Materials and Structures, Paris, Vol. 1, Paper No. A-17 January 1954, 137. RILEM Bull., 19(Part 2), Nov. 1954.

Kaplan, MF, Compressive strength and ultrasonic pulse velocity relationships for concrete in columns, ACI J., 29(54-37), 675, 1958.

Kaplan, MF, The effects of age and water to cement ratio upon the relation between ultrasonic pulse velocity and compressive strength of concrete, Mag. Concr. Res., 11(32), 85, 1959.

Koehler, B, Hentges, G, Mueller, W, Improvement of ultrasonic testing of concrete by combining signal conditioning methods, scanning laser vibrometer and space averaging techniques, NDT E Int. 31 (4) (1998) 281–287.

Krautkramer J & Krautkramer H, in Ultrasonic testing of materials, (Springer, Berlin), 1990, 522–524.

Liang, MT, Wu, J, Theoretical elucidation on the empirical formulae for the ultrasonic testing method for concrete structures, Cem. Concr. Res. 32 (2002) 1763–1769.

Malhotra V, (Ed.), Testing Hardened Concrete: Nondestructive Methods, ACI, monograph No. 9, Detroit, US, 1976.

Mannan MA, Basri HB, Zain MFM, Islam M.N., Effect of curing conditions on the properties of OPS-concrete. Building and Environment, 2002, 37, 1167–1171.

Mehta PK, Concrete: structure, properties, and materials. Englewood, NJ: Prentice-Hall; 1986.

Meloleepszy J, Deja J, The effect of variable curing conditions on the properties of mortars with silica fume. ACI SP-132, 2; 1992, p. 1075–87.

Naik, TR, Malhotra, VM and Popovics, JS, Nondestructive Testing of concretes, part:8, The Ultrasonic Pulse Velocity Method, 2004.

Neville, AM and Brooks, JJ, Concrete Technology, Longman Scientific and Technical, Harlow, 1987.

Neville, AM, Properties of Concrete, Longman Group UK, London, 1995.

Ould Naffa, S, Goueygou, M, Piwakowski, B, Buyle-Bodin, F, Detection of chemical damage in concrete using ultrasound, Ultrasonics 40 (2002) 247–251.

Popovics, S, Rose, JL, and Popovics, JS, The behavior of ultrasonic pulses in concrete, Cem. Concr. Res., 20, 259, 1990.

Qasrawi, HY, Concrete strength by combined nondestructive methods simply and reliably predicted, Cement and Concrete Research, 30 (2000), 739–746.

Ramamurty, K, Narayanan, N, Factors influencing the density and compressive strength of aerated concrete, Magn. Concr. Res. 52, (3), (2000), 163–168.

Ramezanianpour AA, Malhotra VM, Effect of curing on the compressive strength, resistance to chloride-ion penetration and porosity of concretes incorporating slag, fly ash or silica fume. Cement and Concrete Composites 1995, 17(2):125–33.

Robson, TD, High-Alumina Cements and Concretes, Wiley, New York, 1962.

Sayers, CM, Dahlin, A, Propagation of ultrasound trough hydrating cement pastes at early ages, Adv. Cem. Based Mater. (1) (1993) 12–21.

Sturrup, V, Vecchio, F, Caratin, H, Pulse velocity as a measure of concrete compressive strength, in: V.M. Malhotra (Ed.), In situ/Nondestructive Testing of Concrete, ACI SP-82, Detroit, 1984, pp. 201–227.

Sturrup, VR, Vecchio, RJ, and Caratin, H, Pulse Velocity as a Measure of Concrete Compressive Strength, ACI SP 82-11, American Concrete Institute, Farmington Hills, MI, 1984, 201.

Swamy, NR and Al-Hamed, AH, The use of pulse velocity measurements to estimate strength of air-dried cubes and hence in situ strength of concrete, Malhotra, V.M., Ed., ACI SP 82, American Concrete Institute, Farmington Hills, MI, 1984, 247.

Tan, KS, Chan, KC, Wong, BS, Guan, LW, Ultrasonic evaluation of cement adhesion in wall tiles, Cem. Concr. Compos. 18 (1996) 119–124.

Tharmaratnam, K, Tan, BS, Attenuation of ultrasonic pulse in cement mortar, Cem. Concr. Res. 20 (1990) 335–345.

Toutanji AH, Bayasi Z, Effect of curing procedures on properties of silica fume concrete. Cement and Concrete Research 1999; 29(4):497–501.

Turgut, P, Research into the correlation between concrete strength and UPV values, Journal of Nondestructive Testing, Vol.12, No.12, 2004.

Whitehurst, EA, Soniscope tests concrete structures, J. Am. Concr. Inst. 47 (1951 Feb.), 443–444.

Ye, G, Lura, P, Van Breugel, K, Fraaij, ALA, Study on the development of the microstructure in cement-based materials by means of numerical simulation and ultrasonic pulse velocity measurement, Cement and Concr. Comp., Vol. 26, Issue 5, 2004, 491–497.

Excellence in Concrete Construction through Innovation – Limbachiya & Kew (eds)
© 2009 Taylor & Francis Group, London, ISBN 978-0-415-47592-1

The effects of the curing technique on the compressive strength of Autoclaved Aerated mortar

T. Ungsongkhun
Matrix Co. Ltd., Bangkok, Thailand

V. Greepala
Kasetsart University Chalermphrakiat Sakonnakhon Province Campus, Thailand

P. Nimityongskul
Asian Institute of Technology, Pathumthani, Thailand

ABSTRACT: The purpose of this research is to study the effects of the curing technique on the compressive strength of autoclaved aerated lightweight mortars. The proportions of quick lime, cement and aluminum powder were kept constant at 10%, 90% and 0.4% by weight of binder respectively throughout the study. The experimental investigation involved the determination of appropriate curing techniques of autoclaved aerated mortars based on the compressive strength of the mortars. Test results showed that the optimum curing condition based on compressive strength was using curing temperature of 160°C for duration of 8 hours. There was a long-term retrogression of strength of lightweight mortar investigated in this study. Most of the curing conditions showed higher compressive strength at early age and the strength gradually decreased at later age.

1 INTRODUCTION

The aerated concrete manufacturing process consists of the creation of macro-porosity (called introduced porosity) in a micro-mortar matrix made of cement, lime, sand and water with the help of an expansive agent. The agent, generally aluminum powder, reacts with the water and the lime liberated by the hydration of the binder (Wittman 1983). The gaseous release generated by this chemical reaction causes the fresh mortar to expand and leads to the development of pores, which give aerated concrete its well known characteristics: low weight and high thermal performance (Narayanan & Ramamurthy 2000). The widespread utilization of aerated concretes in building as lightweight-bearing elements requires the use of ever more mechanically efficient materials (Cabrillac & Malou 1996). Moreover, the high porosity of aerated concretes, essential to their main function, which is thermal insulation, leads to very poor mechanical strength compared to normal concrete. The quantity of pores and the pores' distribution mainly influence the mechanical properties (Alexanderson 1979). The most common technique to make up for this lack of strength is an autoclave treatment performed under high pressure and high temperature to create Autoclaved Aerated Concrete (ACC) (Narayanan &

Ramamurthy 2000), but this is economically and environmentally costly (Cabrillac et al. 2006).

In Thailand, demand for ACC products is increasing rapidly due to the increasing volume of concrete construction in the homebuilding industry and the outstanding properties of AAC. The lower weight of AAC products results in faster building rates and is easier to handle than conventional materials. In addition, AAC products can replace normal weight concrete and result in more economic design due to the reduction of dead loads. As a result, it seems necessary to conduct research and apply appropriate technology to produce good quality AAC products for construction purpose and other needs.

The aim of this research is to study the effect of autoclaving technique on the compressive strength of lightweight mortars. The investigated parameters are curing temperature and duration. The experimental study was conducted using $50 \times 50 \times 50$ mm standardized samples. Nine samples were cast for each autoclaving condition in order to have a representative average of measured characteristics. Aluminum powder was used as an expansion agent. Water to cementitious ratio was varied for all mixes in order to maintain the same consistency. After casting and rising of the slurry, the test samples were removed from the mould and cured in an autoclave under the investigated conditions. Subsequently, the test samples were retained in a climatic room. Compressive strength tests (TIS 1998) were conducted to investigate the mechanical properties when the test samples had ages of 1, 3 and 7 days.

2 EXPERIMENTAL INVESTIGATION

The purpose of this research is to determine the suitable high pressure steam curing condition. Although high pressure steam curing involves three basic factors, which are temperature, time and pressure, the main parameters investigated were only the curing temperature and time. The pressure used in curing was kept constant at 2 bars throughout. The experimental study was carried out in four steps: specimen preparation, curing in autoclave, testing of compressive strength, and comparison of test results as shown in Figure 1. In order to obtain the suitable high pressure steam curing cycle and make a comparison for each parameter, the eight curing conditions were experimentally conducted under the same mix proportion as shown in Table 1. Two levels of curing temperature, are 160°C and 180°C, were conducted to investigate the influence of curing temperature at the same curing duration. The influences of curing duration at each level of curing temperature were investigated by varying the curing time from 4, 6, 8 and 10 hours. To achieve the appropriate curing condition for this study, the compressive

strengths at 1, 3 and 7 days were investigated and compared for all of the curing conditions.

2.1 Materials used

The binder consisted of Ordinary Portland Cement (OPC) Type I and Quick Lime or calcium oxide (CaO), which was used in the slurry to increase the rate of hardening. The chemical composition of the binder is shown in Table 2. Natural river sand passing ASTM sieve No. 50 and retained on sieve No. 100 (ASTM 2004b) was used as fine aggregate. The physical properties of sand used are shown in Table 3. A fine silvery aluminum powder was used as an expansion agent which will react with hydrated lime ($Ca(OH)_2$) and produce hydrogen to form air bubbles in the matrix. Ordinary tap water was used throughout the experiment. During mixing, water to cementitious ratio was varied for all mixes in order to keep the same consistency according to ASTM C124-71 (ASTM 1971). All of the mixes were cast in formworks having dimensions of $50 \times 50 \times 50$ mm moulds according to ASTM C109 (ASTM 2004a).

2.2 The autoclaved aerated concrete manufacturing process

Under laboratory conditions, sand, cement, aluminum powder, quick lime, and water were weighed by a digital weighing machine.

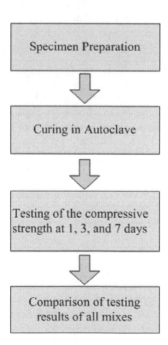

Figure 1. Flow chart of experimental processes.

Table 1. Curing conditions and mix proportion to determine the optimum curing process.

Curing destination	Maximum temperature (°C)	Time at maximum temperature (hours)	Composition			Aluminum powder (% of binder)	Water/cementitious ratio
			Dry solid (%)				
			Binder (55%)		Sand (45%)		
			Quick lime	Cement			
C160H4	160	4	10	90	45	0.4	0.5
C160H6	160	6					
C160H8	160	8					
C160H10	160	10					
C180H4	180	4					
C180H6	180	6					
C180H8	180	8					
C180H10	180	10					

* Note: Pressure used in autoclaving is 2 bars

The sand was mixed with quick lime which was combined with cement. Next, all of the dry solid was introduced in a mixer, then water and the expansion agent were added. Subsequently, the slurry was cast into 50-mm cube moulds to about 3/4 full and vibrated. The slurry was left to rise to exceed the top of the moulds and set, and then the excessive part was cut to a desirable shape while it was still soft. After removing the moulds, the 50-mm cube specimens were cured in an autoclave under high-pressure steam curing, which consists of several temperature envelopes as shown in Figure 2. Finally, the specimens were moved from the autoclave and were ready for compressive strength testing.

2.3 *Testing of the compressive strength of Autoclaved Aerated Mortar*

After the period of curing in an autoclave under investigated condition, the test samples were allowed to dry at room temperature. The investigations consist of determining the compressive strength at 1, 3 and 7 days according to TIS 1505-2541 (TIS 1998).

The 50-mm cube test samples were measured and recorded the dimensions according to TISI Standard: TIS 1505-2541 (TIS 1998), then the test sample was placed in the testing machine and load was applied perpendicularly to the direction of rising during manufacture at a constant rate of 25–35 s. The maximum load

Table 2. Chemical composition of binder used to produce lightweight mortars.

| Oxide | Concentration (% wt) | |
	Cement	Quick lime
Na_2O	0.09	–
MgO	0.88	1.83
Al_2O_3	4.64	0.21
SiO_2	21.45	1.39
SO_3	2.47	0.19
K_2O	0.59	–
CaO	64.99	93.44
TiO_2	–	–
Fe_2O_3	3.05	0.01
Traces	1.84	2.93
Total	100.00	100.00
Loss of Ignition (LOI)	1.33	15.26

Table 3. Physical properties of sand.

Properties	Sand
Specific Gravity	2.55
Absorption (%)	0.98
Dry Rodded Unit Weight (kg/m^3)	1635
Moisture Content (%)	4.75

at which the specimens fail was recorded. The compressive strength is the failure load of each specimen divided by the area over the load.

3 RESULTS AND DISCUSSION

The compressive strengths at 1, 3 and 7 days of 50-mm cube test samples subjected to eight curing conditions, determined from the average of 3 specimens, are summarized and shown in Table 4. It can be observed that the curing at temperature of 160°C for 8 hours (C160H8) showed the highest compressive strength at every age of testing. Curing No. C160H4, which had a temperature of 160°C for 4 hours, resulted in the lowest compressive strength at the same ages. The effects of curing temperature and duration on the compressive strength of lightweight mortar are as follows:

3.1 *Effect of temperature*

The effects of curing temperature on compressive strength at different ages of lightweight mortars under the same mix proportion are considered for each curing

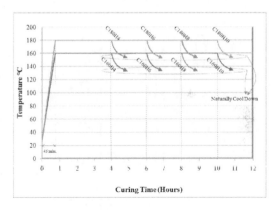

Figure 2. Temperature envelopes used in curing.

Table 4. Compressive strength at different ages of lightweight mortars subjected to various curing conditions.

| Curing No. | Compressive strength (ksc) | | |
	1 Day	3 Days	7 Days
C160H4	15	19	18
C160H6	29	29	25
C160H8	36	37	30
C160H10	33	29	26
C180H4	28	23	24
C180H6	25	23	21
C180H8	30	28	24
C180H10	33	33	29

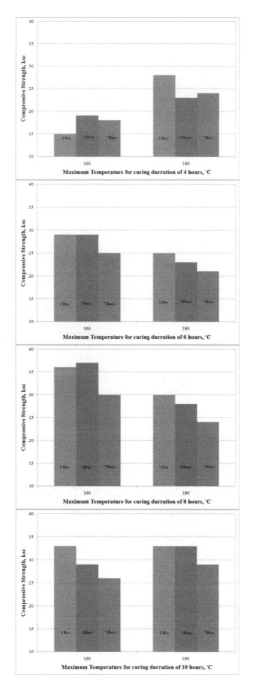

Figure 3. Effects of curing temperature on compressive strength at different ages of lightweight mortars.

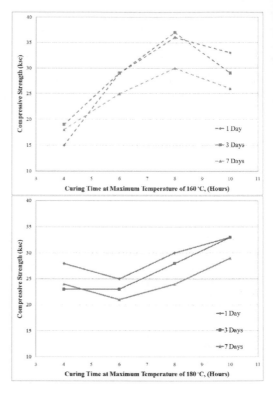

Figure 4. Effect of curing duration on compressive strength at different ages of lightweight mortars.

the use of a curing temperature of 180°C resulted in higher compressive strength at every age. However, for the medium curing duration (6–8 hours), the use of a curing temperature of 160°C produced higher compressive strength. It was also found that at the long curing duration (10 hours), the compressive strength of specimens which were cured at temperature of 180°C was slightly higher than that of the specimens cured at a temperature of 160°C.

3.2 *Effect of curing duration*

The effects of curing duration on compressive strength at different ages of lightweight mortars under the same mix proportion are considered at the same level of curing temperature. The relation between compressive strength and curing time for temperature of 160°C and 180°C are summarized and shown in Figure 4. It was found that at the same curing temperature of 160°C, the compressive strength of lightweight mortar significantly increased as the curing time was increased until 8 hours; subsequently an increase in curing duration slightly decreased the compressive strength. At the curing temperature of 180°C, the compressive strength of lightweight mortar slightly increased as the curing

time. The relation between compressive strength and curing temperature for curing times of 4, 6, 8 and 10 hours are summarized and shown in Figure 3. It can be seen that for the short curing duration (4 hours)

time was increased. However the curing period of 6 hours induced the lowest compressive strength.

Moreover it can be observed from the comparison of the compressive strength of lightweight mortar subjected to eight curing conditions that using a longer period of curing at a lower temperature leads to a higher strength than using a higher temperature for a shorter time, as proposed by Neville (1997). Neville (1997) also concluded that for any one period of curing there is a temperature which leads to an optimum strength. Hence, the curing at a temperature of 160°C for duration of 8 hours, which gave the highest compressive strength, was the optimum curing condition.

It is interest to observe that most of curing conditions showed higher compressive strength at early age than that at later age. The compressive strength at 1 day seemed to be the highest and thereafter there was a fluctuation in compressive strength for later age. However the compressive strength at 1 day was higher than that at 7 days. In other words, there was a decrease of the compressive strength of the test samples. This is due to the fact that a rise in the curing temperature speeds up the chemical reactions of hydration which appears to form products of a poorer physical structure, probably more porous, so that a proportion of the pores will always remain unfilled (Neville 1997).

The adverse effects of a high early temperature on later strength has been proven by Verbeck and Helmuth, who suggest that the rapid initial rate of hydration at higher temperatures retards the subsequent hydration and produces a non-uniform distribution of the products of hydration within the paste. The reason for this is that, at the high initial rate of hydration, there is insufficient time available for the diffusion of the products of hydration away from the cement particles and for a uniform precipitation in the interstitial space (as is in case at lower temperatures). As a result, a high concentration of the products of hydration is built up in the vicinity of the hydrating particles, and this retards the subsequent hydration and adversely affects the long-term strength (Neville 1997).

4 CONCLUSIONS

Based on the results obtained from this study, the following conclusion can be drawn:

1. The optimum temperature and duration of autoclave curing based on compressive strength and density were 160°C and 8 hours respectively.

2. The use of a longer period for curing at a lower temperature leads to a higher strength than using a higher temperature for a shorter time.
3. There was a long-term retrogression of strength of lightweight mortar investigated in this study. Most of the mixes showed higher compressive strength at an early age and the strength gradually decreased at a later age.

REFERENCES

Alexanderson, J. 1979. Relations between structure and mechanical properties of autoclaved aerated concrete *Cement Concrete Composite*, 9.

ASTM. 1971. Method of Test for Flow of Portland-Cement Concrete by Use of the Flow Table, ASTM C124-71, *Annual Book of ASTM Standards*, American Society for Testing and Materials, West Conshohocken, United States.

ASTM. 2004a. Standard Test Method for Compressive Strength of Hydraulic Cement Mortars (Using 2-in. or [50-mm] Cube Specimens), ASTM C109. *Annual Book of ASTM Standards*, American Society for Testing and Materials, West Conshohocken, United States.

ASTM. 2004b. Standard Test Method for Sieve Analysis of Fine and Coarse Aggregates, ASTM C136-95a. *Annual Book of ASTM Standards*, American Society for Testing and Materials, West Conshohocken, United States.

Cabrillac, R., Fiorio, B., Beaucour, A.-l., Dumontet, H.l.n., & Ortola, S. 2006. Experimental study of the mechanical anisotropy of aerated concretes and of the adjustment parameters of the introduced porosity. *Construction and Building Materials*, 20, 286–295.

Cabrillac, R., & Malou, Z. Problems about optimisation of porosity and properties of aerated concretes. *International congress concrete in the service of the mankind*, Dundee, Scotland.

Narayanan, N., & Ramamurthy, K. 2000. Structure and properties of aerated concrete: a review. *Cement Concrete Composite*, 22.

Neville, A.M. 1997. *Properties of Concrete*, New York: John Wiley & Sons.

TIS. 1998. Autoclaved Aerated Lightweight Concrete Elements, TIS 1505–2541. Thai Industrial Standards Institute.

Wittman, F. 1983. *Development in civil engineering. Autoclaved aerated concrete, moisture and properties*, Netherlands: Elsevier.

Excellence in Concrete Construction through Innovation – Limbachiya & Kew (eds)
© 2009 Taylor & Francis Group, London, ISBN 978-0-415-47592-1

Mechanical properties of concrete encased in PVC stay-in-place formwork

K.G. Kuder, C. Harris-Jones, R. Hawksworth, S. Henderson & J. Whitney
Seattle University, Seattle, WA, USA

R. Gupta
Octaform Systems, Inc, Vancouver, BC, Canada (Currently at British Institute of Technology)

ABSTRACT: Stay-in-place formwork can be an attractive alternative to traditional formwork (steel or wood) and is known to improve constructability and produce a more durable final product. However, the influence of the formwork on mechanical performance is not completely understood. Recently, the effect of a patented polyvinyl chloride (PVC) stay-in-place forming system on the compressive and flexural performance of the concrete that it encases was examined. The results indicate that the PVC encasement enhances both compressive and flexural performance. Compressive strength is increased by the confining action of the PVC and flexural performance is improved due to the increased tensile capacity of the sections under flexural loading.

1 INTRODUCTION

Stay-in-place formwork can be used to improve the constructability and durability of the concrete that it encases. The formwork components are assembled on-site, braced and the concrete is then poured. After the concrete has gained sufficient strength, the bracing is removed and the stay-in-place formwork remains, providing an exterior shell.

Octaform Systems, Inc. has developed a stay-in-place formwork technology that consists of interlocking extruded polyvinyl chloride (PVC) components that can be assembled to form straight or curved walls. Figure 1 presents a retaining wall constructed using the PVC stay-in-place formwork, which can be used for any type of vertical wall structure. The PVC panels are assembled on-site, as shown in Figure 2, and allow for steel reinforcement to be used as needed, as depicted in Figure 3.

The PVC stay-in-place formwork is typically used for barns and holding tanks, where a protective barrier is needed to protect the concrete and/or when curved walls are present, but is also used for retaining walls and in residential construction. In addition to being easy to construct, the walls are easy to clean, bacteria-, insect- and rodent- resistant and meet the two hour fire rating (Intertek Testing Services NA Limited 2000).

The PVC encasement consists of a variety of interlocking parts, some of which are shown in Figure 4.

Figure 1. Retaining wall encased with PVC stay-in-place formwork.

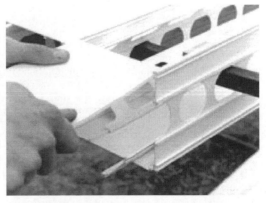

Figure 2. PVC stay-in-place formwork cell assembly.

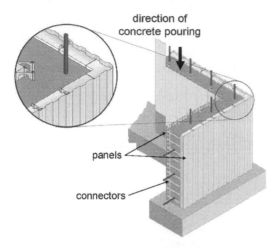

Figure 3. Isometric view of PVC-encased wall with steel reinforcement.

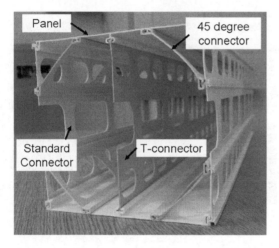

Figure 4. Interlocking PVC components (top view).

The most basic Octaform configuration consists of six inch wide panels that form the exterior of the concrete wall, as well as standard connectors that run through the cross-section of the concrete wall and come in varying widths depending on the wall depth. In addition to this basic configuration, 45 degree braces and T-connectors can be inserted into the wall, depending on the application. The T-connector and 45 degree connectors are designed to strengthen the formwork and carry the lateral pressure of the concrete during pouring. Concrete is poured from the top of the walls, vertically through each cell, as shown in Figure 3.

The PVC components are assembled into a variety of configurations based on the construction circumstances and intended application of the system. Figure 5 presents four common configurations and indicates

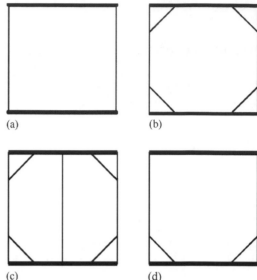

Figure 5. PVC configurations: (a) Configuration 1, Present in all Octaform Walls, (b) Configuration 2, Stabilizes formwork during concrete pouring in thick walls or when concrete not used, (c) Configuration 3, Configuration with all components, used to resist additional lateral movement during erection and concrete pouring phases of construction and (d) Configuration 4, Insulated walls, custom insulation piece added to side of wall opposite 45° braces.

the situations in which each would be used. These configurations represent a single cell that could be replicated many times to create a wall.

The research presented herein investigates the effect of the PVC stay-in-place formwork on the compressive and flexural performance of the concrete that it encases. The influence of different PVC formwork configurations on composite performance is determined experimentally. In addition, an analytical model is developed to explain the effect of the PVC on the flexural behavior of steel reinforced sections.

2 EXPERIMENTAL PROGRAM

The influence of the PVC stay-in-place formwork on compressive and flexural performance was evaluated for the four configurations shown in Figure 5. The performance of these specimens was compared to control specimens that were cast using wooden forms. Compression testing was conducted on cubic specimens, while flexural performance was evaluated for unreinforced and steel reinforced beams. In the field, concrete is poured vertically through the cells, as is shown in Figure 3, allowing the cement paste to penetrate between the interlocking components before hardening, likely increasing the bond between the PVC elements and the concrete. To replicate this final-state condition, specimens were cast by pouring the concrete through the connectors of an assembled section.

2.1 Materials

The concrete mix design typically used in the field was selected for this study. The mix design was as follows: cement – $350 \, kg/m^3$; coarse aggregate – $1150 \, kg/m^3$; fine aggregate – $700 \, kg/m^3$; water $170 \, kg/m^3$; superplasticizer – $600 \, ml/m^3$; air entrainer – $200 \, ml/m^3$. Lafarge Type I cement was used. The coarse aggregate had a maximum particle size of $10 \, mm$ and the fine aggregate was river sand. The superplasticizer was Glenium 3000 NS (poly-carboxylate-based) and the air entrainer was MB-VR Standard. Both of these admixtures are produced by BASF Admixtures, Inc.

The reinforcing steel was a #3 deformed reinforcing bar, with a tensile yield strength of $690 \, MPa$ and an elastic modulus of $200 \, GPa$. The extruded PVC had an inelastic response that is characterized by an ultimate tensile strength of $43 \, MPa$ and an elastic modulus of $2.7 \, GPa$ (Octaform Systems Inc. Accessed August 2006).

2.2 Mixing procedure

The concrete was mixed in a Goldblatt rotary drum mixer. First, the dry ingredients (cement, coarse aggregate, and fine aggregate) were placed in the drum and mixed for five minutes. Then the wet ingredients (superplasticizer, air entrainer, and water) were added to the mixer and combined for two more minutes. The sides of the drum and the turning blades were scraped to prevent adhesion to the drum mixer and mixing was continued to achieve a more homogeneous mixture.

To help ensure mix consistency, a slump test was performed according to ASTM C-143 "Standard Test Method for Slump of Hydraulic Cement Concrete (ASTM C143-03 2003)". Slump measurements ranged from 210–235 mm.

2.3 Specimen preparation

2.3.1 Compression

Compression specimens were $152.4 \times 152.4 \times 152.4 \, mm$. In addition to the four PVC configurations (Fig. 5), a plain concrete sample (no PVC) was prepared. Wooden molds were prepared that were slightly larger than the PVC configurations to avoid using bracing, as is done in the field. The PVC components were assembled and then placed into the molds. The concrete was cast through the connectors in two lifts, with each lift consolidated by rodding the concrete. Specimens were covered with wet burlap for 24 hours and then demolded and immersed in water. After 48 days the specimens were removed from the water and were

Figure 6.   Casting procedure for configurations 1, 2 and 4 – concrete poured through the connectors.

tested at 49 days. Six replications were made for each configuration as well as for the control.

2.3.2 Flexure

The influence of the PVC encasement on flexural performance was examined for both unreinforced and reinforced beams. For the unreinforced specimens, configurations 1, 2 and 3 as well as a plain specimen (no PVC) were examined. Steel reinforced flexural specimens were cast with the four PVC configurations (Fig. 5) as well as with a plain concrete (no PVC). Six replications were made for each configuration as well as for the control.

The beam size selected was 152.4×152.4 with a length of 609.6 mm for easy handling and to facilitate laboratory testing. For the reinforced specimens, #3 rebar was used as longitudinal reinforcement with approximately 38.1 mm clear cover and hooked ends to provide sufficient development length. In addition, transverse reinforcement was provided at a spacing of 63.5 mm to avoid a brittle shear failure.

Flexural specimens were cast to simulate field casting as best as possible. For configurations 1, 2 and 4, the concrete was poured through the connectors, as is shown in Figure 6. However, for configuration 3, this process was not effective due to the presence of the T-connector, which prevented concrete flow to the bottom half of the specimen. Therefore, for configuration 3, the bottom layer was first poured in from the side of the mold and then the next layer poured through the connectors. Note that this consolidation issue does not exist in the field since the concrete is poured vertically, as shown in Figure 3. After the beams were cast, they were covered with wet burlap for 24 hours and then submerged in water. After seven days the beams were removed from the water and left to cure in ambient laboratory conditions. Beams were tested between 41–43 days after casting.

Table 1. Compression strength of concrete encased in PVC stay-in-place formwork.

| Configuration | Compression strength (MPa) | |
	Average	Std. Dev.
Control	27.6	4.1
1	39.3	3.4
2	35.2	2.1
3	28.9	4.1
4	35.2	1.4

2.4 Mechanical Testing

2.4.1 Compression

A Riehle hydraulic testing machine, instrumented with a 1,334 kN load cell was used for the compression testing. The load was applied through a spherical ball bearing steel plate so that uniform loading could be applied. Compression loading was applied in piston-displacement control at a rate of approximately 0.085 mm/sec and the load was recorded. Compressive strength was defined as the maximum axial load sustained divided by the cross-sectional area.

2.4.2 Flexure

Three-point bending was applied using a Riehle hydraulic testing machine with a 1,334 kN capacity load cell. Beams were simply supported with a span of 508 mm. Center-point displacement was measured with a linear variable displacement transducer (LVDT) that had a stroke of ±25.4 mm. Loading was applied at a piston-displacement control rate of 0.085 mm/sec. Load and center point deflection were recorded. Beams were tested to a center-point deflection of 7 mm. Flexural toughness was determined as the area under the load versus center-point-deflection curve up to the 7 mm deflection.

3 RESULTS

3.1 Compression

Table 1 summarizes the average compressive strength of the four different PVC-encased systems and the control specimen (without PVC). In general, the PVC encasement increases the compressive strength when compared with the control. Based on the average of the six replications, configuration 1 has the highest compressive strength, configuration 3 has the lowest compressive strength (although still high than the control) and configuration 2 and 4 have similar strengths.

The enhancement in compressive strength is believed to be due to the confining action of the PVC

Figure 7. Control specimen after peak compressive load reached.

Figure 8. PVC-encased specimen after peak compressive load reached.

cell. Confinement is known to improve the compressive behavior of concrete (Considère 1902, Richart, et al. 1928). As microcracks begin to form, normal strength concrete expands laterally. Assuming deformation compatibility between the confining material and the concrete, lateral stresses develop in the confining material and act against the expansion of the concrete. For normal strength concrete, the effectiveness of the confining material depends on the geometry of the cross section, the deformation compatibility between the two materials and the properties of the confining material (Mirmiran & Shahawy 1997, Mirmiran, et al. 1998). Researchers have shown steel jackets and ties, as well as fiber reinforced polymer (FRP) spirals, wraps and jackets to be extremely effective at enhancing compression strength and ductility under loading (Nanni & Bradford 1995, Richart, et al. 1928).

Typical failures for the control and PVC-encased specimens are shown in Figures 7 and 8, respectively.

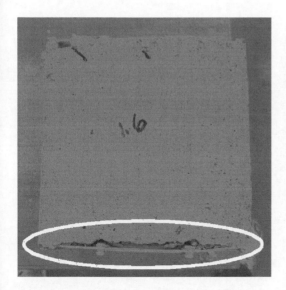

Figure 9. Top-view of PVC-encased specimen after compression testing with connector-debonding failure.

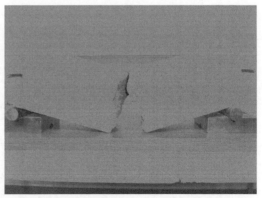

Figure 10. Plain, unreinforced beam after failure.

Figure 11. PVC-encased specimen after 7 mm deflection.

The presence of the PVC-encasement helps to contain the concrete once failure has occurred, which could be useful in the case of extreme loading. Visual inspection of the concrete during and after the compression tests suggests that failure occurs due to the debonding of the connector from the concrete. In all tests, the connector debonded from the concrete before the panel. This trend is expected since, if the PVC is resisting lateral expansion of the concrete, the weakest point would be at the connectors where the concrete is not evenly covered by the PVC and can expand freely between the connector openings. This issue would not arise in the field since as multiple cells would be constructed and, therefore, concrete would cover the connectors on both sides.

The results presented in Table 1 suggest that the specific PVC configurations affect compressive behavior, with configurations with more components having lower compressive strengths. However, it is not clear whether or not this trend is due to testing artifacts or due to material behavior for a number of reasons. First, as the number of PVC components increases, consolidation of the concrete becomes more difficult. This issues was observed in the laboratory specimens and might not be a concern for field application. Second, specimens were not ground or capped before testing so the compression loading might not have been applied uniformly. Finally, an uneven state of confinement pressure is known to exist with prismatic sections because of the stress concentrations that exist at the corners (Campione 2006). However, the data do suggest that compressive strength increases with the presence of the PVC.

3.2 Flexure

3.2.1 Unreinforced concrete

Figures 10 and 11 show an unreinforced plain and PVC-encased specimen, respectively, after flexural testing. Due to the brittle nature of unreinforced concrete, the plain beams fail suddenly once the peak load is reached, losing all load carrying capacity. All PVC-encased specimens, however, remained in tact up to 7 mm of deflection even without the presence of reinforcement.

Figure 12 presents the load versus deflection behavior for configuration 1. The initial load-deflection relationship is characterized by a steep ascent, after which a dramatic decrease in the load carrying capacity occurs due to the failure of the concrete. Subsequently, the beams sustain more load as the capacity of the beam increases up to the 7 mm deflection. This increase in capacity is most likely due to the PVC components that pick up the load. Good reproducibility among the beams is seen in Figure 12. Similar repeatability was observed for the other configurations.

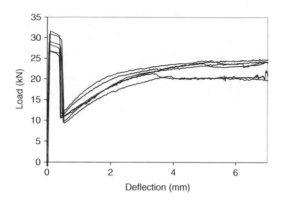

Figure 12. Load versus center-point deflection for configuration 1.

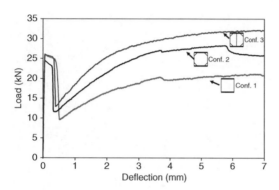

Figure 13. Representative load versus deflection curves for the unreinforced beams with PVC encasement.

Table 2. Maximum flexural load and toughness up to 7 mm center-point deflection.

Configuration	P_{max} (kN)		Toughness (kN-mm)	
	Average	Std. Dev.	Average	Std. Dev.
1	30.29	5.35	112.77	9.01
2	30.59	2.98	171.84	4.58
3	34.18	2.66	189.01	6.15

Figure 13 presents representative load versus deflection curves for configurations 1, 2 and 3. In addition, Table 2 summarizes the flexural performance of the three configurations. Due to the brittle and catastrophic nature of the failure for the plain composites, no further analysis was possible. However, it is clear that the PVC encasement enhances the flexural performance of the concrete specimens. Little effect is seen on the initial portion of the load-deflection curve due to the low modulus of the PVC. However, in the post-peak, the type of PVC configuration appears to

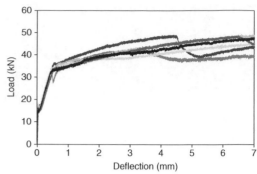

Figure 14. Load versus deflection for control specimens (no PVC).

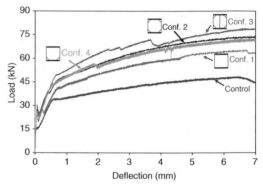

Figure 15. Representative load versus deflection behavior for PVC-encased and control specimens.

influence performance with the composites with more PVC having enhanced capacity at higher deflections and larger toughness values.

3.2.2 Reinforced concrete
3.2.2.1 Experimental results
Figure 14 presents the load versus deflection behavior for the control beams (no PVC). Good repeatability is observed with similar repeatability trends noted for the other configurations tested.

Failure modes of the composites suggest a flexure-initiated failure. For the control specimens, the first observed crack was always a flexure crack. After further loading was applied, either flexure or shear cracks appeared. With the PVC components, observing the cracking pattern of the concrete was more difficult. However, it did appear that the failure patterns were similar to the control, with initial cracking due to flexural loading and subsequent cracking due to either flexure or shear. These observations for the PVC-encased composites are based on the cracking observed through the connectors.

Figure 15 presents representative flexural load versus center-point deflection curves for the four

Table 3. Flexural performance of PVC-encased and control specimens.

| Configuration control | P_{max} (kN) | | Toughness (kN-mm) | | Increase Over Control (%) | |
	Average	Std. Dev.	Average	Std. Dev.	P_{max}	Toughness
	48.7	3.1	277.8	16.9	–	–
1	67.8	2.4	390.3	14.9	39	41
2	77.9	7.2	422.6	13.4	60	52
3	80.7	3.0	444.5	15.5	66	60
4	78.0	9.9	411.5	10.4	60	48

Figure 16. PVC-encased specimen after 7 mm deflection with concrete still contained.

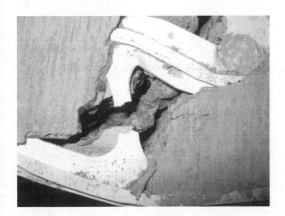

Figure 17. Weakest-point of PVC-encased specimen after 7 mm deflection.

configurations tested as well as the control. In addition, Table 3 summarizes the average data obtained for all the specimens tested. Presence of the PVC clearly increases both flexural strength and toughness. Configuration 1 has the smallest increase in load capacity and toughness, 39 and 41%, respectively, when compared to the control (no PVC). Configuration 3 shows the greatest increase in load capacity and toughness compared to the control, 60 and 66%, respectively. Configuration 2 and 4 have similar performances that fall between the two.

During the test, the control specimens began to lose concrete at higher loads, while the PVC-encasement contained the concrete. Figures 16 and 17 show a PVC-encased specimen loaded beyond 7 mm center-point deflection with the concrete still contained and after the connector has failed at the weakest cross section, respectively. This containment of concrete is advantageous in the event of impact, seismic or blast loading.

3.2.2.2 Analytical modeling

The flexural data suggest that PVC encasement improves flexural performance of the concrete systems and that the extent of the improvement depends on the PVC-configuration. To predict the effect of PVC encasement on flexural performance, an analytical model was developed based on limit state analysis. Using this model, the moment capacity of each of the PVC configurations was determined. Chahrour and colleagues (Chahrour, et al. 2005) examined the effect of a PVC-encasement system using a similar approach and found good agreement between experimental results and analytical predictions. However, they did not examine the influence of different PVC configurations. Similar approaches have also been taken to model the flexural performance of reinforced concrete with externally bonded FRP.

Since the beams investigated in this study were relatively deep beams, a non-linear analysis would be required to accurately model flexural performance (Nawy 2003). Therefore, analytical results are not compared directly with the experimental results, but rather, are used to help describe the influence of the PVC encasement on the flexural performance of the composite system.

The moment capacity of the configurations is modeled using limit state analysis. The following assumptions are made:

- Euler-Bernoulli beam theory applies.
- Perfect composite action is assumed, meaning that no slip occurs between the PVC and the concrete or between the steel and the concrete up to a deflection of 7 mm.
- Tensile forces carried by the cracked concrete (below the neutral axis) and compressive forces carried by the PVC are neglected.

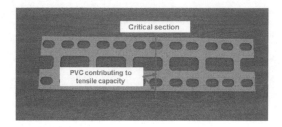

Figure 18.   PVC connector with critical section.

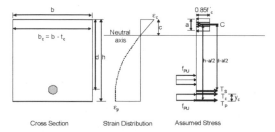

Figure 19.   General PVC-Encased Cross Section with Strain Distribution and Assumed Stress.

- The confining action of the PVC in the compression zone is neglected and the compressive forces are modeled using the Whitney Stress Block.
- The critical section occurs where the PVC connectors have the least amount of PVC (at the location of the openings – Fig. 18).
- The contribution of the PVC component between the top ellipse and the large center opening (Fig. 18) is neglected in the moment capacity, as the strain in this section is relatively small

Figure 19 presents a general cross section of the beams, with the strain distribution and the assumed stresses. The width of the cross section is b (mm); the width of the concrete section is b_c (mm); the thickness of the standard connector is t_c (mm); the height is h (mm); the depth to the reinforcing bar is d (mm); the depth to the neutral axis is c (mm); the strain in the concrete is ε_c (mm/mm); the strain in the polymer is ε_p (mm/mm); the depth of the Whitney stress block is a (mm); the compressive strength of the concrete is f'_c (MPa); the tensile strength of the polymer is f_{pu} (MPa); the compressive force in the concrete is C (kN); the tensile force in the PVC panel is T_p (kN) and the tensile force in the steel is T_s (kN). The tensile force in the connector is T_c (kN) and the centroid of the connector measured from the bottom panel is y_c (mm). The values for T_c and y_c vary for each configuration.

The forces acting in the cross section are:

$$C = 0.85f'_c ab_c \tag{1}$$

$$T_s = A_s f_y \tag{2}$$

Table 4.   Limit state analysis of flexural behavior of different PVC-encased concrete specimens and control.

	A_c (mm^2)	A_p (mm^2)	y_c (mm)	a (mm)	M_n (kN-mm)	Increase M_n over control (%)
Control	–	–		13.7	5249.9	–
Panel only	–	181.9	–	16.2	6325.8	20.5
Conf. 1	83.4	181.9	31.5	17.2	6706.3	27.7
Conf. 2	146.5	181.9	28.2	17.9	7002.1	33.4
Conf. 3	181.9	181.9	29.7	18.4	7180.4	36.8
Conf. 4	146.5	181.9	28.2	17.9	7002.1	33.4

$$T_c = A_c f_{pu} \tag{3}$$

$$T_p = A_p f_{pu} \tag{4}$$

Where A_s is the area of the steel (mm^2), f_y is the yield strength of the concrete (MPa), A_c is the area of the concrete in compression (mm^2) and A_p is the area of the PVC in tension (mm^2).

By equilibrium, the depth of the concrete stress block is:

$$a = \frac{(A_c + A_p)f_{pu} + A_s f_y}{0.85f'_c ab_c} \tag{5}$$

The moment capacity is then calculated by:

$$M_n = A_s f_y (d - \frac{a}{2}) + A_p f_{pu}(h - \frac{a}{2}) + A_c f_{pu}(h - y_c - \frac{a}{2}) \tag{6}$$

The results from the limit state analysis are present in Table 4. A_c and y_c are calculated for each cross section given the assumptions stated above. In addition, the contribution of the panel alone is considered. The model predicts that the PVC encasement will enhance flexural performance and that the extent of the improvement depends on the amount and location of the PVC components. The highest moment capacity is given by configuration 4, followed by configurations 2 and 5 and, finally, configuration 1. These trends are similar to those observed in the experimental part of work and suggest that the PVC-encasement enhances the flexural performance by increasing the tensile capacity of the cross section.

4   CONCLUSION

The influence of various PVC component configurations on compression and flexural behavior was evaluated. Based on this work, the following conclusions can be drawn:

- PVC encasement increases compressive strength. The increase in strength is attributed to the confining action of the PVC. The maximum load appeared to be reached when the connectors debonded from the concrete.

- The PVC encasement significantly increases the flexural performance of both unreinforced and reinforced concrete. The data suggest that the type of PVC configuration influence the extent of the enhancement.
- Limit state analysis indicates that composites containing more PVC in the tension regions will have a higher moment capacity. Data from the experimental work support these findings.
- PVC encapsulations provide significant reduction in spalling under compressive and flexural loading.

ACKNOWLEDGEMENTS

This work was sponsored by Octaform Systems Inc and the Seattle University Senior Design Project Center. Cement, aggregates and admixtures were supplied by Lafarge, Glacier Northwest and BASF Admixtures, Inc.

REFERENCES

ASTM C143-03 2003. Standard Test Method for Slump of Hydraulic Cement Concrete. 04.02

G. Campione 2006. Influence of FRP Wrapping Techniques on the Compressive Behavior of Concrete Prisms. *Cement and Concrete Composites* 28(5): 497–505.

A. H. Chahrour, K. A. Soudki and J. Straube 2005. RBS Polymer Encased Concrete Wall Part I: Experimental Study and Theoretical Provisions for Flexure and Shear. *Construction and Building Materials* 19(7): 550–563.

M. Considère 1902. Etude Theorique de la Resistance a la Compression du Beton Frette (The Resistance of Reinforced Concrete to Compression) (in French). *Comptes Rendus de l'Academie des Sciences*

Intertek Testing Services NA Limited 2000. Report of a Pilot Scale Fire Test Program Conducted on Vinyl Encompassed Concrete Wall System. 491–7930.

A. Mirmiran and M. Shahawy 1997. Behavior of Concrete Columns Confined by Fiber Composites. *Journal of Structural Engineering* 123(5): 583–590.

A. Mirmiran, M. Shahawy, M. Samaan, H. E. Echary, J. C. Mastrapa and O. Pico 1998. Effect of Column Parameter on FRP-Confined Concrete. *Journal of Composites for Construction* 2(4): 175–185.

A. Nanni and N. M. Bradford 1995. FRP Jacketed Concrete Under Uniaxial Compression. *Construction and Building Materials* 9(2): 115–124.

E. G. Nawy 2003. Reinforced Concrete: A Fundamental Approach.

Octaform Systems Inc. Accessed August 2006. http://www.octaform.com.

F. E. Richart, A. Brandtzaeg and R. L. Brown 1928. Study of Failure of Concrete Under Combined Compressive Stresses. *University of Illinois – Engineering Experiment Station – Bulletin* 26(12): 102.

Excellence in Concrete Construction through Innovation – Limbachiya & Kew (eds)
© 2009 Taylor & Francis Group, London, ISBN 978-0-415-47592-1

Study and comparison between numerical and mathematical method for optimizing structures

S.A.H. Hashemi & E. Hashemi
Islamic Azad University of Qazvin, Iran

ABSTRACT: The "Optimization Theory" presents how to find the best. First "Optimization" is defined, and how it is recognized or measured. It is very difficult to achieve the "Best", because too many parameters such as complicated analyze and design equations and their reciprocal effects, implementation realities, increase in expenses and etc, prevent finding "The Best". A better answer compared with previous ones has been scrutinized here. In the present paper, three appropriate methods have been investigated. These methods are "Calculation", "Computational" & "Random". In addition Optimization purpose, Optimization different methods, Differences between Random and other methods, the relation between Random method & Nature visions, the History & Procedure of Random methods have been studied and analyzed, and finally their Preferences have been mentioned.

1   INTRODUCTION

Since optimizing theory studies how to acquire the "most fitness" ones, it should be define optimization and how to measure it and tis goodness and badness recognition. It may be said that optimizing is the science of searching manner of advancement methods towards the optimum points. Note that the former definition includes two parts: 1- advancement process 2- optimum points and there is a clear difference between these two parts. When assessment of an optimizing method only its convergence is considered (whether the method acquired the optimum point) and its first part, that's advancement; is often forgotten. This is while advancement is important for nature.

2   WHAT'S THE GOAL OF OPTIMIZING?

It may be said, however, that acquiring the "most fitness" isn't the only assessment Criterion in many life affairs but movement towards the "most fitness" is operation criterion. Goal of optimizing in complicated issues, then, isn't the best solution but closeness to optimum solution accepts it. In structure engineering it's fitfully impossible to acquire the "most fitness" solution. This is because some various parameters result in giving up the "most fitness" solution, such as complicated relations governing the analysis & designing and their interactions, executive realities, raising the costs due to the more complicated calculations and etc. Advancement and acquiring a more

fitness solution than previous ones however may be considered (Shirani, 1996).

3   VARIOUS OPTIMIZING METHODS

There are overall three optimizing methods including the calculus based methods, enumerative methods and random methods (Shirani, 1996). Calculus based methods are more well-known than others. In calculus-based methods, optimizing math model is defined as below:

Cost finding a n-variables vector as $x = x(x_1, x_2, \ldots, x_n)$ from designing Parameters which minimize the f(x) function

$$f(x) = f(x_1, x_2, \ldots, x_n) \tag{1}$$

Under stipulations (2-2) and (3-2)
   K is a function as:

$$h_i(x) = h_i(x_1, x_2, \ldots, x_n) = 0 \quad \text{k i}= 1 \text{ through} \tag{2}$$

m is a Non-function as:

$$g_j(x) = g_j(x_1, x_2, \ldots, x_n) \leq 0 \quad \text{j}=1 \text{ through m} \tag{3}$$

There are two methods to minimize the function f(x) including the direct and indirect methods (Kulaterjari, 2001). Local optimum Points in indirect method are determined through Solution of some non-linear functions when function equals to zero. A primary hypothesis from x-value is speculated in direct method and

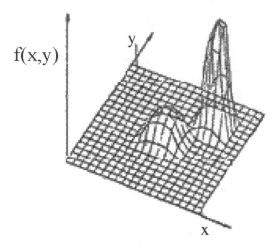

Figure 1. Function f (x,y) has two peak paints while one of them is real maximum point of function (Shirani, 1996).

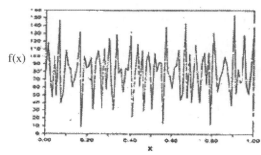

Figure 2. Function F(x) is non-continuous with many variations.

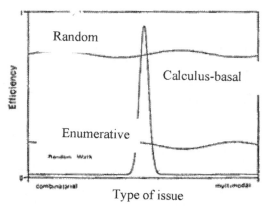

Figure 3. Efficiency diagram of different optimizing methods.

x-value, then, is calculated through error and trial method in which x should be minimize function f(x). note that these two methods require an even & continuous function and Optimum points may produce , locally. For in stance, look at the figure (1) and you understand that there are two answers for question if it is calculated by calculus-based methods of optimum points while there is only one optimum point for function overlay and this is a weakness for these methods.

Calculus-based methods, on the other hand, require a continuous and derivable function while many issues in the nature have non-continuous and fully-variations functions. For instance, Figure (2) illustrate that function f(x) isn't derivable and has many variations. Calculation of optimum point through calculus-based methods there fore is impossible. It many be said then that calculus-based methods aren't "able" to calculate the optimum points of various functions and available issues in the nature.

Enumerative methods are frequently applied in daily life for instance, suppose that you want to find the tallest student of the school you should measure height of all students of the school and then choice the tallest student amongst them. As you see the range of variables should be short in enumerative method so that this method may be used. There are, however, many issues in the nature with extensive uncountable range. Enumerative methods, therefore, may not be used for them and it may be said that enumerative methods are also "unable" ones.

Regarding to weakness Point of calculus-based and enumerative methods, notoriety of random methods is raised day after day. These methods apply a random model to acquire the best answer and their answers are more reliable than enumerative methods in long executions and they may have non-continuous range of variations. Note that there is difference between these methods and "randomized acquisition of answer". These methods apply random selection as a tool to achieve the goal.

Random selection may look surprising in the first view but nature proves its contrast. Genetic algorithm is an example of random methods which apply random selection tool. It should be note that random selection in these methods never leads a certain route and it is just as a tool. Random methods may be example of an "able" method.

As mentioned above, both calculus-based & enumerative methods are unable methods. This never means that they are unsafe methods. In contrast, they are the best selections to acquire the answer in some issues. In an overall selection, able methods (may be a combination of aforesaid methods) are more efficient. Figure 3 illustrates efficiency diagram of different methods. As it can be seen, enumerative methods have less efficiency. Calculus-based methods have high efficiency in a small scope of various issues and haven't efficiency in many other issues. Able methods are a relatively high efficiency in all issues.

4 DIFFERENCE OF GENETIC ALGORITHM AS A RANDOM METHOD AGAINST OTHER OPTIMIZING METHODS

Genetic algorithm, thereinafter referred as GA, is different from other optimizing methods in the four flowing cases:

1. GA doesn't deal with issues' parameters but use their converted forms into acceptable codes.
2. GA use a group of points, nor a unique point.
3. GA deals with the value of function and there are no requirements to derivative of function derivative and the math information's.
4. GA uses the rule of probable change, nor the rule of decisive change.

5 GA AND INTUITION OF NATURE

Phrase of struggle for existence often associates us with its negative idea value. This is true for the phrases of jungle rule and existence rule of more power being.

More power binges were not always the winner. For example, while dinosaurs were giant & powerful, they defeated in the play of struggle for existence and many weaker animals continue their life. It's clear that the nature doesn't select the "most fitness" in the basis of their body! Indeed, it's better to say that the nature select the fittest, nor the strongest one.

According to natural selection rule only those species of a population may reproduce which have the most fitness characteristics and those who haven't these characteristics will be destroy gradually.

For example, suppose that a particular species of individuals are more intelligent than other ones in an ideal society or colony. In the completely natural conditions, these individuals advance more and acquire more convenience and convenience itself also results in long life and better reproduction. If this characteristic (intelligence) is heritage, subsequent generation of the same population naturally will be contained more intelligent individuals.

If this procedure is kept it will be seen that our sample society will be include more and more intelligent individual during consequent generations.

Thereby, simple natural mechanism successes to delete less intelligent individuals of a population within a few generations as well as increase mean intelligence level of a population constantly.

There by, it may be seen that nature could constantly enhance various characteristics of each generations using a very simple method (gradual deletion of inappropriate species and the same time higher reproduction of optimum species).

What mentioned above, of course, may not alone describe what occur really through evolution in nature. Optimization and gradual evolution alone may not assist the nature to create the fittest samples. Let to describe this issue by an example.

Better automobiles with higher speeds and applications than primary ones were produced gradually some years after its first invention. Naturally, these new automobiles are result of struggles of design engineers to optimize the previous models. Note that optimization of an automobile result in a "better automobile".

But, if are may say that this procedure is true for invention of aero plane? Or if we may say that spaceship are results of optimization of primary aero planes design.

Reply is that though aero plane invention surely affected by attainments of automobile industry, it may never says that aero plane is simply result of automobile optimization or spaceship is result of aero plane optimization. This procedure is true in the nature. There are more evolve species that we may not say that they are simply result of gradual evolution of previous species. Concept of "random" or "mutation" may help us to recognize this issue.

In other words, aero plane design is a mutation comparing to automobile design, nor attainment of a gradual evolution. This is true in the nature. Some characteristics change completely random in every new generation. They, then, are kept through gradual evolution mentioned above only if this new random characteristic may meet needs of nature. They, otherwise, will be automatically deleted from nature cycle.

6 HISTORY OF GENETIC ALGORITHM

Genetic algorithm is derived from nature. Nature evolution or Darvin theory was formation basis of genetic algorithm. Darvin believed that nature has an evaluative route. He believed that recent intelligent human is evaluative monkey of thousands years ago (Kulaterjari, 2001). While his theory isn't proven yet and there is any fossil evidence confirming this reality, it may be seen, on the other hand, evolution process in the nature. For example, skull volume of primitive man was less than recent one and this fact was confirmed by fossils acquired from men skulls. Figure 4 schematically shows evdution process of men.

Increasing the brain volume during the history, of course, doesn't mean that there was no intelligent but indicates that there was no intelligent man at past or there is no low intelligent man at present but indicates that intelligence factor in man was enhanced through changing the generations. Another example in this field is insects' resistance against the poisonous against. Insects were killing using these poisonous agents at past but these old Poisonous agents have no effects on insects at present, any more. There're thousands other examples in nature indicating the compatibility of beings with environmental conditions.

 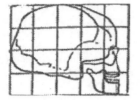

Figure 4. Evolution process of men skulls during various periods of history.

Table 1. String score table.

No	String	Score	Score percent
1	01101	169	14.4
2	11000	576	49.2
3	01000	64	5.5
4	10011	361	30.9
		1170	100.0

As we know chromosomes are distributed completely random when reproduction. Which factor may result in that new generation of beings have more compatibility with environmental conditions?

Biologists reply that higher-degree chromosomes in random combination have more chance than other ones (this chance may be factor of factor and other factors which result in more compatibility with environment for next generation). Briefly, we may say that genetic algorithm is also founded in the basis of random combination of chromosomes and raising the combination chance for choromosoms with higher degrees.

7 OPERATIONAL PROCEDURE OF GENETIC ALGORITHM

While operational mechanism of genetic algorithmic simple, it has so effect on advancement towards the optimum solution. Description of string concept is necessary to review the genetic algorithm Astringes asset of consequent zero & one figures. (Other strings are also defined- in genetic algorithm but zero & one strings are more common). Number of consequent zero & one depends on type of is sue and its various Parameters. For example, 01101 is a string with length 5. Note that genetic algorithm only deals with strings, nor issue's parameters. First step to use of genetic algorithm, therefore, is changing related data of issue into associated strings (issue coding). As said before, genetic algorithm deals with a set of points. This means that at first genetic algorithm forms a set of strings. This set may be includes 5, 10, 15, 20, strings which are determined regarding to type of issue. Here suppose that our set includes four following strings:

$A_1 = 01101$

$A_2 = 11000$

$A_3 = 01000$

$A_4 = 10011$

Above strings are formed randomly in genetic algorithm for example, suppose that a coin is fifed 4*5 times.

We consider figure one for tail and figure zero for head. Thereby, a random set of strings is formed which

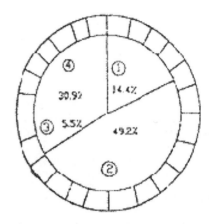

Figure 5. A sample of roulette wheel.

is referred as "first generation" set in genetic algorithm (Kulaterjari, 2001).

Here, we describe three main operators in genetic algorithm:

1. Natural selection operator
2. Mutual combination (cross over) operator
3. Mutation operator

These operators are discussed later.

Natural selection is a procedure in which a string with higher score increases its lasting chance in the system. Please, look at table (1-2). Suppose that the score of each selected string in the last section is according to column 3 of this table. Score percents of string is calculated by dividing these scores into total scores. Column 4 of table 1 presents percents of scores.

Natural selection has different methods. Simplest method of natural selection is selection based on roulette wheel and its division regarding to score percent of each string. Figure 5 illustrates a sample of roulette wheel.

Parents of next generation are selected using roulette wheel. This wheel is turned in terms of strings of available set (4 strings in this example) and a point is, selected randomly. Desired string, then, is selected regarding to position of this point. A string with higher score percent, therefore, has more chance to selection.

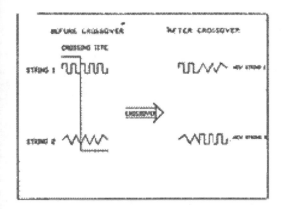

Figure 6. A scheme of cross over operator. As the figure shows, strings 1 and 2 are combined together and make a new string.

Task of natural selection operator is finished after selection of first generation parents. These parents, then, are delivered to mutual combination (cross over) operator. Mutual combination operator is also a random one. Figure 6 describes operational procedure of this operator.

Parents of new generation, then, are paired. Each string is combined with another random selected string in order to produce offspring's.

Combination procedure is through random selection of a figure (such as k) more than one and less than (L-1) regarding to string length (L). both string of k are divided into two parts. Resulted cross over then is produced as shows in Figure (6-2). Note that probability of mutual combination operator equals to Pc (Pc ranges 0.8 through 1). This means that all parents don't combine together and a few of them are transferred into next generation without change. For instance, string A1 and A2 (offspring 5 members) are produced as below:

$$\begin{cases} A_1 = 0110[1 \\ A_2 = 1100[0 \end{cases} \Rightarrow \begin{cases} A_1' = 01100 \\ A_2' = 11001 \end{cases}$$

The last operator is mutation operator. This operator may be deleted from genetic algorithm but its existence improves results. Its task, in fact, is a mechanism in which an organized & completely random change is produced in a string. Result of change may improve nature of string. It should be note that variation size should not be large since this may raise probability of changing the string operation. This operator is available because mutual combination operation

of set string may often divert from the best absolute operation towards a better relative operation. In other words, function will incline towards the relative optimum instead of absolute optimum. Mutation increases probability of acquiring the absolute optimum. It should be remembered that raise the mutation percent (through raise the variation), decreases the algorithm convergence process. If positive mutation is available, natural selection operator will accept it but operator will reject negative mutations.

Operational procedure of mutation operator is through changing a particular position (such as J) in some strings from zero into one or vice versa. Mutation probability of genetic algorithm ranges 0.001 through 0.05 (Shirani, 1996).

Intelligent GA method successfully find global optimum design without regarding to restricting hypothesises such as continuity of search space or existence of derivatives. This method changes a stipulated optimizing issue, with definition of a particular penalty function and its combination to objective and making a modified objective function, into an unstipulated optimizing issue.

As mentioned before, genetic algorithm is derived from rules of natural genetic and natural selection in which only the fittest beings have right to existence. GA operates through coded design variables which are a string of character with finite length.

8 RESULTS

We described that calculus-based and enumerative methods are unable ones. This doesn't mean that these are unseful methods while they are the best selection in some issues to acquire the optimum solution. Overall able methods (a combination of aforesaid methods), however, are more efficient than enumerative methods. Calculus-based methods are high in a small scope of various issues and are inefficient ones in many other issues. Able random methods are more efficient in all issues.

REFERENCES

Kulatejari, V, 1380/2001, optimizing the roof-trusses using genetic algorithm an algebric method of forces. PHD thesis, science & industry university.
S. Ragasenaran and G.A. Vijayalakshmi pai, 2004, Neural Networks, fazzy logic and Genetic Algorithms synthesis and Applications. Published by A. Soke K. Ghosh, New Dehli, pp. 253–261.
Shirani, K, 1375/1996, intelligent designing & optimizing using genetic algorithm. MS. Thesis, Tehran university.

Excellence in Concrete Construction through Innovation – Limbachiya & Kew (eds)
© 2009 Taylor & Francis Group, London, ISBN 978-0-415-47592-1

The mathematical explanation of genetics algorithm method for optimizing structures

S.A.H. Hashemi & E. Hashemi
Islamic Azad University of Qazvin, Iran

ABSTRACT: Genetics Algorithm is very capable for optimizing structures using disconnected variations. Many scientists have investigated optimization in many fields related to "Structure Engineering" using Genetics Algorithm methods. In the present paper the Mathematical explanation of Genetics Algorithm & the reason why less degree, shorter and more point chromosomes can develop more in optimizing structures, have been investigated. In addition, an equation is optimized by Genetics Algorithm with its mathematical logic. Finally the mathematical proof for "structural blocks theory" is been explained.

1 INTRODUCTION

Application of genetic algorithm as a random optimizing method doubts us whether there is a rational to guide the GA towards the relative optimization. This paper also shows that GA is directed in the rational basis of probable change towards the relative optimization.

2 SCHEMA CONCEPT

A schema is used to describe similarity among two or more string. Schema is defined using three marks including {0, 1,*}. Mark "*" (or any other the same mark) indicates that if mark * is available, them zero or one may be considered in the string. For instance, consider schema *111*. This schema then includes one of the following strings:

{01110, 11111, 01111, 11110}

There is of course, difference between different schemas. For instance, consider two schemas 1***0 and **11*. First schema will disappear with high probability by cross over. This is because if points 2, 3, 4 are placed in the cutting place of cross over operator, the schema will be disappear. Second schema, on the other hand may disappear only in one cross over position. It may be said however that a schema with short length and high score has more chance after cross over operation. Schemas with short length and high score are referred as "building blocks". They develop in consequent generation. Here, we discuss mathematical description of optimizing issue in genetic algorithm and operational procedure of schemas.

3 WHICH SCHEMA IS DISAPPEAR ELAND WHICH ONE IS STABILIZED

As mentioned before, schemas are different. In order to more familiarity with them them it's necessary to determine following items for every schema.

1. Order of a schema
2. Length of a schema

Order of schema (H) which is illustrated by symbol O (II) describes number of fixed points in a schema number of figures zero & one in strings with base2.

O (011*1**) = 4

Length of schema (H) which is illustrated by symbol $\delta(H)$ describes length between the first and the last fixed point in schema. For instance, length of schema in schema H = 011*1** equals to

$$\delta(H) = 5 - 1 = 4$$

Effect of natural selection schemas is determined easily. Suppose that there are m similar schemas in generation *tth* in population A (t) which is illustrated as m (H,t). A (t) indicates number of available strings in the population of the generation. As we know, each string in natural selection operator is selected regarding to probability of $P_i = f_i / \sum f \cdot P_i$ indicates the string score and $\sum f$ indicates the sum of available string in the population. Now, we expect that generation $(t+1)$ includes following schemas:

$$m(H,t+1) = m(H,t).nf(H)/\sum f \qquad (1)$$

Or

$$m(H,t+1) = m(H,t).f(H)/\bar{f} \qquad (2)$$

$\bar{f}$ and n indicate scores mean of strings number of and available strings, respectively. Schemas, in other words, develop in terms of their score against to the population scores mean.

Relations (1) and (2) may be presented in another form for schemas with scores higher than scores mean:

$$f(H) = \bar{f} + c\bar{f} \qquad (3)$$

$$\Rightarrow m(H,t+1) = m(H,t)\frac{(\bar{f}+c\bar{f})}{\bar{f}} = (1+C).m(H,t) \qquad (4)$$

C is a fixed value and is calculated as below:

$$C = \frac{f(H)-\bar{f}}{\bar{f}} \qquad (5)$$

If $t=0$ is first generation, relation (4) may rewrite as below:

$$m(H,t) = m(H,0).(1+C)^t \qquad (6)$$

This is a compound interest function indicating the excellent development of schemas with high scores. For example, if $C = 0.5$, then after 5 generations:

$$m(H,t) = m(H,0).(1.5)^5$$
$$m(H,t) = 7.6m(H,0)$$

Above relation indicating 7.6 times development of schemas in the population. An interesting note here, of course, backs to variability of C value within different generations since scores mean of a population is increased by changing the generations and C value will be decreased.

Now, we study crossover operator after natural selection operator.

As we know crossover operator results in information exchange between the strings and produces new strings.

Crossover function may disappear some schemas. Look at the example below:

Suppose that we consider two following schemas of the schema A in length 7.

A=0111000

$H_1 = *1****0$

$H_2 = ***10**$

Both above schemas are divided in to two parts due to crossover operator and division point is selected randomly. Division point for above schemas is figure 4. Then:

$A = 011[\ 1000$

$H_1 = *1*[***0$

$H_2 = ***[10**$

As you see, two fixed point s of schema (H1) are separated from each other but two fixed points of schema (H2) aren't separated. Schema H1, thereby, is disappearing but schema H2 is stabilized in next generation. Regarding to these facts, it may be said that disappearance probability of scheme H1 equals to:

$$L-1 = 7-1 = 6$$

$$\delta(H_1) = 5 \Rightarrow P_d = \delta(H_1)/(L-1) = 5/6$$

Then stability profanity of schema H1 equals to:

$$P_S = 1 - P_d = 1/6$$

Thereby, disappearance probability of schema H2 equals to:

$$P_d = \delta(H_2)/(L-1) = \frac{1}{6}$$

And its stability probability equals to:

$$P_S = 1 - P_d = \frac{5}{6}$$

If P_c indicates crossover Probability, then stability Probability of each schema will be equal to:

$$P_S \geq P_c \frac{\delta(H)}{L-1} \qquad (7)$$

Combination of relations (2) and (7) result in:

$$m(H,t+1) \geq m(H,t).\frac{f(H)}{\bar{f}}[1-P_c.\frac{\delta(H)}{L-1}] \qquad (8)$$

Above relation considers synchronous effect of both natural selection and crossover operators. This relation indicates that the schema will develop in which scores are higher than scores mean of population with shorter length.

Last operator is mutation operator. As we know in order to a schema doesn't change, it's fixed points should be stabilized. Since mutation probability for each of these points is P_m and probability of mutation absence is $(1 - P_m)$, following relation may be used to

80

Table 1. Function procedure of schemas in GA.

String no.	1st generation	X value	Function value (X^2)	Proportion of score to total scores	Proportion of score to scores average	Random selection from roulette wheel
1	01101	13	169	0.14	0.58	1
2	11000	24	576	0.49	1.97	2
3	01000	8	64	0.06	0.22	0
4	10011	19	361	0.31	1.23	1
Total			1170	1.00	4.00	4.0
Average			293	0.25	1.00	1.0
Max. value			576	0.49	1.97	2.0

determine stability probability of a schema owing to mutation:

$$P_1 = (1 - P_m)^{O(H)} \qquad (9)$$

Than P1 approximately equals to: $P_m \ll 1$ If

$$P_1 = 1 - O(H).P_m \qquad (10)$$

With combination of relations (8) and (10) then:

$$m(H,t+1) \geq m(H,t).\frac{f(H)}{\bar{f}}(1 - P_c \frac{\delta(H)}{L-1} - O(H)P_m) \qquad (11)$$

Owing to above relation, a schema will develop in which scores are higher than scores mean of population with shorter length and lower order.

4 STUDY OF SCHEMA AUGMENTATION PROCEDURE

For example, if we aims to minimaize the function $f(x) = x^2$:

We consider three schemas H1, H2, H3 which are defined as below:

$$H_1 = 1****$$
$$H_2 = *10**$$
$$H_3 = 1***0$$

As it can be seen from tables (1) and (2), string 2 & are subset of schema H1 and strings 2&3 are subset of schema H2 and string 2 is a subset of schema H3.

Scores mean of schema are determined as below:

$$f(H_1) = \frac{576 + 361}{2} = 468.5$$

$$f(H_2) = \frac{576 + 64}{2} = 320$$

$$f(H3) = 576$$

Table 2. Selected schemes.

		Similar string in 1st generation	Scores mean of string associated to schema
H1	1****	2.4	469
H2	*10**	2.3	320
H3	1***0	2	576

Table 3 presents results of natural selection & crossover operators on strings & schemas.

As it can be seen from table (2), there are two dependent strings for schema H1, that's:

$$m(H_1,t) = 2$$

On the other hand, mean value scores of schema H1 equals to:

$$f(H_1) = 468.5$$

Thereby, regarding to presented theories it's expected that there are following schema H1 in population after actions of natural selection operator:

$$m(H_1,t+1) = \frac{f(H_1)}{\bar{f}} \times m(H_1,t) = 3.2$$

(3) is acceptable regarding to it's real value.

Actions of mutation & cross over operator don't change the number of schemas H1. Since in the schema, H1, cross over operator may not change this schema. On the other hand, number of bits which may change in this schema due to mutation operator equals to: (if mutation probability is $P_2 = 0.001$)

$$m.P_m = 3 \times 0.001 = 0.003$$

Bits

Regarding to its no considerable value, so schema H1 will not be changed.

Table 3. Results of natural selection & cross over operators on string & schemas.

Random position of crossover actions	Selected pair	Actions position of crossover operator	New generation of offspring	X value	Function value X^2
0110[1	2	4	01100	12	144
1100[0	1	4	11001	25	625
11[000	4	2	11011	27	729
10[011	3	2	10000	16	256
Total					1754
Mean					439
Maximum					729

Table 4. Results on schemas.

After natural selection operator				After actions of LL OPERATORS		
Schema	Calculus	Real number	Dependent strings	Calculus number	Real number	Dependent strings
H_1	3.2	3	2,3,4	3.2	3	2,3,4
H_2	2.18	2	2,3	1.64	2	2,3
H_3	1.97	2	2,3	0.0	1	4

As it can be seen from table (2) the number of schemas H1 after operation of all operators will be fixed.

There are two dependent strings for schema H2, that's:

$$m(H_2,t) = 2$$

On the other hand, scores mea value of H2 equals to:

$$f(H_2) = 320$$

It's expected, therefore, that following numbers of schema H2 are available after calculated action of natural selection operator:

$$m(H_2,t+1) = \frac{f(H_2)}{\bar{f}} \times m(H_2,t) = 2.18$$

(2) is acceptable regarding to it's real value.
It's expected that number of schemas H2 are changed as below after calculated actions of mutation & crossover operator:

$$\delta(H_2) = 1$$

$$L-1 = 5-1 = 4 \qquad \Rightarrow P_s = 1 - \frac{1}{4} = 0.75$$

$$m(H_2,t+) \geq 0.75 \times 2.18 = 1.635$$

Real number of H2 after operations of all algorithm operators equals to 2 which is compatible with above relation.

There is one dependent string for schema H3, that's:

$$m(H_3,t) = 1$$

On the other hand, scores mean of H3 equals to:

$$f(H_3) = 576$$

It's expected, therefore, that following numbers of schema H3 are available in population after calculated action of natural selection operator:

$$m(H_3,t+1) = \frac{f(H_3)}{\bar{f}} \times m(H_3,t) = 1.97$$

Figure (3) is acceptable regarding to its real value.
It's expected that number of schemas H3 are changed as below after operations of mutation & cross over operators.

$$\delta(H_3) = 4$$

$$\Rightarrow P_s = 1 - \frac{4}{4} = 0$$

$$L-1 = 4$$

$$m(H_3,t+1) \geq 0 \times 1.97 = 0$$

Real number of H3 after actions of algorithm operators equals to one which is compatible with above relation. Above example clearly shows difference amongst the different schemas. A schema with higher score, therefore, (such as H1, H2, H3) will develop in natural selection operator but schemas with longer lengths have less chance to pass cross over operator (one Point) and schemas with shorter lengths also have more chance to pass cross over operator (one point).

Natural selection operator increases chance of schemas with higher scores exponentially from one generation to another generation.

5 RESULTS

GA is guided towards the relative optimization in the basis of probable change rational. A schema will develop during the optimization process which includes shorter length and lower degree and higher score than score mean. These schemas are building blocks. Numbers of schemas in the algorithm are function of n2.

REFERENCES

A.M. Raich and J. Ghaboussi, 2000. Evolving structural design solutions using an implicit redundant Genetic Algorithm.

D. Hales. Introduction to Genetic Algorithms, www.davidhales.com.

Larsen, C and Sindholt, J. 2003. Optimization of compliant mechanisms using genetic algorithms. Department of Mathematics, Technical University of Denmark.

Malcolm I. Heywood. Holland's GA Schema Theorem, CSCI6506 – Genetic Algorithm and Programming.

S. Ragasenaran and G.A. Vijayalakshmi pai, 2004, Neural Networks, fazzy logic and Genetic Algorithms synthesis and Applications. Published by A. Soke K. Ghosh, New Dehli, pp. 253–261 ibid. p. 227–281.

Shirani, K.H., 1375/1996, intelligent and optimum designing of structure using GA. MS thesis, Tehran university.

S-Y. Chen and S.D. Rajan. Using Genetic Algorithm as An Automatic Structural Design Tool, Department of Civil Engineering, Arizona State University.

Wei Lu. 2003. Optimum Designe Of Cold-Forme Steel Purlins Using Genetic Algorithm. Helsinki University of Technology Laboratory of Steel Structures Publications.

Excellence in Concrete Construction through Innovation – Limbachiya & Kew (eds)
© 2009 Taylor & Francis Group, London, ISBN 978-0-415-47592-1

Optimising 3D structure frames using GA

S.A.H. Hashemi

Islamic Azad university of Qazvin, Iran

ABSTRACT: As it is known the first parameter for decreasing structure's expenses is the weight of the structure therefore, the designing process continues until the weight of the structure is reduced and strains & stresses are within allowable limit. Of course not only the weight of the structure but also its Topology plays a determining role.To design a structure usually after designing its form and the position of elements and loads we make the first guess for sections, and then analyze them to find the strains and stresses. The structure will be designed using the new stresses, and then the sections will be chosen to be analyzed and designed again. The process continues until the stains & stresses are allowable and minimum. The whole process can be done automatically using Genetics Algorithm. And all different sections are designed and chosen with the least weight (minimum) and allowable strains and stresses. In the present paper the procedure in which Genetics Algorithm relates with 3D frames considering joist vector, has been investigated.

1 INTRODUCTION

Direction of joists arrangement in roofs is a major parameter which is by engineers through experience or through suspicion & opinion. This parameter directly affects the weight of building. We, therefore, decided to study a method to optimize the 3-D frames regarding to direction of joists. Don't remember that not only this paper focuses on joists arrangement but also optimize the profiles of elements (beams, columns, wind brace). Following states may be study in optimization process of joists direction and profiles 3-D frame's elements using GA regarding to capacities of prepared software: 1-static analysis 2-dynamic analysis 3-making the bending 3-D frame system 4-making the joint 3-D frame system as well as wind brace system 5-making the 3-D compound bending and joint frame system 6-ability to make any type of extensive & tie load 7-analysis and designing under any desired load.

2 OPTIMIZATION PROGRAM

Computer programs conformably to GA were prepared to optimize the joists arrangement in building frame and minimize the weight of 3-D frame. This software was based on visual basic language (VB6). Since optimization of building isn't possible without building analysis, It was clear that we need a credit building analysis program. Amongst the various building analysis programs such as SAP90, SAP2000, Etabs 90, Etabs 2000 and etc, we selected SAP90 program

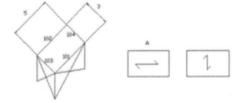

Figure 1. 3-D frame of sample no. 1.

due to its many applications and easy link into every unfamiliar program.

SAP 90 (Habiballah, 1995) is retrieved by analyzer to analyze the building when necessary in optimization process. SAPSTL processor program is used to design the building. This program may retrieve by analyzer when necessary.

3 OPTIMIZATION SAMPLE OF 3-D FRAME'S JOISTS DIRECTION AND COMPARING THEM WITH TRADITIONAL DESIGNING

3.1 Sample no.1

This sample describes functional procedure of genetic algorithm to optimize the direction of joists and 3-D profiles in Figure 1.

Loading includes vertical and lateral ones. Designing variables include direction of joists profiles of frame's elements, a live load 0.3 t/m^2 and dead load

0.2 t/m². Lateral loading is determined according to code 2800. This sample aims to find the best direction of joists and the best profiles of elements (beams, columns, braces). Mathematical formulation is described as below.

Weight objective functions f(x) is minimized under stipulations $g_i(x) \leq = 1, 2, \ldots, m$. M indicates the number of stipulations.

Since we aim to minimize the weight, f(x) for the above frame will be as following relation (1).

$$f(x) = \sum_{i=1}^{J} \omega A_j L_j + a \qquad (1)$$

Here, *Aj, Lj, W and a* indicate profile area, length of *i*th element, special weight of materials and total weight of joists round steels, respectively. Stipulations of problem include stress (2) and translocation (3).

$$\delta_j \leq \delta_a \qquad j = 1, 2, 3\ldots m \qquad (2)$$

$$2U_j \leq U_a \qquad j = 1, 2, 3\ldots \qquad (3)$$

U_j and U_a indicate floor translocation and authorized translocation, respectively.

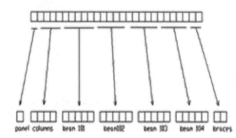

Figure 2.   Chromosome sample 1.

If designing variables are supposed discrete, their values should be looked up in available profiles lists. This paper, therefore, uses profiles available in Iran. Since GA deals with coded designing variables, it's necessary to encode the designing variables of profile area as a string. Each designing variable may select one of the profiles in the list. A binary string in length 27 therefore is sufficient for this sample each bit indicates a panel (Figure 2).

Table 1 presents values F (value of competence function), $\Phi(x)$ (modified objective function) and Cg (sum of structure Violations).

These chromosomes are divided into substrings related to panels, columns, beams and wind braces.

These substrings, then, are decoded by Analyzer program and it determines ratio of stresses and translocations and their violation value. Values in third column of Table 1, therefore, indicate sum of the structure violations. Building weight that's building weight of every chromosome is calculated and is located in columns (4) of Table 1. Value of modified objective function, then, is calculated using the relation of "rajiv and kreshnamourti". (4) and (5). (5th column of Table 1).

$$\sum_{q=1}^{\ell} \max\left[0, g_q(x)\right] \quad c= \quad \text{fpenalty=f(x).k.c,} \qquad (4)$$

$$(x)\Phi = f(x) + p_{penalty} \qquad (5)$$

$f_{penalty}$, $f(x)$ and $g_q(x)$ indicate penalty function, objective function or weight and value of building violation in terms of stipulations, respectively Φ and X indicating the stipulations governing the problem and vector of designing variables, respectively. K is constant value and equals to 10. Value of

Table 1.   1st generation of sample frame 1.

0.067 $F/\sum F$	F	$\phi(x)$	W	Cg	Chromosome number
	11723	11567.6	1051.6	1	1
8.03E-03	1444	24736	912.8	2.61	2
8.03E-03	1444	24736	912.8	2.61	3
8.03E-03	1444	24736	912.8	2.61	4
0.035	6075	17216	1076.6	1.50	5
0.035	6075	17216	1076.6	1.50	6
0.035	6075	17216	1076.6	1.50	7
0.025	4360	18931	1113.6	1.6	8
0.12	21908	1382.56	1382.56	0	9
0.12	21895	1395.6	1395.6	0	10
0.12	21846	1444.56	1444.56	0	11
0.12	21846	1444.56	1444.56	0	12
0.037	6475	16816	1051.6	1.50	13
0.119	20645	2646	2646	0	14
0.116	20228	3063.96	3063.96	0	15

competence function for every, than, is calculated from recommended relation by "Rajiv and Kreshnamourti".

$$F = [\Phi_{max}(x) + \Phi_{min}(x)] - \Phi(x) \qquad (6)$$

F and $\Phi(x)$ indicates competence value and modified objective function for every chromosome, respectively. Φmax (x) and Φmin (x) also indicate maximum and minimum value of modified objective function in current population, respectively. According to above relation, a chromosome with minimum modified objective function has maximum value of competence. Figure 3 and 4 clearly illustrates procedure of optimization.

Optimum weight is achieved (2439 kg). Figure 5 illustrates profiles of optimum design as a file.

3.1.1 Confirmation of optimization results

Now, we manually analyze the direction of joist in above sample which was optimized by prepared Analyzer program and compare them with results of the program. Directions of joist in 3-D frame of sample 1 include two states which are shown in figure (1). Frame, at first, is shown in the state of joist arrangement direction which is analyzed in figure (A-1). Frame, then, is shown in the state of joist arrangement direction which was analyzed in figure (B-1). We, finally, compare their weights.

3.1.2 Study of frame in state A

Building weight in this state along with weight of panel round steels equals to 2391 kg, shown in Table 2.

3.1.3 Study of frame in state b

Building weight as well as weight of joists round steel in this state equals to 2453kg. We At last found that suspicion of analyzer program to optimum direction of joists is correct, as shown in Table 3.

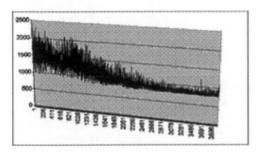

Figure 3.   Diagram of building weight loss against the GA process development. Optimum direction of joist is the same as figure 4 below.

3.2   Sample no. 2:

This sample describes functional procedure of genetic algorithm to optimization of direction of joist and profile of frame in Figure 6. Since there is no instruction on the direction of joists in the recommended plan, design may consider any desired direction of joists in the plan.

Design is acceptable only if stress of building elements and translocations don't extreme authorized levels. Analyzer program therefore optimizes direction of joists and profiles of frame element. Frame in four states then, is shown and is analyzed traditionally and its weight is calculated and is compared with results of program.

At last optimum direction of joists is acquired according to Figure 7.

Optimum weight of building equals to 2751 Kg and optimum profiles are presented as a file which is shown in Figure 8.

3.2.1   Confirmation of optimization results

Now, we manually analyze the direction of joist in above sample which was optimized by prepared

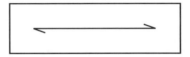

Figure 4.   Direction of optimum joist sample 1.

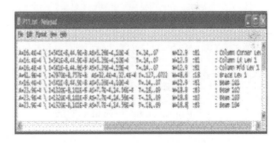

Figure 5.   Optimum profile of sample 1.

Table 2.   Presents elements designing for this state.

Element no.	4 Profiles
1	IPE 14
2	IPE 14
3	IPE 14
4	IPE 14
101	IPE 14
102	IPE 14
103	IPE 16
104	IPE 16
105	L18

Table 3. Presents frame designing in state B.

Element no.	Profiles
1	2IPE 14
2	2IPE 14
3	2IPE 14
4	2IPE 14
101	IPE 18 + PL7*ITOP
102	IPE 18 + PL7*ITOP
103	IPE 14
104	IPE 14
105	L8

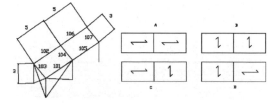

Figure 6. 3-D frame in sample 2.

Table 4. Frame designing in state A.

Element no.	Profiles
101	I 14
102	I 14
103	IPE 16
104	IPE 20
105	IPE 14
106	IPE 14
107	I 16
108	L8

Analyzer program and compare them with results of program. Frame, at first, is analyzed in four different states which are shown in figures (A-4), (B-4),(C-4) and (D-4) and then we compare their weights.

3.2.2 *Study of frame in state A:*
Table 4 presents analysis of frame. Frame weight in this state along with weight of joist round steels is estimated 2712 Kg.

3.2.3 *Frame designing in states B and C:*
Since B and C are symmetry, we just study state C which is presented in Table 5. Building weight in this state equals 2752 Kg.

3.2.4 *Study of frame in state D:*
Table 6 briefly presents analysis and designing of frame. Frame weight in this state equals 2795 Kg.

As it can be seen, structure has minimum weight in state (A) of joist arrangement and suspicion of analyzer program on direction of joists is correct.

Table 5. Frame designing in states B and C.

Element no.	Profiles
101	IPE 20
102	IPE 20
103	IPE 14
104	IPE 14
105	IPE 20
106	IPE 20
107	IPE 14
108	L8

Table 6. Frame designing in sate D.

Element no.	Profiles
101	IPE 14
102	IPE 14
103	IPE 16
104	IPE 16
105	IPE 20
106	IPE 20
107	IPE 14
108	L8

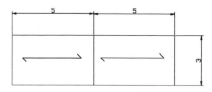

Figure 7. Optimum direction of joist in sample 2.

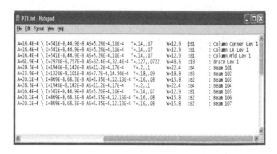

Figure 8. Optimum profiles in sample 2.

4 OPTIMIZATION OF JOISTS DIRECTION IN FRAME NO. 3:

Here, we optimize the frame in Figure 9 which is a two floors 3-D frame and each floor includes 6 panels. This frame totally includes 12 panels. Frame tolerates lateral earthquake load in two X and Y directions by wind bracing system. Earthquake load is as tie load in mass joint.

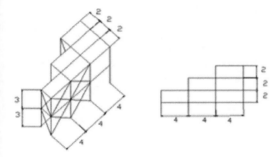

Figure 9.   3-D frame NO. 3.

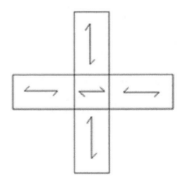

Figure 12.   Direction of joists in 3-D frame no. 4.

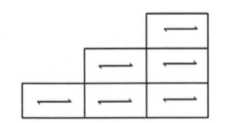

Figure 10.   Optimum direction of joists in frame no. 3.

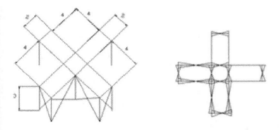

Figure 11.   Frame no. 4.

Since structure in both X and Y directions is as a simple building frame and is aligned with wind bracing system, its behavior index and authorized translocation are considered 6 and 0.015 m, respectively, program optimized the building frame with mutation and cross over (0011). Figure 10 illustrates optimum direction of joists in both floors.

5   DIRECTION OPTIMIZATION OF JOISTS IN FRAME NO. 4

Here, direction of joists in frame of Figure 11 is optimized which includes a one floor 3-D frame with 5 panels. This frame tolerates lateral load of earthquake in two directions X and Y using wind bracing system.

Loading includes vertical and lateral loads and live & dead loads are considered 0.5 t/m^2 and 0.5 t/m^2, respectively.

Authorized translocation is considered according to 2800 code criteria (0.03* H/R). H and R are floor height and building behavior index, respectively. Since structure in both X and Y directions is as a simple building frame and is aligned with wind bracing system , its behavior index and authorized translocation are considered 6 and 0.015 m, respectively.

Since structure has UN undisciplined plan during optimization process, dynamic analysis was used to analyze the structure and acquire the values of stress violation and translocations. Analyzer program in this sample was used to dynamic analysis of GA optimization process.

Dynamic load was considered as an earthquake item (north-south) occurred in 1912, may at "Alsentro" which answer spectrum in direction of 30 degree affects the building axis.

Since frame totally has 5 panels, according to error and trial method there are 2^5 possible state for direction of joists (Figure 12). In this sample corner columns are considered as type 1. Wind braces are also considered as type 1.

Designing specifications include: Elasticity module   E = 2.1e6 kg/cm^2; steel volume unit weight W = 7.85e − 3 kg/cm^3; loading includes the vertical and lateral loads. We suppose that this is a general building. According to code criteria, minimum live and dead load in technical buildings & construction are considered 0.5 t/m^2 and .5 t/m^2, respectively.

Since this frame in clouds 12 panels, according to error and trial method there are 2^12 possible states for direction of joists. In this sample corner columns are considered as type, lateral columns as type 2 and middle columns as typer 3.

Wind braces of first and second floors are considered as type1 and type 2, respectively. According to 2800 code criteria, relative authorized translocation is considered 0.03H/R which H and R are floor height and structure behavior index, respectively.

6 RESULTS

Regarding to optimized samples which were compared with analysis results we concludes that this program determines direction of joists correctly and optimization is acceptable. Regarding to results of "optimization of joists, direction in steel frames using GA" the best direction of joists arrangement to decrease the weight loss of 3-D joint steel frame is long direction.

REFERENCES

Habiballah, ovilson, general program of static & dynamic analysis of structures SAP90, A, Hoda press, fourth 1995.
Habiballah, ovilson, designing program of steel structure SAP90, A, Hegradi press, fourt 1999.

Excellence in concrete construction – through innovation

S.K. Manjrekar

Sunanda Speciality Coatings Pvt. Ltd., Mumbai, Maharashtra, India

ABSTRACT: The placement of conventional concrete mixtures in underwater construction results in a high percentage of material loss owing to washout of cement paste. The current paper presents the indigenous development of antiwashout admixture (AWA) and its influence on various properties of concrete. The concrete mixtures were tested for slump, slump flow, washout resistance and compressive strength. Test results indicated that the use of an AWA ensures lower washout of cement and facilitates the production of flowable concrete mixtures. The self-flowing underwater concretes developed 28-day compressive strengths ranging from 20 MPa.

1 INTRODUCTION

In mega hydro power structures and for mass underwater concreting projects like construction of Weir, concreting for dams, erection of cassions, concrete placed under water is inherently susceptible to cement washout, laitance, segregation, cold joints, and water entrapment. It must posses some unique workability characteristics. The essential workability requirements are that the concrete must

(a) Flow easily,
(b) Retain adequate cohesion against washout and segregation,
(c) Possess self-consolidating characteristics (because it is impractical to consolidate concrete under water by using mechanical vibration).

Underwater concrete is also required to remain cohesive while being placed under water. A high degree of cohesiveness improves homogeneity and strength of the underwater concrete by minimizing cement washout. The required degree of concrete cohesion, however, depends on many variables such as the thickness and configuration of placements, flow distance, required in-place strength, and exposure to flowing water during placements.

Concrete placed underwater typically undergoes a wide range of Kinetic states i.e. concrete falls through a tremie pipe at a high velocity, it mixes and flows out of the tremie pipe at slower speeds and finally consolidates under pseudostatic condition.

Underwater concrete must be able to easily flow out of the tremie pipe, completely fill the placement area and consolidate under its own weight.

It is reported that the workability of underwater concrete should be higher than 175 mm slump which helps in self consolidation under its own buoyant weight.

The workability of underwater concrete includes additional requirements such as self-leveling and high anti-washout characteristics.

2 WEIR PROJECT AT SRISALEM, HYDERABAD

M/s Patel Engineering Ltd, Mumbai who are the pioneers in Heavy Civil engineering construction especially in hydro power structures since 1950's were awarded the job as turnkey contractors for the construction of a weir project at Sri Salem, Hyderabad.

2.1 Project details

Project Title:
Weir Project at Srisalem, Dam, Srisalem.
Clients:
Andhra Pradesh Power Generation Company (APGENCO)
Clients Consultants:
SNC Lavelin Limited, Canada: SNC Lavelin Limited, Canada
Contractors:
M/s. Patel Engineering Limited, Hyderabad.
Concreting for Weir:
30,000 cum: 30,000 cum
Dredging quantity:
40,000 cum: 40,000 cum.
Estimated Project Cost:
App. USD 62 Million.

2.2 Mix design detail of underwater concrete as recommended

- Desired Strength: 20 MPa grade of concrete
- Slump requirements: 170 mm
- The use of Anti-washout admixture for tremie concrete.
- The effectiveness of Anti-washout admixture shall be measured in accordance with the United States Corps of Engineers specification CRD-C 61— "Test method for determining the resistance of freshly mixed concrete to washing out in water". issued on 1st DEC 1989.
- The maximum washout shall not exceed 8% cumulative mass loss.
- The tremie concrete mix shall include water reducing retarding and Anti-washout admixtures. Also 15% of the cementitious material shall consist of a natural pozzolan or fly ash. Alternatively 8 to 10% silica fume may be used instead of a natural pozzolan or fly ash.
- The slump shall vary between 170 and 230 mm with admixtures.

Table 1. Successful trial using anti washout admixture.

Successful trial	
W/C	0.45
Cement	7.980 kg
Fly ash	2.660 kg
C. A. – 40 mm	20 kg
20 mm	10 kg
Sand	9.580 kg
Dust (0–5)	9.580 kg
Water	4.880 l
Polytancrete NGT	70 ml
SUNPLAST AWA®	10 ml
Slump	170 mm

3 NEW BEGINNING

M/s Patel Engineering, Mumbai and the consultants started searching for the Anti-washout admixture in the Indian subcontinent and had approached all the leading construction chemical manufacturing companies.

All the companies based in India concluded that the materials are required to be imported. They were also not sure about their suitability and stated that the effectiveness can be ascertained only after laboratory and site trials. The product could have been sourced from outside India which would have meant delay in procurement of material for already awarded time bound contract. Besides, it would also mean the heavy cost due to exchange rate, freight and various duties etc.

M/s Patel Engineering approached the Concrete Materials Consultancy Division of Sunanda Speciality Coatings Pvt. Ltd, Mumbai in January 2004 and the Research & Development efforts for the advent of an Anti-washout system for the first time in the country had begun.

4 R & D EFFORTS

On global scene development of Anti-washout admixture for underwater concrete is comparatively a newer development. Though the initial works are reported to be done in early seventies more "State of the Art" works are reported in early nineties.

The project was divided into two parts:

- Product development, mix design trials and sitetesting from January 2004 to April 2004.
- Execution of 30,000 cu.m of underwater concreting in May-June 2004.

Sunanda's R&D division is recognized by ISO 9001-2000 and has in-house capabilities to develop newer materials. As a first step, extensive literature survey was conducted from international journals and other reported case studies. Soon after that the experimental programme started at our R&D centre. Almost for two months several experiments were conducted which looked at the desired requirements of Slump, Anti washout properties, Strengths after seven days, simultaneous compatibility with fly ash and cement.

After about 100 plus trials proper materials were developed which were satisfactory in terms of Anti-washout properties as well as compatibility and desired development of strength of concrete. It was also one of our ongoing R&D projects in the past and hence those results and past works became useful in developing the technology further.

Subsequent to this, various trials were conducted at the concrete laboratory of Jawaharlal Nehru Technological University, Hyderabad (JNTU). These tripartite trials were conducted in the presence of our R&D Engineers, Engineers from Patel Engineering Ltd and Concrete Technologist from JNTU.

Following is the finally developed Mix Design i.e the representative successful trial using Anti-washout mixture SUNPLAST AWA and Compatible Super plasticizer POLYTANCRETE NGT conducted at Jawaharlal Nehru Technical University in Hyderabad.

The entire focus was to develop user friendly material which can be easily handled on difficult sites in difficult terrains. Also the designed material was tested for skin irritation test after conducting trials on humans which was found satisfactory. After getting certificate of analysis of Anti-washout admixture – SUNPLAST AWA® conducted by the Govt. Approved Institution from India.

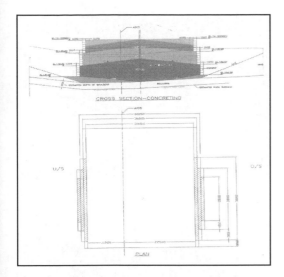

Figure 1.   Plan & Cross section of under-water concreting.

After satisfactory test results form JNTU and approval from the consultants around 10,000 ltrs Anti-washout admixture and 78,000 ltrs of Compatible Superplasticizer – POLYTANCRETE NGT worth around Rs. 45 lakhs was dispatched on the project site in various batches. These tailor made materials were manufactured and transported on the project site at Srisalem. A.P. on an average of 10,000 liters per day in eight consignments.

5   QUALITY ASSURANCE

The material was needed to be dispatched to site in quick succession sometimes at the interval of two days only. Prompt dispatches were necessary in order to maintain the heavy pace of concreting work. However as per ISO norms and in house practice, no consignment was dispatched unless quality control reports were obtained both from chemical laboratory as well as from concrete laboratory.

As per ISO guidelines quality control at all the stages was strictly adhered to which included

– Quality control of raw materials
– Process control
– Quality control of finished goods
– Quality control of Packaging.

After the actual laboratory tests, various on site tests were also conducted for monitoring and checking the behavior of concrete parameters after admixing the designed Anti-washout admixture, SUNPLAST AWA® and compatible superplasticizer – POLYTANCRETE NGT.

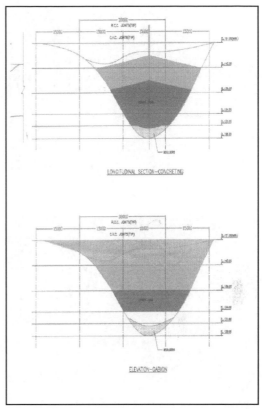

Figure 2.   Elevational & Longitudinal section of under-water concreting.

Figure 3.   Floating Pontoon.

To monitor the quality control of site, a team comprising of R&D experts, Concrete Technologists, and Quantity Surveyors were deputed from CMCD of SSCPL from Mumbai to Srisalem site. Following is the sequence of execution of the weir project at Srisalem (Refer pictorial preview).

93

Figure 4.   Gabion Shuttering.

Figure 7.   Dosing of Sunplast AWA to batching plant.

Figure 5.   Placing of Gabion Shuttering in progress.

Figure 8.   Gabion Shuttering.

Figure 6.   Concrete Batching plant erected on site.

6   CONCLUSION

The development of SUNPLAST – AWA® has had a revolutionary impact on the performance of underwater concrete. Underwater concrete can now achieve high flowability at a low water-cement ratio, and yet retain adequate cohesion to resist cement washout and concrete segregation and bleeding.

Primarily this breakthrough in the anti-washout technology was for the first time in India and will

make the construction of hydraulic structures easier. Hydraulic structures are of paramount importance for building up of industrialization and self sufficient India in terms of Power, Irrigation, Drinking Water, etc.

Use of such Anti-washout admixtures has opened an avenue for many applications in the construction industry in India wherever "Under Water Concrete" is necessary like

- Seismic Retrofit design and construction of bridges.
- Tremie Concreting to connect precast concrete pile caps.
- Underwater repair of stilling basins.
- Concreting for cassions.
- Concreting for jetties, weirs, dams, canals etc.

REFERENCES

Detwiler, R.J.; Bhatty, J.I.; and Bhattacharja, S., "Supplementary Cementing *Materials for use in Blending Cements*," Portland Cement Associations, 1996.

Gerw.ck, B.C.; Holland, T.C.; and Komendant, G.J., "Tremie Concrete for Bridge Piers and other Massive Underwater Placements," *Report No*. FHWA/RD-81/153, 1981.

Gerwick, B.C., "Placement of Tremie Concrete," *Symposium on Concrete Construction in Aqueous Enviroments*, SP-8, American Concrete Institute, Farmington Hills, MI, 1964, pp. 9–20.

Halloran, P.J. and Talbot, K.H., "The properties and Behaviour of Underwater Plastic Concrete," ACI *Journal*, Proceedings V. 39, June 1943, pp. 461–492.

Hasan, N.; Faerman, E.; and Berner, D., "Advances in Underwater Concreting: St. Lucie Plant Intake Velocity Cap Rehabilitation," High *Performance Concrete in Severe Enviroments*, SP-140, P. Zia, ed., American Concrete Institute, Farmington Hills, MI, 1993, pp. 187–214.

Kanazawa, K.; Yamada, K.; and Sogo, S., "Properties of Low Heat Generating Concrete Containing Large Volume of Blast Furnace Slag And Fly Ash," *Fly Ash, Silica Fume, Slag and Natural Pozzolans in Concrete, Proceedings*, Fourth International Conference, Sp-132, V. 1, V.M. Malhotra, ed., American Concrete Institute, Farmington Hills, MI, 1992 pp. 97–118.

Khayat, K.H.; Yahia, A, and Sonebi, M., "Applications of Statistical Models For Proportioning Underwater Concrete," ACI Materials Journal, V. 96, No. 6, Nov–Dec. 1999, pp. 634–640.

Nagataki, S., "Use of Antiwashout Underwater Concrete for Marine Structures," Tokyo Institute Of Technology publication, 1992.

Netherlands Committee for Concrete Research, "Underwater Concrete," HERON, 1973.

Tasillo, C.L., "Lessons Learned from High Performance Flowable and Self-Consolidating Concrete," U.S. Army Corps of Engineers 2003 Infrastructure Systems Conference, 2003.

U.S. Army Corps of Engineers Waterways Experiment Station, "Test Method for Determining the Resistance of Freshly Mixed Concrete to Washing Out in Water," CRD-C 61-89A, 1989.

Yao, S.X., and Gerwick, B.C., "Underwater Concrete Part-1: Design concepts Practises," *Concrete International*, V. 26, No. 1, Jan. 2004, pp. 78–83.

Yao, S.X., Berner, D.E, and Gerwick, B.C., "Assessment of Underwater Concrete Technologies for In-the Wet Construction of Navigation Structures," TR INP- SL-1, U.S. Army Corps of Engineers, 1999.

Development of eco earth-moist concrete

G. Hüsken & H.J.H. Brouwers
University of Twente, Faculty of Engineering Technology, Department of Civil Engineering, The Netherlands

ABSTRACT: This paper addresses experiments on eco earth-moist concrete (EMC) based on the ideas of a new mix design concept. Derived from packing theories, a new and performance based concept for the mix design of EMC is introduced and discussed in detail. Within the new mix design concept, the grading line of Andreasen & Andersen (1930), modified by Funk & Dinger (1994), is used for the mix proportioning of concrete mixtures. The innovative part of this approach is to consider the grading of the entire mix as an optimization problem. The formulation and solution of the optimization problem will be explained and validated by experiments on EMC. Therefore, mixes consisting of a blend of slag cement and Portland cement, gravel (4–16), granite (2–8), three types of sand (0–1, 0–2 and 0–4) and a polycarboxylic ether type superplasticizer are designed using the new mix design concept and tested on lab sale.

1 INTRODUCTION

The mass production of concrete products like concrete pipes, slabs, paving blocks and curb stones is based on the advantageous working properties of earth-moist concrete (EMC). So, in comparison to normal-weight concrete, the dry consistency of EMC allows for direct stripping of concrete products after filling and vibrating the molds. As a result, short processing times of the production process can be realized.

Traditional EMC mixes are characterized by high cement contents between 350 kg and 400 kg per m³ concrete and low content of fine inert particles. Besides, they show low water/cement ratios (w/c < 0.4) combined with stiff consistency and high degree of compactibility. As a result of low water/cement ratios, high compressive strength values are achieved by EMC mixes and the durability is improved compared to ordinary concrete.

However, despite these positive properties there is potential for improvements regarding the workability, green strength, packing density and reduction of cement content by using secondary waste materials. An optimization of the particle packing and the ratio of water in voids, in consideration of the material properties of the available raw materials, can lead to an increase in the green strength of the fresh concrete and also to higher compressive strength values of the concrete in hardened state.

This enormous potential for optimizing EMC mixes is the starting point for the development of a new and performance based mix design concept.

2 PARTICLE PACKING IN CONCRETE

The packing of the solids is of essential importance for the properties of fresh and hardened concrete. By the help of the fundamental understanding of the working mechanism of particle packing it is possible to control the behavior and the characteristics of products based on granular materials. Therefore, a lot of research into the field of particle packing was carried out in the last century. Based on the work of Fuller & Thomsen (1907), Andreasen & Andersen (1930) studied the packing of continuously graded particles and led them to the semi-empirical equation for the cumulative volume fraction $F(D)$:

$$F(D) = \left(\frac{D}{D_{max}} \right)^q \qquad \forall D \in [0, D_{max}] \qquad (1)$$

where D = particle size; D_{max} = maximum particle size; and q = distribution modulus.

Numerous publications afterwards refer to Equation 1 using a distribution modulus of 1/2 as 'Fuller curve' or 'Fuller parabola'. Also Hummel (1959) referred to the Fuller curve for composing aggregates used in standard concrete. The Fuller curve is also used by most of the European standards for composing concrete and is located within the recommended area between sieve line A and B of the Dutch standard NEN 5950 (1995) (cp. Figure 1).

Analyzing the mix design concepts it becomes obvious that the Fuller curve is applied for the grading of aggregates bigger than 250 μm only. This limitation has as a consequence that maximum amounts of fine

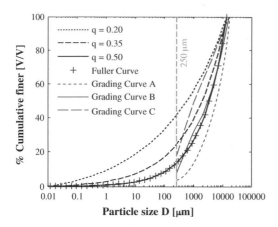

Figure 1. Equation 2 for varying distribution moduli using $D_{max} = 16$ mm, $D_{min} = 0.01$ μm and grading curves A, B and C according to NEN 5950 (1995) as well as Fuller curve.

material are prescribed but without considering the granulometric properties of the fine material. This will not be the case under practical conditions as there will always be a minimum particle size depending on the grading of the ingredients used. Accordingly, a modified version of Equation 1 was introduced by Funk & Dinger (1994) which prescribes the grading for continuously graded aggregates considering a minimum and maximum particle size in the mix. This modified equation for the cumulative volume fraction $F(D)$ reads:

$$F(D) = \frac{D^q - D_{min}^q}{D_{max}^q - D_{min}^q} \qquad \forall D \in \left[D_{min}, D_{max}\right] \qquad (2)$$

where D = particle size; D_{min} = minimum particle size; D_{max} = maximum particle size; and q = distribution modulus. According to Figure 1, it can be assumed that the Fuller curve and Equation 2 using $q = 0.5$ seem to be concordant if both graphs are compared only visual. But this is not the case as the difference of the functional values becomes more relevant the closer D approaches D_{min}.

In Equation 2, the distribution modulus q influences the ratio between coarse and fine particles and allows therefore the composition of ideal graded mixtures for different types of concrete having special requirements on the workability by using one equation for the aimed grading line. Higher values of the distribution modulus ($q > 0.5$) are leading to coarse mixtures whereas smaller values ($q < 0.25$) are resulting in mixtures which are rich in fine particles. Own experiments showed that appropriate values for q are in the range between 0.35 and 0.40 for EMC.

3   MIX DESIGN CONCEPT

In concrete technology, ingredients with a wide spread range of particle size distributions (PSDs) are combined and therefore different approaches can be found for the composition of concrete mixtures. The idea of the new mix design concept is to design a performance based concrete mix considering the granulometric properties of all solid raw materials. This idea is realized by the formulation of an optimization problem using the modified equation of Andreasen and Andersen (Equation 2).

The positive influence of the modified A&A equation on the properties of self-compacting concrete (SCC) was already shown by Brouwers & Radix (2005), and Hunger & Brouwers (2006). Furthermore, the suitability of the modified A&A equation was shown by Schmidt et al. (2005) for EMC. However, an aimed optimization of the grading line of the composed concrete mixes in consideration of Equation 2 was not carried out by Schmidt et al. (2005). The application of the modified A&A equation is constricted in this case to a comparison in the shape of the curve of the various composed mixes and Equation 2 using different distribution moduli.

Hence, it appears that an aimed composition of the concrete mix considering the grading line given by Equation 2 can result in concrete that meets the required performance properties. Therefore, an algorithm was developed which helps to compose the concrete mix according to the PSD given by the modified A&A equation based on m ingredients ($k = 1, 2, \ldots, m$), including the non-solid ingredients air and water.

For proportioning the aggregates using Equation 2, as particle size D in Equation 2 is taken the geometric mean of the upper and lower sieve size of the respective fraction obtained by sieve analysis or laser diffraction analysis as follows:

$$D_i^{i+1} = \sqrt{D_i D_{i+1}} \qquad for\ i = 1, 2, \ldots, n-1 \qquad (3)$$

The sizes of the fractions vary in steps of $\sqrt{2}$ starting from 0.01 μm up to 125 mm. Consequently, 44 sizes ($i = 1, 2, \ldots, n+1$) are present and 43 fractions (n) are available for the classification of the m-2 solid ingredients.

Taking this wide range in the PSDs of the granular ingredients into account, the entire grading of all aggregates, binders, and filler materials will be considered in the mix design to obtain an optimized packing. As mentioned in the beginning, the mix design concept will result in the formulation of an optimization problem and requires therefore the definition of the optimization problem via:

– Target value
– Adjustable values
– Constraints

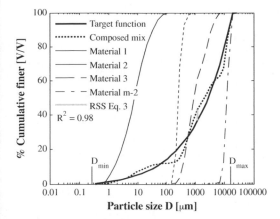

Figure 2. Composed aggregate mix; $D_{max} = 16\,mm$, $D_{min} = 0.275$, $q = 0.35$.

3.1 Target value

The target value represents the objective of the optimization problem. This value shall either be minimized or maximized. In the considered case, the deviation between the desired grading of the composed mixture $F_{mix}(D)$ (Equation 8) and the grading given by the target function $F_{tar}(D)$ (Equation 2) shall reach a minimum and results, in other words, in a curve fitting problem. To solve this curve fitting problem, the least squares technique is commonly used. Thereby, the sum of the squares of the residuals RSS is minimized. Equation 4 expresses the least squares technique mathematically.

$$RSS := \sum_{i-1}^{n} e_i^2 = \sum_{i-1}^{n} \left(F_{mix}\left(D_i^{i+1}\right) - F_{tar}\left(D_i^{i+1}\right) \right)^2 \rightarrow min! \quad (4)$$

$$\text{with } F_{tar}\left(D_i^{i+1}\right) = \frac{\left(D_i^{i+1}\right)^q - D_{min}^q}{D_{max}^q - D_{min}^q} \qquad \forall D_i^{i+1} \in \left[D_{min}, D_{max}\right]$$

$$D_{min} = D_i \qquad \text{for } F(D_{i-1}) = 0 \wedge F(D_i) > 0$$
$$D_{max} = D_i \qquad \text{for } F(D_{i-1}) < 100 \wedge F(D_i) = 100$$

As criterion for evaluating the quality of the curve fit, the coefficient of determination R^2 is used. Figure 2 shows as an example the grading of a composed aggregate mix based on 4 solid ingredients (m-2).

This figure illustrates that by combining four ingredients only, that each have their own PSD, a mix can be designed that closely follows the specified target PSD given by Equation 2.

3.2 Variables

The variables are adjustable values on the system. Changeable values are used by the optimization algorithm to approach the target value. Values which can be changed in the system are the total volume of solids

V_{sol}^{tot} and the volumetric proportion $v_{sol,k}$ of each solid ingredient.

$$V_{sol}^{tot} = \sum_{k=1}^{m-2} V_{sol,k} \qquad (5)$$

$$v_{sol,k} = \frac{V_{sol,k}}{V_{sol}^{tot}} \qquad for\ k = 1, 2, \ldots, m-2 \qquad (6)$$

The volumetric proportion $v_{sol,k}$ of each solid component influences the grading (computed sieve residue $P_{mix}(D_i)$) of the composed mix via:

$$P_{mix}(D_i) = \frac{\sum_{k=1}^{m-2} \dfrac{v_{sol,k}}{\rho_{sol,k}^{spe}} P_{sol,k}(D_i)}{\sum_{i=1}^{n} \sum_{k=1}^{m-2} \dfrac{v_{sol,k}}{\rho_{sol,k}^{spe}} P_{sol,k}(D_i)} \qquad (7)$$

where $P_{sol,k}(D_i) = $ sieve residue of solid k on the sieve with the mesh size D_i; and $\rho_{sol,k}^{spe} = $ specific density of solid k. And finally, the computed cumulative finer fraction of the composed mix is given by:

$$F_{mix}\left(D_i^{i+1}\right) = \begin{cases} F_{mix}\left(D_{i-1}^i\right) - P_{mix}(D_i) & \text{for } i = 1, 2, \ldots, n-1 \\ 1 & \text{for } i = n \end{cases} \quad (8)$$

As mentioned before, the total volume of solids V_{sol}^{tot} per m³ fresh concrete is also changed by the optimization algorithm. This value is not directly connected with the target value. Here, a connection exists via the constraints.

3.3 Constraints

Constraints are used to express real-world limits (physical constraints) or boundary conditions (policy and/or logical constraints) of the formulated optimization problem. Physical constraints are determined by the physical nature of the optimization problem whereas policy constraints are representing requirements given by standards and logical constraints are reflecting particular requirements on the designed concrete. In the considered case the following constraints have to be accounted for:

– Non-negativity constraint; $v_{sol,k} > 0$
– Volumetric constraint; $\Sigma v_{sol,k} = 1$
– Minimum cement content
– Maximum cement content
– Water/cement ration (w/c)
– Water/powder ration (w/p)
– Ratio between binder 1 and binder 2 (B₁/B₂)

Moreover, the total concrete volume V_{con} of all concrete ingredients (including air and water) per m³ fresh concrete cannot be higher or lower than 1 m³ and follows from:

$$V_{con} = V_{sol}^{tot} + V_{wat} + V_{adm} + V_{air}$$
$$= V_{agg} + V_{cem} + V_{fil} + V_{wat} + V_{adm} + V_{air} = 1\,m^3 \qquad (9)$$

Table 1. Mix proportioning of composed concrete mixes in kg per m³ concrete.

Mix	CEM I 52.5 N	CEM III/B 42.5N LH/HS	Sand 0-1	0-2	0-4	Gravel 2-8	4-16	8-16	Water	w/c	w/p	SP*
Mix 1	–	310.0	482.8	–	475.4	584.8	366.0	–	139.5	0.45	0.41	–
Mix 2	–	310.0	482.8	–	475.4	584.8	366.0	–	139.5	0.45	0.41	–
Mix 3	–	310.0	88.6	–	594.1	818.7	448.8	–	124.0	0.40	0.39	–
Mix 4	–	310.0	227.8	–	604.7	605.0	512.7	–	124.0	0.38	0.37	–
Mix 5	–	310.0	92.0	–	599.5	826.2	452.9	–	116.2	0.38	0.37	–
Mix 6	–	310.0	92.0	–	599.5	826.2	452.9	–	116.2	0.38	0.37	0.63
Mix 7	–	310.0	92.0	–	599.5	826.2	452.9	–	116.2	0.38	0.37	0.94
Mix 8	–	310.0	92.0	–	599.5	826.2	452.9	–	116.2	0.38	0.37	0.93
Mix 9	–	310.0	400.2	–	522.7	599.2	448.6	–	116.3	0.38	0.35	0.63
Mix 10	–	310.0	400.2	–	522.7	599.2	448.6	–	116.3	0.38	0.35	0.94
Blend 1	130.0	245.0	–	698.0	–	–	–	356.0	131.3	0.35	0.34	–
Blend 2	112.7	212.3	–	602.0	–	–	–	396.9	113.7	0.35	0.34	–
Blend 3	112.7	212.3	–	602.0	–	–	–	396.9	113.7	0.35	0.34	1.63
Blend 4	112.7	212.3	–	602.0	–	–	–	396.9	113.7	0.35	0.34	0.98

* SP: Superplasticizer

where V_{agg} = aggregate volume; V_{cem} = cement volume; V_{fil} = filler volume; V_{wat} = water volume; V_{adm} = admixture volume; and V_{air} = air volume.

4 CONCRETE EXPERIMENTS

To demonstrate the suitability of geometric packing for EMC and the application of the modified A&A equation, several concrete mixes have been designed and tested in the lab. The composition of the designed mixes is given in Table 1. For the mix design, three types of sand (0–1, 0–2 and 0–4), two types of gravel (4–16 and 8–16) and granite 2–8 are used. Slag cement or a blend of 65% slag cement and 35% Portland cement are used as binder. The air content per m³ fresh concrete is estimated a priori to be 0.04 m³ (=4.0%), this value needs to be verified later as it is depending on the packing fraction, the water content and the compaction efforts.

The designed concrete mixes are tested both in fresh and hardened state. The consistence of the concrete mix in fresh state is assessed by the degree of compactibility c (compaction index) according to DIN-EN 12350-4 (1999). The tested EMC mixtures resulted in values between 1.40 and 1.65 for the degree of compactibility and can therefore be classified as zero slump concrete.

Based on the mix proportioning, the packing fraction PF of the fresh concrete mixes in loose as well as dense state can be computed as follows:

$$PF = \frac{V_{sol}}{V_{ves}} = \frac{\sum_{k=1}^{m-2} \frac{1}{\rho_{sol.k}^{spe}} \frac{M_{sol.k}}{M_{con}} M_{ves}}{V_{ves}} \quad (10)$$

where V_{sol} = solid volume; M_{con} = concrete mass; $M_{sol,k}$ = mass of solid k; and M_{ves} = mess of the measuring vessel.

The computed packing fraction uses the density of the fresh concrete densely and loosely packed in a round vessel having a fixed volume of 8 liters and a diameter of 205 mm.

Furthermore, the degree of saturation of the void fraction can be computed and reflects the ratio of the total volume of water to the total volume of voids using the values of the packing fraction as well as the water content of the mix as follows:

$$S = \frac{V_w}{V_{void}} \quad (11)$$

where V_w = water volume; and V_{void} = void fraction.

The designed mixes have been poured in standard cubes, cured sealed during the first day, demolded, and subsequently cured for 27 days submersed in a water basin at $20 \pm 2°C$. After 28 days, the cubes are tested for compressive strength. The compressive strength f_c of each cube is determined according to the standard DIN-EN 12390-4 (2002). The mean values of the compressive strength of each tested series, based on three cubes per series, are depicted in Figure 3.

As the amount of cement per m³ fresh concrete as well as the applied w/c ratio is varying marginally, it can be assumed that the increase in the compressive strength for mixes having higher packing fractions is caused by an improved granular structure. The positive relation between higher values of the packing fraction due to an improved packing of the solids and improved mechanical properties is one of the basic features of the new mix design concept.

But not only the compressive strength of a mixture can be improved by an optimized and dense packing

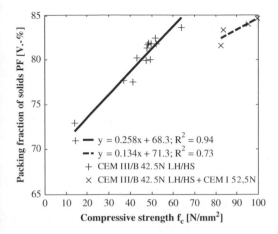

Figure 3. Measured packing fraction versus compressive strength.

Table 2. Compressive strength (28 days), packing fraction and computed cement efficiency for tested concrete mixtures.

Specimen	Compr. Strength f_c [N/mm²]	Packing Fraction [V−%]	Cement Efficiency x_c [Nm³/kgmm²]	Saturation S [%]
Mix 1	41.1	77.5	0.133	62.0
Mix 2	36.9	77.7	0.119	62.6
Mix 3	48.8	81.9	0.157	68.5
Mix 4	52.3	81.8	0.169	68.1
Mix 5	48.2	81.7	0.156	63.5
Mix 6	47.3	79.9	0.153	57.8
Mix 7	49.9	81.8	0.161	63.8
Mix 8	63.7	83.6	0.205	70.9
Mix 9	43.0	80.2	0.139	58.7
Mix 10	48.0	81.3	0.155	62.2
Blend 1	82.6	81.6	0.216	71.4
Blend 2	83.5	83.5	0.235	68.9
Blend 3	95.3	84.0	0.279	71.1
Blend 4	100.2	84.7	0.308	74.3

of all granular ingredients, but also the cement can be used more efficiently.

This is indicated by the compressive strength cement efficiency x_c or the flexural strength cement efficiency x_f of the concrete mixture and described by:

$$x_c = \frac{f_c}{M_{cem}} \text{ and } x_f = \frac{f_f}{M_{cem}} \qquad (12)$$

where f_c = compressive strength; f_f = flexural strength; and M_{cem} = mass of cement per m³ concrete. The computed values for the compressive strength cement efficiency are given in Table 2 combined with the obtained packing fractions and compressive strength

values. Based on the values given in Table 2, the cement efficiency could be increased for mixtures using only slag cement, from 0.13 up to 0.21 Nm³/kgmm². For mixtures using a blend of 65% slag cement and 35% Portland cement the cement efficiency could even be improved from 0.22 to 0.31 Nm³/kgmm² by means of an optimized particle packing.

Considering both the high compressive strength values of mixes having high packing fractions and improvements regarding a more efficient use of cement, reducing the cement content of the designed concrete mixtures is one of the major interests for the further research.

Following the idea of eco EMC, the application of stone waste materials should be preferred as they are produced in large quantities.

5 STONE WASTE MATERIALS

During the production of washed rock aggregates, high amounts of fine stone waste powders with particle size $<125\,\mu m$ are generated in slurry form throughout the washing process. The material characterization, particularly with respect to particle size distribution at present, showed that this type of waste material can be used as cement replacement in order to turn the production process of the broken rock aggregates as well as the concrete production to a more sustainable and environmental friendly process.

Depending on the origin and the generation of the stone waste materials, different approaches are possible for the application of stone waste materials in concrete. In this part of the research, the direct use of the untreated and unwashed intermediate product is favored. In this case, the stone waste material will not be generated as the original product allows a direct use of the material in special types of concrete (e.g. EMC or SCC). This method will result in higher financial and environment-friendly aspects as an intermediate step in the production of broken rock aggregates is eliminated. Therefore, the direct use of this untreated product, its sand fraction here named Premix 0–4, was focused.

Four different EMC mixes having varying cement contents are designed by means of the ideas of the new mix design concept and tested on mortar base as mortar tests permit a quick and handy preliminary evaluation of the designed mixtures. Table 3 shows the composition of the tested concrete mixes serving as starting point for the mortar tests.

The designed mixes are using premixed sand (Premix 0–4), containing both fine aggregate and inert stone powder, in combination with varying cement contents. The distribution modulus q is chosen to be 0.35 for all mixes. Considering a distribution modulus of $q = 0.35$ and a w/p ratio of 0.35, the necessary

Table 3. Mix proportioning of composed concrete mixes containing stone waste material.

Mix	CEM I 52.5 N	CEM III/B 42.5N LH/HS	Premix 0–4	Granite 2–8	Gravel 8–16	Water	w/c	w/p
Premix C275	95.3	179.7	999.6	377.4	583.3	118.3	0.43	0.33
Premix C250	75.0	175.0	1034.5	346.8	591.8	121.6	0.49	0.33
Premix C200	60.0	140.0	1140.3	283.6	626.6	109.1	0.55	0.33
Premix C175	52.5	122.5	1182.9	252.0	638.8	108.7	0.62	0.35

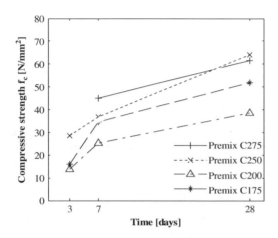

Figure 4. Mean compressive strength of tested mortar samples.

cement content amounts to 235 kg per m^3 fresh concrete to follow the given target line with the lowest deviation.

For the mortar experiments, the cement content is reduced starting from 275 kg down to 175 kg per m^3 fresh concrete and is compensated by higher amounts of Premix 0–4 and gravel as well. For designing the concrete mixtures, a blend of 65% slag cement (CEM III/B 42.5N LH/HS and 35% Portland cement (CEM I 52.5 N) was used.

Due to the high content of fines and low cement contents in the mixtures, the amount of water is maintained as low as possible in order to achieve w/c ratios around 0.50. Therefore, the use of plasticizers is necessary to match the industrial compaction efforts with available compaction efforts under laboratory conditions.

The mortar samples are tested regarding their compressive strength as well as flexural strength. For testing the compressive strength, cubes with dimensions of $50 \times 50 \times 50$ mm have been produced using constant compaction efforts, cured sealed during the first day, demolded, and subsequently cured submersed in a water basin at $20 \pm 2°C$. The produced cubes have been tested after 3, 7 and 28 days. The mean values of the compressive strength tests are depicted in Figure 4.

It seems that the compressive strength after 28 days of the mortar samples having a cement content of 275 kg and 250 kg per m^3 fresh concrete is not influenced by the cement content of the designed mixtures. Here, the mix having a cement content of 250 kg cement per m^3 fresh concrete achieves the same compressive strength after 28 days as the mix having 275 kg. The optimum cement content for following the given target line with the lowest deviation amounts to 235 kg. This amount of cement results from the granulometric properties (PSD) of the used materials in combination with a distribution modulus of $q = 0.35$ (Equation 2).

It appears that a reduction in the cement content is not influencing the compressive strength when the original cement content is already higher than actually needed. In this case, the additional cement only acts as a kind of filling material. The reduction in the cement content shows higher effect on the compressive strength if the cement content is already below the necessary amount needed for optimum packing and a further reduction in the cement content is then influencing the granular structure in a negative way.

Considering the data of the mix proportioning given in Table 3, the w/p ratio and the workability is constant for mixes having a cement content of 275, 250 and 200 kg per m^3 fresh concrete.

But the w/p ratio was increased from 0.33 to 0.35 for the mix having 175 kg per m^3 fresh concrete. This slightly increase in the water content improved the workability properties of the mixtures and the granular structure of the hardened concrete. Therefore, the mix containing 175 kg cement per m^3 fresh concrete achieved higher compressive strength values than the mix using 200 kg cement per m^3 fresh concrete. The highest cement efficiency for the compressive strength test, considering a constant w/p ratio, has been achieved for the EMC mix using 250 kg cement per m^3 fresh concrete. The present results regarding the compressive strength show therefore that cement can be used more efficient when the mix design results in an optimized particle packing.

Significant variations on the flexural strength in dependence on the varying cement contents are only recognizable for the strength development up to seven days. To show the effect of cement reduction clearly on

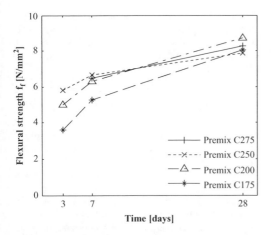

Figure 5. Mean flexural strength of tested mortar prisms.

Table 4. Test results and computed cement efficiency for tested mortar mixtures (28 days).

Specimen	Compr. Strength Cement Efficiency x_c [Nm^3/kgmm^2]	Flexural Strength Cement Efficiency x_f [Nm^3/kgmm^2]
Premix C275	0.225	3.02E-02
Premix C250	0.255	3.16E-02
Premix C200	0.192	4.35E-02
Premix C175	0.297	4.63E-02

the results of the flexural strength tests after 28 days is hardly possible as the standard deviation for each particular series is higher than the difference of the mean values among each other.

All tested series using Premix 0–4 showed high flexural strength values in the range between 7.9 and 8.1 N/mm^2. These high values of the flexural strength are caused by the angular shape of the broken material (Premix 0–4).

Due to the marginal difference between the values of the flexural strength, the cement efficiency regarding flexural strength is increasing with decreasing cement contents. This points out clearly that the cement can also be used in a more efficient way when the flexural strength is considered.

6   DISCUSSION

The tests on EMC learned that a general relation between the distribution modulus of the modified A&A equation and the packing fraction can be derived. Based on the data obtained from the concrete experiments, the relation between distribution modulus and packing fraction is shown in Figure 6. It becomes

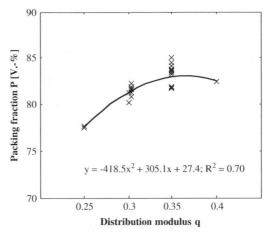

Figure 6. Influence of various distribution moduli on the packing fraction.

clear from Figure 6 that highest packing fractions have been achieved for a distribution modulus of q = 0.35. The same fact applies for the compressive strength of the hardened concrete. Here, mixes with a distribution modulus of q = 0.35 and high packing fractions achieved higher compressive strength values than the mixes with lower packing fractions, caused by their lower distribution moduli.

Besides the use of suitable distribution moduli for composing concrete aggregates, some further properties of the fresh concrete mix are also important for workable concrete mixes and partly influenced by the chosen distribution modulus. Based on the analysis of the obtained test results, the following values are advisable for the characterization of workable EMC mixes:

– Distribution modulus (q): 0.35–0.40
– Paste content (<125 μm): 0.225–0.250 m^3 per m^3 fresh concrete
– w/p ratio (<125 μm): 0.30–0.35

Traditional EMC mixes are characterized by low contents of fine inert or reactive filler materials. Therefore, the difference between the w/c ratio and the w/p ratio is small or nonexistent for these mixes when using a classical design approach. Decreasing cement contents and increasing amounts of filler materials, however, augments the difference between the w/c and the w/p ratio. Considering the desired workability of the concrete as a function of the w/p ratio, it is more appropriate for the mix design to take the w/p ratio into account than the w/c ratio.

This fact is also reflected by the strength development of the tested mortar samples depicted in Figure 4. The mix having a cement content of 175 kg per m^3 fresh concrete and a w/c ratio of 0.62 (w/p = 0.35) achieved higher compressive strength values after 28 days than the mix containing $200\,kg$ cement per m^3 fresh concrete and the corresponding lower w/c ratio of 0.55 (w/p $= 0.33$). The higher compressive strength of the mix containing less cement is caused by a better workability due to a slightly higher w/p ratio. The difference in the w/p ratio improved the packing of the mix using constant compaction efforts. Considering the w/c ratio of the two mixes, this is in contrast to the observations made by Locher (1976) on cement stone regarding cement hydration and strength development of the hardening cement stone.

According to Locher (1976), an increase in the w/c ratio is resulting in higher values of the capillary porosity. Caused by the remaining water content, capillary pores are formed which will be filled with hydration products in a progressed hydration state. As a result of increasing w/c ratios, also the capillary porosity of the cement stone is increasing, which makes the cement stone weak. Therefore, an increase in the w/c ratio will result in a decrease in the compressive strength. This is not affirmed by the conducted experiments as a possible reduction in the compressive strength of the cement stone is compensated by a positive influence on the compressive strength of an optimized and denser granular structure. So, it appears that the absolute water content in a mix is more relevant than the w/c ratio.

Furthermore, it is of great interest for EMC mixes to consider also the degree of saturation of the fresh concrete mix as the formation of the capillary forces is connected to the air/water saturation in the void fraction. This value was also computed for the designed mixtures and is given in Table 2. The designed mixes are having values between 58% and 75% and can therefore be classified as 'moist' according to soil mechanical definitions.

Nevertheless, the use of low w/c ratios is still important for the durability properties of the concrete as the capillary pores are influencing the impermeability as well as the durability of the hardened concrete. But the application of high contents of fine materials (e.g. stone waste materials) seems to be hindering for this purpose as the water demand and therewith the w/c ratio is increasing with increasing amounts of fine materials. Here, the application of plasticizers showed a positive effect. Due to the use of plasticizers, the workability of EMC mixes has been improved for mixes containing high amounts of fine materials and low w/c ratios. Packing fractions between 84 and 85% have been achieved in the lab by using paste content of $0.246\,m^3$ per m^3 concrete with a w/p ratio of 0.33. These mixes have been resulted in high 28 days compressive strength values between 95.7 and $100.3\,N/mm^2$.

Using the Premix 0–4, the cement content can be reduced without changes in the entire PSD of the designed mix since a suitable filler material is provided by this premixed and unwashed aggregate. This allows a more efficient use of the binding material by means of an optimized granular structure and suitable filler materials.

7 CONCLUSIONS

This paper demonstrates that the packing of all solids is of fundamental importance for the aimed optimization of the properties of earth-moist concrete, both in fresh and hardened states. An optimization in the packing improves the properties of the fresh concrete (water demand, compactibility, green strength) and therefore the properties of the hardened concrete as well.

The new mix design concept has proved its potential for a performance based composition of concrete mixes considering the granular properties of the granular ingredients such as particle size, particle shape and surface texture. Thereby, the improvement of the concrete properties is based upon an aimed optimization of the entire particle size distribution of all solids in the mix. For the optimization of the composed PSD, the geometric packing of continuously graded particles following the modified equation of Andreasen and Andersen, according to Funk and Dinger (1994), is employed.

The results show that hardened concrete achieves higher compressive strength values if an improved granular structure can be achieved. This includes both an increase in the grain-to-grain points of contact and a better packing of the aggregates used.

Considering both the enhancement in the compressive strength as well as the improved cement efficiency, the study reveals how the optimum packing of all concrete ingredients is influencing the mechanical properties positively. Particular attention should not only be paid to high compressive strength values, but also to the fact that a part of the cement seems to be unused if the granular structure of the hardened concrete is very porous and weak due to poor packing of the solids. This causes the addition of cement in order to increase the compressive strength of the hardened concrete. Whereas the added cement is used as a filler material instead of a binding material which contributes primarily to the strength development of the hardened concrete.

In view of the market prices for cement as well as the energy consumption involved with cement production, this is hardly acceptable for the mass production of concrete products. For the production of cost efficient and environmental friendly concrete products, the cement content should be reduced to a minimum amount necessary for fulfilling the demands on the mechanical and durability properties. This can be achieved by means of an optimum packing and the application of fine stone waste materials as substituent for primary raw materials.

Furthermore, the use of superplasticizers showed good results for EMC mixes containing high powder contents. By means of plasticizers, the amount of fine inert particles could be increased without increasing the water content of the mix or a loss in the workability of the concrete mix.

Within the conducted research the new mix design concept showed its suitability for the optimization of the PSD of the composed concrete mix, both for the optimization of already used mixtures and the composition of new mixes containing fine inert filling materials in form of stone waste powders. Moreover, this approach allows for a more performance based mix design of EMC concrete mixes.

ACKNOWLEDGEMENTS

The authors wish to express their sincere thanks to the European Commission (I-Stone Project, Proposal No. 515762-2) and the following sponsors of the research group: Bouwdienst Rijkswaterstaat, Rokramix, Betoncentrale Twenthe, Betonmortelcentrale Flevoland, Graniet-Import Benelux, Kijlstra Beton, Struyk Verwo Groep, Hülskens, Insulinde, Dusseldorp Groep, Eerland Recycling, Enci, Provincie Overijssel, Rijkswaterstaat Directie Zeeland, A&G Maasvlakte (chronological order of joining).

REFERENCES

Andreasen, A.H.M. & Andersen, J. 1930. Ueber die Beziehungen zwischen Kornabstufungen und Zwischenraum in Produkten aus losen Körnern (mit einigen Experimenten). *Kolloid-Zeitschrift* 50: 217–228.

Brouwers, H.J.H. & Radix, H.J. 2005. Self-Compacting Concrete: Theoretical and experimental study. *Cement and Concrete Research* 35: 2116–2136.

DIN Deutsches Institut für Normung e. V. 1999, *DIN-EN 12350-4: Testing fresh concrete – Part 4: Degree of compactibility; German version EN 12350-4:1999*. Berlin: Beuth Verlag GmbH.

DIN Deutsches Institut für Normung e. V. 2002, *DIN-EN 12390-4: Testing hardened concrete – Part 3: Compressive strength of test specimens; German version EN 12390-4:2000*. Berlin: Beuth Verlag GmbH.

Dutch Normalization-Institute. 1995. *NEN 5950: Voorschriften Beton Technologie – Eisen, vervaardiging en keuring*. Delft: Nederlands Normalisatie Instituut.

Fuller, W.B. & Thompson S.E. 1907. The laws of proportioning concrete. *Transactions of the American Society of Civil Engineers* 33: 222–298.

Funk, J.E. & Dinger, D.R. 1994. *Predictive Process Control of Crowded Particulate Suspensions: Applied to Ceramic Manufacturing*. Boston: Kluwer Academic Press.

Hummel, A. 1959. *Das Beton-ABC – Ein Lehrbuch der Technologie des Schwerbetons und des Leichtbetons*. Berlin: Verlag von Wilhelm Ernst & Sohn.

Hunger, M. & Brouwers, H.J.H. 2006. Development of Self-Compacting Eco-Concrete. In H.B. Fischer (ed.), *Proc. 16th Ibausil, International Conference on Building Materials, Weimar, 20–23 September 2006*, Weimar: F.A. Finger-Institut für Baustoffkunde.

Locher, F.W. 1976. Die Festigkeit des Zements. *Beton* 26(8): 283–286.

Schmidt, M., Bornemann, R. & Bilgeri, P. 2005. Entwicklung optimierter hüttensandhaltiger Zemente für den Einsatz in der Betonwarenindustrie. *Betonwerk International* 2005(3): 62–72.

Excellence in Concrete Construction through Innovation – Limbachiya & Kew (eds)
© 2009 Taylor & Francis Group, London, ISBN 978-0-415-47592-1

Contaminated soil concrete blocks

A.C.J. de Korte & H.J.H. Brouwers
University of Twente, Enschede, The Netherlands

ABSTRACT: According to Dutch law the contaminated soil needs to be remediated or immobilised. The main focus in this article is the design of concrete blocks, containing contaminated soil, that are suitable for large production, financial feasible and meets all technical and environmental requirements. In order to make the design decision on the binder composition, binder demand and water demand needed to be made. These decisions depend on the contaminations present and their concentration. Two binder combinations were examined, namely slag cement with quicklime and slag cement with hemi-hydrate. The mixes with hemi-hydrate proved to be better for the immobilization of humus rich soils, having a good early strength development. Based on the present research, a concrete mix with a binder combination of 90% blast furnace cement and 10% hemihydrate, a binder-content of 305 kg/m^3 and water-binder factor of 0.667 gave the best results.

1 INTRODUCTION

In the Netherlands, there is a large demand for primary construction materials. At the same time, many locations in the Netherlands are contaminated and need to be remediated according to the national environmental laws (WBB, 1986) (BMD, 1993). Since the amendment to the National Waste Management plan in 2005, immobilization is considered to be equivalent to remediation of waste (LAP, 2005). Immobilization of contaminated soil can be a partial solution for both needs. Immobilization also fits the sustainable building concept, because waste materials are re-used, so less primary construction material is needed.

The Netherlands Building Material Degree (BMD, 1993), which applies to stony materials, distinguishes two categories of construction materials: shape retaining and non-shape retaining materials. The determination of the leaching, the leaching limits and the composition limits differ for these two categories. The successful production of a non-shape retaining building material using contaminiated dredging sludge and the binders slag cement and lime was presented by Brouwers et al. (2007). Shape-retaining materials are defined as sustainable shape-retaining and which have a volume of at least 50 cm^3. Sustainable shape-retaining implies a limited weight loss of 30 gr/m^2 during the diffusion test (BMD, 1993). An additional problem with the immobilization of soil is the possible presence of humus with the soil. Humus can retard the hydration of cement and can have a negative influence on the characteristics of a mix.

The immobilisates need to be able to replace products which are made from primary raw material.

Therefore, the immobilisates need to fulfil, besides the leaching limits, the same requirements as products based on primary materials. In this case, where a construction block is produced, at least a compressive strength of 25 N/mm^2 is required. In addition, the immobilisate needs to represent a financially feasible solution, which means that the profit on producing the immobilisate must be the same or better than that of the primary product. Furthermore, this production of the immobilisates should be possible on a large scale. So, financially feasible solutions and production on a large scale are both important criteria for the design of the mix.

The purpose of the present research is the development of financially feasible mixes for the immobilization of contaminated soil by producing a shaped construction materials using cement, lime and additives, such that mixes can be applied on a large scale in accordance with national law. This objective was furthered through theoretical and laboratory research.

The research consists of two parts: the main experiment and an additional experiment. The main experiment was focused on the immobilization of two soils.

2 MATERIALS AND METHODS

This section describes the used materials and methods.

2.1 *Materials*

Two soils were used within this research. The physical characteristics of both soils, henceforth named D- and

Table 1. Physical and chemical characterization of soils.

Parameter	D-soil	J-soil
Dry matter (dm)	94.8% m/m	63.6% m/m
Organic matter (H)	2.4% dm	19.0% dm
Lutum (L)	7.9% dm	2.4% dm
CaCO3	1.6% dm	17.0% dm

Table 2. Standard strength development of CEM III/B 42.5 N LH.

Cement	Standard strength after N days			
	N/mm^2			
	1	2	7	28
CEM III/B 42,5 N LH	–	12	36	59

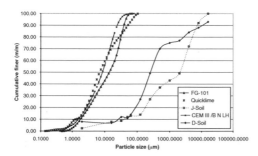

Figure 1. Particle size distribution of the applied soils and binders.

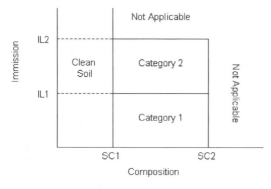

Figure 2. Classification of soil for the applicability as building material (SC = standard composition, IL = immision level).

J-soil, are presented in Table 1 and the particle size distribution is presented in Figure 1. The D-soil is poor in humus, clay containing and sandy soil. The J-soil is a humus rich, clay containing and sandy soil and comes near to a peat soil. Both soils are common soil types in the Netherlands.

Besides the physical characteristics, the environmental characteristics are important. In the Netherlands, soil is considered as a non-shaped material, while concrete is considered as a shaped material. The methods for the determination, when a material may be used, are different for both categories.

For non-shaped material, the leaching is determined using a column test, which is described in the standard NEN 7343. The BMD distinguishes four categories of non-shaped material, Figure 2 shows the distinction

between these categories. The distinction is made based on two parameters: immission and composition.

In this research a shaped material is produced. For a shaped material, the leaching is measured by the diffusion test, which is described in standard NEN 7345. For the mix design, only the composition is used, because the composition is a measure for the availability of heavy metals for leaching. Since the calculation methods differ for both categories of materials, a comparison of the immission of soil (non-shaped) and immission of product (shaped) is not possible. Therefore immission is very difficult to use for the mix design.

As can be seen from Figure 2, two limits (SC1 and SC2) are available for the composition of non-shaped material. Both limits depend on the physical parameters of the soil. This means that the two used soils have different limits. The D-soil only contains one pollutant, which renders it Not Applicable: cadmium is above the so called SC2-level. The J-soil contains arsenic, chromium, copper, lead, zinc and mineral oil levels that are above the SC2-level. These pollutants render this soil Not Applicable as well.

In this research a third soil is composed which consists of half J-soil and half 0-2 sand. This soil has a lower humus level and is used for the additional experiments in order to measure the effect of humus on the hydration and immobilization. This soil is named the J½-soil.

Besides the soils, some other materials were used like binders and aggregates. The used binders were slag cement (CEM III/B 42.5 N LH from ENCI), quick-lime (from Lhoist) and hemihydrate(FG-101 from Knauf). The strength development of the cement is shown in Table 2. Figure 1 shows the particle size distribution of the different binders besides the particle size distribution of the soils.

2.2 Methods

A description of the test procedures for slump-flow, V-funnel, capillary water absorption, compressive strength and tensile splitting strength, can be found in Brouwers & Radix (2005).

The consistency of the mixes is assessed by the relative slump flow using the Abrams cone (concrete) or Haegermann cone (mortar). The relative slump (Γ) is determined with Eq. (2), whereby d_1 and d_2 are the maximum diameters, rounded off at 5 mm, and d_0 is the base diameter of the Haegermann cone.

$$\Gamma = \left(\frac{d_i}{d_0}\right)^2 - 1 \qquad \text{with } d_i = \frac{d_1 + d_2}{2} \qquad (1)$$

Besides the consistency, also the required quantity of water for the cement and the filler can be determined by the slump flow. The spread flow test is a common way to assess the water demand of pastes and mortars. This yields a relation between relative slump flow (Γ) and water/powder ratio (V_w/V_p). The powders are defined here as all particles smaller than 125 μm. The test is executed analogously to Domone & Wen (1997). Ordinary tap water is used as the mixing water in the present research. This relation is described by;

$$\frac{V_w}{V_p} = E_p \cdot \Gamma + \beta_p \qquad (2)$$

This method was originally developed for powders only (Okamura & Ouchi, 2003). But the same procedure can be applied on mortar mixes with the use of the same Heagermann cone. Besides the determination of relation between relative slump flow (Γ) and water/powder ratio, it is also possible to do this for the water-/solid ratio. Solids means in this case the powders and sand in de mix.

Finally the leaching of the hardened product is measured by the diffusion test (NEN 7375). The cubes are places in 1 (in case mortar cubes) and 7 litres (in case of concrete cubes) of acid water of pH 4. The acid water is replenished after 0.25, 1, 2.25, 4, 8, 16, 36 and 64 days. This water is analysed on the concentration of heavy metals. From this leached amount the immission can be calculated according to NEN 7375 and BMD. Further details of the followed calculation procedure can be found in de Korte (2006).

3 PREVIOUS RESEARCH

In this section, previous research will be recapitulated. This information will serve as a basis for the new mix designs developed and tested here, presented in the next Section.

3.1 The influence of contaminants on immobilisation

The contaminants' characteristics influence the degree to which immobilization is possible. Arsenic, lead, chromium and cadmium are solvable in acid environments. Arsenic and lead are amorphous, which mean that they are soluble in both acidic and base environments. Immobilisates which are produced using cement have a high pH. This means that heavy metals are soluble and available for leaching. The leaching behaviour strongly depends on the valence of the metal. Both arsenic and chromium have more than one valence. Chromium (III) is for instance easier to retain than chromium (VI) (Mattus & Gilliam, 1994).

Heavy metals also influence the hydration of cement. Copper, lead and zinc will retard the hydration of cement (Mattus & Gilliam, 1994). Chromium shortens the gel fibres and increases the matrix porosity (Palomo & Palacios, 2003).

The way heavy metals are incorporated in the hardened product differs from case to case. Cadmium, zinc and arsenic can replace calcium within CSH (Pomies et al., 2001). Chromium and lead are absorbed within the CSH-binding, but nickel cannot be absorbed within the CSH binding (Bonen & Sarkar, 1995). Chromium (III) can replace aluminium within the CAH-binding (Duchesne & Laforest, 2004). The different binders have a different oxide composition and therefore they have a different level of bindings. This means that the most suitable binder can be selected based on the required bindings.

3.2 Possible binder combinations

In this section the feasible binder combinations are described. The first and most known binder is ordinary Portland cement (OPC). Portland cement is suitable for immobilization of most heavy metals. Pure blast furnace slag is more suitable for the immobilization of heavy metals in humus rich soils. The use of slags results in a lower porosity and permeability compared to the use of Portland cement. A lower porosity normally results in a lower level of leaching. However, a major disadvantage of the use of slag is the slower reaction rate. This reaction rate decreases further due to the presence of heavy metals and humic acid. Hence, an iniator could be needed when slags are deployed. The main reason is the absence of a calcium source within slag (Chen, 2007). Possible iniators are quicklime, anhydrite and hemi-hydrate. The advantage of the use of calcium sulphates is the possible formation of ettringite. Ettringite can fill the pores between the soil particles and so decrease the porosity and permeability. A lower porosity will result in a lower level of leaching (Mattus & Gilliam, 1994).

Portland cement is also an initiator for slag. The combination of Portland cement and slag, i.e. slag blended cement, results in a higher compressive strength and better immobilization than when Portland cement is used only. The combination of Portland cement and slag has the same effect as when blast furnace slag cement is used. For instance, the combination of 25% Portland cement and 75% blast furnace slag has the same composition as many available blast furnace slag cements.

Another possible binder is pulverized fly ash (PFA), although it is less suitable than blast furnace slag. For the immobilisation of cadmium and copper, PFA is less suitable (Peysson et al., 2005). For chromium, PFA is completely unsuitable, because it appeared that no strength development took place at all (Palomo & Palacios, 2003). PFA combined with Portland cement is suitable for the immobilization of copper but unsuitable for lead (Thevenin & Pera, 1999). Besides, as PFA reacts slowly, the strength development is slow too. So fly ash can better not be used for the immobilization of heavy metals.

The combination of calcium sulfoaluminate cement (CSA) and hemihydrate can be used instead of blast furnace slag. In a ratio of 70/30 CSA/Hemihydrate it is suitable for all heavy metals except six valence chromium. For six valence chromium, a ratio of 80/20 is suitable. The combination of CSA with hemihydrate can result in the formation of ettringite. Ettringite can fill the pores between soil particles and therefore results in lower porosity and permeability, and also a lower level of leaching (Peysson et al, 2005). A disadvantage of the combination of CSA with hemihydrate is the introduction of more sulphate into the mix. Ettringite and gypsum are dissolved at low pH values, which results in the release of sulphate. The leaching of this sulphate is also regulated in the Building Material Decree. This problem also exists with the combination of blast furnace slag and hemihydrate. However, in the case of blast furnace slag cement and hemihydrate the problem is smaller due to a lower amount of sulphate in the binder.

3.3 Required binder amount

In this section, the determination of the amount of binder per m^3 of concrete is described. An amount of 250 kg binder per m^3 concrete mix is currently used for the production of concrete blocks by Dusseldorp groep. According to Axelsson et al (2002), between 100–200 kg/m^3 is needed for the immobilisation of mud, 150–250 kg/m^3 for peat and 70–200 kg/m^3 for hydraulic filling. Nijland et al. (2005) used 250 kg/m^3 for the immobilisation of contaminated Gorinchems clay. Based on these finding, here also a binder level of 250 kg/m^3 is included. The binder amount of 350 kg is selected as well to overcome the possible negative effects of heavy metals and humus. A binder amount of 500 kg is introduced to investigate if the addition of extra binder can neutralize the possible negative effect of large quantities of humus. Hence, in this research, binder amounts of 250, 350 and 500 kg/m^3 are selected. These amounts correspond with 13.6, 21.9 and 26.7% (m/m) dry matter of D-soil. While the

amount of 350 kg for the J-soil corresponds to 38.4% and 500 kg with 58.2% dm.

3.4 The composition of the binder

This section summarizes possible binder combinations that will be used in this research. The first binder combination is slag cement and quicklime. This ratio is set to 90/10. Brouwers et al. (2007) researched the immobilization on heavily contaminated (Class 4) dredging sludge. The ratio of 90/10 slag cement/quicklime gave good results. This finding is compatible with Janz and Johansson (2002), who point out that the optimal mix lies between 60–90% slag cement and 40–10% quicklime.

The choice of a ratio of 60/40 slag cement/hemi-hydrate is based on the research of Huang (Huang, 1997). This ratio was confirmed by the research of Peysson et al. (2005). Peysson et al. (2005) indicated that 70% CSA and 30% gypsum is a suitable binder for the immobilization of most heavy metals. CSA itself also contains calcium oxide and sulphate. Because of that, the levels of calcium and calcium sulphate are higher than at the same ratio of slag cement and hemi-hydrate. In order to compensate this, here the proportion of hemi-hydrate is increased to 40%.

4 EXPLORATORY TESTS

As a preparation for the main-research, some exploratory tests were done to identify possible problems which could arise with the application of soil in concrete mixes. The main findings of the exploratory tests were;

- Both mixtures had a too low 28 days compressive strength, caused by a too high water/binder ratio and non-optimal particle size distribution
- The leaching of the mixtures was within the limits of the Dutch law. But the leaching of sulphates for the binder combination with gypsum approaches the limit. The leaching of heavy metals is better retained with blast furnace slag cement and quicklime.

5 MORTAR EXPERIMENTS

In this section the results of the experiments on mortar are described. The experiments on mortar are divided into two parts. The first part is the determination of the water demand for flowability, but at the same time the water content should be as low possible in order to achieve a high compressive strength and low leaching properties. The second part is the production of mortar cubes (50*50*50 mm^3). The mix used for casting these mortar cubes was based on the results of the

Table 3a. Composition D-mixes mortar cubes (in kg/m^3).

	Present Mix	D-HK-250m	D-HG-250m	D-HK-350m	D-HK-500m
Slag cement		195.3	130.2	303.3	360.5
Portland cement	250				
Quicklime		21,7		33.7	40.1
Hemihydrate			86,8		
D soil (dry)		1590.2	1598.8	1538.2	1499.5
Sand	1950	0	0	0	0
Water D soil		87.2	87.7	84.4	82.2
Superplastifizer		4.7	4.9	6.4	7.2
Mix water	125	214.2	206.6	195.3	190.3

Table 3b. Composition J-mixes mortar cubes (in kg/m^3).

	J-HK-350m	J-HG-350m	J$^1/_2$-HK-350m	J$^1/_2$-HG-350m	J-HK-500m
Slag cement	298.4	198.2	408.4	259.7	431.7
Portland cement					
Quick lime	33.2		45.0		48.0
Hemihydrate		132.1		173.1	
J soil(dry)	863.2	854.4	590.7	559.7	823.8
Water J soil	494.7	489.7	338.5	320.8	472.2
Sand 0–2			590.8	559.6	
Sand + gravel					
Superplastizer	8.4	8.5	9.0	8.9	10.5
Mix water	44.6	47.8	36.9	78.5	32.8

water demand part. The compressive strength, capillary absorption and diffusion of these mortar cubes will be determined. The results of the main and additional experiment are incorporated in this section. Section 5.3 will address the main findings of the additional experiment.

5.1 Water demand determination

The water demand determination was carried out using the slump flow test for the mortar mix. This mortar mix includes binders, soil (fraction that passes the 4 mm sieve) and sand. The soil was sieved in order to make it possible to use a small mortar mixer. The D-soil could be sieved wet, but for the J-soil this was not possible. The J-soil is therefore dried during 24 houres at 105 +/− 5°C. Before using this soil for the mortar mixes, the amount of water evaporated during drying was re-added, and mixed with the soil. These soil-water mixes stood for 30 minutes, so the soil could absorb the water. By doing so a wet soil could be simulated, which is closer to the practice since in practice a wet soil will be used in the immobilization process.

The mixes are displayed in Tables 3a, b. The slump flow was measured for different water/powder ratios (m/m) and with differed amounts of superplastizer (Glenium 51), based on the mass of powders in the mix. The mass of powders is the sum of all particles smaller than 125 μm present in the mix. The function

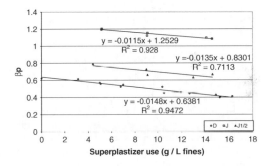

Figure 3. β_p versus SP dosage for D-soil (215 < binder < 400 kg/m^3), J-Soil (330 < binder < 480 kg/m^3) and J½-Soil (430 < binder < 455 kg/m^3).

of a superplastizer is to reduce the quantity of water while maintaining the same workability.

The relative slump flow is plotted against the water/powder ratio to construct the spread-flow line. The relative slump flow is computed with Eq. (1) with d_1 and d_2 as the diameters of the slump flow and d_0 the base diameter of the Haegermann cone.

The water demand (β_p) of a mix is the interception point of the linear regression function based on these results (Okamura & Ouchi, 2003), (see also Section 2.2 and Eq. (2)). In Figure 3, the water demands of mixes for different amounts of superplastizer are shown.

111

Table 4a. Measured properties of mortar cubes D-soil.

Property		D-HK-250m	D-HG-250m	D-HK-350m	D-HK-500m
Slump flow	*[mm]*	108–109	107–108	139–140	107–110
Relative Slumpflow (Γ)		0.177	0.156	0.946	0.177
Compressive strength 7 days	*[N/mm^2]*	1.91	3.09	8.31	10.05
Compressive strength 28 days	*[N/mm^2]*	4.77	4.57	8.55	23.62
Density 7 days	*[kg/m^3]*			1773	1941
Density 28 days	*[kg/m^3]*	1647	1603	1761	1968

Table 4b. Measured properties of mortar cubes J-Soil.

Property		J-HK-350m	J-HG-350m	J½-HK-350m	J½-HG-350m	J-HK-500m
Slump flow	*[mm]*	103–104	100–103	148–149	109–110	152–146
Relative Slumpflow (Γ)		0.071	0.030	1.205	0.188	1.220
Compressive strength 7 days	*[N/mm^2]*	–	0.75	2,27	3,81	–
Compressive strength 28 days	*[N/mm^2]*	–	–	–	–	5.68
Density 7 days	*[kg/m^3]*	0.68	1.64	7.33	6.51	6.64
Density 28 days	*[kg/m^3]*	1640	1600	1868	1799	1707

The different binder types all had their specific water demand. The mixes with slag cement and hemihydrate had a lower water demand (β_p measured as water/powder ratio) than the mixes with slag cement and quicklime. Also, the mixes with 350 kg binder had a lower water demand (β_p) than mixes with 250 kg binder. This lower water demand for higher binder amounts is partly caused by the chosen definition of water demand. Water demand is defined as the amount of water in de mix divided by the powder amount. Mixes with a higher binder content also have a higher powder content and hence, at same water content a lower water/powder ratio. But this effect can not explain the difference completely, because the total amount of water in the mixes is lower at higher binder contents. A possible explanation could be that the soil absorbed some of the mix water, so it is not available for enabling flowability. The mixes with higher binder content have namely a lower soil content. The mixes with J½ had a lower water demand than the normal J-mixes, which could be expected as J½ contains less fines. Section 6 contains a more detailed analysis of this effect/phenomenon.

5.2 *Mortar cubes results*

The mixes for mortar cubes were based on the results from the water demand study. A relative slump flow of 0.2 and a superplastizer use of 15 g/l powder formed the two constraints used for the mix designs. The mix compositions are presented in Tables 3a, b.

The hardened mortar was tested for compressive strength, density, leaching and capillary absorption. The last two properties could only be measured for the

Table 5. Performance overview of binders on most important aspects.

Aspect	Quicklime	Hemihydrate
Early Strength		+
Final Strength	+	
Leaching Sulphate		–
Retaining heavy metals	+	++
Sustainable shape retaining	+	+
Humus neutralisation		+
Legenda: ++ very suitable, + suitable, – unsuitable		

mixes containing slag cement and quicklime, because hemi-hydrate and gypsum readily dissolve when they come into contact with water.

The results of the experiments are presented in Tables 4a, b. A difference in the flowability was visible during the mixing. First, the mix was very dry and after a few minutes the mix became flowable. This time gap can be explained by the time the superplastizer needs to form a thin layer around the particles, which is a commonly known phenomenon.

In Table 5, a comparison between the quicklime and hemi-hydrate mixes is presented. The mixes containing quicklime had a lower early strength than comparable mixes with hemi-hydrate. The mix with quicklime could be crushed manually. These effects became very clear when the humus content of the soil was increased. The 500 kg variant of the J-soil with 19% humus could be crumbled manually after 1 day, but achieved a good compressive strength after 28 days. After 28 days, almost every mix with quicklime

Table 6. Leaching results of mortar D-soil (in mg/m^2).

	i_{max}	i_{bv}		
	$[mg/m^2]$	$[mg/m^2]$		
		D-HK-250m	D-HK-350m	J-HK-500m
Sulphate	100,000	30,352	52,080	10,158
Cadmium	12	0.04	0.05	0.07
Chromium	1500	0.04	0.04	1.40
Copper	540	0.11	0.10	14.00
Nickel	525	0.11	0.11	3.50
Lead	1275	0.45	0.45	3.50
Zinc	2100	0.37	0.38	1.40
Cobalt	300	0.07	0.07	1.40
Arsenic	435	0.84	0.84	3.50

had a higher compressive strength than the comparable mix with hemi-hydrate. In general, the compressive strength increased when the binder amount increased. This effect was partly caused by a decrease of the water/powder ratio, which has a direct relation on the compressive strength (Hunger & Brouwers, 2006) and partly to the binder as such.

A high leaching of sulphate is considered negative due to the limitations for the leaching of sulphate according to the Building Material Degree. Mixes containing gypsum, hemi-hydrate and anhydrite have a high sulphate leaching level. This was already shown in the exploratory tests. The solubility of gypsum, hemi-hydrate and anhydrite is relatively high, which results in a high leaching of sulphate. This means that mixes with hemi-hydrate are less suitable than mixes with quicklime.

The retention capability of heavy metals is also important for the immobilisates. Hemihydrate mixes can retain heavy metals better than the quicklime mixes. But here it appeared that both hemi-hydrate and quicklime were suitable for retaining heavy metals within the immobilisates (Table 6). The metal leaching was less than 5% of the limits specified by the Building Material Decree. The possibility that leaching of sulphate was mask by the leaching of the heavy metals has to be taken into account.

Both the hemi-hydrate and quicklime mixes had a sustainable shape retention during exploratory tests. The tested quicklime mortar cubes also retained their shape during the diffusion test. The mixes with hemi-hydrate were not tested for this aspect, as they would dissolve.

5.3 Additional experiment

The mixes with hemi-hydrate appeared to be more suitable for immobilisation of humus rich soils. This effect becomes apparent at humus levels of 9.5 and 19%.

Table 7. Comparison between different alternatives for immobilization of soil with high humus content.

	Mix with 0-2 sand	Extra binder
Compressive strength	+	+
Capillary water absorption	+	−
Financial feasibility	−	+

Legend: ++ very good, + good/better, − bad/less

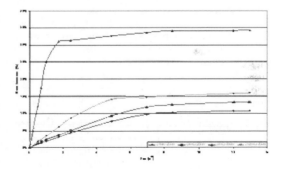

Figure 4. Capillary water absorption of the mortar cubes.

There was a threshold visible for the mixes with a high humus level. The compressive strength of the 350 kg variants is lower than 1.7 N/mm^2, while the 500 kg variant has a compressive strength of 6.7 N/mm^2. This was also the compressive strength for a mix containing half soil and half 0-2 sand (J½) and 350 kg binder. So an alternative to reducing the humus content by mixing with sand was the use of more binder (Table 7). But humus also increases the capillary absorption, and this was not reduced by adding extra binder (Figure 4). From these capillary absorption tests it follows that capillary absorption increases when the level of humus is increased. From the financial point of view, adding extra binder is more desirable. This is because the soil

Table 8. Specific densities of employed materials.

Component	Specific density
Slag cement	$2950 \, \text{kg/m}^3$
Quicklime	$3345 \, \text{kg/m}^3$
Hemihydrate	$2700 \, \text{kg/m}^3$
D-soil	$2736 \, \text{kg/m}^3$
J-soil	$2679 \, \text{kg/m}^3$
Sand	$2650 \, \text{kg/m}^3$
Organic Matter	$1480 \, \text{kg/m}^3$
Mineral fraction	$2700 \, \text{kg/m}^3$

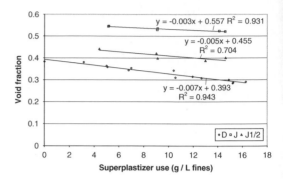

Figure 5. Void fraction versus SP dosage for the three employed soils.

used in the mix does not need to be remediated, which generates revenues as remediation costs of soil can be avoided. By the application of the soil in the mix, there is no need for this remediation, so the saved cost of remediation can be seen a revenue. On the other hand, the addition of sand will lead to extra costs.

6 ANALYSIS OF THE WATER DEMAND

In Section 5.1, it was observed that the water demand of a mix decrease when the binder contents increase. This is contrary to what would be expected, namely that a higher powder content results in a higher water demand. A possible explanation could be that the soil was finer than binder, but this was not the case (Figure 1). In this section the relationship between water demand and the properties of the mixes is examined in more detail.

6.1 Spread-flow analysis

As discussed in Section 2, there is a relation between relative slump flow (Γ) and water-/solid. The β_p of the different mixes at different amount of superplastizer (SP) is shown in Figure 3. Having a closer look at Figure 3, it can be noticed that the β_p of three soils result each in one line independently of the used mix design, which differs in amount of binder and binder combination. It appears that for each soil β_p depends linearly on the applied superplastizer dosage only.

6.2 Void fraction

For every soil a spread-flow line can be drawn. In order to analyse the lines the soil volume in the calculation is splitted into mineral part and organic matter, since these two parts have a different specific density. The mineral phase has a density between 2650 and $2750 \, \text{kg/m}^3$, while organic matter has a density of $1480 \, \text{kg/m}^3$. Table 8 shows the used specific densities for the calculation.

It is also possible to assess the void fraction based on the β_p of the mixes. According to Brouwers and Radix (2005) this can be done according to Eq. (3).

$$\phi_{water}(\Gamma = 0) = \frac{V_w}{V_{total}} = \frac{V_w}{V_w + V_s} = \frac{\beta_p}{\beta_p + 1} \quad (3)$$

In which β_p is the interception of the spread flow line with the abscissa. It appears that for each soil the void fraction depends linearly on the applied superplastizer dosage. This is shown in Figure 5. For the D-soil this relation reads:

$$\beta_p = -4.6 \cdot 10^{-3} \cdot SP + 0.46 \quad (4)$$

With SP in grams per liter fines

6.3 Particle packing theory

The packing of a granular mix is closely related to the particle size distribution. Continuously graded granular mixtures are often based on the Fuller parabola. The cumulative finer fraction is given as

$$F(d) = \left(\frac{d}{d_{max}} \right)^{0.5} \quad (5)$$

Where d is the sieve term and d_{max} represents the maximum sieve size (i.e. where 100% passing takes place). The introduction of a distribution modulus q by Andreasen and Andersen and a minimum particle size by Funk and Dinger (1994) led to an alternative equation, which reads as follows

$$F(d) = \frac{d^q - d_{min}^q}{d_{max}^q - d_{min}^q} \quad (6)$$

It is believed that values of q that range from 0 to 0.28, lead to optimum packing (Hunger & Brouwers,

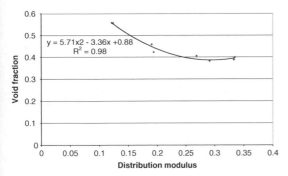

Figure 6.   Derrived void fraction ($\beta_p/(\beta_p - 1)$) and distribution modulus at $\Gamma = 0$ and SP $= 0$.

2006). Hummel (1959) mentioned an optimum distribution modulus of 0.4 for round shaped aggregates and 0.3 for more angular shape aggregates. According to Brouwers (2006) several researchers refer to a distribution modulus of 0.37 for spatial grain distribution in order to obtain optimal packing and therefore minimum void fraction.

Using the particle size distribution of the mixes, the distribution modulus is assessed using fitting minimizing the sum of the squares of the residuals (RSS).

The distribution moduli is calculated for all mixes, for which holds $d_{max} = 2.8$ mm and $d_{min} = 1$ μm for all solids. The distribution modulus versus the void fraction is shown in Figure 6. The relation between both characteristics can be described according to a quadratic function. The void fraction is minimal at a distribution modulus of about 0.3. This is line with the range which is mentioned by previous authors (Hummel, 1959; Hunger & Brouwers, 2006; Brouwers, 2006).

6.4 Prelimary conclusions

It is found that there is a linear relationship between superplastizer content and void fraction/packing for each soil depending for every relative slump flow, independent from the amount of binder and the binder composition. Furthermore, it is concluded that;

- The finer the mix, the higher the water demand and, because of this higher water demand, a higher void fraction.
- The D, J and J½ mixes have a distribution modulus of a proximally 0.3, 0.1 and 0.2.
- According to measurements a distribution modulus of 0.29 would minimize the void fraction. This is in line with the volume mentioned by previous authors (Hummel 1959; Brouwers 2006; Hunger & Brouwers 2006).

- Superplastizer additions reduces linearly water demand of the mixes.
- The amount of binder showed a limited effect on the water demand within one soil. But the variance in amount of binder, used in this research, was limited, so the conclusion can be different when a binder amount differs substantially. The amount of binder used here is in range with the amounts used in practice.

Furthermore, the relationship between the SP dependency lines of the different soils is investigated. This relationship is typical for each soil on which property this depends is so far unknown. From the present research we could exclude dry matter, organic matter, lutum and $CaCO_3$.

7   CONCRETE MIX RESULTS

The results of the mortar research were used for the preparation of the concrete mixes. In this part of the research, only D-Soil is used. A combination slag cement and quicklime forms the basis of the mixes. This binder combination performed well on all aspects during the mortar tests and does not have major drawbacks. Better results are to be expected from this combination compared to the binder combination slag cement with hemi-hydrate, for instance in regard to leaching (Sections 4 and 5).

Concrete is distinct from mortar because of the presence of bigger aggregates. The concrete mix consisted of 70–75% mortar and 25–30% coarse aggregates (Brouwers & Radix, 2005). The concrete mixes are based on the mortar mixes D-HK-350m and D-HK-500m (Tables 3a, b).

Based on these two mix definitions, a preliminary mix was designed. This mix was optimized to meet two objectives. The first objective was the optimisation of the particle size distribution. This means a minimization of the sum of absolute deviations from the modified Andreasen and Andersen line with $q = 0.35$, $d_{min} = 1$ μm and $d_{max} = 16$ mm. The second target was to design a mix which is more cost effective than the present one.

Table 9 shows the composition of the final mixes. In Figure 7, the particle size distribution of the final mixes is shown. These mixes were selected (out of a number of possible mixes) because they had an acceptable deviation from target function (the modified AA-line) and lowest costs.

7.1 Tests on fresh concrete

The fresh concrete tests can be divided into slump flow and V-funnel tests. Two batches were made of each mix. The results of these tests are presented in Table 10. The mixes were designed for a relative slump flow of 0.2. The relative slump flow and V-Funnel time differed considerably between the batches. The second batch of D-HK-500e was almost self-compacting, whereas the first batch was completely unflowable, this is due to fluctuations in the soil composition. These differences can be explained by the heterogeneity of the soil. For instance the amount of powder (all particles smaller than 125 μm) in the soil differs, resulting in a change in the water/powder ratio and the superplastizer content on powder. But differences in the sulphate level can also result in a different flowablity and workability. Fluctuations in the flowablility and workability of a mix can cause problems, when the mix is used in a production line. A solution for this problem is homogenizing of the soil prior to treatment, in order to reduce fluctuations in the composition of the soil and so fluctuations in the flowability and workability.

Table 9. Mix design final mixtures.

		D-HK-350e	D-HK-500e
Slag cement	kg/m³	209.5	274.9
Quick lime	kg/m³	23.3	30.6
Hemihydrate	kg/m³		
Dry DD ISM soil	kg/m³	1104.0	1143.7
Water DD ISM soil	kg/m³	60.6	62.7
Gravel 4-16	kg/m³	740.7	621.8
Water Extra	kg/m³	138.0	141.2
SP-solution	kg/m³	4.4	5.4

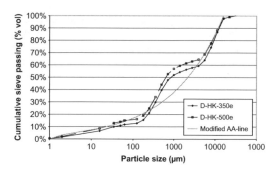

Figure 7. Cumulative finer function of final mixes (q = 0.35, d_{min} = 1 μm and d_{max} = 16 mm).

7.2 Results on hardened concrete

This section deals with the results of the hardened concrete tests, which can be divided into compressive strength, flexural strength, density, capillary absorption and leaching.

The 28 days compressive strength of the D-HK-350e mix did not fulfil the requirement of 25 N/mm², but D-HK-500e did fulfil this requirement. The measured compressive strength values were lower than the expected values based on general equations for the relation between compressive strength and the water/cement ratio. The compressive strength (f_c) of a concrete, with an uncertainty up to 5 N/mm², can be assessed by

$$f_c' = \alpha N_n + \frac{\beta}{wcf} - \gamma \qquad (7)$$

The α, β and γ in Eq. (7) are depending on the cement that is used. For slag cement this value are 0.75, 18 and 30 (ENCI, 2006). The N_n is the standard strength of the used cement after N days. In Table 2 the standard strength of the used slag cement (CEM III/B 42,5 N LH) is shown.

The flexural strength of the mixes was between 1.5 and 2.7 N/mm². These values were in line with the expectations for the flexural strength based on the measured compressive strength.

The density of the mixes was lower than the expected value. This can mean a higher air content of the mixes than expected. The air content can be calculated from the measured and calculated densities (Table 11). For D-HK-350e follows an air content of 9.2% and for D-HK-500e 4.2%.

The capillary absorption of both mixes was lower than the requirement for self compacting concretes

Table 11. Density and air content of final mixtures.

	Mix design		Measured	
	Density	Air content	Density	Air content (calculated)
	kg/m³	% V/V	kg/m³	% V/V
D-HK-350e	2280	1	2093	9.2
D-HK-500e	2280	1	2206	4.2

Table 10. Results fresh concrete of the final concrete mixtures.

		D-HK-350e Batch 1	Batch 2	D-HK-500e Batch 1	Batch 2
Slump flow	mm	280–280	220–220	200–200	510–520
Relative Slump flow		0.96	0.21	0	5.63
V-Funnel	Sec.	13–14	–	–	15–13
V-Funnel after 5 min	Sec.	25 sec.	–	–	16

(less than $3 \, \text{mm/h}^{0.5}$). D-HK-350e had a sorption-index of $0.77 \, \text{mm/h}^{0.5}$ and D-HK-500e had a sorption-index of $1.21 \, \text{mm/h}^{0.5}$.

The leaching of final mixes was very low compared to the limits of the BMD. In Table 12, the results of the test are displayed for D-HK-350e. The mixes fulfil the requirements of the BMD regarding leaching.

8 CONCLUSIONS

In this section, some conclusions are presented based on the research. The research consisted of a main experiment (Sections 4, 5.2 and 7) and an additional experiment (Section 5.3). First, the conclusions of the main experiment will be given and next the conclusions of the additional experiment will be described.

8.1 *Main experiment*

The results of the experiments have been examined for their financial feasibility, feasibility for production on large scale, shape retaining and strength. The following conclusions can be drawn;

- All mixes within the experiment were more cost effective than the current mixes with primary material.
- The mixes are suitable for production of immobil-isates on large scale. The design of these real mixes has been adapted to the use of wet soil, instead of the dried material within the 'normal' laboratorium concrete production.
- The leaching of the mixes containing D-soil was tested according to the limitations of the Building Material Decree i.e. the diffusion test. The leaching of sulphate was near the limit for the mixes with hemi-hydrate during pre-research, due to the solving of gypsum when in contact with water.
- The J and D mixes were sustainable shape retaining. This means that the products can categorised

Table 12. Leaching of D-HK-350e.

	Measured immission	Maximum Immission
	mg/m^2	mg/m^2
Sulphate	3,171	100,000
Cadmium	0.04	12
Chromium	0.16	1500
Copper	0.08	540
Nickel	0.11	525
Lead	0.42	1275
Zinc	0.35	2100
Cobalt	0.06	300
Arsenic	0.84	435

as a shaped material, which also implies that the diffusion (leaching) tests are indeed applicable.
- The strength of the final mix D-HK-500e was higher then the required compressive strength of $25 \, \text{N/mm}^2$. The other mixes had a compressive strength of less than $25 \, \text{N/mm}^2$. When a slightly lower compressive strength is acceptable, e.q. $20 \, \text{N/mm}^2$, then already $350 \, \text{kg/m}^3$ would have been sufficient.

Given these results, the final mixes of D soil, quick-lime and slag cement (500 kg binder) fulfilled all the objectives. So the objective of the main experiment fulfilled both the technical and financial requirements.

8.2 *Additional experiment*

Humus strongly influences the hydration of cement. The mixes with hemi-hydrate were more suitable for the immobilization of humus rich soils. This was due to the better strength development of these mixes compared to mixes containing quicklime.

Two possible ways to reduce the effects of humus were considered during this research. The first method is the mixing of J-soil with 0-2 sand to achieve a soil with a reduced humus level. The second method is the use of more binder.

Given the results, it seems that it is not possible to immobilize soil with a humus content of 19% with $350 \, \text{kg/m}^3$ binder only. Based on the present research, two possible solutions are available to immobilize such soils. The first method is the reduction of humus content by replacing half of the soil with 0-2 sand. But from a financial point of view, it is more attractive to increase the proportion of binder. The increase to $500 \, \text{kg/m}^3$ of binder is a way to achieve the required compressive strength, without reduced the high capillary absorption, which is a result of the present of humus.

ACKNOWLEDGEMENTS

The authors would like to thank Ir. S.J. Dijkmans from Jaartsveld Groen en Milieu (Steenbergen, The Netherlands) and Ing. E.M.H. Schildkamp from Dusseldorp Groep (Lichtenvoorde, The Netherlands) for providing the soils, their practical advice and providing the chemical and leaching tests. This research was financially supported by the Delta Marine Consultants, Civil Engineering Division of the Directorate-General for Public Works and Water Management, Jaartsveld Groen en Milieu, Senter-Novem Soil+, Rokramix, Betoncentrale Twenthe, Betonmortelcentrale Flevoland, Graniet-Import Benelux, Kijlstra Beton, Struyk Verwo Groep, Hülskens, Insulinde Recycling & Milieu, Dusseldorp Groep, Eerland Recycling Services and ENCI.

REFERENCES

Axelsson, K., Johansson, S., & Andersson, R. (2002). Stabilization of organic soils by cement and pozzolanic reactions—feasibility study. Swedish Deep Stabilization Research Center Rep. No, 3.

BMC. (2002). Aanvulling brl 1801, aanvulling op de nationale beoordelingsrichtlijn betonmortel (brl 1801), hoogvloeibare, verdichtingsarme en zelfverdichtende betonmortel. Gouda, The Netherlands: Certificatie Instelling Stichting BMC.

BMD. (1993). Bouwstoffenbesluit (Building Material Degree). from http://www.wetten.overheid.nl

Bonen, D., & Sarkar, S. L. (1995). The effects of simulated environmental attack on immobilization of heavy metals doped in cement-based materials. Journal of Hazardous Materials, 40(3), 321–335.

Brouwers, H. J. H. (2006). Particle-size distribution and packing fraction of geometric random packings. Physical Review E, 74, 031309.

Brouwers, H. J. H., Augustijn, D. C. M., Krikke, B., & Honders, A. (2007). Use of cement and lime to accelerate ripening and immobilize contaminated dredging sludge. Journal of Hazardous Materials, 145(1), 8–16.

Brouwers, H. J. H., & Radix, H. J. (2005). Self-compacting concrete: Theoretical and experimental study. Cement and Concrete Research, 35(11), 2116–2136.

Chen, W. (2007). Hydration of slag cement: Theory, modelling and application. University of Twente, Enschede, The Netherlands.

CUR. (2002). Cur-aanbeveling 93, zelfverdichtende beton. Gouda, The Netherlands: Stichting CUR.

de Korte, A. C. J. (2006). Koude immobilisatie van verontreinigde grond; koude anorganische immobilisatie van verontreinigde grond met behulp van cement, kalk en gips. Master Thesis, University of Twente, Enschede.

Domone, P., & HsiWen, C. (1997). Testing of binders for high performance concrete. Cement and Concrete Research, 27(8), 1141–1147.

Duchesne, J., & Laforest, G. (2004). Evaluation of the degree of Cr ions immobilization by different binders. Cement and Concrete Research, 34(7), 1173–1177.

ENCI. (2006). Productinformation of slag cement 42,5 n. 's Hertogenbosch, the Netherlands: ENCI.

Funk, J. E., & Dinger, D. R. (1994). Predictive process control of crowded particulate suspension, applied to ceramic manufacturing: Kluwer Academic Press.

Huang, X. (1997). On suibatility of stabilizer based on chemical analysis of the liquid from stabilized soil. Proceedings Fourteenth international conference on soil mechanics and foundation engineering, Hamburg, Published by A.A. Balkema, Rotterdam/Brookshield. 1613–1616.

Hummel, A. (1959). Das beton-abc; ein lehrbuch (12th ed.). Berlin: Verlag von Wilhelm Ernst & Sohn.

Hunger, M., & Brouwers, H. J. H. (2006). Development of self-compacting eco-concrete. Proceedings 16th IBausil, International Conference on Building Materials (Internationale Baustofftagung), Weimar, 2-0189-2-0198, Ed. H.B. Fischer, F.A. Finger-Institut für Baustoffkunde, Weimar, Germany.

Janz, M., & Johansson, S. E. (2002). The function of different binding agents in deep stabilization. Swedish Deep Stabilization Research Center, Linkoping: SGI, 9.

LAP. (2005). Landelijk afvalbeheer plan 2002-2012. from http://www.wetten.overheid.nl

Mattus, C. H., & Gilliam, T. M. (1994). A literature review of mixed waste components: Sensitivities and effects upon solidification/stabilization in cemnet-based matrices. Tennessee, US: Oak Ridge National Laboratory.

Nijland, T. G., van der Zon, W. H., Pachen, H. M. A., & van Hille, T. (2005). Grondstabilisatie met hoogovencement: Verharding en duurzaamheid; stabilisatie voor de boortunnel van randstadrail. Geotechniek(Oktober), 44–52.

Okamura, H., & Ouchi, M. (2003). Self-compacting concrete. Journal of Advanced Concrete Technology, 1(1), 5–15.

Palomo, A., & Palacios, M. (2003). Alkali-activated cementitious materials: Alternative matrices for the immobilisation of hazardous wastes: Part ii. Stabilisation of chromium and lead. Cement and Concrete Research, 33(2), 289–295.

Peysson, S., Pera, J., & Chabannet, M. (2005). Immobilization of heavy metals by calcium sulfoaluminate cement. Cement and Concrete Research, 35(12), 2261–2270.

Pomies, M.-P., Lequeux, N., & Boch, P. (2001). Speciation of cadmium in cement: Part i. Cd2+ uptake by c-s-h. Cement and Concrete Research, 31(4), 563–569.

Thevenin, G., & Pera, J. (1999). Interactions between lead and different binders. Cement and Concrete Research, 29(10), 1605–1610.

WBB. (1986). Wet bodembescherming. from http://www.wetten.overheid.nl

118

Excellence in Concrete Construction through Innovation – Limbachiya & Kew (eds)
© 2009 Taylor & Francis Group, London, ISBN 978-0-415-47592-1

Who is the key decision maker in the structural frame selection process?

H. Haroglu, J. Glass & T. Thorpe
Civil and Building Engineering Department, Loughborough University, UK

C. Goodchild
The Concrete Centre, UK

ABSTRACT: Selecting the correct structural frame is crucial to a project's feasibility and success, but this decision can have profound implications for the future performance of a building project. In practice, the eventual choice of a frame may involve various parties including client, project manager, cost consultant, structural engineer, architect, main contractor, etc. This paper presents research findings on the levels of influence of these project team members on the structural frame selection process. It describes the results of a two-year study in which various research methods were undertaken including a state-of-the-art literature review, semi-structured interviews and a postal questionnaire survey. The interviews showed that cost consultants, project managers and clients were found to be the most influential people in the structural frame decision-making process, so a postal questionnaire survey was sent to a sample of UK companies operating in these areas to further examine their priorities and views in detail. The data collected was subsequently analyzed and produced a rank ordering of project team members in relation to the influence they have on the choice of frame type at each stage of design process. In fact, they agreed that the structural engineer was the most influential decision-maker in the structural frame selection process. So, this paper asks the question 'who really is the key decision maker?' The conclusions will be of interest to all those concerned with project teams, structural frame design and selection and effective leadership in decision making.

1 INTRODUCTION

The framed structure market cuts across several traditionally defined sectors such as residential, education, commercial, health, retail, leisure etc. The UK has a tradition of in-situ concrete construction and in the past in-situ concrete frame construction dominated the frame market. Over the past 20 years concrete has lost significant market share to structural steel in the framed structure market (BRE, 2005). However, concrete's range of structural frame solutions, its thermal efficiency, inherent fire resistance, acoustic and vibration performance, durability and low maintenance ensure that it performs well in a number of UK markets such as commercial and residential buildings (TCC, 2005). Nevertheless, the concrete market has remained steady over the past 18 months, with the exception of reinforcement prices, which are still volatile (Bibby, 2006).

Selecting the correct structural frame is crucial to a project's feasibility and success but this decision on the structural frame type can have profound implications for the future performance of a building project (Soetanto et al, 2006a). Furthermore, the project stakeholders' requirements should be captured and taken into consideration so as to ensure apt decisions in the design stage (Soetanto et al, 2006b). Therefore, we tend to make an assumption that the choice of an appropriate structural system during the design stage will lead to a successful project outcome. It is therefore essential to recognize the decision makers in the structural frame selection process. In practice, the eventual choice of a frame may involve various parties including client, project manager, cost consultant, structural engineer, architect, main contractor, etc. So who is the key person to influence what structural frame type is used, and any changes to the design of building project.

This paper describes the results of a two-year study in which various research methods were used including a state-of-the-art literature review, semi-structured interviews and a postal questionnaire survey. As a result of these interviews, cost consultants, project managers and clients were found to be the most influential people. A postal questionnaire, aimed at these three disciplines, to address the influence of project team members upon choosing appropriate frame type for building projects. The results were analysed using Statistical Package for the Social Sciences (SPSS), and through *frequency analysis*, confirmed that all project members, perceived by these respondents to the survey, have a great deal of influence in the choice of frame type. The severity index has been further used to rank the project team members (decision makers) for the degree of influence they have in the structural frame selection process. Lastly, *Spearman's rho* (ρ) analysis has been calculated to establish a measure of agreement between cost consultants, project managers and clients in the rankings of these decision makers at each stage of design process. The study presents findings of a questionnaire survey to establish a ranking of the decision makers (or project team members) at each stage of the design process and to investigate the degree of agreement among cost consultants, project managers, and clients with regards to the rankings. The aim is to provide a view of the different professions, decision makers involved in choosing the structural frame at each key step of the design process.

2 PROJECT TEAM MEMBERS

Although the precise contractual obligations of the project participants vary with the procurement option adopted, the project participants must carry out certain essential fundamental functions. The project team consists mainly of client, architect, project manager, structural engineer, cost consultant and main (principal) contractor (CIOB, 2002). Each member of the project team is described below:

Client: A client is a person or organisation paying for the services and can be represented by others, such as clients' representative, employer's agent, project manager, etc. Their chief interest would be to satisfy themselves that the contractor(s) were performing in accordance with the contract and to make sure they are meeting their obligations to pay all monies certified for payments to the consultants and the contractor(s) (CIOB, 2002).

Architect and Structural Engineer: The architect is in charge of the architectural issues, whereas the engineer is concerned with more technical issues. The design should be developed with the involvement of both sides: architect and engineer. There are different driving forces: technical for the engineer whose main aim is to make things "work" without compromising the architects' concept. The architect deals with the appearance of the structure which needs to be true to the concept and fit the context and use (Larsen and Tyas, 2003).

Project manager: Construction and development projects involve the coordinated actions of many different professionals and specialists to achieve defined objectives. The task of project management is to bring the professionals and specialists into the project team at the right time to enable them to make their possible contribution, efficiently. Effective management requires a project manager to add significant and specific value to the process of delivering the project. The value added to the project by project management is unique: no other process or method can add similar value, either qualitatively or quantitatively. The project manager in the main has a role which is principally that of monitoring the performance of the main contractor and the progress of the works (CIOB, 2002).

Cost Consultant (quantity surveyor): The cost consultant has responsibility to advise on building cost and estimating, which can have two distinct roles (Morrison, 1984):

– Part of the design team for cost advice but not management of budget.
– Appointed separately by the client as a cost consultant.
– Main contractor: The principal management contractor has a duty to (CIOB, 2002):
– Mobilize all labour, subcontractors, materials, equipment and plant in order to execute the construction works in accordance with the contract documents.
– Ensure the works are carried out in a safe manner
– Indemnifying those working on site and members of the public against the consequences of any injury resulting from the works.

The extent to which the above-mentioned roles are likely to influence the choice of frame type for a building project depends on various matters such as the procurement route adopted, existing attitudes within the organisations involved, type of the building project, project value etc. Nevertheless, a study by Haroglu *et al.* (2008) identified several issues perceived to be the most important to the structural frame decision-making process and established an agreement between cost consultants, project managers and clients over the significance of these issues influencing the choice of a frame type for a building project. Therefore, it is also important to appreciate the common approach adopted by the members of a typical building project to the structural frame selection process. As a result, this paper examines project team members' influence on the choice of frame type at each stage of the design process.

3 DATA COLLECTION

Although a few research studies have been carried out in this field, a state-of-the-art literature review was first completed in order to understand the process in which the structural frame is normally selected as well as identifying the decision makers in this process. Semi-structured interviews were then conducted with structural engineers to determine the most influential people in the structural frame selection process with the intention of capturing their perceptions in the postal questionnaire survey.

The work stages of the RIBA Plan of Work (2007) are used in this research as the stages are well-known and widely recognized throughout the UK construction industry. We can therefore acknowledge that the design stage consists of three parts: Stage C (Concept), Stage D (Design Development) and Stage E (Technical Design).

3.1 Semi-structured interviews

Nine interviews were arranged with structural engineers in selected consultancies to retrieve information about structural frame options and by whom they are evaluated. The core topics discussed during these interviews included: the frame types applied in their projects, influential people in selecting the frame type, and the rationale behind the preferred frame type of their current project. Consequently, cost consultants, project managers and clients were found to be the most influential people in the structural frame decision-making process. These interviews were carried out in total over a two-month period at the interviewees' work places, each lasting approximately 30 minutes. Each interview was tape recorded and subsequently transcribed verbatim and analysed.

3.2 Questionnaire survey

As a result of the interviews, cost consultants, project managers and clients were surveyed in an attempt to better understand their views of the relative influence of each project team member on the choice of frame type. The respondents were asked to rate the influence of the project team members on a 4-point Likert scale ranging from 0 for 'lowest level' to 3 for 'highest level' as by using an odd number of response points, respondents may be tempted to 'opt-out' of answering by selecting the mid-point (Fellows and Liu, 2003). Having developed the questionnaire, a pilot study was carried out with a sample of nine people from both industry and academia to see how they understand the questions and the response options. Having made a few alterations to the questionnaire as a result of the pilot study, the questionnaire survey was distributed amongst cost consultants, project managers and construction clients to establish the significance and ranking order of the project team members.

The individual respondents were selected randomly from a database of professional companies held by The Concrete Centre (TCC), irrespective of the size of the company. As shown below in Table 1, 239 postal questionnaires were sent to selected names, working for cost managers, project managers and client bodies, in the public and private sectors. As a result, 70 questionnaires were received in total, giving an overall response rate of 29.29% which is considered sufficient enough

Table 1. Questionnaire distribution and response rate.

Respondent group	Number of Questionnaires		Response rate %
	Distributed	Returned	
Cost Consultant	86	20	23.26
Project Manager	74	25	33.78
Client	79	25	31.65
Total	239	70	29.29

Table 2. Issues ranked in Concept Design.

Concept (Stage C of RIBA)

Project Team Members	Frequency of responses for score of				No. of responses	Severity Index %	SPSS Rank
	0	1	2	3			
Structural Engineer	1	6	17	42	66	85.23	1
Architect	0	9	28	29	66	79.17	2
Cost Consultant	2	13	25	27	67	73.13	3
Project Manager	1	19	31	12	63	63.49	4
Client	4	22	20	18	64	60.16	5
Main Contractor	21	14	19	11	65	44.23	6

to meet the research reliability level compared with the norm of 20–30% with regard to questionnaire surveys in the construction industry (Akintoye and Fitzgerald, 2000). Of the responses received, 20 were from cost consultants, 25 from project managers and 25 from clients (Table 2).

The respondents were also asked about their influence over the choice of frame type for a building project in order to appreciate the value of each individual's response to this survey. Below Figure 1 shows that 44% of the respondents had a great deal of influence over the choice of frame type for a building project whereas only 9% had none, which suggests that the respondents were generally influential in the structural frame selection, and possessed an immense understanding in the structural frame selection process.

The results confirmed that all of the project team members included in the survey were considered to be influential, proving the validity of the decision makers of a typical building project as a basis for consideration in the choice of frame type. Because of this, and the considerable degree of influence the respondents have on the choice of frame type, the returned sample was considered to be representative of the actual decision-making population. The next section illustrates some of the results in detail.

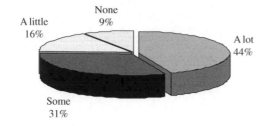

None
9%

A little
16%

A lot
44%

Some
31%

Figure 1. The influence of the respondents on the choice of frame type.

4 ANALYSIS AND RESULTS

The questionnaire was designed to provide predominantly descriptive data. An ordinal scale was used to rank the responses in this survey that there was no indication of distance between scaled points or commonality of scale perceptions in the Likert scale by respondents. It essentially provided a hierarchical ordering. Therefore, non-parametric tests were used in the analysis because non-parametric statistical tests are available to treat data which is inherently in ranks (Siegel and Castellan, 1956; Johnson and Bhattacharyya, 1996); the analysis was then carried out on the ranks rather than the actual data. The non-parametric procedures adopted for this study were frequency, severity index analysis, and Spearman's rho (ρ) test.

First of all, frequency analysis was applied to examine the degree of influence for each project team member. The severity index was used to rank the project team members for the degree of influence. The results of the frequency analysis and the ranking (severity index) have been based on analyses of all the completed responses. Individuals within these three disciplines were asked to provide information based on their own experiences from one of their projects that had recently started on site. However, these experiences were gained from distinct disciplines at each part of design stage, so it was essential to conduct a comparative analysis to distinguish between their responses. Since the variables are at the ordinal level, there are two prominent methods for examining the relationship between pairs of ordinal variables namely, *Spearman's rho* (ρ) (or Spearman rank correlation rs) and *Kendall's tau* (τ) – the former being more common in reports of research findings (Brymer and Cramer, 2005). Kendall's tau usually produces slightly smaller correlations, but since Spearman's rho is more commonly used by researchers, it was decided to be applied in this case. The Spearman's rho correlation coefficient is produced by using the rank of scores rather than the actual raw data (Brymer and Cramer, 2005; Hinton et al., 2004; Kinnear and Gray, 2006). The Statistical

Package for the Social Sciences (S.P.S.S.) was used to compute and run these statistical analyses.

4.1 Ranking the project team members: frequency and severity index analysis

This stage of the statistical analysis ranked the project team members in order of influence for each part of design process. In this case, frequency analysis was first carried out to obtain the frequency of the respondents, using the Statistical Package for the Social Sciences (S.P.S.S.). The frequencies of responses were therefore used to calculate severity indices for each project team member via Equation 1 (Ballal, 2000):

$$\text{S.I.} = [\sum_{i=1}^{i=n} \omega i * fi] * 100\% / n \qquad (1)$$

where: S.I. = severity index; fi = frequency of responses; ωi = weight for each rating; n = total number of responses

Since the 4-point Likert scale ranging from 0 for 'lowest level' to 3 for 'highest level', was used for the survey in order for the respondents not to be tempted to 'opt-out' of answering by selecting the mid-point, the weight assigned to each rating and is calculated by the following Equation 2 (Ballal, 2000):

$$\omega i = (\text{Rating in scale}) / (\text{number of points in a scale}) \qquad (2)$$

Therefore, $\omega 0 = 0/4 = 0$; $\omega 1 = 1/4 = 0.25$; $\omega 2 =$ No mid-point in the scale; $\omega 3 = 3/4 = 0.75$; $\omega 4 = 4/4 = 1$

Example: An example of the calculation for the severity index is given below:

Influence of "Architect" at the Stage D:

	Not imp = 0	Of little imp = 1	Quite imp = 2	Extremely imp = 3	Total (n)
Frequencies (fi)	0	11	30	24	65

S.I. = ((0*0 + 11*0.25 + 30*0.75 + 24*1)/65)*100 = 75.77%

The project team members were then ranked in order of value of severity index, the highest value having a rank of 1, and the lowest value assigned a rank of 6. Tables 2, 3 and 4 present the project team members ranked in terms of influence for each stage of the design process. In addition to that, Figure 2 displays the respondents' view of the degree of influence of the project team members on the choice of frame type during the design process.

Figure 2 shows the opinions of the respondents of the level of influence that the project team members have on the choice of frame type at the three stages of design process. 'Structural Engineer' appeared to be

Table 3.   Issues ranked in Design Development.

Design Development (Stage D of RIBA)

Project Team Members	Frequency of responses for score of				No. of responses	Severity Index %	SPSS Rank
	0	1	2	3			
Structural Engineer	0	6	15	45	66	87.50	1
Cost Consultant	1	8	29	29	67	78.73	2
Architect	0	11	30	24	65	75.77	3
Project Manager	2	17	26	18	63	66.27	4
Main Contractor	10	9	27	19	65	63.85	5
Client	5	18	27	14	64	60.55	6

Table 4.   Issues ranked in Technical Design.

Technical Design (Stage E of RIBA)

Project Team Members	Frequency of responses for score of				No. of responses	Severity Index %	SPSS Rank
	0	1	2	3			
Structural Engineer	0	3	22	40	65	88.08	1
Main Contractor	4	10	14	38	66	77.27	2
Architect	5	14	26	19	64	65.63	3
Cost Consultant	4	17	26	19	66	64.77	4
Project Manager	3	18	27	14	62	62.50	5
Client	6	24	19	14	63	54.37	6

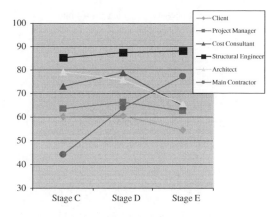

Figure 2.   Respondents' view of the influence of the project team members at all design stages.

Table 5.   Comparison of severity index and ranking for each group at Concept Design.

Concept (Stage C of RIBA)

Project Team Members	Cost Consultant		Project Manager		Client	
	S.I %	SPSS Rank	S.I %	SPSS Rank	S.I %	SPSS Rank
Cost Consultant	84.21	1	65.22	3	72.00	3
Structural Engineer	83.82	2.5	86.46	1	85.00	1
Architect	83.82	2.5	78.13	2	77.00	2
Project Manager	66.18	4	60.87	4.5	64.13	4
Client	58.82	5	60.87	4.5	60.42	5
Main Contractor	35.29	6	34.78	6	59.00	6

the most influential at all stages. Note the increasing influence of the 'Structural Engineer' and especially the 'Main Contractor', with the influence of the other members decreasing.

4.2   Investigating agreement: Spearman's rho (ρ) test

To examine the agreement, if there is any, between three disciplines on the ranking of the project team members in relation to the influence they have in the structural frame selection process, Spearman's rho (ρ) test was employed. The frequency of responses and severity indices were again computed for each group to generate a separate ranking of the project team members, as shown in Tables 5, 6 and 7. Additionally, the

Figures 3, 4 and 5 were used to display the results of the analyses for the readers of this paper to assimilate more readily.

As a result of this, Spearman's rho (ρ) (or Spearman rank correlation rs) test was computed using the Statistical Package for the Social Sciences (S.P.S.S.). The three groups are compared statistically by applying Spearman Rho test. Table 8 presents all of the Spearman Rho correlations computed, using SPSS, as shown below.

The level of significance was set by SPSS both at 0.05 and 0.01 levels, which indicated the degree of relationship amongst the three rankings. $p < 0.05$ means that there is less than a 5 per cent chance that there is no relationship between the two rankings, whereas $p < 0.01$ means that there is less than a 1 percent chance, and can be accepted at the 99% confidence level (Bryman and Cramer, 2005; Fellows

123

Table 6. Comparison of severity index and ranking for each group at Design Development.

| Project Team Members | Design Development (Stage D of RIBA) | | | | | |
| | Cost Consultant | | Project Manager | | Client | |
	S.I %	SPSS Rank	S.I %	SPSS Rank	S.I %	SPSS Rank
Structural Engineer	83.82	1	87.50	1	90.00	1
Cost Consultant	81.58	2	73.91	2	81.00	2
Architect	80.88	3	72.83	3	75.00	4
Project Manager	61.76	4	66.30	4	69.57	5
Main Contractor	60.29	5	51.09	6	78.00	3
Client	55.88	6	59.78	5	64.58	6

Table 7. Comparison of severity index and ranking for each group at Technical Design.

| Project Team Members | Technical Design (Stage E of RIBA) | | | | | |
| | Cost Consultant | | Project Manager | | Client | |
	S.I %	SPSS Rank	S.I %	SPSS Rank	S.I %	SPSS Rank
Structural Engineer	86.76	1	91.30	1	86.00	2
Cost Consultant	73.68	2	57.95	5	64.00	4
Main Contractor	72.06	3	70.83	2	87.00	1
Project Manager	64.71	4	62.50	4	60.87	5
Architect	61.76	5	63.64	3	70.00	3
Client	50.00	6	52.27	6	59.38	6

and Liu, 2003; Field, 2000). From Table 8, most of the correlations written with asterisks did achieve statistical significance at either $p < 0.05$ or $p < 0.01$ which confirmed that there are strong relationships amongst the rankings of three groups, assuring that agreements amongst the three rankings was much higher than it would occur by chance. As a result, it may be concluded that the rankings obtained from the three groups, as given by the severity index analysis, was consensual amongst the respondents.

5 FINDINGS AND DISCUSSION

With regard to the results of frequency and severity index analyses, all of the project team members were ranked by the respondents to the survey in order of

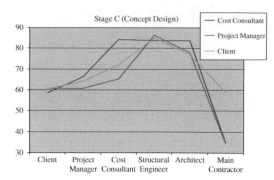

Figure 3. The views of the three sets of respondents on the degree of influence the project team members have at stage C.

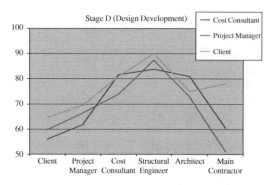

Figure 4. The views of the three sets of respondents on the degree of influence the project team members have at stage D.

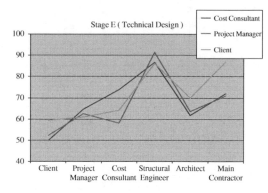

Figure 5. The views of the three sets of respondents on the degree of influence the project team members have at stage E.

influence they have at each stage of the design process. The Spearman's rho test was then applied to establish the consensus between the three sets of respondents in relation to the rankings of the project team members at each stage.

Table 8. Spearman's Rho (r) test results between the rankings of three groups.

Design Stages of RIBA	Cost Consultant vs. Project Manager	Cost Consultant vs. Client	Project Manager vs. Client
Correlations			
Correlation Coefficient			
Stage C	0.794	0.812*	0.986**
Stage D	0.943**	0.829*	0.657
Stage E	0.600	0.600	0.886*

Note: ** , * denotes 'strong' with $p < 0.01$ and 'some' with $p < 0.05$ statistical evidence of significant similarities.

As the design develops, note the increasing influence of the structural engineer and especially the contractor, with the influence of the other members decreasing, as shown in Figure 2. It is evident that 'Structural Engineers' influence was perceived to be far more than the other project team members at all times during the design process. However, the structural engineers interviewed indicated that they were not the most influential party in the choice of frame type, citing cost consultants, project managers and clients as more influential. This may be because structural engineers are not aware of their influence, or because they do not want to pronounce that they are powerful in the structural frame selection process. On the other hand, 'Clients' were perceived to be unexpectedly the least influential decision-maker for the duration of design process in general. 'Architect' and 'Cost Consultant' were perceived to be very influential during stages C and D when the structural frame of a building project is generally selected (Ballal and Sher, 2003). In addition, regarding the magnitude of the severity indices, there appears to be a relatively large gap separating the 'Structural Engineer', 'Architect' and 'Cost Consultant' as the top three decision makers from the rest at the stages C and D, as shown in Tables 2 and 3. 'Project Manager's influence is highest at stage D where it was ranked the fourth by the respondents which indicates that 'Project Manager' is not considered with the same degree of influence as are 'Structural Engineer', 'Architect' and 'Cost Consultant'. In addition, 'Main Contractors' influence rises to be number two at stage E. However, it may well be too late for the main contractor to influence the choice of frame type at this stage.

From the results of the Spearman's rho (ρ) test, there appeared to be a significant agreement in the rankings of project team members amongst the three groups and the degree of agreement was higher than would have occurred by chance, as shown in Table 8. The degree of agreement amongst the three groups is higher at stages C and D than it is at stage E as regards the correlations written with asterisks in Table 8. 'Structural Engineer' is generally agreed upon to be the most influential decision-maker in the selection of a frame type. Although the cost consultants, project managers and client were in good agreement with each other in relation to the degree of influence of the decision makers (or project team members), they differ in some places, particularly the degree of influence of 'Main Contractor' at stages D and E. 'Main Contractor' was considered to be very influential by clients in the structural frame selection process at stages D and E, whereas cost consultants and project managers did not consider 'Main Contractor' very influential at stage D (it was ranked the least influential decision-maker by project managers). A possible reason for this is that contractor involvement in a building project at stage D is perceived to be higher or more effective by clients than it is in reality. In addition, not surprisingly whilst 'Cost Consultant' was considered to be the second most influential at stage E by cost consultants, it was ranked by project managers and clients to be the fifth and fourth respectively.

As in any research based on a questionnaire survey, this study is subject to some biases and limitations. Firstly, with regard to the use of The Concrete Centre's database; although it may not necessarily represent the whole UK construction industry, it is large (25,000 names), up to date and nationwide. Secondly, since the postal questionnaire was sent through the post from The Concrete Centre to the respondent, it may have been presumed that the main thrust of this survey was about concrete frames rather than structural frames in general.

None-the-less, it can be said that the ranking of the six decision makers obtained from the respondents to the survey are representative of the views of the UK construction industry in relation to the structural frame selection process. Since selecting the correct structural frame is crucial to a project's feasibility and success, the assumption made earlier on in this paper was that the choice of an appropriate structural system will lead to a successful project outcome. The rank ordering at each stage of design process can therefore be of much interest to all those concerned with project teams, structural frame design and selection and effective leadership in decision making. Above all, the findings can give useful insights into the frame industry. For instance, it is evident that 'Main Contractor' appeared to have a significant input at both stages D and E which means that contractors should be a major audience in the frame market.

6 CONCLUSION

The decision on the choice of frame has significant short- and long-term implications for the building's function and its client's needs (Soetanto et al., 2007). Having undertaken a literature review and semi-structured interviews, cost consultants, project managers and clients were found to be the most influential decision makers in the selection of structural frame process. So this study asked these people the question 'who really was the key decision maker?' through a postal questionnaire survey. The respondents to the survey were requested to base their answers on one of their projects that had recently started on site. So, as project participants moved through the design stages, their influence was evaluated by the respondents. A total of 70 detailed responses were received and analyzed, providing a number of useful insights into the view of professionals about the decision makers in the structural frame selection process.

As a result of the questionnaire survey, the structural engineer was evidently found to be the most influential decision-maker in the choice of frame at each stage of design process. This is an outstanding contrast to the results of semi-structured interviews carried out with the structural engineers earlier on in this research. Further research in this field might examine how the key decision makers in the choice of frame for a building project vary by sector, project value, type of procurement route, etc. Furthermore, it was found that the contractor's influence is particularly high, as perceived by the respondents, at stages D and E which indicates that contractors could make quite an impact on the choice of frame type for a building project.

In conclusion there were some areas of disagreement amongst the three sets of respondents, such as the main contractor. This warrants specific research in this field. It is not known yet whether the main contractor could exert influence to change the frame type or any specifications of a building project after being involved. Hence there confirms to be a gap in knowledge about who the key decision maker is and while this paper has offered some key insights, the role of the contractor now appears to be next area of focus for research, particularly if we are seeking a clear model for how this area of decision making works in practice.

ACKNOWLEDGEMENT

The research project reported in this article was funded by the Engineering and Physical Sciences Research Council (EPSRC) along with The Concrete Centre (TCC) and this support is gratefully acknowledged. The authors are grateful to all of those involved with the project. The authors would also like to give their special thanks to all the people that had spent their valuable time to complete and return the questionnaire.

REFERENCES

Akintoye, A., and Fitzgerald, E. 2000. A survey of current cost estimating practices in the UK, *Construction Management and Economics*, No: 2, 18: 161–172.

Ballal, T. M. (2000). *The Use of Artificial Neural Networks for Modelling Buildability in Preliminary Structural Design*. Ph.D. thesis, Loughborough University of Technology, UK.

Ballal, T.M.A., and Sher, W.D. 2003. Artificial Neural Network for the Selection of Buildable Structural System. *Engineering, Construction and Architectural Management*, 10 (4): 263–271.

Bibby, G. 2006. Structures update, *Building Magazine*, 3: 64–69.

BRE Report 2005. *Towards a successful future for concrete frame construction in the UK*, prepared for DTI, December 2005. London: Building Research Establishment Ltd.

Bryman, A., and Cramer, D. 2005. *Quantitative Data Analysis with SPSS 12 and 13: A Guide for Social Scientists*. London: Taylor & Francis Group.

Fellows, R., and Liu, A. 2003. *Research Methods for Construction*. Oxford: Blackwell Science Ltd.

Field, A. 2000. *Discovering Statistics Using SPSS for Windows: Advanced Techniques for Beginners*. California: SAGE Publications Inc.

Haroglu, H., Glass, J., Thorpe, T. and Goodchild, C. 2008. *Critical Factors Influencing the Choice of Frame Type at Early Design*. CSCE 2008 Annual Conference June 10-13, 2008, Canada.

Hinton, P.R., Brownlow, C., McMurray, I., and Cozens, B. 2004. *SPPS Explained*. London: Routledge.

Johnson, R. A. and Bhattacharyya, G. K. 1996. *Statistics: Principles and Methods*. New York: Wiley.

Kinnear, P.R. and Gray, C.D. 2006. *SPSS 14 MADE SIMPLE*. New York: Psychology Press.

Larsen, O.P., and Tyas, A. 2003. *Conceptual Structural Design: bridging the gap between architects and engineers*. London: Thomas Telford.

Morrison, N. 1984. The accuracy of quantity surveyors' cost estimating, *Construction Management and Economics*, No: 1, 2: 57–75.

RIBA 2007. The RIBA Plan of Work Stages 2007. See http://www.snapsurveys.com/learnandearn/ribaoutlineplanofwork2007.pdf

Soetanto, R., Dainty, A. R. J., Glass, J. and Price, A. D. F. 2006a. A Framework for Objective Structural Frame Selection, *Structures & Buildings*, 159: 45–52.

Soetanto, R., Dainty, A.R.J., Glass, J. and Price, A.D.F. 2006b. Towards an explicit design decision process: the case of structural frame, Construction Management and Economics, 24: 603–614.

Soetanto, R., Glass, J., Dainty, A.R.J., and Price, A.D.F. 2007. Structural Frame Selection: Case Studies of Hybrid Concrete Frames, *Building Research & Information*, 35(2): 206–219.

The Chartered Institute of Building (CIOB). 2002. *Code of Practice for Project Management for Construction and Development*. Oxford: Blackwell Publishing.

The Concrete Centre (TCC) 2006. Benefits of Concrete. Surrey: The Concrete Centre. Available See http://www.concretecentre.com/main.asp?page=13.

126

Excellence in Concrete Construction through Innovation – Limbachiya & Kew (eds)
© 2009 Taylor & Francis Group, London, ISBN 978-0-415-47592-1

Performance of surface permeability on high-performance concrete

A. Naderi, A.H.H. Babei & N.S. Kia
Azad University of Navab, Qazvin, Iran

ABSTRACT: Concrete is the one of the important materials which people use in their buildings. It has many qualifications that you recognize it better. 1. tension strength 2. compressive strength 3. permeability. With a glance at materials in concrete and spice of concrete, it has been understood that permeability is the most important parameter that will provide the long life of your massive concrete and it can omit the effect of the two other parameters in long life it has been tried to study on a high performance concrete that has low absorption.

1 INTRODUCTION

The durability control and lifetime of concrete is one of the most concrete industry sections. Concrete with the least level permeability and on chemical attack of sulfates-carbonates and chlorine has shown the best resistance and prevents steel corrosion in hydraulic structures like dams.

Channels and drainages the permeability of concrete is the most factor in structure design. To access the least is the most factors in structure design. To access the least concrete level permeability and with regarding its pressure resistance about too samples and more than 20 fusion plan were totally prepared in concrete and building research center of Azad university.

The samples are experimented under level water absorption test and under-pressure water absorption test and pour. It was tried to use ultra-lubricants and different cements exit in Iran to preview their reactions in front of level permeability and volume water absorption and also suitable combination plans in experiment concrete placing.

2 CEMENT CHOICE

It was chosen cement (type Π), SAVEH cement factory, and (type Π) with higher percentage of ABYEK and POZZOLAN cement related to ABYEK cement factory for sampling was done under complete hydration and Pressure resistance experiments.

The POZZOLAN cement adding's because of POZ-ZOLANIES.

3 COMPLETE HYDRATION EXPERIMENT

The cement (type Π) with higher tiny percentage and (type C) need the more water rates for hydration. The water volume in addition to 0.2 W_W that are considered for cement complete hydration in reference was previewed.

4 PRESSURE RESISTANCE EXPERIMENT

With equal combination, the cement (type Π) with higher tiny percentage and type Π show the most resistance. And form samples, the built sample with SAVEH cement (type Π) showed the most suitable water absorption. The reason can be finding in little production of level bubble and accumulation of sample level caused by less produced heat in hydration process.

5 THE AGGREGATE CHOICE WITH THE BEST RESISTANCE AND WATER ABSORPTION

From exist aggregate in combination plans, it was used from sedimentary (river). Granite and quartz are experimented by water absorption, the form of aggregates after break and pressure resistance in concrete. The granite and quartz aggregates were approximately similar and with middle absorption less than 1% in relation to sedimentary aggregates has less water absorption. From point of geometry shape, the quartz and sedimentary were more suitable in relation to granite and more circle corner. With similar conditions the granite and quartz showed higher resistance relation to sedimentary aggregates. Finally it was used from the quartz and granite in combination plan.

6 THE GRAPH OF AGGREGATE GRACING

In normal carves, the material distribution was not considered less than 0–125 mm.

The kind of gracing and tiny bits make water decreasing to cement and water maintenance and leakage decreasing in concrete.

In general, by using the fine-grained stone material in the main plan make decreasing level capillary pipes and the caused cracks by concrete shrinkage and finally with higher density, makes decreasing %90–60 level for every concrete combination plan, it was considered two gracing graphs one, fuller * Thompson and the other, the modification graph proportional.

With desired conditions in concrete combination plan. In general, in comparison whit normal graphs like ASTM C_{33}, the designed sample with the method from the point of volume water absorption has %90–60 decreasing and pressure resistance (28 days) %25–30 increasing (Potter, 1998 & ASTM, 1998).

7 THE RATE OF WW/ WC IN COMBINATION PLAN

The used water rate in concrete has important role in producing empty spaces after concrete complete setting prepared level cracks, level capillary pipes and finally complete hydration of the cement. By using a fix amount of W_W/W_C in concrete is related completely to environmental circumstances of concrete detail, kind of cement, and absorption of water by aggregates in a plan.

In more general conditions by preventing from concrete seepage and suitable mold and also using from ultra-lubricants, in 200 samples to get determine water rate in experimental conditions with optimal amount of 0.25–0.35 W_W/W_C was chosen.

8 PREVIEW OF ULTRA-LUBRICANT

To access to mixes of self-compacting and zero energy concretes and decreasing of í to ultra-lubricants like MRWR (ASTM c 404) is necessary. As using of POZ-ZOLANS which has feller role in combined plan like micro silica, calcareous rock, makes decreasing concrete slump and need to water, then using from ultra-lubricants is inevitable.

It was tried to preview all produced lubricants in Iran. So, it was used from PARSA-production lubricants and built 220 samples by them totally.

The ultra-lubricants showed the best response to high resistance and low permeability of concrete level. Huge concrete placing by using cold water in concrete mixing doesn't need to late setting material because the necessary stop is made by cold water.

ultra-lubricant	Efficiency (except mental increasing and concrete slump)	consumption interval	Interval with high output	The best response on the kind of cement
p-116	To decrease volume permeability 50%–120% to increase resistance of 28 days concrete	1.2%–2.4% W_C	1.2%–3.5% W_C	type Π
p-117	To decrease volume permeability 50%–120%	2%–3% W_C	2%–2.5% W_C	type Π

In used POZZOLANS, by previewing micro silica, the volume permeability and concrete level is decreased In addition to it, made relative density between 50%–80% in the previewed plans.

Additive	Consumption output	Interval with high output	Permeability decreasing	The rate of increasing of 40 days resistance
Micro silica	9%–20% W_C	11% W_C	50–80%	50%

Decreasing of concrete cracking by control the combining plan and curing.

The kind of concrete cracking in hydraulics structures is a main factor of structure demolition. The cracks are the main factor of water absorption and chemical ions in concrete and decreasing of concrete structure resistance. The signs of demolition type in concrete structure like cracking, to shell, to spell, steel corrosion and changing forms or structure clamping is very much.

With correct design, control of curing conditions like suitable cover of steel in front of carbonate or chlorate attack (the last cover in standard Austria with 18mm. and the most cover steel in standard UK with 56mm. the environmental circumstances are determinable). Tensile stresses that makes concrete be cracked are not necessary by external loads.

Concrete extraction and also thermal changes make concrete volume changing. If a structure is in effect of the changes, it will crack in front of the changes. In the laboratory, the cracks by using PP fiber with 12mm, 19mm dimensions and in interval 0.001%–0.003% of cement weight and using of the above ultra-lubricants M.V.A was completely removed.

Cracking (except for load)	The first factor of cracking	The lateral factor of cracking	Necessary arrangements in redesign to omit cracking
Plastic settlement cracks	To make water unlimited	To become dry fast of concrete level	Using from micro silica to prevent bleeding, making air with calm and repeat shaking
Plastic contraction cracking		Low rate of making water and quick evaporation	Using from suitable ultra-lubricants like quicker curing and suitable floating
Thermal contraction cracking quickly	Making much heat unsuitable W_W/W_C	Quick getting cold of curing by cold water	Decreasing temperature or isolation, correct rate of W_W/W_C, using from filler like micro silica instead of parts of cement
Cracks made by steel corrosion	To spell concrete made by little cover and more calcium chloride in environmental	Concrete with low cement content and high moisture in environment	Decreasing of level permeability by new decreasing of water in capillary pipes by suitable POZZOLAN correct cover of steel

text, the received results are executable with some changes in huge concrete especially in precast concrete. The cement type hast the best efficiency in decreasing permeability and increasing pressure resistance (in condition that there is necessary water for cement hydration). In previewing aggregates, the one with circle angle with high density and compatible with cement has direct role in initial water consumption of the concrete block like quartz. The choice of the graph is the important concrete design factor and has direct role in concrete density and decreasing permeability. In addition, it makes developing the pressure resistance and more safety in front of chemical attacks.

Finally, the results of large applications of ultra-applications and POZZOLANS were described completely. But the ultra-lubricants with high viscosity and darker color that has hot smell act more quick and with high output and has the combination of POZZOLAN like the experiments and give excellent results to decrease concrete permeability (like micro silica). Even the best concrete designs to prepare cracks were previewed some factors and strategies which by sapling and permeability increasing and also decreasing of pressure resistance were faced. The above strategies save concrete highly in front of cracks.

By using the plans with suitable pressure resistance and least permeability rate band then low electrical resistance in foundation of soil dams and sheet pilling of dams and also drainages and bodies of curve dams and other concrete structures increase the life and safety factor of structure in long times.

REFERENCES

American society for testing & materials, standard test method for half-cell potentials of uncoated reinforcing steel in concrete, ASTM C876-87, 1988.

American society of testing & materials, standard test method for bleeding of concrete, ANSI/ASTM c232-71.

British standard institute, analysis of hardened concrete, BS 1881: pan6, 1971.

Emerson. M, mechanisms of water absorption by concrete T & R, lab, UK, 1990.

Potter, R.H., Quality of cover & its in fluency on durability, SP100 International conference on concrete Durability, vol. 1. Aci.

9 CONCLUSIONS

According to the researches, to decrease concrete water absorption which was described in detail in the

Application of nano composites in designing and manufacturing of cement and concrete

A. Bahari, J.R. Nasiri & O.J. Farzaneh
Islamic Azad University-Tabriz Branch, Iran

ABSTRACT: Today, by introducing nano composites, strengthening and nano reinforcements in building materials a new ware has been created by rapid speed in building industry. Carbon nano tube, empty cylinders are graphite mono sheet wrapped cylindrical have extraordinary mechanical and electrical structural properties and their cylindrical Symmetry. Research have been conducted in carbon, carbon nano tubes, carbon fibers, their properties, applications and methods of their diverse production and modifying them based on requirements. Since carbon nano tubes are produced form graphite carbon, they impose very good resistance against chemical attacks and they have good thermal resistance. In this paper, the result of an investigational plan for manufacturing concrete by using nano composites have been proposed and exhaustive research has been conducted about the method of producing nano tubes. Then the application of nano composites and in general nano technology in building industry especially in concrete and cement has been evaluated.

Keywords: Nano composite, cement, concrete, nano technology, nano tube, carbon

1 INTRODUCTION

Nanotechnology or materials control in molecular scale. Is a gate opening to nature's secrets in ale fields from engineering to medicine? In near future, in newly-made buildings, bricks may repair themselves facing with a crack inside. Also, machines may be covered with layers as strong as diamond to be protected against scratches. Physicians will be able to diagnose against scratches. Physicians will be able to diagnose hundred types of diseases, too, just by putting one blood drop in a machine and getting the result after a few seconds. Nanotechnology is controlled in a very smau world. Its purpose is to make objects, atom by atom, molecule by molecule, and from bottom to top, a way followed by molecule, and from bottom to top, a way followed by nature for million years. Nano is a scientific-prefix means "one-in-billion", and its domain is "a billion in a meter", a dimension in which all atoms are combined with one another and molecules impose crass effects on each other. The aim is that if human being can tell the atoms how to do alignment themselves and how to behave, he can control many features of a material. As we see in nature, carbon atoms in coal are turned in to diamond by changing their arrays, so that characteristics like color, strength, and breaking can be determined in atomic level. Scientists believe that if they can make a brick atom by atom, they can treat its molecules in a way to repairer a possible crack, or to comply with wet weather by decreasing or increasing their porosity. Therefore, nanotechnology hopes to make all imaginable things from smallest cranes and motors to self-assembling metallic or plastic layers. For the first time in scientific imagination, because of recent progresses to see the world in nano scale, these stories will be reasonable. Many kinds of new microscopes and strong simulating computer programs, developed in recent ten years, made a revolution in nano technology; microscopes let scientists not only to see atoms, but also to replace them. Likewise scientists in Alemaden Research center relied no IBM, IN 1990, did a famous experiment through which they could write the word "IBM" BY 35 "Zenon" atoms. Nowdays, a team of IBM physicists announced another progress which makes circuits in atomic scale. This progress, named "Quantum Mirage", shows that in formation can wirelessly move among solids. The new teel is: "eyes", "Fingers", and "pincers" that can work in nano world. Engineering Assistant, Evgene wang, in America science National Association reported on nanotechnology to parliament members.

"Nanotechnology promises to attract increasing number of people, those in terested in science, government and private industries". Dr. Tom Schaeider, a mathematician biologist in cancer National Anistitute said "the reason why people accept this is its real scientific backing". He added "we can make whatever we want in future". Lost year, leader scientists announced in America science National Association that nanotechnology will effect basically on world peoples' healthy state, security and economical position, at least will be important in making anti-biotics, ICS, and polymers in 20th century. In 1998, the white House Technology and science Council established a working team among IWGN required by government industrial and scientific parts. The team had to developed U.S outlook on nanotechnology during 10–20 years in future. U-S government had invested about 260 million in this technology during 1990. President Clinton, also, suggested increasing of nano-technology budget up to $500 million in 2001. Group IWGN predicts that nanotechnology will lead to progresses on fields such as it, medicine, biology, engineering, automobile industry, energy, and national security. According to quick advances of scientific and applied researches of nano sciences in au fields of industry and science, application of this phenomenon in building industry and in general making buildings has not been paid much attention, Recently, however, according to nano-related reinforcers and consolidators in building making materials, a new wave has quickly included buildings industry. Nano – tubes Carbon (CNT) is one of the most important research fields in nanotechnology, including their especial features and unigue potentials in useful business application, a wide range from eldctronics to Chemical processes Control. As it was said, Nano – tubes Carbon are hollow Cylenders made up of graphitic Sigle-layers and wrapped in to Cylender Shape. These materials have extraordinary Structural, electrical, and mechanical features due to thir especial characteristics of carbon bonds and cylenderical symmetry. On carbon, nano-tubes Carbon, thir different features, applications and production way, also about thir modification systems according to customers needs, many researches have been done, however, there is no room heve to discuss about them. Young's modulus and stress-to-strain ratio, i.e material resistance consumption against flexibility varies for different materials from a few GPA to 600 GPA for the strongest one lide diamond. This quantity Changes from 0.4 to 4.15 TPA in nano tubes. Its average for SWNT is about 1 TPA, and for MWNT is nearly 1/28 Tpa-While nano-tubes diameter increases, its Strain rate decreases and youngs modulus goes wp. Strength to denisity ratio is, also, very important in building materials designing. This amount for nano tubes carbon is 100 times more than steel. This quantity for current carbonic fibers is 40 times more than steel. Since carbon nano tubes are made up of graphit carbon, they have a very well resistance against chemical attacks and show high heating resistance.

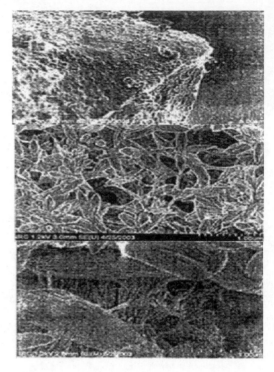

Figure 1.

2 TUBE STRUCTURE

One can imagine nano-tubes carbon the modified and shape-changed from of graphit. Graphit consists of many carbon atoms layers Connected to one another in hexagonal from and make flat sheets. The bond amony layers are cery weak and the ones amony atoms are strong. CNT can be assumed graphity sheet or sheets turned into cylinder from. CNT can be available in froms of SWNT single-wall nano-tubes, like a cylenderical sheet, or MWNT multi-wall like a few sheets truned together into a cylinder.

2.1 Concrete in nano-scale

Concrete is among porous materials, that their porousity size is in nano scale made by chemical reaction between water and cement. These nano porousities Control features of calcium-cilicahydrated product, so concrete is a nano – material in some aspects Chemical attacks take place Through concrete holes and influence on its inside steel, lead to its xidation, which is one reason of concrete rupture and breaks. Therefore we should pay a lot of attention to concrete micro-structure. A Concrete mixture which includes fumed silca nano-structure (Sub-product of industrial glass productiony is known as a great reform in concrete structure resistance against salinity. Generally, by adding fumed silca, which has addative role in nano scale, we can make highly – resistive concrete, but if it is added so much (more than need), it will make it to be fragile, so we should be careful 0 (-its amount).

2.2 Active cement of nano structure

During the researches, done in FHWA Anistitute, the researchers used nuclear resonance reaction analysis (NRRA) TO Study cement hydratation in nano scale, therefore, one can get useful information from events happened in cement hydrated particles level. A light of nitrogene atoms is radiated to cement full of water, and the results are drawn as a graph, called hydrogelle vertical cut in depth, this grogh shows water in fluence speed and various surface layers alignment achieved through the reaction. Surface layer of 20 nm thickness acts as a semi-osmosic dam, which allows water inter cement particles and react with calcium ions. Silica ions, which are bigger than, are stopped behind this layer. When the reaction continues a layer of silicat-gel is made under the surface-layer which is cement inflation factor, and suddenly layer is broken down. This rupture makes those stopped surface silicats free and they make reaction with calcium ions, cakiam-silicat hydrated gel production results in concrete stiffness. Gradual completion of hydrogen uertical cut shows the time of surface lager rupture. This information is used to study concrete stiffing process as a function of temperature, heating, chemistry, cement, etc. for instance by the use of NRRA one can determine hydrated cement breaks in 86 of for 1/5 hours after adding water.

2.3 Alkali-silica gel reactions (ASR)

Among cement alkali's and silica's active form in materials, there is an event called silica-Alkali Reaction. This reaction produces silica-Alkali gel. If there is enough moisture, the gel will extend and make cracks in concrete. ASR causes concrete weakness and its damage in some points against external forces. Therefore it is called concrete Aids. ASR gel expansion includes transformation and changing of Morph-gel in nano-scale which is studied by transformation notronic diffraction, taken place in nano scale, as a function of gel chemistry, temperature, and relative moisture.

2.4 Fly ash features

The Reaction between fly ash and nano scale gel of Portland cement has affection on concrete strength and endurance. Through notronic diffraction, scientists study quality and the way of changes happening as a function of time and fly ash combination.

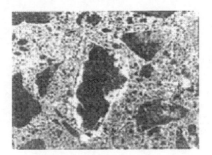

Figure 2. Alkali-Silica.

Figure 3. The Reaction between fly ash and nano scale gel of Portland cement.

2.5 Cement hydratation speed

It is very important to achieve a correct model of water and cement reaction as a function of temperature, water to cement ration, and cement particle size. However it is difficult to get these basic in formation by current analysis ways, as reactions take place in nano holes of cement gel. Notronic diffraction procedure is very useful to measure water movement and reactions. Also, researchers study uarious factors' affects on cracks progress speed in cement.

3 CONCLUSIONS

Nano technology is a modern field which will have lots of effects on other technologies and human's life quality in future. Iran has, also, started to do some researches in this field.

Carbon nano tubes are one of the most important materials surueyed and discussed for nano technolog applications. Their especial features include a wide spectrum from their extraordinary strength to electronical unusual behaviour, high heat conductivity, and save ability of nano particles inside the tubes, potential of lots os application of nano tubes carbon and their especial features encouraged many researches in different scientific and engineering courses. CNT usage in building industry includes a wide range of material complsite to combinations and parts with high resistance of structure and heat conductivity technology.

REFERENCES

"Nano technology, creation technology image", presidentary technological coworking office, Nanotechnology policy suruey committee.

Dai, H., "Nano tube Growth and cterization" in carbon Nano tubes:

Engines. f creation / X.Eric Drexler. Chor book seditions. USA.

WWW. Nano tech news.com

WWW. Smalltimes. Com

www.nano.ir

Excellence in Concrete Construction through Innovation – Limbachiya & Kew (eds)
© 2009 Taylor & Francis Group, London, ISBN 978-0-415-47592-1

Lightening and strengthening of building using structural lightweight concrete

M. Mohammadi, S. Nanpazi & M. Ghassabi K.
Civil Engineering Faculty of Islamic Azad University – Marand Branch

D.B. Zadeh
Islamic Azad University, Iran

ABSTRACT: Using lightweight natural materials has been considered as a suitable way in order to decrease the dimensions of supporting structure, minimize the force of earthquake on buildings, facilitate the execution and economizes the project. In order to decrease the losses and damages resulting from earthquake, besides observing correct principles and criteria of designing, the structures should be designed and executed as light as possible. In this paper we have investigated the tests results of a research project executed in East Azerbaijan Science and Technology Park. In this research project 16 mix designs in concrete laboratory of Sahand University of Technology and 9 mix designs in concrete laboratory of Tabriz University have been used and the structural lightweight concrete with pumice has been designed and made according to ACI213R02 standard.

1 INTRODUCTION

Providing suitable housing for the public and fulfilling this basic need of Iranian society is one of the main and fundamental problems, which has not been solved despite all discussions and actions taken in this regard. Those who are involved in construction technology undertake the execution of these projects. Among the main problems in construction is to provide materials with possible low price and minimize the expenses of building construction. So, besides having the required strength, the construction materials should be available with lowest prices and the expenses of transportation and application should also be low. According to the researches carried out by the experts of construction industry, concrete is one of the most proper construction materials with advantages like being strong and resistance against earthquake, soundproof and heatproof. Unfortunately, using ordinary concrete has some disadvantages including: higher amount of cement needed and heaviness of its components which both cause the concrete to be expensive. The aim of structural lightweight concrete project is to offer a solution to decrease the disadvantages of ordinary concrete and keep its advantages.

2 SITUATION OF MINE

The mine is located in 45 km of southeast of Tabriz and the road connecting the mine Tabriz is an asphalted road.

3 QUALITY & QUANTITY OF MINERALS

The minerals excavated from then mine is pumice and pumicite, which have separate special applications. Pumicite excavated from this mine is applied as a natural pozzolan. According to the tests of pozzolanic activity carried out by the laboratory of Housing & Construction Research Center, the pozzolanic activity of this product is 99% and considering that the standard index of this activity is 75%, its is obvious that this product is of high quality. Pumice, which is a natural lightweight product with good disadvantages and applications, is another product of this mine, proposed for preparation of structural lightweight concrete.

4 FIRST STAGE

By starting the laboratory activities of project, with the aim of observing the amount of cement as $350\,kg/m^3$,

Table 1. Specifications of concrete laboratories designs of Sahand University of Technology [Cubic samples $10 \times 10 \times 10$ cm].

Super plasticizer Kg/m^3	Water Kg/m^3	Silica fume Kg/m^3	Cement Kg/m^3	L.W.A 8 to 15 mm Kg/m^3	L.W.A 4 to 8 mm Kg/m^3	L.W.A 2 to 4 mm Kg/m^3	L.W.A 0 to 2 mm Kg/m^3	Mix No.
0	322	0	356	109	358	275	377	S1
0	324	36	322	109	360	276	379	S2
1.75	314	35	312	106	349	268	367	S3
0	320	0	354	256	356	178	323	S4
0	316	35	314	252	351	176	318	S5
1.75	316	35	314	252	351	175	318	S6

Table 2. Results of concrete laboratories designs of Sahand University of Technology [Cubic samples $10 \times 10 \times 10$ cm].

Slump cm	Compressive strength of 28 days kg/cm^2	Compressive strength of 14 days kg/cm^2	Density of fresh concrete kg/m^3	Mix No.
0	230	190	1795	S_1
0	234	205	1804	S_2
4	233	222	1751	S_3
0	215	196	1787	S_4
0	232	202	1762	S_5
4	225	211	1763	S_6

Table 3. Specifications of workshop mix designs [Cubic Samples $10 \times 10 \times 10$ cm].

Super plasticizer Kg/m^3	Water Kg/m^3	Cement Kg/m^3	Lightweight aggregates Kg/m^3 4 to 15 mm	Lightweight aggregates Kg/m^3 2 to 4 mm	Lightweight aggregates Kg/m^3 0 to 2 mm	Mix No.
1.8	270	370	484	286	392	S1
1.75	265	355	666	193	358	S2

six mix designs were prepared and tested in concrete laboratory of Sahand University of Technology, the specifications and results of which have been mentioned in Tables 1 and 2.

In these mix designs, we have tried to take into account the effect of silica fume and the amount cement and super-plasticizer. We have also tried to use the grading of the mine in order for the results to have the least problems in practical application. Lightweight aggregates were dampened with mix water for 30 minutes before mixing in order to prevent absorption of much water by the aggregates during the first stage of mixing. The cement used here is type 2 Soufian cement. The amount of plasticizer is about 0.5% of adhesive materials like cement and silica fume and the replacement percentage of cement with silica fume is 10%. Considering the obtained results, we decided to omit silica fume powder, which was an additive, in order to economize the project and in order to increase the slump and plasticity of the concrete as it was used in construction workshop besides being used in ready-mixed concrete factories, we used low amounts of plasticizer, the expenses of which is even negligible.

5 SECOND STAGE

By analyzing the results of the first stage, two mix designs were selected for workshop laboratory tests. In order to analyze the effects of workshop conditions on laboratory results of first stage, the samples of concrete with two mix designs were prepared in the mine, the specifications and results of which are mentioned in tables 3 and 4.

Table 4. Results of workshop mix designs [Cubic Samples 10 × 10 × 10 cm].

Slump cm	Compressive strength of 28 days kg/cm²	Compressive strength of 14 days kg/cm²	Density of fresh concrete kg/m³	Mix No.
7	192	169	1804	S7
9	200	176	1839	S8

Table 5. Specifications of final workshop mix designs [Cubic Samples 10 × 10 × 10 cm].

Super plasticizer Kg/m³	Water g/m	Cement Kg/m³	Lightweight aggregates 8 to15 mm Kg/m³g/m³	Lightweight aggregates 4 to 8 mm Kg/m³	Lightweight aggregates 2 to 4 mm Kg/m³	Lightweight aggregates 0 to 2 mm Kg/m³	Mix No.
2	254	406	203	205	205	205	S9
1.75	244	259	275	406	193	263	S10

Table 6. Results of final workshop mix designs [Cubic samples 10 × 10 × 10 cm].

Slump cm	Compressive strength of 28 days kg/cm²	Compressive strength of 14 days kg/cm²	Density of fresh concrete kg/m³	Mix No.
5.5	233	203	1380	S9
5	216	189.5	1843	S10

Here it is necessary to mention that the materials used in the workshop were damp enough when used and this can be observed by comparing the amount of consumed water in tables 1 and 3.

6 THIRD STAGE

Analyzing the results of workshop mix designs and the obtained low strengths resulting from high amount of cement in proportion to water and non-conformity of grading with laboratory grading. In this stage, two mix designs with 350 and 4000 kg/m³ were considered taking into account the results of two previous stages and the samples of concrete were prepared in the workshop. The specifications and results of these designs are mentioned in Tables 5 and 6.

It should be mentioned that the grading of lightweight aggregates used in these two designs is rather different from laboratory grading. The dimensions of the concrete samples were 10 × 10 × 10 cm and in order to convert their strength to standard cylindrical strength we should apply the coefficient of 15 × 30 on them. Researches show that this coefficient

is variable from 0.75 to 0.95 according to the strength and weight of the concrete and in higher strengths this ratio approaches 1. Considering the ordinary strength of concretes and their relative lightness, 0.85 was selected as coefficient of converting cubic strength to cylindrical strength. The final results of these two designs are mentioned in Table 7. Taking into account the results in the table we can see that increasing the amount of cement from 350 to 400 had not an important role in increasing the strength. This can be explained that: we can describe a strength limit for each type of lightweight aggregates, called strength capacity. Up to a certain amount of cement, the increase rate of mortar and the correspondent concrete is the same. In other words, increasing the amount of cement in mortar and the lightweight concrete made with it increases the strength of mortar and concrete made with this mortar. But at ratios higher than this amount, the increase rate of mortar strength is more than the increase rate of the strength of its correspondent concrete. In order to determine the strength capacity of lightweight aggregates being investigated, six mix designs were prepared and the samples of lightweight concrete and the related mortar were made, the results of which are shown in tables 8 and 9.

Table 7. Conversion of the results of final designs to standard results.

Slump cm	Compressive strength of 28 days kg/cm^2	Compressive strength of 14 days kg/cm^2	Density of fresh concrete kg/m^3	Super plasticizer	Water Kg/m^3	Cement Kg/m^3	Lightweight aggregates	Mix No.
5.5	198	173	1840	2	254	406	1168	S9
5	184	161	1843	1.75	244	259	1238	S10

Table 8. The specifications of designs for determining the strength capacity with lightweight sand [Cubic Samples $10 \times 10 \times 10$ c].

Compressive strength of 7 days kg/cm^2	Density Kg/m^3	Super plasticizer	Water Kg/m^3	SILICA FUME Kg/m^3	Cement Kg/m^3	Lightweight aggregate passed through sieve number 6	Lightweight sand [between sieve number 6 and 1/2 inches]	Mix No.
96	1797	–	240	–	288	519	750	S11
121	1787	–	245	–	330	496	716	S12
130	1746	3.1	275	52	304	455	657	S13
145	1827	–	250	–	380	490	707	S14
147	1811	3.9	231	–	380	490	706	S15
153	1810	–	278	–	140	460	664	S16

Table 9. Specifications of mortar designs offered for determining the strength capacity with lightweight aggregate [Cubic Samples $5 \times 5 \times 5$ cm].

Compressive strength of 7 days kg/cm^2	Density Kg/m^3	Super plasticizer	Water Kg/m^3	SILICA FUME Kg/m^3	Cement Kg/m^3	Lightweight aggregate passed through sieve number 6	Mix No.
236	2000	–	204	–	608	1088	S11
248	2016	–	226	–	672	1008	S12
252	1872	6.8	226	96	576	956	S13
268	2016	–	192	–	926	728	S14
276	2040	6.08	192	–	952	728	S15
292	1944	–	176	–	848	760	S16

By increasing the amount of cement from 288 to 380, no considerable change was seen in curve but when the amount of cement was increased to 380, the curve went higher. This shows that when the amount of cement is 380, increasing the amount of cement increase the strength of the mortar, while it has less effect on increasing the strength of the correspondent concrete. So we can say that increasing the amount of cement more than 380 kg/m^3 to lightweight concretes made with this type of aggregate is not economical and will have considerable expenses. As there are some disagreements about applying the coefficients in order to convert compressive strength of cubic samples to standard cylindrical samples (15×30) and as all valid codes of concrete have expresses the standards and requirements in accordance with standard cylindrical samples, we decided to follow the test carried out in concrete laboratory of Tabriz University and the mix designs were sampled and using cylindrical molds and tested to ensure the results. It should be mentioned that no special grading was considered for concreted molded in the laboratory and the materials were used with the range of grading in the mine in dimensions of 0–2 mm, 2–4 mm, 4–8 mm and about 15–8 mm in

Table 10. Specifications of concrete laboratories designs of Tabriz University (Standard cylindrical samples).

Super plasticizer	Water Kg/m^3	Cement Kg/m^3	Pumice aggregate				Mix No.
			8–15 mm	0–8 mm	0–4 mm	0–2 mm	
59	8.7	8	4	7	6	6	T1
56	2.2	8	2	7	4	5	T2
65	4.5	9	2	8.5	4.2	7.1	T3
67	4.5	7.15	6	8.2	4.43	8.23	T4

Table 11. Results of concrete laboratories designs of Tabriz University (Standard cylindrical samples).

Tensile strength of 28 days kg/cm^2	Compressive strength of 28 days kg/cm^2	Density Kg/m^3	Mix No.
16	–	1765	T11
–	142	1770	T12
–	125	1762	T13
16	–	1785	T21
–	156	1787	T22
–	165	1760	T23
17	–	1795	T31
–	140	1790	T32
–	144	1809	T33

order to economize the mass production. The specifications of designs and results are shown in Tables 10 and 11.

The amounts of plasticizer were 0.73, 0.675, 0.71, 0.82, respectively, for the content of cement.

– Three samples were molded for each design, two of which will be used for compressive strength test and one for tensile strength test on 28th day.
– As there was no standard mold to be used in the laboratory and because of religious holidays, we only made four mix designs.

As the minimum compressive strength for structural lightweight concrete with cylindrical sample should be 170 kg/cm^2, so the above-mentioned designs were optimized in the next stage and we used silica fume gel, which is a new and cost-effective product in Iran, and mixture of silica fume powder and concrete plasticizer in order to increase the compressive strength of the samples and reach higher standard of structural lightweight concrete and the plasticizer used in designs was omitted, instead.

– Two samples from each design were sampled in cylindrical molds in order to test the compressive strength.

– The amount of silica fume gel for each design was 10% of the cement weight, which will be decrease in later designs.

According to the obtained results, the working team made a structural lightweight concrete in accordance with Iranian and American Concrete Codes. The requirements of structural lightweight concrete of these codes are mentioned below: According to Iranian Concrete Code (ABA), the structural lightweight concrete should have the 28-day compressive strength of more than 160 kg/cm^2 (standard cylindrical sample). According to American standard (ASTM- No. C330 and American Concrete Development Standard ACI 313), the air-dried specific gravity of 28-day standard cylindrical sample in structural concrete containing lightweight aggregates was about 1440–1850 kg/m^3 and compressive strength was more than 17.2MP. According to the American structural lightweight concrete requirements and standards code (Aci213R-03), the specific gravity of 28-day air-dried sample (standard cylindrical sample) should be between 1120–1920 kg/m^3 and the minimum compressive strength at 28-day should be more than 170 kg/cm^2, while the specific gravity and compressive strength of the made concretes were within the mentioned ranges.

7 SPECIFICATIONS OF BUILDING BEING STUDIES FOR LIGHTENING WITH PUMICE

A five-storey residential building + parking and ridge, located in Tabriz, with a very high relative line Code of earthquake lateral force distribution (UBC94) (According to 2800 third edition).
Code of concrete structure design (ACI318-99)
Type of Land: III

Type of Land	T_0	T_s	S
III	0.15	0.7	1.75

$$T = 0.07 \times H^{3/4} => T = 0.07 \times 18.8^{3/4} = 0.63 \, S$$
$$1.25 \, T \rightarrow 1.25 \times T = 0.79 \, S$$

139

Table 12. Specifications of optimized designs of Tabriz University (Standard cylindrical samples).

Silica fume gel gr	Water Kg/m^3	Cement Kg/m^3	Pumice aggregate				Mix No.
			8–15 mm	4–8 mm	2–4 mm	0–2 mm	
890	3.2	8.9	0	5	5	9	T5
930	3.5	9.3	0	2	5	12	T6
930	4	9.3	0	4	5	11	T7
390	4.3	9.4	2	6	5	8	T8
930	4.2	9.3	2	7	5	5.5	T9

Table 13. The results of designs optimized in concrete laboratory of Tabriz University (Standard cylindrical samples).

Compressive strength kg/m^2	Age of samples	Density Kg/m^3	Mix No.
305	11	1925	T51
278	11	1920	T52
303	11	1928	T61
297	11	1919	T62
250	10	1905	T71
177	10	1910	T72
264	10	1865	T81
261	10	1860	T82
250	9	1820	T91
200	9	1809	T92

$T > 0.79$ S $\rightarrow$ for correct distribution of earthquake force and whiplash effect of it, that is (ft), we should use UBC94.

$$B = (\varepsilon + 1)(\frac{T_s}{T})^{2/3} \Rightarrow B = 2.75(\frac{0.7}{0.79})^{2/3} = 2.54$$

$$C = \frac{ABI}{R} = \frac{0.35 \times 2.54 \times 1}{7} = 0.127$$

In order to compare the results we need to model two structures with similar dimensions.

7.1 Building A

Ordinary concrete with specific gravity of 2500 kg/m^3 and compressive strength of 210 kg/cm^2

7.2 Building B

Structural lightweight concrete with natural pumice aggregated, specific gravity of 1900 kg/m^3 and compressive strength of 210 kg/cm^2

– Both structures are calculated with exact dynamic analysis.

The spectrum used is the standard spectrum for soil type III and all the regulations of ABA Concrete Code and ACI318-99 and 2800 of earthquake, third edition in Iran, have been observed.

Elasticity modulus with calculated weigh relations and weight of lightweight concrete in calculations were considered as 1900 kg/m^3.

As it is clear from dynamic analysis:

– The amount of basic shearing in a structure made with lightweight concrete is 226780 kg and the weight of structure is 178791 kg.
– The amount of basic shearing in structure made with ordinary concrete is 257376 kg and weight of structure is 2028792 kg.

The results show that by the using lightweight concrete in structure, the weight of the structure was decrease up to 12%.

So 12% of earthquake force applied on the structure was decreased directly. These results can be clearly seen in steels needed in the structure.

Statistically, the percentage of decrease in steels of columns is decrease up to 14.26%. (The constant dimensions are considered for the purpose of comparison).

Dimensions of Columns	Percentages of Steels Decrease	Storeys
60 × 60	16.8% Decrease	1st Storey
55 × 55	12.8% Decrease	2nd Storey
50 × 50	12.1% Decrease	3rd Storey
45 × 45	15.8% Decrease	4th Storey
45 × 45	13.8% Decrease	5th Storey
40 × 40	14.3% Decrease	6th Storey
	14.26	Average

This amount in beams is 74% of that mentioned for columns.

On the whole, the total weight of the structure decrease up to 12% and the amount of longitudinal steel in beams and columns decrease up to 12.4%, which is a considerable amount in consumed steels and its expenses in a reinforced concrete structure with medium height.

Obviously, good results can be offered in higher buildings, even in higher dynamic analyses the results increase up to 14%.

Of course this type of lightening has been used only in the skeleton of the structure and coatings of floors and if it is applied on internal walls, partitions and external walls, the results will be very desirable.

8 CONCLUSIONS

Considering the workshop tests carried out in the place of mine we can mention the following points:

– We can prepare structural lightweight concrete with the investigated materials according to the definition of ACI-R-03.
– When the amount of the prepared concretes is about 380 kg/m^3, we can expect the following specifications from the concretes made with light sand and gravel in the workshop.
– Specific gravity of fresh concrete is about 1850 kg/m^3 and the weigh of dried concrete is about 1800 kg/m^3. It should be explained that according to definitions of (ACI213R-03) code, the specific gravity of structural lightweight concrete is about 1120 to 1920 kg/m^3 and the specific weight of lightweight concretes are within this range.
– 28-Day compressive strength of standardized cylindrical samples is about 300 kg/cm^2. It should be mentioned that according to (ACI213R-03) code, the minimum of 28-days compressive strength of structural lightweight concrete is higher than this amount.
– The slump of fresh concretes will be about 5 cm and it is obvious that this efficiency will increase in case of using lots of super-plasticizer.

REFERENCES

Concretology, translated by Hormoz Famili, University of science and technology, 1996.
Famili, Hormoz, Research project of lightweight aggregate concrete, *University of science and technology, 2000.*
Josefh J. Waddell, Joseph A. Dobrowolski, *Concrete construction handbook*, Third edition, 2001.

Excellence in Concrete Construction through Innovation – Limbachiya & Kew (eds)
© 2009 Taylor & Francis Group, London, ISBN 978-0-415-47592-1

The role of nano particles in self Compacting concrete

O.J. Farzaneh, J.R. Nasiri & A. Bahari
Islamic Azad University - Tabriz Branch, Iran

ABSTRACT: The aim of this study was to investigate the relationship between nano technology and constituent materials of new concrete called self Compacting Concrete in both fresh self Compacting concrete and hardened self Compacting concrete starts as applied method In this paper, by designing and manufacturing self Compacting concrete according international criteria and application of nano silica (silica colloidal), the properties and effects of this material have been investigated in type 5 concrete and mechanical properties like compression strength, tensile strength, expansion and condensation for curing conditions and different ages in concrete have been tested and the results were compared with control self Compacting concrete (without nano particle). The results showed that the mechanical properties and resistance of self Compacting concrete with nano silica (silica colloidal) particles are better than control samples because of filling the voids of concrete in nanometer scale.

1 INTRODUCTION

The idea of self consolidating concrete was expressed by a Japanese scientist, Okamura, in 1980 for the first time, and today's many research centers throughout the world are doing surveys about it. In Iran, survey, show that researches on this kind of concrete took place, Since our country should apply nano technology in different fields, present paper has evidently done some surveys first with Concrete Type V of Kerman cement Factory to design and produce self – consolidating concrete and has studied its mechanical features by the help of nano colloidal until 28-day age, and then has compare it with control self-consolidating ones (without those colloidal). Results indicate that those including nano colloidal have better performance than the control group.

2 APPLIED MATERIALS FEATURES

In this research, the researcher has used cement Type V of Kerman cement factory, which is Stronger in front of Sophistic attacks than other Portland cements. Micro silica, used in this experiment, includes about 94% of sio_2 with especial weight of $2.3\,g/cm^2$. The used nano-silica is on the basis of silicate particles of 5–50 nm. This kind of silica is a solution with cream color (10–50% of solid amount) with especial weight of 1.03 and ph of 10. In one measure of 3–5%, it may decrease, and increase resistance against separation Due to especial surface area, that is very much (80–1000 m^2/g),

and because of colloidal silicate particles circle shape, self-consolidating concrete has high endurance, especially when there is little fulfiller (Porro, 2005, Maghsoudi & Hoornahad, 2005 and Maghsoudi & Hoornahad, 2006).

3 EXPERIMENTS EXPLANATION

3.1 *Fresh self-consolidating concrete*

To do survey and comparison of nano-particles' effects on self-consolidating concrete including cement Type V, the researcher selected two mixture designs among the various types designed and made previously. One of them was used by adding nano-particles in self-consolidating (scc1 mn) (as the experimental group) and the other one was applied without any additions of nano particles (SCC 1 m) to self-considering concrete (as the control group) Table 1 indicates mixing pattern in details.

3.2 *Experiments results on fresh concrete*

In this paper, the researcher has done four experiments, taken place in paste phase of self-consolidating concrete, and results have been compared in table 2 with the values recommended by world valid regulations. To see more explanation on using these experiments refer to [12]. The results of this table indicate similarity of experiments in paste phase of self-considering concrete with these described internationally, which shows successful production of this concrete in this research.

Table 1. Mixing pattern in details.

	SCC1M	SCC1MN
Cement type V (Kg)	270	270
Micro-silica (Kg)	30	30
Nano-silica (Liter)	–	1.8
Water (Kg)	188	188
LSP (Kg)	225	225
Coarse Aggregate (Kg)	750	750
Fine Aggregate (Kg)	870	870
PCE (Liter)	3.675	3.675
W/CM 1	0.626	0.626

1 CM: Cement + Micro-silica   2 P: Cement + Micro-silica + Lime Stone Powder.

Table 2. Fresh properties of SCC.

Type of test	SCC1M	SCC1MN	Recommended values [8]
Slump Flow (mm)	640	630	600 to 700
L-Box (h2/h1)	0.78	0.83	Up 0.75
J-Ring (mm)	620	600	Same as Slump Flow
V-Funnel, 5 (min, s)	4.43	6.5	6 to 12

3.3 Experimental results of stiffed concrete

In this paper, three curing conditions for concrete samples have been considered.

Dry curing condition (D): In this state, samples are kept for 7 days in saturated slacked lime, and then are kept in lab in dry condition until experiment stage. Wet curing condition (w): In this state, samples are kept for 28 days in saturated slacked lime.

Sophistic curing condition (s): In this state, samples are kept for 7 days in saturated slacked lime are kept for 7 days in saturated slacked lime.And then in sulphuric acid with pH 1.5 until experiment stage.

3.4 Pressure strength

According to U.K standard regulation, we prepared cubic samples of 10*10* 10cm to determine pressure strength of them in various stages. For each curing condition in each stage (age), at least three samples were considered and experiments were determined for ages of 3,7,14, and 28 days. Figure 1 shows the experimental results on pressure strength. As it is seen in this figure, for cements type V, Concretes pressure strength including nano-silica is higher than ones lack of nano – silicate. Also, figures convergence in concretes including nano-particles, in various curing conditions of wet and dry, indicates more complete hydration in concrete's lower ages (Fig. 1). Figure 2

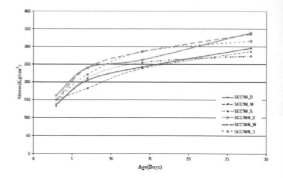

Figure 1. Comparison of results of three curing conditions for pressure strength of SCC1MN & SCC1M samples.

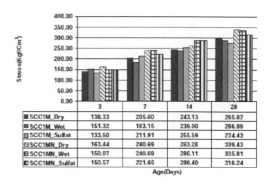

Figure 2. Comparison of pressure samples with and without nano particles for various curing conditions.

represents comparison of results of samples with and without nano particles.

Strength increasing rhythm. In wet condition of curing (w), at age of 28 days and in comparison with a mixture without nano, the value of strength increase was 17%, for curing in dry condition (D), at the age of 28 days, it was 14.7% and for curing and sulphat (s), at the age of 28 days, it was 15.2%. Rhythm of achieving pressure strength for samples for determined time intervals indicates quality of the way of achieving this strength, represented in tables 3 & 4. As it is clear in Table 3, rhythm of achieving strength in ages of 3–7 days in concrete mixing pattern with nano-particles has significantly grown – Also, in a condition where corroding sulphate is active, concretes full of nano-particles indicate much better strength than those ones without nano-particles. This fact shows that using nano- particles leads to concrete's porosity decrease.

3.5 Bending strength

Results of bending experiments are shown in figure 3.

Results show that curing conditions have little effect on getting samples' bending strength. The achieved

Table 3. Achieving pressure strength rhythm of samples in various time intervals and curing conditions.

Mixes label	Curing conditions	3.7	7.14	14.28
SCC1M	(D)	32.7	15.4	17.8
SCC1M	(W)	17.4	23.4	16.7
SCC1M	(S)	37	17.1	6.86
SCC1MN	(D)	32.1	5.56	24.9
SCC1MN	(W)	37.6	15.8	14.8
SCC1MN	(S)	32	22.6	9.4

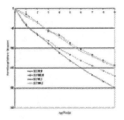

Figure 4. Swelling deformation.

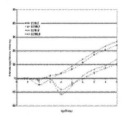

Figure 5. Average readings.

As it is observed, samples with nano particles have the least shrinkage value.

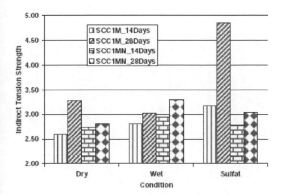

Figure 3. Comparison of bending samples with and with out nano particles for various curing conditions.

bending strengths are close to one another for all curing conditions.

3.6 *Swelling and shrinkage*

To determine swelling and shrinkage values of self-consolidating concrete samples, for each pattern, two prismatic samples by dimensions of 45 X 10X 10 cm for swelling and two others for shrinkage were made. 24 hours after making, samples were taken off from frames and, at the age of one – day, were installed. In two sides against the samples in 5 cm intervals from samples center (10 cm from each other), and swelling samples were put inside the saturated slacked lime. Shrinkage samples for dry curing condition were in slacked lime for 7 days. For wet curing condition this time was 28 days. At the end of these periods, the samples were taken off from slacked lime and kept in normal weather. Therefore the only shrinkage value in dry curing condition has been shown. Samples swelling value was read by mechanical strain gauge. Figure 4 indicates average values of readings in two opposite sides. As it is seen, samples including nano particles have the least swelling value. Average values of readings in two opposite sides are shown in figure 5.

4 CONCLUSIONS

– Applying nano-silica in self – consolidating concrete has several results as: pressure strength increase – swelling and shrinkage decrease due to fulfillment of many porous and holes inside concrete because of using nano particles in comparison with control self-consolidating concrete (without nano- particles application).
– Pressure strength increase and swelling and shrinkage decrease in samples with nano particles than samples lack of them are two important steps towards using little comment in mixing self – consolidation concrete. In other words, if we use little comment and apply nano particles, we will be able to get the same strength with control group. Little usage of comment in concrete leads to decrease hydration heat and premature cracks (in early hours and first week of concreting) in early ages of P-to concreting that is very important.
– By the help of nano particles in self–consolidating concrete, we can decrease swelling and shrinkage. In other words this decrease is known as a significant and positive step towards decreasing cracks and swelling in reinforced concrete structures, and leads to increase their life span and endurance especially in inappropriate weather conditions (in Persian gulf region).
– Swelling and shrinkage decrease in self ansolidating concretes with nano particles is considered positive

145

in terms of per-load concrete structures, as in these kinds of structures the value of force lose in steels decreases and results in economical usage of them a long with their life span and endurance increase.

REFERENCES

Engineering, University of Tarbiat Modares, Tehran, Iran, May 2006, p. 12.

Maghsoudi, A. A. and Hoornahad, H., "Investigation of Engineering Properties of SCC with Colloidal Silica", Proceeding of the 3rd International Conference on Construction Materials, University of British Columbia, Canada, 2005, p. 11.

Maghsoudi, A. A. and Hoornahad, H., "Workability and Engineering Properties of Self-Compacting Concrete", Proceeding of the 3rd International Conference on Civil.

Maghsoudi, A. A. and Hoornahad, H., "Self Compacting Concrete by Use of Kerman's Materials", Proceedings of the 2nd International Conference on Concrete & Development, Tehran, Iran, April–May, 2005.

Porro, A., Dolado, J., S., Camillo, I., Erkizia, E., De Miguel, Y., Saez de Ibarra, "Effects of Nanosilica Additions on Cement Pastes", Application of Nanotechnology in Concrete Design, International Conference, University of Dundee, Scotland, UK, 2005.

Theme 2:   Composite materials in concrete construction

Excellence in Concrete Construction through Innovation – Limbachiya & Kew (eds)
© 2009 Taylor & Francis Group, London, ISBN 978-0-415-47592-1

Strength properties of high-volume fly ash (HVFA) concrete incorporating steel fibres

R. Siddique
Department of Civil Engineering, Thapar University, Patiala (Punjab), India

J.M. Khatib
School of Engineering and the Built Environment, University of Wolverhampton, UK

İ. Yüksel
Zonguldak Karaelmas University, Department of Civil Engineering, Zonguldak, Turkey

P. Aggarwal
Department of Civil Engineering, National Institute of Technology (Deemed University), Kurukshetra, India

ABSTRACT: Properties of high-volume fly ash (HVFA) concrete incorporating steel fibres are presented in this paper. Three concrete mixtures were prepared with 35%, 45%, and 55% of Class F fly ash as partial replacement of cement. Straight-round steel fibres of aspect ratio 80 (0.75% by volume of concrete) were added in each of the fly ash concrete mixtures. Tests were performed for fresh concrete properties, compressive strength, splitting tensile strength, flexural strength, and impact strength at the ages of 28, 91 and 365 days. 28 days test results indicated that steel fibres increased compressive strength of high-volume fly ash concrete by 12 to 19%, splitting tensile strength by 37 to 45%, flexural strength by 15 to 21%, and enhanced impact strength tremendously (4 to 6 times) depending upon the fly ash content. At later age (91 and 365 days) continuous increase in strength properties of high-volume fly ash concrete was observed. Paper also presents the correlation between impact strength with compressive strength, splitting tensile strength, and flexural strength. It has been found that impact strength of high-volume fly ash concrete has very good correlation with compressive strength (R^2 between 0.9816 and 0.99), splitting tensile strength (R^2 between 0.9432 and 0.9838) and flexural strength (R^2 between 0.9846 and 0.9970) depending upon the fly ash content and testing age.

1 INTRODUCTION

Cement is the most costly and energy intensive component of concrete. Concrete cost could be substantially reduced by the partial replacement of cement with fly ash. Accordingly, the replacement of cement with fly ash reduces concrete cost, and conserves energy. In concrete, the replacement of cement with fly ash has its beneficial effects of lower water demand for similar workability, reduced bleeding, and lower evolution of heat. It has been used particularly in mass concrete applications and large volume placement to control expansion due to heat of hydration and also helps in reducing cracking at early ages.

High-volume fly ash concrete has emerged as a construction material in its own right. This concrete normally contains more than 35 to 40% fly ash by mass of total cementitious materials. Many researchers have reported their findings on the various aspects of high-volume fly ash concrete, but here, in this paper only few are listed [1–13]. Ravina and Mehta (1986) reported that by replacing 35 to 50% of cement with fly ash, there was 5 to 7% reduction in the water requirement for obtaining the designated slump, and rate and volume of the bleeding water was either higher or about the same compared with the control mixture.

Malhotra and others (Langley et al. 1989; Giaccio and Malhotra 1988; Malhotra 1990; Alsali and Malhotra 1991; Bilodeau and Malhotra 1992; Bilodeau et al. 1994; Langley et al. 1992) have reported extensively on high-volume fly ash concrete. Concrete containing high volumes of Class F fly ash exhibited excellent mechanical properties, good durability with regard to repeated freezing and thawing, very low permeability to chloride-ions (Langley et al. 1989; Giaccio and Malhotra 1988; Malhotra

1990), and showed no adverse expansion when reactive aggregates were incorporated into concrete Malhotra (1990). Alasali and Malhotra (1991) reported that high-volume of Class F fly ash in concrete has proved to be highly effective in inhibiting the alkali-silica reaction. Superplasticized high-volume fly ash concrete containing up to 60% of fly ash of total cementitious materials, had poorer abrasion resistance than concrete without fly ash (Bilodeau and Malhotra, 1992). Air-entrained high-volume fly ash concrete exhibited excellent characteristics regardless of the type of fly ash (eight fly ashes from U. S.) and cements (two Portland cements from U. S.) (Bilodeau et al., 1994). For concrete blocks containing high volumes of low-calcium (ASTM Class F) fly ash, ratio of the 42 days core compressive strength to the 28 days laboratory-cured compressive strength ranged from 78% for the control concrete to 120% for the high-volume fly ash concrete. At 365 days, these ratios were 78 and 92 percent, and at 730 days, the respective ratios were 88 and 98 percent (Langley et al., 1992). Maslehuddin (1989) reported that addition of fly ash as an admixture increased the early age compressive strength and long-term corrosion-resisting characteristics of concrete. Tilasky et al. (1988) concluded that the abrasion resistance of concrete made with Class C fly ash was better than both concrete without fly ash and concretes containing Class F fly ash.

Papadakis (1999) proposed a model for performance prediction of concrete made with low-calcium fly ash as an additive. 28 days compressive strength of 80 MPa could be obtained with a water-to-binder ratio of 0.24, with a fly ash content of 45% (Poon, 2000). Ravina (1997) reported the effects of high volume of fly ash (two ASTM Class F and two Class C fly ash) as partial fine sand replacement on the fresh concrete properties.

Siddique (2003a) reported the abrasion resistance of concrete incorporating four levels of fine aggregate replacement (10, 20, 30 and 40%) with Class F fly ash. Test results indicated that abrasion resistance and compressive strength of concrete mixtures increased with the increase in percentage of fine aggregate replacement with fly ash. Abrasion resistance of concrete was improved approximately by 40% over control mixture with 40% replacement of fine aggregate with fly ash, and concrete with fine aggregate replacement could be suitably used.

Siddique (2003b) reported the mechanical properties of concrete mixtures in which fine aggregate (sand) was replaced with five percentages (10, 20, 30, 40, and 50%) of Class F fly ash by weight. Test results indicate significant improvement in the strength properties of plain concrete by the inclusion of fly ash as partial replacement of fine aggregate (sand), and can be effectively used in structural concrete. Siddique (2004) presented the properties of concrete incorporating high volumes of Class F fly ash. Portland cement was replaced with three percentages (40, 45 and 50%) of Class F fly ash. Test results indicated that the use of high volumes of Class F fly ash as a partial replacement of cement in concrete decreased its 28 days compressive, splitting tensile, and flexural strengths, modulus of elasticity, and abrasion resistance of the concrete. However, all these strength properties and abrasion resistance showed continuous and significant improvement at the ages of 91 and 365 days, which was most probably due to the pozzolanic reaction of fly ash. Based on the test results, it was concluded that Class F fly ash can be suitably used up to 50% level of cement replacement in concrete for use in pre-cast elements and reinforced cement concrete (R.C.C.) construction.

The use of fly ash in concrete is found to affect strength characteristics adversely. Loss in concrete strength can be retrieved to a large extent by incorporating steel fibres, which have proved their worth in enhancing the strength characteristics of concrete. Extensive work has been reported on the steel fibre reinforced concrete, but, in this paper, only few references are quoted (Roumaldi and Batson 1963a; Roumaldi and Batson 1963b; Snyder and Lankard 1972; Swamy and Mangat 1974a; Swamy and Mangat 1974b). An effort was made to study the effects of straight round steel fibres on the properties of high-volume fly ash concrete. The effects of steel fibres on fresh concrete properties, compressive strength, splitting tensile strength, flexural strength, and impact strength of the high-volume fly ash concrete are presented in this paper.

2 EXPERIMENTAL PROGRAMME

2.1 Materials

Ordinary Portland (43 grade) cement was used. It was tested per Indian Standard Specifications IS: 8112 (1989) and results are given in Table 1. Class F fly ash was used, and was tested per ASTM C 311. Its chemical composition is given in Table 2. Natural sand (4.75 mm maximum size) and gravel (12.5 mm maximum size) were used as fine and coarse aggregates, respectively.

Their properties were determined per Indian Standard Specifications IS: 383 (1970). Their physical properties and sieve analysis results are shown in Tables 3 and 4, respectively. Straight-round steel fibres (0.75% by volume of concrete) having an aspect ratio of 80 (length 32.8 mm, diameter 0.41 mm) were used. Physical and mechanical properties of steel fibres are given in Table 5. A commercially available superplasticizer Centriplast FF90, based on melamine formaldehyde was used in all mixtures.

Table 1. Physical properties of Portland cement.

Physical test	Results	IS: 8112 (1989) Specifications
Fineness (retained on 90 μm sieve	7.8	10 max
Fineness: specific surface (air permeability test) (m²/kg)	345	225 min
Normal consistency	31%	–
Vicat time of setting (minutes)		
Initial	110	30 min
Final	260	600 max
Compressive strength (MPa)		
3 days	24.5	23.0 min
7 days	37.2	33.0 min
28 days	48.0	43.0 min
Specific gravity	3.13	–

Table 2. Chemical composition of fly ash.

Chemical analysis	Class F fly ash (%)	ASTM requirement C 618 (%)
Silicon Dioxide, SiO_2	56.2	–
Aluminum Oxide, Al_2O_3	26.8	–
Ferric Oxide, Fe_2O_3	5.9	–
$SiO_2 + Al_2O_3 + Fe_2O_3$	88.9	70.0 min
Calcium Oxide, CaO	5.8	–
Magnesium Oxide, MgO	2.3	5.0 max
Titanium Oxide, TiO_2	1.1	–
Potassium Oxide, K_2O	0.7	–
Sodium Oxide, Na_2O	0.6	1.5 max
Sulfur trioxide, SO_3	1.6	5.0 max
LOI (1000°C)	1.4	6.0 max
Moisture	0.6	3.0 max

Table 3. Physical properties of aggregates.

Property	Fine aggregate	Coarse aggregate
Specific gravity	2.63	2.62
Fineness modulus	2.30	6.62
SSD absorption (%)	0.88	1.13
Void (%)	34.2	38.2
Unit weight (kg/m³)	1685	1635

2.2 Mixture proportions

Initially, three high-volume fly ash concrete mixtures containing 35%, 45%, and 55% of Class F fly as replacement of cement were proportioned per Indian Standard Specifications IS: 10262 (1982). After this, steel fibres (0.75% by volume of concrete) were added in each of the fly ash concrete mixtures. Mixtures proportions are given in Table 6.

2.3 Preparation, casting, and testing of specimens

Initially, cement and fly ash were dry mixed properly. After all the constituent materials were mixed, about 1\5 of the required water was added to the mix. Small quantities of fibres were released manually and gradually taking care that the fibres were not mixed in bundles. After adding about 1/3 of the quantity of fibres, some more water (about 1/3 of the remaining quantity) was added to the mixer, and the remaining quantity of fibres was added again slowly and in small quantities. Finally, the remaining water was added, and the mixing was done till good homogeneous mixture, as visually observed, was obtained. If any lumping or balling was found at any stage, it was taken out, loosened and again added manually.

Following the mixing, fresh concrete properties tests were performed. Standard 150 mm cubes for compressive strength, 153×305 mm cylinders for splitting tensile strength, $101.6 \times 101.6 \times 508$ mm beams for flexural strength were cast in accordance with the specifications of Indian Standard Specifications IS: 516 (1959). For impact strength, concrete sheets of size $500 \times 500 \times 30$ mm were cast.

The specimens were covered immediately for complete moisture retention. The specimens were demoulded after 24 hours of casting, and were then placed in a water-curing tank at temperature of $26 \pm 10C$.

Compressive strength, splitting tensile strength, and flexural strength were determined at the ages of 28, 91 and 365 days per Indian Standard Specifications IS 516 (1959). For impact strength measurement, a set-up was designed. Impact strength test was carried out by a falling weight method. In this test a cylindrical metallic piece of weight 40 N was dropped from a constant height (1000 mm). The number of blows required to fail the specimen gives the impact strength of the sheets. Since damage inflicted by the blows of impact load stays during the subsequent blows, it was assumed that impact energy imparted by the drop of load is absorbed by the sheets. The cumulative energy imparted to the slab in kN-m to cause failure is expressed as mgh x average number of blows. Impact strength test was also performed at the ages of 28, 91 and 365 days.

Table 4. Sieve analysis of aggregates.

Fine aggregates			Coarse aggregates		
Sieve No.	Percent passing	Requirement IS: 383-1970	Sieve size	Percent passing	Requirement IS: 383-1970
4.75 mm	96.2	90–100	12.5 mm	95	90–100
2.36 mm	95.6	85–100	10 mm	73	40–85
1.18 mm	80.2	75–100	4.75 mm	8	0–10
600 μm	60.2	60–79			
300 μm	37.8	12–40			
150 μm	4.8	0–10			

Table 5. Physical and mechanical properties of steel fibres.

Property	Value
Diameter (mm)	0.41
Length (mm)	32.8
Aspect Ratio	80
Specific gravity	7.84
Tensile Strength (MPa)	845–890
Elongation (%)	2.06–2.54

Table 6. Concrete mixture proportions.

Mixture number	M-1	M-2	M-3	M-4	M-5	M-6
Fly ash, %	35	45	55	35	45	55
Fibres, %	–	–	–	0.75	0.75	0.75
Cement, kg/m^3, C	260	220	180	260	220	180
Fly ash, kg/m^3, FA	140	180	220	140	180	220
Fibres, kg/m^3	–	–	–	58	58	58
Water, kg/m^3, W	148	148	148	148	148	148
W/(C+FA)	0.37	0.37	0.37	0.37	0.37	0.37
SSD Sand, kg/m^3	620	620	620	620	620	620
Coarse aggregate, kg/m^3	1190	1190	1190	1190	1190	1190
Superplasticizer, L/m^3	2.6	2.7	2.8	6.1	6.2	6.3
Slump, mm	85	90	100	80	75	85
Air content, %	3.1	3.2	3.3	3.4	3.6	3.8
Air temperature, °C	23	24	24	24	23	25
Concrete temperature, °C	25	25	26	25	25	27
Concrete density, kg/m^3	2360	2360	2360	2422	2422	2422

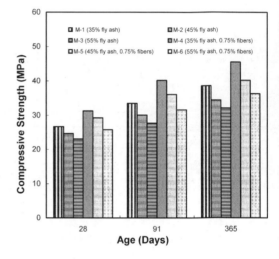

Figure 1. Compressive strength versus age.

3 RESULTS AND DISCUSSION

3.1 *Compressive strength*

Compressive strength results of concrete mixtures containing 35, 45, and 55% fly ash, and the effects of steel fibres on the compressive strength of high-volume fly ash concrete are shown in Fig 1.

Concrete Mixtures M-1 (35% fly ash), M-2 (45% fly ash), and M-3 (55% fly ash) achieved compressive strengths of 26.7, 24.7 and 23.1 MPa, respectively at the age of 28 days. Compressive strength of concrete Mixtures M-1, M-2, and M-3 significantly increased at 91 and 365 days. Mixture M-1 (35% fly ash) achieved strength of 33.5 MPa at the age of 91 days, and 38.6 MPa at the age of 365 days, an increase of 25.5 and 44.5%. Similarly, Mixtures M-2 (45% fly ash) and M-3 (55% fly ash) achieved strengths of 30.1 MPa and 27.7 MPa at the age of 91 days, and 34.4 MPa and 32.1 MPa at the age of 365 days. Increase in compressive strengths of fly ash concrete mixtures was probably due to significant pozzolanic reaction of fly ash.

It is evident from Figure 1 that for a particular fly ash percentage, compressive strength of high-volume fly ash concrete mixtures increased with the incorporation of 0.75% of steel fibres at all ages (28, 91, and 365 days). At the age of 28 days, increase in compressive

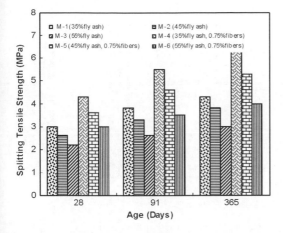

Figure 2.   Splitting tensile strength versus age.

strength, increase in splitting tensile strength at later ages is due to the pozzolanic reaction of fly ash.

The effects of steel fibres on the splitting tensile strength of high-volume fly ash concrete are shown in Fig 2. It is clear from this figure that for a particular fly ash percentage, splitting tensile strength of fly ash concrete mixtures increased with the addition of 0.75% of steel fibres at the age of 28, 91, and 365 days. At 28 days, increase in the splitting tensile strength was 45% for Mixture M-1 (35% fly ash), 36% for Mixture M-2 (45% fly ash), and 37% for Mixture M-3 (55% fly ash).

At later ages (91 and 365 days), splitting tensile strength of concrete mixtures continued to increase. At 91 days, it was 44% for Mixture M-1 (35% fly ash), 38% for Mixture M-2 (45% fly ash), and 35% for Mixture M-3 (55% fly ash). At 365 days, it percentage increase in strength was 45% for Mixture M-1 (35% fly ash), 39% for Mixture M-2 (45% fly ash), and 34% for Mixture M-3 (55% fly ash). Increase in strength at later ages was probably due to the pozzolanic action of fly ash, leading to more densification of the concrete matrix, and development of more effective bond between fibres and fly ash concrete matrix.

It is clear from these results that maximum percentage increase in splitting tensile strength of high-volume fly ash concrete occurred with 35% fly ash content, and then there was slight reduction in the percentage increase in strength compared of high-volume fly ash concrete as in concrete made without fly ash with that of 35% fly ash content. However, these results indicated that steel fibres could be as effective in enhancing the splitting tensile strength.

3.3   Flexural strength

Flexural strength test results of concrete mixtures containing 35, 45, and 55% fly ash, and the effects of steel fibres on the flexural strength of high-volume fly ash concrete are shown in Figure 3. At the age of 28 days, Mixtures M-1 (35% fly ash), M-2 (45% fly ash), and M-3 (55% fly ash) achieved flexural strengths of 2.9, 2.5 and 2.3 MPa, respectively. Mixture M-1 (35% fly ash) achieved strength of 4.5 MPa at the age of 91 days, and 5.0 MPa at the age of 365 days. Similarly, Mixtures M-2 (45% fly ash) and M-3 (55% fly ash) achieved strengths of 3.9 MPa and 3.1 MPa at the age of 91 days, and 4.2 MPa and 3.3 MPa at the age of 365 days. Like compressive strength and splitting tensile strengths, increase in flexural strength at later ages is also due to the pozzolanic reaction of fly ash.

It is clear from Figure 3 that for a particular fly ash percentage, flexural strength of fly ash concrete mixtures increased with the addition of 0.75% of steel fibres at the ages of 28, 91, and 365 days. At 28 days, increase in the flexural strength was 21% for Mixture M-1 (35% fly ash), 19% for Mixture M-2 (45%

strength was 17% for Mixture M-1 (35% fly ash), 19% for Mixture M-2 (45% fly ash), and 12% for Mixture M-3 (55% fly ash). At 91 days, increase in compressive strength was 20% for Mixtures M-1 & M-2, and 14% for Mixture M-3. At 365 days, the increase in compressive strength was 18% for Mixture M-1 (35% fly ash), 17% for Mixture M-2 (45% fly ash), and 13% for Mixture M-3 (55% fly ash).

It is evident from these results that percentage increase in compressive strength for 55% fly ash concrete was less than that of 35% and 45% fly ash concrete. This could be probably due to larger number of spherical particles of fly ash present in the concrete in comparison with those in 35% and 45% fly ash concrete, resulting in comparatively weaker bond between steel fibres and concrete matrix. The effects of addition of steel fibres on the properties of high-volume fly ash concrete was similar to that of its effects on the properties of concrete (without fly ash) as reported by several authors including (Roumaldi and Batson 1963a; Roumaldi and Batson 1963b; Snyder and Lankard 1972; Swamy and Mangat 1974a; Swamy and Mangat 1974b).

3.2   Splitting tensile strength

Splitting tensile strength of high-volume fly ash concrete mixtures, and effect of steel fibres on its splitting tensile strength is shown in Figure 2.

At 28 days, Mixtures M-1 (35% fly ash), M-2 (45% fly ash), and M-3 (55% fly ash) achieved splitting tensile strengths of 3.0, 2.6 and 2.2 MPa, respectively. Mixture M-1 (35% fly ash) achieved strength of 3.8 MPa at the age of 91 days, and 4.3 MPa at the age of 365 days. Similarly, Mixtures M-2 (45% fly ash) and M-3 (55% fly ash) achieved strengths of 3.3 MPa and 2.6 MPa at the age of 91 days, and 3.8 MPa and 3.0 MPa at the age of 365 days. Like compressive

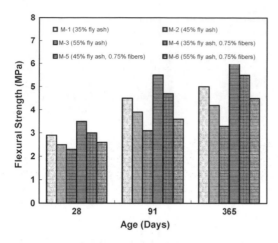

Figure 3. Flexural strength versus age.

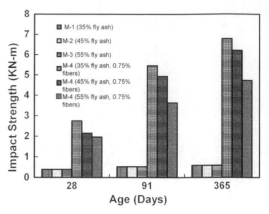

Figure 4. Impact strength versus age.

fly ash), and 15% for Mixture M-3 (55% fly ash). At later ages (91 and 365 days), flexural strength of mixtures further increased. At 91 days, increase was 22% for Mixture M-1 (35% fly ash), 20% for Mixture M-2 (45% fly ash), and 16% for Mixture M-3 (55% fly ash). At 365 days, percentage increase in strength was 23% for Mixture M-1 (35% fly ash), 21% for Mixture M-2 (45% fly ash), and 17% for Mixture M-3 (55% fly ash).

Pattern of increase in flexural strength of high-volume fly ash concrete with fibre addition was similar to that of splitting tensile strength results. In this case, maximum percentage increase in flexural strength of high-volume fly ash concrete occurred with Mixture M-1 (35% fly ash), and then there was very slight reduction in the percentage increase in strength compared with that of 35% fly ash content. These results indicated that steel fibres could be as effective in enhancing the flexural strength of high-volume fly ash concrete as in concrete made without fly ash [17-23].

3.4 Impact strength

Impact strength test results of steel fibre reinforced high-volume fly ash concrete mixtures, at ultimate failure are shown in Figure 4. It is clear from this figure that for all fly ash concrete mixtures, addition of steel fibres enhanced the impact strength significantly. For concrete Mixtures M-1 (35% fly ash), improvement in impact strength was 6 times at the age of 28 days, whereas it was 9.5 and 10.5 times at 91 and 365 days. For concrete Mixtures M-2 (45% fly ash), improvement in impact strength was 4.5 times at the age of 28 days, whereas it was 8.5 at 91 days, and 9.5 times at 365 days. For concrete Mixtures M-3 (55% fly ash), improvement in impact strength was 4 times at the age of 28 days, whereas it was 6 times at 91 days and 7 times at 365 days.

The maximum improvement in impact strength of high-volume fly ash concrete mixture, with the addition of fibres, was observed with Mixture M-1 (35% fly ash), and then there was marginal reduction in the improvement of impact strength of Mixtures M-2 (45% fly ash) and M-3 (55% fly ash). The lesser improvement in impact strength in 45% and 55% fly ash concretes in comparison with 35% fly ash concrete is probably be due to the comparatively weaker bond between fibres and concrete matrix due the spherical nature of fly ash particles.

3.5 Relationship of Impact strength with compressive strength, splitting tensile strength and flexural strength

Figures 5–6 show the relationship between compressive strength, splitting tensile strength, and flexural strength with that of impact strength. These figures have been plotted with 28, 91, and 365-day values of compressive strength, splitting tensile strength and flexural strength on X-axis and Impact strength on Y-axis, for 35%, 45%, and 55% fly ash concrete with 0.75% fibre content. The equation and Correlation coefficients are presented in Table 7. In the equation, 'X' represents that particular strength property and 'Y' is the impact strength.

It as clear from Figure 5 that impact strength is closely related with compressive strength, and the correlation (value of R2) ranged from 0.9816 to 0.99. Fig. 6 makes abundantly clear that impact strength of concrete is closely related with splitting tensile strength as well, and in this case, correlation (value of R2) was very good; ranged from 0.9432 to 0.9838. Figure 7 presents the relationship between impact strength with that of flexural strength. In this case, the correlation (value of R2) ranged from 0.9846 to 0.9970. It can be observed from this analysis that the best correlation (value of R2) occurred with 35% fly ash concrete

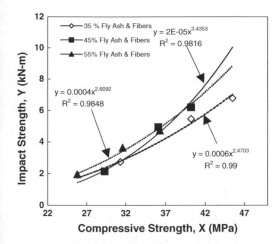

Figure 5. Relationship between compressive strength and impact strength.

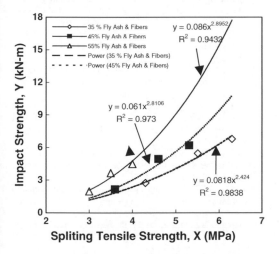

Figure 6. Relationship between splitting tensile strength and impact strength.

and fibres, and then there was very small reduction in correlation value. Though the correlation coefficient is very good with all the three strengths with that of the impact strength, but it is better with reference to compressive strength and flexural strength rather than splitting tensile strength.

4 CONCLUSIONS

The following conclusions are drawn from the present investigation:

1. Compressive strength, splitting tensile strength and flexural strength of high-volume fly ash (HVFA)

Table 7. Correlation coefficient of relationship of impact strength with compressive strength, splitting tensile strength and flexural strength.

Compressive strength (X) versus Impact Strength (Y)			
Equation	35% Fly Ash and Fibres	45% Fly Ash and Fibres	55% Fly Ash and Fibres
	$Y = 0.0006 x^{2.4703}$	$Y = 2E-05 x^{3.4353}$	$Y = 0.0004 x^{2.6092}$
Correlation Coefficient (R^2)	0.9900	0.9816	0.9848
Splitting tensile strength (X) versus Impact Strength (Y)			
Equation	$Y = 0.0818 x^{2.424}$	$Y = 0.061 x^{2.8106}$	$Y = 0.086 x^{2.8952}$
Correlation Coefficient (R^2)	0.9838	0.9730	0.9432
Flexural strength (X) with Impact Strength (Y)			
Equation	$Y = 0.3676 x^{1.5992}$	$Y = 0.3097 x^{1.7706}$	$Y = 0.4263 x^{1.624}$
Correlation Coefficient (R^2)	0.9970	0.9978	0.9846

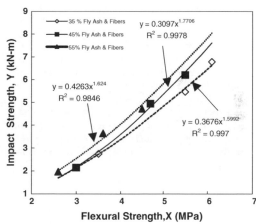

Figure 7. Relationship between flexural strength and impact strength.

concrete increased with age in comparison with its 28-day strength. Increase in the strengths of HVFA concrete clearly indicated the pozzolanic reaction of fly ash.

2. Steel fibres marginally increased the compressive strength of high-volume fly ash concrete. At the age of 28 days, increase in compressive strength was between 12 and 20% depending up on the fly ash content. At 91 and 365 days, compressive strength

of steel fibre reinforced high-volume fly ash concrete increased, which is clearly due to pozzolanic reaction of fly ash.

3. Steel fibres significantly enhanced the 28 days splitting tensile strength of high-volume fly ash concrete mixture by 36 to 45% depending upon fly ash content. At later ages (91 and 365 days), there was further improvement in the splitting tensile strength of high-volume fly ash concrete. This was probably due to the pozzolanic reaction of fly ash, leading to more densification of the concrete matrix, and development of more effective bond between fibres and fly ash concrete matrix.

4. Addition of steel fibres marginally increased the 28 days flexural strength of high-volume fly ash concrete by 15 to 21% depending upon the fly ash content. Flexural strength of high-volume fly ash concrete also increased with age.

5. Use of steel fibres substantially enhanced the 28-day impact strength (4 to 6 times) of high-volume fly ash concrete. With the increase in age, improvement in impact strength was between 6 and 9.5 times at the age of 91 days, and between 7 and 10.5 times at the age of 365 days, depending upon the fly ash content.

6. The findings of this investigation suggested that fibres have similar effects in improving the strength properties of high-volume fly ash concrete as it does is for control concrete made without fly ash.

7. Impact strength of high-volume fly ash concrete has very good correlation with compressive strength ($R2$ between 0.9816 and 0.99), splitting tensile strength ($R2$ between 0.9642 and 0.9838) and flexural strength ($R2$ between 0.9846 and 0.9970) depending upon the fly ash content and testing age. Conclusions should state concisely the most important propositions of the paper as well as the author's views of the practical implications of the results.

REFERENCES

Alasali, M.M. & Malhotra, V.M. 1991. Role of concrete incorporating high volumes of fly ash in controlling expansion due to alkali-aggregate reaction, *ACI Materials Journal* 88(2): 159–163.

Barr, B. & Noor, M.R.M. 1985. The toughness index of steel fibre reinforced concrete, *American Concrete Institute Journal* 82(5): 622–629.

Bilodeau, A. & Malhotra, V.M. 1992. Concrete incorporating high volumes of ASTM class F fly ashes: mechanical properties and resistance to deicing salt scaling and to chloride-ion penetration, *Proceedings of the Fourth CAN-MET/ACI International Conference on the Use of Fly Ash, Silica Fume, Slag, and Natural Pozzolans in Concrete, Istanbul, Turkey, ACI SP-132, Editor V.M. Malhotra, American Concrete Institute*, 319–349.

Bilodeau, A., Sivasundaram, V., Painter, K.E., & Malhotra, V.M. 1994. Durability of concrete incorporating high volumes of fly ash from sources in the USA, *ACI Materials Journal* 91(1): 3–12.

Giaccio, G.M. & Malhotra, V.M. 1988. Concrete incorporating high volumes of ASTM Class F fly ash, *Cement, Concrete and Aggregates* 10(2): 88–95.

Hughes, B.P. & Fattuhi, N.I. 1976. The workability of steel fibre reinforced concrete, *Magazine of Concrete Research* 28(96): 157–161.

IS: 10262-1982, Recommended guidelines for concrete mix design, Bureau of Indian Standards (BIS), New Delhi, India.

IS: 383-1970, Specifications for coarse and fine aggregates from natural sources for concrete, Bureau of Indian Standards (BIS), New Delhi, India.

IS: 516-1959, Indian standard code of practice- methods of test for strength of concrete, Bureau of Indian Standards, New Delhi, India.

IS: 8112-1989, Specifications for 43-grade portland cement, Bureau of Indian Standards (BIS), New Delhi, India.

Langley, W.S., Carette, G.G. & Malhotra, V.M. Strength development and temperature rise in large concrete blocks containing high volumes of low-calcium (ASTM Class F) fly ash, *ACI Materials Journal* 89(2): 362–368.

Langley, W.S., Carette, G.G. & Malhotra, V.M. 1989. Structural concrete incorporating high volumes of ASTM Class F fly ash, *ACI Materials Journal* 86(5): 507–514.

Malhotra, V.M. 1990. Durability of concrete incorporating high-volume of low-calcium (ASTM class F) fly ash, *Cement & Concrete Composites* 12(4): 271–277.

Maslehuddin, M. 1989. Effect of sand replacement on the early-age strength gain and long-term corrosion-resisting characteristics of fly ash concrete, *ACI Materials Journal* 86(1): 58–62.

Papadakis, V.G. 1999. Effect of fly ash on Portland cement systems: Part 1. low-calcium fly ash, *Cement and Concrete Research* 29(11): 1727–1736.

Poon, C.S. 2000. Study on high strength concrete prepared with large volumes of low- calcium fly ash, *Cement and Concrete Research* 30(3): 447–455.

Ravina, D. & Mehta, P.K. 1986. Properties of fresh concrete containing large amounts of fly ash, *Cement and Concrete Research* 16(2): 227–238.

Ravina, D. 1997. Properties of fresh concrete incorporating a high volume of fly as partial fine sand replacement, *Materials and Structures* 30(202): 473–479.

Romualdi, J.P. & Batson, G.B. 1963a. Mechanics of cracks arrest in concrete, *Journal of Engineering Mechanics Division, ASCE* 89: 147–160.

Romualdi, J.P. & Batson, G.B. 1963b. Behavior of reinforced concrete beams with closely spaced reinforcement, *Journal of the American Concrete Institute Proceedings* 60(3): 775–789.

Siddique, R. 2003a. Effect of fine aggregate replacement with class F fly ash on the mechanical properties of concrete, *Cement and Concrete Research* 33(4): 539–547.

Siddique, R. 2003b. Effect of fine aggregate replacement with class F fly ash on the abrasion resistance of concrete, *Cement and Concrete Research* 33(11): 1877–1881.

Siddique, R. 2004. Performance characteristics of concrete containing high-volumes of class f fly ash, *Cement and Concrete Research* 34(3): 487–493.

Snyder, M.J. & Lankard, D.R. 1972. Factors affecting the flexural strength of steel fibrous concrete, *Journal of the American Concrete Institute Proceeding* 69(2): 96–100.

Swamy, R.N. & Mangat, P.S. 1974a. A theory for the flexural strength of steel fibre reinforced concrete, *Magazine of Cement and Concrete Research* 4(2): 313–335.

Swamy, R.N. & Mangat, P.S. 1974b. Flexural strength of steel fibre reinforced concrete, *Institution of Civil Engineers, Proceedings Part 2, Research and Theory* 57: 701–707.

Tikalsky, P.J., Carrasquillo, P.M. & Carrasquillo, R.L. 1988. Strength and durability considerations affecting mix proportions of concrete containing fly ash, *ACI Materials Journal* 85(6): 505–511.

Excellence in Concrete Construction through Innovation – Limbachiya & Kew (eds)
© 2009 Taylor & Francis Group, London, ISBN 978-0-415-47592-1

Permeability of high strength concrete

S.M. Gupta, P. Aggarwal, Y. Aggarwal & V.K. Sehgal
Civil Engineering Department, N.I.T. Kurukshetra, India

R. Siddique
Civil Engineering Department, Thapar University, Patiala, India

S.K. Kaushik
Civil Engineering Department, I.I.T. Roorkee, India

ABSTRACT: Permeability of concrete is of particular significance in structures which are intended to retain water or which come in contact with water. Permeability is intimately related to durability of concrete and it influences the resistance of concrete to sulphate and alkali attack, rate of corrosion of reinforcement, fire and frost. In this paper, permeability of High Strength Concrete is presented. In the High Strength Concrete the water cement ratio is reduced to great extent, therefore the concrete becomes denser in comparison of Normal Strength Concrete. Thus due to increased denseness of high strength concrete, the permeability of concrete is reduced to great extent. The High Strength Concrete is made with concrete mix 1:0.8:2.2 with water-cement ratio as 0.30. The desired degree of workability of concrete at this low water cement ratio is obtained by adding superplasticizer dose at the rate of 2% by weight of cement. In the present investigation the permeability of concrete with 53 grade cement, 12.5 mm size granite as coarse aggregate and Yamuna and Badarpur sand as fine aggregate is studied at 28 and 180 days. Further the results of permeability of high strength concrete are compared with the normal strength concrete. The permeability of High Strength Concrete without fly ash and silica fume and with partial replacement of cement by equal amount of flyash; silica fume and flyash and silica fume in varying percentage is studied. Test results indicate that with the replacement of cement by equal amount of flyash, silica fume and flyash and silica fume both in varying percentage (as indicated above) in concrete reduces the coefficient of permeability of concrete by 8 to 12% and 25 to 50% respectively at 28 days and 180 days. Further, the coefficient of permeability of concrete with flyash is 15 to 20% less than the coefficient of permeability of concrete with replacement of cement by equal amount of silica fume. The type of the fine aggregate also reduces the coefficient of permeability of concrete. From the test results of above investigation, it can be concluded that the coefficient of permeability high strength concrete with Badarpur Sand as fine aggregate is 8 to 10% less than the coefficient of permeability high strength concrete with Yamuna Sand as fine aggregate, the coarse aggregate being same as sand stone of 12.5 mm size. Thus from the results of above investigation, it can be concluded that the coefficient permeability of High Strength Concrete with flyash, silica fume and granite, Badarpur and Yamuna sand as aggregate reduces the coefficient of permeability to the great extend.

1 INTRODUCTION

Permeability of concrete may be defined as the ease with which a fluid can pass through concrete under a pressure difference and is measured in terms of coefficient of permeability. For steady state flow, the coefficient of permeability (k) is determined form Darcy's expression

$$K = Q / [ATH/L] \qquad (1)$$

where K = coefficient of permeability (m/s); Q/T = rate of flow (m³/sec); H/L = ratio of pressure head to thickness of specimens both expressed in same units; and A = Area of cross section of the specimen (m²).

In the High Strength Concrete the water cement ratio is reduced to great extent, therefore the concrete becomes denser in comparison of Normal Strength Concrete. Thus due to increased denseness of high strength concrete, the permeability of concrete is reduced to great extent.

2 REVIEW OF PREVIOUS WORK

The coefficient of permeability of concrete to gases or water vapour is much lower than the coefficient of permeability for water, therefore, the coefficient of permeability is determined using water free from dissolved air (Constantine et al. 1992).

Watson & Oyeka (1981) studied the permeability of both hardened cement paste and concrete to oil, using a specially constructed permeameter and applying a maximum pressure of oil as $10 \, kg/cm^2$. The size of specimens was 150 mm diameter $\times 50$ mm thickness with w/c ratio varying from 0.4 to 0.8. It was found that the co-efficient of permeability decreased during early hours of curing and with an increase in cement content. But permeability increased linearly with increases in applied pressure and w/c ratio.

According to Mehta (1985), w/c ratio determines the size, volume and continuity of capillary voids, which ultimately influence the permeability of concrete. The permeability of concrete is more than that of cement paste because the transition zone between aggregates and cement paste contains micro-cracks, which are more wide than capillary cavities present in cement mortar paste. These cracks allow inter connectivity of pores, which increase the permeability of cement concrete as compared to cement mortar paste.

Kaushik (1991) observed that the air-entrained concrete was more permeable than plain concrete at low w/c ratios and vice-versa. The reason being that when water is forced into air-entrained concrete, the water also enters into the entrained air and the path of water gets shortened. It was also concluded from the study that to make the concrete water tight, continuous moist curing and early age curing were very important.

Permeability of Silica fume and Fly ash High Strength Concrete:

Pozzolanic materials such as silica fume, Fly ash, blast furnace slag etc. are increasingly being used to produce dense and impermeable cement concrete. Due to increased denseness of concrete, damage due to sulphate attack, reinforcement corrosion and alkali-aggregate reaction is reduced (Vaish & Nautiyal, 1995). Vaish & Nautiyal (1995) made permeability studies by replacing 5% cement by silica fume and found that permeability was reduced to 20% to that of plain cement concrete at same curing age conditions. Austin & Robins (1997) carried out the studies on performance of silica fume concrete cured in temperate and hot climates. The 10% silica concrete responded well to hot climate. Four days moist curing reduced the permeability of silica fume concrete in hot climate by a factor of four compared to wet one day curing. This factor however, was 2 to 3 in temperate climate. The silica fume concrete was found gaining strength and becoming less permeable at later ages (28 to 180 days)

due to continuing pozzalanic reaction, while the plain cement concrete showed little change after 28 days.

Khatri & Yu (1997) also studied the importance of curing for concrete with silica fume. They cast cylindrical specimens with height to depth ratio 0.5 and subjected them to a pressure of $7 \, kg/cm^2$. From the results, they observed that 28 days continuous moist curing gave the lowest permeability as compared to intermittent curing. They also concluded that the addition of silica fume to plain concrete considerably reduced the permeability.

Bamforth (1987) also reported similar finding that the use of Fly ash and slag had no significant effect on permeability at 28 days when compared to equal strength of OPC concrete. Establishing relationship between w/c ratio, air-entraining and permeability, Bamforth revealed that the air-entrained concrete was less permeable than the plain one. This reduction was not due to air itself but due to reduced w/c ratio. Murata found that at higher w/c ratio (>0.6) air-entertainment reduced the permeability.

3 EXPERIMENTAL PROGRAMMES

Experimental programme consist of casting of Cylindrical specimens of 150 mm dia and 150 mm height were cast and tested for the determination of co-efficient of permeability at 28 and 180 days of water curing as per IS: 3085-1965. Briefly stating the samples were placed in a specially designed cell subjected to a hydrostatic pressure of $9 \, kg/cm^2$. The quantity of water thus entering and leaving the samples was measured in given time interval at steady state condition. The coefficient of permeability was computed using Darcy's Law. The permeability test results as average of two specimens are given in Tables 1 to 2.

4 RESULTS AND DISCUSSION

The results of the permeability of various concrete mixes with Yamuna and Badarpur sand and 12.5 mm size Sandstone aggregate with replacement of 5, 10 and 15 per cent cement with equal amount of fly ash, silica fume, fly ash and silica fume are given Tables 1 and 2. These results are discussed in detail as under:

Permeability of Concrete with Yamuna Sand and 12.5 mm size Sandstone Aggregate:

The coefficient of Permeability of concrete with Yamuna sand and 12.5 mm size Sandstone aggregate without fly ash and silica fume (HSC-01) at 28 and 180 days is $3.88 \times 10-11$ m/s and $2.69 \times 10-11$ m/s respectively. Thus the coefficient of permeability of concrete mix HSC-01 at 180 days is 0.693 times the coefficient of permeability of concrete at 28 days. The coefficients of permeability of concrete mixes at

Table 1.   Permeability of concrete.

Water Head H = 90 meter   Fine Aggregate: Yamuna sand
Dia and Length of Specimen Coarse Aggregate: 12.5 mm size sandstone
D = L = 150 mm

Sr. No.	Mix designation	Cube compressive strength of concrete (N/mm^2) at		Water flow in 7 hours after saturation Q × 10^{6}m^3 at		Coefficient of permeability K × 10^{11} m/sec. at		Mean temp of 7 hours in °C at		Corrected value of K × 10^{11} m/sec. at 27 ± 2°C at	
		28 days	180 days	28 days	180 days	28 days	180 days	28 days	180 days	28 days	180 days
1	HSC-1 FA & SF = 0%	75.00	78.80	10.40	7.20	3.88	2.69	21.00	30.50	3.57	2.77
2	HSC-2 FA = 5%	76.50	80.60	9.50	4.20	3.54	1.57	20.50	30.00	3.22	1.60
3	HSC-3 FA = 10%	76.85	80.95	9.00	3.75	3.36	1.40	20.00	31.00	3.02	1.46
4	HSC-4 FA = 15%	77.10	81.25	9.30	3.50	3.47	1.31	21.00	29.50	3.19	1.35
5	HSC-5 SF = 5%	73.90	78.40	10.00	5.40	3.73	2.02	20.50	30.00	3.39	2.28
6	HSC-6 SF = 10%	73.45	77.75	9.60	4.80	3.58	1.79	20.00	30.50	3.22	1.91
7	HSC-7 SF = 15%	73.15	77.35	9.80	5.20	3.66	1.94	21.00	31.00	3.37	2.02
8	HSC-8 FA + SF = (5 + 5)%	76.50	80.40	9.50	4.40	3.54	1.64	21.00	30.00	3.26	1.67
9	HSC-9 FA + SF = (10 + 5)%	76.35	80.95	9.20	3.80	3.43	1.42	19.50	29.50	3.05	1.43
10	HSC-10 FA + SF = (5 + 10)%	74.10	78.30	9.40	4.10	3.51	1.53	20.00	30.50	3.16	1.58
11	HSC-11 FA + SF = (10 + 10)%	74.60	78.75	9.60	4.50	3.58	1.68	21.00	30.00	3.30	1.71
12	NSC-01 0	37.45	41.50	32.50	24.00	12.12	8.95	19.50	31.00	10.78	8.38

Note: FA = Flyash
SF = Silicafume

Table 2.   Permeability of concrete.

Water Head H = 90 meter Fine Aggregate: Badarpur sand
Dia and Length of Specimen Coarse Aggregate: 12.5 mm size sandstone
D = L = 150 mm

Sr. No.	Mix designation	Cube compressive strength of concrete (N/mm^2) at		Water flow in 7 hours after saturation Q × 10^{6}m^3 at		Coefficient of Permeability K × 10^{11} m/sec. at		Mean temp of 7 hours in °C at		Corrected value of K × 10^{11} m/sec. at 27 ± 2°C at	
		28 days	180 days	28 days	180 days	28 days	180 days	28 days	180 days	28 days	180 days
1	HSC-12 FA & SF = 0%	79.30	83.30	10.00	6.80	3.74	2.54	21.00	31.50	3.44	2.67
2	HSC-13 FA = 5%	80.70	85.05	9.20	4.00	3.44	1.49	21.50	32.00	3.20	1.55
3	HSC-14 FA = 10%	81.00	85.40	8.70	3.60	3.25	1.34	21.00	31.00	3.00	1.39
4	HSC-15 FA = 15%	81.40	85.70	9.00	3.20	3.36	1.19	21.50	30.50	3.12	1.22
5	HSC-16 SF = 5%	77.75	82.50	9.80	5.20	3.66	1.94	22.00	31.50	3.40	2.23
6	HSC-17 SF = 10%	77.35	82.05	9.30	4.80	3.47	1.72	22.50	32.00	3.30	1.89
7	HSC-18 SF = 15%	76.90	81.70	9.50	5.00	3.55	1.87	21.00	31.00	3.26	1.95
8	HSC-19 FA + SF = (5 + 5)%	80.95	85.20	9.20	4.20	3.44	1.57	21.50	31.50	3.20	1.65
9	HSC-20 FA + SF = (10 + 5)%	81.35	85.65	9.00	3.60	3.36	1.34	22.00	30.50	3.16	1.38
10	HSC-21 FA + SF = (5 + 10)%	78.40	82.85	9.20	4.00	3.44	1.50	21.00	31.00	3.17	1.56
11	HSC-22 FA + SF = (10 + 10)%	78.80	83.50	9.40	4.30	3.51	1.61	21.00	31.50	3.23	1.69
12	NSC-12 0	39.50	43.85	31.50	22.50	11.77	8.41	21.00	31.00	10.83	8.75

Note: FA = Flyash
SF = Silicafume

28 days with replacement of 5, 10 and 15 per cent cement with equal amount of fly ash (HSC 02, 03, 04) are 0.912, 0.886 and 0.894 times respectively the coefficient of permeability of concrete mix HSC 01. At 180 days the coefficient of permeability of concrete mixes (HSC-02, 03, 04) are 0.584, 0.520 and 0.487 times respectively the coefficient of permeability of concrete mix HSC-01. The coefficient of permeability of concrete mixes at 28 and 180 days with replacement of 5, 10 and 15 per cent cement with equal amount of silica fume (HSC 05, 06 and 07) are 0.961, 0.923, 0.943 and, 0.833, 0.688, 0.721 times respectively the coefficient of permeability of concrete mix HSC-01 at 28 and 180 days. Further, the coefficients of permeability of concrete mixes at 28 and 180 days with 10, 15, 15 and 20 per cent replacement of cement by equal amount of fly ash and silica fume (HSC-08, 09, 10 and 11) are 0.912, 0.884, 0.905, 0.923 and 0.610, 0.528, 0.569, 0.624 times respectively the coefficient of permeability of concrete mix HSC-01 at 28 and 180 days. The coefficients of permeability of concrete mix HSC-01 at 28 and 180 days are 0.320 and 0.334 times respectively the coefficient of permeability of NSC-01 at 28 and 180 days Thus, the coefficient of permeability of above concrete mixes decreases at 180 days in comparison to the coefficient of permeability of the same mixes at 28 days. Further, the coefficients of permeability of concrete mixes with varying percentage replacement of cement by equal amount of fly ash, silica fume and fly ash and silica fume at 28 and 180 days are less than the coefficients of permeability of concrete without fly ash and silica fume (HSC-01) at 28 and 180 days.

Permeability of Concrete with Badarpur Sand and 12.5 mm Sandstone Aggregate:

To study the effect of type of fine aggregate on the coefficient of permeability of concrete, the results of the coefficient of permeability of concrete mixes with Badarpur sand and 12.5 mm size Sandstone aggregate are given in Table 2 and are compared with the corresponding results of coefficient of permeability of concrete with Yamuna sand and 12.5 mm size Sandstone aggregate (Table 1).

The coefficient of permeability of concrete with Badarpur sand and 12.5 mm size Sandstone aggregate without fly ash and silica fume (HSC-12) at 28 and 180 days is $3.74 \times 10\text{--}11$ m/s and $2.54 \times 10\text{--}11$ m/s respectively. Thus the coefficient of permeability of concrete mix HSC-12 at 180 days is 0.679 times the coefficient of permeability of concrete at 28 days. Further, the coefficients of permeability of concrete mix HSC-12 at 28 and 180 days are 0.964 and 0.944 times the corresponding values of coefficient of permeability of concrete mix HSC-01. The coefficient of permeability of mixes with replacement of 5, 10 and 15% cement by equal amount of fly ash at 28 and 180 days are 0.972, 0.967, 0.968 and 0.949, 0.957,

0.908 times respectively the corresponding values of fly ash concrete mixes with Yamuna sand and 12.5 mm size Sandstone aggregate. The coefficient of permeability of concrete mixes with replacement of 5, 10 and 15% cement by equal amount of silica fume at 28 and 180 days are 0.981, 0.969, 0.970 and 0.960, 0.961, 0.964 times respectively the corresponding values of silica fume concrete mixes with Yamuna sand and 12.5 mm size Sandstone aggregate. Further, the coefficients of permeability of concrete mixes at 28 and 180 days with 10, 15, 15 and 20% replacement of cement by equal amount of fly ash and silica fume both are 0.979, 0.979, 0.980, 0.981 and 0.957, 0.944, 0.980, 0.958 times respectively the corresponding values of fly ash-silica fume concrete mixes with Yamuna sand and 12.5 mm size Sandstone aggregate. The coefficient of permeability of concrete mix HSC-12 at 28 and 180 days are 0.318 and 0.302 times respectively the coefficient of permeability of NSC-12 at 28 and 180 days. Further, the coefficients of permeability of NSC-12 at 28 and 180 days are 0.971 and 0.939 times the corresponding values of NSC-02. Thus the coefficients of permeability of concrete mixes with Badarpur sand and 12.5 mm size Sandstone aggregate decrease by 2 to 7 per cent in comparison to the corresponding values of the concrete mixes with Yamuna sand and 12.5 mm size Sandstone aggregate.

5   CONCLUSIONS

- The coefficient of permeability of concrete at 180 days is about 0.68 times the coefficient of permeability at 28 days.
- With partial replacement of cement by fly ash and silica fume, the coefficient of permeability of concrete at 28 days is slightly less (8 to 12%) compared to the coefficient of permeability of concrete without fly ash and silica fume. At 180 days, the coefficient of permeability of fly ash and silica fume concrete reduces by 25 to 50% compared to the coefficient of permeability of concrete without fly ash and silica fume.
- Further, the coefficient of permeability of fly ash concrete is less (15 to 25%) than the coefficient of permeability of silica fume concrete.
- The coefficient of permeability of concrete with Badarpur sand is less (5 to 8%) than the coefficient of permeability of concrete with Yamuna sand.
- The coefficient of the permeability of high strength is very less in comparison of normal strength concrete.

REFERENCES

Austin, S. A. & Robins, P. J. 1997. Influence of Early Curing on the Surface and Permeability and Strength of Silica fume Concrete, *Magazine of Concrete Research*. 49(178): 23–24.

Bamforth, P. B. 1987. Relationship Between Permeability Coefficient for Concrete Obtained Using Liquid and Gas, *Magazine of Concrete Research* 39(138): 3–11.

Constantine, A. M., Tia, M. & Bloomquist, 1992. Development of a Field Permeability Task Apparatus and Method of Concrete, *ACI Material Journal*, 89(1): 83–89.

Kaushik, S.K. 1991. Permeability and Durability of High Strength Concrete, *National Seminar on HSC, ICI Bangalore* December 5–6: 1–68.

Khatri, R. P. & Yu, L. K. 1997. Effect of Curing on Water Permeability of Concrete Prepared with Normal Portland Cement and Slag and Silica Fume, *Magazine of Concrete Research* 49(180): 167–172.

Mehta, R. K. 1985. Concrete Structure, Properties and Materials, *Prentice Hall, INC, Englewod Cliffs*, New Jersey.

Murata, J. Studies on Permeability of Concrete, *RILEM* 29: 47–54.

Vaish, M. K. & Nautiyal, B.D. 1995. Permeability of Concrete and Factors Affecting It-A Review, *ICI Bulletin* 52(July–September): 25–29.

Waston, A. I. & Oyaka, C. C. 1981. Oil Permeability of Hardened Cement Pasts and Concrete, *Magazine of Concrete Research* 33(115): 85–95.

Excellence in Concrete Construction through Innovation – Limbachiya & Kew (eds)
© 2009 Taylor & Francis Group, London, ISBN 978-0-415-47592-1

Formulation of Turonien limestone concrete of the Central Saharian Atlas (Algeria)

Z. Makhloufi & M. Bouhicha

Laboratory of Civil Engineering, University of Laghouat (Algeria)

ABSTRACT: This contribution consists in making concretes containing limestone aggregates coming from rocks laminated in decametric benches of Turonien which belongs to the cretaceous of mesozoïc age of the Algerian central Saharian Atlas, to study their need on water and mechanical performances, and to work out graphs and an abacus of formulation of these concretes. Three particle size distributions of aggregates were the subject of a series of normative tests in order to lead to a characterization, more or less detailed, that allow to assess the quality of these aggregates in comparison with the standards in vigour, the crushed limestone sands 0/5 and gravels 5/15, 15/25 mm.

1 INTRODUCTION

In this study, we will present the formulation of the Turonien limestone concrete made from a mixture of aggregates showing three particle size distributions and which were the subject of a series of normative tests in order to lead to a characterization, more or less detailed, that allow to appreciate the quality of these aggregates in comparison with the standards in vigour (AFNOR), it is about crushed limestone sands 0/5 and gravels 5/15, 15/25 mm. We use, for the preformulation of the concrete, the Dreux-Gorisse method (Dreux & Groisse, 1993, Dreux & Festa, 1995) based on a graphic process, to determine in approximate way the ponderal proportions of the various components of a concrete. Then Baron-Lesage method to determine the optimal formula of the mixture thus giving the best compactness.

The properties of hardened concrete were appreciated by the compressive and the tensile strength which were measured on standardized test-tubes (cylinder Ø 16 × 32 cm). The various tests carried out at the laboratory led to a series of experimental results which enable to establish curves and an abacus which the use appears easy.

2 CHARACTERISATION OF MATERIALS USED

Cement used is the (CPJ-CEMII/ A) of Aïn Touta (table 1), its absolute density is $3.00 \, g/cm^3$, its specific surface is $3678 \, cm^2/g$, the true class of cement is 37 MPa.

Table 1. Chemical composition of cement (%).

SiO_2	21,04
Al_2O_3	4,97
Fe_2O_3	3,91
CaO	63,80
MgO	1,08
K_2O	0,62
Na_2O	0,08
SO_3	0,96
Perte au Feu	0,57

Table 2. Physical characteristics of sands and gravels crushed limestones.

Tests	Symbol	Unit	Sand 0/5	Gravel 5/15–15/25
Densities:				
absolute	ρs	Kg/m^3	2700	2680
apparent	ρ	Kg/m^3	1530	1270
real	ρr	Kg/m^3	2610	2590
Absorption	Abs	%	2.5	1.42
Porosity	η	%	3.33	3.36

Results of characterization tests of studied materials are summarized in table 2.

The particle size distribution curve of the limestone sand 0/5, showed in figure 1, is continuous; it is normally spread out and entirely located in the grading envelope of a natural aggregate for ordinary concrete.

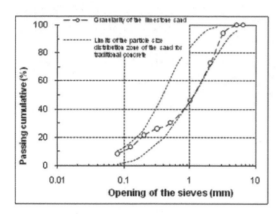

Figure 1. The particle size distribution curve of the limestone sand 0/5 of Laghouat.

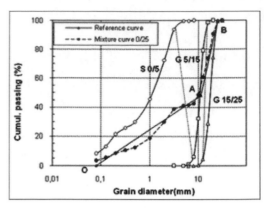

Figure 2. Application of the Dreux-Gorisse method to determine the composition of the limestone concrete.

The percentage of the fine fraction (fillers) and the fineness modulus of sand, normally, will to give a better compromise strength-workability of the mixture homogeneity. The crushed limestone gravels 5/15 and 15/25 mm are whitish; relatively compact, clean not very porous.

Los Angeles coefficient of 20%, absolute density and porosity of these materials allows classifying them among hard limestones. These aggregates are primarily made up of calcium carbonate (97,5%).

3 THE MIXTURE FORMULATION PROCESS

3.1 Principle of the Dreux-Gorisse method

The pragmatic methods of formulation depend generally on the granular curves of reference, the tables and abacuses established from the practical observations. Currently, the Dreux-Gorisse method seems to be largely employed for the industry of the traditional concretes containing aggregates in conformity with the specifications in vigour.

Some morphological and petrophysical properties of the aggregates must be known initially, it is fixed, then, the average properties of the concrete to be formulated, in particular its consistency and its mechanical strength to compression at 28 days.

3.2 Application of the method

We must to verify the results, and to readjust them to really obtain the unit volume of the concrete, having the properties which approach those fixed at the beginning. The selected average properties of the concrete are fixed equal to those of a plastic consistency concrete, that is to say a slump with the characteristic strength and slump test A = 6 cm to 28 days Fc28 = 25 MPa. The general properties of

the aggregates intended for the manufacture of the concrete, are enumerated as follows:

Diameter of the most coarse grains: D = 25 mm.
Granular coefficient: G = 0,5.
Fineness modulus: FM = 2,7.
True class of cement: $\sigma C = 37$ MPa.
Mixing water is water of the tap.

3.2.1 Proportioning of cement and water

In Bolomey formula, the C/W ratio used allows to express strength in a linear function of C/W. Cement dosage will be deduced from this formula which is written (Bolomey, 1996):

$$\overline{\sigma}_{28} = G.\overline{\sigma}_C \left(\frac{C}{W_{us}} - 0,5 \right) \tag{1}$$

With $\overline{\sigma}_{28}$ or RC28: average compressive strength at 28 days, estimated according to Dreux-Gorisse at FC28 + 5%fc28 is 28,75 MPa. After calculation we obtain a ratio C/W = 2. It is of use to adopt the opposite factor W/C = 0,5 for the current concretes. By using an abacus (Fleming, 1952), for a C/W ratio = 2 and a slump A = 6 cm, we obtain cement dosage C = 400 Kg/m³.

Useful water dosage (Wus) of dry material is thus: Wus = 200l.

3.2.2 Aggregate proportioning

According to Dreux-Gorisse, the S/G ratio = 0,71.

These proportions enabled us to calculate the real granular composition which would be deferred on figure 2, it would thus correspond to the real curve of the mixture 0/25. The curve of mixture 0/25 is not superimposed exactly on the grading curve of reference (GCR), but notably it approaches it, except in the part whose grains diameter lies between 0,2 and 2 mm where a light spacing is noted, in this zone, which can have a small influence on the compactness of the mixture.

To evaluate the masses of the aggregates allowing to manufacture one cubic meter of concrete we must determine the coefficient of compactness $\gamma_{D25} = 0,822$ (Fleming, 1952). As that the sand is crushed, it is thus advisable to make a correction which is $-0,03$, the coefficient of compactness will be thus $\gamma_{D25} = 0,792$. The absolute total volume of the solid components $V_T = V_C + V_S + V_{G1} + V_{G2} = 100\,\gamma = 792$ liters.

As the volume of cement introduced into the formula is: $V_C = 133l$ The real volume of the aggregates (V_G) is then:

$$V_G = V_S + V_{G1} + V_{G2} = 658,67l \qquad (2)$$

Thus for $1\,m^3$ of freshly-mixed concrete, the pre-formulation by the method of Dreux-Gorisse led to the ponderal quantities of the following dry materials:

Cement CPJ-CEMII/ A):	400 kg
Useful water (W_{us}):	200 kg
Sand 0/5 (S):	736,26 kg
Gravel 5/15 (G_1):	398,94 kg
Gravel 15/25 (G_2):	635,48 kg
Total mass:	2371 kg.

3.2.3 Effect of hygroscopy aggregates

Within this framework we define the following types of water (Dupain et al., 1995):

- Absorptive water (W_{Abs}) or retained by the pores, also called internal water.
- Brought water (Wbr): It is the quantity of water brought by the aggregates.
- Surface Water or remaining water (Wrem): It is the water retained on the surface of the grains.

$$W_{rem} = W_{br} - W_{abs} \qquad (3)$$

To take into account what precedes, we define for a mixture of concrete:

- Effective Water also called useful water (Wus): It is the sum of the water of manufacture (Wman) added to the moment of mixing and of remaining water. It is this water which ensures the plasticity of the concrete and the hydration of cement.

$$W_{us} = W_{man} + W_{rem}. \qquad (4)$$

Total Water (Wtot): Quantity of water presents in a mixture of freshly-mixed concrete.

$$W_{tot} = W_{us} + W_{abs} = W_{man} + W_{br} \qquad (5)$$

The quantity of water brought is:
W_{br} = mass of wet material − mass of dry material
Sand $W_S = 1,40$ L,

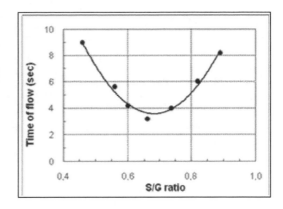

Figure 3. Variation of the time of flow according to S/G ratio.

Gravel $W_{G1} = 0,32$ L,
Gravel $W_{G2} = 0,44$ L.
$W_{br} = W_S + W_{G1} + W_{G2} = 2, 16$ L
The quantity of water of manufacture is thus:
$W_{man} = W_{us} - W_{br} + W_{abs}$ $W_{man} = 231\,l$

3.2.4 Optimization of the mixture

The principle of this method initially consists in carrying out several mixtures of the same volume by keeping the Wus/C ratio constant (Dreux-Gorisse method) and while varying S/G ratio (Baron & Lesage, 1976). The mixture thus optimized, corresponds to the best proportioned mixture. A light increase in the Wus/C ratio = 0,52 whose workability with the slump test is equal to slump initially considered (Slump = 6 cm).

Figure 3 shows evolution of the time of flow at LCPC maniabilimeter in function S/G ratio by maintaining the Wus/C ratio = 0,52 constant. Optimum of the parabola of figure 2 representing the optimal S/G ratio located at 0,68 which is slightly lower than that found by the method of Dreux-Gorisse (0,71).

3.2.5 Variation of the w_{us}/c ratio

Three cement proportionings were selected: 250, 350 and 450 kg/m³. For each considered proportioning, several mixtures of concrete were made, while varying the W_{us}/C ratio, in order to obtain current workabilitys. During this stage S/G ratio was of course kept constant (S/G = 0,68). The corrected quantities of the aggregates are as follows:

Sand 0/5: 704,41 kg;
Gravel 5/15: 399,32 kg;
Gravel 15/25:635,44 kg.

Figure 4 shows, for a constant S/G ratio, that the workability of the freshly-mixed concrete depends on useful water dosage, because it grows with the increase in the W_{us}/C ratio. This increase all the more significant as cement dosage is weak, like is indicated by

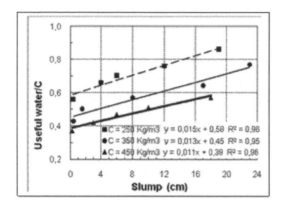

Figure 4. Evolution of the Wus/C ratio according to slump.

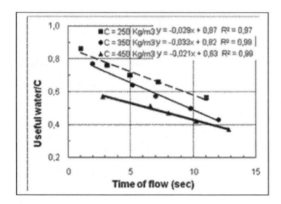

Figure 5. Evolution of the Wus/C ratio according to the time of flow.

Table 3. Density of the freshly-mixed concrete and coefficient of correction.

C (kg)	Wus/C	Dens. meas. (kg/m^3)	Dens. théor. (kg/m^3)	Coef.cor.
250	0.56	2393	2129	1.124
	0.66	2371	2154	1.101
	0.70	2359	2164	1.090
	0.76	2342	2179	1.075
	0.86	2298	2204	1.043
350	0.43	2392	2240	1.068
	0.50	2368	2264	1.046
	0.57	2339	2289	1.022
	0.64	2315	2313	1.001
	0.77	2241	2359	0.950
450	0.37	2403	2356	1.020
	0.42	2407	2378	1.012
	0.47	2352	2401	0.980
	0.51	2320	2419	0.959
	0.57	2280	2446	0.932

theoretical density, in particular for weak cement proportionings, or for the mixtures low in water, i.e. for the firm concretes.

Corrected masses of the mixture obtained by multiplying the coefficient of correction by the initial masses are shown in table 4. Cement proportionings are not equal any more to proportionings initially considered, it is thus appropriate to carry out a readjustment of the components of the mixture to obtain a formula of one real cubic meter concrete, having the same quantity of cement as that fixed at the beginning.

The representation of current state of a freshly-mixed concrete, led us to develop the experimental laws evolution of ingredients of the mixture according to three slumps of reference:

To = 3 cm: firm concrete, A = 9 cm: plastic concrete, A = 15 cm: fluid concrete.

Table 5 obtained by interpolation, summarizes the data of the adjustment model. The evolution of the W_{us}/C ratio according to cement which follows a hyperbolic law of type $W_C/C = a/c + B$ is shown in figure 6 for three selected slumps of reference. The equations of the curves of this figure obtained by regression are as follows:

$A = 3\ cm$: $W_{us}/C = 147/C + 0,105$ $R^2 = 0,99$
$A = 9\ cm$: $W_{us}/C = 166/C + 0,116$ $R^2 = 0,99$ (7)
$A = 15\ cm$: $W_{us}/C = 181/C + 0,117$ $R^2 = 0,99$

An experimental law of adjustment of the W_{us}/C ratio according to cement proportioning C and slump A, may be obtained by construction a linear regression on the coefficients A and B of the equations (7), this law is given by the following expression:

$$W_{us}/C = (2{,}83.A + 139)/C + 0{,}001.A + 0{,}104 \qquad (8)$$

the slopes of the straight regression lines. In addition, for the same workability (slump), the W_{us}/C ratio decreases with the increase in cement proportioning. In figure 5, we observe a reduction in the time of flow in LCPC maniabilimeter with the increase in the W_{us}/C ratio for three cement proportionings.

3.2.6 *Readjustment with the real cubic meter*
The volume of the concrete for each mixture differs, in general, slightly of the volume initially estimated by the Dreux-Gorisse method. The results were adjusted to the real cubic meter. The coefficient of correction which allows adjusting initial formula is given by (Count et al., 1995):

Coefficient of correction = measured Density / theoretical Density   (6)

The coefficients of correction for each cement proportioning are given in table 3. The measured density, is in the majority of the cases, higher than the

Table 4. Initial formulas restored to m³.

C (Kg)	Wus. (Kg)	Wtot. (Kg)	Cor.	C (Kg)	Wus. (Kg)	Wtot. (Kg)	Sdry (Kg)	G1dry (Kg)	G2dry (Kg)
250	140	173	1.124	281	157	194	790	448	714
	165	198	1.101	275	182	218	774	439	699
	175	208	1.090	273	191	227	766	435	692
	190	223	1.075	269	204	240	755	429	682
	215	248	1.043	261	224	259	733	416	662
350	151	184	1.068	374	161	197	751	426	678
	175	208	1.046	366	183	218	735	417	664
	200	233	1.022	358	204	238	718	408	649
	224	257	1.001	350	224	257	704	399	636
	270	303	0.950	333	257	288	668	379	603
450	167	200	1.020	459	170	204	717	407	648
	189	222	1.012	455	191	225	711	404	643
	212	245	0.980	441	208	240	689	391	622
	230	263	0.959	432	221	252	675	383	609
	257	290	0.932	420	240	270	656	372	592

Table 5. Data for model of adjustment the limestone concrete.

A = 3 cm		A = 9 cm		A = 15 cm	
C	Wus/C	C	Wus/C	C	Wus/C
277	0,63	271	0,73	266	0,80
364	0,52	357	0,58	352	0,62
455	0,42	434	0,50	424	0,55

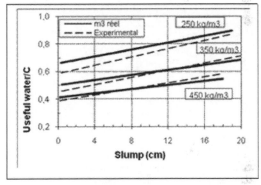

Figure 7. Evolution of the Wus/C ratio according to slump.

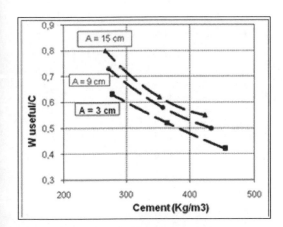

Figure 6. Evolution of the Wus/C ratio according to cement proportioning, for slumps of reference.

For a wished dosage of cement and a slump, the experimental law given previously allows to obtain the ratio W_{us}/C necessary to manufacture one real m³ of concrete. On a purely comparative basis, as seen in figure 7, the evolution of the ratio W_{us}/C according to slump given by the developed law of adjustment, and that obtained from the experimental results not adjusted.

The examination of figure 8 enables to draw the following observations: For the high dosages of cement, the adjustment of the formulas involves a decrease of the W_{us}/C ratio, in particular for slumps more or less large, which leads in the majority of the cases to a reduction in the quantity of cement fixed at the beginning. In addition, in the formulas at moderated dosage of cement, the adjustment leads to an increase in the W_{us}/C ratio for the range of the firm and plastic concretes and to a decrease in this ratio for the fluid concretes.

The real total volume of ingredients is:

$$V_T = \frac{S}{\rho_S} + \frac{G_1}{\rho_{G1}} + \frac{G_2}{\rho_{G2}} + \frac{C}{\rho_C} + \frac{W_{ut}}{\rho_w} \qquad (9)$$

Table 6 gives the real total volume, which slightly exceeds the volume of one cubic meter.

Average total volume for all dosages is however $V_T = 1007$ L.

From the equation (8) we obtain:

$$V_G = 868 - 2{,}83.A - (0{,}001.A + 0{,}104).C - \frac{C}{\rho_o} \quad (10)$$

The mass of the aggregates to be introduced: $G = \%G.V_G.\rho_G$, with G, %G and ρ_G are respectively the masses, percentages and real densities of the aggregates.

Figures 9 allow to calculate the quantities of ingredients necessary to the manufacture one cubic meter of concrete according to C and A.

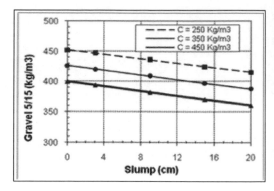

Figure 9a.   Quantities of sand 0/5 mm to be introduced in the concrete to obtain a given slump.

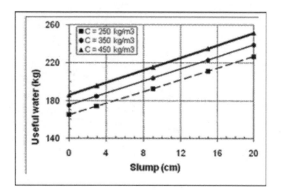

Figure 8.   Quantity of useful water to bring to the limestone concretes to obtain a given workability.

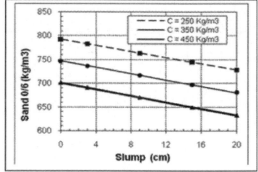

Figure 9b.   Quantities of fine gravels 5/15 mm to be introduced in the concrete to obtain a given slump.

Table 6.   Formulas restored with the m 3 with initial cement proportionings.

C (kg)	Wus/C	Wus. (kg)	Wtot. (kg)	S.dry (kg)	G1 dry (kg)	G2 dry (kg)	S+G+C γ	Vt (l)
250	0.66	166	202	791	452	726	837	1003
	0.71	177	213	779	445	716	832	1009
	0.73	183	219	773	441	710	827	1011
	0.81	202	237	753	430	692	810	1012
	0.89	224	257	731	417	671	785	1008
350	0.50	177	211	745	425	684	823	1000
	0.51	180	214	741	423	681	824	1004
	0.57	201	234	720	411	661	806	1007
	0.66	229	261	690	394	634	781	1010
	0.71	249	279	670	382	615	754	1003
450	0.41	186	219	700	400	643	819	1005
	0.43	196	228	690	394	634	824	1020
	0.46	205	237	680	388	625	804	1010
	0.49	219	249	666	380	612	787	1006
	0.54	245	274	639	365	587	758	1003

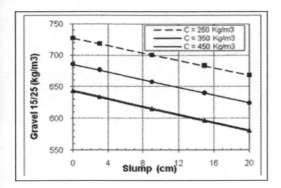

Figure 9c. Quantities of gravel 15/25 mm to be introduced in the concrete to obtain a given slump.

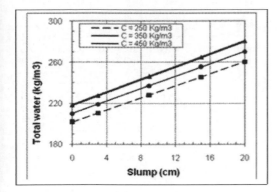

Figure 10. Variation of total water according to slump.

The quantity of total water (W_{tot}) is given by the following relation

$$W_{tot} = W_{us} + S.Abs_S + G_1.Abs_{G1} + G_2.Abs_{G2} \qquad (11)$$

Figure 10 allows to determine the quantity of total water necessary for the mixture of concrete.

4 MECHANICAL BEHAVIOR

Table 7 gathers the results of the strength tests to compression at 28 days R_{C28} and the tensile strength to 28 days R_{t28}.

4.1 Compressive strength

Figure 11 shows the variation of the compressive strength R_{C28} according to the W_{us}/C ratio .This variation can be comparable with a linear form whose equation is as follows:

$$R_{C28} = 16,69.\frac{C}{W_{us}} - 12,24 \qquad R^2 = 0,95 \qquad (12)$$

Table 7. Mechanical performances at 28 days of the limestone concretes.

Cement (Kg/m³)	S/G	Slump (cm)	Rc28 (MPa)	CV %	Rt28 (MPa)	Cv %
250	0,68	0,3	14,45	4,94	2,78	6,42
	0,68	4	13,27	4,63	2,51	5,21
	0,68	6	11,21	10,10	2,31	7,22
	0,68	12	10,49	11,12	2,19	1,80
	0,68	19	6,36	6,54	1,63	1,42
350	0,68	0,4	25,40	5,74	3,55	3,11
	0,68	1,5	19,27	3,48	3,26	6,14
	0,68	8	17,10	4,26	2,93	6,83
	0,68	17	16,30	4,11	2,50	5,31
	0,68	23	7,68	4,97	1,71	10,04
450	0,68	0,2	32,75	8,87	3,84	3,40
	0,68	3	27,10	10,57	3,43	8,24
	0,68	6	25,69	9,37	3,27	6,97
	0,68	10	21,06	4,45	3,01	2,04
	0,68	18	19,98	9,87	3,00	7,46

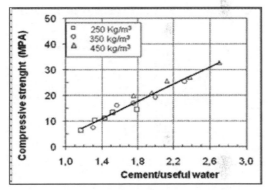

Figure 11. Evolution of the compressive strength according to the C/W useful ratio.

The application of the equation (12) also allows to correct strengths, so as to take into account cement proportionings initially envisaged. The results of this adjustment which consists in introducing into the equation (12) the W_{us}/C ratio readjusted for obtain corrected R_{C28} are gathered in table 8.

Figure 12 shows the evolution of strength according to slump. We notice a reduction in the compressive strength R_{C28} with the increase in the slump which is due to the influence of the Wus/C ratio. It should be noted that the values of measured and corrected strengths are close to each other for considered cement proportioning.

In addition to a given slump and a discounted strength, the formulation of the limestones concretes seems to lead to cement proportionings as shown in figure 13.

171

Table 8. Values of the compressive strengths corrected according to the law of regression.

Cement (kg)	Wus/C	Wut/Ccor	Slump (cm)	Rc28 mes (MPa)	Rc28 cor (MPa)
250	0.56	0.66	0.3	14.45	12.91
	0.66	0.71	4	13.27	11.29
	0.70	0.73	6	11.21	10.50
	0.76	0.81	12	10.49	8.42
	0.86	0.89	19	6.36	6.43
350	0.43	0.50	0.4	25.40	20.82
	0.50	0.51	1.5	19.27	20.18
	0.57	0.57	8	17.10	16.85
	0.64	0.66	17	16.30	13.22
	0.77	0.71	23	7.68	11.26
450	0.37	0.41	0.2	32.75	28.04
	0.42	0.43	3	27.10	26.15
	0.47	0.46	6	25.69	24.31
	0.51	0.49	10	21.06	22.12
	0.57	0.54	18	19.98	18.44

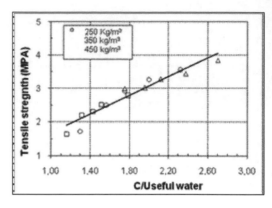

Figure 14. Variation of the tensile strength according to the C/Wus ratio.

Table 9. Calculation of the Rt/Rc ratio.

Cement (kg)	Wus/C	Wus/C cor	Rc28 cor (MPa)	Rt28 cor (MPa)	Rt/Rc
250	0.56	0.66	12.91	2.38	0.18
	0.66	0.71	11.29	2.24	0.20
	0.70	0.73	10.50	2.18	0.21
	0.76	0.81	8.42	2.00	0.24
	0.86	0.89	6.43	1.84	0.29
350	0.43	0.50	20.82	3.04	0.15
	0.50	0.51	20.18	2.99	0.15
	0.57	0.57	16.85	2.71	0.16
	0.64	0.66	13.22	2.41	0.18
	0.77	0.71	11.26	2.24	0.20
450	0.37	0.41	28.04	3.65	0.13
	0.42	0.43	26.15	3.49	0.13
	0.47	0.46	24.31	3.34	0.14
	0.51	0.49	22.12	3.15	0.14
	0.57	0.54	18.44	2.84	0.15

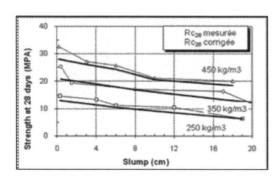

Figure 12. Evolution of the compressive strength at 28 days according to slump.

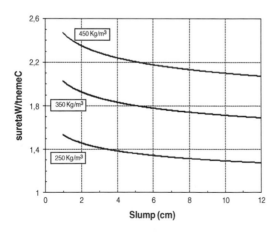

Figure 13. Evolution of the compressive strength at 28 days and C/W useful ratio according to slump.

4.2 Tensile strength

The values of tensile strength by splitting grow from 1, 63 to 3,84 MPa (figure 14) and follow the linear of equation:

$$R_{t28} = 1,40. \frac{C}{W_{us}} + 0.27 \qquad R^2 = 0,92 \qquad (13)$$

The tensile strengths develop less quickly than the compressive strengths. The ratios R_t/R_C vary from 0,13 to 0,29 (table 9).

5 FORMULATION ABACUS

Finally, the model suggested previously allows to elaborate an abacus of formulation intended for the production of the limestone concretes. This abacus

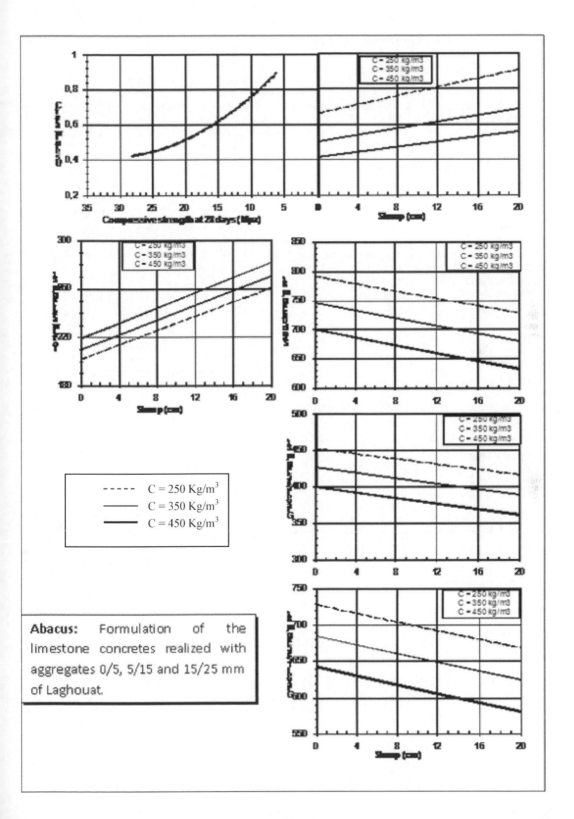

Abacus: Formulation of the limestone concretes realized with aggregates 0/5, 5/15 and 15/25 mm of Laghouat.

173

envisages the quantities of ingredients entering in the manufacture of one cubic meter of concrete produced with limestone aggregates.

Example of application

We want to make $1 \, m^3$ of concrete whose desired properties are as follows:

$R_{C28} = 20 \, MPa$;
$A = 13 \, cm$.
From curves 1 and 2 proportioning necessary is:

C: 450 kg;
W_{tot}: 264 kg;
S: 656 kg;
G_1: 375 kg;
G_2: 603 kg.
Let us suppose that during the manufacture of the concrete, the aggregates are wet and that the water content of sand is 3% and the fine gravels is 1%.
The masses of the wet aggregates are:

S_h: 676 kg;
G_{1h}: 379 kg;
G_{2h}: 609 kg.
The water of manufacture

$$W_{man} = W_{us} - W_{br} + W_{abs} \quad W_{man} = 231 \, l$$

The quantities of materials to waste to manufacture $1 \, m^3$ of the concrete are:

C: 450 kg;
W_{man}: 235 kg;
S: 676 kg;
G_1: 379 kg;
G_2: 609 kg.

6   CONCLUSIONS

This contribution to the formulation of a concrete made from limestone sands and gravels has allow drawing the following conclusions:

The physical, geometrical, morphological and mechanical characteristics of the Turonien crushed limestone aggregates of the Central Saharian Atlas are satisfactory for the manufacture of the hydraulic concretes. Since they carried out the requirements demand of the conformity criteria necessary for the aggregates constituting the current concretes.

The concretes manufactured by these aggregates, have a mechanical behaviour which approaches that of the current concretes, since for plastic workability of the freshly-mixed concrete, and after 28 days of conservation in an ambient environment, the compressive strength can reach 25 MPA.

Possibility to elaborated a simple law which allows to adjust the cubic meter of the mixture carried out with unit volume (real $1 \, m^3$) and to propose graphs of formulation for the limestone concretes, which allows the quantification of the formulas to manufacture $1 \, m^3$ of concrete, the workability of which extends from the firm to the fluid state.

BIBLIOGRAPHY

BARON J. and LESAGE R. (1976) "the composition of the hydraulic concrete, the laboratory to the building site", Research report LCPC, n°64, Paris.

BOLOMEY (July–August 1966) Bulletin of connection of the laboratories of the Highways Departments (LCPC), n°20.

COUNT A., MASSON L. REMILLON A. (1995) "Formulation of the current concretes with aggregates LORRAINE limestones 0/6 and 6/20 Annals of the Institute of the Building industry and Public works (ITBTP), N°: 539, concrete series 326, pp 1–24.

DREUX G. and FESTA J. (1995) "concrete guide". Seventh Edition Eyrolles Edition.

DREUX G. and GORISSE F. (1983) "Composition of the concretes: Dreux-Gorisse". Annals of the ITBTP, n°414, concrete series 214, pp 86–105.

DUPAIN R., HOLY LANCHON R. and ARROMAN J.C. (1995) "Aggregates, grounds, cements and concretes, Characterization of materials of civil engineering by the laboratory tests", Editions CASTEILLA, Paris.

FLEMING J. (1952) "The Atlasique chains and the Northern edge of the Sahara". Publications of the service of the geological map, regional monograph. First series, n°14.

French association of standardization AFNOR.

Excellence in Concrete Construction through Innovation – Limbachiya & Kew (eds)
© 2009 Taylor & Francis Group, London, ISBN 978-0-415-47592-1

Effect of cumulative lightweight aggregate volume in concrete on its resistance to chloride-ion penetration

X.M. Liu & M.H. Zhang
National University of Singapore, Singapore

ABSTRACT: This paper presents an experimental study on the resistance of lightweight aggregate concretes to chloride-ion penetration in comparison to that of normal weight concrete of similar w/c. Salt ponding test (based on AASHTO T 259), rapid chloride permeability test (ASTM C 1202) and rapid migration test (NT Build 492) were carried out to evaluate the concrete resistance to the chloride-ion penetration. Results indicate that in general the resistance of the LWAC to the chloride-ion penetration was in the same order as that of NWAC of similar w/c. However, the increase in cumulative LWA volume and the incorporation of finer LWA particles led to higher charge passed, migration coefficient, and diffusion coefficient. Since the LWACs had lower 28-day compressive strength compared with that of the NWAC of similar w/c, the LWACs may have equal or better resistance to the chloride-ion penetration compared with the NWAC of equivalent strength. The trend of the resistance of concretes to chloride-ion penetration determined by the three test methods was reasonably consistent although there were some discrepancies due to different test methods.

1 INTRODUCTION

Lightweight aggregate concretes (LWAC) are generally used to reduce the unit weight of concrete and hence the self-weight of structures. They have been used for structural applications in long-span bridges, high-rise buildings, buildings where soil conditions are poor, and floating structures. It has also been proposed that lightweight concrete may be used in steel-concrete-steel composite plates for ship structures (Eyres, 2007). By reducing unit weight of concrete with a given strength, the strength/weight ratio of the concrete can be increased.

In addition to the strength/weight ratio, the other consideration that is of significance for structures exposed to severe environments is their long-term durability. When exposed to severe environments, e.g. marine environment, the resistance of concrete to chloride-ion penetration is a great concern.

Mechanisms for chloride-ion penetration in concrete generally include (1) diffusion, driven by concentration gradient, (2) permeation, driven by pressure gradients, and (3) absorption, driven by moisture gradients under capillary force. If the concrete is in a saturated condition, the principal mechanism for the penetration of chloride-ions into the concrete to reach steel reinforcing bars is diffusion.

Concrete has highly heterogeneous and complex microstructures. It may be considered as a three-phase composite material consisting of inert aggregate particles of various sizes embedded in a uniform matrix of hydrated cement paste, and the interface transition zone (ITZ) between the aggregate and cement paste. It may also be considered as a composite consisting of coarse aggregates embedded in a mortar matrix and an ITZ between these two components. Concrete resistance to the chloride-ion penetration will therefore be affected by the matrix, aggregate, and interface transition zone between them. Lightweight aggregates are generally more porous compared with normal weight aggregates (NWA). The ITZ is generally considered a potential weak link in NWAC due to its high porosity compared with the bulk cement paste. However, the ITZ around the LWA is generally denser and more homogenous due to the absorption of LWA (Zhang & Gjørv, 1990). With more porous LWA and denser ITZ, the resistance of LWAC to the chloride-ion penetration depends on which of these factors has more dominant effect.

Published research on permeability and resistance to chloride ion penetration of LWA concretes and mortars showed contradictory findings. According to Nyame (1985), mortars incorporating lightweight sand and with a water-to-cement ratio (w/c) of 0.47 were about twice as permeable as those made with natural sand. Chia and Zhang (2002) evaluated the resistance of LWAC to chloride-ion penetration in comparison to that of NWAC with

Table 1. Mix proportion of the concretes (w/c = 0.38).

			Natural aggregate				Dry LWA					
		Mixing	0–1.18	1.18–2.36	2.36–4.75	4.75–9.5	0–1.18	1.18–2.36	2.36–4.75	4.75–9.5		Slump
#	Cement	water*	mm	mm	mm	mm	mm	mm	mm	mm	SP**	mm
NC	500	191	514	171	76	850	–	–	–	–	2.9	200
LC1	500	193	514	171	76	–	–	–	–	395	2.7	125
LC2	500	193	514	171	–	–	–	–	40	395	2.3	70
LC3	500	191	514	–	–	–	–	83	40	395	2.3	100
LC4	500	188	–	–	–	–	321	83	40	395	–	85

Mix proportion, kg/m^3

* Water absorbed by LWA was not included here.
** Naphthalene-based superplasticizer (SP), l/m^3.

water-to-cementitious material ratios (w/cm) of 0.35 and 0.55. The LWAC included coarse LWA but natural sand. They found that the resistance of the LWAC to the chloride-ion penetration was similar to that of the corresponding NWAC with the same w/cm. Al-Khaiat and Haque (1999) found that the concentrations of chloride penetrated into the LWAC with both coarse and fine LWA were somewhat higher than that in NWAC of the same 28-day design compressive strength of 50 MPa. However, Thomas (2006) found that the use of LWA (both coarse and fine aggregates) substantially reduced the electrical conductivity and chloride penetrability (determined by ASTM C 1202 and the bulk diffusion test) of high-performance LWAC with w/cm of 0.30 and made with blended cement with silica fume.

These published information indicates that cumulative LWA content in the concrete, whether lightweight sand was used, and test method may have influence on the results. In many applications, lightweight sand needs to be used in order to reduce the unit weight of concrete. This research was thus carried out to evaluate the effect of cumulative LWA content (with and without lightweight sand) in concrete on its resistance to chloride-ion penetration and the results was presented in this paper. The results were compared with those of the corresponding NWAC with the same w/c of 0.38. Salt ponding test (based on AASHTO T 259 (2002), rapid chloride permeability test (based on ASTM C 1202 (2005) and rapid migration test (based on NT Build 492 (1999) were carried out to evaluate the concrete resistance to the chloride-ion penetration.

2 EXPERIMENTAL DETAILS

2.1 Concrete mixtures and materials

Five mixtures were included in this study which involved one NWAC and four LWACs. The volumes of

Table 2. Properties of lightweight aggregates.

Type of aggregate	Grain size (mm)	Dry particle density (g/cm^3)	24 hours absorption
F6.5	>4.75	1.14	13.0%
F6.5	2.36–4.75	1.29	11.6%
F4.5	1.18–2.36	1.04	12.8%
LW sand	<1.18	1.60	~30.0%

coarse and fine aggregates of the concrete were kept the same. For the LWACs, different size fractions of the LWA were included in the concretes so that cumulative volume of the LWA increased from 50% by volume of total aggregate in Mix LC1 to 100% in LC4. In Mix LC1, only coarse LWA was included, whereas in Mixes LC2 and LC3 and LC4, different amounts and size fractions of lightweight sand were also included. Table 1 summarizes the mixture proportions of the concretes and the slump of the concretes.

ASTM Type I normal Portland cement was used for the concretes. A naphthalene based superplasticizer was used in concretes for workability purpose. The superplasticizer was a dark brown solution with a specific gravity of 1.2, and contained about 40% solids. The superplasticizer conforms to the requirement of ASTM C494 (2005) – Type F high range water-reducing admixture.

Lightweight aggregates for this study were expanded clay with maximum size about 9.5 mm commercially available, and their properties are shown in Table 2. The lightweight aggregates of the size fractions >1.18 mm had round shape and similar 24-h water absorption of about 11.6 to ~13.0%. However, the lightweight sand with size <1.18 mm was crushed particles, thus had more open pores and much higher water absorption than others. Because of the nature of the crushed particles and more open pores, it is difficult to obtain saturated surface dry condition of the aggregate and to accurately determine the absorption of the aggregate. Therefore, the absorption value given in Table 2 was, at most, an estimate.

For the NWAC, granite aggregate with a maximum size of 9.5 mm was used. Natural sand was used in the concretes except for Mix LC4. Specific gravity of the coarse and fine normal weight aggregates was 2.61 and 2.56, respectively.

In order to have a good control for grading, both normal weight and lightweight aggregates were separated into four fractions of <1.18, 1.18–2.36, 2.36–4.75, and >4.75 mm, and then recombined to satisfy ASTM C 33 grading requirements.

2.2 Concrete preparation and curing

The concretes were mixed in a pan mixer with a mixing speed of 50 rounds per minute (rpm) at ambient temperature of about 30°C. Slump of the concrete was determined according to ASTM C143 (2005). The LWAC was prepared with prewetted lightweight aggregate soaked in water for 24 h before the concrete mixing. Prior to soaking, the lightweight aggregate was oven dried and cooled to room temperature.

For each mixture, three $100 \times 100 \times 100$ mm cubes, three $300 \times 300 \times 70$ mm slabs and two $\varnothing 100 \times 200$ mm cylinders were prepared for compressive strength test, salt ponding test, and ASTM C 1202 and NT Build 492 test, respectively. After demould at 24 hours, the specimens were transferred to a fog room and cured at a temperature of about 28°C until 7 days. The specimens were then exposed to the lab air at similar temperature for 21 days.

After 7 days fog curing, approximately 20 mm each from the top and bottom part of the cylinders were removed and, each middle part were then cut into three small cylinders with a dimension of $\varnothing 100 \times 50$ mm which were used for ASTM C 1202 and NT Build 492 tests.

The density and compressive strength of the concretes were determined according to BS EN 12390 (2000 & 2002).

2.3 Test methods to determine the resistance of concrete to chloride penetration

The resistance of the concrete to the chloride-ion penetration was evaluated by the following three methods.

2.3.1 Rapid chloride permeability test
The rapid chloride permeability test (RCPT) was carried out at 28 days according to ASTM C 1202-05 and the test set up is shown in Fig. 1. The total charge passed after 6 hrs was obtained from integration of current over the time duration.

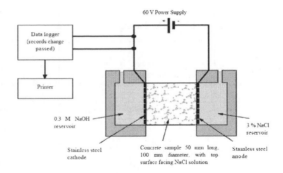

Figure 1. ASTM C 1202 test set-up (Stanish et al., 1997).

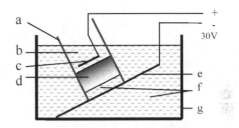

Figure 2. Setup of NT Build 492 test. (a. Plastic tube; b. 0.3 N NaOH; c. Anodic stainless steel plate; d. Concrete specimen; e. Cathodic stainless steel plate; f. 10% NaCl; g: Glass container).

2.3.2 Rapid migration test (Tang & Nilsson, 1991)
Migration coefficient (also referred to as apparent diffusion coefficient) was determined according to NT Build 492 test method (Fig. 2). The principle of the test setup is similar to that of ASTM C 1202, but the external potential applied was 30 V instead of 60 V and sodium chloride concentration was 10% instead of 3% compared with ASTM C 1202 test.

The concrete specimen was exposed to a 10% NaCl solution on one side and a 0.3 M NaOH solution on the other side. The external potential of 30 V was applied across the specimen for 24 hours. After that the specimen was split into two halves across its circular cross section. The split open surfaces were sprayed with 0.1N AgNO$_3$ solution to determine the chloride penetration depth, which was then used to calculate the migration coefficient according to Eq. (1).

$$D_{nssm} = \frac{RT}{zFE} \cdot \frac{x_d - \alpha\sqrt{x_d}}{t} \qquad (1)$$

where

$$E = \frac{U-2}{L} \qquad (2)$$

$$\alpha = 2\sqrt{\frac{RT}{zFE}} \cdot \mathrm{erf}^{-1}\left(1 - \frac{2c_d}{c_0}\right) \qquad (3)$$

177

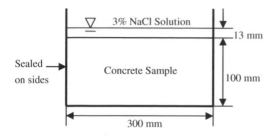

Figure 3. Salt ponding test setup.

where D_{nssm} = non-steady-state migration coefficient (m^2/s); z = absolute value of ion valence, for chloride $z = 1$; F = Faraday constant $(9.648 \times 10^4 \, J/(V \cdot mol))$; U = absolute value of the applied voltage (V); R = gas constant $(8.314 \, J/(V \cdot mol))$; T = average value of the initial and final temperatures in the anolyte solution (K); L = thickness of the specimen (m); x_d = average value of the penetration depths (m); t = test duration (s); erf^{-1} = inverse of error function; c_d = chloride concentration at which the color changes, $c_d \approx 0.07 \, N$ for OPC concrete; c_0 = chloride concentration in the catholyte solution, and $c_0 \approx 2 \, N$ (NT BUILD 492).

2.3.3 Salt ponding test

The resistance of concrete to chloride ion penetration was also determined according to AASHTO T259 method with some modifications. One of the modifications was 3 day water ponding of the specimens prior to the salt ponding. The purpose was to create a nearly-saturated condition to simulate a diffusion process for chloride ions rather than a combination of diffusion and absorption as the original AASHTO test. The other modification was the curing age. AASHTO T 259 specifies 14 days moist curing followed by 28 days drying before the test. In this research, the slabs were moist cured for 7 days followed by exposure in lab air for 21 days.

At 28 days, the sides and bottom of the slabs were sealed by epoxy, and dams around the top edges of the specimens were built (Fig. 3). The epoxy was allowed to dry overnight. After that, the specimens were ponded by water for 3 days before the ponding with a 3% of sodium chloride solution for 90 days. The top of the ponding dams was covered with plastic sheet to minimize evaporation, and additional solution was added periodically to keep the 13 mm depth of solution specified by AASHTO method.

After 90 days of exposure, the ponding solution was removed and the specimens were allowed to dry. The surfaces were brushed to remove salt crystal buildup. Concrete samples were taken at various depths of the slabs by drilling at four locations of each slab and then combined. The samples were dried at 105°C and ground to pass a 150 μm sieve. Acid soluble chloride content of the concrete was determined according to

BS 1881-124. The chloride profile was plotted where the chloride content was expressed as percentage of cement in dry samples and the depth was the mid-point of each interval.

The diffusion of chloride ions in concrete may be described by Fick's second law as

$$\frac{\partial C}{\partial t} = K_d \frac{\partial^2 C}{\partial x^2} \tag{4}$$

where C = concentration (mol/m^3); t = time (s); K_d = diffusion coefficient (or diffusivity) (m^2/s); and x = depth (m). The solution for the ponding test of a semi-finite slab to this equation is given by Equation (5) with boundary conditions $C_{(x=0, \, t>0)} = C_0$ and $C_{(x=\infty, \, t>0)} = 0$ and initial condition $C_{(x=0, \, t=0)} = 0$.

$$C(x,t) = C_0 \left[1 - erf\left(\frac{x}{2\sqrt{K_d t}} \right) \right] \tag{5}$$

where $C(x, \, t)$ = the chloride concentration at depth x and time t; C_0 = the chloride concentration at the surface; and erf = error function.

Based on the chloride profile after the ponding test and Fick's second law, chloride diffusion coefficient K_d of the concretes was obtained.

3 RESULTS AND DISCUSSION

Results of the concretes obtained from the tests described in Section 2.3 were summarized in Table 3. The density of the concretes after 1 day and compressive strength at 28 days were also given in Table 3. With the increase in LWA content and reduction in NWA in concrete, the density and compressive strength of the concrete decreased. The results reported are the averages of three specimens for each test.

3.1 Resistance to chloride-ion penetration determined by ASTM C 1202 test

The results in Table 3 indicate that the average charge passed through the LWAC and NWAC was all within the range from 2000 to 4000 coulombs, classified as "moderate" chloride penetrability according to ASTM C 1202. The LWAC with LWA size >2.36 mm (LC1 and LC2) was similar to that of the NWAC with similar w/c ratio. However, the charge passed through the LWAC with LWA size ≤2.36 mm (LC3 and LC4) was higher than that of the NWAC.

3.2 Chloride migration coefficient determined from NT BUILD 492 test

The penetration depth determined by AgNO₃ spray and migration coefficients of the LWAC calculated was in

Table 3. Test results of different mixtures.

#	Density at 1 day (kg/m^3)	28-day compressive strength (MPa)	Total Coulombs*	Cl^- penetration depth (mm)	Migration coefficient $\times 10^{-12}$ (m^2/s)	Diffusion coefficient $\times 10^{-12}$ (m^2/s)
NC	2360	67.1	2437 ± 112	19.4	10.1 ± 1.2	5.6 ± 0.5
LC1	1900	50.3	2385 ± 141	12.0	6.5 ± 2.2	5.3 ± 1.1
LC2	1860	46.7	2496 ± 175	15.5	7.6 ± 0.1	5.9 ± 1.5
LC3	1740	42.3	3278 ± 273	17.8	8.8 ± 0.8	6.4 ± 1.9
LC4	1620	33.5	3621 ± 529	18.3	8.9 ± 0.6	NA

* According to ASTM C 1202 (2005).

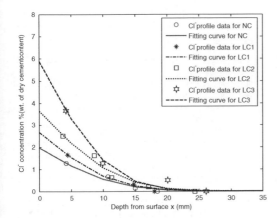

Figure 4. Chloride profiles and the fitting curve according to Fick's 2nd Law of the concrete (w/c = 0.38).

the same order as that of the NWAC with similar w/c. However, with the increase in the cumulative LWA content and incorporation of finer LWA particles in the concrete, the migration coefficient of the concrete increased somewhat. This trend seems to be consistent with that based on the charge passed through the concrete discussed above.

3.3   Chloride diffusion coefficient from AASHTO T259 ponding test

Figure 4 shows the typical chloride profiles of the mixtures after 90 days of the ponding test except for Mix LC4. The test of LC4 is still on going. The diffusion coefficients presented in Table 3 are the averages from three concrete slabs. The diffusion coefficient of LWAC was in the same order as that of the NWAC with similar w/c.

The incorporation of lightweight coarse aggregate did not affect the chloride-ion penetration, and the chloride profile of the concrete with only coarse LWA (LC1) was similar to that of the NWAC as shown in Fig. 4. This was consistent to the findings of Chia and Zhang (2002). The resistance to the chloride-ion

penetration of the LWAC with LWA size ≤ 4.75 mm was somewhat lower than that of the NWAC. This may be related to the cumulative volume of the LWA and combined pore structure of LWA and cement paste. The diffusion coefficient determined by the ponding test showed similar trend as those from the other two tests.

3.4   Discussion

The above results were reasonably consistent although there were some discrepancies due to different test methods. In general the resistance of the LWACs was in the same order as that of NWAC of similar w/c. However, the increase in cumulative LWA volume and the incorporation of finer LWA particles led to higher charge passed, migration coefficient, and diffusion coefficient.

For NT Builder 492 test, the chloride penetration depth was determined by the whitish color due to the precipitation of AgCl. In the calculation of migration coefficient according to the method, chloride concentration at which the color change occurred was approximately 0.07 N for OPC concrete. According to Otsuki et al. (1992), this whitish color will only be visible when the chloride content is $>0.15\%$ by weight of cement with the spray of the Ag NO_3 solutions. The actual chloride penetration depth was probably greater than that presented in Table 3. This may influence the migration coefficient determined.

For the ponding test, the chloride diffusion coefficient was calculated based on the acid soluble chloride content in concretes. Thus the trend established by the diffusion coefficient would be more accurate than that by the migration coefficient although the latter took shorter time and was easier to determine from test point of view.

For ASTM C 1202 test, the results were based on the total charges passed, which are affected by the pore structure of the concrete and all ions in the pore solutions.

Since the LWACs had lower 28-day compressive strength compared with that of the NWAC of similar w/c, the LWACs may have equal or better resistance to the chloride-ion penetration compared with the NWAC of equivalent strength.

4 CONCLUSIONS

Based on the results, the following conclusions may be drawn.

1. In general the resistance of the LWAC to the chloride-ion penetration was in the same order as that of NWAC of similar w/c. However, the increase in cumulative LWA volume and the incorporation of finer LWA particles led to higher charge passed, migration coefficient, and diffusion coefficient.
2. The trend of the resistance of concretes to chloride-ion penetration determined by the three test methods was reasonably consistent although there were some discrepancies due to different test methods.

ACKNOWLEDGEMENTS

Authors acknowledge the contribution of Lui Wing Fai for his assistance in experimental work. Support and assistance from staff of Structural and Concrete Laboratory, Department of Civil Engineering, National University of Singapore is appreciated. Dr. Chia Kok Seng's review is appreciated as it allowed this paper to be improved.

REFERENCES

AASHTO T 259-02, Standard Method of Test for Resistance of Concrete to Chloride Ion Penetration.
AASHTO T 260-94, Sampling and Testing for Chloride Ion in Concrete and Concrete Raw Materials.
Al-Khaiat, H. & Haque, N. 1999. Strength and durability of lightweight and normal weight concrete. *Journal of Materials in Civil Engineering.* 11(3): 231–235.
ASTM C 143/C 143M-05, Standard Test Method for Slump of Hydraulic-Cement Concrete, *Annual Book of ASTM Standards*, V 04.02, ASTM International.
ASTM C 494/C 494M-05a, Standard Specification for Chemical Admixtures for Concrete, *Annual Book of ASTM Standards*, V 04.02, ASTM International.
ASTM C 1202-05, Standard Test Method for Electrical Indication of Chloride's Ability to Resist Chloride, *Annual Book of ASTM Standards*, V 04.02, ASTM International.
British Standards Institution, Testing Concrete, BS 1881-124, 1988.
British Standards Institution, Testing Hardened Concrete, BS EN 12390: Part 3, 2002.
British Standards Institution, Testing Hardened Concrete, BS EN 12390: Part 7, 2000.
Chia, K.S. & Zhang, M.H. 2002. Water permeability and chloride penetrability of high-strength lightweight aggregate concrete. *Cement and Concrete Research*, 32(4): 639–645.
Eyres, D.J. 2007. Ship Construction. Oxford: Butterworth-Heinemann.
Nyame, B.K. 1985. Permeability of normal and lightweight mortars, *Magazine of Concrete Research*, 37(130): 44–48.
NT BUILD 492, 1999. Concrete, Mortar and Cement-Based Repair Materials: Chloride migration coefficient from non-steady-state migration experiments.
Otsuki, N., Nagataki, S. & Nakashita, K. 1992. Evaluation of $AgNO_3$ solution spray method for measurement of chloride penetration into hardened cementitious matrix materials, *ACI Materials Journal*, 89(6): 587–592.
Stanish, K.D., Hooton, R.D. & Thomas, M.D.A. 1997. Testing the Chloride Penetration Resistance of Concrete: A Literature Review, FHWA Contract DTFH61-97-R-00022.
Tang, L. & Nilsson, L. 1992. Rapid determination of the chloride diffusivity in concrete by applying an electrical field. *ACI Materials Journal*, 89(1): 49–53.
Thomas, M.D.A. 2006. Chloride diffusion in High-performance lightweight aggregate concrete. *Durability of Concrete, Proceedings 7th CANMET/ACI International Conference*, SP234, Montreal, Canada.
Zhang, M.H. & Gjørv, O.E. 1990. Microstructure of the interfacial transition zone between lightweight aggregate and cement paste. *Cement and Concrete Research*, 20(4): 610–618.

Excellence in Concrete Construction through Innovation – Limbachiya & Kew (eds)
© 2009 Taylor & Francis Group, London, ISBN 978-0-415-47592-1

Effect of bagasse ash on water absorption and compressive strength of lateritic soil interlocking block

P. Khobklang, K. Nokkaew & V. Greepala
Kasetsart University Chalermphrakiat Sakonnakhon Province Campus, Thailand

ABSTRACT: The effects of the proportion of bagasse ash-blended Portland cement in mixtures of lateritic soil interlocking block were investigated. The bagasse ash was sieved through a No. 200 sieve. Different replacement percentages of 0, 15, 30 and 40% of Portland cement were mixed with river sand, water and lateritic soil (from Kasetsart University Chalermphrakiet Sakon Nakhon Province Campus area). The results revealed that an increase in the amount of bagasse ash significantly increased water absorption. The interlocking block using bagasse ash in a ratio of more than 15% to binder provided higher compressive strength, compared to those using lower bagasse ash content, at the testing age of 90 days. Moreover it was found that an increase in the water to binder ratio significantly decreased water absorption. However, this ratio had no significant influence on the compressive strength of interlocking block specimens.

1 INTRODUCTION

Ordinary Portland cement is recognized as a major construction material throughout the world (Bentur 2002). Industrial wastes, such as blast furnace slag, fly ash and silica fume are being used as supplementary cement replacement materials (Ganesan et al. 2007). In addition to these, agricultural wastes such as rice husk ash and wheat straw ash are used as pozzolanic materials, and hazel nut shell is used as cement replacement material (Biricik et al. 1999; Cook 1986; Demirbas & Asia 1998; Mehta 1977; Mehta 1992). When pozzolanic materials are added to cement, the silica (SiO_2) present in these materials react with free lime released during the hydration of cement and forms additional calcium silicate hydrate (CSH) as new hydration products (Boating & Skeete 1990), which improve the mechanical properties of concrete formulation. Bagasse ash is an agricultural by-product of sugar manufacturing. When juice is extracted from the cane sugar, the solid waste material is known as bagasse. When this waste is burned under controlled conditions, it also produces ash containing amorphous silica, which has pozzolanic properties. Therefore it is possible to use bagasse ash as a cement replacement material to improve quality and reduce the cost of construction materials such as mortar, concrete pavers, concrete roof tiles and soil cement interlocking block.

The aim of this research is to investigate the effects of the proportion of bagasse ash-blended Portland cement on the compressive strength and water absorption of lateritic soil interlocking block mixtures. The investigated parameters are the replacing content of bagasse ash in binder and the water to binder ratios.

This experimental study examines the compressive strength and water absorption of $250 \times 120 \times 110$ mm interlocking block samples. The main ingredients were Portland cement type I, bagasse ash, lateritic soil and river sand. After mixing, the interlocking block samples were cast and retained in a climatic room for 24 hours. Subsequently all test samples were cured in water until the compressive strength and water absorption tests (TIS 1987) were conducted.

2 EXPERIMENTAL INVESTIGATION

This research investigated the effect of the proportion of bagasse ash-blended Portland cement on the compressive strength and water absorption of lateritic soil interlocking block mixtures. The main parameters were the bagasse ash to binder ratio and the water to binder ratio. The experimental study was carried out in three steps: materials preparation, water absorption test, and compressive strength test. In order to effectively test the main parameters, altogether eight mix proportions were produced. Determination of the influence of bagasse ash content was achieved by varying the bagasse ash to binder ratio (BA/b) from 0, 15, 30 and 40% by weight. The two levels of the water to binder ratio (w/b) investigated were 1.14 and

Table 1. Experimental program and details of mix proportion.

| Mix No. | Mix destination | Mix proportion (kg/m³) | | | | |
		Cement	Bagasse ash	Sand	Lateritic soil	Water
1	SC114	210.0	–	730	730	240
2	SCB114-15	178.5	31.5	730	730	240
3	SCB114-30	147.0	63.0	730	730	240
4	SCB114-40	126.0	84.0	730	730	240
5	SC124	210.0	–	730	730	260
6	SCB124-15	178.5	31.5	730	730	260
7	SCB124-30	147.0	63.0	730	730	260
8	SCB124-40	126.0	84.0	730	730	260

Table 2. Atterberg's limit of Lateritic soil used in this study.

Atterberg's properties	
Liquid limit (LL)	39.17
Plastic limit (PL)	26.55
Plastic index (PI)	12.61

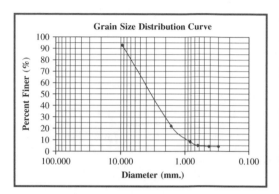

Figure 2. Grain size distribution of Lateritic soil used in this study.

a) Original

b) Passing sieve No.200

Figure 1. Bagasse ash used in this study.

1.24 by weight. The details of the mix proportions are summarized and shown in Table 1.

2.1 Materials

Ordinary Portland cement (OPC) and bagasse ash passing sieve No .200 (ASTM 2004b) obtained from "Rerm-Udom Sugar Factory, Udontani, Thailand"

(Figure 1) were used as blending components. Lateritic soil from Kasetsart University Chalermphrakiet Sakon Nakhon Province Campus area was used as an aggregate ingredient. The lateritic soil has an average specific gravity of 2.88 (ASTM 2004c) and its Atterberg's limit and grain size distribution curve (ASTM 2004a) are given in Table 2 and Figure 2 respectively. River sand passing sieve No. 8 (ASTM 2004b) with a specific gravity of 2.71 was also used as an aggregate ingredient. The mix ratio between the lateritic soil and sand is 1:1 by weight.

The materials were mixed in a laboratory drum mixer for a total of five minutes. Twenty $250 \times 120 \times 110$ mm interlocking blocks, as shown in Figure 3, were cast from each mix for compressive strength testing at 14, 28 and 90 days and water absorption testing at 90 days. After casting, all specimens were left covered in the casting room for 24 hours. The specimens were then demolded and transferred to the curing bath until the time of testing.

2.2 Water absorption test

Percentage of water absorption is a measure of the pore volume or porosity in a hardened specimen, which is occupied by water in saturated condition. Water absorption values of interlocking block specimens were measured in accordance with TIS 58-2530 (TIS 1987) after 90 days of water curing. The difference between the saturated mass and oven dry mass

Figure 3. Interlocking block specimen.

Table 3. Compressive strength and water absorption of interlocking block.

Mix destination	Average compressive strength (ksc)			Average water absorption at 90 Days (%)
	14 Days	28 Days	90 Days	
SC114	73.51	81.56	82.26	11.29
SC114-15	70.74	78.35	81.93	11.38
SC114-30	62.73	77.20	90.75	11.83
SC114-40	62.14	72.00	88.53	11.92
SC124	71.54	80.92	81.46	10.54
SC124-15	70.02	78.16	82.29	10.64
SC124-30	62.66	75.44	90.37	10.92
SC124-40	62.13	71.04	88.27	10.92

Table 4. Standard properties of interlocking block given by TIS 57-2533 and TISTR.

Properties	TIS 57-2533	TISTR
Compressive strength (ksc)	≥ 70.00	≥ 70.00
Water absorption (%)	≤ 8.00	≤ 15.00

expressed as a fractional percentage of oven dry mass gives the water absorption. For each mix proportion, the water absorption value was obtained by taking the average of five specimens.

2.3 Compressive strength test

Compressive strength testing of the lateritic soil interlocking block specimens was conducted after 14, 28 and 90 days' water curing in accordance with TIS 58-2530 (TIS 1987). The interlocking block samples were measured and the dimensions recorded according to TIS 58-2530 (TIS 1987), then the test sample was placed in the testing machine and a load was applied perpendicular to the direction of rising during manufacture at a constant rate of 1 kN/s. The maximum load at which the specimens failed was recorded. The compressive strength is the failure load of each specimen divided by the net area over the load.

3 RESULTS AND DISCUSSION

The compressive strengths at 14, 28 and 90 days and the water absorption at 90 days of $250 \times 125 \times 105$ mm interlocking block samples determined from the 5 specimens are summarized in Table 3. Besides being useful as a cement replacement material, bagasse ash was found to be satisfactory for producing interlocking block due to its compressive strength and water absorption, compared with a standard hollow load-bearing concrete masonry unit provided by TIS 57-2533 (TIS 1990) and a standard interlocking block provided by Thailand Institute of Scientific and Technological Research (TISTR), as shown in Table 4. Moreover it was found that a BA/b ratio less than 15% resulted in a satisfactory interlocking block within 14 days. However, a BA/b ratio of 30% resulted in the

highest compressive strength at 90 days. The influence of bagasse ash content and the water to binder ratio (w/b) on compressive strength and water absorption of interlocking block samples can be summarized as follows:

3.1 Effect of bagasse ash content

An experimental investigation was conducted to investigate the influence of the bagasse ash content by varying the bagasse ash to binder ratio (BA/b) from 0, 15, 30 and 40% by weight. The effects of bagasse ash content on compressive strength and water absorption at different ages of interlocking block under two levels of water to binder ratio (w/b), w/b = 1.14 and 1.24, are summarized and shown in Figures 4 and 5. It can be seen that the use of BA/b higher than 15% led to lower compressive strength at early stages; however, the compressive strength at 90 days is 10% higher than those with 0% bagasse ash replacement.

This difference can be explained by the fact that the reduction of calcium silicate hydrate (CSH) from hydration reaction due to decreased cement content through replacement by bagasse ash resulted in lower compressive strength in the first 56 days. However the silica (SiO_2) in bagasse ash reacts with free lime released during the hydration of cement and forms additional calcium silicate hydrate (CSH) at later stages of curing, resulting in more CSH content at 90 days. Therefore the interlocking block using

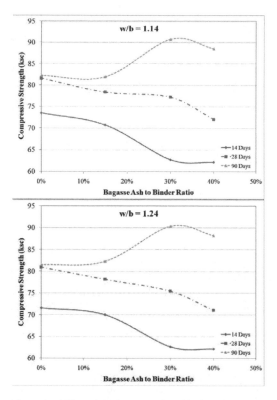

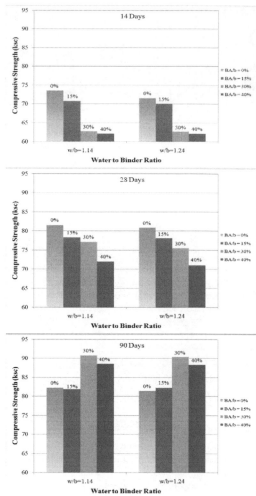

Figure 4. Effect of the bagasse ash to binder ratio (BA/b) on compressive strength of interlocking block specimen.

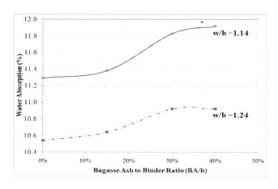

Figure 5. Effect of the bagasse ash to binder ratio (BA/b) on water absorption of interlocking block specimen.

Figure 6. Effect of water to binder ratio (w/b) on compressive strength at 14, 28 and 90 days of interlocking block specimen.

3.2 Effect of water to binder ratio (w/b)

An experimental investigation was conducted to investigate the influence of the water to binder ratios of 1.14 and 1.24 by weight. The effects of water content on compressive strength at different ages of interlocking blocks are summarized and shown in Figure 6. It was found that an increase in water content very slightly decreased compressive strength of interlocking block specimens. Therefore, it could be said that the water to binder ratio had no significant influence on the compressive strength of interlocking block specimens.

Regarding water absorption, it was found that an increase in water to binder ratio from 1.14 to 1.24 led to a reduction in water absorption by around 6–8%.

higher bagasse ash content showed higher compressive strength at 90 days.

Regarding water absorption, it was found that an increase in bagasse ash content significantly increased the water absorption of interlocking blocks, regardless of water to binder ratio.

4 CONCLUSIONS

On the basis of the results obtained from this study, the following conclusion can be drawn:

1. Besides being useful as a cement replacement material, bagasse ash was found to be satisfactory for producing interlocking blocks due to its compressive strength and water absorption compared with the standard hollow load-bearing concrete masonry unit provided by TIS 57-2533 (TIS 1990) and the standard interlocking block given by Thailand Institute of Scientific and Technological Research (TISTR).
2. An increase in the amount of bagasse ash significantly increased water absorption. The interlocking block using a bagasse to binder ratio of more than 15% provided higher compressive strength, as compared to those using lower bagasse ash content at the testing time of 90 days.
3. An increase in the water to binder ratio significantly decreased water absorption. However, this ratio had no significant influence on the compressive strength of interlocking block specimens.

ACKNOWLEDGEMENTS

The authors wish to express their profound gratitude to Dr. S. Sujjawanit for the invaluable advice, extraordinary guidance and consistent encouragement provided throughout the duration of this research. A grateful acknowledgment is extended to P. Wongsearam and E. Pocapanich for providing support to carry out this research work.

REFERENCES

ASTM. 2004a. Standard Test Method for Particle-Size Analysis of Soils, ASTM D422-63. *Annual Book of ASTM Standards*, American Society for Testing and Materials, West Conshohocken, United States.

ASTM. 2004b. Standard Test Method for Sieve Analysis of Fine and Coarse Aggregates, ASTM C136-95a. *Annual Book of ASTM Standards*, American Society for Testing and Materials, West Conshohocken, United States.

ASTM. 2004c. Standard Test Method for Specific Gravity of Soils, ASTM D854-98. *Annual Book of ASTM Standards*, American Society for Testing and Materials, West Conshohocken, United States.

Bentur, A. 2002. Cementitious materials – nine millennia and a new century: past, present and future. *ASCE Journal Materials in Civil Engineering*, 14(1), 1–22.

Biricik, H., Akoz, F., Berktay, I., & Tulgar, A.N. 1999. Study of pozzolanic properties of wheat straw ash. *Cement and Concrete Research*, 29, 637–643.

Boating, A.A., & Skeete, D.H. 1990. Incineration of rice hull for use as a cementitious materials; The Guyana experience. *Cement and Concrete Research*, 20, 795–802.

Cook, J.D. 1986. Rice husk ash. *Concrete technology and design cement replacement material*, S. R.N., ed., Surrey University, London, 171–95.

Demirbas, A., & Asia, A. 1998. Effect of ground hazel nutshell, wood and tea waste on the mechanical properties of cement. *Cement and Concrete Research*, 28(8), 1101–1104.

Ganesan, K., Rajagopal, K., & Thangavel, K. 2007. Evaluation of bagasse ash as supplementary cementitious material. *Cement and Concrete Composites*, 29, 515–524.

Mehta, P.K. 1977. Properties of blended cement made from rice husk ash. *ACI Materials Journal*, 74(9), 440–442.

Mehta, P.K. Rice husk ash – A unique supplementary cementing material. *Proceeding of the international symposium on advances in concrete tech*, Athens, Greece, 407–430.

TIS. 1987. Standard for Hollow Non-Load-Bearing Concrete Masonry Unit, TIS 58-2530. Thai Industrial Standards Institute.

TIS. 1990. Standard for Hollow Load-Bearing Concrete Masonry Unit, TIS 57-2533. Thai Industrial Standards Institute.

Light-weight TRC sandwich building envelopes

J. Hegger & M. Horstmann

Institute of Structural Concrete, RWTH Aachen University, Aachen, Germany

ABSTRACT: Textile Reinforced Concrete (TRC) is a composite material made of open-meshed textile structures and a fine-grained concrete. In order to increase the efficiency compared to glass-fiber reinforced concrete (GFRC) the fibers are aligned as two-dimensional fabrics in the direction of the tensile stresses. The application of TRC leads to the design of filigree concrete structures with high durability and high quality surfaces allowing economic savings in terms of material, transport and anchorage costs. Thus, ventilated façade systems have become the state-of-the-art-application for TRC in recent years. Large panel sizes and spans can only be realized in combination with bracing substructures. In current projects light-weight sandwich structures with large spans and compact sections of thin TRC-facings and sustainable insulation cores applied for self-supporting façades and for modular buildings concepts have been developed at RWTH Aachen University. In the paper production methods, developed connectors between the thin concrete shells, test results of sandwich members under bending and shear loading as well as deduced calculation models and first applications are presented.

1 INTRODUCTION

Structural concrete has been an economic building material for façade constructions like ventilated façade panels and sandwich elements. Numerous applications in recent decades induced a decreasing acceptance of the material for façades due to the insufficient architectural design range, the clumsy appearance and the corrosion damages.

Thus, non-corrosive reinforcement materials have gained importance in the last 3 decades to achieve the goal of precast, filigree and lightweight concrete structures with high durability, high quality surfaces and a wide-spread design range. Since the cheap and capable asbestos fibres were discovered to be carcinogen chopped AR-glass fibres have been widely used in Glassfibre Reinforced Concrete (GFRC) for the production of non-structural building elements.

The development and application of Textile Reinforced Concrete (TRC) incorporates the advantages of GFRC adding a structural load-bearing capacity in arbitrary directions. The used textile fabrics can be customized as 2D or 3D reinforcements to the production method and load-bearing behavior of the structure. Thus, TRC complements and broadens the design and application range opened up by GFRC.

2 VENTILATED FAÇADE SYSTEMS

TRC allows economic savings in terms of material, transport and anchorage costs and thus has been

Figure 1. Ventilated façades (DIBt 2004, Engberts 2006).

severally used for thin-walled and light-weight ventilated façade systems in recent years (Hegger et al. 2006, Brameshuber 2006). At present, small panel sizes of 0.5–$3\,m^2$ (Fig. 1) are state-of-the-art in application of TRC in Germany. Panel sizes of up to $7\,m^2$ can only be realized in combination with bracing studframe systems (Engberts 2006, Hegger & Voss 2005).

Due to the missing design codes the application of TRC façade elements in Germany requires either an individual approval for each construction or a general approval for defined boundary conditions of the German building inspection (DIBt 2004).

3 SANDWICH ELEMENTS WITH TRC

The application of sandwich panels for façades of factory and industrial buildings has gained importance in the past 50 years due to the prefabrication irrespective the weather conditions as well as the reduced time effort during mounting.

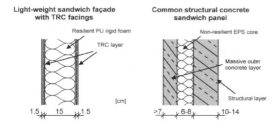

Figure 2. Comparison of light-weight self-supporting sand-wich panels made of TRC and common structural panels.

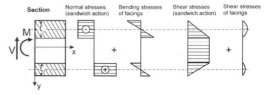

Figure 3. Stress distribution of thick facings and core due to bending and shear forces.

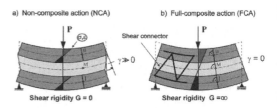

Figure 4. NCA and FCA of sandwich panels (Seeber 1997).

Common structural sandwich elements made of concrete consist of a structural, load-bearing layer ($h = 10$–14 cm), a heat insulation layer and an outer facing ($h \sim 7$ cm). Although in standard non-composite action panels the outer facing has no structural function, a steel reinforcement is necessary to bear constraint forces caused by constricted deformations induced by temperature and shrinkage. In load-bearing structures as well as in façades a concrete cover of about 35 mm complying with current design codes (DIN 1045-1, MC 90) has to be provided to avoid corrosion of the steel reinforcement. If the massive outer layer of usual structural concrete panels is replaced by a thin-walled TRC-layer, the overall thickness of the panel can be reduced about 5–6 cm and the number of required connectors between the concrete layers diminishes. In case the inside facing is also produced with TRC in combination with a sustainable insulation foam light-weight sandwich structures with large spans can be obtained (Fig. 2).

Either the overall panel thickness can even be more reduced or, keeping to the overall thickness, the size of the insulation layer is increased obtaining a superior heat transfer coefficient exceeding the requirements of current regulations by far.

3.1 Load-bearing behaviour

The load-bearing behaviour of sandwich panels with rigid facings depends primarily on the thicknesses of the layers, the overall height and the (shear) stiffness of the core (Stamm & Witte 1974).

In contrast to panels with flexible facings the concrete facings are not only stressed by diaphragm forces but also by bending and shear forces according to their flexural stiffness related to the panel stiffness (Fig. 3).

The magnitude of the flexural and diaphragm forces follows the theory of the elastic composite. Facings connected by a core with a low stiffness react decoupled to the loading (non-composite action, NCA, Seeber 1997). The relative shear deformations of the layers cause large deformations and a non-validity of the Bernoulli-Hypothesis (Fig. 4).

With an increasing shear modulus of the core the composite action of the upper and lower facing is more activated. This leads to decreasing deformations and for an infinite core stiffness to a full composite action (FCA, Seeber 1997). The relating stress and strain distributions for both NCA and FCA panels is illustrated in.

3.2 Sandwich components

3.2.1 Textile Reinforced Concrete

The development of TRC is based on the experiences with Glassfibre-Reinforced Concrete (GFRC). A drawback of the reinforcement with chopped strands is the partial unoriented distribution of the fibers over the total cross section, reducing their effectiveness. In contrast in TRC the fibers are aligned as two-dimensional fabrics in the direction of the tensile stresses. This leads to a higher utilization of the reinforcement material and better load-bearing properties of the component.

Textile reinforcements are made out of rovings which themselves consist of some hundreds to thousands of single fibers (filaments). The currently most favorable fiber materials are alkali resistant glass (AR-glass), carbon and aramid. Due to the multifilament assembly of the rovings and the small filament diameters ($\varnothing = 9$–$30 \, \mu m$) compared to the matrix particles varying bond conditions of filaments determine the load-bearing behaviour of the fibers and lead to an inhomogeneous utilization of the roving. A detailed survey of the investigations concerning the materials, the bond characterization and the load-bearing behaviour as well as their influence on the design of TRC is given in Brameshuber (2006) and Hegger & Voss (2006).

The applied fabrics have been produced out of AR-glass rovings, respectively, having a cross-sectional area of $0.44 \, mm^2$ (fabric 1) and $0.89 \, mm^2$

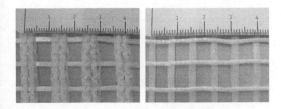

Figure 5. Used AR-glass fabrics: uncoated fabric 1 (left) and fabric 2 with prepreg epoxy resin (right).

Table 1. Geometric design of applied AR-glass fabrics.

		Fabric 1	Fabric 2
Type of Roving	(–)	Vetrotex LTR 5325, 1200 tex (0.44 mm²)	Vetrotex LTR 5325, 2400 tex (0.89 mm²)
Impregnation		–	Prepreg epoxy resin
Section of fabric			
0°-direction	(mm²/m)	71.5	108.0
90°-direction	(mm²/m)	36.0	105.0
Tensile strength in concrete	(MPa)	780	1395
Elastic Modulus	(MPa)	63000	59000

Table 2. Properties of used concrete mixtures (mean values at age of 28 days).

		Concrete 1	Concrete 2
Compressive strength	(MPa)	67.0	86.0
Tensile strength	(MPa)	2.5	4.0
Elastic Modulus	(MPa)	22700	34000

(fabric 2) each. Fabric 1 is used without a coating, fabric 2 is impregnated with a thermoset prepreg-epoxy resin. Figure 5 and Table 1 depict geometric and mechanical properties of the used fabrics.

The properties of the textile reinforcement lead to special demands on the concrete mixtures. In order to enable the penetration of the fabric mesh the maximum grain size is limited to about 2 mm and a high flowability is necessary.

Concrete 1 is a premix concrete with an amount of 2% per weight AR-glass short fibers which has been widely used as a GFRC for the fabrication of ventilated façade systems (Engberts 2006, Tab. 2).

Concrete 2 is the standard mixture (PZ-0899-01, Brockmann 2005) of the Collaborative Research Center 532 at RWTH Aachen University. Table 2 lists the major mechanical properties.

3.2.2 Core materials

Polyurethane (PU) foam is a highly crosslinked, low-density, cellular thermoset plastic composed of closed cells. PU is advantageously and widely used for

Table 3. Properties of applied PU cores (mean values, manufacturer information).

Property		PU 32	PU 40	PU 50	PU 200
Density	(kg/m³)	32	39	49	195
Compr. strength	(MPa)	0.19	0.29	0.38	2.9
Shear strength	(MPa)	0.15	0.19	0.23	1.5
Tensile strength	(MPa)	0.08 [1]	0.25	0.40	–
Elastic Modulus	(MPa)	4.5–5.5	6.5–8.5	10.0 [1]	70–80
Shear Modulus	(MPa)	1.4	2.0	–	17
Therm. conductivity	(W/mK)	0.023	0.023	0.024	0.038

[1] Derived from own measurements.

Table 4. Properties of applied XPS core (mean values, manufacturer information).

Property		XPS 32
Density	(kg/m³)	32
Compr. strength	(MPa)	0.25
Shear strength	(Mpa)	0.2
Tensile strength	(Mpa)	0.45
Elastic Modulus	(MPa)	10.0
Shear Modulus	(MPa)	7.0
Thermal conductivity	(W/mK)	0.036

the continuous production of sandwich panels with metal facings where the insulation material is foamed directly between the thin metal sheets. PU also can be obtained as prefabricated rigid foams as used in the conducted studies. Due to the cellular composition PU is a material with visco-elastic and anisotropic properties which have to be regarded for the design of elements with dead loads. The load-bearing capacity increases as well as the shear and elastic modulus with the density of the foam. In contrast the thermal conductivity and the sound insulation are affected adversely. For the conducted bending and shear tests rigid polyurethane foams (PU-slabstocks) of different densities as shown in Table 3 were applied.

Another suitable core material is extruded polystyrene (XPS), which is different from expanded polystyrene (EPS) and commonly known as Styrofoam. The combination of a low density, a sufficient thermal conductivity and a high load-bearing capacity (Tab. 4) makes it also a proper construction material for sandwich panels made with TRC.

3.3 Self-supporting sandwich panels

3.3.1 Production methods

For the experimental program prefabricated PU rigid foams as well as XPS foams ($h_c = 150$ mm) were

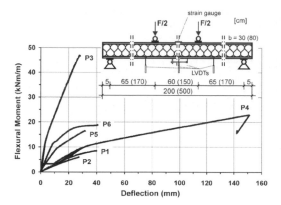

Figure 6. Load-deflection curves of panels P1 to P6.

Table 5. Selected results of test on sandwich beams.

Panel	Core	Interface	Span L_s (m)	Failure Load F_u (kN/m)	Deflection w (mm)
P1	PU 32	Notched	1.9	27.6	43.0
P2	PU 32	Glued	1.9	18.1	29.2
P3	PU 200	Notched	1.9	151.2[1]	28.0
P4	PU 40	Notched	4.9	23.2	99.8
P5	PU 50	Notched	1.9	49.0	32.0
P6	XPS 32	Notched	1.9	58.0	40.0

[1] tensile failure of fabrics 1 (5 layers) in lower facing.

used as a core being attached to the TRC-facings ($h_f = 15$ mm) either by gluing or by pressing a notched core into a fresh concrete layer. The notches were oriented perpendicular to the beam axis with an interspace of 5 cm. The concrete facings were produced in a lamination process in which concrete (concrete 1) and three fabrics (fabric 1) are alternately placed in the formwork.

3.3.2 Results of bending and shear tests

In four-point-bending tests on sandwich panels with spans of 1.90 m and 4.90 m (Fig. 6) a satisfactory load-bearing behavior was determined which mainly depends on the (shear) stiffness of the core material and the joint quality between core and facings.

All used PU cores were cut out of slabstocks and a fine dust of PU-cells covered the cutting edges. The dust could not be removed with compressed air nor be brushed off the surfaces. If the PU core was pressed into the fresh concrete the particles were easily bounded and had no noticeable influence on the bond quality. Compared to panels with direct bond, the inferior joint quality of the glued panels resulted in at least 30% lower ultimate loads (Fig. 6, Tab. 5).

The panel with the XPS core (P6) failed more ductile and at a 20% higher load lever than the panel with the soonest comparable PU 50 core.

Figure 7. Shear failure of panel P4.

The ultimate load in the tests was determined by a shear failure of the core (Fig. 7) except for panel P3 with a high density PU core failing by tensile rupture of the textile reinforcement in the lower facing.

For the panels P1 to P6 no connectors were used. The sandwich action was only established by the bond between the foam and the concrete layers.

3.3.3 Fire resistance and sound/heat insulation

A first examination of the fire resistance of the sandwich panels with PU core materials was conducted with a SBI (single burning item) test according to DIN EN 13823 (2002). The panels were categorized in the second highest class according to DIN EN 13501-1 (2002) as *A2/B, S1, d0* making them suitable for façades of office buildings and factory floors. The airborne sound insulation was determined in a testing facility according to DIN EN 140-3 (2005). The measured sound reduction index of $R'_w = 43$ dB is sufficient for factory floors and office buildings. The heat transfer coefficient U for the homogeneous section was assessed to 0.22 W/m^{2}K which complies with the limit ($U = 0.35$ W/m^{2}K) of the current German Energy Saving Regulation (EnEV 2001).

3.3.4 Design model

The test data was used to deduce a design model for the TRC-facings and the core in ultimate limit state.

Stamm & Witte (1974) deduced for sandwich beams with an arbitrary transverse loading $q(x)$ the following decoupled universally valid differential equations for the displacement w and the transverse strain γ

$$-\frac{B_u + B_o}{A} w^{VI} + \left(\frac{B}{B_S}\right) w^{IV} = \frac{q}{B_S} - \frac{q''}{A}, \quad (1)$$

$$-\frac{B_u + B_o}{A} \gamma^{IV} + \left(\frac{B}{B_S}\right) \gamma'' = -\frac{q'}{A}, \quad (2)$$

where

$$A_c = G_c \frac{b\,c^2}{h} \quad \text{and} \quad B = B_s + B_{f,u} + B_{f,l} \quad (3)$$

are the shear rigidity of the core (index c) and the bending stiffness of the panel. B is the total of

$$B_{f,u/l} = E_{f,u/l}\,I_{f,u/l} \quad \text{and} \quad B_S = \frac{E_{f,u}A_{f,u} \cdot E_{f,l}A_{f,l}}{E_{f,u}A_{f,u} + E_{f,l}A_{f,l}}c^2 \quad (4)$$

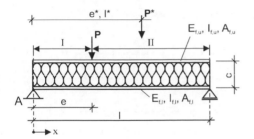

Figure 8. Static system for derivation of model.

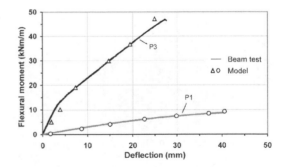

Figure 9. Comparison of derived model and tested beams P1/P3.

which are the bending rigidity of the facings (index f) and the sandwich action (index s) of the facings with distance c.

The differential equations (1) and (2) can be solved for the boundary conditions of a simple beam loaded with a single transverse load P at a distance e from support A (Fig. 8) to obtain the flexural moments and normal forces in the facings and the shear stress in the core. Due to the unsteadiness at the load point Stamm 1974 distinguished two fields I and II divided by the load P.

The normal and bending internal forces for field I evoked by P as a function of x are

$$N_{f,u,I} = \pm \frac{M_{sI}}{c} = \frac{Pl}{c} \frac{1}{1+\alpha} \psi , \quad M_{f,u,I} = P\,l\,\frac{\alpha_{u,o}}{1+\alpha} \psi \quad (5)$$

where

$$\psi = \left[(1-\varepsilon)\,\xi - \frac{sinh(\lambda(1-\varepsilon))}{\lambda\,sinh\,\lambda}\,sinh(\lambda\,\xi) \right], \quad (6)$$

$$\varepsilon = \frac{e}{l}, \quad \xi = \frac{x}{l}, \quad \lambda = \sqrt{\frac{1+\alpha}{\alpha\,\beta}}, \quad \alpha = \frac{B_o + B_u}{B_S}, \quad \beta = \frac{B_S}{A\,l^2}. \quad (7)$$

The associated deflection is

$$w_I = \frac{P\,l^3}{B}\left[\frac{1}{6}(1-\varepsilon)\,\xi\,(2\varepsilon - \varepsilon^2 - \xi^2) + \frac{1}{\alpha\,\lambda^2}\psi \right]. \quad (8)$$

To derive the deflection and the maximum internal forces of the conducted four-point bending tests, the internal forces and deformations caused by a second single load P^* with distance e^* have to be superposed to those of P e.g. with the aid of a spreadsheet. The superposition is valid due to the linear elastic calculation model.

With N_f and M_f the stresses of the facings are determined and the structural design of facings is accomplished. If stresses exceed the tensile strength of the matrix the cracking leads to a loss of rigidity and to an underestimation of the deflection. The reduced rigidity of the facings with bending cracks is assessed to

$$I_{ef} = \kappa\,\frac{b\,d^3}{12}, \quad (9)$$

where d is the distance of the concrete reinforcement to the top of the compression zone and κ is calculated as

$$\kappa = 4\,\xi^3 + 12\,\alpha_e\,\rho\,(1-\xi)^2, \quad \rho = \frac{A_{fabric}}{b\,d}, \quad (10)$$

$$\xi = -\alpha_e\,\rho + \sqrt{(\alpha_e\,\rho)^2 + 2\,\alpha_e\,\rho}, \quad \alpha_e = \frac{E_{fabric}}{E_{concrete}}. \quad (11)$$

Figure 9 illustrates the sufficient calculation accuracy of the model for the tested short beams P1 and P3.

The shear stress of the core evoked by a single load P is denoted by Stamm & Witte (1974) as

$$\tau_c = \frac{P}{c \cdot b}\,\frac{1}{1+\alpha}\,(1 - \varepsilon - \frac{sinh\lambda(1-\varepsilon)}{sinh\,\lambda}\,cosh\lambda\,\xi). \quad (12)$$

The cracking of the concrete facings causes a shear transfer from the facings to the core which has to be considered for a secure design of the core.

4 SANDWICH PANELS FOR MODULAR BUILDINGS

The advantages of the sandwich technology are also applied to the design study of a modular building consisting of load-bearing and demountable sandwich panels for walls and roofing. Based on a basic grid of 1 m 12 wall (clear height: 2.82 m) and 4 roof elements (span 4.73 m) are assembled to a small prototype building (Figure 10).

4.1 Design of modular panels

The sandwich panels were designed with the theory of the elastic composite (section 3.3.4) and a comparative linear finite element analysis. The inner TRC layer (concrete 2, Tab. 2) of the roof elements was profiled and a PU 50 (Tab. 3) was chosen to reduce the

191

Figure 10. Assembly of modular prototype sandwich building.

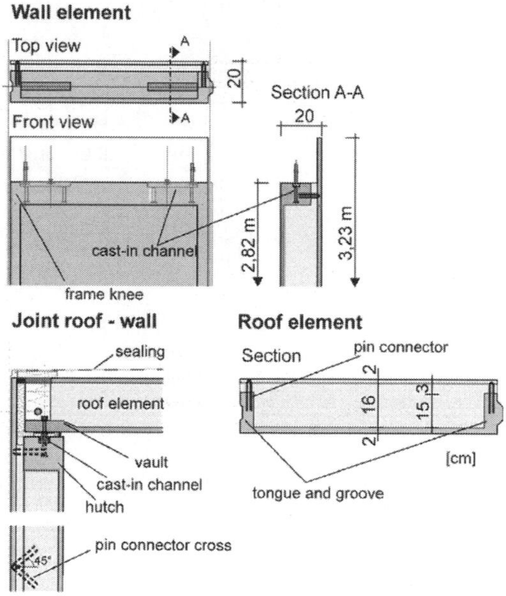

Wall element

Top view

Section A-A

Front view

cast-in channel

frame knee

2,82 m

3,23 m

Joint roof - wall

sealing

roof element

vault

cast-in channel

hutch

pin connector cross

Roof element

Section

pin connector

tongue and groove

[cm]

Figure 11. Sections, mounting devices and connectors of modular panels.

shear portion and the thus induced creep deformations of the visco-elastic core material (Fig. 11).

The section of the inner layers of the wall elements (concrete 2, Tab. 2) was designed similar to the roof elements but forming a hutch with additional horizontal beams at the top and the bottom. In the horizontal beams cast-in channels connecting the wall elements to the foundation and the roof elements were integrated.

The sides of the vertical webs of the inner layers were profiled as tongue and groove for both roof and wall elements.

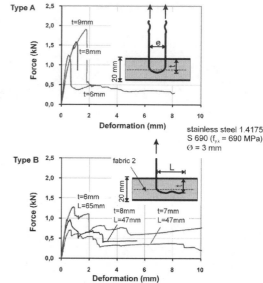

Figure 12. Pullout tests on pin connectors with different anchorage systems (concrete age 7d).

4.2 Connecting devices

Temperature, shrinkage and wind suction apply normal stresses to the bond area of the concrete layers and the core. Thus, connecting devices are necessary to ensure a durable connection between the sandwich layers and a sustainable sandwich action. To avoid a fatigue of the bond action and to bear the normal forces 8–10 pin connectors ($\emptyset = 3$ mm) made of stainless steel with a low bending and shear stiffness are applied in wall and roof elements (Figs. 11–12). In the wall elements the dead load of the outer thin concrete layer is transferred by the foam and a pin connector cross to the profiled inner structural layer.

The load-bearing capacity of pin connectors with different anchorage geometries and small anchorage lengths was determined in pullout-tests after 7 days of curing ($f_c = 59{,}8$ MPa, Fig. 12). Based on the test results type A was chosen for the production of test specimen and the modular elements for the prototype building.

4.3 Tailoring of 3D textile reinforcement

For a simple cast process a capable and rigid 3D AR-glass reinforcement (fabric 2, Tab. 1) was tailored (Fig. 13).

The fabrics were laminated with a thermoset prepreg epoxy resin (Hexion, Germany) and cured on a metal form in an oven to shape them to e.g. rectangular reinforcement cages. These were fixed to the cnc-milled cores and in the knee points the transverse

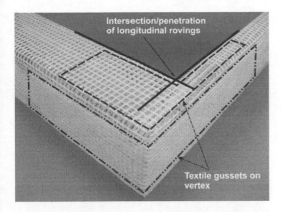

Figure 13. Tailored AR-glass reinforcement for frame knees.

Figure 14. Casting of modular roof panels in precast plant.

rovings of the crossing cages were removed to enable a penetration of the longitudinal rovings. Additionally placed textile gussets supported a sufficient frame knee action (Fig. 11) in the vertexes.

The elements were cast upside down in a three step production method: (1) lamination of the upper plain layer with concrete 2, an additional amount of 1% per weight short fibers and one layer of fabric 2 (Tab. 2), (2) positioning of the core assembled with the tailored reinforcement and the lateral formwork and (3) casting of the lower profiled layer with concrete 2 (Tab. 2, Fig. 14).

4.4 *Bending tests on modular roof and wall elements*

In Figure 15 the setup of the bending tests on a roof element (P7) and a wall element (P8) is illustrated.

The roof element was tested with a positive flexural load and supported true-to-detail, the wall element was loaded by a negative flexural load and supported on the horizontal beam of the inner hutch (Fig. 15). The outside concrete facing thus was shortened and able to deform without any constraint.

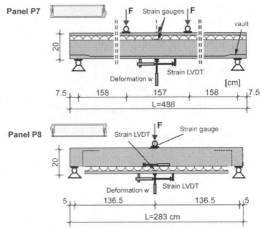

Figure 15. Setup for bending tests on roof/wall element (P7/P8).

Table 6. Results of bending tests on sandwich panels P7/P8.

Panel	Core	Span L_s (m)	Failure load F_u (kN/m)	Deflection w (mm)
P7	PU 50	4.73	28.0	108.0
P8	PU 50	2.73	34.6	38.9

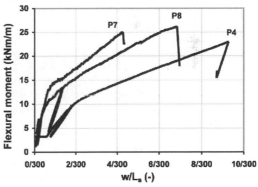

Figure 16. Comparison of load-bearing behaviour of panels P4, P7 and P8.

Table 6 and Figure 16 show the results obtained by the bending tests on panels P7 and P8.

In comparison to panel P4 the profiled inner concrete layer and the slightly higher core density led to a much stiffer load-deflection curve (Fig. 16) of panels P7 and P8.

The ultimate load of the roof element calculated in the design stage was exceeded due to (a) a shear block action caused by the vault (Fig. 15) of the inner facing

193

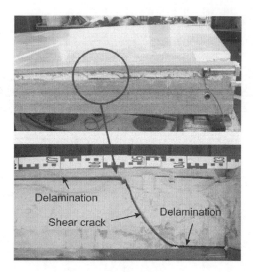

Figure 17.   Shear failure of panel P7.

and the compressive stresses at the supports and (b) a non negligible influence of the non-linear material properties of the foam encaged by the profiled lower concrete facing.

Panels P7 and P8 both failed due to shear rupture of the PU core leading to a subsequent delamination between concrete facings and the core (Fig. 17).

5   SUMMARY AND CONCLUSION

The presented investigations proved TRC to be a proper and capable construction material with a high adaptivity to the requirements of light-weight and filigree building components. The potential of TRC compliments the utilization of GFRC and broadens the application of load-bearing structures of complex geometry. In addition to simple joining techniques and static dimensioning models the basis is formed for the development of future constructions with optimized concrete sections, sharp edges and excellent concrete surfaces. Combined with the manufacturing as precast elements and the entailed simple assembly and disassembly of buildings also the demand for a sustainable method of construction is fulfilled.

ACKNOWLEDGMENTS

The authors gratefully acknowledge the financial support of the Deutsche Forschungsgemeinschaft (DFG) within the Collaborative Research Center (SFB) 532 "Textile Reinforced Concrete – Development of a new technology" and thank BAYER MaterialScience AG as well as HEXION, Germany, for the financial and technical support.

REFERENCES

Brameshuber, W. 2006. Textile Reinforced Concrete. State-of-the-Art Report of RILEM Technical Committee 201-TRC. RILEM Report 36, RILEM Publications, Bagneux, France, 2006.

Brockmann, T. 2005. Mechanical and fracture mechanical properties of fine grained concrete for Textile Reinforced Composites. PhD thesis, Institute of Building Materials Research (Institut für Bauforschung), RWTH Aachen University, 2005.

CEB-FIP Model Code 1990. 1993. Thomas Telford Ltd., London, 1993.

DIBt 2004. Deutsches Institut für Bautechnik (DIBt): General approval: Concrete façade panel with backside mounting devices for the utilization as ventilated cladding. Approval-No. Z-33.1-577, 2004.

DIN 1045-1. 2001. Concrete, reinforced and prestressed concrete structures – Part 1: Design and construction. Beuth Verlag, Berlin, 2001.

DIN EN 13501-1. 2002. Fire classification of construction products and building elements – Part 1: Classification using test data from fire reaction to fire tests; German version EN 13501-1:2002.

DIN EN 13823. 2002. Reaction to fire tests for building products – Building products excluding floorings exposed to the thermal attack by a single burning item; German version EN 13823:2002.

DIN EN ISO 140-3. 2005. Acoustics – Measurement of sound insulation in buildings and of building elements – Part 3: Laboratory measurements of airborne sound insulation of building elements (ISO 140-3:1995 + AM 1:2004); German version EN 20140-3:1995 + A1:2004, 2005.

EnEV. 2001. German Energy Saving Ordinance, 2001.

Engberts, E. 2006. Large-size façade elements of Textile Reinforced Concrete. In: Textile Reinforced Concrete. Proceedings of the 1st International RILEM Symposium (ICTRC), Aachen, 05.-09.09.2006, RILEM Proceedings 50, pp. 297–307.

Hegger, J. & Voss, S. 2006. Dimensioning of textile reinforced concrete structures. Proceedings of the 1st International RILEM Symposium (ICTRC), Aachen, 05.-09.09.2006, RILEM Proceedings 50, pp. 151–160.

Hegger, J. & Voss, S.. 2005. Thin facade panels made of textile reinforced concrete. In: JEC-Composites, No.15, February-March 2005, S. 36–37.

Hegger, J., Will, N. et al. 2006. Applications of Textile Reinforced Concrete Textile. In: Brameshuber, W. State-of-the-Art Report of RILEM Technical Committee 201-TRC. RILEM Report 36, RILEM Publications, Bagneux, France, 2006.

Seeber, K. et al. 1997. State-of-the-Art of Precast/Prestessed Sandwich Wall Panels. PCI Committee Report, PCI Journal 04/1997, pp. 93–134.

Stamm, K. & Witte, H. 1974. Sandwichkonstruktionen: Berechnung, Fertigung, Ausführung. Springer-Verlag, 1974.

Excellence in Concrete Construction through Innovation – Limbachiya & Kew (eds)
© 2009 Taylor & Francis Group, London, ISBN 978-0-415-47592-1

Thermo-mechanical properties of HSC made with expanded perlite aggregate

M.B. Karakoç & R. Demirboğa
Ataturk University, Eng. Faculty, Civil Eng. Dept., Erzurum, Turkey

ABSTRACT: The effects of curing conditions and expanded perlite aggregate (EPA) on the high strength concrete (HSC) properties were studied at 28 days. Time depended dry and wet curing effects on the thermal conductivity (TC) were compared. Five different concrete mix designs with the same mix proportions and different EPA replacements of aggregate were used: 0% (control), 7.5%, 15%, 22.5% and 30% EPA. The effects of EPA replacement and curing conditions upon concrete properties were examined. The properties examined included compressive strength (CS), TC, ultrasound pulse velocity (UPV) and oven dry density (ODD) properties of HSC containing EPA. CS, TC, UPV and ODD of HSC were decreased with increasing of EPA. Dry curing reduced the CS compared to the wet curing condition in this study.

1 INTRODUCTION

TC has been defined as 'The rate of flow of heat per unit area per unit temperature gradient when heat flow is under steady state conditions' (Tennent, 1997). The TC of porous materials depends upon various parameters, such as the thermal properties of the constituent phases and the microstructural parameters, which include the volume fractions of the constituent phases, geometrical distribution of the phases, the size and size distribution of the particles and the geometry of the pore structure (Zumbrunnen et al. 1986, Loeb 1954, Francl & Kingery 1954, Cheng & Vachon 1969, Prakouras et al. 1978).

Accordingly, TC values have been chosen with the abovementioned range without much consideration. However, TC of concrete is greatly affected by mix proportioning (Morabito 1989, Neville 1995, Lanciani et al. 1989, Shin et al. 2002), aggregate types (Morabito 1989, Neville 1995, Lanciani et al. 1989, Khan 2002, Marshall 1972, Zoldners 1971, Cambell-Allen & Thorne 1963, Harmathy 1970, Khan & Bhattacharjee,1995, USBR, 1940, Bhattacharjee & Krishnamurthy, under revision, Ashworth 1991, FIP 1978), mineralogical character of the aggregate (Neville 1995, Zoldners 1971, Harmathy 1970) and aggregate sources (Morabito 1989, Neville 1995, Lanciani et al. 1989), as well as moisture status (Morabito 1989, Neville 1995, Lanciani et al. 1989, Shin et al. 2002, Khan 2002, Marshall 1972, Chambell-Allen & Thorne 1963, Harmathy 1970, Khan & Bhattacharjee, 1995, USBR 1940, Ashworth 1991, FIP 1978, Short & Kinniburg 1978, Xu &

Chung 2000, Bomberg & Shirtliffe, 1978, Vries 1987, Sandberg 1983, Sandberg 1995, Schnider 1982, cement content (Gül et al. 1997, Örüng 1996) porosity (Marshall 1972, Cambell-Allen & Thorne 1963, Harmathy 1970, Khan & Bhattacharjee,1995, USBR, 1940) and unit weight in the dry state (Morabito 1989, Neville 1995, Lanciani et al. 1989, Shin et al. 2002, Steiger & Hurd 1978, Brewer 1967). Lu-Shu et al. (1980) mathematically modeled a relationship between density and TC.

There are many studies related to the UPV; for example, Demirboğa et al. (2004) reported relationship between UPV and CS of mineral-admixtured concrete, Tharmaratnam & Tan (1990) provided the empirical formula of the combined UPV and ultrasonic pulse amplitude (UPA). Liang & Wu (2002) studied theoretical elucidation of the empirical formulae for the UPV and UPA and combined methods. Ye et al. (2004) determined the development of the microstructure in cement-based materials by means of numerical simulation and UPV.

2 MATERIALS AND METHODS

ASTM C150 Type I Portland cement (PC), with the chemical composition shown in Table 1, was utilized in preparing the concrete specimens. Silica fume (SF) and EPA were obtained from Antalya Electro Metallurgy Enterprise and Etibank Perlite Expansion Enterprise in Izmir, Turkey, respectively. The chemical composition of the materials used in this study was summarized in Table 1. The physical and mechanical

Table 1. Chemical composition of PC, SF and EPA (%).

Component	PC	SF	EPA
SiO_2	19.94	93.7	71–75
Al_2O_3	5.28	0.3	12–16
Fe_2O_3	3.45	0.35	–
CaO	62.62	0.8	0.2–0.5
MgO	2.62	0.85	
SO_3	2.46	0.34	
C	–	0.52	
Na_2O	0.23	–	2.9–4.0
K_2O	0.83	–	
Chlor (Cl^-)	0.0107	–	
Sulphide (S^{-2})	0.17	0.1–0.3	
Undetermined	0.08		
Free CaO	0.51		

Table 2. The physical and mechanical properties of PC.

Specific gravity (g/cm^3)		3.13
Specific surface(cm^2/g)		3410
Remainder on 200-mm sieve (%)		0.1
Remainder on 90-mm sieve (%)		3.1
Setting time initial (min)		2.10
Setting time final (min)		3.15
Volume expansion (Le Chatelier, mm)		3
Compressive strength (MPa)	2 days	23.5
	7 days	35.3
	28 days	47.0
Flexural strength (MPa)	2 days	5.0
	7 days	6.2
	28 days	7.7

properties of PC summarized similarly in Table 2. The specific gravity of SF and EPA were 2.18 and 0.28, respectively. The percentages of EPA that replaced fine aggregate in this study were: 0, 7.5, 15, 22.5 and 30 percent. The binder dosage, water to cement ratio, and the dosage of SF was kept constant at 500 kg/m³, 0.25 and 7%, respectively, throughout the study. Sulphonate naphthalene formaldehyde was used as a superplasticizer, it conformed to Type F of ASTM C494 F (high-range water reducer) at a dosage of 2.0 ml/kg of cement.

The cement, sand, and coarse aggregates were normally blended first and the SF, superplasticizer, and water were then added. The lightweight aggregate (EPA) was first mixed, with water needed for dry surface-saturated for half an hour before blending, with cement, sand and coarse aggregate. Mixing was done in a revolving drum type mixer for approximately 3–5 min to obtain uniform consistency. After mixing, the concrete was filled in the cylindrical moulds in two layers and consolidated by vibrator to remove entrapped air. For each mixture, 70 mm × 70 mm × 280 mm prisms for TC

and ODD, and 100 mm diameter × 200 mm height cylinders molds were used for CS and UPV. After casting, the specimens were covered with wet burlap and cured in the laboratory at a temperature of 20 C for 24 h prior to demolding. A group of samples exposed to the air curing and the other cured under calcium hydroxide solution till the time of test. The maximum size of coarse aggregate was 16 mm.

The samples were tested at 28-day for UPV, CS and ODD in accordance with ASTM C 597-83, ASTM C 39 and ASTM C 332 respectively. In addition, TC of both air and wet cured samples were determined, at 3, 7, 28 and 90 days, by a Quick Thermal Conductivity Meter based on ASTM C 1113-90 Hot Wire Method and detailed elsewhere (Moore et al. 1969, Daire & Downs 1980, Willshee 1980, Sengupta et al. 1992, Demirboğa 2003).

Dry curing is defined as samples were air cured at uncontrolled temperature and relative humidity until the test age and similarly, the wet curing is defined as samples were immersed in 20 ± 3°C lime-saturated water until the test age.

3 TEST RESULTS AND THEIR EVALUATIONS

3.1 Oven dry density

Tables 3–4 show that ODD decreased with increasing EPA ratios. The lowest ODD value was observed at 30% replacement of EPA and it was 2108 kg/m³. Maximum reduction in the ODD due to EPA was 12 percent. It can be attributed to the lower density of EPA, grading of the aggregate and their moisture content and compaction of the concrete (Neville 1995).

Reduction in the ODD densities due to the dry curing condition was higher than that of wet curing. The maximum difference was around 5 percent. It may be due to the evaporation of the mix water from samples cured in dry curing condition.

3.2 Compressive strength

Tables 3–4 present the test results of CS due to the changing of EPA ratios and curing conditions. It was observed that, CS decreased with increasing EPA ratios. Reductions at 28-day for both wet and dry curing periods were 12, 13, 17 and 33, and 16, 19, 30 and 30 percent for 7.5%, 15%, 22.5% and 30% EPA, respectively. This was owing to the porous and weak structures of EPA (2001). Ramamurty & Narayanan (2000), Gül et al. (1997), Akman & Tasdemir (1977) and Faust (2000) reported that the decrease in density causes reduction of the CS. Zhou et al. (1998) reported that the replacement of the normal aggregate in HSC with the lightweight aggregate results in a 22% reduction in the CS. Goble & Cohen (1999) have concluded that the sand surface area has

Table 3. Hardened concrete properties of samples in wet curing.

Samples	EPA ratios (%)				
	0	7.5	15	22.5	30
Compressive strength, (MPa)	80.77	70.72	70.52	67.20	53.97
ODD, (kg/m³)	2388	2341	2316	2251	2210
UPV, (m/s)	4681	4599	4507	4436	4410
3 days TC, (W/mK)	1.3947	1.2601	1.2089	1.0526	1.0403
7 days TC, (W/mK)	1.4211	1.3351	1.2918	1.1994	1.1711
28 days TC, (W/mK)	1.5189	1.3511	1.3429	1.2437	1.2204
90 days TC, (W/mK)	1.5412	1.3856	1.3753	1.2529	1.2335

Table 4. Hardened concrete properties of samples in dry curing.

Samples	EPA ratios (%)				
	0	7.5	15	22.5	30
Compressive strength, (MPa)	56.77	47.62	46.10	39.64	39.62
ODD, (kg/m³)	2325	2278	2247	2133	2108
UPV, (m/s)	4377	4263	4193	4080	4020
3 days TC, (W/mK)	1.2241	1.1537	1.0537	0.9954	0.9606
7 days TC, (W/mK)	1.2309	1.1787	1.1151	1.0539	1.0260
28 days TC, (W/mK)	1.2408	1.1886	1.1468	1.0606	1.0394
90 days TC, (W/mK)	1.2604	1.2100	1.1570	1.0699	1.0482

a significant influence on the mechanical properties of Portland cement mortar. Demirboğa et al. (2001) reported results of an extensive laboratory study evaluating the influence of EPA and mineral admixtures on the CS of low density concretes. They concluded that the addition of mineral admixtures increased the CS of concrete produced with EPA. They reported that the CS decreased because the density decreased with increasing lightweight aggregate ratio instead of the normal aggregate.

The moisture conditions in the concrete affect both the mechanical properties and the hydration of the cement. Thus, the CS of the concrete will be affected. As it can be seen from Tables 3–4 that, CS of samples cured in dry curing condition were reduced 30, 33, 35, 41 and 27 percent for 0%, 7.5%, 15%, 22.5% and 30% EPA, respectively, when compared to the wet cured ones. This observation indicates that in dry curing, a

significant part of cement remains unhydrated due to unavailable water to develop the hydration products because evaporation of water from sample is very high.

3.3 Ultrasonic pulse velocity

The UPV results of the high strength concretes made with EPA were determined at 28 days. Tables 3–4 show that EPA reduced UPV values of the concretes at all levels of replacement at 28 days in wet and dry curing conditions. However, reduction in UPV values due to EPA replacement was much lower than that of CS. In other words, the UPV took a shorter time to reach a plateau value for HSC containing EPA when compared to the CS (Demirboğa 2004). Jones (1954) reported the relationship between the pulse velocity and age. He showed that velocity increases very rapidly initially but soon flattens. UPV values changed between 4681 and 4020 m/s. Maximum reduction occurred at 30% EPA replacement and it was 6 and 8 percent at the 28 days at wet and dry curing conditions, respectively. The difference in the reduction of the UPV values for wet and dry curing conditions was 2 percent. Jones (1954) reported that a 4 to 5% increase in pulse velocity can be expected when dry concrete with high w/c ratio is saturated. In this study, w/c ratio was rather low and wet cured samples were unsaturated. Thus, reduction percent in UPV due to dry curing was lower than that of reported by Jones. Kaplan (1958), also, found that the pulse velocity for laboratory-cured specimens were higher than for site-cured specimens.

3.4 Thermal conductivity

As it can be seen from Tables 3–4 that TC results of the high strength concretes made with different EPA ratios were determined at 3, 7, 28 and 90 days.

3.4.1 Effect of EPA on TC
The variation of TC of concrete with EPA is shown in Tables 3–4. This table shows that the highest value of TC of concrete is obtained for specimens produced without EPA. The maximum reduction in TC of concrete occurred at the maximum EPA (30%) replacement of fine aggregate. For 7.5%, 15%, 22.5% and 30% EPA replacement, keeping other conditions constant; the reductions were 10, 13, 25 and 25 percent, 6, 9, 16 and 18 percent, 11, 12, 18 and 20 percent, and 10, 11, 19 and 20 percent, for 3, 7, 28 and 90 days, respectively, compared to the corresponding control specimens at wet curing. For 7.5%, 15%, 22.5% and 30% EPA replacement, keeping other conditions constant; the reductions were 6, 14, 19 and 22 percent, 4, 9, 14 and 17 percent, 4, 8, 15 and 16 percent, and 4, 8, 15 and 17 percent, for 3, 7, 28 and 90 days, respectively, compared to the corresponding control specimens at dry curing. This is because the density decreased with increasing EPA content. The reduction in density of

concrete by means of EPA is probably related to the increasing of porosity due to the addition of EPA in fine aggregate (Ramamurty & Narayanan (2000). Losiewicz et al. (1996), Demirboğa & Gül (2003), Gül et al. (1997), Demirboğa (2003), Uysal et al. (2004), Akman & Tasdemir (1977) and Blanco et al. (2000) also reported that the TC decreased due to the density decreasing of concrete, which results in an increase void content. Demirboğa & Gül (2003) reported that EPA reduced the TC of samples up to 43.5%. Wang & Tsai (2006) reported that TC of the concrete is related to the apparent density: the lower the aggregate density, the smaller the TC coefficient will be.

3.4.2 Effect of curing time and conditions on TC

TC values of HSC made of EPA increased with increasing curing periods. Comparing 3 days samples' TC with 7, 28 and 90 days, the increment in TC in wet curing were 2%, 9% and 11%; 6%, 7% and 10%; 7%, 11% and 14%; 14%, 18% and 19%; and 13%, 17% and 19% due to 0, 7.5, 15, 22.5, and 30 percent EPA, respectively. Comparing 3 days samples' TC with 7, 28 and 90 days, the increment in TC in dry curing were 1%, 1% and 3%, 2%, 3% and 5%, 6%, 9% and 10%, 6%, 7% and 7%, and 7%, 8% and 9% due to 0, 7.5, 15, 22.5, and 30 percent EPA, respectively. Thus, we can be seen that in this study with increasing curing period, TC values of samples increased little. TC of control samples were due to the curing period was negligible for both dry and wet curing periods. Kim et al. (2003) reported that thermal conductivities of cement, mortar, and concrete mixtures were revealed independent of curing age. Blanco et al. (2000) reported that TC remained almost constant after only 5 to 28 days of curing. However, increasing with EPA ratios differences in TC due to the curing period increased up to 19 percent. The higher TC of EPA concretes in comparison with those of control samples were probably a consequence of the slow hydration at short curing time. Increment in TC for 28 and 90 days of curing time were nearly the same for all replacement levels of EPA. It can be concluded that TC for EPA concrete independent of days of curing after 28 days. The curing condition was also investigated. The TC values measured in wet conditions were higher than those of dry conditions for all level of EPA replacement. Kim et al. (2003) reported that, TC is dramatically increasing as the status changes from fully saturated to fully dried. This is attributed to changes in air voids filled with water, whose TC is superior to that of air, therefore accounting for the effect of the degree of saturation on TC.

4 CONCLUSIONS

The EPA replacement of fine aggregate decreased CS, UPV, TC and ODD values of samples. Results showed that the CS, UPV, TC and ODD of samples at both wet and dry curing conditions were decreased %33, %6, %20 and %7, and %30, %8, %16 and %9 for 28 days, respectively. The increasing of curing periods improved TC. TC values samples in wet curing condition were also higher than those of dry curing condition. It also can be concluded that TC for concrete containing EPA independent of days of curing after 28 days.

REFERENCES

A. M. Neville, Properties of concrete. UK: Longman, 1995.
A. L. Marshall, The thermal properties of concrete, Building Science, 7 (1972) 167–174.
A. Lanciani, P. Morabito, P. Rossi, Measurement of the thermophysical properties of structural materials in laboratory and in site: methods and instrumentation, High Temp. High Press. 21 (1989) 391–400.
A. Short, W. Kinniburg, Lightweight Concrete, Galliard (Printers), Great Yormouth, Great Britain, 1978, p. 113.
ASTM C 1113-90, Test method for thermal conductivity of refractories by hot wire (Platinum Resistance Thermometer Technique).
Bhattacharjee B, Krishnamurthy S. A model for thermal conductivity of porous building materials, under revision.
Bomberg M and Shirtliffe C J 1978 Thermal Transmission Measurements of Insulation ASTM STP 660 211.
Boulder Canyon Project Report. Thermal properties of concrete. Final Report by USBR, Bulletin No. 1, Part VII, 1940.
Brewer, H.W., 1967. General relation of heat flow factors to the unit weight of concrete. Journal of Portland Cement Association, Research and Development Laboratories 9 (1), 48–60.
C. Goble, M. Cohen, Influence of aggregate surface area on mechanical properties of mortar, ACI Mater. J. 96 (6) (1999) 657–662.
Cheng S G and Vachon R L 1969 Int. J. Heat Mass Transfer 12 537
D. Cambell-Allen, C.P. Thorne, The thermal conductivity of concrete, Magazine of Concrete Research, 15 (43) (1963) 39-48.
De Vries D A 1987 Int. J. Heat Mass Transfer 30 1343.
F. Blanco, P. Garcia, P. Mateos, J. Ayala, Characteristics and properties of lightweight concrete manufactured with cenospheres, Cem. Concr. Res. 30 (2000) 1715–1722.
F.P. Zhou, R. V. Balendran and A. P. Jeary, Size effect on flexural, splitting tensile, and torsional strengths of high-strength concrete, Cement and Concrete Research, 28 (1998) 1725–1736.
FIP State of Art Report, Principles of Thermal Insulation with respect to Lightweight Concrete, FIP/8/1, C&CA, Slought, England, 1978.
Francl J and Kingery W D 1954 J. Am. Ceramic Soc. 37 99.
G. Ye, P. Lura, K. van Breugel, A.L.A. Fraaij, Study on the development of the microstructure in cement-based materials by means of numerical simulation and ultrasonic pulse velocity measurement, Cement and Concrete Composites, Vol. 26, 2004 (5), 491–497.
H. Uysal, R. Demirboğa, R. Şahin and R. Gül, The effects of different cement dosages, slumps, and pumice aggregate

ratios on the thermal conductivity and density of concrete, Cement and Concrete Research 34 (2004) 845–848.

H.Y. Wang, K.C. Tsai, Engineering properties of lightweight aggregate concrete made from dredged silt, Cement and Concrete Composites, 28, 2006 (5), 481–485.

İ. Örüng, A research on usage possibilities of ground lightweight aggregate in agricultural buildings, Atatürk Univ. Ziraat Fakultesi Dergisi, Turkey 26 (1) (1996) 90–111.

J.C. Willshee, Comparison of thermal conductivity methods, Proc. Br. Ceram. Soc. 29 (1980) 153.

Jones, R., Non-Destructive Testing of Concrete, Cambridge University Press, London, 1962.

Jones, R., Testing of concrete by an ultrasonic pulse technique, RILEM Int. Symp. on Nondestructive Testing of Materials and Structures, Paris, Vol. 1, Paper No. A-17 January 1954, 137. RILEM Bull., 19(Part 2), Nov. 1954.

K. Lu-shu, S. Man-qing, S. Xing-Sheng, L. Yun-xiu, Research on several physicomechanical properties of lightweight aggregate concrete, Int. J. Lightweight Concr. 2 (4) (1980) 185–191.

K. Sengupta, R. Das, G. Banerjee, Measurement of thermal conductivity of refractory bricks by the nonsteady state hot-wire method using differential platinum resistance thermometry, J. Test. Eval. 29 (6) (1992) 455–459.

K. Tharmaratnam, B.S. Tan, Attenuation of ultrasonic pulse in cement mortar, Cem. Concr. Res. 20 (1990) 335–345.

Kaplan, M.F., Compressive strength and ultrasonic pulse velocity relationships for concrete in columns, ACI J., 29(54–37), 675, 1958.

KH Kim, SE Jeon, JK Kim, S Yang, An experimental study on thermal conductivity of concrete, Cement and Concrete Research 33 (2003) 363–371.

Khan MI, Bhattacharjee B. Relationship between thermal conductivities of aggregate and concrete. In: Reddy RR, editor. Civil engineering materials and structures. Osmania University Hyderabad, India: 1995. p. 162–6.

KY Shin, SB Kim, JH Kim, M Chung, PS Jung, Thermophysical properties and transient heat transfer of concrete at elevated temperatures, Nuclear Engineering and Design 212 (2002) 233–241.

Loeb A L 1954 J. Am. Ceram. Soc. 37 96.

M. I. Khan, Factors effecting the thermal properties of concrete and applicability of its prediction models, Building and Environment, 37 (2002) 607–614.

M. Losiewicz, DP. Halsey, SJ Dews, P. Olomaiye and FC. Harris, An investigation into the properties of micro-sphere insulating concrete, Construction and Building Materials, Vol. 10, No. 8, 583–588, 1996.

M.S. Akman, M.A. Taşdemir, Taşıyıcı Malzeme Olarak Perlit Betonu (Perlite concrete as a structural material), 1st National Perlite Congress, Ankara, Turkey, 1977.

M.T. Liang, J. Wu, Theoretical elucidation on the empirical formulae for the ultrasonic testing method for concrete structures, Cem. Concr. Res. 32 (2002) 1763–1769.

Moore, J.P., Stradley, J.G., Graves, R.S., Hanna, J.H., McElroy, D.L., 1969. Some Thermal Transport Properties of a Limestone Concrete. ORNL-TM-2644, Oak Ridge National Laboratory.

N. G. Zoldners, Thermal properties of concrete under sustained elevated temperatures, ACI Publication SP-25, 1971, pp. 1–31.

P. Morabito, Measurement of thermal properties of different concretes, High Temp., High Press. 21 (1) (1989) 51–59.

Prakouras A G, Vachon R I, Crane R A and Khader M S 1978 Int. J. Heat Mass Transfer 21 1157.

R. Demirboğa, and R. Gül, The effects of expanded perlite aggregate, silica fume, and fly ash on the thermal conductivity of lightweight concrete, Cement and Concrete Research, 33 (5) (2003) 723–727.

R. Demirboga, I. Orung, R. Gul, Effects of expanded perlite aggregate and mineral admixtures on the compressive strength of low-density concretes, Cem. Conc. Res., 31 (11), (2001), 1627–1632.

R. Demirboğa, I. Türkmen, M. B. Karakoc, Relationship between ultrasonic velocity and compressive strength for high-volume mineral-admixtured concrete, Cement and Concrete Research 34 (2004) 2329–2336.

R. Demirboğa, Influence of mineral admixtures on thermal conductivity and compressive strength of mortar, Energy build. 35 (2003), 189–192.

R. Gül, H. Uysal, R. Demirboğa, Kocapınar Pomzası ile Uretilen Hafif Betonların ısı Iletkenliklerinin Araştırılması (Investigation of the Thermal Conductivity of Lightweight Concrete Made with Kocapınar's Pumice Aggregate), Advanced in Civil Eng.: III. Technical Congress, vol. 2, METU, Ankara, Turkey, 1997, pp. 553–562 (In Turkish).

R. Gül, R., Şahin, R. Demirboğa, Investigation of The Compressive strength of Lightweight Concrete Made with Kocapınar's Pumice aggregate (Kocapınar Pomzası ile Üretilen Hafif Betonların Basınç Dayanımlarının Araştırılması), Advanced in Civil Eng. III. Technical Congress, Vol. 3, METU, Ankara, Turkey, 1997, pp. 903–912. (In Turkish).

R.W. Steiger, M.K. Hurd, Lightweight insulating concrete for floors and roof decks, Concr. Constr. 23 (7) (1978) 411–422.

Ramamurty, K. and Narayanan, N., Factors influencing the Density and Compressive Strength of Aerated Concrete. Mag. Concr. Res. (52) (3), 2000, 163–167.

Sandberg P I 1983 Report Swedish National Testing Institute.

Sandberg P I 1995 J. Thermal Insul. Bldg. Env. 18 276.

T. Ashworth, E. Ashworth, in: R.S. Graves, D.C. Wysocki (Eds.), Insulation Materials: Testing and Applications, vol. 1116, ASTM STP, Philadelphia, 1991, pp. 415–429.

T. Faust, Properties of different matrix and LWAs their influences on the behavior of structural LWAC, Second International Symposium on Structural Lightweight Aggregate Concrete, 18–22 June. Kristiansand, Norway, 2000, 502–511.

T. Z. Harmathy, Thermal properties of concrete at elevated temperatures, Journal of Materials, (1970) 5:47–74.

Tennent RM, editor. Science Data Book. Edinburgh: Oliver and Boyd; 1997.

U. Schnider, Behavior of concrete at high temperatures, Deutscher Aussehuss fürstahlbeton, Heft 337, Berlin, 1982.

W.R. Daire, A. Downs, The hot wire test—a critical review and comparison with B 1902 panel test, Trans. Br. Ceram. Soc. 79 (1980) 44.

Y. Xu, D.D.L. Chung Cement of high specific heat and high thermal conductivity, obtained by using silane and silica fume as admixtures, Cement and Concrete Research 30 (2000) 1175–1178.

Zumbrunnen D, Viskanta R and Incropera F P 1986 Int. J. Heat Mass Transfer 29 75.

Composition and microstructure of fly ash geopolymer containing metakaolin

K. Pimraksa
Department of Industrial Chemistry, Chiang Mai University, Thailand

T. Chareerat
Department of Civil and Environmental Engineering, Kasetsart University, Thailand

P. Chindaprasirt
Department of Civil Engineering, Khon Kaen University, Thailand

N. Mishima & S. Hatanaka
Department of Architecture, Mie University, Japan

ABSTRACT: This article reports a study of composition and microstructure of fly ash geopolymer containing metakaolin which is obtained at different firing temperatures. The mixtures of fly ash and metakaolin are activated by alkali with SiO_2/Al_2O_3 molar ratio of 3.83, Na_2O/Al_2O_3 of 1.26 and W/B of 0.27. The glassy components of these starting materials are chemically transformed into glassy alumino-silicate network that can create very strong solid. The highly reactive metakaolin is attained at the temperature at which the hydroxyl groups in octahedral sites of kaolinite are totally removed. XRD-pattern of metakaolin burnt at 600°C shows the presence of highly amorphous phases which are the origins of the glassy geopolymer matrix as seen by XRD and SEM while the patterns of metakaolin burnt at 400°C and 800°C contain crystalline phases identified as kaolinite and spinel respectively. Those crystalline phases exit in the matrix of glassy geopolymer cement indicating no dissolution of the crystalline phases with alkali activation. It is found that mechanical strength of fly ash geopolymer cement using metakaolin fired at 600°C is the highest. This suggests that in order to achieve the high strength geopolymer, the optimum firing temperature of kaolin that hence the highly glassy phase should be controlled.

1 INTRODUCTION

1.1 *General in geopolymer*

Recently, attempt in making environmentally friendly cementitious material is now ongoing in the research and development of inorganic alumino-silicate polymer, namely "Geopolymer". The term of geopolymer was firstly initiated by Prof. Davidovits in 1970s (Davidovits 1970). Geopolymer is the new material that does not need the presence of Portland cement as a binder. Instead, the silica- and alumina-source raw materials such as fly ash are activated by alkali liquids to produce the binder (Davidovits 1999, Hardjito et al, 2004). Hence, concrete with no Portland cement could be an alternative construction material in the future with similar appearance, higher strength and superior durability.

It is well known that kaolin contains a large amount of silica and alumina. When burnt at optimum temperature, metakaolin (MK) can be obtained and also used as raw material for making geopolymer. In Thailand, there are various sources of kaolin for use in ceramics industries. A study on the use of fly ash incorporating with MK to produce geopolymer so far was researched and developed by various researchers (Davidovits 1970, Davidovits 1999, Hardjito et al, 2002, Chareerat et al, 2006). Ongoing study in the composition and microstructure after the chemical reaction of polymerization has also been proposed and investigated in this paper. The knowledge of the use of high calcium lignite fly ash together with MK to produce geopolymer, therefore, would be beneficial to the correctly understanding and to the future use of these materials in the field of construction.

Table 1. Chemical compositions of OFA and kaolin (%).

Binder	SiO$_2$	Al$_2$O$_3$	CaO	Na$_2$O	K$_2$O	LOI
OFA	38.72	20.76	16.55	1.19	2.70	0.10
Kaolin	60.30	30.70	0.21	0.00	1.23	1.01

2 MATERIALS AND METHODOLOGY

2.1 Materials

In this experiment, fly ash (OFA) and metakaolin (MK) were used as the starting materials. Original fly ash namely "OFA" comes from Mae Moh electric power station plant in the north of Thailand. Kaolin is from the south of Thailand. Sodium silicate solution (Na$_2$O = 15.32 %, SiO$_2$ = 32.87 % and water = 51.81% by mass) and 10 Molarities of NaOH solution were used as the alkaline activators. This lot of fly ash contains a reasonably large amount of calcium oxide and is rather coarse with Blaine fineness of 2,100 cm^2/g. The fly ash has mean particle size of 65 μm and percentage of retain on sieve #325 of 50%. The chemical compositions of the fly ash as determined by X-Ray Fluorescence (XRF) analysis are given in Table 1. Kaolin was calcined at various temperatures under oxidation atmosphere to determine the optimum firing condition of 400, 600 and 800°C. The chemical compositions of kaolin are also given in Table 1.

2.2 Methodology

In this test, the symbols of "POFA" stands for the pure fly ash geopolymer paste while "PMK" stands for fly ash geopolymer with 20% MK replacement at different firing temperature. All geopolymer mortars were made with sand to binder ratio of 2.75. The SiO$_2$/Al$_2$O$_3$ and Na$_2$O/Al$_2$O$_3$ molar ratio of 3.79 and 0.88 with W/B of 0.21 was obtained for POFA mix. While PMK mix, SiO$_2$/Al$_2$O$_3$ of 3.83, Na$_2$O/Al$_2$O$_3$ of 1.26 and W/B of 0.27 was obtained. In order to obtain the workable geopolymer mortar, a minimum base water content of 5% by mass of the geopolymer paste (fly ash (and MK), NaOH, sodium silicate and base water) was used for all mixes. For example, a typical POFA with 10M NaOH and 1.00 sodium silicate/NaOH ratio mix consisted of 503 kg fly ash, 1,382 kg. sand, 127.5 kg. NaOH, 127.5 kg. sodium silicate and 40 kg. base water. The mixing was done in an air conditioned room at approximately 25°C. The mixing procedure started with mixing of NaOH solution, base water, fly ash (and MK) for 5 minutes in a pan mixer. Sand was then added and mixed for 5 minutes. This was followed by the addition of sodium silicate solution and followed by a final mixing of another 5 minutes. Right after the mixing, the flow

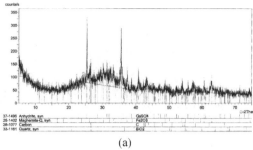

(a)

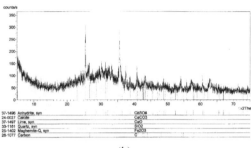

(b)

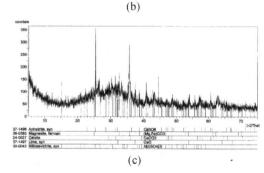

(c)

Figure 1. XRD patterns of OFA.

test was controlled within 110–125% and followed by the casting of cylindrical moulds (Ø5 cm with 10 cm height). Three cylinders were prepared for each test variable. After casting, the samples were immediately covered by a vinyl sheet to avoid the loss of water. After being left in room temperature for 1 hour, specimens were cured in an oven at 75°C for 24 hours. Specimens were then cooling down and left in room-temperature until the age meets 7 days. The fractures of specimens after strength test were then collected for the further composition and microstructure investigations.

3 RESULTS AND DISCUSSION

3.1 X-ray diffraction patterns (XRD)

XRD as seen in Fig. 1 shows that the original fly ash (OFA) basically composes of glassy phase and some minor crystalline phases (quartz, mullite,

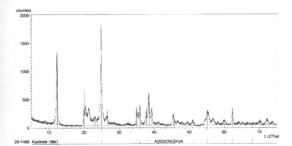

Figure 2. XRD pattern of MK after firing at 400°C.

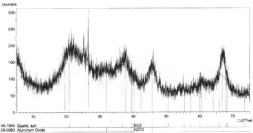

Figure 4. XRD pattern of MK after firing at 800°C.

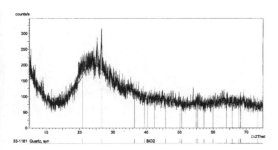

Figure 3. XRD pattern of MK after firing at 600°C.

hematite, magnetite, calcite, anhydrite, magnesite, CaO, $Al_2(SO_4)_3$ and unburned carbon). The elements existing in glassy phase of fly ash are estimated by XRF cooperating with XRF indicating the presence of Si, Al and some alkalis such as Ca and K. Silicon and aluminium ions generated from OFA and MK are bonded as tetrahedral structure to be a framework of geopolymer matrix. Those K and Ca ions from OFA together with the additional Na ions from basic solutions are located at the sites at which the charge balancing is acquired.

Figures 2–4 show XRD patterns of kaolin burnt at different firing temperatures. On heating under oxidation atmosphere, non-crystalline phase of metakaolin is obtained primarily consisting of short range ordered structure of alumino-silicate phase as shown by a reflection of board peak between 20 and 30° of 2θ as shown in Fig. 3. With low firing temperature (400°C) as shown in Fig. 2, the kaolinite structure has not been changed with calcination due to strong bonds of hydroxyl groups with aluminium ions in octahedral sites of 2-layer kaolinite. Therefore, kaolinite structure associated with small amount of quartz is still found in 400°C calcined kaolin. With an increase of firing temperature to 600°C (Fig. 3), the kaolinite structure becomes more short range ordered or more chemically reactive due to more complete removal of hydroxyl groups at higher calcination temperatures. With calcination at 800°C (Fig. 4), the pattern of short range alumino-silicate phase becomes again more long range

ordered containing quartz and γ-alumina which is kind of spinel. A removal temperature of the chemically combined water in kaolinite structure is therefore an important factor to design the reactivity of kaolin.

Figs. 5–8 show XRD patterns of fly ash geopolymer paste and fly ash geopolymer paste containing 20% kaolin burnt at different temperatures. All patterns appear with low and scattered bands of low degree of crystalline structures which overlap partially with those of the original fly ash and MK. Using pure fly ash, the geopolymer material contains iron silicate which is newly developed after the chemical reaction. Mullite and maghemite containing in OFA are still found in the matrix due to their inert phases as seen in Fig. 8.

Using MK burnt at 400°C, the produced geopolymer contains maghemite and quartz as found in OFA and kaolinite as found in 400°C MK as seen in Fig. 5. Furthermore, there is new phase developed at the same time of geopolymerization reaction, called "heulandite $(Ca_{1.23} (Al_2Si_7) O_{18} \cdot 6H_2O)$" which is zeolite material containing framework aluminosilicates with an infinitively extending 3-dimensional network of AlO_4 and SiO_4 tetrahedral linked to each other by sharing of all of the oxygen atoms, quite similar to the structure of geopolymer material. The difference of those materials is the degree of crystallinity in that the geopolymer is essentially amorphous while the zeolite material possesses much higher degree of crystallinity. Another kind of new phase developed is "tilleyite $(Ca_5Si_2O_7 \cdot (CO_3)_2)$" which can be found when hydrated calcium silicate is subjected to carbonate ions (Medvescek 2006). Here, the hydrated calcium silicate is contributed by fly ash when exposed to water.

New phase development of geopolymer material produced using 20% of 600°C MK replacing fly ash can be seen in Fig. 6. The new phase composition is known as "thaumasite $(Ca_3Si(CO_3)(SO_4)(OH)_6 \cdot 12H_2O)$". Normally, the formation of thaumasite eventually results in a softening within the cement matrix causing disintegration of the concrete (Stark 2003). Its occurrence in the geopolymer is therefore required to be investigated in more detail.

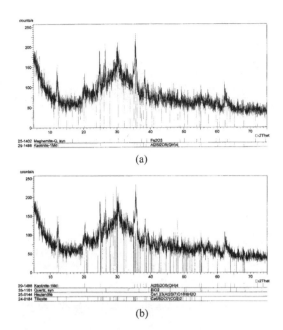

(a)

(b)

Figure 5.   XRD patterns of PMK 400°C.

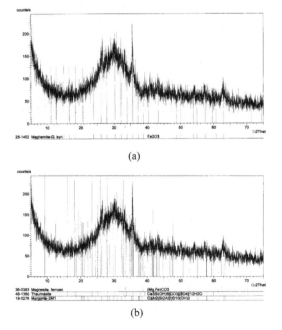

(a)

(b)

Figure 6.   XRD patterns of PMK 600°C.

Using 800°C calcined kaolin in the geopolymer synthesis, gismondine ($CaAl_2Si_2O_6 \cdot 4H_2O$) which is kind of zeolite-material can be developed as shown in Fig. 7. This finding is similar to Bakharev's work

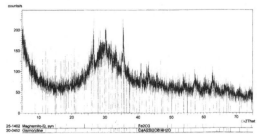

Figure 7.   XRD patterns of PMK 800°C.

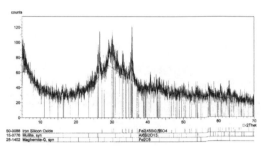

Figure 8.   XRD patterns of POFA paste.

(Bakharev 2005) saying that gismondine was developed in geopolymer matrix at the elevated curing temperature during 75–95°C. Normally, zeolite type used for ion exchange can be synthesized using 1–4 of SiO_2/Al_2O_3 ratio. In this experiment, the addition of MK in the fly ash mixture creates the ratio of 3.83 which is suitable for the zeolite development. It should be noted here that those minor phases developed from kaolin calcined at different temperatures are not the same although the SiO_2/Al_2O_3 ratios used at the beginning are similar. It is a result of the different solubility of starting kaolin burnt at different temperature when attacked by alkali solution that hence the varying of SiO_2/Al_2O_3 ratios in the geopolymer structures. It can be concluded that kaolin burnt at 600°C is the most suitable starting powder for geopolymer production in this case as all aluminium ions leached from both fly ash and metakaolin can play a role as the framework of geopolymer. However, geopolymer incorporating with all those minor phases should be observed in more detail so that their applications will be revealed at the highest potential.

3.2   Photomicrographs

Figs. 9–12 demonstrates morphological aspects of the geopolymer pastes. Microstructures of geopolymer pastes known as POFA, PMK400°C, PMK600°C and PMK800°C show considerable glassy dense matrixes with no grain boundary incorporating with partially

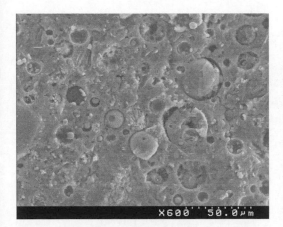

Figure 9.   Photomicrograph of PMK 400°C.

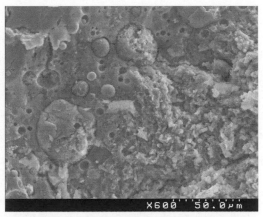

Figure 11.   Photomicrograph of PMK 800°C.

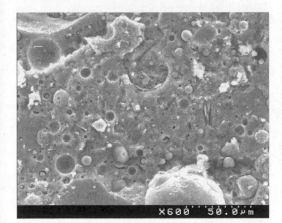

Figure 10.   Photomicrograph of PMK 600°C.

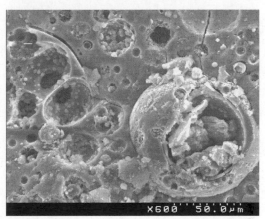

Figure 12.   Photomicrograph of POFA.

unreacted fly ash particles. It is pointed out that the alkaline activated fly ash and calcined kaolinite mixtures produce alkaline aluminosilicate glassy structures that strongly bind the entire solid together. It is the matter of molecular bonding of geopolymer structure that results in very high strength material, not similar to particle-particle attractive force so called "van der Waals force" resulting from the hydration of ordinary Portland cement.

3.3   Mechanical strength

The result of strength is shown in Table 2. The addition of 20% MK in the mix makes strength lower in every firing temperature of kaolin when compared to the mix containing pure fly ash. Good strength is obtained when fly ash is replaced by 20% of kaolin burnt at 600°C. As the matter of fact that the addition of MK increases the SiO_2/Al_2O_3 molar ratio, mechanical property is degraded when compared to pure fly

Table 2.   Mechanical strength of geopolymer paste.

Mix	POFA	PMK400	PMK600	PMK800
Str.(MPa)	68.0	51.5	60.5	45.5

ash geopolymer material. The optimum firing temperature of kaolin is thus 600°C to develop good strength. The results suggest that 600°C calcined kaolin gives the proper ratio of SiO_2/Al_2O_3 that hence the stronger bonds of alumino-silicate structures than those mixes using 400°C and 800°C calcined kaolins.

4   CONCLUSIONS

Compositions and microstructures of fly ash geopolymer containing 20% of metakaolin compose mainly of a highly amorphous phase incorporating with some un-reacted crystalline phases which exist in the original fly ash particles and minor phases which are newly developed after geopolymerization. Those minor phases are heulandite and tilleyite for PMK using 400°C calcined kaolin, thaumasite for PMK using 600°C calcined kaolin and gismondine for PMK using 800°C calcined kaolin. The optimum burning temperature of kaolin is 600°C to develop good strength geopolymer material.

ACKNOWLEDGEMENTS

This research was partial financially supported by the Sustainable Infrastructure Research and Development Center (SIRDC) and Kasetsart University, Chalerm phrakiat Sakon Nakhon Province Campus. Special thanks are also due to Dr. Akihiro Maegawa and Dr. Yukihisa Yuasa from Mie Prefecture Science and Technology Promotion Center, Japan for their kind assistance.

REFERENCES

Chareerat, T., Lee-Anansaksiri, A. and Chindaprasirt, P., Synthesis of High Calcium Fly Ash and Calcined Kaolin Geopolymer Mortar, the 1st International Conference on Pozzolan, Concrete and Geopolymer, Thailand, May 24–25, 2006, pp. 327–335.

D.C. Stark, Occurrence of thaumasite in deteriorated concrete, Cem. Concr. Comp. 25 (2003), pp. 1119–1121.

Davidovits, J., Chemistry of Geopolymeric Systems, Terminology, in Geopolymer, International Conference, eds. Joseph Davidovits, R. Davidovits & C. James, France, 1999, pp. 9–40.

Davidovits, J., Geopolymer, man-made rock geosysthesis and the resulting development of very early high strength cement, Journal of Material Education, 16, 2–3, 91–137.

Hardjito, D., Wallah, S. E., J. Sumajouw D. M., and Rangan B. V., 2005, On the Development of Fly Ash–Based Geopolymer Concrete, ACI Mat. Jour, Vol. 101(6), Nov. 2004, pp. 467–72.

S. Medvescek, R. Gabrovsek, V. Kaucic, A. Meden, Hydration products in water suspension of Portland cement containing carbonates of various solubility, Acta Chim. Slov., 53 (2006), pp. 172–179.

T. Bakharev, Geopolymeric materials prepared using class F fly ash and elevated temperature curing, Cem. Concr. Res., 35 (2005), pp. 1224–1232.

Excellence in Concrete Construction through Innovation – Limbachiya & Kew (eds)
© 2009 Taylor & Francis Group, London, ISBN 978-0-415-47592-1

A computed-based model for the alkali concentrations in pore solution of hydrating Portland cement paste

W. Chen & Z.H. Shui
School of Materials Science and Engineering, Wuhan University of Technology, Wuhan, China

H.J.H. Brouwers
Department of Civil Engineering, University of Twente, Enschede, Netherlands

ABSTRACT: A computed-based model for the alkali concentrations in pore solution of hydrating Portland cement paste is proposed. Experimental data reported in different literatures with thirteen different recipes are analyzed. A 3-D computer-based cement hydration model CEMHYD3D is used to simulate the hydration of these pastes. The models predictions are used as inputs for the alkali partition theory, which is used to derive the alkali binding capacity of C-S-H in hydrating Portland cement paste. A linear relation between the amount of bound-alkali Na^+ in C-S-H and its concentration in the pore solution is found, whilst a non-linear relation should be employed for the amount of bound-alkali K^+ in C-S-H. New methods for predicting the alkali concentrations in the pore solution of hydrating Portland cement pastes are proposed based on the computer model CEMHYD3D, which is also validated with experimental results.

1 INTRODUCTION

The cement hydration process can essentially be considered as the interactions between the solid compounds and the liquid phase in the paste. The liquid phase contains some ions and is therefore called "pore solution". The pore solution normally contains ions like OH^-, K^+, Na^+, Ca^{2+}, SO_4^{2-}, $Al(OH)_4^-$, $H_3SO_4^-$, etc. with the alkali ions (K^+ and Na^+) and OH^- being the most dominant ones. The hydroxyl concentration in the pore solution of concrete is important due to its dominant effect on the likelihood of alkali-silica reaction (ASR) in concrete, the hydration of cement mixture containing contaminants for immobilization and stabilization (Van Eijk, 2001), stabilization of the oxide film on the surface of steel bar inhibiting further corrosion and the reactivity of some supplementary materials (slag, fly ash, etc.) in concrete.

The hydroxyl concentration is mainly controlled by the alkali concentrations in the pore solution, because the alkali ions are the most abundant cations in the solution and the hydroxyl ion is generated to maintain the charge neutrality.

Number of research on the alkali concentrations in the pore solution of hardening cement paste and concrete has been rising rapidly in the past decades. Most of the research is focused on the development of the alkali concentrations in the pore solution with

the aids of experimental design. Effects of different factors on the alkali concentrations are investigated, like the use of supplementary materials (Diamond, 1981; Longuet, 1976), addition of mineral salt (Page & Vennesland, 1983; Schäfer & Meng, 2001), alkalinity of cement (Dehwah et al., 2002) and carbonation of concrete (Anstice et al., 2004). Most of the analysis is based on the pore solution compression method firstly introduced by Longuet et al. (1973).

Methods for predicting the alkali concentrations in the pore solution of hydrating cement paste are useful, although experimental setup is widely used to investigate the pore solution composition in hydrating cement paste. First, it takes lots of efforts to measure the alkali concentrations in mature concrete. Special care should be taken to minimize the environmental effect (Glasser, 2003). Second, the experiments cannot obtain quick results because no reliable methods are available to accelerate the hardening process of cement, the principal binder in concrete (Taylor, 1987). Third, the different factors can hardly be considered in a single batch of experiments. A large set of experiments need to be carried out to clarify the effect of each factor. Therefore, models for the pore solution composition can on one hand help to design the experimental scheme, and on the other hand predict the effect of various factors without the need of carrying out the time-consuming and costly experiments.

In this study, a new computer-based model is proposed for the alkali concentrations in the pore solution of hydrating Portland cement. They are based on the methods developed by Taylor (1987) and Brouwers & Van Eijk (2003), and a large set of experimental data taken from literature. The alkali-binding capacity of the hydration product C-S-H is determined with the theories and experimental data. Two binding models are proposed for Na^+ and K^+, respectively.

2 MODELS FOR ALKALI CONCENTRATIONS IN PORE SOLUTION

2.1 Pore solution composition

The evolution of pore solution composition is firstly discussed. Immediately after mixing with water, remarkable amounts of alkali ions are released into the water, together with sulfate ions. After about 6 hours, the sulfate concentration starts to decline, accompanied by a rapid increase of the hydroxyl concentration. The drop of the sulfate ion is most likely caused by the formation of ettringite, which is proven by the rapid drop of aluminum in the pore solution. The alkali concentrations increase steadily during the first 24 hours, and keep increasing to maxima at about one week. Then, they start to decline and reach constant values. The calcium concentration remains always very low, although it is relatively higher in the first 20 hours than that in the later ages. The hydroxyl concentration in the pore solution is very low in the early ages, and it remains almost constant as long as sulfate presents in a large amount in the pore solution. Immediately after the sulfate ion becomes depleted, the hydroxyl concentration is increased greatly. The overall trend for the change of hydroxyl concentration is very similar to those of the alkali ions. The aluminum and silicon concentrations in the pore solution remain very low throughout the whole hydration process, less than 0.5 mmol/L.

Alkali salts are normally very soluble, i.e. they have high solubility in water; hence, considering the situation in the pore solution of hydrating cement paste, equilibrium between the solid compounds of alkali salts and the pore solution can hardly be established. Thus, the alkali ions are always dissolved into the solution when they are available. Therefore, their concentrations are determined by the amount of available alkalis and the volume of pore solution. For predicting the alkali concentrations, the methods proposed by Taylor (1987) and further developed by Brouwers & Van Eijk (2003) are first introduced.

2.2 Taylor's method

Taylor (1987) proposed a method for predicting alkali ion concentrations in the pore solution of hydrated cement paste from more than one day old. The method takes use of the total contents of Na_2O and K_2O in the components and the water available for the pore solution. The alkali ions are partitioned between the pore solution and the hydration products.

The amount of alkali ions taken up by the hydration products is assumed proportional to its concentration in the solution and the amount of products as adsorbent. The concentration of ions in the solution is calculated from the remaining amount of alkali ions and the volume of solution. An empirical constant, namely the binding factor, is defined and derived both for Na^+ and K^+. In the computations, the amounts of alkali ions released by the cement hydration and the hydration products are estimated using some empirical equations. The volume of pore solution is computed from the total water content in the paste and that combined in the products. The detailed procedure for predicting the alkali concentrations is as following.

The alkali oxides in cement are divided into two groups according to their state, namely "soluble" and "insoluble" alkalis. Pollitt & Brown (1968) found part of alkali oxides in cement present as sulfate, which is instantly soluble after contact with water. The proportion of this part of alkali oxide depends on the sulfate content of cement. A detailed routine for computing this proportion is given by Taylor (1990). The rest of the alkali oxide is often found in solid solution in the alite, belite, aluminate and ferrite. It is released into the pore solution simultaneously as the hydration proceeds. Both these two parts of alkali oxides are available for the pore solution. The amount of alkali ions (Na^+ or K^+) in the cement can be computed as:

$$n_i^T = \frac{2x_i}{M_i} \cdot m^p \tag{1}$$

where n_i^T = the moles of alkali ion i (K^+ or Na^+) in the cement; n_i = the mass fraction of alkali oxide i in cement; m^p = the mass of Portland cement. The moles of alkali ions existing as sulfate are calculated as:

$$n_i^{sul} = f_i^{sul} \cdot n_i^T \tag{2}$$

where n_i^{sul} = the moles of alkali ion i in the sulfate form; f_i^{sul} = the fraction of alkalis as sulfates. The procedure for determining f_i^{sul} is given by Taylor (Taylor, 1990).

In the case that the sulfate content in clinker is not known, estimated values for f_i^{sul} may be used, for example, 35% of Na_2O and 70% K_2O in the soluble sulfate form (Taylor, 1987). The amount of the other alkali ions bound in the clinker phases is calculated as:

$$n_i^c = f_i^c \cdot n_i^T = (1 - f_i^{sul}) \cdot n_i^T \tag{3}$$

where f_i^c = the fraction of alkalis as solid solution in the clinker. The moles of alkali ions in each clinker

Table 1. Fraction of alkali ions in individual clinker phases to the amount of non-sulfate alkali ions in clinker (Taylor, 1987).

Alkali	Alite	Belite	Aluminate	Ferrite	Total
Na_2O	0.44	0.17	0.36	0.03	1
K_2O	0.29	0.41	0.27	0.03	1

phase (C_3S, C_2S, C_3A, C_4AF, etc.) are calculated by using n_i^c and the distribution of alkali ions in clinker phases given in Table 1 (Taylor, 1987).

The amount of alkali ions which exist as non-sulfate and are released by the clinker hydration is calculated as:

$$n_i^{r.c} = n_i^T \cdot \sum_{j=1}^{4}(f_{i,j} \cdot f_i^c \cdot \alpha_j) = n_i^T \cdot \sum_{j=1}^{4}[f_{i,j} \cdot (1-f_i^{sul}) \cdot \alpha_j] \quad (4)$$

where i = the clinker phase (alite, belite, aluminate and ferrite); $f_{i,j}$ = the fraction of alkali ion i (to the amount of non-sulfate ions in clinker) in the phase j; α_j = the hydration degree of phase j. Hence, the total amount of alkali ions released by the cement hydration is:

$$n_i^r = n_i^{sul} + n_i^{r.c} = n_i^T \cdot \{f_i^{sul} + \sum_{j=1}^{4}[f_{i,j} \cdot (1-f_i^{sul}) \cdot \alpha_j]\} \quad (5)$$

Obvious partition of alkali ions between the solid and aqueous phases takes place. The alkalis bound in the solid phases of cement are continuously released into the aqueous phase in the paste as the hydration proceeds. Parts of these alkali ions are absorbed by the hydration products (Taylor, 1987) and are immobilized, which are not available for the pore solution. The amount of alkali ions released by the cement hydration is accordingly divided into two parts: those in the solution and those bound in hydration products, yielding:

$$n_i^r = n_i^b + n_i^s \quad (6)$$

where n_i^b = the amount of alkali ions bound in products; n_i^s = s the moles in the solution, which reads:

$$n_i^s = C_i \cdot V_w \quad (7)$$

where C_i = the concentration of alkali ion i (mol/L); V_w = the volume of pore solution (L).

Taylor (1987) assumed that the amount of bound alkali ion is proportional to the concentration in the pore solution and the amount of hydration products, yielding:

$$n_i^b = ba_i \cdot C_i \cdot F \quad (8)$$

where ba_i = the binding factor (L) of alkali i; F = fraction between the quantity of hydration products (dimensionless), which are able to take up alkali cations in the paste and that after complete hydration of cement.

Substituting Equation 8 and 7 into Equation 6, the alkali concentration in the pore solution is calculated as:

$$C_i = \frac{n_i^r}{V_w + ba_i \cdot F} \quad (9)$$

Note that the computations of Taylor (1987) are based on 100 g cement. In the Taylor's method, most of the parameters necessary for the computation are estimated using empirical equations. Thus, uncertainties are inevitable because those equations can hardly be valid for all cements under investigation. Furthermore, a constant value for ba_i is used, implying a constant binding capacity of the hydration products. As stated by Taylor (1987), the assumption of this linear dependency had no theoretical basis. Experimental results by Hong & Glasser (2002), Stade (1989) and computations in this study prove that the relation is not necessarily linear.

2.3 Brouwers & Van Eijk's method

Brouwers & Van Eijk (2003) further developed the method proposed by Taylor (1987). The concepts of alkali release and adsorption are taken over. Furthermore, the hydration degree of cement, the amount of alkali ions released by the cement hydration, the amount of C-S-H and volume of pore solution are computed from the outputs of one computer-based cement hydration model CEMHYD3D (Van Eijk, 2001). Uncertainties in the theory of Taylor (1987) induced by using some empirical equations are minimized by distinguishing the main hydration product C-S-H and all others. Hence, it is expected to give more accurate predictions over a wide range of cements. The authors took use of the results of the experiments by Hong & Glasser (1999) to compute the C-S-H binding factors of alkalis and compared them to those used by Taylor (1987). Results from the experiments by Larbi et al. (1990) were used to validate the improved model. The model predictions agreed fairly well with the experimental results.

Brouwers & Van Eijk (2003) proposed the amount of alkali ions bound in the products is calculated as

$$n_i^b = Rd_i \cdot C_i \cdot m_{C-S-H} \quad (10)$$

where m_{C-S-H} = the mass of C-S-H in the solids and Rd_i = the distribution ratio of alkali i (Na^+ or K^+), which is defined as (Hong & Glasser, 1999):

$$Rd = \frac{\text{alkali in solid C-S-H}}{\text{alkali concentration in solution}} \quad (mL/g) \quad (11)$$

In this method, the alkali-binding capacity of C-S-H is assumed linearly proportional to the alkali concentrations in the solution because constant values of Rd_i are used. Only C-S-H in the products is considered as adsorbent because it is the most abundant phase in the products and is concluded to be the main binder of alkali ions (Brouwers & Van Eijk, 2003). Substituting Equation 7 and 10 into Equation 6 yields:

$$n_i^r = C_i \cdot V_w + Rd_i \cdot C_i \cdot m_{C-S-H} \qquad (12)$$

Hence, the concentration of alkali ion is solved from Equation 12 as:

$$C_i = \frac{n_i^r}{V_w + Rd_i \cdot m_{C-S-H}} \qquad (13)$$

where n_i^r = computed from Equation 5. The parameters in Equation 5 and 13 (α_i, m_{C-S-H}, and V_w) is computed by using the CEMHYD3D.

3 NEW MODELS

The alkali-binding capacity of hydration products (or the main product C-S-H) is an essential factor in the methods proposed by Taylor (1987) and Brouwers & Van Eijk (2003). In both methods, the alkali-binding factors are set to be constant based on the experimental results by using the synthetic C-S-H, or constant values largely based on assumptions. However, a constant alkali-binding capacity of C-S-H is not supported by the experimental results of Hong & Glasser (2002) and Stade (1989). In this study, new non-linear methods for determining the binding factors of C-S-H to Na^+ and K^+ in hydrating Portland cement pastes are proposed, which are derived from a large set of experimental results reported in literature.

3.1 Alkali-binding capacity of C-S-H

Stade (1989) studied the incorporation of alkali hydroxides in synthetic C-S-H and C-A-S-H gels. It is found that the amount of alkali hydroxide incorporated in C-S-H gels increases with decreasing C/S ratio in it. The alkali-binding capacity of the Al-containing C-S-H gel is smaller than the Al-free gel at equal C/S ratios. No obvious differences between the binding capacities for Na^+ and K^+ are observed in the experiments.

Hong & Glasser (1999, 2002) studied the alkali binding in synthetic C-S-H and C-A-S-H gels as well. For the alumina-free C-S-H, the alkali-binding capacity is found to increase linearly with increasing alkali concentrations in the solution. This linear relation is concluded from the approximately constant distribution ratios for C-S-H with fixed C/S ratios. However, if

alumina is incorporated into C-S-H, which takes place in real hydrating cement paste, the obtained C-A-S-H gels have obviously enhanced alkali-binding capacity, which is in contrast to the conclusion of Stade (1989). This enhancement is more obvious for C-A-S-H gels with low C/S ratios. On the one hand, for all C-S-H and C-A-S-H gels, with increasing alkali concentrations in the solution, more alkali ions are held in the gels. On the other hand, the distribution ratio decreases with increasing alkali concentrations in the pore solution, which indicates that a linear relationship between the binding capacity and the alkali concentrations is questionable. Similar to the conclusion of Stade (1989), there is no significant difference between the alkali-binding capacity of C-S-H for Na^+ and K^+.

A possible explanation for the different observations found in the experiments of Stade (1989) and Hong & Glasser (1999, 2002) is the way of preparing the C-S-H gel. Stade (1989) made the C-S-H gel at 150°C by the autoclave reaction with CaO and silica, and at 80°C by precipitation from sodium silicate solutions with calcium chloride. Hong & Glasser (1999, 2002) prepared the C-S-H gel by mixing $Ca(OH)_2$ and a very reactive, high surface area silica gel in double-distilled, CO_2-free water, sealed for 12 months at $20 \pm 2°C$ with regular agitation. The difference in the way of preparation and temperature may have a significant influence on the heterogeneity and structure of the C-S-H. The influence of different preparation temperature is already observed in the experiments of Stade (1989) as well.

Furthermore, the alkali concentrations used by Hong & Glasser (1999, 2002) are between 0.015 to 0.3 mmol/L. The alkali concentrations in the hydrating cement pastes under investigation are frequently outside this range because normally they are evaluated due to their high alkali oxide contents, which can potentially induce ASR in concrete. Therefore, higher concentrations of alkali ions in the pore solution are more relevant. Furthermore, the distribution ratio in real hydrating cement pastes can be significantly different from that of synthetic ones, which is of primary importance in modeling the pore solution composition.

4 ALKALI-BINDING IN PRODUCTS

Thirteen cement pastes using different cements or recipe tested in experiments are taken as basis. All the pastes are cured in sealed environment at various temperatures. Pore solutions are collected at the planned ages using the liquid compression method (Longuet et al., 1973). The oxide compositions of these cements are listed in Table 2 together with the recipe of the paste and the curing temperatures.

The hydration of these cement pastes are first simulated by using the computer model CEMHYD3D

Table 2. Properties of the Portland cements and pastes used in experiments*.

Num	C_3S	C_2S	C_3A	C_4AF	SO_3	Na_2O	K_2O	w/c	Temp (°C)
1[a]	56.8	19.7	6.9	11.3	3.25	0.43	1.23	0.42	23
2[a]	70.0	8.8	7.4	11.3	3.0	0.2	0.47	0.42	23
3[b]	54.5	21.3	10.9	9.2	3.1	0.21	0.82	0.4	22
4[b]	54.5	21.3	10.9	9.2	3.35	0.21	0.82	0.45	22
5[b]	54.5	21.3	10.9	9.2	2.7	0.21	0.82	0.56	22
6[c]	55.0	15.0	7.9	8.1	0.88	0.08	1.12	0.5	20
7[d]	58.0	16.0	7.0	12.0	0.91	0.16	0.51	0.35	22
8[e]	63.8	12.4	11	8.8	3.21	0.14	0.95	0.5	20
9[e]	69.8	6.8	9.6	9.7	2.59	0.19	1.22	0.5	20
10[e]	65.8	15.2	8.7	10.2	3.21	0.64	0.78	0.5	20
11[f]	64.3	17.3	8.5	9.9	3.1	0.32	0.6	0.4	23
12[g,†]	55.1	25.0	9.7	10.2	0.35	0.25	1.27	0.5	20
13[g,‡]	55.1	25.0	9.7	10.2	0.35	0.25	1.27	0.5	20

*. Mineral compositions are computed with the Bogue method.
a: taken from Bérubé et al. (2004); b: taken from Larbi et al. (1990); c: taken from Lothenbach & Winnefeld (2006); d: taken from Rothstein et al. (2002); e: taken from Schäfer (2004); f: taken from Diamond (1981); g: taken from Longuet (1976);
†: 4.5 m/m% gypsum added;
‡: 8.6 m/m% gypsum added;

(Van Eijk, 2001). For cements whose fineness is unknown, a value of $380\,m^2/kg$ is assumed, corresponding approximately to CEM I 32.5R produced in the Netherlands. For facilitating the discussion, a similar concept as the "distribution ratio" is used, called "Molality" (Ma) of alkalis in C-S-H. The molality physically represents the moles of alkali ions adsorbed by unit mass of the solid C-S-H gel. It is thus independent of the alkali concentrations in the pore solution and allows deriving the non-linear binding relationship. According to its definition, the molality is calculated as:

$$Ma = \frac{n_i^b}{m_{C-S-H}} \text{ (mmol/g)} \tag{14}$$

Substituting Equation 14 and 7 into Equation 6 gives:

$$Ma_i = \frac{n_i^r - C_i \cdot V_w}{m_{C-S-H}} \tag{15}$$

If the molality is known, the alkali concentration is calculated as:

$$C_i = \frac{n_i^r - Ma \cdot m_{C-S-H}}{V_w} \tag{16}$$

Similarly, the distribution ratio is calculated as:

$$Rd_i = \frac{n_i^r - C_i \cdot V_w}{C_i \cdot m_{C-S-H}} \tag{17}$$

The parameters n_i^r, V_w and m_{C-S-H} in Eqs (16) and (18) are obtained with the CEMHYD3D. Therefore, the Ma and Rd_i for alkalis in C-S-H can now be computed. The calculated Ma for Na^+ and K^+ are plotted as a function of the alkali concentration in Figure 1 and the calculated Rd_i in Figure 2.

It can be seen in Figure 2a that there is a linear relationship between the molality of Na^+ in C-S-H and its concentration, implying a linear binding capacity of C-S-H for Na^+. This linear binding model is in agreement with the hypothesis by Taylor (1987) and Brouwers & Van Eijk (2003).

However, for the molality of K^+ in C-S-H (Fig. 2b), this linear relationship cannot be discerned. On the contrary, when observing the distribution ratio for K^+ (Fig. 3b), it can be seen that it decreases with increasing K^+ concentrations, indicating a non-linear binding of C-S-H to K^+ ions. Most likely a non-linear binding model should be applied here.

A linear regression analysis for the linear relation in Figure 2a gives:

$$Ma_N = 0.45C_i \text{ (mg/g)}; \quad Rd_N = 0.45 \text{ (mL/g)} \tag{18}$$

The non-linear relation in Figure 2b can be fitted with:

$$Rd_K = 0.20C_i^{-0.76} \text{ (mL/g)} \tag{19}$$

The Equation 19 complies with the Freundlich isotherm, which is widely used to describe the adsorption of solutes in solution by solid phases. It can be seen in Figure 2 that the distribution ratios of Na^+

211

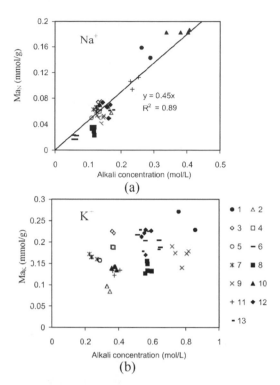

(a)

(b)

Figure 1. Molality of alkalis versus concentrations in solution calculated with Equation 16. Numbers in the legend correspond to the cement numbers in Table 2.

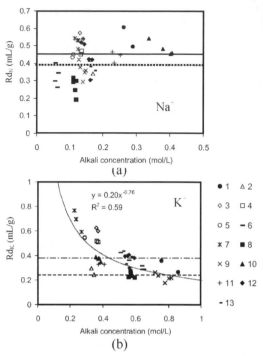

(a)

(b)

Figure 2. Distribution ratio of alkalis versus concentrations in solution calculated with Equation 18. Numbers correspond to the cement numbers in Table 3. "–": values for Na^+ used by Taylor (1987) and Brouwers & Van Eijk (2003); "——": values for K^+ used by Brouwers & Van Eijk (2003); "---": values for K^+ used by Taylor (1987), "——": values suggested in the present work.

and K^+ for hydrating Portland cement paste have similar values. A detailed comparison shows that for low alkali concentrations (about 0–400 mmol/L) the distribution ratio of K^+ is slightly higher than that of Na^+. With increasing alkali concentrations, the later gradually surpasses the former.

The constant values of Rd_N and Rd_k used by Brouwers & Van Eijk (2003) and those derived from the study of Taylor (1987) are included in Figure 2 as well. It can be seen that the values of Rd in these two studies (Brouwers & Van Eijk, 2003; Taylor, 1987) are in line with the calculations in this study. The used values of Rd_N are the same in the two studies, because (a) the linear relation used are indeed valid for Na^+ and (b) the used values are very close to the predictions in this study. However, two remarkably different values are used for Rd_K because of the differences in the concentrations of K^+ measured in the different experiments. The Rd_K value used by Taylor (1987)—0.25 mL/g—is valid for the high concentrations (e.g. 400–600 mmol/L in Figure 1 of Taylor (1987), while the value used by Brouwers & Van Eijk (2003)—0.38 mL/g—is valid for relative lower concentrations (370 mmol/L). Therefore, for low K^+ concentrations,

the predictions with Taylor's value are obviously higher than the measurements.

Furthermore, it is illustrated in this figure that C-S-H in hydrating cement paste can bind more K^+ than Na^+, due to the generally lower concentration of Na^+ than that of K^+. Therefore, it is essential to distinguish the alkali types in the cement while evaluating the alkali-binding capability of C-S-H in hydrating cement paste. If the cement is low in both Na_2O and K_2O, relatively more K^+ is immobilized in the solid phases. If the levels increase for both alkalis, the binding capability of Na^+ is much more enhanced than that of K^+, and surpasses the latter at a certain level.

5 VALIDATIONS OF THE NEW MODEL

Taylor (1987) determined the value of binding factor from measurements in nine laboratories and validated it with four batches of samples in these laboratories. A similar idea is used here as well, by validating the new method with four of the 13 recipes in Table 2.

These four recipes cover cements from different literatures and with varying alkali contents. The proposed method for determining Rd value of Na^+ and K^+ in hydrating cement paste (Equation 19 and 20) is used to predict the alkali concentrations in the pore solution. The hydration of these four cements is simulated with the CEMHYD3D. The results are included in Figure 3, together with the experimental measurements with Cement 4-7.

It can be seen that concentrations are correctly predicted for both Na^+ and K^+, to a better extent for Na^+. Therefore, it can be concluded that the proposed method can accurately predict the alkali concentrations in a wide range. The method takes different factors into account, for example, the mineral composition of cement, its fineness, the w/c ratio, the alkali contents in cement, and the curing temperature. For modeling the Na^+ concentration, the linear model by Taylor (1987) and Brouwers & Van Eijk (2003) is indeed valid, and for modeling the K^+ concentration, the non-linear method given in Equation 20 should be followed.

6 CONCLUSIONS

A new method is proposed in this study for determining the distribution ratio of alkalis in hydrating Portland cement pastes. It is derived from experimental results taking use of a large number of cement pastes selected from literature. The adsorption of alkali ions by the solid product C-S-H in hydrating Portland cement pastes complies with the Freundlich isotherm. The higher alkali concentrations, the more alkali ions are bound in the solid products. The linear binding model used by Taylor (1987) and Brouwers & Van Eijk (2003) is valid for Na^+, but a non-linear method should be followed for K^+.

A new computed-based model is proposed as well, which can precisely predict the alkali concentrations in the pore solution of hydrating Portland cement paste. Together with computer modeling, the models established in this study can be used to investigate the long-term changes of alkali concentrations in the pore solution.

ACKNOWLEDGEMENT

The authors wish to thank the following institutions for their financial support of the research of Dr. Wei Chen during his stay at University of Twente: Dr. ir. Cornelis Lely Foundation, Delta Marine Consultants, Betoncentrale Twenthe, Rokramix, Dutch Ministry of Infrastructure, SenterNovem Soil+, Jaartsveld Groen en Milieu.

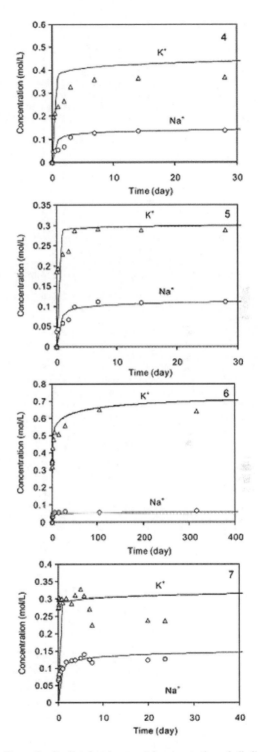

Figure 3. Predicted and measured concentration of alkali ion in the pore solution for Cement 4–7 from Table 2.

REFERENCES

Anstice, D. Page, C.L. Page, M. 2004. The pore solution phase of carbonated cement pastes. *Cement and Concrete Research* 35(2): 377–388.

Bérubé, M.A. Tremblay, C. Fournier, B. Thomas, M.D. Stokes, D.B. 2004. Influence of lithium-based products proposed for counteracting ASR on the chemistry of pore solution and cement hydrates. *Cement and Concrete Research* 34 (9): 1645–1660.

Brouwers, H.J.H. Van Eijk, R.J. 2003. Alkali concentrations of pore solution in hydrating OPC. *Cement and Concrete Research* 33: 191–196.

Dehwah, H. Malslehuddin, M. Austin, S. 2002. Effect of cement alkalinity on pore solution chemistry and chloride-induced reinforcement corrosion. *ACI Materials Journal* 99 (3): 227–233.

Diamond, S. 1981. Effect of two Danish fly ashes on alkali contents of pore solutions of cement-flyash pastes. *Cement and Concrete Research* 11: 383–394.

Glasser, F.P. 2003. The pore fluid in Portland cement: its composition and role. In: *Proc. 11th ICCC:* 341–352. Durban, South Africa.

Hong, S.Y. Glasser, F.P. 1999. Alkali binding in cement pastes: Part I, The C-S-H phase. *Cement and Concrete Research* 29: 1893–1903.

Hong, S.Y. Glasser, F.P. 2002. Alkali sorption by C-S-H and C-A-S-H gels: Part II. Role of alumina. *Cement and Concrete Research* 32: 1101–1111.

Larbi, J.A. Fraay, A.L. A. Bijen, J.M. 1990. The chemistry of the pore fluid of silica fume-blended cement systems. *Cement and Concrete Research* 20: 506–516.

Longuet, P. 1976. La protection des armatures dans le béton armé élaboré avec des ciments de laitier. *Silicates Industriels* 7/8: 321–328.

Longuet, P. Burglen, L. Zelwer, A. 1973. The liquid phase of hydrated cement. *Rev. Mater. Constr.* 676: 35–41.

Lothenbach, B. Winnefeld, F. 2006. Thermodynamic modelling of the hydration of Portland cement. *Cement and Concrete Research* 36(2): 209–226.

Page, C.L.Vennesland, O. 1983. Pore solution composition and chloride binding capacity of silica-fume cement pastes. *Materials and Structures* 16(91): 19–25.

Pollitt, H. Brown, A.W. 1968. The distribution of alkalis in Portland cement clinker. In: *Proc. 5th ISCC*: 322–333. Tokyo.

Rothstein, D. Thomas, J.J. Christensen, B.J. Jennings, H.M. 2002. Solubility behavior of Ca-, S-, Al-, and Si-bearing solid phases in Portland cement pore solutions as a function of hydration time. *Cement and Concrete Research* 32: 1663–1671.

Schäfer, E. 2004. Einfluss der Reaktionen verschiedener Zementbestandteile auf den alkalihaushalt der Porenlösung des Zementsteins. Ph.D. thesis, Clausthal University of Technology, Clausthal-Zellerfeld, Germany.

Schäfer, E. Meng, B. 2001. Influence of cement and additions on the quantity of alkalis available for an alkali-silica reaction. *Tech. rep. VDZ*, Düsseldorf, Germany.

Stade, H. 1989. On the reaction of C-S-H(di, poly) with alkali hydroxides. *Cement and Concrete Research* 19(5): 802–810.

Taylor, H.F.W. 1987. A method for predicting alkali ion concentrations in cement pore solutions. *Advances in Cement Research* 1(1): 5–16.

Taylor, H.F.W. 1990. Cement chemistry (1st Ed.), London: Academic press.

Van Eijk, R. J. 2001. Hydration of cement mixtures containing contaminants. Ph.D. thesis, University of Twente, Enschede.

Research on the absorbing property of cement matrix composite materials

B. Li & S. Liu

School of Material Science & Engineering, Dalian University of Technology, Dalian, China

ABSTRACT: A new kind of cement matrix composite material used for absorbing electromagnetic wave was studied in this paper. Based on cement, single or double layer samples were made by packing hollow spheres (EPS, e.g.) and absorbents. Tested by the arching method in an anechoic chamber, reflectivities of single-layer samples are all better than 8 dB in the frequency range 8 ~ 18GHz and its of double-layer samples have 12.7 GHz better than 10 dB in the frequency range 2 ~ 18GHz. The impedance matching theory and energy conservation theory were used to explain the factors' impact on absorbing property and the connection among these factors. The resonant theory was used to analyse the rule of absorbing peak and absorbing width.

1 INTRODUCTION

Electromagnetic interference (EMI) preventing is particularly needed for underground vaults containing transformers and other electronics that are relevant to electric power and telecommunication. It is also needed for deterring any electromagnetic forms of spying (Xiang et al. 2001, Cao & Chung 2003). It is in this sense that the cement based building composite material which is not only a structural material, but also can have some electromagnetic wave absorbing properties has caused more and more attention.

Cement based material which has rich resources and good environmental adaptability is one of the most common structural materials used in engineering constructions. Cement is slightly conductive and its wave absorbing property is very low, but it is a simple and practical method to increase the cement composite's absorbing effectiveness by introducing conductive fillings and loadings (Guan et al. 2006). There have been many studies on the reflection loss of cement matrix composite materials, and most of the fillings are metal powders (Xiong et al. 2004, Xiong et al. 2005), fibers (Yang et al. 2002, Ohmi et al. 1996, Kimura & Hashimoto, 2004) or ferrites (Li et al. 2003, Morimoto et al. 1998, Oda 1999, Yamane et al. 2002, Kobayashi et al. 1998a,b). With these fillings, cement matrix materials can get a high reflection loss of 8~20 dB in the frequencies tested. In the field of cement matrix wave absorbing materials, studies on microwave-transparent materials filling composites are few (Guan et al. 2006). The weight of the cement matrix absorbing material is still a concerning problem and few studies have dealt with it.

Expanded polystyrene has a series of excellent properties such as low density, high specific strength, low water absorption, and has been used in concrete to produce lightweight concrete or in cement for thermal insulation (Cook 1983, Chen & Liu 2004). Besides, EPS has low electromagnetic parameters (Bandyopadhyay, 1980) and it can be used to adjust the parameters of the cement composite. In this paper, the electromagnetic characteristic of the cement matrix material filling with absorbents such as CB and EPS beads was studied.

2 EXPERIMENTS

2.1 Materials

The cementitious starting material used in this study was Portland cement of Type P·O 32.5R, which was produced by Dalian-Onoda Cement Co., Ltd., China. The specific area and ignition loss are $3300\,cm^2 \cdot g^{-1}$ and 0.6%, respectively. EPS beads, were provided by Dalian Hongyu Foam Plastics Co., Ltd., China. Carbon black N234 was produced by Fushun Dongxin Chemical Co. Ltd., China.

2.2 Sample preparation

For preparation of the samples, a UJZ-15 mortar mixer was used to mix water and cement and the ratio of water to the total cementitious material was 0.34. After 10 min, the EPS beads were added to the cement paste and mixed for another 10 min. After pouring the mixture into the oiled moulds with the size of $200\,mm \times 200\,mm$, the moulds were vibrated on a vibration table for 1 min and then smoothed with a float to facilitate compaction and decrease the amount of air bubbles. The specimens were demolded after 24 hours and then cured at the room temperature for 28 days.

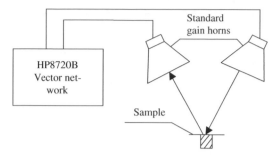

Figure 1. Set-up for arched test method.

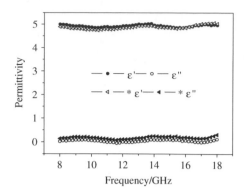

Figure 2. Permittivity of cement and cement after curing for 28 days (noted by *).

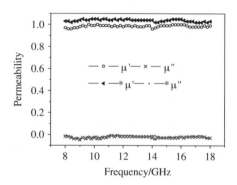

Figure 3. Permeability of cement and cement after curing for 28 days (noted by *).

Generally, CB or other absorbents were added and mixed for 10min before adding EPS beads. For double-layer samples, the thickness proportion of matching layer and absorbing layer is 2 to 1.

2.3 Testing method

The electromagnetic parameter of cement was tested by coaxial flange method. The electromagnetic absorbing effectiveness of wave absorbing material was denoted with the reflectivity R, which was expressed as $R = 20\lg|E_r/E_i|$ (dB), where E_i and E_r referred to the electric field strength of the incident and reflective electromagnetic wave, respectively (Kaynak 2000). The electromagnetic reflection loss of the composite material was tested in an anechoic chamber using the arched testing method [19], with an Agilent 8720B vector network analyzer (VNA), shown in Figure 1. The VNA was first calibrated before testing and then the sample was put on the support structure for reflection loss testing. The frequency tested is 2~18 GHz.

3 RESULTS AND DISCUSSION

3.1 Absorbing mechanics of cement

The absorptive efficiency of absorbing material depends upon the complex dielectric permittivity and magnetic permeability of the material. The complex dielectric permittivity and magnetic permeability of pure cement and cement after curing for 28 days are shown in Figure 2 and Figure 3, respectively. It can be seen that, for pure cement, its imaginary part and real part of complex magnetic permeability is close to 0 and 1, respectively, but its imaginary part of complex dielectric permittivity is positive value, and the real part is about 5. It is obviously that the electromagnetic absorptive capacity of cement depends upon the dielectric loss by some metal oxides and minerals in cement and there is little magnetic loss for cement (Cao & Chung 2004). Compared to pure cement, the complex dielectric permittivity and

magnetic permeability of the cement after curing for 28 days don't have obviously modification and the imaginary part increases a little because some silicates are generated by complex reaction after adding water in cement. Since the dielectric constant of cement is low, absorbing agents are needed to improve absorbing capacity of cement, and in order to improve impedance matching with free space, addition agents with little absorbing ability is needed. In this research, CB, manganese dioxide and ferrite are used as absorbing agent and EPS is used as addition agent.

For a good absorbing material, it firstly should have a good impedance matching with free space and has a low reflectivity of electromagnetic wave. In other words, the absorber should provide transmission paths for electromagnetic wave. It secondly should have a high absorptive capacity. For single layer absorber, this two terms is a contradiction and it had to satisfy together. But for double layer absorber, it is easy to reach this request. By a matching layer with little absorbing capacity, the absorber can has a good impedance matching, and by an absorbing layer, the absorber can have a good absorbing capacity.

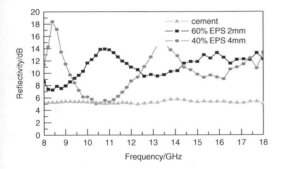

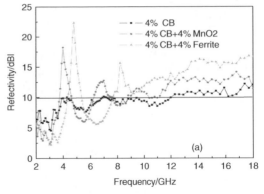

Figure 4. Reflectivities of samples with different microwave-transparent materials.

3.2 Effect of microwave-transparent materials on absorbing characteristic

It can be seen from Figure 4 that reflectivities of samples with EPS are obviously better than its of pure cement. The average reflectivities of samples with EPS are all better than 10dB.

As is analyzed above, because of the unmatched impedance cement has a limited absorbing ability. If there are enough transmission paths for electromagnetic, absorbing property can be improved. In these samples, EPS beads were added as a kind of wave-transparent material, which form a wave-transparent network and ensure electromagnetic wave transmitting in the absorber. In one side, the cement packs these paths and the electromagnetic wave would be weakened when it transmitted through the paths; in another side, the wave would be scattered, refracted and even interfered when it transmits from one bead to another. Because of these the absorbing property can be increased a lot by adding microwave-transparent materials which has a absorbing peak of 18.4 dB.

Besides, for the transmission paths for electromagnetic are free, absorbing property can be improved by adding thickness. Samples in Figure 4 has a thickness of 10 mm and if adding its thickness, the length of transmission path for electromagnetic wave would be increased and absorbent increases too, and this can absorb more electromagnetic wave and improve the absorbing property of sample. It is predicted that the absorbing property would be better if the absorber thickness is more than 10 mm.

3.3 Effect of absorbents on absorbing characteristic

Figure 5 shows the reflectivities of samples filled with different absorbents. It can be seen that most of the reflectivities are better than 10dB and the peak value can reach 23dB. The reflectivities of samples with complex absorbents are better than that of samples with single absorbent. EPS beads form a

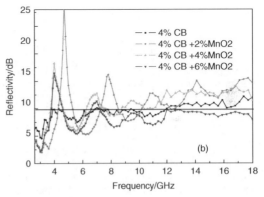

Figure 5. Reflectivities of samples with different absorbents.

wave-transparent network and absorbents distribute along it, which together composed a resonant-cavity group through which the incident electromagnetic wave would be depleted to its best ability. For a rectangle resonant cavity, its resonant frequency can be marked as follows:

$$f = \frac{\sqrt{\left(\frac{m}{n}\right)^2 + \left(\frac{n}{b}\right)^2 + \left(\frac{p}{d}\right)^2}}{2\sqrt{\mu_r \mu_0 \varepsilon_r \varepsilon_0}} \tag{1}$$

where $a = b = 200\,mm > d = 20\,mm$, m, n, p are the modules along the three coordinate axises. The reflectivities in Figure 5(b) show that the resonant frequencies drift toward to the lower with the absorbents' increase, which is accord to the equation.

Carbon black, manganese dioxide and ferrite are three kinds of absorbents. When the used together different absorbing principles work cooperative and the reflectivities would be improved as Figure 5(a) showed. The ratio between the absorbents has a optimal value. Take the CB and manganese dioxide as an example. A sample with 4% CB can forms a developed electric conduction network without agglomerated particles and it has a well adsorbing ability. When a little manganese dioxide added, resistance depletion

217

Table 1. Design of samples (%vol.).

Sample	1#	2#	3#	4#
Matching layer	60%EPS	60%EPS	60%EPS	60%EPS
Absorbing layer	6%CB+ 60%EPS	2.5%CB+ 60%EPS	6%CB	2.5%CB

(CB) and dielectric depletion (manganese dioxide) woke together and the reflectivities increase obviously with increasing content of the absorbents which get a peak value at 4% manganese dioxide. When the content of manganese dioxide keep added then the glomeration phenomenon aggravates and the electric conduction network is broken causing the decrease of the absorbability.

3.4 Effect of conformation on absorbing characteristic

Figure 6 shows the reflectivities of samples 1#∼4# with double layers. It can be seen from Fig.6 that average reflectvities are better than 10dB and the peak value can about 20dB. The reflectivities of samples 1# is obviously better than others.

In pursuit of zero reflection of absorbing materials, samples with double layers were designed as Table 1. The surface layer is matching layer formed by cement and EPS beads, and the bottom layer is absorbing layer formed by cement, EPS beads and CB. The matching layers are all the same and have enough paths for wave's transmission. The absorbing property is different because of the different content of CB in absorbing layer. For double-layer absorber, the electromagnetic wave is reflected partly when it transmits from matching layer to absorbing layer, which can weaken absorbing capacity. When the CB content is 6%, the reflection of electromagnetic wave by the conductive network or partial conductive network formed by CB particles is strong and the absorbing property is worse. Sample 1# has the same CB content with sample 3#, but by adding EPS beads in absorbing layer, the absorbing property of 1# is better than 3#. This is because the transmission paths in absorbing layer of 1# are freer than that of 3# due to the EPS beads and the reflection between the two layers is less. The reflectivities of samples 2# and 4# are not optimal, but the reasons are different. On the one hand, the CB content in absorbing layer is low and the absorber can not attenuate electromagnetic wave effectively; on the other hand, the impedances of the two layers are not matching and too much wave is reflected on the interface.

As is known, each absorbing material has certain frequency selectivity and has its own characteristic absorbing peak in certain frequency, so in a wide frequency band it can have one or several absorbing peak. Sample 1# shows the frequency selectivity obviously.

Of this five samples, the reflectivities of 1# are lower than others and 1# has a good absorbing property in the frequencies tested.

4 RESULTS

By introducing EPS into cement, the cement matrix composite plates have well matching impedance with the air. By compounding absorbents, the cement plates have a better absorbing property with sharp absorbing peak and width frequencies. By designing matching layer and absorbing layer, the cement matrix composite absorbing plates have a good absorbing property.

The composite plate materials have rich resources, simple production process, light and high absorption, and can be used to build anechoic chamber.

REFERENCES

A. J. Simmons, W. H. Emerson. An anechoic chamber making use of a new broadband absorbing material. IRE International Convention Record. 1953; 1 (2): 34–41.

Akif Kaynak. Electromagnetic shielding effectiveness of galvanostatically synthesized conducting polypyrrole films in the 300–2000 MHz frequency range. Mater. Res. Bull. Vol. 31(7), 1996, p 845–60.

B. Chen, J. Y. Liu. Properties of lightweight expanded polystyrene concrete reinforced with steel fiber. Cement & Concrete Research. 2004; 34 (7): 1259–63.

D. J. Cook. Expanded polystyrene beads as lightweight aggregate for concrete. Precast Concrete. 1983; 45 (12): 691–93.

D. Xiang, L. Xu, L. Jia. The design of basement rebuilding into radial shielded room. J. Shandong Institute of Arch & Eng. 2001; 16 (3): 64–68. (in Chinese)

G. Xiong, L. Xu, M. Deng. Research on absorbing EMW properties and mechanical properties of nanometric TiO_2 and cement composites. J. Functional Mater. & Devices. 2005; 11 (1): 87–91. (in Chinese)

G. Xiong, M. Deng, L. Xu. Absorbing electromagnetic wave properties of cement-based composites. J. Chinese Ceramic Society. 2004; 32 (10): 1281–84. (in Chinese)

H. Guan, S. Liu, Y. Duan. Cement based electromagnetic shielding and absorbing building materials. Cement & Concrete Composites. 2006 (in press).

H. Yang, J. Li, Q. Ye. Research on absorbing EMW properties of steel–fiber concrete. J. Functional Mater. 2002; 33 (3): 341–43. (in Chinese)

J. Cao, D.D.L. Chung. Coke powder as an admixture in cement for electromagnetic interference shielding. Carbon. 2003; 41 (12): 2433–36.

Jingyao Cao, D.D.L. Chung. Use of fly ash as an admixture for electromagnetic interference shielding [J]. Cement & Concrete Research. 2004, 34 (10): 1889∼92.

K. Kimura, O. Hashimoto. Three-layer wave absorber using common building material for wireless LAN. Electronics Letters. 2004; 40 (21): 1323–24.

M. Kobayashi, Y. Kasashima, H. Nakagawa. Anti-radio wave transmission curtain wall used in buildings. J. Japan Soc. Composite Mater. 1998; 24 (1): 32–34. (in Japanese)

M. Morimoto, K. Kanda, H. Hada, et al. Development of electromagnetic absorbing board for wireless communication environment. In Proceedings of the Conference of Architectural Institute of Japan. 1998, D-1. p. 1069–70.

M. Oda. Radio wave absorptive building materials for depressing multipath indoors. Electromagnetic Compatibility, 1999 International Symposium on. 1999. p. 492–95.

M. Kobayashi, Y. Kasashima. Anti-radio wave reflection curtain wall used in tall buildings. J. Japan Soc. Composite Mater. 1998; 24 (3): 110–13. (in Japanese)

N. Ohmi, Y. Murakmi, M. Sbibayama, et al. Measurements of reflection and transmission characteristics of interior structures of office buildings in the 60 GHz band. PIMRC'96, 7th International Symposium on. 1996. p. 14–18.

P. B. Bandyopadhyay. Dielectric behavior of polystyrene foam at microwave frequency. Polymer Eng. & Sci. 1980; 20 (6): 441–46.

T. Yamane, S. Numata, T. Mizumoto, et al. Development of wide-band ferrite fin electromagnetic wave absorber panel for building wall. Electromagnetic Compatibility, 2002 International Symposium on. 2002, vol. 2. p. 799–804.

X, Li, Q. Kang, C. Zhou. Research on absorbing properties of the concrete shielding material at 3mm wave bands. Asia-Pacific Conference on Environmental Electromagnetics. Hangzhou, China. 2003. p. 536–40.

Excellence in Concrete Construction through Innovation – Limbachiya & Kew (eds)
© 2009 Taylor & Francis Group, London, ISBN 978-0-415-47592-1

Visual examination of mortars containing flue gas desulphurisation waste subjected to magnesium sulphate solution

J.M. Khatib
School of Engineering and the Built Environment, University of Wolverhampton, Wolverhampton. UK

L. Wright
Pick Everard, Halford House, Charles Street, Leicester, UK

P.S. Mangat
Centre for Infrastructure Management, MERI, Sheffield Hallam University, Sheffield, UK

ABSTRACT: This work forms part of a wide ranging research project on the use of waste from the flue gas desulphurisation (FGD) processes in concrete. It examines the resistance to magnesium sulphate, using visual examination, of mortars containing simulated desulphurised waste (SDW). The mortar consists of 1 part binder to 3 parts sand and the water to binder ratio was 0.55. The binder consists of cement, SDW and slag. The cement was partially replaced (by mass) with 0% to 70% SDW and 0% to 90% slag. Mortar specimens were exposed to 2.44% magnesium sulphate (MgSO$_4$) solutions for up to 450 days. Visual examination of mortar specimens suggest that replacing cement with increasing levels of SDW improved sulphate resistance in each sulphate solution. This was attributed to the dilution of C$_3$A and CH due to the reduction in cement, pore refinement, and the formation of ettringite during early periods of hydration. Mortars underwent some deterioration due to the reaction with the C-S-H, C-S and aluminate phases. At SDW contents above 20%, resistance to magnesium sulphate, based on visual examination of specimens, was satisfactory. The use of slag increased the sulphate resistance of mortar.

1 INTRODUCTION

If concrete is to be used in practice it is necessary that it performs as designed throughout its service life, i.e. maintains its designed strength and serviceability. However, during the service life of concrete structures it is common that elements are subjected to external factors that cause deterioration and wear which may lead to an unserviceable structure if left untreated. Hence, the durability of concrete determines the concrete's ability to resist such attacks and subsequent deterioration. One of the most common forms of concrete deterioration is from sulphate attack, which occurs due to the presence of sulphates in surrounding ground water (Wild et al. 1997, Mangat & El-Khatib 1992, Matthews 1995). Sulphate attack occurs through the reaction of aggressive sulphate ions with constituents of the hydrated material such as tricalcium aluminate hydrate (C-A-H) and calcium hydroxide (CH) (Bureau 1970, Mehta 1986, Lea 1998, Collepardi 2001). These reactions generally result in the formation of gypsum (CaSO$_4$·H$_2$O) and ettringite

(C$_3$A·3C$\bar{S}$·H32). This can cause excessive expansion, which can lead to cracking and strength loss. Magnesium sulphate reacts with the calcium aluminate hydrate (C-A-H) phases to form ettringite as shown in Equation 1.

$$2(3CaO.Al_2O_3.12H_2O) + 3(MgSO_410H_2O) \rightarrow$$

$$3CaO.Al_2O_3.3CaSO_4.31H_2O + 2Al(OH)_3 + 3MgOH + 17H_2O \quad (1)$$

In addition, magnesium sulphate reacts with the CH formed during hydration, and deposits gypsum as shown in Equation 2

$$Ca(OH)_2 + MgSO_4.10H_2O \rightarrow CaSO_4.2H_2O + MgOH + 8H_2O \quad (2)$$

Magnesium sulphate (MgSO$_4$) does not only react with the CH and C-A-H phases but also the calcium silicate hydrate (C-S-H) phases, this is shown in Equation 3.

$$3CaO.2SiO_2.aq + MgSO_47H_2O \rightarrow$$

$$CaSO_4.2H_2O + Mg(OH)_2 + SiO_2.aq \quad (3)$$

Given that magnesium sulphate reacts with all phases of cement-based materials, the attack is more severe than other types of sulphate attack. It is also clear that the presence of CH and C-A-H aggravate the attack due to sulphates. Therefore, the reduction of lime (CaO) and tricalcium aluminate (C_3A) in the cement minimises the risk of sulphate attack. Santhanam et al. (2001) indicated that decreasing the C_3A content might improve sulphate resistance when exposed to sodium sulphate. However, when exposed to magnesium sulphate the use of low C_3A cements maybe inadequate because it attacks the C-S-H formed.

The sulphate resistance of concrete depends mainly on the amount of minerals present within the concrete that contribute to the attack. It also relies on its ability to resist the transport of sulphate ions throughout the body of the material that initiates the chemical attack process (Manmohan & Metha 1981, Huges 1985, Gollop & Taylor 1996). In normal concretes, the C_3A in cement is an important factor in the sulphate resistance of cement. Generally, the sulphate resistance is improved by reducing the amount of C_3A present. One other way of reducing the amount of C_3A in the system is to replace the cement with pozzolanic materials such as fly ash (Santhanam et al. 2001, Dunstan 1980, Bilodeau & Malhotra 1996). The pozzolanic reactions occurring from the inclusion of fly ash consumes CH, which also increases sulphate resistance. Reducing porosity and permeability of the concrete matrix improves the resistance to sulphate by reducing the flow of water through the concrete (Yang et al. 1996). In concrete, permeability can be improved by reducing the water to cement ratio and by providing adequate curing, which produces a much denser structure by reducing voids and pores present in the concrete. Materials such as fly ash also improve durability by increasing pore refinement and reducing permeability (Huges 1985, Dunstan 1980, Bilodeau & Malhotra 2000, Khatri et al. 1997, Tikalsky & Carrasquillo 1992). Therefore, in this paper, the sulphate resistance of mortars containing cement, a typical simulated desulphurised waste (SDW) and ground granulated blastfurnace slag was studies by visual examination.

2 EXPERIMENTAL

2.1 Materials and mix proportions

A standard 42.5N cement (C), conforming to BS12: 1996, ground granulated blastfurnace slag (S) conformed to BS6699: 1992, and BS EN 196-2: 1995, fly ash (FA) conformed to BS3892: 1997 were used. The gypsum (G) was wallboard grade quality with $aCaSO_4 \cdot 2H_2O$ purity of 95%, and the fine aggregate was a class M sand conforming to BS 882:1983.

Table 1. Details of mixes containing cement and a typical simulated desulphurised waste (C-SDW blends).

Mix No	Mix ID	Proportions (% weight of binder)	
		Cement (C)	SDW
1	REF (100$_C$)	100	0
2	90$_C$10$_{SDW}$	90	10
3	80$_C$20$_{SDW}$	80	20
4	70$_C$30$_{SDW}$	70	30
5	60$_C$40$_{SDW}$	60	40
6	30$_C$70$_{SDW}$	30	70

Table 2. Details of mixes containing cement, ground granulated blastfurnace slag and a typical simulated desulphurised waste (C-S-SDW blends).

Mix No	Mix ID	Proportions (% weight of binder)		
		Cement (C)	Slag (S)	SDW
7	10C90S0SDW	10	90	0
8	10C80S10SDW	10	80	10
9	10C70S20SDW	10	70	20
10	10C60S30SDW	10	60	30
11	10C50S40SDW	10	50	40
12	10C20S70SDW	10	20	70

Table 1 gives details of the mortar mixes. All mixes had a constant proportion of 1 (binder): 3 (sand) with water to binder ratio of 0.55. The binder consists of cement and cement replacement materials which include slag and a typical simulated desulphurised waste (SDW). The composition of the SDW was 85% fly ash and 15% gypsum. Mix 1 represents a reference mortar mix of 100% cement binder. Mixes 2 to 6 contain different blends of cement (C) with a typical simulated desulphurised waste (SDW). The cement was replaced with 0 to 90% SDW. The mix ID (column 2) represents the constituents of the binder. For example, mix 70$_C$30$_{SDW}$ represents a binder containing 70% cement and 30% of SDW by weight of binder. Columns 3 and 4 show the cement (C) and SDW content respectively.

Table 2 gives proportions of mixes 7 to 12 which contain different blends of cement, ground granulated blast furnace slag and a typical simulated desulphurised waste (C-S-SDW blends). A cement content of 10% was provided to initiate the reactivity of the slag and to maximise waste content. The slag content ranged from 90 to 20%, and the SDW content ranged from 0 to 70%. The mix ID (column 2, Table 2) represents the constituents of the binder. For example mix $10_C 70_S 20_{SDW}$ represents a binder containing 10% cement, 70% slag and 20% SDW by weight of binder. Columns 3 to 5 show the cement (C), ground granulated blastfurnace slag (S) and simulated desulphurised waste (SDW) content respectively.

2.2 Mixing, casting, curing and testing

Fly ash and gypsum were mixed by hand to produce the typical simulated desulphurised waste (SDW). All binder constituents were then mixed by hand until homogeneity was achieved. The fine aggregate was placed in a mixer followed by the binder. The dry materials were mixed for three minutes. The water was added over a period of 30 seconds while mixing continued until a homogenous mixture was achieved. The mortar was placed in steel moulds of dimensions $40\,\text{mm} \times 40\,\text{mm} \times 160\,\text{mm}$ prisms. After casting, all specimens were placed in a mist curing room at $20°C \pm 1°C$ and $95\% \pm 5\%$ relative humidity until demoulding. For most specimens this initial mist curing was for 24 hours, however, several specimens required a longer initial period due to an increase in setting times. After demoulding, all cubes were cured in water at $20°C$ for 28 days.

After 28 days of water curing, mortar specimens were immersed in 5% magnesium sulphate solution ($MgSO_4 \cdot 7H_2O$ or 2.44% $MgSO_4$). The sulphate solution was renewed every month and specimens were visually examined at different periods in order to assess the attack due to the exposure to sulphate solution.

3 RESULTS AND DISCUSSION

Figures 1 and 2 show the appearance of C-SDW mortars immersed for respectively 365 and 450 days in magnesium sulphate. The effects of sulphate attack are shown on the top face of the specimens with respect to SDW content of the mix. The reference specimen is on the right of the plate and is labelled SA1. During the first 60 days of immersion, a white layer forms on the surface of the reference specimens. The formation of the white layer became thicker with time. No cracking was observed with the reference mix after 150 days, however, the presence of the white layer may have masked the start of cracking. After 180 days, the edges and corners of the reference mix had started to soften, which resulted in loss of material on handling. Slight cracking was observed, and was more prominent around the edges and corners.

During the first 365 days the mortar containing 10% SDW follow a similar deterioration process to the reference mix, i.e. white discoloration, softening of edges and corners, slight cracking leading to loss of material. The process of deterioration appeared to be delayed by approximately 30 days, with the formation of the white

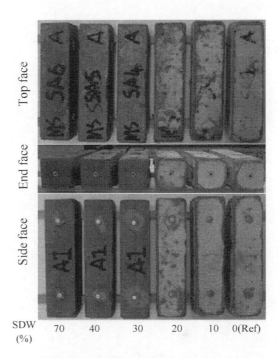

Figure 1. Sulphate attack of C-SDW mortars immersed in magnesium sulphate for 365 days.

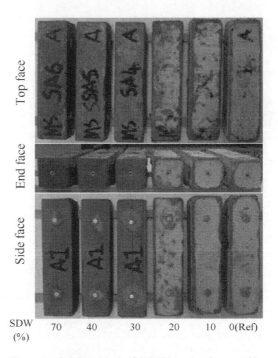

Figure 2. Sulphate attack of C-SDW mortars immersed in magnesium sulphate for 450 days.

223

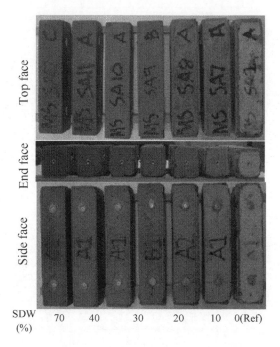

SDW (%)   70   40   30   20   10   0(Ref)

Figure 3. Sulphate attack of C-S-SDW mortars immersed in magnesium sulphate for 365 days.

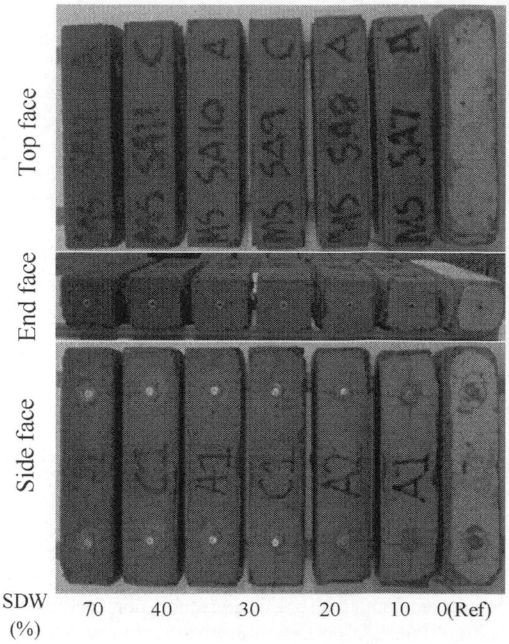

SDW (%)   70   40   30   20   10   0(Ref)

Figure 4. Sulphate attack of C-S-SDW mortars immersed in magnesium sulphate for 450 days.

layer on the surface of the specimens occurring after 90 days of immersion. Increasing the SDW content above 10% seemed to reduce the deterioration process. The formation of the white layer was slower, and during the first 365 days was almost non-existent for mixes containing 30% SDW and above. The softening, cracking and loss of material was minimised as the level of SDW increased, and as with the formation of the white layer, cracking was almost non-existent at 365 days.

Between 365 and 450 days, the reference mix and the mortar containing 10% SDW continue to exhibit a softening and deterioration of the edges and corners, and a clear distortion of the specimens was observed. At 450 days, large white deposits on he surface of the specimens masked any apparent cracking present, and made the specimens unrecognisable from the reference specimens cured in water for the same duration. After 450 days, the mixes containing SDW contents above 10% showed little signs of deterioration, with minimal softening, cracking and deterioration. Mixes containing SDW contents above 20% showed no discoloration or crystal growth on the surface of the specimen, therefore, making surface cracking visible. The deterioration of mixes containing SDW contents above 20% appears to be different to that of the reference mix, and the mix containing 10% and 20% SDW.

Figures 3 and 4 show the appearance of C-S-SDW mortars immersed for respectively 365 and 450 days in

magnesium sulphate. The effects of sulphate attack are shown for the top face of the specimens with respect to SDW content (%) and slag content (%) of the mix. The reference specimen is on the right of the plate and is labelled SA1.

During the first 365 days of exposure, the reference mix underwent considerable deterioration in the form of softening of the edges and corners, which resulted in a loss of material on handling. The formation of the white layer on the reference specimens after 60 days of exposure was not observed for the C-S-SDW mortars. During the first 90 days of exposure, the C-S-SDW mortars containing 0% and 10% SDW ($10_C90_S0_{SDW}$, $10_C80_S10_{SDW}$) start to exhibit advanced deterioration in the form of friable surfaces and cracking around the edges and corners. This resulted in flaking of material on handling. This sort of deterioration was not present in the reference mix until 180 days of exposure, indicating that the white layer formed on the surface of the reference specimens inhibits sulphate attack. As the SDW content increased beyond 10%, the amount of cracking observed decreased until almost no signs of deterioration were observed in the mix containing 70% SDW ($10_C20_S70_{SDW}$).

The replacement of cement with SDW significantly improved the sulphate resistance of mortars immersed in magnesium sulphate solution. The sulphate resistance of C-SDW mortars increased as the SDW content in the mix increases. The improvement in sulphate resistance manifests itself in the form of reduced cracking and deterioration. The composition of the SDW is predominantly fly ash (85% FA, 15% G). The sulphate resisting properties of fly ash in cement is well documented, and generally leads to improved resistance, when exposed to sulphate environments (Mangat & El-Khatib 1992, Metha 1986, Collepardi 2001, Jueshi 2001, Poon et al. 2000).

Sulphate attack occurs when Calcium Aluminate Hydrate (C-A-H) phase from cement hydration and calcium hydroxide (CH), formed during cement hydration, react with sulphates in the solution (i.e. $MgSO_4$). This results in the formation of gypsum and/or ettringite that leads to the deterioration and destruction of the specimens. The improved sulphate resistance due to the replacement of cement with SDW can be attributed to its chemical and mineralogical composition. Unfortunately, no attempts were made to determine the mineralogical composition of the SDW. The replacement of cement with SDW firstly reduced the C_3A and CaO content of the mix available for reaction. Secondly, the fly ash in the SDW undergoes long-term pozzolanic reactions, which consumes CH, one of the main reaction products in sulphate attack, to form additional cementing C-S-H and low calcium C-A-H phases (Neville 1995).

Gollop and Taylor (1996) reported that by increasing the slag content in cement mortars from 69% to 92%, sulphate resistance could be significantly improved when exposed to magnesium sulphate. However, the magnitude of the attack was much greater when exposed to magnesium sulphate. It was reported that the Al_2O_3 available for reaction was an important factor in the sulphate resistance of slag cements. For mixes containing the same amount of slag, if the Al_2O_3 content of the slag increased the sulphate resistance worsened. However, if the same slag type was used, increasing the slag content (increasing the Al_2O_3) improved the sulphate resistance. It was suggested that the improvement in sulphate resistance by increasing the slag content resulted from the reduction in the available Al_2O_3 for reaction, which was taken up in the formation of additional C-S-H. When exposed to magnesium sulphate, similar reactions associated with sodium sulphate attack occurred, however, the decomposition of the C-S-H was much more severe.

4 CONCLUSIONS

Replacing cement with increasing levels of a typical simulated desulphurised waste (SDW) resulted in mortars with superior sulphate resistance compared to reference mortars when immersed in magnesium sulphate solutions. The improved sulphate resistance of

mortars containing SDW was attributed to the dilution of C_3A and CH due to the reduction in cement, and the formation of expansive products, such as ettringite, during early periods of hydration. An increase in the SDW content led to the formation of a white surface layer and loss of material.

Mortars containing cement, ground granulated blastfurnace slag, and a typical simulated desulphurised waste (SDW) exhibited superior sulphate resistance compared to reference mortars immersed in magnesium sulphate.

All C-S-SDW mortars immersed in magnesium sulphate underwent some form of deterioration. The reason being that the magnesium sulphate reacts with the C-S-H phases as well as aluminate and CH phases. When immersed in sulphate, increasing the SDW content from 0 to 70% improved sulphate resistance. An increase in SDW caused softening and deterioration of the edges and corners.

REFERENCES

Bilodeau A, Malhotra VM, High-volume fly ash system: concrete solution for sustainable development, ACI Materials Journal, No. 97-M6, Jan–Feb 2000

Collepardi M, Ettringite formation and sulphate attack on concrete, Fifth CANMET/ACI International Conference, Recent Advances in Concrete Technology, Ed. Malhotra VM, SP 200-2, pp. 21–37, 2001

Gollop RS, Taylor HFW, Microstructural and microanalytical studies of sulphate attack: IV. Reactions of a slag cement paste with sodium and magnesium sulphate solutions, Cement and Concrete Research, Vo. 26, No. 7, pp. 1013–1028, 1996

Hughes DC, Sulphate resistance of OPC, OPC/fly ash and SRPC Pastes: Pore structure and permeability, Cement and Concrete Research, Vol. 15, pp. 1003–1012, 1985

Khatri RP, Sirivivatnanon V, Yu LK, Effects of curing on water permeability of concretes prepared with normal Portland cement and with slag and silica fume, Magazine of Concrete Research, Vol. 49, No. 180, pp. 167–172, Sept 1997

Jueshi Q, Caijun S, Zhi W, Activation of blended cements containing fly ash, Cement and Concrete Research, Vol. 31, pp. 1121–1127, 2007

Lea FM, LEA'S Chemistry of Cement and Concrete, Forth Edition, Arnold, ISBN 0 340 56589 6, Chapter 10 – Pozzolana and Pozzolanic Cements, pp. 471–633, 1998

Mangat PS, EL-Khatib JM, Influence of initial curing on sulphate resistance of blended cement concrete, Cement and Concrete Research, Vol. 22, pp. 1089–1100, 1992

Matthews JD, Performance of PFA concrete in aggressive conditions: 1. Sulphate resistance, Building Research Establishment Laboratory Report, 1995

Mehta PK, Effect of fly ash composition on sulphate resistance of cement, ACI Journal, pp. 994–1000, Nov–Dec 1986

Neville AM, Properties of Concrete, Fourth edition, Longman Group Ltd, ISBN 0-582-23070-5, Chapter 10-Durability of Concrete, pp. 482–537, 1995

Poon CS, lam L, Wong YL, A study on high strength concrete prepared with large volumes of low calcium fly ash, Cement and Concrete Research, Vol. 30, pp. 447–455, 2000

Santhanam M, Cohen MD, Olek J, Sulphate attack research – whither now? Cement and Concrete Research, 31, pp. 845–851, 2001

Tikalsky PJ, Carrasquillo RL, Influence of fly ash on the sulphate resistance of concrete, ACI Materials Journal, Vol. 8, No. 1, pp.69–75, Jan–Feb 1992

Wild S, Khatib JM, O'Farrell M, Sulphate resistance of mortar, containing ground brick clay calcined at different temperatures, Cement and Concrete Research, Vol. 27, No. 5, pp. 697–709, 1997

Yang S, Zhongzi X, Mingshu T, The process of sulphate attack on cement mortars, Advanced Cement Based Materials, No. 4, pp. 1–5, 1996

Effect of cement type on strength development of mortars containing limestone fines

J.M. Khatib
School of Engineering and the Built Environment, University of Wolverhampton, UK

B. Menadi & S. Kenai
Geomaterials Laboratory, Civil Engineering Department, University of Blida, Algeria

ABSTRACT: There has been an increasing interest in the use of crushed limestone fines (LSF) in the production of concrete where river sand is not widely available. This paper reports the compressive strength results of crushed sand mortar made with different types of cements and containing different amounts of LSF. Fine aggregates was partially replaced with 0, 5, 10, 15, 20% LSF. The binder to cement and the water to binder ratios were maintained constant at 1:3 and 0.55 respectively. Compressive strength testing was conducted at 2, 7, 28 and 90 days of curing. Generally, the use of LSF to replace crushed sand up to at least 20% (by mass) did not have detrimental effect on compressive strength.

1 INTRODUCTION

In recent years, there has been an increasing interest in limestone fines (LSF) from limestone quarries in concrete construction to overcome inherent deficiencies in river sand in particular regions of North Africa. Crushed limestone fine sand is a by-product of the quarry process and typically does not have a significant demand due to its high content of small particles whose diameters are less than 80 μm. The introduction of high content of crushed limestone fines to concrete mixes is limited due to its negative effects on water demand and strength of concrete. However, the use of these fines at a reasonable percentage can improve the fresh and hardened concrete properties. In some countries, the content of these small particles exceeds the standard allowable limit of 5% and most of the LSF are disposed of in landfill sites.

The annual production in Algeria, one of the North African countries, is 68 million tonnes of aggregates (fine and coarse). Nearly one half of the aggregate production goes to the building sector and one third to road construction. The fine aggregate consists of 22% (15 million tonnes) of the total aggregate produced. Between 8–25% of the fine aggregate contains particles whose diameters are below 80 μm (i.e limestone fines) and they are mainly used in road construction and usually sold at a lower price (about the third of other types of aggregate). One option to meet the increasing demand and shortage of fine aggregate is to use more LSF in construction as it is widely available

(Kenai et al., 1999). This will lead to conservation of finite natural resources.

Other countries in the world including Spain, France and Argentina are experiencing similar shortage in natural fine sand. Standards in these countries made greater usage of LSF in excess of 15% in cement manufacture. Standards include European Standards (EN197-1).

There is a number of research papers on the use of LSF in the production of concrete. Replacing sand with 12–18% LSF does not cause any harmful effect on the physical and mechanical properties of mortar and concrete (Kenai et al. 2006, Nehdi et al. 1996, Bonavetti & Irassar 1994, Ramirez et al. 1987, Bertrandy & Charbernoud, Chi et al. 2004, Aitcin & Mehta 1999, Donza et al. 2002, Donza et al. 1999, Poitevin 1999, Ramirez et al 1990). The main objective of this paper is to further examine the influence of replacing sand with LSF on strength of mortar. Three types of Portland cement were used and sand was partially replaced with 0, 5, 10, 15, and 20% LSF. The compressive strength was determined at 2, 7, 28 and 90 days. In addition shrinkage of mortar containing 0 and 15% LSF was determined during the first 180 days.

2 EXPERIMENTAL

The chemical and physical properties of the cement and limestone fine (LSF) used in this investigation

Table 1. Composition of the various cements.

Cement type	Cement A	Cement B	Cement C
Chemical properties			
SiO_2	21.33	21.50	20.32
CaO	64.65	65.52	64.34
MgO	1.21	1.05	0.56
Fe_2O_3	3.07	2.84	3.20
Al_2O_3	5.74	5.13	4.71
Loss on ignition	0.21	2.02	4.26
SO_3	2.30	2.53	1.26
Insoluble residue	0.06	1.60	–
Free CaO	0.30	0.39	0.21
Mineralogical composition			
C_2S	19.88	15.45	20.30
C_3S	53.44	56.58	50.59
C_3A	9.96	8.79	7.7
C_4AF	9.31	8.63	9.55

Table 2. Chemical properties of crushed limestone fines.

Oxide	SiO_2	CaO	MgO	Fe_2O_3	Al_2O_3	H_2O	Loss on ignition
%	1.78	54.3	0.20	0.34	0.79	2.8	42.5

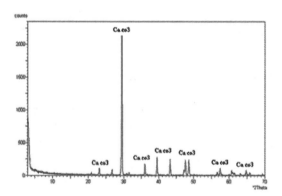

Figure 1. X-ray diffraction of crushed limestone fines.

are shown in Tables 1 and 2 respectively. The X-ray diffraction of the crushed sand limestone fine is shown in Figure. 1. Three types of cements were used, cement A, cement B and cement C. The LSF used is obtained by sieving the crushed sand through 80 μm sieve. The fine aggregates used were crushed sand with maximum particle size of 4 mm. The grading curve of the fine aggregate is presented in Figure 2. The mix proportions, by mass, for mortars were 1 (cement): 3 (sand): 0.55 (water). The binder content was kept constant at 450 kg/m³ for all mortars.

Specimens were placed in prismatic moulds $40 \times 40 \times 160$ mm. They were left covered at 20°C until demoulding. After 24 hours, demoulding took

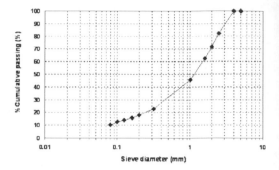

Figure 2. Grading curve of the fine aggregate.

place and for compressive strength determination, specimens were placed in water at 20°C until the time of testing. Compressive strength was determined at 2, 7, 28 and 90 days of curing according to AFNOR (1990). For the determination of shrinkage, specimens were placed in a controlled chamber at 20°C and 60% RH and their length was monitored up to 180 days of curing.

3 RESULTS AND DISCUSSION

3.1 *Compressive strength*

Experimental results of the effect of limestone fines as sand replacement on the compressive strength development of crushed sand mortar for the three types of cements used in this investigation can be seen in Figures 3 to 5. Higher compressive strength can be observed for cement B and C compared to cement mortar with and without fines. This may be attributed to the class type of cement. During the first 2 days and 7 days, the compressive strength increases with the increase of fines up to 10% for mortar with cement B and C and up to 15% for mortar with cement A. The increase of strength at early ages may be due to filler effect, acceleration of hydration of C_3A and C_3S, changes in the morphology of the C-H-S gel, and formation of carboaluminates by the reaction of $CaCO_3$ with C_3A. These results are in accordance with several studies in the case of limestone filler in cement (Nehdi et al 1996, Tsivillis et al. 2000). The same trend was observed at long-term for mortar with cement A and B. However, for mortar with cement C, the compressive strength decreases sharply between 5% and 15% of fines at the age of 90 days. The decrease in strength for all mortars with different cement types at higher level of LFS may be due the dilution effect. By comparing the different compressive strength results from these figures, it can be concluded that the use of crushed limestone fines in the range of 10% to 15% could be used without affecting the compressive strength.

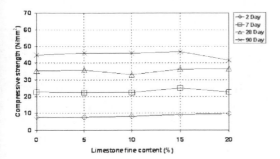

Figure 3. Compressive strength of cement-A mortars containing varying amounts of limestone fines.

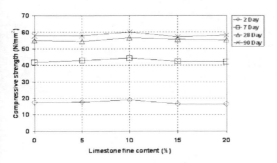

Figure 4. Compressive strength of cement-B mortars containing varying amounts of limestone fines.

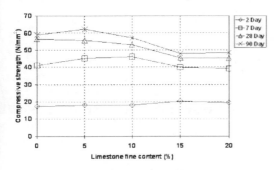

Figure 5. Compressive strength of cement-C mortars containing varying amounts of limestone fines.

The effect of cement type on relative strength to 28 days strength of crushed sand mortar with and without limestone fines as sand replacement material are illustrated in Figures 6 to 10. It can be seen from these figures that the relative compressive strength increases with the age for all cement types with and without limestone fines. This can be attributed to the development of the hydration of cement. The rate development of relative strength of cement mortar B and C was greater initially (early ages) and decreased at the age of 90 days in comparison with cement mortar A for all limestone fines content (0, 5, 10, 15 and 20%).

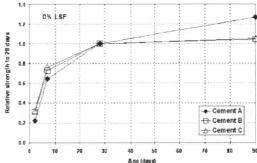

Figure 6. Effect of cement type on relative strength to 28 days strength of mortar without limestone fines.

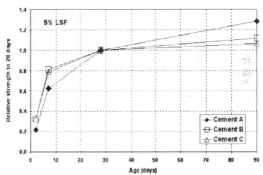

Figure 7. Effect of cement type on relative strength to 28 days strength of mortar containing 5% limestone fines.

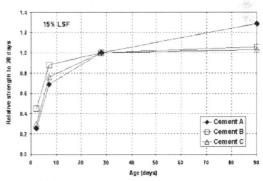

Figure 8. Effect of cement type on relative strength to 28 days strength of mortar containing 10% limestone fines.

The incorporation of 10% of limestone fines as sand replacement gives the higher difference in relative compressive strength between cement C and A. The lower value in relative strength was obtained for 20% of LFS at the age of 90 days.

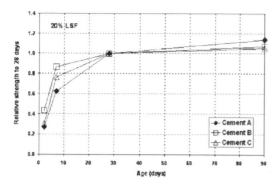

Figure 9. Effect of cement type on relative strength to 28 days strength of mortar containing 15% limestone fines.

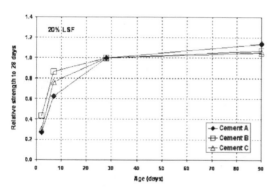

Figure 10. Effect of cement type on relative strength to 28 days strength of mortar containing 20% limestone fines.

3.2 *Drying shrinkage*

On the basis of the optimum limestone fines found previously in the experimental study of cement mortar compressive strength, the amount of crushed limestone fines LSF was fixed at 15% as replacement of crushed sand for all cement mortars types.

Results of drying shrinkage of crushed sand mortars containing 0% and 15% of limestone fines for the three types of cement used in this investigation are shown in Figure 11. It can be seen from this figure that mortar with 15% of limestone fines as replacement of crushed sand, increases the drying shrinkage at all ages and for all cements used. The increase of shrinkage of mortars with limestone fines may be attributed to the formation of carboaluminates. Cement mortar (C) with 15% of crushed limestone fines exhibits higher shrinkage compared to mortars A and B. Whereas, the comparable values of shrinkage were obtained for mortars B0 and C0. The increase of shrinkage from 0% to 15% is about 8%, 11% and 22% for cement A, B and C respectively at 180 days. The results are in agreement with those obtained by Bonavetti et al. (1994).

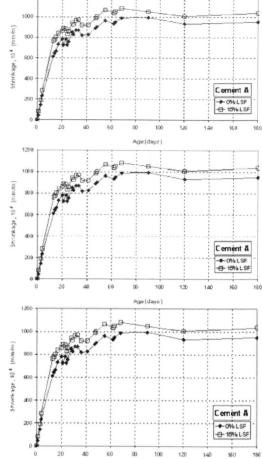

Figure 11. Effect of LSF on shrinkage for different types of cement.

4 CONCLUSIONS

The results of this study lead to the following conclusions:

- From this experimental study on cement mortar, substitution of 10% to 15% of crushed limestone fines as sand replacement improved the compressive strength at all curing ages and for all cement types used.
- An increase of the crushed limestone fines content resulted in an increase of the drying shrinkage irrespective of the type of cement. The shrinkage of cement mortar (C) with and without LSF was lower than that of cement mortar A and B and remains in the limit proposed by standards.

REFERENCES

AFNOR, European Committee for Standardization, NF EN 196-1, 'Méthodes d'essais des ciments : Détermination des résistances mécaniques', AFNOR, Paris Mars 1990.

C. Chi, Y. Wu, C.Riefler; The use of crushed dust production of self-consolidating concrete, Recycling Concrete and other materials for sustainable developement, Editors Tony C.Liu Christian Meyer; ACI International SP-219, 2004.

European Committee for Standardization, prEN 197-1, Draft September, 1996.

H. Donza, M. Gonzalez, and O. Cabrera, Influence of fine aggregate mineralogy on mechanical properties in high strength concrete, Second international conference on high performance concrete, Gramado, Brazil, (1999).

H. Donza, O. Cabrera, E.F. Irassar, High-strength concrete with different fine aggregate, Cem. Conc. Res. 32 (2002), 1719–1729.

J.L. Ramirez, J.M Barcena, and J.I Urreta, Proposal for limitation and control of fines in calcareous sands based upon their influences in some concrete properties, Materials and Structures, 1990; 23, pp. 277–288.

J.L. Ramirez, J.M. Barcena, and J.I. Urreta, Sables calcaires à fines calcaires et argileuses : influence et nocivité dans les mortiers de ciment, Matériaux et Construction, 1987, 20, 202–213.

M. Nehdi, S. Mindess, and P.C. Aitcin, Optimisation of high strength limestone filler cement mortars, Cem.concr. Res. 26 (6) (1996) 883–893.

P. Aitcin, P.K. Mehta, Effect of coarse aggregate characteristics on mechanical properties of high-strength concrete, ACI Mater. J. 87 (2) (1999) 103–107.

P. Poitevin, Limestone aggregate concrete, usefulness and durability, Cem. Conc. Com, 21(11), (1999), pp. 99–105.

R. Bertrandy, and J.L. Chabernaud, Study of the influence of calcareous fillers on concrete. Travaux; N 437-438; 38–52.

S. Kenai, B. Menadi, and M. Ghrici, Performance of limestone cement mortar, Eight CANMET/ACI Internatioanl Conference on Recent Advances in Concrete Technology, Montréal-Canada, 2006, pp. 39.

S. Kenai, Y. Benna, and B. Menadi, The effect of fines in crushed calcareous sand on properties of mortar and concrete, Int. Conf. on Infrastructure regeneration and rehabilitation, Sheffield, Editor R.N. Swamy, 1999, pp. 253–261.

S. Tsivillis, G Batis, Chaniotakis E, Grigoriadis Gr, Theodossis D. Properties and behavior of limestone cement concrete and mortar. Cem Concr Res 2000; 30: 1679–1683.

V.L. Bonavetti, E.F. Irassar, The effect of stone dust content in sand, Cem. Conc. Com. 24 (3)(1994) 580–590.

Excellence in Concrete Construction through Innovation – Limbachiya & Kew (eds)
© 2009 Taylor & Francis Group, London, ISBN 978-0-415-47592-1

Adiabatic temperature rise of metakaolin mortar

J.M. Khatib
School of Engineering and the Built Environment, University of Wolverhampton, Wolverhampton, UK

S. Wild
School of Technology, University of Glamorgan, Pontypridd, Mid Glamorgan, UK

R. Siddique
Department of Civil Engineering, Thapar University, Patiala (Punjab), India

S. Kenai
Geomaterials Laboratory, Civil Engineering Department, University of Blida, Algeria

ABSTRACT: Using metakaolin (MK) in concrete has gained momentum in recent years. This paper investigates the temperature rise in mortar containing MK during the first 48 hours of hydration. Mortar specimens in the form of 300 mm cubes were prepared. The Portland Cement (PC) was partially replaced with with between 0 and 25% MK (by mass) and between 0 and 25% fine sand. The mortar had a proportion of one part binder to 3 parts aggregate sand. The water to binder (PC + MK) and (PC + FS) ratios were kept constant at 0.55 for all mortar mixes. The results show that the peak temperature increases with increase in MK content up to between 10% and 15% MK, beyond which the peak temperature starts to decrease. However the peak temperature still remains higher than that of the control (0% MK) at 20% MK but drops below the control peak temperature at 25% MK. Also the time taken to reach the peak temperature is reduced in the presence of MK indicating acceleration of hydration.

1 INTRODUCTION

In recent years, there has been an increasing interest in the utilisation of metakaolin (MK) as a supplementary cementitious material in concrete (Zhang & Malhotra, 1995, De Silva & Glasser, 1990, Ambroise et al, 1994, Basher et al, 1999). MK is an ultra fine pozzolan and is produced by calcining kaolin at temperatures between 700 and 900°C and it consists predominantly of combined silica and alumina. The presence of MK increases the strength of concrete especially during the early ages of hydration (Wild et al, 1996, Wild et al, 1997). After 14 days of curing the contribution that MK provides to concrete strength is reduced (Wild et al, 1996). It was suggested (Wild et al, 1996) that the increase in compressive strength of MK concrete relative to the equivalent PC control concrete is caused by the combined effects of: (i) the ultra-fine MK particles filling the void space between the cement particles, (ii) the acceleration of cement hydration, and (iii) the pozzolanic reaction of the MK. Although the total pore volume of PC-MK paste has been shown (Khatib & Wild, 1996) to slightly increase with increase in the MK content, the pore structure of paste becomes much finer when cement is partially replaced with MK (Khatib & Wild, 1996, Frias & Cabrera, 2000). The improvement in pore structure occurs for MK levels up to least 20% (Khatib & Wild, 1996). The incorporation of MK in cement mortar also results in an increase in sulphate resistance and this increase is systematic as the MK content in the mortar increases (Khatib & Wild, 1998) although at very low MK levels (up to 5%) sulphate resistance may be reduced. This increased sulphate resistance in mortars containing MK is attributed to the reduction in portlandite content (which reduces the gypsum and ettringite formation) and the refinement in pore structure (which hinders the ingress of sulphate ions). Also Zhang & Malhotra, 1995 found that the resistance of 10% MK-90% PC concrete to the penetration of chloride ions was significantly higher than that of the control and Bai et al., 2003 showed that for ternary MK-PFA-PC concrete immersed in seawater , increasing MK content generally leads to reduced penetration of chloride. In addition Zhang & Malhotra, 1995 found that MK concrete showed excellent performance under freezing and thawing with a durability factor of about 100. With respect to concrete durability one aspect of concrete production that needs to

Table 1. Composition and properties of cement and metakaolin.

		PC^1	MK^2
SiO_2	%	20.2	52.1
Al_2O_3	%	4.2	41.0
Fe_2O_3	%	2	4.32
CaO	%	63.9	0.07
MgO	%	2.1	0.19
SO_3	%	3	–
Na_2O	%	0.14	0.26
K_2O	%	0.68	0.63
Insoluble Residue	%	0.37	–
Loss on Ignition	%	2.81	0.6
Free Lime	%	2.37	–
Specific Surface Area	m^2/kg	367.8	12,000
Residue Retained on 45 μm Sieve	%	15.16	–
Initial Set	min	115.0	–

1. Portland cement. 2. metakaolin.

be taken into consideration, particularly when pouring large volumes of concrete, is the temperature rise which, if not controlled can lead to thermal cracking. Metakaolin is known to be a highly active pozzolan and to make a significant contribution to the heat of hydration of PC-MK blends (Curcio et al., 1998). Therefore a knowledge of the temperature rise of PC-MK concrete and mortar provides a useful indication of what will occur in practice. In the current work the temperature rise of PC-MK mortars with a wide range of MK contents is determined over the first 48 hours of hydration. The data are compared with the temperature rise of equivalent PC-fine sand (FS) mortars.

2 EXPERIMENTAL

2.1 Materials

The mix constituents were Portland cement, metakaolin (MK) supplied by Immerys, fine sand (FS), water and aggregate sand. The aggregate sand used complied with class M of BS 882, 1992 and the fine sand (FS) used was silica sand (>97% silica) with particle size below 300 μm. The compositions of the cement and MK are given in Table 1.

2.2 Mix proportions

A total of eleven mortar mixes were employed to examine the influence of MK on temperature rise during the early period of hydration. The control mix (M1) had a proportion of 1 (cement): 3 (sand) and did not include MK. In mixes M2, M3, M4, M5 and M6, the cement was partially replaced with 5%, 10%, 15%, 20% and 25% MK respectively. In mixes M7-M11, the cement was partially replaced with 5%, 10%, 15%, 20% and

Table 2. Details and composition of mixes.

	Proportions (% by mass of binder[1])		
Mix No.	PC	MK	FS^2
M1	100	0	0
M2	95	5	0
M3	90	10	0
M4	85	15	0
M5	80	20	0
M6	75	25	0
M7	95	0	5
M8	90	0	10
M9	85	0	15
M10	80	0	20
M11	75	0	25

1. PC + MK 2. Fine sand.

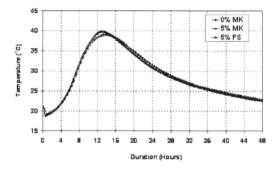

Figure 1. Effect of replacing cement with 5% MK or FS on the temperature profile.

25% fine sand (FS). The water to binder ratio was kept constant at 0.55. The binder consists of cement and MK. Details of the binder composition of the mixtures are given in Table 2.

2.3 Casting, curing and monitoring

Cubes of 300 mm in size were used for the monitoring of temperature during the early stages of hydration. The mortar mixtures were cast in timber cube moulds in a room kept at 20°C and 70% r.h. with a thermocouple inserted in the centre of the cube. The temperature was continually monitored during the first 48 hours after casting.

3 RESULTS

The temperature profiles during the first 48 hours for the mortar mixes containing 0–25% MK and 0–25% FS are presented in Figures 1–6 respectively. Generally, the temperature continues to rise reaching a

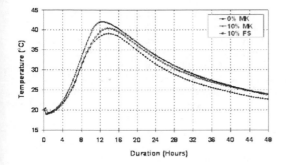

Figure 2. Effect of replacing cement with 10% MK or FS on the temperature profile.

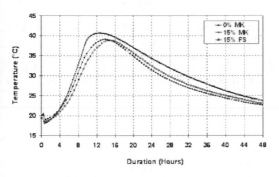

Figure 3. Effect of replacing cement with 15% MK or FS on the temperature profile.

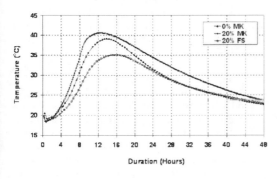

Figure 4. Effect of replacing cement with 20% MK or FS on the temperature profile.

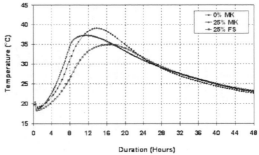

Figure 5. Effect of replacing cement with 25% MK or FS on the temperature profile.

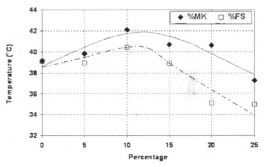

Figure 7. Effect of MK and FS content on peak temperature of mortar.

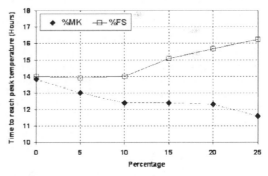

Figure 8. Effect of MK and FS on time taken to reach peak temperature of mortar.

maximum temperature between 11.8 and 13.8 hours. Beyond that, the temperature starts to slowly decrease. Using MK levels of up to 10% as partial cement replacement tends to increase the peak temperature. Beyond 10%, the peak temperature begins to reduce. This is clearly illustrated in Figure 7 where the peak temperature is plotted against MK content. Despite the diminishing quantity of PC, increasing the MK level from between 5 and 20% yields maximum temperatures higher than that of the PC control mix (i.e. 0% MK) indicating a significant contribution to the heat output, from the pozzolanic activity of the MK. However at the 25% MK level the maximum temperature drops below that of the control. As might be expected the maximum temperature reached by the PC-FS mortars is consistently below that of the equivalent PC-MK mortars (Figure 7) and the difference increases as the PC replacement level increases. Somewhat surprisingly however at 10% replacement of PC with FS the peak temperature achieved is slightly greater ($\sim 1.3°C$) than that for the PC control. Figure 8 shows the time taken to reach peak temperature versus replacement level for the PC-MK and PC-FS mortars. For the PC-MK mortars increasing MK levels produce decreasing times to reach peak temperature. This indicates acceleration of the hydration reactions particularly for the compositions from 0 to 10% MK. In contrast for PC-FS mortars increasing FS levels produce increasing times to reach peak temperature. This indicates retardation of hydration reactions, particularly for compositions from 10% to 25% FS.

4 DISCUSSION

It has been demonstrated by a number of authors (Snelson et al., 2008, Bai & Wild, 2002, Frias et al., 2000) that MK, when partially replacing PC, enhances the overall hydration process, due to the onset of rapid pozzolanic action. That is, the reaction of MK with CH and H contributes to the cumulative heat evolved by the hydrating systems, to the extent that with increasing MK content (up to a specific limit) the total heat evolved is greater than that from 100% PC. However the MK-CH-H reaction is controlled by the availability of CH which itself is provided by the hydrating PC. Thus, as the proportion of MK increases and that of PC decreases, the supply of CH from the hydrating PC will decrease but the demand for CH, from the increasing proportion of MK, will increase. Hence there will be a limiting MK : PC ratio above which the reaction of MK with CH will be curtailed due to insufficient supply of CH. This will decrease the rate of heat output and thus the maximum temperature reached by the PC-MK mortar. This is clearly illustrated in Figure 7 where as the percentage of MK increases the maximum temperature achieved increases up to an MK level between 10 and 15% and then at higher MK levels subsequently decreases.

Surprisingly the FS, which is a fine silica sand and not expected to be pozzolanic, shows a similar peak-temperature – composition profile to that of the PC – MK mortar. Although as the replacement level increases, the peak temperatures do drop substantially below those of the equivalent PC – MK mortars, although at 10% FS the peak temperature slightly exceeds that of 100% PC mortar. However it has been demonstrated (Gutteridge & Dalziel, 1990, Kadri & Duval, 2002) that materials blended with PC do not necessarily have to be pozzolanic in order to influence the hydration kinetics of the cement blend. For example Kadri & Duval (2002) have shown that mortar with a binder comprising 90% PC and 10% fine quartz sand, gives off more heat in the first six days of hydration than does the 100% PC control mortar. They attributed this to the fine quartz particles acting as heterogeneous nucleation sites at which the primary cement hydrates could precipitate, thus driving the PC hydration reaction forward and increasing the rate of hydration. However they do suggest also that the quartz particles exhibit some pozzolanic activity, although their evidence for this is unconvincing. It is in fact highly unlikely that the quartz is pozzolanic as crystalline quartz is very inert and even when highly strained reacts only very slowly.

However Snelson et al., (2008) have concluded, from calorimetry and heat evolution measurements, that during the early stages of hydration of Portland cement – PFA blends the PC hydration is significantly enhanced even though at this stage the PFA does not show any significant pozzolanic reaction. They attribute this enhanced hydration to the effective increase in w/c ratio as the PFA to PC ratio increases. In addition the PFA particles, which have a similar particle size range to that of the PC increase the separation of the cement particles allowing easier access of water to cement particle surfaces. It is suggested that the FS is behaving in a similar manner in the current work.

5 CONCLUSIONS

- Partial replacement of PC with MK in mortar (up to a particular MK level), produces increases in peak temperatures above that for PC mortar. This is attributed to the contribution of the pozzolanic activity of the MK to the overall heat of hydration of the PC-MK blend. However above this level (between 10 and 15% MK) the peak temperature declines. This is attributed to the reduced supply of CH from the decreasing amount of PC, which inhibits the MK-CH-H reaction.
- Partial replacement of PC with FS in mortar also produces increases in peak temperatures at low FS levels but substantially lower than those for equivalent PC-MK mortars. The peak temperatures (other than at 10% FS) are below those for PC mortar. The behaviour is attributed to the effective increase in w/c ratio which enhances PC hydration rates and the increased separation of PC particles which allows easier access of water to the cement particle surfaces. However at higher FS levels ($>10\%$) the reduction in PC content becomes the dominant effect and thus the peak temperatures achieved fall rapidly.
- The time to peak temperature for PC-MK mortars (Figure 8) shows a significant decrease with increase in MK level, particularly at between 0 and 10% MK. This is attributed to the contribution of the pozzolanic reaction of the MK in accelerating the overall hydration rate of the system.
- The time to peak temperature for PC-FS mortars (Figure 8) shows a significant increase in time to

peak temperature, particularly at FS levels above 10% FS. This is attributed to an overall reduction in hydration rate as a result of the diluting effect of the FS at higher FS levels (>10%).

ACKNOWLEDGEMENTS

The authors would like to thank Mr Mansell for conducting the experimental programme.

REFERENCES

Ambroise, J., Maxmilien, S. and Pera, J., "Properties of MK blended cement", Advanced Cement Based Materials, Vol. 1, pp. 161–168, 1994.

Bai J., Wild S. and Sabir B. B. "Chloride ingress and strength loss in concrete with different PC-PFA-MK binder compositions exposed to synthetic seawater". *Cement and Concrete Research*, **33**, (2003), pp 353–362.

Bai, J. and Wild, S. "Investigation of the temperature change and heat evolution of mortar incorporating PFA and metakaolin", Cement & Concrete Composites, Vol. 24, pp. 201–209, 2002.

Basheer, P. A. M., McCabe, C. C. and Long, A. E., "The influence of metakaolin on properties of fresh and hardened concrete", Proc. Int. Conf. Infrastructure Regeneration and Rehabilitation Improving the Quality of Life through Better Construction, Swamy, R. N. (Ed). pp 199–211, 1999.

Curcio, F., DeAngelis, B. A. and Pagliolico, S., "Metakaolin as a pozzolanic microfiller for high-performance mortars", Cement and Concrete Research, Vol. 28, No. 6, pp. 803–809, 1998.

De Silva, P. S., and Glasser, F. P., "Hydration of cements based on metakaolin: thermochemistry", Advances in Cement Research, Vol. 3, pp 167–177, 1990.

Frias, M. and Cabrera, J., "Pore size distribution and degree of hydration of metakaolin-cement pastes", Cement and Concrete Research, Vol. 30, pp. 561–569, 2000.

Frias, M. Sanchez de Rojas, Cabrera, J. "The effect that the pozzolanic reaction of metakaolin has on the heat evolution in metakaolin-cement mortars", Cement and Concrete Research, Vol. 30, pp. 209–216, 2000.

Gutteridge, W.A. and Dalziel, R. "Filler cement: The effect of the secondary component on the hydration of Portland cement", Cement and Concrete Research, Vol. 20, pp. 778–782, 1990.

Kadri, E.H. and Duval, R. "Effect of ultrafine particles on heat of hydration of cement mortars", ACI Materials Journal, Vol. 99, No. 2, pp. 138–142, 2002.

Khatib, J. M., and Wild, S., "Pore size distribution of metakaolin paste", Cement and Concrete Research, Vol. 26, pp. 1545–1553, 1996.

Khatib, J. M., and Wild, S., "Sulfate resistance of metakaolin mortar", Cement and Concrete Research, Vol. 28, pp 83–92, 1998.

Snelson, D. G., Wild, S. and O'Farrell, M. "Heat of hydration of Portland Cement-Metakaolin-Fly ash (PC-MK-PFA) blends". Cement and Concrete Research (2008).

Wild, S. and Khatib, J. M., "Portlandite consumption in metakaolin cement pastes and mortars", Cement and Concrete Research, Vol. 27, No. 1, pp 127–146, 1997.

Wild, S., Khatib, J. M., and Jones, A., "Relative strength pozzolanic activity and cement hydration in superplasticised MK concrete", Cement and Concrete Research, Vol. 26, pp 1537–1544, 1996.

Zhang, M. H., and Malhotra, V. M., "Characteristics of a thermally activated alumino-silicate pozzolanic material and its use in concrete", Cement and Concrete Research, Vol. 25, pp 1713–1725, 1995.

Excellence in Concrete Construction through Innovation – Limbachiya & Kew (eds)
© 2009 Taylor & Francis Group, London, ISBN 978-0-415-47592-1

Cement-based composites for structural use: Design of reactive powder concrete elements

G. Moriconi & V. Corinaldesi

Dept. of Materials and Environment Engineering and Physics, Università Politecnica delle Marche, Italy

ABSTRACT: Reactive Powder Concrete (RPC), with compressive strength higher than 200 and up to 800 MPa as well as flexural strength higher than 60 and up to 150 MPa, at the moment potentially represents a new material for structural use in building and engineering in general, even though its application fields have not yet been well defined. RPC can be also considered as the ultimate step in the development of high performance concrete, even though its classification as a concrete material may be not quite proper, based on its microstructure and mechanical behaviour. Also, its production technology, by pressure moulding as well as extrusion, takes it even further from a common concrete. The wide range of achievable strengths for RPC requires careful design of the material, strictly related to the structural design and appropriate to the specific project, with maximum cooperation between materials engineering and structural engineering. For this, RPC can be used at best by developing new shapes and structural types specially designed for it. In this paper potential application of RPC for high span roofing structural elements is exploited and discussed in comparison with other materials typically used for this particular structural application. By applying external pre-tensioning to RPC elements, traditional limits of cement-based materials as inability to bear tensile stress and fragile behaviour can be overcome. Moreover, new structural characteristics like slenderness and lightness can make RPC able to compete with steel in the fabrication of very high span durable elements.

1 INTRODUCTION

Historically, new materials are related to the shape and development of new structural concepts. One need only think of the megalithic structures, in which stone prevented a span higher than 5 m, until the introduction of pozzolanic cement, used to join bricks and stones, which allowed the building of high spanning arch structures, covering up to about 50 m, like the Pantheon's dome in Rome.

Further evidence is provided by steel as a structural material, which became available between the eighteenth and the nineteenth century and whose high tensile strength permitted sweeping changes in building technology by allowing higher span beams, frame structures, truss girders, suspended structures, tall buildings, and so on.

In actual construction technology, structures are mainly built by concrete, timber or steel; however, new composite materials, reinforced by polymer, metal, glass, or carbon fibres, are in prospect of appearing, giving rise to new interesting practical applications owing to their improved mechanical performance (Collepardi et al. 2001).

A new general category of so-called CBC (Chemically Bonded Ceramics) materials resulted from recent research aimed at the attempt to reduce micro-porosity of cementitious materials. The term CBC attributed by Roy (1987) to this new class of cementitious materials points out, beyond the chemical nature of the involved bond, the inorganic, non-metal character of the material, which turns ceramic because of the particular processes involved in its manufacturing.

The CBC materials (Fig. 1) can be grouped in two large categories (Roy 1992): MDF (Macro Defect Free) and DSP (Densified with Small Particles) materials, the main difference being the role played by the polymeric component in the manufacturing process.

In MDF materials (Birchall et al. 1981) fully hydro-soluble polymers play a very important role in order to significantly change the rheology of the cement paste and so to obtain a dough material, able to be extruded or rolled.

In DSP materials, instead, sulphonated or acrylic polymers make possible either the compressive moulding of wet powders or the soft casting of flowable mixtures. Among the DSP materials, RPC (Re-active Powder Concrete), with compressive strength higher than 200 and up to 800 MPa as well as flexural strength higher than 60 and up to 150 MPa, at the moment represents potentially a new material for structural use in building and engineering in general, even though its application field has not yet been well defined. RPC can be also considered as the ultimate step in the

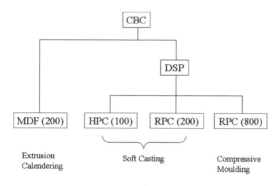

Figure 1. Outline of innovative cementitious materials and their related manufacturing process. Numbers enclosed in brackets, expressed as MPa, stand for compressive strength of HPC or RPC and flexural strength of MDF.

Table 1. RPC mixture proportions, processing treatment and related mechanical performance.

Ingredients (kg/m^3)	RPC	RPC 200	RPC 600	RPC 800
Portland Cement	955	1000	1000	1000
Silica Fume (18 m^2/g)	229	230	230	230
Fine Aggregate (150–400 µm)	1051	1100	500	–
Very Fine Quartz Sand (diameter 10 µm)	–	–	390	390
Amorphous Silica (35 m^2/g)	10	–	–	–
Superplasticizer	13	19	19	19
Steel Fibres (L = 13 mm, L/d 72)	191	175	–	–
Micro-Fibres (L = 3 mm)	–	–	630	630
Metal Aggregates (diameter < 100 µm)	–	–	–	490
Water	153	190	190	190
Treatment				
Compressive Stress (on fresh mixture, MPa)	–	–	50	50
Curing Temperature (°C)	20	90	250–400	250–400
Mechanical Performance				
Compressive Strength (MPa)	200	230	680	810
Flexural Strength (MPa)	50	60	45	140
Elastic Modulus (GPa)	50	60	65	75
Fracture Energy (kJ/m^2)	20	40	12	20

development of HPC (High Performance Concrete), even though its classification as a concrete material may be not quite proper based on its microstructure and mechanical behaviour.

Also, its production technology, by pressure moulding as well as extrusion, takes it even further from a common concrete. The wide range of achievable strengths for RPC requires careful design of the material, strictly related to the structural design and appropriate to the specific project, with maximum co-operation between materials engineering and structural engineering. For this, RPC can be used at best by developing new shapes and structural types specially designed for it.

As for any new building material, one of the main issues in RPC initial use is represented by its high production cost, even if economy can be achieved in the long term by lower maintenance cost and longer service life, as a consequence of RPC extraordinary durability.

Another obstacle to remove is to consider RPC as an ordinary concrete by measuring its performance on traditional structures in which RPC strength levels are not required. This means that new shapes and structural typologies must be developed for this material in order to maximize its performance.

Within this frame work, the paper presents the experimental results obtained by the mechanical characterisation of RPC prepared in the laboratory, and, based on these data, exploits its use for high span roofing applications in comparison to other typical structural materials employed for this purpose.

2 RPC MIXTURE PROPORTIONS AND EXPERIMENTAL APPROACH

The achievement of DSP materials is based on combined use of water-soluble polymers and ultra-fine

(≤ 0.1 µm) solid particles, which mainly consist of amorphous silica. The role of water-soluble polymers is to improve the rheological behaviour of cement mixtures with a very low amount of water. The role of ultra-fine silica particles is to reduce interstitial porosity among cement grains and to ensure the formation of calcium hydro-silicates by reaction with hydrolysis lime from cement hydration.

The ultimate goal is to produce easily formable materials through the soft casting technique in addition to the compressive moulding technique. By this method, even large sizes and complicated shapes may be produced, also by using extremely flexible reinforcing fibres (polymeric or amorphous cast-iron-based), instead of ordinary steel fibres.

In Table 1 typical mixture proportions of differently prepared RPC (Richard & Cherezy 1994) are reported together with the achievable mechanical performance. However, in this work a different aim was pursued: to obtain typical performance of RPC 200 by using in the mixture easily available raw materials as in common practice for precast concrete. In this way, a cement type CEM II/A-L 42.5 R was used instead of CEM I

Table 2. Mixture proportions, processing treatment and related mechanical performance of laboratory prepared RPC materials.

Ingredients (kg/m^3)	RPC 200-a	RPC 200-b
CEM II/A-L 42.5 R Cement	960	960
Silica Fume (18 m^2/g)	250	250
Limestone Aggregate (0.15–1 mm)	960	960
Acrylic-based Superplasticizer	96	96
Brassed Steel Fibres (L/d = 72)	192	192
Water	240	240
Treatment		
Curing Temperature, °C	20	160
Mechanical Performance		
Compressive Strength (MPa)	150	170
Flexural Strength (MPa)	33	34
Tensile Strength (MPa)	14	15
Fracture Energy (J/m^2)	44000	45000
Secant Elastic Modulus (GPa)	36	40
Tangent Elastic Modulus (GPa)	63	77
Poisson Modulus	0.19	0.17
Bond Strength with Steel (MPa)	32	34

Table 3. Characteristics and performance data of different construction materials.

	R.C.	Glulam	Steel	RPC 200	RPC 800
Elastic modulus (GPa)	25	12	210	60	75
Compressive strength (MPa)	30	32	360	200	800
Tensile strength (MPa)	3	15	360	45	100
Flexural strength (MPa)	5	32	360	60	130
Unit weight (kN/m^3)	25	5	78.5	23	28
Specific elasticity (10^{6}m)	1.0	2.4	2.7	2.6	2.7
Specific strength (10^{3}m)	1.2	6.4	4.6	8.7	28.6
Elastic strain (%)	0.15	0.25	0.18	0.33	0.80
Ultimate strain (%)	0.30	0.25	14	2	2
Ductility (%)	2.0	1.0	77	6.1	2.5
Fracture energy (kJ/m^2)	0.3–0.4	–	–	20–40	20

52.5 R as usual in RPC mixtures. Moreover, a limestone instead of quartz aggregate was used, which was also coarser (0.15–1 mm) than usual (150–600 μm). Finally, a lower quality black type silica fume was added.

According to this approach, the influence of an easily attainable thermal treatment, such as 24 hours air curing at 160°C, on the mechanical performance of this mixture was also evaluated. The thermal treatment was applied on demoulded H-shaped specimens after 1 day's casting. The two RPC materials in this way obtained are later on labelled RPC 200-a (without thermal curing) and RPC 200-b (thermally cured) respectively, notwithstanding that a compressive strength of 170 MPa was achieved instead of 200 MPa because of the change in the specification of the raw materials.

The mixture proportions of the RPC materials prepared for this work are reported in Table 2 together with the experimental results of the tests performed on them.

3 COMPARISON OF STRUCTURAL CHARACTERISTICS OF DIFFERENT CONSTRUCTION MATERIALS

In Table 3 a comparison is made, in terms of characteristics and performance, between five different structural materials usable for challenging structures, like for high span roofing structural elements.

These materials are: reinforced concrete (R.C.), glued laminated timber (Glulam), steel, RPC 200

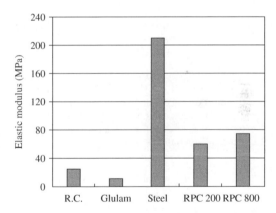

Figure 2. Comparison of elastic modulus of different construction materials.

(made by soft casting) and RPC 800 (made by pressure moulding and high temperature curing).

In Figures 2–3 also a comparison is made, in terms of elastic modulus and strength respectively, between these materials.

4 RPC STRUCTURE DESIGN TRIAL

In the absence of a precise reference standards frame, calculations of RPC elements have been carried out

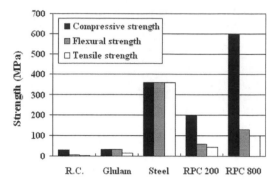

Figure 3. Comparison of mechanical strength of different construction materials under compression, bending and tension.

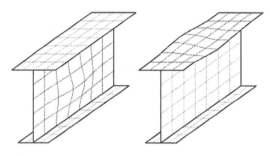

Figure 4. Schematic representation of local instability phenomena of slender parts in structural elements, like the web (left) and the flanges (right) of a beam.

by way of reference to Eurocode 2 (Parts 1–1, 1–3, 1–5), Document UNI/CIS/SC4-SFRC n°29 (Design of structural elements made of fibre reinforced concrete), AFGC (Association Française du Génie Civil) Recommendations on "Ultra High Performance Fibre-Reinforced Concretes".

4.1 An unusual problem for concrete beams

Due to RPC high strength and consequently to high slenderness attainable for RPC elements, a new problem rises, unusual for concrete: local instability of thin parts making up the RPC beam, analogously to steel beams. This issue compels to verify the equilibrium stability of compressed parts in the element section, as for instance the web of a H-shaped beam subjected to normal and/or shear stress, or its compressed flange (Fig. 4).

4.2 Preliminary dimensioning of a long span beam

Firstly, dimensioning of structural members made of the different construction materials reported in Table 3 was carried out. Seven beams with 14 m span were designed according to EC2, EC3, EC5 in order to bear the same bending stress with the same deflection, equal to 1/200 of the span length.

The dimensions for each beam resulting from calculations are compared in Figure 5. The seven beams were made of:

- C 30/35 concrete reinforced with FeB44k steel;
- C 40/45 concrete reinforced with pre-stressed tendons (2 ducts containing 6 strands each);
- a truss-girder with members made of steel Fe 360;
- glued laminated timber according to EC5;
- steel Fe 360 (full cross section beam);
- RPC 200 reinforced with FeB44k steel only at the lower edge in tension;
- RPC 800 reinforced with FeB44k steel only at the lower edge in tension.

4.3 High span beam design

Five beams with 25 m span were designed according to EC2, EC3, EC5 in order to bear the same bending stress with the same deflection, equal to 1/200 of the span length. The five beams were made of:

- C 30/35 concrete reinforced with FeB44k steel;
- steel Fe 360 (full cross section beam);
- RPC 200 reinforced with FeB44k steel only for tension and not for shear and compression;
- RPC 200 and two external pre-stressed tendons (RPCP 200), according to the scheme in Figure 6;
- RPC 800 and two external pre-stressed tendons (RPCP 800), according to the scheme in Figure 6.

The dimensions for each beam resulting from calculations are compared in Figure 7. When the designed span was increased to 35 m, reinforced concrete could not be used anymore in order to bridge this span in safe and serviceable conditions. The dimensions for the remaining beams resulting from calculations are compared in Figure 8.

At last, when the span was further increased to 50 m, also steel was no longer able to be advantageously used, so that only RPC beams were able to bridge this very high span (Fig. 9).

5 CONCLUSIONS

RPC material shows very high compressive and tensile strength as well as high toughness according to its high fracture energy.

This excellent behaviour, which takes RPC further from a common concrete, is due to accurate mixture proportioning and processing with selected raw materials. However, even using more easily available ingredients in order to make RPC more affordable, this work shows that very high mechanical performance can be usefully achieved, allowing one to avoid steel reinforcement for compression and shear

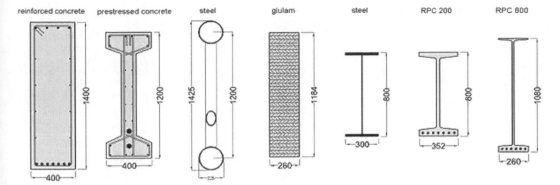

Figure 5. Comparison of equivalent strength beam cross sections obtained from calculations with seven different structural materials (all dimension are in mm).

Figure 6. Geometry of the external prestressed tendon applied to RPC beams.

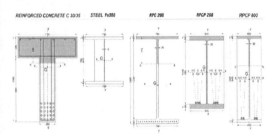

Figure 7. Comparison of equivalent strength cross sections for a 25 m span.

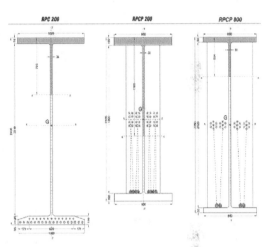

Figure 9. Comparison of equivalent strength cross sections for a 50 m span.

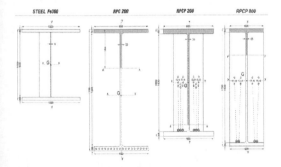

Figure 8. Comparison of equivalent strength cross sections for a 35 m span.

and use it only for tension. This approach makes the girder cross section much more slender, which gives rise to unusual structural issues for cementitious elements, as high strain and equilibrium stability of the

beam web. These problems can be in any case overcome by externally pre-stressing the beams. In this way, external pre-stressing, which completely avoids traditional steel reinforcement, suits extremely well calendered or extruded RPC beams. Further, external pre-stressing disallows any cracking under the service conditions, and significantly increases the durability of the structural member.

In conclusion, RPC proves to be an innovative material able to outrun traditional limits of cementitious materials, as well as to compete with structural steel in challenging structures.

REFERENCES

Birchall, J.D., Howard, A.J. and Kendall, K., 1981. Flexural Strength and Porosity of Cements. *Nature* 289: 388–390.
Collepardi, M., Corinaldesi, V., Monosi, S., Moriconi, G., 2001. DSP Materials Applications and Development

Progress. In M. Cerný (ed.), *CMSE/1 Composites in Material and Structural Engineering, Proc. Intern. Conf., Prague, June 3–6 2001, Czech Republic*, 49–52.

Richard, D. and Cherezy, M.H., 1994. Reactive Powder Concrete with High Ductility and 200–800 MPa Compressive Strength. In P.K. Mehta (ed.) *Concrete Technology: Past, Present and Future, Proc. Intern. Congress, San Francisco, USA*, 507–518.

Roy, D.M., 1987. New Stronger Cement Materials: Chemically Bonded Ceramics. *Science* 6: 651–658.

Roy, D.M., 1992. Advanced Cement Systems Including CBC, DSP, MDF. *Proc. 9th Intern. Congress on the Chemistry of Cement, New Delhi, India,* Vol.1, 357–380.

Excellence in Concrete Construction through Innovation – Limbachiya & Kew (eds)
© *2009 Taylor & Francis Group, London, ISBN 978-0-415-47592-1*

Biomass ash and its use in concrete mixture

V. Corinaldesi, G. Fava, G. Moriconi & M.L. Ruello

Dept. of Materials and Environment Engineering and Physics, Università Politecnica delle Marche, Italy

ABSTRACT: Chemical and physical characterization of biomass ash coming from burning of paper mill sludge was carried out in order to evaluate the possibility of its use as a cement replacement in concrete manufacturing. Several cement pastes were prepared with water to cement ratios of 0.4 and 0.6 by partially replacing cement with biomass ash at a percentage of 0, 5, 10, 15, 20, 30, 40, 50% by weight of cement. Concrete mixtures were also prepared with the same fluid consistency and different water to cement ratios (0.4 and 0.6). In the case of concretes, cement was replaced with biomass ash at percentages of 0, 10, 20, 40% by weight of cement. Cement paste and concrete were characterized from a mechanical point of view by means of compression tests and also their leaching behaviour was monitored by means of dynamic leaching test, which is a suitable method for estimating release of the most representative soluble ions concerning cementitious materials.

1 INTRODUCTION

Europe produces about 1.3 billion tons per year of sludge which comes principally from the manufacturing industries and mineral processing. This huge amount produces a severe impact on the environment and drives European Governments to promoting any process that will enable its reduction and/or possible valorisation. At this purpose, waste can be submitted to a high temperature sintering process in order to convert it into more environmentally friendly products, which will still be classified as waste but with little dangerous effect on the environment and public health. A number of researchers have demonstrated, for example, a reduced leaching from sintered products as compared to the original waste (Wunsch et al. 1996, Ozaki et al. 1997, Wiebusch et al. 1998, Favoni et al. 2005).

Paper mill sludge is a by-product of paper production. About 6 kg of sludge are produced per ton of paper. In the year 2004, the production of paper mill sludge in Italy was around $6 \cdot 10^5$ tons (Asquini et al. 2008). Paper mill sludge is composed of mineral fillers, inorganic salts, small cellulose fibres, water and organic compounds. The composition of mineral fillers depends on the type of paper produced. Paper mill sludge is often burnt in order to reduce the waste disposal and sometimes to recover heat. This process is achieved first by de-watering (i.e. evaporation) at low temperature ($<200°C$), followed by incineration at high temperature ($>800°C$). During incineration, paper and organic compounds are burned out at temperatures of around $350 \div 500°C$, whereas mineral fillers and inorganic salts are transformed

into the corresponding oxides at higher temperatures ($>800°C$). CaO, Al_2O_3, MgO and SiO_2 are the most abundant oxides in incinerated paper mill sludge (Liaw et al. 1998). The obtained paper mill sludge ash is classified as waste, and at present it is mainly conferred to landfill at high costs. Recycling it would have beneficial effects for paper producers, and the environment as well.

A possible reuse of paper mill sludge is its blending with natural raw materials extracted from the ores in the production of bricks or cements (Marcis et al. 2005, Ernstbrunner 2007, Liaw et al. 1998), since the main constituent elements of paper mill sludge are Al, Mg, Si and Ca, whose oxides are largely used in the ceramic industry.

In this work the characterization of a biomass ash coming from burning of paper mill sludge was carried out in order to evaluate the possibility of its use as a cement replacement in concrete manufacturing.

2 EXPERIMENTAL PART

2.1 Materials

A commercial portland-limestone (20% maximum limestone content) blended cement type CEM II/A-L 42.5 R according to EN-197/1 was used. The Blaine fineness of cement was $410 \, m^2/kg$, and its specific gravity was $3.05 \, kg/m^3$.

A biomass ash was used coming from burning of paper mill sludge. Its Blaine fineness was $635 \, m^2/kg$, and its specific gravity was $1.72 \, kg/m^3$. In order to characterize the biomass ash, thermal analysis and

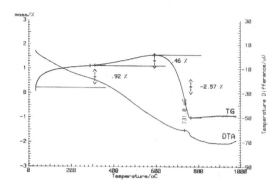

Figure 1. Results of the thermogravimetric (TG) and differential thermal analysis (DTA) of the biomass ash.

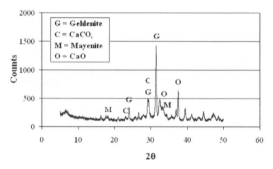

Figure 2. X-ray diffraction of the biomass ash.

X-ray diffraction were carried out. Thermal analysis shows that the examined material contains about 6% of calcite, $CaCO_3$. As a matter of fact, as it can be observed in Fig. 1, a sharp weight loss, corresponding to the flex of the differential thermal analysis (DTA) curve occurs from about 600°C to 800°C while strong heat absorption is detected. This is due to the decomposition reaction of calcite, which is endothermic. X-ray diffraction analysis (Fig. 2) shows also the presence of calcium oxide (CaO), gehlenite ($Ca_2Al_2SiO_7$) and mayenite ($Ca_{12}Al_{14}O_{33}$). The remaining part of the biomass ash consists of amorphous silica. Biomass ash was observed at increasing magnifications by means of scanning electron microscope (SEM, Fig. 3).

Natural sand (range of particle size: 0–5 mm) and fine gravel (range of particle size: 2–12 mm) were used in each concrete mixture. Their main physical properties, determined according to EN 1097-6 and EN 933-1, are reported in Table 1.

A common water-reducing admixture based on polycarboxylate polymers (30% aqueous solution) was used for preparing concrete mixtures.

2.2 Cement paste mixture proportions

Cement pastes were prepared with two different water to binder ratios: 0.4 and 0.6.

Biomass ash was added to the cement paste mixtures at different percentages: 0, 5, 10, 15, 20, 30, 40 and 50% by weight of cement. Cement paste mixture proportions are reported in Table 2.

2.3 Concrete mixture proportions

Natural sand and fine gravel were combined at 40% and 60% respectively in order to optimize grain size distribution within the concrete mixtures according to the Bolomey particle size distribution curve (Collepardi 1991).

The dosage of superplasticizer was always equal to 1.0% by weight of cement in order to reduce the water dosage of about 30%. The concrete workability was the same for all the concrete mixtures and equal to 180–190 mm (measured according to EN 12350-2).

Eight concrete mixtures were prepared with two diffent water to binder ratios (0.4 and 0.6) and four different biomass ash percentages: 0, 10, 20 and 40% by weight of cement. Biomass ash was always added as a partial substitution of cement. Concrete mixture proportions are reported in Table 3.

2.4 Preparation and curing of specimens

Twelve cylindrical specimens (diameter of 25 mm and height of 50 mm) for each cement paste mixture were prepared for cement paste mechanical characterization.

Fifteen concrete prismatic specimens (40 × 40 × 160 mm) for each concrete mixture were prepared for mechanical characterization of concrete and leaching tests.

All the specimens were wet cured at a temperature of about 20°C.

3 RESULTS AND DISCUSSION

3.1 Compression tests on cement pastes

The compressive strength of cement pastes was evaluated after 1, 3, 7 and 28 days since their preparation. The results obtained are reported in Table 4.

Cement replacement by 10–20% biomass ash seems to improve the mechanical performance of cement pastes independently of the water to binder ratio. As a matter of fact, with higher biomass contents the strength development seems to slow down, particularly with the higher water to cement ratio (0.60).

On the basis of the results reported in Table 4 the activity index of biomass ash can be evaluated (CEN 2005): it is equal to 1.46 and 1.36 for the water to binder ratios of 0.4 and 0.6 respectively.

3.2 Flexural strength of concretes

The flexural strength of concretes was evaluated after 1, 7 and 28 days since their preparation.

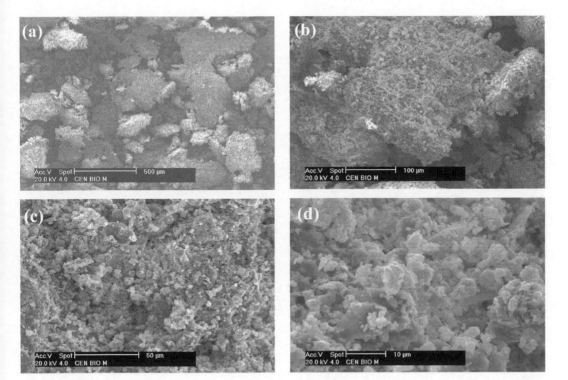

Figure 3.   SEM observations of biomass ash at different magnifications: 500 μm (a), 100 μm (b), 50 μm (c) and 10 μm (d).

Table 1.   Physical properties of aggregate fractions.

Fraction	Water absorption (%)	Specific gravity in SSD* condition (kg/m^3)	Passing 75 μm sieve (%)
Natural sand	4.1	2530	0.7
Fine gravel	3.4	2570	0.0

$^{(*)}$SSD = saturated surface-dried.

Results obtained are reported in Fig. 4. In this case the optimum percentage of replacement seem to be 10% for both water to binder ratios (0.4 and 0.6). In fact, for the mixture prepared with 10% biomass ash and a water to cement ratio of 0.6 (0.6–10%ba), even 30% higher flexural strength was obtained with respect to the reference mixture (0.6–0%ba).

3.3   *Compressive strength of concretes*

The compressive strength of concretes was evaluated after 1, 7 and 28 days since their preparation.

The results obtained are reported in Fig. 5, and they seem to confirm data collected in terms of flexural strengths. In fact good performances were obtained with biomass ash replacing 10% of cement, particularly with the higher water to binder ratio (0.6).

Table 2.   Mixture proportions of cement pastes prepared with different percentages of biomass ash (ba).

Mixture	Water/ Binder	Ingredient dosage (g)		
		Water	Cement	Biomass Ash (ba)
0.4–0%ba	0.4	550	1370	0
0.4–5%ba	0.4	550	1300	70
0.4–10%ba	0.4	550	1235	135
0.4–15%ba	0.4	550	1165	205
0.4–20%ba	0.4	550	1095	275
0.4–30%ba	0.4	550	960	410
0.4–40%ba	0.4	550	820	550
0.4–50%ba	0.4	550	685	685
0.6–0%ba	0.6	660	1100	0
0.6–5%ba	0.6	660	1045	55
0.6–10%ba	0.6	660	990	110
0.6–15%ba	0.6	660	935	165
0.6–20%ba	0.6	660	880	220
0.6–30%ba	0.6	660	770	330
0.6–40%ba	0.6	660	660	440
0.6–50%ba	0.6	660	550	550

Further studies are needed in order to clarify the reasons of the excellent behavior detected for the concrete mixtures prepared with 10% biomass ash replacing cement. Probably, biomass ash can give an active

Table 3. Mixture proportions of concretes prepared with different percentages of biomass ash (ba).

Mixture	Ingredient dosage (kg per m³ of concrete)					
	Water	Cement	Biomass ash	Natural sand	Fine gravel	SPA*
0%ba(0.4)	154	386	0	1080	720	3.9
10%ba(0.4)	154	347	39	1080	720	3.9
20%ba(0.4)	154	308	78	1080	720	3.9
40%ba(0.4)	154	230	156	1080	720	3.9
0%ba(0.6)	154	257	0	1140	760	2.6
10%ba(0.6)	154	230	27	1140	760	2.6
20%ba(0.6)	154	206	51	1140	760	2.6
40%ba(0.6)	154	154	103	1140	760	2.6

(*)SPA = Superplasticizing admixture.

Table 4. Compressive strength values (MPa) measured on cement pastes prepared with different percentages of biomass ash.

Mixture	Curing time (days)			
	1	3	7	28
0.4–0%ba	4.1	14.0	18.3	18.3
0.4–5%ba	8.5	14.7	24.0	26.9
0.4–10%ba	7.7	15.6	18.4	21.6
0.4–15%ba	7.3	16.7	18.8	27.0
0.4–20%ba	11.1	21.5	22.2	26.6
0.4–30%ba	9.4	15.2	14.9	21.5
0.4–40%ba	4.9	9.9	16.5	19.0
0.4–50%ba	3.6	9.2	13.5	18.3
0.6–0%ba	4.6	9.9	13.3	14.4
0.6–5%ba	5.2	10.0	9.0	18.8
0.6–10%ba	6.3	13.1	13.4	22.2
0.6–15%ba	5.1	8.1	10.4	16.1
0.6–20%ba	4.5	9.4	8.2	19.6
0.6–30%ba	2.0	7.2	5.4	14.8
0.6–40%ba	1.7	3.1	6.6	9.6
0.6–50%ba	0.4	7.3	9.7	15.5

contribution to cement paste hardening (confirmed by the positive value of the activity index, see §3.1), but if its dosage becomes too high the mechanical strength of the mixture turns lower due to its high fineness (and consequently high water absorption).

3.4 Leaching tests

Dynamic leaching tests were conducted following Italian regulations for reuse of no-toxic waste materials as by-products (D.M.A. 1998). Examples of similar international standard test include ISO 6961:82 and ASTM C1220:92. According to the extraction protocol of the Italian regulations, a sample is placed in contact with a precise amount of deionised water (CO_2 free) for predetermined extent of time, at $20°C \pm 4°C$.

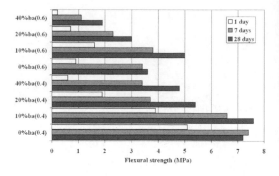

Figure 4. Flexural strengths vs. curing time of concretes.

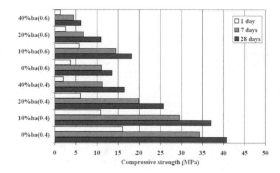

Figure 5. Compressive strengths vs. curing time of concretes.

Sample is placed in a polyethylene test vessel, with enough space around, and leachant water is added and allowed to wet the sample all around (minimum level: at least 2 cm above the specimen). The solid to liquid ratio, expressed as the ratio of volume of solid to volume of the leachant, is 1:5. The test vessel must be hermetically closed to prevent acidification of the water for CO_2 diffusion. The leachant solution is renewed to drive the leaching process. The renewing sequence is: 2, 8, 24, 48, 72, 102, 168, and 384 hours. At each renewing sequence, the fluid is collected for analysis. Because the physical integrity of the sample matrix is maintained during the test, the concrete-specimen/leachant properties affect how much material can be leached out as a function of time. In particular, the surface area reactivity of the test sample, more than the extraction force of the leachant, provides leached contaminants in the water, and the kinetic information about the dissolution process. Metal released cumulated values were then compared with the Italian code requirements (D.M.A. 1998) in order to verify the Italian environmental standards. Additionally, in order to follow the release mechanisms, pH, electrical conductivity, and temperature of the leachant solution were recorded. Total alkali cumulated curves are plotted as a function of time and

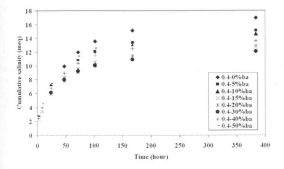

Figure 6. Cumulative salinity vs. time of cement pastes with water to cement ratio of 0.4.

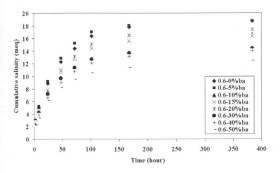

Figure 7. Cumulative salinity vs. time of cement pastes with water to cement ratio of 0.6.

related to both specific surface area and bulk density of the test specimens (see Figures 6 and 7).

Curves are also compared with the pores distribution of specimens. Heavy metal leachates result fully conform to the Italian legal standards. The leachates, alkaline throughout the testing period with pH values greater than 10, indicate that the interstitial pore fluid in contact with hydrated cementitious materials is buffered by the presence of alkaline ions. So far as the local condition of the leachant remain unchanged, the heavy metals leaching remains very low, as found in previous works (Sani et al. 2004, Carrescia et al. 2004, Naik et al. 2001). The resistance to the leaching is the most important factor for the evaluation of the immobilised matrices, because water would be the primary liquid available for the potential dispersion for the heavy-metal ions. The Italian waste control regulation recognizes the concrete as a technology for the safe disposal of wastes providing that leachates satisfy the enforced standards. The concrete specimen manufactured with biomass ashes containing heavy metals did not manifest a release higher than the Italian standards and the pH leachates remained alkaline throughout the testing period.

4 CONCLUSIONS

On the basis of the data collected within this experimental work it can be concluded that biomass ash, particularly if replacing 10% cement, show a positive effect on time development of concrete mechanical performance.

Probably, biomass ash can give an active contribution to cement paste hardening (positive value of the activity index). On the other hand, due to its high fineness, and consequently high water absorption, its dosage should not be too high (the upper limit appears to be 10% by weight of cement).

Finally, the concrete specimen manufactured with biomass ashes containing heavy metals did not manifest a release higher than the Italian standards and the pH leachates remained alkaline throughout the testing period.

REFERENCES

Asquini, L., Furlani, E., Bruckner, S., Maschio, S. 2008. Production and characterization of sintered ceramics from paper mill sludge and glass cullet. *Chemosphere* 71, 83–89.

ASTM C1220:1992, Static Leaching of Monolithic Waste Forms for Disposal of Radioactive Wastes. *American Society for Testing and Materials, Annual Book of ASTM Standards*, 04.02. West Conshohocken, Pennsylvania.

Carrescia, P.G., Fava, G., Sani, D. 2004. Regulation aspects and a dynamic leaching test for non-ferrous slag recycling in concrete manufacturing. *Proc. VII Congresso Nazionale AIMAT*. Ancona, Italy.

Collepardi, M., 1991. *Concrete Science and Technology* (in Italian), Third Edition, Hoepli, Milano, Italy.

D.M.A. 5 febbraio 1998. Individuazione dei rifiuti non pericolosi sottoposti alle procedure semplificate di recupero ai sensi degli articoli 31 e 33 del Decreto Legislativo 5 febbraio 1997 n°22 (Identification of non dangerous wastes undergoing recovery simplified procedure according to articles 31 and 33 in Legislative Decree 5 February 1997 n°22, in Italian). *Ordinary Supplement G.U.*, n°88, April 16, 1998.

EN 197-1. Cement – Part 1: Composition, specifications and conformity criteria for common cements, 2000.

EN 450-1. Fly ash for concrete, 2005.

EN 933-1. Tests for geometrical properties of aggregates – Determination of particle size distribution – Sieving method, 1997.

EN 1097-6. Tests for mechanical and physical properties of aggregates – Determination of particle density and water absorption, 2000.

EN 12350-2. Testing fresh concrete – Slump test, 1999.

Ernstbrunner, L., 1997. Rejects from paper manufacture utilized in the cement works, *Papier 51* (6), 284–286.

Favoni, C., Minichelli, D., Tubaro, F., Bruckner, S., Bachiorrini, A., Maschio, S., 2005. Ceramic processing of municipal sewage sludge and steelwork slag. *Ceram. Int.* 31, 697–702.

ISO 6961:1982, Long-term leach testing of solidified radioactive waste forms.

Liaw, C.T., Chang, H.L., Hsu, W.C., Huang, C.R., 1998. A novel method to reuse paper sludge and cogeneration ashes from paper mill. *J. Hazard. Mater.* 58, 93–103.

Marcis, C., Minichelli, D., Bruckner, S., Bachiorrini, A., Maschio, S., 2005. Production of monolithic ceramics from incinerated municipal sewage sludge, paper mill sludge and steelworks slag, *Ind. Ceram.* 25 (2), 89–95.

Naik, T.R., Singh, S.S., Ramme, B.W. 2001. Performance and Leaching Assessment of Flowable Slurry. *Journal of Environmental Engineering*, 127(4).

Ozaki, M., Watanabe, H., Wiebusch, B., 1997. Characteristics of heavy metal release from incinerated ash, melted slag and their re-products. *Water Sci. Technol.* 36 (11), 267–274.

Sani, D., Moriconi, G., Fava, G., 2004. Estimation of alkali diffusivity from dynamic leaching test. *Computational Methods in Materials Characterisation*, WIT press, 307–316.

Wiebusch, B., Ozaki, M., Watanabe, H., Seyfried, C.F., 1998. Assessment of leaching tests on construction material made of incinerator ash (sewage sludge): investigations in Japan and Germany. Water Sci.Technol. 38 (7), 195–205.

Wunsch, P., Greilinger, C., Bieniek, D., Kettrup, A., 1996. Investigation of the binding of heavy metals in thermally treated residues from waste incineration. *Chemosphere* 32, 2211–2218.

Excellence in Concrete Construction through Innovation – Limbachiya & Kew (eds)
© 2009 Taylor & Francis Group, London, ISBN 978-0-415-47592-1

Properties of lightweight concretes made from lightweight fly ash aggregates

N.U. Kockal & T. Ozturan

Bogazici University, Department of Civil Engineering, Bebek, Istanbul, Turkey

ABSTRACT: The study presents the effects of aggregate characteristics on the lightweight fly ash aggregate concrete (LWAC) properties. The influence of characteristics of four aggregate types (two sintered lightweight fly ash aggregates, cold-bonded lightweight fly ash aggregate and normalweight crushed limestone aggregate) such as strength, microstructure, porosity, water absorption, bulk density and specific gravity on the behaviour of concrete mixtures were discussed. The results of this study revealed the achievement of manufacturing high-strength air-entrained lightweight aggregate concretes using sintered and cold-bonded fly ash aggregates. In order to reach target slump and air content, less amount of chemical admixtures was used in lightweight concretes than in normalweight concrete, leading to reduction in production cost. The use of lightweight aggregates (LWA) instead of normalweight aggregates in concrete production slightly decreased the strength. As it was expected, the lightweight aggregate concretes had lower compressive strength and modulus of elasticity values than the normal weight concrete due to the higher porosity and lower strength of the aggregate included in the concrete. The models given by codes, standards and software and equation derived in this study gave close estimated values to the experimental results.

1 INTRODUCTION

Large quantity of fly ash is still disposed of in landfills and storage lagoons leading to environmental damage by causing air and water pollution on a large scale. As large quantities of the fly ash remain unutilized in most countries of the world, the manufacture of lightweight fly ash aggregates seems to be an appropriate step to utilize a large quantity of fly ash.

Recently, some attempts have been made to use the waste materials such as fly ash for the production of lightweight aggregate by two different methods. Cold bonding and sintering are the main production processes in order to obtain lightweight fly ash aggregates which are generally used for the hardening of fly ash pellets. The researchers (Yun et al. 2004) indicated that by using fly ash (FA), furnace bottom ash (FBA) and Lytag, it was possible to manufacture lightweight concrete with density in the range of 1560–1960 kg/m³ and strength of medium range. Lo & Cui (2004) presented the mechanical properties of a structural grade lightweight aggregate made with fly ash and clay. The findings indicated that water absorption of the green aggregate was large but the crushing strength of the resulting concrete could be high.

The concrete made with fly ash lightweight aggregates obtained by crushing the sintered fly

ash briquettes was 25% stronger than that made using pelletized fly ash based lightweight aggregates (Kayali, in press). The use of sintered fly ash aggregate in concrete as a partial replacement of granite aggregate was examined (Behera et al. 2004). The concrete so produced was light in nature (up to 14% reduction in unit weight) and the development of such concrete with sintered aggregate minimised the consumption of granite rock, resulting in protection of the natural environment. Experimental study was carried out to obtain the compressive strengths and elastic moduli of cold-bonded pelletized lightweight aggregate concretes (Chi et al. 2003). Different types of aggregates were made with different fly ash contents. Test results showed that the properties of lightweight aggregates and the water/binder ratio were two significant factors affecting the compressive strength and elastic modulus of concrete.

When these lightweight aggregates are used, we can expect not only the reduction of selfweight and other positive effects for concretes but also the reduction of environmental pollution through recycling of waste resources. Thus, high-strength air-entrained lightweight fly ash aggregate concrete can be an attractive construction material which possesses required properties for structural materials used in the applications mentioned. However, there is not sufficient number of published papers related with the properties of high-strength LWC made with sintered and cold bonded fly ash pellet aggregates.

This paper presents the results of one part of an investigation with a much larger scope. The aim of this paper is to study the effects of characteristics of different lightweight fly ash aggregates on concrete properties. Regression and graphical analysis of the data obtained for concrete properties were also performed. Analysis of variance (ANOVA) is a useful technique for analyzing the experimental data. Thus, a software TableCurve 3D V4.0 was used for analyzing the hardened concrete properties. The software offers a large number of graphical and numerical tools to enable the user to choose the best equation for particular needs. The validity of predicted values of different models specified by standards and literature, derived in this study and suggested by software was compared and discussed.

2 MATERIALS AND METHODOLOGY

2.1 Properties of materials used in lightweight aggregate and concrete production

Chemical composition of cement, fly ash, bentonite, glass powder and silica fume are exhibited in Table 1. Physical properties and the results of compressive strength and pozzolanic activity tests of cement and fly ash determined according to TS 639 are shown in Table 2 and Table 3, respectively. TS 639 specifies that the strength of test specimen must not be less than compressive strengths of control specimens in 7 and 28 days for compressive strength test and must have a minimum of 70% of the 28-day strength of the control specimen for pozzolanic activity test. As can be seen in Table 3, the results met the requirements specified in TS 639 for both compressive strength and pozzolanic activity tests. Quantachrome-NOVA 2200e BET (Brunauer Emmett–Teller) method was used to determine the specific surface area from N_2 adsorption isotherms. Fly ash used in this study conforms to Class F as specified in ASTM C 618. Particle size distribution of raw materials is illustrated in Figure 1. Mastersizer was used to determine the grading curves of materials by laser diffraction. It can be seen that fly ash was much coarser than Portland cement and glass powder was finer than fly ash, but much coarser than bentonite and cement.

Among the 18 different sintered fly ash lightweight aggregates produced in this research project, the two sintered aggregates (aggregate containing 10% bentonite sintered at 1200°C (10B1200) and aggregate containing 10% glass powder sintered at 1200°C (10G1200)) were selected due to their lower density and higher strength in order to produce lightweight high-strength concrete. Cold-bonded lightweight fly

ash aggregate (CB) and normalweight calcerous crushed stone aggregate (NW) were also used in making concrete for the comparison of their properties.

The superplasticizer and the air-entraining agent were used in this study. The superplasticizer was based on modified polycarboxylic ether complying with EN

Table 1. Chemical composition of cement (C), fly ash (FA), bentonite (BN), glass powder (GP) and silica fume (SF) (% by weight).

Oxide (%)	C	FA	BN	GP	SF
SiO_2	20.55	59.00	57.84	70.62	94.12
Al_2O_3	4.78	19.58	13.77	1.38	0.57
Fe_2O_3	3.64	7.23	6.14	0.82	0.55
CaO	63.94	0.54	3.75	8.75	0.50
MgO	1.50	4.64	3.04	3.54	1.30
SO_3	2.77	0.69	1.34	1.85	0.42
Na_2O	0.25	0.48	2.80	10.85	0.54
K_2O	0.77	5.95	2.80	1.53	1.75
Cl^-	0.035	0.011	0.004	0.003	0.0049
Loss on ignition	1.24	0.49	8.40	0.12	0.19
$CaCO_3 +$ $MgCO_3$	–	–	6.50	0.75	–
Insolubleresidue	1.36	–	–	–	–
Free CaO	1.25	–	–	–	–
Specific Gravity(gr/cm^3)	3.16	2.06	2.40	2.50	2.28

Table 2. Physical properties of cement and fly ash.

		CEMI 42.5R	Fly ash
Retention on sieves (%)	45 μm	12.0	45.2
	90 μm	1.7	25.3
	200 μm	0.1	6.5
Specific gravity (gr/cm^3)		3.16	2.06
Specific surface area (m^2/gr)		3.27	2.88
Setting time (min)	Initial	139	–
	Final	188	–

Table 3. Compressive strength and Pozzolanic activity tests.

		Contr. spec.	Compr. st. test	Pozz. act. test
Materials (gr)		C	FA + C	FA + C
Cement content (gr)		450	450	293
Fly ash content (gr)		–	112	104
Sand content (gr)		1350	1233	1350
Water content (gr)		225	225	199
Compr. strength (N/mm^2)	2 days	26.9	28.8	–
	7 days	40.9	43.9	–
	28 days	54.0	59.8	42.6
2 – day percentage			107.06	–
7 – day percentage			107.33	–
28 – day percentage			110.74	78.89

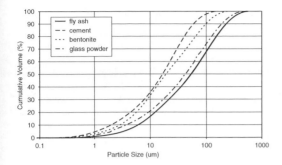

Figure 1. Particle size distribution of fly ash, cement, bentonite and glass powder.

934 part 2. The air-entraining agent was an ultra stable air entraining admixture for use in all types of concrete based on aliphatic alcohol and ammonium salt. It has been formulated to meet the requirements of TS EN 206-1 and relative specifications.

2.2 Test performed on aggregates

To determine the strength of lightweight aggregates, individual particles were placed between parallel plates and were crushed by being loaded diametrically. A minimum of 20 pellets with a diameter of approximately 10 mm were tested to calculate the average crushing strength at each aggregate type. Aggregate crushing value test (ACV) was also conducted on the normal and lightweight coarse aggregates according to BS 812-110.

Pore structure (total porosity and pore size distribution) of aggregate particles was estimated by mercury porosimetry with pressure up to 220 MPa. Quantachrome-PoreMaster Mercury porosimetry was used to determine total porosity. The porosity and critical pore diameter were determined from the figures representing pore size distribution curves of each aggregate.

Specific gravity and water absorption of fine and coarse aggregates were determined according to ASTM C 128 and ASTM C 127, respectively. The specific gravities of natural sand and crushed sand were 2.62 and 2.63, respectively. The coarse aggregates were tested in oven-dry condition utilizing the shoveling and rodding procedure to determine the unit weight (loose and rodded) and void content according to ASTM C 29-97.

Table 4 shows the results of tests performed on aggregates used in concrete production. Cold-bonded fly ash aggregate had the lowest unit weight and the highest aggregate crushing value among other aggregates. The crushing value of the artificial aggregates was lower than that of the natural aggregates because of their porous structure (Jo et al. 2007). Unit weight and crushing value of aggregates are

Table 4. Aggregate characteristics used in concrete production.

Characteristics	Aggregates			
	CB	10B1200	10G1200	NW
Specific gravity based on SSD condition	1.63	1.57	1.60	2.71
Apparent specific gravity	1.89	1.58	1.60	2.71
Specific gravity based on OD condition	1.30	1.56	1.59	2.70
28-hour water absorption (%)	25.5	0.7	0.7	0.8
Total porosity (%)	31.1	8.4	6.2	–
Critical pore diameter (nm)	398.4	28.03	12.49	–
Rodded bulk density (kg/m^3)	842	993	999	1537
Loose bulk density (kg/m^3)	789	933	936	1433
Voids (%) loose	39.2	40.1	41.0	46.8
Voids (%) rodded	35.1	36.2	37.0	43.0
ACV (%)	35	28	30	23
Pellet crushing strength (MPa)	3.7	12.0	9.6	–

rated in the same order with specific gravity and particle crushing strength results, respectively. The interparticle void space percentages of lightweight aggregates were similar because they all had round shape and smooth surface. However, this value for normalweight aggregate was relatively larger than those of other lightweight aggregates, because grain shape of normalweight aggregates was more angular.

2.3 Grading of aggregates

Grading of aggregates used in concretes was determined in accordance with ASTM C 136-06. The grading of lightweight and normalweight aggregates for structural concrete conformed to the requirements given in ASTM C 330-05 and ASTM C 33-03, respectively. The same passing percentages were selected for lightweight fly ash and normalweight crushed limestone aggregates considering both standards to eliminate the effect of grading difference on concrete properties.

2.4 Production of normalweight and lightweight concretes

The concretes were coded as LWGC (Lightweight glass powder added fly ash aggregate concrete), LWBC (Lightweight bentonite added fly ash aggregate

Table 5. The proportions of the concrete mixtures (based on SSD condition) (kg/m³).

| Concr. | C | SF | W | Fine agg | | CA | AEA[a] (%) | SP[b] (%) |
				NS	CS			
LWCC	551	55	158	318	318	592	1.2(0.2)	6.7(1.1)
LWBC	548	55	157	316	317	567	1.2(0.2)	6.7(1.1)
LWGC	549	55	157	317	317	580	1.2(0.2)	6.7(1.1)
NWC	551	55	158	317	319	981	1.6(0.265)	7.3(1.2)

[a] AEA: Air-entraining agent.
[b] SP: Super-plasticizing admixture.

concrete), LWCC (Lightweight cold-bonded fly ash aggregate concrete) and NWC (Normalweight concrete) referring to the type of coarse aggregate used. Initial trial mixes were made to obtain a slump of 150 mm, air content of 4% and desired compressive strength and unit weight using a pan-type mixer of 50 dm³ capacity.

Lightweight concrete specimens were cast using ordinary Portland cement (PC 42,5), superplasticizer, air entraining agent, water, natural sand, crushed sand and lightweight coarse aggregate with a water/binder ratio of 0.26. The density of aggregates affects the density of concretes considerably, since the volume of aggregate is significant in the mixture. Four different coarse aggregates with a volume ratio of 0.60 (volume of coarse aggregate/volume of total aggregate) were considered in the mix proportions.

Trial batches were made to obtain the desired strength, slump and air content, therefore proportions of the superplasticer and air entraining agent were adjusted. The content of these admixtures were kept constant for LWCs, but not for NWC. The amount of other constituents in NWC and LWCs were same. The only difference was the type of coarse aggregate among four concretes.

The specimens were cast according to ASTM C 192. The moulded specimens were covered with a plastic sheet to prevent water loss. The specimens were demoulded at 24 hours after casting and cured in lime-saturated water at 22°C for 28 and 56 days before testing. The mix proportioning for the concretes are exhibited in Table 5.

2.5 Tests performed on fresh concretes

The slump test was performed immediately after production according to ASTM C 143. The fresh density and air content were measured in general compliance with the standard procedures in ASTM C 138-01a and ASTM C 231-04. The results of air content measured by both ASTM C 138 and ASTM C 231 were close to each other.

2.6 Tests performed on hardened concretes

The 100 × 200 mm cylindrical specimens were used to determine the density in hardened concrete. The compressive strength and modulus of elasticity of the 100 × 200 mm cylinder specimens were measured according to ASTM C 39 and ASTM C 469-02, respectively, at the age of 28 and 56 days. Also, modulus of elasticity values of concrete specimens were calculated by the equations (1) and (2) given by ACI 318 Building Code and BS 8110, respectively.

$$E_c = w_c^{1.5} x 0.043 x \sqrt{f_c} \qquad (1)$$

$$E_c = w_c^{2} x 0.0017 x f_c^{0.33} \qquad (2)$$

where E_c = modulus of elasticity, w_c = air dry density of concrete, f_c = cylinder compressive strength.

3 TEST RESULTS AND DISCUSSION

3.1 Slump, air content and density of concretes

As can be seen in Table 6, the slump and air content values for all concrete types were 16 ± 1 cm and 4 ± 0,3%, respectively. It is clear that in order to obtain similar slump and air content value for lightweight concretes as compared to normal weight concrete, a lower amount of superplasticizer and air entraining agent was needed which contributes to reduction in cost (Table 5). This could be due to the spherical particle shape of lightweight aggregates compared to the angular particles of crushed stone causing a "ball-bearing effect" and resulting in decrease of the water demand of the fresh concrete which was also reported by Yun et al. (2004). However, it was also reported that the lighter the mix, the less the slump, because the work done by gravity was lower in the case of lighter aggregate (Hossain 2004). Due to lower aggregate density, structural low-density concrete did not slump as much as normal density concrete with the same workability. The recommended range (ACI Committee 213 1987) of total air content for lightweight concrete is given as 4–8% for 19 mm maximum size of aggregate which was consistent with this study.

The concretes were workable, cohesive and segregation was not observed. It can be seen that when natural coarse aggregate was replaced with lightweight aggregate, there was a significant reduction in the density of fresh and hardened concrete (Table 6). All of the lightweight concretes had fresh and oven-dry densities lower than 2000 kg/m³. Although LWCC had a greater fresh density than the other lightweight concretes, its oven dry density was lower than the others because the cold-bonded aggregates had much higher water absorption due to their high volume of open pores. A reduction in the self-weight in oven dry condition

Table 6. The properties of the fresh and hardened concretes.

Concr.	Slump (cm)	Fresh Density (kg/m^3)	Air content (%)	28-day density (SSD) (kg/m^3)	28-day density (OD) (kg/m^3)	56-day density (OD) (kg/m^3)
LWCC	15	1991	3.9	2025	1860	1868
LWBC	15.5	1960	4.3	1979	1915	1922
LWGC	16.5	1975	4.1	1997	1943	1963
NWC	17	2381	3.8	2387	2316	2323

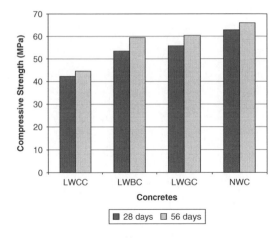

Figure 2. The compressive strength values of 28-day and 56-day concrete specimens.

between 19% and 25% was obtained with the replacement of normal aggregates by lightweight fly ash aggregate with 10% glass powder and cold bonded fly ash aggregate, respectively. In the case of fresh density and density in SSD condition, the maximum reductions in the selfweight were 21.5% and 20.6%, respectively, with concrete made from aggregates with 10% bentonite. This suggested that the low densities of both cold bonded and sintered fly ash aggregates was beneficial to produce lightweight concrete. In this study, the 28-day OD density in the range of 1860–1943 kg/m^3 was achieved. As expected, 56-day density (OD) values were relatively higher than the 28-day density (OD) values. Lightweight concrete density class LC 2.0 is achieved with high strength values based on oven-dry density according to BS EN 12390–7. Much lighter concretes can be obtained by partial or total replacement of fine aggregate as well as full replacement of coarse aggregate by using combined fine and coarse lightweight aggregate following the grading requirement suggested by ASTM C 330. In this study, only the coarse aggregates were replaced fully with lightweigth aggregates. However, it is obvious that replacement of fine aggregates would further reduce the compressive strength of concretes.

3.2 Compressive strength and modulus of elasticity

The values of compressive strength and modulus of elasticity of concretes are shown in Figures 2 and 5. As it was expected, the lightweight aggregate concretes had lower compressive strength and modulus of elasticity than the normal weight concrete due to the higher porosity and lower strength of the lightweight aggregate. LWCC, on the other hand, had lower compressive strength and modulus of elasticity compared to LWBC and LWGC which had comparable strength and modulus values. Lower density and crushing strength of cold bonded fly ash aggregates are most probably responsible for this. The replacement of the normal-weight aggregate in high strength concrete with the lightweight aggregate resulted in a 13%, 18% and 49% reduction in the 28-day compressive strength for LWGC, LWBC and LWCC, respectively, whereas this

reduction ratios decreased in 56-day strength to 9%, 11% and 48%, respectively. The highest strength gain from 28 days to 56 days (11.2%) of LWBC could be attributed to the higher pozzolanic activity of 10B1200 aggregates, while NWC showed the lowest strength gain (4.9%).

Considering the 28-day strengths of LWBC and LWGC, although 10B1200 aggregate had higher crushing strength, the concrete prepared by 10G1200 aggregate exhibited higher compressive strength. Wasserman and Bentur (1997) also noted that higher strength aggregate does not necessarily lead to higher, or proportionally higher, strength concrete. In their study, they revealed that at early age lower strength was obtained with aggregates of smaller absorption and the closure of the gap in strength at later age between the concrete from untreated aggregate and the heat treated aggregates might be attributed to the higher pozzolanic activity of the latter. However, in our study, water absorptions of aggregates were same for sintered aggregates used in concrete. Also, the closure of the gap in strength was observed between LWBC and LWGC at 56 days. Nevertheless, as can be seen in Table 6, this relatively higher compressive strength of LWGC as compared to LWBC was attributed to the difference in density between these two concretes having similar air content, 4.3% for LWBC and 4.1% for LWGC. The 28-day compressive strength and the dry concrete density of lightweight concretes varied from 42.3 to 55.8 MPa and from 1860 to 1943 kg/m^3, respectively. LWGC was heavier than LWBB. Heavier concrete resulted in stronger concrete. Nemes and Jozsa (2006) indicated that the achieved maximum compressive strength depended on the particle density of LWA. This result was also confirmed by Figure 3 which exhibits the relationship between the compressive strength and the oven-dry density.

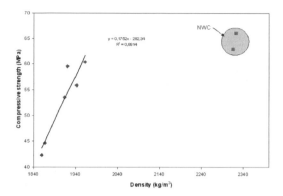

Figure 3.   Compressive strength vs density.

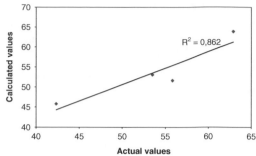

Figure 4.   Calculated values vs. actual values of compressive strength of 28-day lightweight concretes.

However, there is no general relationship between density and strength classes for different types of lightweight aggregates. In this study, regression analysis was performed on the compressive strength and density among the lightweight aggregates using linear regression models which gave the best fit against other regression models.

The linear model showed the linearity between the compressive strength and density. The compressive strength was directly proportional to the OD density of hardened concrete. Large R-square value of 0.88 for linear model showed that the regression was established under good fit, which means that there was no remarkable variation in the compressive strength and density of concrete results. However, this regression analysis did not include normalweight concrete test results (Figure 3). The reason is the different mechanisms of ligtweight and normalweight concretes. The influence of lightweight aggregates on concrete strength is greater than that of normal aggregates (Lijiu et al. 2005). The strength of aggregate of common concrete is 1.5–2.0 times higher than the strength of concrete by itself, while the strength of lightweight aggregate is far lower than that of concrete, so they have different destructive forms if compressed. When the values of normalweight concrete were imported, the R-square values decreased. Strength of lightweight concrete was closely associated with the density. An increase in compressive strength was observed with the increase in the oven dry density. Lo et al. (2004) also demonstrated that the compressive strength increased with increasing concrete density. The slight scattering of strength results indicated a minor variation in the quality of the lightweight aggregate. The particle strength and stiffness of a lightweight aggregate depends on pore-size distribution, shape and the total volume of pores in the individual particles; large, irregularly shaped cavities will weaken the aggregate particles leading to weaker strength in the LWAC.

Equations were proposed for calculating the compressive strength of lightweight aggregate concrete (Lijiu et al. 2005). In these equations, the parameters affecting the compressive strength of lightweight concrete were effective water–cement ratio of lightweight aggregate concrete, dry apparent specific gravity of lightweight coarse aggregate, actual consumption of lightweight aggregate, actual strength of cement and compressive strength of lightweight aggregate. Also, in another study (Hossain 2004), it was concluded that the concrete strength depends on the strength, stiffness and density of coarse aggregates. In the current study, effective water–cement ratio of lightweight aggregate concrete, actual consumption of lightweight aggregate and actual strength of cement were same for all concrete types. However, the specific gravity and the strength of aggregates were the varying parameters.

From the regression analysis between the compressive strength of concrete and specific gravity of aggregate and aggregate crushing value, the following equations could be derived:

$$\sigma = -1{,}6392\mu + 101{,}16,\ R^2 = 0{,}9078 \qquad (3)$$

$$\sigma = 11{,}699\delta + 32{,}713,\ R^2 = 0{,}7255 \qquad (4)$$

By multiplying the equations 3 and 4, the following equation for strength in terms of both specific gravity of aggregate and aggregate crushing value could be yielded:

$$\sigma = 1183{,}471\ \delta + 53{,}623\ \mu - 19{,}177\ \mu\ \delta + 3309{,}247 \quad (5)$$

where σ = compressive strength of concrete (MPa), δ = aggregate crushing value (%), μ = specific gravity. Figure 4 shows the relationship between experimental results and calculated values obtained by equation 6 for compressive strength which had an R-square value of 0.86 meaning that there was no remarkable variation in the actual and calculated values of compressive strength of the concretes.

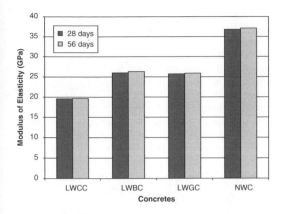

Figure 5. The modulus of elasticity values of 28-day and 56-day concrete specimens.

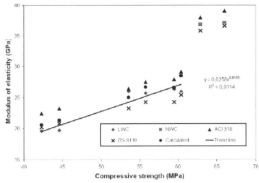

Figure 6. Relationship between modulus of elasticity and compressive strength.

The modulus of elasticity for LWAC ranged from 22.4 to 28.6 GPa which was comparatively less than that for normal-weight concrete with similar strength having 38.0 and 38.9 GPa for 28 and 56 days, respectively. The modulus of elasticity is affected by the compressive strength of concrete, stiffness and volume of the LWA used, interfacial zone between the aggregates and the paste and the elastic properties of the constituent materials (Hossain 2004). In this study, only the aggregate types were different as stated previously.

In high-strength concrete, the modulus of elasticity of the hardened cement paste is high and the difference between the modulus of elasticity of the aggregates and the hardened cement paste becomes small enough to result in higher bond strength and monolithic behaviour (Kayali et al. 2003). Similarly, lightweight aggregate concrete is expected to manifest a clearer monolithic behaviour than normal weight aggregate concrete. This is due to the lower modulus of elasticity of the lightweight aggregate, resulting in a smaller difference between its value and that of the hardened cement paste and to the increased aggregate–paste bond due to the porous surface of the lightweight aggregates. LWGC showed slightly lower modulus of elasticity than LWBC, whereas LWGC had slightly higher strength. For almost similar strength values, the E value of the NWC was ∼42% higher than the E value of the LWBC and LWGC. Also, the 28-day E values of LWBC and NWC were 33 and 88% higher than that of the LWCC. The modulus of elasticity depends not only on the density, but also on the pore structure and the surface texture of the ligtweight aggregate. Therefore, an aggregate with a dense structure and evenly distributed pores gives a higher modulus of elasticity and more concrete stiffness than a more porous aggregate.

The three parameters, namely density, total porosity and critical pore diameter of lightweight aggregates,

supported the results of modulus of elasticity. 10G1200 and 10B1200 had close and the highest specific gravity and bulk density values and lowest total porosity and critical pore diameter while CB had the lowest specific gravity and bulk density values and highest total porosity and critical pore diameter (Table 4). The smaller critical pore diameter means a finer pore structure. It is clear that sintered aggregates showed a denser structure with a high density and low total porosity and exhibited an evenly distributed finer pore structure with a small critical pore diameter and pore size distribution than cold-bonded aggregates. As compared to normal weight concrete, a much lower modulus of elasticity would be expected for lightweight concrete (Lo et al. 2004). The modulus of elasticity increase was very small in 56 days for all concretes.

The dependency between the modulus of elasticity and the compressive strength was analyzed; the best regression being obtained with an expression of the power type (Figure 6).

$$E = 0{,}6258\sigma^{0{,}9185}, R^2 = 0{,}9114 \tag{6}$$

for LWC in 28 and 56 days

$$E = 0{,}1155\,\sigma^{1{,}3558}, R^2 = 0{,}8237 \tag{7}$$

for LWC and NWC in 28 and 56 days where the modulus of elasticity (E) being expressed in GPa and the compressive strength (σ) in MPa.

Also, regression analysis was performed on the modulus of elasticity and density (OD) among the lightweight aggregates using power regression model which gave the best fits against the other regression models with the following equations (Figure 7).

$$E = 3\text{E-}19\gamma^{6{,}0735}, R^2 = 0{,}8091 \tag{8}$$

for LWC in 28 and 56 days

$$E = 2\text{E-}07\,\gamma^{2{,}4498}, R^2 = 0{,}8729 \tag{9}$$

for LWC and NWC in 28 and 56 days where E = modulus of elasticity in GPa, γ = density in OD condition in kg/m^3.

By multiplying Equations 6 and 8 and Equations 7 and 9, modulus of elasticity was calculated in terms of strength and density as below:

$$E = 4.33E\text{-}10 * \sigma^{0.459} * \gamma^{3.037} \qquad (10)$$

for LWC

$$E = 4.62E\text{-}08 * \sigma^{0.678} * \gamma^{1.2249} \qquad (11)$$

for LWC and NWC where E = modulus of elasticity in GPa, σ = compressive strength in MPa, γ = density in od condition in kg/m^3.

All models given by the standards are also valid for lightweight concretes as specified earlier in Equations 2 and 3. BS 8110 and ACI 318 suggest that the modulus of elasticity could be estimated from the compressive strength and density of concretes. Table 7 shows the experimental results and the estimated modulus of elasticity values by the relevant standards and the equation suggested in this study while the relationship between experimental, predicted and calculated modulus of elasticity and compressive strength of lightweight concretes are exhibited in Figure 6. Predicted values of modulus of elasticity by ACI 318

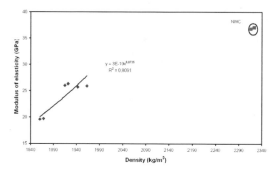

Figure 7. Relationship between modulus of elasticity and density.

overestimated the experimental results whereas BS 8110 underestimated the results except for LWCC. However, both models and equation derived in this study gave close estimated values to the experimental results. All calculated values were between the predicted values proposed by two standards. The models given by codes and standards were also reported to overestimate the modulus of elasticity value of high-strength structural lightweight concretes by a study (Nassif 2003). However, for the concrete reported in a study (Al-Khaiat & Haque 1998) the calculated E values using code formulae were less than the observed value. In the calculation of modulus of elasticity, γ used was the oven dry density instead of air dry density specified in standards.

Analysis of variance is a useful technique for analyzing experimental data. When several sources of variation are acting simultaneously on a set of observations, the variance of the observations is the sum of the variances of the independent sources (Chi et al. 2003). Thus, a software TableCurve 3D V4.0 was used for regression and graphical analysis of the data obtained. TableCurve 3D is the first and only program that combines a powerful surface fitter with the ability to find the ideal equation to describe three dimensional empirical data.

In analyzing the variations, the first step is to compute the sum of squares (SS) and then the mean squares (MS). The total variation includes two parts: the variation within the columns and the variation between the columns. Each of the variations can be reduced to the mean square when they are divided by the corresponding degree of freedom. The ratio of any two of mean squares provides basic information for the F test of significance. Statistical method was applied to analyze the experimental data of the compressive strengths, densities and elastic moduli of concretes. X represented compressive strength, Y represented density and Z represented modulus of elasticity. The test of multivariate analysis of variance with 95% confidence level ($\alpha = 0.05$) was applied.

The application of response surface methodology yielded the following regression equation which is an empirical relationship between the modulus of

Table 7. Experimental, calculated and predicted results of modulus of elasticity.

| Concr. | Modulus of elasticity (GPa) | | | | | | | |
| | Experimental | | ACI 318 | | BS 8110 | | Calculated | |
	28 day	56 day	28 day	56 day	28 day	56 day	28 day	56 day
LWCC	19.6	19.7	22.4	23.2	20.2	20.8	20.5	21.3
LWBC	26.0	26.3	26.4	27.9	23.2	24.2	25.0	26.5
LWGC	25.7	25.9	27.5	29.1	24.2	25.4	26.6	28.5
NWC	36.8	37.1	38.0	39.1	35.8	36.6	–	–

Table 8. Coefficients of the model given by software.

Parm	Value	Std error	t-value	P > |t|
a	−36.6289312	4.57095901	−8.0134018	0.00049
b	0.284671999	0.07737466	3.6791373	0.01431
c	0.023802136	0.00341565	6.9685579	0.00094

Table 9. ANOVA for response surface fit to the experimental results.

Source	Sum of squares	DF	Mean square	F value	P > F
Regr	304.02988	2	152.015	120.48	0.00006
Error	6.3088735	5	1.26177		
Total	310.33875	7			

Table 10. Actual and predicted values by response surface fit for modulus of elasticity.

X	Y	Z	Z Predicted	Residual	Residual %
66	2323	37.1	37.5	−0.351783	−0.948
62.9	2316	36.8	36.4	0.3973156	1.080
60.4	1963	25.9	27.3	−1.38885	−5.362
59.5	1922	26.3	26.1	0.243242	0.925
55.8	1943	25.7	25.5	0.1966835	0.765
53.5	1915	26	24.2	1.8178889	6.992
44.6	1868	19.7	20.5	−0.82983	−4.212
42.3	1860	19.6	19.7	−0.084667	−0.432

elasticity (Z) and compressive strength (X) and density (Y).

$$Z = a + b*X + c*Y \qquad (12)$$

The significance of each coefficient was determined by Student's t-test and p-values (Table 8). The larger the magnitude of the t-value and the smaller the p-value, the more significant is the corresponding coefficient. This implied that the factor most significant (t-value = 6,96856, p-value < 0.05) for modulus of elasticity was the density and the next was the compressive strength.

The results of the response surface fitting in the form of analysis of variables are shown in Tables 9 and 10. The F-test with a very low probability value (p < 0.0001) showed a very high significance for the regression model. The value of the determination coefficient ($R^2 = 0.98$) verifies the suitable fit of the model, thus indicating a discrepancy of 0.02% for total variation, which is a normally accepted range of experimental error. The value of the adjusted determination coefficient (adjusted $R^2 = 0.96$) is also very high which indicates a high significance for the model.

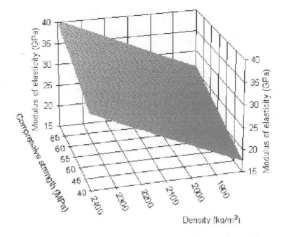

Figure 8. 3D view – response surface plot showing the effect of density and compressive strength on modulus of elasticity.

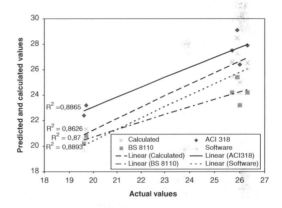

Figure 9. Relationships between actual values and predicted and calculated values of modulus of elasticity of lightweight concretes.

Figure 8 illustrated 3D view of the predicted values obtained by the software.

Relationships between actual values and predicted and calculated values of modulus of elasticity are shown in Figure 9. The large R^2 values were evidences for the good relationships which proved that there was no remarkable variations between the experimental and estimated values of modulus of elasticity of lightweight aggregates. The R-square values were 0.889, 0.887, 0.870 and 0.863 for predicted values obtained by using the software, ACI 318, BS 8110 and the equation derived in this study, respectively. All the estimated values were close to each other and showed small variations with the experimental results. The best fit model was proposed by the software which possessed the largest R-square value of 0.889. All R^2 values

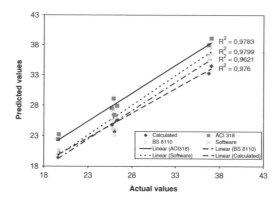

Figure 10. Relationships between actual values and predicted and calculated values of modulus of elasticity of lightweight and normalweight concretes.

of models were large enough to use in predicting modulus of elasticity of concretes. The determination of R-square values was done for lightweight aggregates. Owing to this fact, the R^2 value of the software which analyzed both lightweight and normalweight concrete data, decreased from 0.98 to 0.889 when the normalweight concretes were exported. Figure 10 illustrates the relationships between actual values and predicted and calculated values of modulus of elasticity of lightweight as well as normalweight concretes. Importing the normalweight concrete data into the analysis R^2 values increased as 0.9799, 0.9783, 0.976 and 0.9621 for software, ACI 318, equation in this study and BS 8110, respectively. The best fit model was obtained again by the software with a high R^2 value. All R-square values of models were extremely large enough to apply in estimating the modulus of elasticity of both lightweight and normalweight concretes.

4 CONCLUSIONS

Following conclusions can be drawn from the results of this study:

It is clear that in order to obtain similar slump and air content for lightweight concretes as compared to normal weight concrete, a lower amount of superplasticizer and air entraining agent was needed. All of the lightweight concretes had a fresh and oven-dry density lower than 2000 kg/m^3.

The lightweight aggregate concretes had slightly lower compressive strength than the normal weight concrete due to the higher porosity and lower strength of the aggregate in the concrete. The replacement of the normalweight aggregate in high strength concrete with the lightweight aggregate resulted in a 13%, 18% and 49% reduction in the 28-day compressive strength for LWGC, LWBC and LWCC, respectively, whereas

the reduction was 9%, 11% and 48%, respectively, for 56-day strength.

Heavier concrete resulted in stronger concrete among lightweight concretes when keeping all the constituents same in concrete production. In other words, strength of lightweight concrete was closely associated with its density. When the values of normalweight concrete were imported into the regression analysis between the compressive strength and density, the R-square values decreased.

The relationship between the experimental results and the values calculated by the equation, yielded in this study to find the compressive strength of concrete in terms of specific gravity and strength of aggregates, had large R-square value meaning that there was no remarkable variation in the actual and calculated values of compressive strength.

The modulus of elasticity for lightweight concretes was comparatively less than that for normal-weight concrete.

Models given by standards, the software and the equation derived in this study gave close estimated values to the experimental results. The best fit model was proposed by the software which possesed the largest R-square value. All R^2 values of models were large enough to use in predicting modulus of elasticity of concretes.

ACKNOWLEDGMENTS

The support of Bogazici University Research Fund is gratefully acknowledged through project 07A402.

REFERENCES

ACI Committee 213. 1987. *Guide for Structural Lightweight Aggregate Concrete*. ACI 213R-87.
Al-Khaiat, H. & Haque, M. N. 1998. Effect of initial curing on early strength and physical properties of a lightweight concrete. *Cement and Concrete Research* 28(6): 859–866.
Behera, J. P., Nayak, B. D., Ray, H. S. & Sarangi, B. 2004. Lightweight concrete with sintered fly ash aggregate : a study on partial replacement to normal granite aggregate. *IE (I) Journal* 85: 84–87.
Chi, J. M, Huang, R., Yang, C. C. & Chang, J. J. 2003. Effect of aggregate properties on the strength and stiffness of lightweight concrete. *Cement and Concrete Composites* 25(2): 197–205.
Hossain, K. M. A. 2004. Properties of volcanic pumice based cement and lightweight concrete. *Cement and Concrete Research* 34: 283–291.
Jo, B., Park, S. & Park, J. 2007. Properties of concrete made with alkali-activated fly ash lightweight aggregate (AFLA). *Cement & Concrete Composites* 29: 128–135.
Kayali, O. 2007. Fly ash lightweight aggregates in high performance concrete. *Construction and Building Material. In Press.*
Kayali, O., Haque M. N. & Zhu, B. 2003. Some characteristics of high strength fiber reinforced lightweight

aggregate concrete. *Cement and Concrete Composites* 25(2): 207–213.

Lijiu, W., Shuzhong Z. & Guofan, Z. 2005. Investigation of the mix ratio design of lightweight aggregate concrete. *Cement and Concrete Research* 35: 931–935.

Lo, T. Y. & Cui, H. Z. 2004. Properties of green lightweight aggregate concrete. *International Wokshop on Sustainable Development and Concrete Technology*: 113–118. Beijing: PRC.

Lo, T.Y., Cui H.Z. & Li, Z.G. 2004. Influence of aggregate pre-wetting and fly ash on mechanical properties of lightweight concrete. *Waste Management* 24: 333–338.

Nassif, H. 2003. Development of High-Performance Concrete for Transportation Structures in New Jersey. *Department of Transportation, In cooperation with New Jersey Department of Transportation Division of Research and Technology and U.S. Department of Transportation Federal Highway Administration*, FHWA NJ 2003-016, New Jersey.

Nemes, R. & Józsa, Z. 2006. Strength of lightweight glass aggregate concrete. *Journal of Materials in Civil Engineering* 18(3): 710–714.

Wasserman, R. & Bentur, A. 1997. Effect of lightweight fly ash aggregate microstructure on the strength of concretes. *Cement and Concrete Research* 27(4): 525–537.

Yun, B., Ratiyah, I. & Basheer, P. A. M. 2004. Properties of lightweight concrete manufactured with fly ash, furnace bottom ash, and Lytag. *International Wokshop on Sustainable Development and Concrete Technology:* 77–88. Beijing.

Excellence in Concrete Construction through Innovation – Limbachiya & Kew (eds)
© 2009 Taylor & Francis Group, London, ISBN 978-0-415-47592-1

Experimental studies of the effectiveness of mortar modified with latexes

M.Z. Yusof & M. Ramli
School of Housing Building and Planning, USM, Penang, Malaysia

ABSTRACT: This paper is to investigate the effectiveness latex to modify properties of mortar and ferrocement. A series of test on prism and cubes exposed in NaCl and HCl concentration has been carried out to evaluate the flexural strength, compressive strength, water absorption, carbonation as well as weight loss. For both environmental conditions the test results show that mortar modified with latex have significant improvement in term of durability as compared to the unmodified mortar. The latex modified ferrocement had been tested for their structural performance, which exhibit a superior quality than that of the unmodified ferrocement.

1 INTRODUCTION

The present of dissolved salts in sea water such as sodium chloride, magnesium chloride, magnesium sulphate, calcium sulphate, pottasium chloride and potassium sulphate has significant affect on the durability properties of concrete structures. The deterioration of normal concrete resulting from cyclic exposure in sea water had caused cracking, corrosion of reinforcement. To ensure the durability performance of concrete based material, in practice three types of concrete has been used as alternative to conventional concrete i.e. mortar modified with synthetic or natural rubber latex, resin concrete and polymer impregnated concrete. However in this research, the authors only investigate the durability performance of latex modified mortars which includes Dow latex (DW), carboxylated styrene butadiene copolymer 123 latex (SB123), natural rubber latex (NR), a combination natural rubber – synthetic latexes (Dow latex); NDW, natural rubber-carboxylated styrene butadiene copolymer 123 latex (NSB). These specimen were compared with the unmodified control (CON) for durability performance. All specimens were designed in accordance with method proposed by M.F. Canovas, where maximum diameter of aggregate has been taken into account for determining the quantity of cement as shown in the following equation.

$$C = 700/D^{1/5}$$

where C is the cement proportion in kg/m³

D is the maximum diameter of aggregate

Prior to the test, all mortar specimen were cured in water for 28 days before transferring them to 5% NaCl and 5% HCl for 30, 90, 180 and 365 days for further curing.

2 CONSTITUENT MATERIALS

2.1 Cement and fine aggregates

Latex modified mortars used in the manufacture of ferrocement are comprise of ordinary portland cement (OPC), fine aggregate consisting of 50-50 mixed river-plastering sand with maximum size of 5 mm diameter and fineness modulus of 2.31. Ordinary portland cement had a specific surface of $350 \, m^2/kg$; a R.H. Bogue composition tricalcium silicate, C_3S of 54.1%, dicalcium silicate, C_2S of 16.6%, tricalcium aluminate, C_3A of 10.8% and tetracalcium aluminoferrite C_3AF of 9.1%. In addition to the main R.H. Bogue composition, there exist of minor compounds, such as MgO, alkalis, SO_3, and their percentage composition were 1.5%, 1% and 2%, respectively.

Three different types of latex dispersions had been used which include, natural rubber latex, Dow latex and carboxylated styrene butadiene copolymer 123, and their properties are summarized in Table 1.

Table 1. Properties of latexes.

	Synthetic latex		Natural latex
Properties	Dow*	SB123**	NR
Appearance	white	white	white
Odour	slight	slight	no putrefacting smell
pH	10.4	4.5 ± 1	10.56
Water solubility	soluble	soluble	soluble
Relative density	1.0	1.02	0.94
Solid content	46.6	50 ± 1	0.94

* Dow latex (DW); ** SB123 latex (SB).

2.2 Water

The mixing water used for manufacturing the ferro-cement specimen was taken from portable water with pH number of between 6.5 to 8.5 and comply with BS 3148.

2.3 Reinforcement (wire meshes)

Three layers of 1.0 mm diameter of square welded wire mesh, having 10 mm × 10 mm openings and volume fraction of 0.71% were used as reinforcement for ferrocement. The yield strength and ultimate tensile strength of the wire mesh were 73.2 N/mm² and 122 N/mm², respectively.

3 CASTING OF TEST SPECIMENS

3.1 Latex modified mortar, LMM

The mortar specimens, i.e. CON, DW, SB, NR, N DW, NSB consisting of prisms 100 mm × 100 mm × 500 mm and cubes of 100 mm × 100 mm × 100 mm were cast in steel moulds and compacted in 3 layers using vibrating table.

3.2 Latex modified ferrocement, LMF

The structural behavior of latex modified ferrocements was carried out on the ferrocement test panels consisting of FEDW (Dow latex ferrocement), FESB (SB latex ferrocement), FENR (NR latex ferrocement), FENDW (mixed NR–Dow latexes ferrocement), FENSB (mixed NR-SB ferrocement) and FECON (control). The nominal size of the specimens were 30 mm × 125 mm × 350 mm.

3.3 Design of LMM and LMF

The mix design of the mortar for ferrocement were prepared as follows:

Ordinary portland cement (OPC) of 510 kg/m³, sand of 1275 kg/m³ and having a fineness modulus of 2.31, latex emulsion of 15% solid content by weight of cement, 1.5% anti-forming agent and anti-coagulant agent of 3% by weight of latex solid as shown in Table 2.

3.4 Method of curing

After 24 hours of casting, specimens were demoulded and cured in water for 28 days before exposing them to salt and acid water environments for further curing. For cubes and prisms, the specimens were tested at the following ages; 30, 90, 180, 365 days to determine their flexural strength, compressive strength, water absorption, carbonation and weight loss. Meanwhile, ferrocement specimens were tested for load-deflection characteristics, the ultimate load, cracking load and crack development when exposure to salt and acid environments.

Table 2. Design of mixes for LMM and LMF test specimens.

Setting	Latexes			mortar	
	%	AF%	AC %	W/C	Slump(mm)
CON & FECON*	0	0	0	0.50	50
DW & FEDW*	15	1.5	0	0.31	55
SB & FESB*	15	1.5	0	0.46	60
NR & FENR*	15	1.5	3.0	0.27	80
NDW&FENDW*	15	1.5	3.0	0.35	65
NSB & FENSB*	15	1.5	3.0	0.42	60

* LMF: Latex modified ferocement.
LMM: Latex modified mortar, AF: anti-forming agent, AC: Anti-coagulant agent, W/C: water-cement ratio.

4 PARAMETER MEASUREMENT

4.1 Latex modified mortar

The flexural strength, compressive strength, carbonation depth, water absorption and weight loss of the mortar are determined as follows:

4.2 Flexural strength

The flexural strength of the mortar was determined following the British Standard BS 1881: Part 118: 1983. The specimens were tested at 30, 90, 180 and 365 days using ELE flexural machine.

$$f_{cf} = FL / (d_1 \times d_2{}^2)$$

where F is the breaking load (in N); d_1 and d_2 are the lateral dimension of the cross section (in mm); L is the distance between the supporting rollers (in mm).

4.3 Compressive strength

The compressive strength of 100 mm × 100 mm × 100 mm cubes were tested in accordance with British standard BS 1881: Part 119: 1983 using ELE compression machine.

4.4 Water absorption

Water absorption was calculated in accordance to British Standard BS 8110: part 122: 1983 "Method for determination of water absorption". For a representative sample, a set of three core samples was taken from broken prisms after flexural strength test. The cores having diameter of 75 mm and length 100 mm were taken from broken prisms using coring machine with diamond cutter. The three cores were placed in an oven for a period of 72 ± 2 hours at a temperature of 105°C. On removal from oven the cores were allowed to cool for 24 ± 0.5 hours. Before immersing in tap water for

30 ± 0.5 minutes, the weight of each core was recorded (w_i). After immersing for 30 ± 0.5 minutes, the samples were removed and dried with cloth as rapidly as possible and weighed again (w_a). The water absorption of latex modified mortar and latex unmodified mortar was determined by using the following equations.

water absorption = $\{(w_a - w_i) \times 100 / w_i\} \times$ correction factor

where, correction factor = volume $(mm^3)/\{(surface\ area(mm^2) \times 12.5)\}$

4.5 Carbonation

The carbonation depth was measured by spraying phenolphthalein indicator on a broken surface of the prism after flexural test, in which carbonated surface will changed to colourless and the uncarbonated surface remain pink in colour. This is due to the formation calcium carbonate resulting from reaction of carbon dioxide in the presence of moisture as illustrated by the following equation.

$$Ca(OH)_2 + CO_2 + H_2O \rightarrow CaCO_3 + 2H_2O$$

4.6 Weight Loss of mortar (WL)

To determine the weight loss of mortar, the cube specimens of $100\,mm \times 100\,mm \times 100\,mm$ were dried in an oven for 24 hours at a temperature of $105°C$ before exposed them to hydrochloric acid with 5% concentration. The weight of cube before exposure were recorded (W_i). After exposing in acid water for specified period i.e. 90, 180, 365 days, the specimen were washed in running tap water. Then the specimen were dried in an oven for 24 hours before weighing (W_a). The following expression was used to determined the weight loss of mortar.

$$WL = (W_a - W_i)/W_i \times 100$$

where W_i: Original weight of specimen in grams
    W_a: Final weight after deterioration in grams

4.7 Latex Modified Ferrocement, LMF

The load-deflection characteristics of the latex modified and the unmodified ferrocement; were conducted using a TORSEE testing machine. The specimen was simply supported over a longitudinal span of 300 mm and subjected to four point loading test as shown in Figure 1.
    The load-deflection curves, were recorded manually. Appearance of first crack load and the crack width were measured at the tension face in the longitudinal span when the load was applied at the center of specimen. At each load increment, the number of cracks within the centre of 100 mm in the flexural

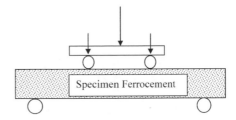

Figure 1. Test set up for deflection and crack.

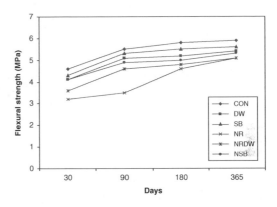

Figure 2. Flexural strength of mortar curing in salt water.

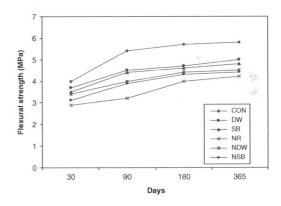

Figure 3. Flexural strength of mortar curing in acid water.

span of specimen was noted and the crack widths for each crack were measured using a hand held microscope, and capable of measuring crack width of about ± 0.01 mm.

5 TEST RESULT

Durability parameter of the latex modified mortar and latex unmodified mortar, i.e. DW, SB, NR, NDW, NSB and CON subjected salt water and acid environments in Figures 2–8.

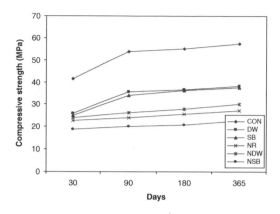

Figure 4.   Compressive strength in salt water.

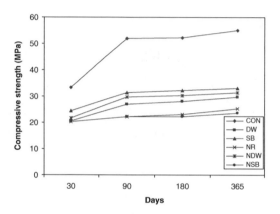

Figure 5.   Compressive strength of mortar in acid water.

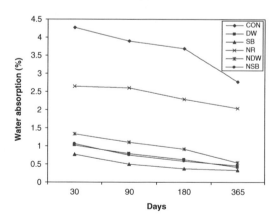

Figure 6.   Water absorption of mortar in salt water.

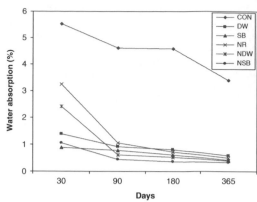

Figure 7.   Water absorption of mortar in acid water.

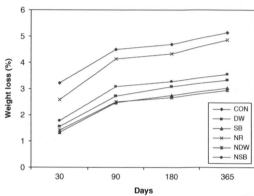

Figure 8.   Weight loss of mortar.

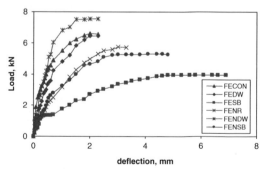

Figure 9.   Load-deflection characteristic of ferrocement in salt water.

6   DISCUSSION

5.1   *Structural performance*

The results of experiments on ferrocement are shown in Figures 9–13 for load-deflection, first crack load, ultimate load and crack spacing, respectively.

6.1   *Latex modified mortar*

6.1.1   *Flexural strength*

Figure 2 showed that, at the age of 365 days, the latex modified mortars subjected to salt water curing

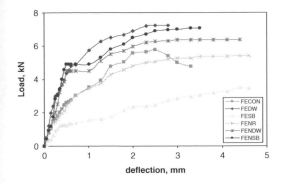

Figure 10. Load-deflection characteristic of ferrocement in acid water.

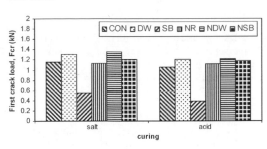

Figure 11. First crack load of ferrocement.

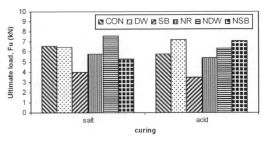

Figure 12. Ultimate load of ferrocement.

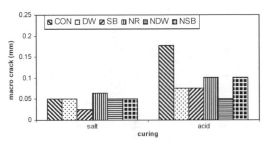

Figure 13. Crack width of ferrocement.

possessing lower flexural strength capacity of 8.5%, 5.1%, 13.6%, 13.6%, 10.2% for DW, SB, NR, NDW and NSB, respectively compared to that of the unmodified, CON. Similarly, the flexural strength of latex modified mortars cured in acid water also showed lower flexural strength, i.e. 73.8% for DW, 17.2% for SB, 27.6% for NR, 22.4% for NDW, 24.1% for NSB compared to that of the unmodified mortar at 365 days of age.

6.1.2 Compressive strength
Figures 4–5 illustrate that latex addition reduces the compressive strength of the modified mortar regardless of type of latex used, curing condition and the age of test. For example, there is also reduction of compressive strength of latex modified mortar even after 365 days of salt water curing i.e 38.1 MPa, 37.7 MPa, 27.2 MPa, 30.3 MPa and 22.7 MPa for DW, SB, NR, NDW and NSB, respectively, compared to that of the unmodified mortar which had compressive strength of 57.4 MPa.

6.1.3 Water absorption
The test result in Figures 6–7 for salt and acid curing had shown that the latex has significantly improvement on the water absorption of the modified mortar. At the age 365 days, specimen cured in salt water exhibit the water absorptions of 0.42%, 0.33%, 2.04%, 0.55% and

0.46% for DW, SB, NR, NDW and NSB respectively as compared to that of the unmodified mortar, CON which had water absorption of 2.77%. Meanwhile, water absorptions for mortar specimen subjected to acid i.e. DW, SB, NR, NDW and NSB were 0.58%, 0.42%, 0.49%, 0.40% and 0.33% respectively, compared to that of the unmodified mortar CON which had the water absorption of 3.39%.

6.1.4 Depth of carbonation
Latex modified mortars DW, SB, NR, NDW, NSB subjected to salt and acidic environments showed no sign of carbonation when the broken surface of prism 100 mm × 100 mm × 500 mm was sprayed with phenolphthalein indicator at the end of 365 days of exposure.

6.1.5 Loss of weight
The test results in Figure 8 show that all latex modified mortar have excellent weight loss resistance compared to the unmodified mortar. At 365 days exposure in acid water showed that mortar modified with mixed natural rubber-synthetic latexes; NDW have the greatest weight loss resistance i.e. 2.95%, followed by mortar modified with latex SB, DW, NSB, NR which had water absorption of 3.04%, 3.34%, 3.55% and 4.85% respectively, and the unmodified mortar which had weight loss of 5.14 %.

267

6.2 Latex modified ferrocement

6.2.1 Load-deflection characteristic

The effect of natural rubber and synthetic latex on load – deflection characteristics of latex modified ferrocement can be seen from Figures 9–10, and be expressed in term of first crack load F_{cr}, ultimate load F_u and crack width W_{av} as show in the following equation.

$$(P\text{-}\delta) = f(F_{cr}, F_u, W_{av})$$

6.2.2 First crack load

The ratios of the cracking load of latex modified ferrocement and the unmodified ferrocement subjected to salt water curing are 1.13 for FEDW, 0.48 for FESB, 0.98 for FENR, 1.17 for FENDW and 1.04 for FENSB at ages of 365 days. Meanwhile for specimen exposured to acidic environment, the ratios were 1.15, 0.37, 1.07, 1.16, 1.12 for FEDW, FESB, FENR, FENDW and FENSB. These results indicate that ferrocement modified with synthetic latex and mixed natural rubber-synthetic latex ferrocement FEDW, FENDW, FENSB and FENR, FENSB gives a significant improvement in the first crack load of ferrocement cured in salt and acid water, respectively.

6.2.3 Ultimate load

For ultimate load, the ratios of latex modified ferrocement as compared to the unmodified ferrocement were 0.98 for FEDW, 0.60 for FESB, 0.87 for FENR, 1.14 for FENDW, 0.80 for FENSB specimens under salt water curing. Furthermore for specimens under acidic environment, the ratio were 1.26, 0.60, 0.94, 1.10, 1.23 for FEDW, FESB, FENR, FENDW and FENSB, respectively compared to that of the unmodified ferrocement. Again, the result of ultimate load for both exposure revealed that Dow latex and mixed natural rubber-synthetic latex has greatly affect the ultimate load of latex modified ferrocement.

6.2.4 Crack width

The result for crack width of latex modified ferrocement and the unmodified ferrocement were presented in Figure 13. It showed that the crack width of latex modified ferrocement was smaller than that of the unmodified ferrocement. For instance, the crack width of latex modified ferrocement and unmodified ferrocement cured in salt water were 0.0762 mm, 0.0762 mm, 0.1016 mm, 0.0508 mm, 0.1016 mm, 0.1778 mm for DW, SB, NR, NDW, NSB and CON respectively. This result clearly indicates that the crack propagation in latex modified ferrocement is much slower than the unmodified ferrocement and would provided sufficient warning before the ferrocement structures would collapses due to excessive loading.

7 CONCLUSION

Based on the experimental study, the following conclusions may be drawn.

1. The latexes; Dow latex, SB123 latex, natural latex has greatly affected the serviceability of ferrocement; load-deflection characteristics, particularly on first crack load, the ultimate load and the crack width. This was achieved by dispersing the solid polymer particles between cement and aggregate in fresh state, bridging them into a monolithic structures, leading to a considerable improvement of the serviceability. However, for ferrocement with mixed NSB latexes show slightly lower first crack load, the ultimate load and the crack width as compared to that of the ferrocement with natural rubber latex only.
2. Latex modified mortar and the unmodified mortar cured in salt and acid environments were not affected by carbonation regardless of curing period.
3. Method of a mixed natural-synthetic latexes; NDW and NSB influenced the durability of modified mortar in terms of water absorption, carbonation and weight loss as compared to the unmodified mortar irrespective of their curing.

REFERENCES

Canovas, M.F. 1988. *Patologia a Terapia do Concrete Armado*, Sao Paulo: PINI Editora

M. Ramli dan M.Z. Yusof. 2000. *Durability properties of mortar modified with natural rubber-synthetic latexes in aggressive environment*, Proc. of the 4th Asia Pacific Structural Engrg. and Construction Conference: 253–261

M.Z. Yusof. 2000, *Durability of polymer modified mor tar in aggressive environment*, PhD thesis, University of Science Malaysia

M.A. Mansur & P. Paramasivam. 1985. *Ferrocement under combined bending and axial loads*, The International Jour nal of Cement Composites and Lightweight Concrete, Vol. 7, Number 3: 151–157

Y. Ohama, T.Kobayashi, K. Takeuchi & K. Nawata. 1986. *Chemical resistance of polymethyl methacrylate concrete*, The International Journal of Cement Composites and Lightweight Concrete, Vol. 8, Number 2: 87–91

Excellence in Concrete Construction through Innovation – Limbachiya & Kew (eds)
© 2009 Taylor & Francis Group, London, ISBN 978-0-415-47592-1

Effect of crystal cement on concrete

H.K. Nezhad & A. Naderi
Islamic Azad University, Qazvin, Iran

ABSTRACT: Concrete is one of the most important materials that people use in construction. It has many qualification that make us to use it concrete has known as a material that has a good construction usages. But never seen it as an architectural material maybe in this new composite concrete a new access has open to the architectures designs the same concrete has known by adding unusual materials to the concrete or by passing light from special parts that use for it but the new cement can pass light easily from side to site and also has a good compressive strength that eager architectures to use concrete more in their designs.

1 INTRODUCTION

Always, in different building designs especially in architecture, the glass has effective presence as a transparent element. Transparency has been interested by ancients from many years ago. As, they have used many kinds of crystals as jewelry on their crown and throne.

Iran, Egypt and Greece kingdoms are included the nations which were interested to transparent crystals according to the remained works. Now, it is mention that the interests in the form of transparent design that are reminded a structure style exist.

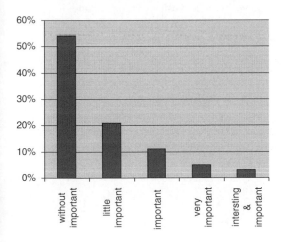

Concrete industry as an important and effective industry has not given the possibility to the architecture designers to make transparent designs. In a questionnaire from architecture engineers in Iran and turkey,

%81 from them believe the necessity to use from a concrete with transparent property.

It's obvious that based on the necessity in 2007 according to many studies and normal reconstruction, it was built the transparent crystal concrete more than 195 times by Ali Naderi and Houra Kazemi Nejad.

2 THE BINDING EFFECT OF CEMENT

The binding effect of cement binding material is in a form that by replacing beside of crystal aggregates, the quartz aggregates are matched with the other aggregates and its nature cement is, if instead of quartz aggregate, it is used from glass, has the silica alkaline reaction capability with glass aggregate and makes chemical contact.

The crystal cement which is made from cement crystallized. If using in concrete, it is given to it, the transparency property, it means that if you move your hands behind a sheet, you can observe the complete hand shadow obviously.

In using %100 from crystal cement aggregate micro silica, the concrete has the below properties.

The most important point that in adding micro silica to the crystal cement, its volume has increased to %8, so the rate of using crystal cement will decrease which makes costs, energy and ... decreasing.

3 THE CRYSTAL CEMENT REACTION WITH USED MATERIALS IN CONCRETE

If the crystal cement is used as the only cement binding in concrete (means: crystal cement aggregate + water)

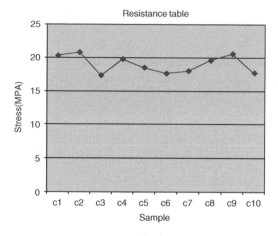

Resistance table

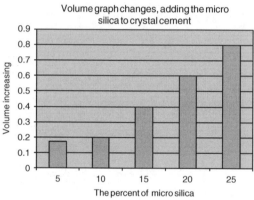

Volume graph changes, adding the micro silica to crystal cement

aggregate, there was no aggregate to be able to change chemical changes on crystal cement as remarkable. But just about the glass aggregate (waste glass), we can say that the glass is effectively captured by crystal cement because of ASR (alkali _silica reaction) that is the benefit of using glass aggregate and crystal concrete jointly because of chemical combination of normal glass, the crystal cement and glass are betrayed.

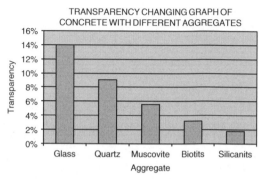

TRANSPARENCY CHANGING GRAPH OF CONCRETE WITH DIFFERENT AGGREGATES

has transparency capability in concrete and as its base is from white cement, if it is added Portland cement or.... to it, the subject will lose its transparency capability but save its resistive properties.

If it is completely used from the cement binding, the concrete texture transparency is guarantee. By using transparent aggregate in concrete in concrete with light color like quartz, waste glass. . . .. This makes concrete to be more transparent, but crystal cement lonely saves light transit capability from itself beside every kinds of aggregate. Also, the kind of aggregate could make the color of concrete darker or lighter.

The crystal cement doesn't effect on aggregate nature, for example, it doesn't decrease from their resistance or change in the color of aggregate. In many case, during more than 60 experiments with different

4 CONCLUSION

As above said, the crystal cement with transparency capability for concrete has made a new operation in concrete application. In addition to more connect (physical and sometimes chemical) with aggregate, it cause to hide from the eyes of viewers. The concrete transparency property is a negligible subject that gives to designers and architectures the probability to prepare a new application from the others and design with a more spread and calmer vision, because the beauty of concrete will be from the other factors which have added to its capabilities.

REFERENCES

BOROSOI, M&S COLLEPARDI, CPPOLA, TROTI, sulfate attack on blended Portland cements CANMET-ACI/International Conference on durability of concrete Barcelona, SPAINGIUGNO 2000.

P. Kumar Mehta, sulfate attack on concrete: separating myths from reality – concrete international, Aug 2000.

Portland cement association, design and control of concrete mixtures, thirteenth edition.

Properties of FRP composite durability

A. Bahari & J.R. Nasiri
Islamic Azad University-Tabriz Branch, Iran

ABSTRACT: FRP composites are a new branch of materials that their endurance is the essential and initial reason for their application in a broad area of structural elements. For this reason, they are used not only in construction building, but also in astronaut, aircraft wing, automobile doors, and containers of liquid gas, ladder and even tennis racket's components. Thus from engineering point of view, not only the resistance and stiffness affair, but also their endurance under expected conditions seem to be thoroughly important. The mechanisms which control the endurance of composites are as follows: Chemical or physical changes in polymer's matrix, loosing the stickiness between fiber and matrix, decreasing in resistance and stiffness of fiber. Environment has a very determinant role in changing properties of composite's matrix polymers. Both matrix and fiber may be affected by moisture, temperature, sun light and especially ultraviolet rays, ozone and presence of some decomposer chemical materials like salts and alkalis. Additionally the repeated changes of temperature may cause some changes in matrix and fiber as freezing and melting cycles. On the other hand, under mechanical loading condition, the repetitious loads may cause fatigue. Also, received loads during a certain time in a constant mode can follow with creep problem. These mentioned problems altogether, affect the endurance of FRP composites. In this article these subjects has been discussed in details.

1 INTRODUCTION

Although plastic materials have some appropriate properties such as low special weight, good formability, simplicity in construction and low price, they don't apply as carrier elements in building because of their low endurance and madulasticiti and essentially they are not appropriate for porter age. For removing the disadvantage of behavior and mechanical properties of plastic materials, the idea of equipping them by fibers with very good mechanical properties developed. So, by equipping plastic materials with fibers, compound materials or composites which called "plastic reinforced with fibers" emerged, that are known shortly as FRP. Inclination to application of composite FRP in building construction and as a constructing material only started two decades ago. Although, in this short period, especially from 1990th, a dazzled development in using of composites FRP in building construction has been obtained, in a way that in hundreds of special and important building, they have been applied as the principal porter age element and considerable research expenses have been spent on them. Perhaps the main reason of the special attention to FRP material in building construction is the very well endurance of these materials in corroding environments. Structural and construction elements' corrosion, especially in alkaline and salty environments, always has been considered as a threatening factor in constructions. Although this problem mostly had been examined in marine areas, in other environments, many alkaline and salty active ions contained in air, hurt the steel parts. This problem is not only an important difficulty for steel constructive components, but they make the reinforced concrete vulnerable too.

2 PHYSICAL AGING OF POLYMER MATRIX

One of the significant aspects which should be considered in the case of composites' endurance, is the role of polymer matrix and it's changes, the primary role of matrix in composite, tension transforming between fibers, protecting fiber surface against mechanical erosion and making an obstacle against inappropriate environment. Matrix plays a deserving role in transforming incision tension in composite plane too. So, if the polymer matrix changes it's properties by time, it should be considered in a special way. It's wholly natural for all of polymers to have a very low change in their chemical (molecular) construction. This change is controlled by environment and especially by temperature and moisture. This gradual change is known as aging. Aging changes in most common thermoset composites is slighter than thermoplastic ones. Affected by physical aging, some polymers may become harder and more brittle; this problem consequently will affect on determinant properties of composite such as incision behavior of composite. In spite of this, often these effects are not critical; because finally the procedure of essential load transformation occurs through fibers, and aging effect on fibers is pretty negligible.

3 INFLUENCE OF MOISTURE

By fast superficial absorption moisture and diffusion of it, most of composites with polymeric matrix in the vicinity of moist weather or humid environments, get the moisture. Usually, percent of moisture by passing of time increases and finally after a few days being in touch with the moist environment, it reaches the saturation (balance) point. The time that is needed for composite to reach to the saturation point, is depended on composite thickness and its amount of environmental humidity. Drying the composite can make this procedure reversed, but it may not result in obtaining the primary properties completely. Absorbing water follows the general Fick's law of emission and is related to time square root. On the other hand, the exact speed of moisture and absorption is depended on some factors such as the amount of pores, fiber type, resin type, fiber direction and construction, temperature, level of incoming tension, and presence of micro cracks. We'll continue our essay with discussing the effect of moisture on composite's components separately.

3.1 The effect of moisture on polymeric matrix

Absorbing water by resin in some cases may change a part of resin's properties. Such changes mainly may occur on temperatures above $120^{\circ}C$ and decrease the composite hardness severely. However such a condition primarily happens in composites uses in civil engineering and building industry rarely and it's not considerable. Additionally moisture absorption has a useful affect on composite too; it caused the resin to be swollen which decreases the hysteresis tensions between matrix and fiber caused by contraction in producing the composite. However, it's reported that in composites which are made inappropriately, as a result of existing some vacuoles in common surface of fiber and matrix may lead to resin flux. This problem can be removed by selecting resin materials or preparing fibers surface appropriately and improving constructing techniques.

3.2 Influence of moisture on fibers

Prevailing doctrine is that if glass fibers place beside water for a long period of time, they will get hurt. The reason is that glass has been made from silica which contains alkaline metal oxides. These alkaline metal oxides are both water absorbent and hydrolysable.

Even though in most uses case in civil engineering, E – glass and S – glass are applied those have only a few amount of alkaline metal oxides and so are resistant against the dangers arising from being in contact with water. However, the composites consisting of glass fibers should have been made well; as they can protect from water permittivity, greatly, because presence of water in glass fibers' surface decreases their superficial energy and may increase the development of cracking. On the other hand, aramid fibers can absorb considerable amounts of water too that results in their swelling and inflammation. In spite of this, most fibers are preserved by a cover that provides the appropriate connection with matrix and retains the water absorption. It should be mentioned that variety of researches indict that moisture has no negative known influences on carbon fibers.

3.3 General action of composites saturated with water

The composites saturated with water usually show some increase in formability as a result of matrix softening. This can be considered as a useful aspect of water absorption in polymeric composites. The limited fall in resistance and module elasticity can take place in composites saturated with water, too. Such changes normally are reversible, so that as soon as drying the composite, maybe the effect of lost properties again compensated. It is noteworthy that increasing in hydrostatic pressure (for example in cases that composites are applied under water or in sea bottom), it doesn't necessarily result in more water absorption by composite and falling its mechanical properties. In this way, it's expected that most polymeric structures have high endurance under water. In fact, under hydrostatic pressure, as a result of obstructing the micro cracks and middle surface damages, water absorption slightly decreases. It should be mentioned that water absorption affects on composites insulation properties. Presence of free water in micro cracks can decrease the insulation property of composite severely.

4 TEMPERATURES – HUMIDITY INFLUENCE

Temperature plays a determinant role in water absorption mechanism in composites and their subsequent irrevocable influences. Temperature affects on water distribution, amount and speed of its absorption. By increasing the temperature, amount and speed of water absorption increases quickly. Researches have shown that damages resulted from placing composite in boiled water for a few hours is equal to separating composite components and cracking it as a result of resting in water with 50°C temperature for 200 days. In normal room temperature, composite patterns have shown no damage and demolition. Such observations have resulted in developing some techniques for accelerated examines of composites aging in high temperature.

5 ALKALINE ENVIRONMENTS

Using composites with glass fibers in alkaline environment, it's necessary to apply the glass fibers with high alkaline resistance; because alkaline solution gives reaction with glass fibers and produces expensed silica jelly. This point is especially important in application of composites with glass fibers as reinforcing rods. Nowadays interest in using FRP rods in glass material in concrete surfaces as a replacement for steel rods, which are corrosive against frozen salts, and in structures in adjacent with water, has increased. Besides in cement hydration process, a solution with high alkaline (PH > 12) emerges. This high alkaline solution can affect the glass fibers and decrease the endurance of FRP rods which have been made from glass fibers. Glass fibers in E – glass type which is mostly applied because of its relatively low price, may have less resistance against alkaline attacks. Using resin and nibble ester by producing an effective obstacle, decreases the alkaline attacks partly. Resistance against alkaline attacks can be improved by designing a structural member for tolerating the lower level of tensions. For improving the endurance it is possible to use the glass fibers with perfect resistance against alkaline too. It is noteworthy that FRPs made from carbon and aramid fibers, essentially show no disadvantage in alkaline environments.

6 LOW TEMPERATURE INFLUENCE

Severe changes in temperature have several main effects on composites. Most of materials expand by increasing temperature. In FRP composites with polymeric matrix, expansion coefficient of matrix temperature is usually in higher rank compared with fibers temperature expansion coefficient. Decreasing the temperature as a result of cooling during building step or composite's function conditions in low temperature, will cause the matrix to be flexed. On the one part, matrix contracting confronts with relatively hard fibers' resistance which have placed in adjacent to matrix. This problem leaves residual tensions in microstructure of material. The residual tensions size is proportionate to temperature difference at composite production and functioning conditions. Even though unless in very cold environments, the leftover tensions will not be considerable. In places where there is severe temperature changes (like areas close to north and south poles), it may emerge some large leftover tension in composite material which will result in micro crack in the material. Such micro cracks decrease composite hardness and increase the permeability and water entering through boundary layer and in this manner participates in composite decomposition process. Another notable effect of low temperatures is corresponding change in resistance and hardness of matrix. Most of resin matrix materials become harder and more enduring by cooling. Such changes affect the fracture condition. For example, it has been shown that squeezing fracture in cylindrical samples of composite with the diameter of 38 mm in 50°C compared with fracture in similar samples in room temperature, accompanies with 17/6 percent increase in pressing resistance and more brittle fracture. So, energy absorption before fraction in lower temperature will be more than room temperature. This special aspect of releasing a lot of energy in fraction moment should be considered in designing the composites which are taken under impact loads in low temperature.

7 THERMO CYCLES EFFECTS IN LOW TEMPERATURE (FREEZING – MELTING)

The effect of freezing – melting phenomenon in the prevailing temperature range (−20°C − +30°C) on resistance, except than cases that composite has considerable percent of connected full of water holes, is insignificant and unimportant. Composites made from glass fibers which are commonly available, have about 0/4 percent holes that don't permit considerable freezing and don't provide serious hurt possibility. In spite of this, thermo cycles in low temperature have some other effects on composites; in composite materials, as a reason of existent differences in thermo expansion coefficients of consisting components in material's micro structure, leftover tensions occur. Such tensions, in very low temperatures, can result in forming micro cracks in matrix resin or in common surface of resin and fiber. Developmental changes of micro crack in prevailing range of exploitation temperature (−20°C − +30°C), are usually insignificant or subordinate. Even though under sever thermo cycle conditions, for example between −60°C − +60°C, the micro cracks may find the possibility of developing and connecting together which leads to formation crack in matrix and distributing it in matrix or around the common surface of matrix and fiber. Such cracks' number and size grows under long term thermo cycles and finally can result in declining the hardness or other properties of matrix.

Also, it has been observed that in very low temperature, the traction resistance of all polymeric composites in fibers direction tends to decreasing; even though the traction resistance in other direction involving orthonormal direction increases. These results are explained with hardening the polymeric matrix in low temperature. On the other hand, in long term thermo cycles between the maximum and minimum temperature follow by resistance and hardness decadence in all directions. Such changes in material properties are considered very important in structural designing in cold areas.

8 THE EFFECT OF ULTRAVIOLET (UV) RAYS' RADIATION

The effect of ultraviolet light on polymeric compositions is thoroughly acknowledged. Under long term sunlight radiation, matrix may become hard or colorless. This problem is generally removable by applying a resistant cover against ultraviolet rays on composite. One of the most considerable points about this subject is declining parts of reinforcing polymeric fibers involving aramids. For example, for an aramid made from thin fibers, after five weeks resting in sunlight in Florida, 50 percent of resistance falling has been reported. However, usually this effect is superficial; so in thicker composites, the effect of this kind declining on structural properties is insignificant. In cases that the superficial properties are important too, it is necessary to consider some notations in order to decrease the shallow cracking under sun rays.

REFERENCES

Aci committee 544, IR – 32, "state-of-the-art report on fiber reinforced concrete", ACI Manual of Concrete Practice, Part 5, PP. 22 – 1984.

Building construction donors, Professor Ahmad Hami.

Dr. ALIREZA KHALOU, MAJID AFSHARI., "Study of bending behavior of concrete slabs reinforced with steel fibers", Technology of Concrete (September 2002), Page 7.

MOHAMMAD REZA SAREBAN, "Internal Reinforcement of Concrete with Wastage of Composite Materials", M.S Course Project, Under Supervision of Dr. MAHMOUD MEHRDAD SHOKRIYEH, Iran University of Science & Technology, Faculty of Mechanical Engineering, Autumn 2004.

Plastic and composite materials, Davood Mostofi nejad.

Ruberts, J.B, Las Casas, E.B. and Oller, S. "An Explicit Finite Element Model for Fiber Reinforced Concrete under Traction". European Congress on Computational Methods in Applied Sciences and Engineering, Barcelona, 11–14 September 2000, pp 11.

SHAPOUR TAHOONI, "Designing Reinforced Concrete Structures", Tehran University Press 1996.

Winter & Nilson, "Design of Concrete Structures" Ninth Editions, Mc Graw Hill, Inc., New York, 1979.

Ultra-high performance concrete

D.B. Zadeh
Islamic Azad University, Iran

A. Bahari & F. Tirandaz
Islamic Azad University – Tabriz Branch, Iran

ABSTRACT: Ultra-High-Performance Concrete (UHPC) with compressive strength higher than 150 N/mm^2 and other perfect properties is a new type of cementations materials. Basic principle to improve concrete properties is advancing a matrix as dense as possible and a good transition zone between matrix and aggregate. In this work an original UHPC was modified through partial replacement of cement or silica fume by fine quartz powder. A self-compacting UHPC with a cylinder compressive strength of 155 N/mm^2 can be produced without heat treatment or any other special measures. Mechanical properties and autogenous shrinkage of the modified UHPC were presented in this work.

1 INTRODUCTION

Performance concrete is a new cementitious material with strength more than 150 N/mm^2 and other perfect properties. Basic principle to improve the concrete properties is the reduction of defect places, such as micro cracks and capillary pores, in concrete. In (Richard & Cheyrezy 1995) some measures have been preferred for the production of UHPC:

- Enhance the homogeneity of the concrete by elimination of coarse aggregate. It is well known that the transition zone between coarse aggregate and matrix is often the source of micro cracks in concrete, due to their different mechanical and physical properties. It was suggested that maximal aggregate size in UHPC should be less than 600 μm (Richard & Cheyrezy 1995)
- Improve the properties of matrix by addition of pozzolanic admixture, i.e. silica fume. The modifying effects of silica fume in concrete are attributed to its pozzolanic reaction with Ca (OH) 2 and filler effect in voids among cement or other components particles. In concrete containing typical Portland cement 18% silica fume, in the weight of cement, is enough for total consumption of Ca (OH) 2 released from cement hydration (Papadakis 1999). However, considering the filler effect the optimal share of the silica fume is about 30% of cement (Richard & Cheyrezy 1995, Reschke 2000). Therefore the silica fume content in UHPC is normally 25–30% of cement

- Improve the properties of matrix by reducing water to binder ratio
- Enhance the packing density of powder mixture. According to the results in (Reschke 2000) a mixture with wide size distribution has a low avoid among their particles. This means powder mixture should be composed of number classes of granular powder
- Enhance the microstructure by post-set heat-treatment since 1994 intensive researches have been carried out in France and Canada. Cement content in these original UHPC ranged between 900–1000 kg/m^3. In this paper a modified UHPC and its mechanical properties and autogenous shrinkage are presented.

2 MATERIALS

Based on the principle for ultra-high performance mentioned above quartz sand with the size of 0.3–0.8 mm was used as aggregate. An ordinary Portland cement CEM I 42.5 R was used as binder. A white silica fume was added as pozzolanic admixture in concrete. Its particle size lies between 0.1–1.0 μm. A quartz powder with a diameter smaller than 10 μm was used as micro filler. Its particle fill the lack between the cement particle and the silica fume and make the grading curve of the mixture composed of cement, silica fume and quartz powder continuous. Super plasticizer on the basis of polyethercarboxylate was used to ensure the concrete flowing ability.

Table 1. Compressive strength of UHPC with quartz powder.

	Cement	Silica cement (%)	Cement replacement (%)	w/c	Compressive strength	
					28d/20°C	14d/20°C + 3d/90°C
1	950	25	0	0.187	149.2	189.7
2	893	26.6	6	0.199	141.6	195.5
3	836	28.4	12	0.213	145.9	186.3
4	779	30.5	18	0.229	151.1	177.5
5	722	32.9	24	0.247	143.0	190.1
6	665	35.7	30	0.268	148.1	201.7

* All of these concretes were cast without vibration.

Table 2. The modified concrete composition and compressive strength.

CEM1 42.5R (kg/m^3)	Silica fume (kg/m^3)	Quartz powder (kg/m^3)	Quartz sand (kg/m^3)	Total water (kg/m^3)	Superplasticiser (kg/m^3)	$F_{c,cy111*300}$ (N/mm^2), 28d/20°C		
						Without vibration	60s vibrated	>90s vibrated
665	200	285	1020	178	23.0	155.5	174.3	197.0

3 MODIFY THE REFERENCE CONCRETE COMPOSITION

In literatures UHPC was characterized with very high cement content about 950 kg/m^3 [4]. Because of the low water to cement ratio, only a part of the cement has hydrated. Unhydrated cement particles lie in matrix as fine aggregate. In this work cement was stepwise replaced by inert quartz powder with same volume. The slump flow of fresh concretes was measured according to DIN 1048 but without shock. It can be seen in Table 1 that even 30% cement was replaced compressive strength was not suffered. Furthermore, with the cement replacement by quartz powder the flowing ability of the concrete was improved. When 30% cement was replaced by quartz powder the slump flow increased from 510 mm (mixing 1) to 620 mm (mixing 6). This can be resulted from that the incorporation of fine quartz powder reduced the voids in the original mixture (in mixing 1) containing only cement and silica fume. Otherwise, with the cement replacement less hydrate have been produced in the first few minutes. There were not enough hydration products to bridge various particles together. Some particles were still free and could move easily.

Finally a part of silica fume was replaced also by quartz powder. The modified composition and compressive strength of self-compacting or vibrated concretes are shown in Table 2.

4 MECHANICAL PROPERTIES OF THE CONCRETE

4.1 Relationship between compressive strength and elastic modulus

Compressive strength and Modulus of elasticity were determined on concretes at ages of 3, 7, 14, 28 and 90 days. Cylindrical specimens, 300 mm high and 100 mm in diameter, were cast as self-compacting concrete or vibrated with a rod vibrator for 60 or 90 sec. Due to the retarding effect of superplsticizer on cement hydration specimens were demoulded 2 days after casting and then immersed in water at 20°C. 3 days before testing they were taken out from water and stored under relative humidity of 80% at 20°C till to test. Elastic Modulus of concrete increases disproportional with compressive strength. For high performance concrete containing quartz coarse aggregate Modulus of elasticity can be estimated from compressive strength as following:

$$E_c = 20.5 \cdot \left(\frac{f_c}{10} \right)^{1/3} \quad (10^3 \, \text{N/mm}^2) \qquad (1)$$

With this equation the E-Modulus of UHPC would be over estimated, because the paste volume in UHPC is much higher than that in conventional high performance concrete. Results in this work would suggest

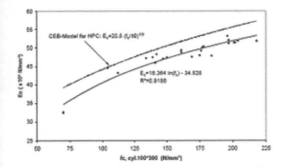

Figure 1. Relation between compressive strength and E-modulus of UHPC cured at 20°C.

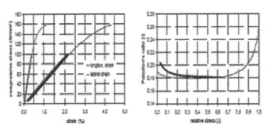

Figure 2. Stress-strain curve of self-compacting UHPC without fibre (6 loading-unloading cycles).

a new relationship between cylinder strength and E-Modulus as following:

$$E_c = 16.364 \cdot \ln(f_c) - 34.828 \quad (10^3 \text{ N/mm}^2) \qquad (2)$$

4.2 Stress-strain curve under axial compressive loading

Stress-strain curves were investigated on 100×300 mm cylinders. After load-unload cycles specimen was loaded to rupture stress. The minimal and maximal stress in the load-unload cycles was 5 N/mm^2 and 65% of compressive strength, respectively. After the first cycle the loading und unloading curves were almost identical. This indicates that only a few new micro cracks were generated during loading cycles. Poisson's ratio of UHPC is about 0,18. The low Poisson's ratio can be resulted from the firm bound between fine quartz sand and matrix. This value kept constant till ca. 70% of compressive strength. After compressive stress exceeded this level, the Poisson's ratio rose abruptly, this indicated a rapid propagation of micro cracks.

5 AUTOGENOUS SHRINKAGE

Autogenous shrinkage is the consequence of chemical volume contraction during cement hydration and

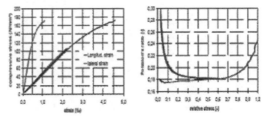

Figure 3. Stress-strain curve of UHPC without fibre (60 sec. vibrated, 4 loading-unloading cycles).

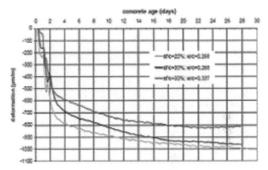

Figure 4. Deformation of sealed self-compacting UHPC.

self-desiccation in concrete. High cement content and low water to binder ratio in UHPC may lead to high autogenous shrinkage, which will induce micro cracks in early ages. Experiments investigating autogenous shrinkage of UHPC were carried out on $150 \times 150 \times 700$ mm beams. In order to prevent the friction between concrete and form a plastic foil had been inserted before casting. Concrete was sealed with over-long foil immediately after casting. Length changes and temperature variation in concrete were recorded every 15 min and saved in computer. Deformations of UHPC with different silica fume contents or water to cement ratios are shown in Figure 4, in which deformation is the sum of autogenous shrinkage and thermal expansion due to temperature increasing during cement hydration.

Autogenous shrinkages shown in Figure 5 were calculated by subtracting thermal expansion from the measured deformations in Figure 4. The Zero-point of the time-axis in Figure 5 corresponds the time, when highest temperature in concrete was reached. Generally UHPC show a higher autogenous shrinkage than conventional HPC. Similar autogenous shrinkages of UHPC with different w/c-ratio or different silica fume content indicate that the both factors have not great influence on autogenous shrinkage at the age of 28 days. However, they have significant influence on the development of autogenous shrinkage. Curves in Figure 5 suggest that autogenous shrinkage of UHPC with

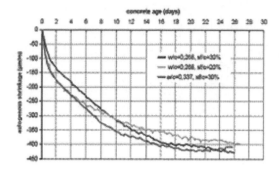

Figure 5. Autogenous shrinkage of self-compacting concrete.

lower w/c-ratio and higher silica fume content could increase continuously.

6 CONCLUSIONS

In this paper some results about the properties of UHPC are presented. Through optimizing the composition of powder mixture in matrix a flow able UHPC has been produced. Its strength depends intensively upon air content in concrete. Compressive strength higher than 200 N/mm^2 can be reached under normal curing condition by reducing the air content less than 1%.

Relationship between compressive strength and elastic Modulus of UHPC is quite different from that in conventional high performance concrete. A new empirical equation predicting E-Modulus of UHPC was suggested.

Because of the high binder (cement and silica fume) content and low water to binder ratio UHPC shows a higher autogenous shrinkage than conventional high performance concrete.

REFERENCES

PIERRE RICHARD, MARCEL CHEYREZY, "Composition of Reactive Powder Concretes", Cement and Concrete Research, Vol. 25, No. 7, 1995, pp. 1501–1511.
PAPADAKIS V., "Experimental Investigation and theoretical Modelling of Silica Fume Activity in Concrete", Cement and Concrete Research, Vol. 29, 1999, pp. 79–86.
RESCHKE T. "Einfluss der Granulometrie der Feinstoffe auf die Gefügeentwicklung und die Festigkeit von Beton", Schriftenreihe der Zementindustrie, Heft 62/2000.

*Theme 3:   Design and construction
in extreme conditions*

Suitability of PTC heating sheets for curing foundation concrete in low temperature environments

M. Sugiyama

Hokkai Gakuen University, Sapporo, Hokkaido, Japan

ABSTRACT: This study evaluated the viability of electric heating sheets for the curing of concrete housing foundations in severe, low-temperature environment. The heating sheets tested use PTC (positive temperature coefficient resistor) ceramic elements. The heating temperature is automatically adjusted by these elements in response to adjacent air temperature. The first series was carried out at $-10°C$ in a thermostatic chamber. The second series involved on-site testing. Under severe low-temperature conditions, it was found that curing was possible at about $+15°C$ through the use of heating sheets. Compressive strength was found to be sufficient. The heating sheet is considered to be extremely useful for curing concrete in cold regions.

1 INTRODUCTION

1.1 *Scope*

This study evaluated the viability of electric heating sheets for the curing of concrete housing foundations in severe, low-temperature environments. The heating sheets tested use PTC (positive temperature coefficient resistor) ceramic elements. The experiment was carried out in two series. The first series, which used small test specimens, evaluated the compressive strength of concrete that had been cured using the heating sheet. The second series involved on-site testing.

1.2 *An outline of the PTC electric heating sheet*

In the past, heating sheets that used nichrome wire displayed erratic temperature changes with the attendant risk of fire and damage to the sheet. Likewise in the case of oil-stove heating, toxic gases were produced and fire was a potential hazard.

The PTC (Positive Temperature Coefficient Resistor) element has many advantages over previous technologies used in heating sheets. The PTC element is extremely durable, safe and flexible and is thus resistant to damage. The PTC element is able to maintain temperature stability by automatically detecting ambient temperature and adjusting if necessary. Heating sheets that utilize PTC elements run on domestic electricity at 100(V), which makes the power source convenient and efficient.

2 BASIC EXPERIMENT

2.1 *Plan and method*

The first series evaluated the compressive strength of concrete that had been cured using the heating sheet. This series was carried out using small test specimens. The experiments were carried out in a thermostatic chamber at $-10°C$ and outdoor in Sapporo in November. First, a box that simulated the footing foundations of the concrete was produced. Next, specimens of the fresh concrete were inserted into this box, and the heating sheet was placed on the upper surface. Compressive strength was tested after the 3rd, 7th and 28th day of curing.

The dimensions of the specimen were $10\varphi \times 20$ cm. The box was made of 10 mm thick wood, and the dimensions were L400×W120×H200. Three pieces of the concrete specimen ($100\varphi \times 200$) were positioned in this box. The box was filled with foam beads, $3 \sim 5$ mm in diameter. Steel plates (L400 × H200 × @4.2 mm) acted as inner walls in three of the test boxes. Because steel reacts more sensitively to outside temperature, we were able to reproduce severe curing conditions. Table 1 shows the design of the experiments. Table 2 shows curing conditions. Table 3 shows the concrete mix proportions.

2.2 *The results*

The temperature results for concrete specimens cured at $-10°C$ are shown in Figure 1. The concrete was

Table 1. Plan of experiment.

Curing condition	In chamber	Outdoor
Temperature	−10 degrees Celsius	Open air
Curing period	4 weeks	4 weeks
Strength	After 3 days, 7 days,	28 days

Table 2. Curing conditions.

Symbol	PTC heating sheet	Conditions
A	With	In wood box
B	With	In steel plates box
C	Without	Out air
D	20 deg. Celsius	water

Table 3. Concrete mix proportions.

W/C %	Slump cm	Air %	Water kg/m³	Cement kg/m³	Sand kg/m³	Gravel kg/m³
50	18	4	165	330	762	1122

*Cement (density 3.16), river sand (density 2.69), river gravel (density 2.75).

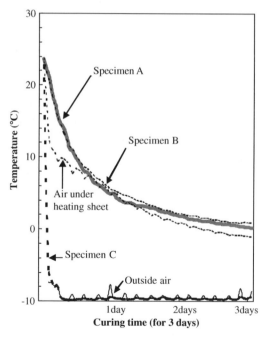

Figure 1. Temperature of concrete specimens (−10 deg.).

mixed at a temperature of 24°C. Specimens A and B (cured under the heating sheet) cooled slowly. However, specimen C (cured without the heating sheet) cooled rapidly and froze at a temperature of −10°C.

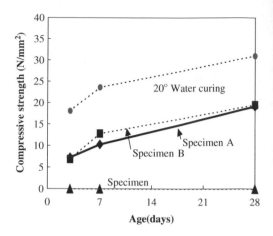

Figure 2. Compressive strength (−10 deg.).

Mean temperatures after 3 days of curing were as follows; Specimen A (wooden frame) cured under the heating sheet was 5.1°C. Specimen B (steel plate) cured under the heating sheet was 5.5°C. Specimen C cured in outside air was −8.9°C. The air under the heating sheet was 3.4°C. The difference in mean temperature between specimen A (with the heating sheet) and specimen C (without the heating sheet) was 14°C.

Compressive strength is shown in Figure 2. Compressive strength after 3 days of curing was as follows; Specimen A (wooden frame) cured under the heating sheet was 7.24 N/mm². Specimen B (steel plate) cured under the heating sheet was 6.85 N/mm². Both specimens reached the target strength of 5 N/mm² for mould removal. The strength of specimen C could not be assessed due to freezing damage in the initial stages.

The temperature results for concrete specimens cured outside in November in Sapporo are shown in Figure 3. Specimen A and B (cured under the heating sheet) once again, cooled slowly. However, specimen C (cured without the heating sheet) cooled rapidly. Mean temperatures after 3 days of curing were as follows; Specimen A (wooden frame) cured under the heating sheet was 12.4°C. Specimen B (steel plate) cured under the heating sheet was 12.3°C. Specimen C cured in outside air was 0.4°C. The air under the heating sheet was 7.7°C. The mean temperature of outside air was 1.0°C. The difference in mean temperature between specimen A (with the heating sheet) and specimen C (without the heating sheet) was about 12°C.

Compressive strength is shown in Figure 4. Compressive strength after 3 days of curing was as follows; Specimen A (wooden frame) cured under the heating sheet was 16.0 N/mm². Specimen B (steel plate) cured under the heating sheet was 15.1 N/mm². Specimen C cured in outside air was 3.44 N/mm². Specimen A and B reached the target strength of 5 N/mm² for mould

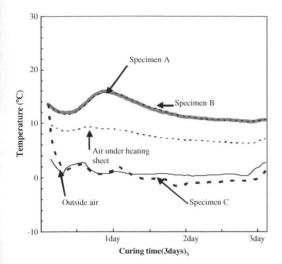

Figure 3. Temperature of concrete specimens (outdoor in November, Sapporo, Japan).

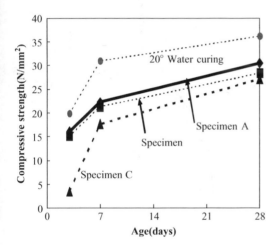

Figure 4. Compressive strength (outdoor in November, Japan).

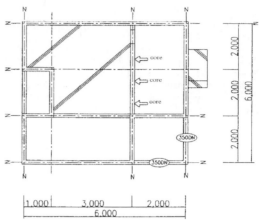

Figure 5. Foundation.

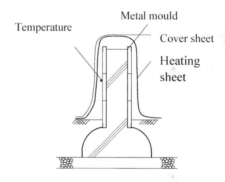

Figure 6. Section of foundation.

removal. Specimen C, cured without the heating sheet, could not reach the required strength target.

The results show that curing, using the heating sheet, produces concrete of sufficient strength, even in severe low temperature conditions.

3 FIELD EXPERIMENT

3.1 Outlines

The second series involved on-site testing. The tests were carried out in Sapporo City, Hokkaido prefecture,

the northern-most region of Japan. Under severe, low-temperature conditions, of up to −11°C outside air temperature, it was found that curing was possible at about +15°C through the use of heating sheets.

3.2 Plan and method

The foundation plan is shown in Figure 5. The experiment was carried out using an actual foundation with metal mould. Foundation dimensions were 6 × 6 m square, wall thickness was 16 cm and the height varied – in sections 1.1 m and in others 1.4 m. The foundation was covered with the heating sheets about 1 hour after the concrete had been poured. The foundation cross section is shown in Figure 6. The heating sheet was activated 4 days after the initial pouring.

The plan of the experiment is shown in Table 4. An outline of the test-pieces used is shown in table 5. A core test-piece was used to examine the foundation strength. The concrete mix proportions are shown in Table 6. The water/cement ratio was 50%, the slump was 15 cm and the temperature was 13°C.

Table 4. Plan of experiment.

Place	Sapporo city, Japan
Period	Feb.4th–Mar.3th.
Curing period	4 days
Compressive strength	After 4 days, 7 days, 28 days

Table 5. Curing conditions.

Symbol	PTC heating sheet	Position	
A	With	Upper foundation	10 cm Core
B	With	Lower foundation	10 cm Core
C	Without	Center foundation	10 cm Core
D	With	–	10 cm Cylinder
O	Outdoor	–	10 cm Cylinder
S	20 deg. Water	–	10 cm Cylinder

Table 6. Concrete mix proportions.

W/C %	Slump cm	Air %	Water kg/m^3	Cement kg/m^3	Sand kg/m^3	Gravel kg/m^3
50	15	4.4	155	310	1)539 2)236	1075

*Cement (density 3.16), sand (density 1)2.69, 2)2.68), gravel (density 2.64).

3.3 The results

3.3.1 Temperature

The temperature results are shown in Figure 7. The temperature of the foundation concrete lowered slowly. Mean temperatures after 4 days of curing are as follows; ①Center of the foundation concrete (cured under the heating sheet) was 16.5°C. ②AT division of the foundation concrete (cured under the heating sheet) was 17.6°C. ③BFoundation concrete without heating sheet was 3.4°C. ④CTemperature of outdoor air was −4.4°(−11°C lowest temperature). ⑤Air temperature in the tent was −0.7°C. ⑥Air temperature under the heating sheet was 12.3°C. ⑦Air temperature between the heating sheet and vinyl sheet was 8.6°C.

In an environment where the lowest outside air temperature reached −11°C, curing at about +16°C was possible for concrete cured under the heating sheet. The temperature of the concrete cured under the heating sheet (① and ②) was between 13°C to 14°C higher than in concrete cured without the heating sheet (③). The temperature of the concrete cured under the heating sheet (① and ②) remained stable when compared to the change in outside air temperature.

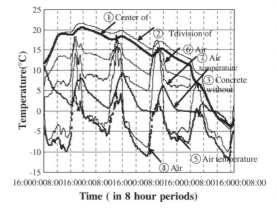

Figure 7. Temperature survey result of field concrete.

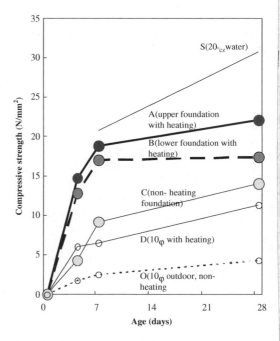

Figure 8. Compressive strength in field concrete.

3.3.2 Compressive strength

Compressive strength is shown in Figure 8. After the 4th day of curing, the compressive strength of the core drilled from the upper part of the concrete (cured with the heating sheet) (A) was 14 N/mm^2, while the core drilled from the lower part of the concrete (B) was 12 N/mm^2. Both A and B exceeded the minimum strength of 5 N/mm^2 required to prevent freezing damage in the initial state. Core strength C (cured without the heating sheet) was 4 N/mm^2 after the 4th day. The effectiveness of the heating sheet is clear.

Table 7. The advantage of the PTC heating sheet.

	The PTC electric heating sheet	The stove (Charcoal or Oil heater)
Temperature irregularity	Nothing	There is temperature variation.
Safety	Little risk of fire. 100 V electric power source.	There is great risk of fire.
Effect on global environment	Nothing	Large amount of carbon dioxide produced.
Total evaluation	Good	No-good

In the case of specimen D, the temperature lowered significantly before the heating sheet was put in place (due to the small size of the specimen) and therefore compressive strength was extremely low.

3.3.3 The merit of the PTC heating sheet for curing of concrete

The advantages of the PTC heating sheet are shown in Table 7.

In regards to the PTC electric heating sheet, there are no temperature irregularities, no risk of fire, and no adverse effects on the environment whereas, in the case of the nichrome wire sheets, a temperature regulator is necessary, there is a substantial risk of fire and damage to the sheet and carbon dioxide is generated in the case of fire.

In the case of the stove heating method (charcoal or oil heater use), there is substantial variation between air temperature close to the heating source and in the corners. A temperature regulator is necessary when using this method and there is a great risk of fire and burn damage to the sheet. A large amount of carbon dioxide gas is produced which has a negative effect on the environment.

The PTC electric heating sheet therefore, has substantial advantages over traditional heating methods.

4 CONCLUSION

This study evaluated the viability of electric heating sheets for the curing of concrete housing foundations in severe, low-temperature environment. The heating sheets tested use PTC (positive temperature coefficient resistor) ceramic elements. The heating temperature is automatically adjusted by these elements in response to adjacent air temperature. The experiment was carried out in two series. The first series was carried out at $-10°C$ in a thermostatic chamber. The second series involved on-site testing. Under severe, low-temperature conditions, of up to $-11°C$ outside air temperature, it was found that curing was possible at about $+15°C$ through the use of heating sheets. Compressive strength was found to be sufficient. The PTC heating sheet is extremely durable, safe and flexible and is thus resistant to damage. It has significant benefits when compared to traditional heating methods. This heating sheet is considered to be extremely useful for curing concrete in cold regions.

REFERENCES

Masashi SUGIYAMA: Suitability of PTC Heating Sheets for Curing Foundation Concrete in Low Temperature Environment, International Congress, Dundee, Scotland, 2002.

Excellence in Concrete Construction through Innovation – Limbachiya & Kew (eds)
© *2009 Taylor & Francis Group, London, ISBN 978-0-415-47592-1*

Improvement of the durability of sand concrete to freezing-thaw and wet-dry cycling by treatment of wood shavings

M. Bederina & M.M. Khenfer
University AmarTélédji, Laghouat, Algérie

A. Bali
ENP, Alger, Algérie

A. Goullieux & M. Quéneudec
Laboratoire des Technologies Innovantes, IUT, France

ABSTRACT: The objective of this work is the valorisation of local materials and the re-use of the industrial wastes. The valorised materials are two local sands: a dune sand and a river sand separately used, and the re-used wastes are fillers and shavings coming respectively from crushing waste and woodwork activities waste. It consists in studying the effect of shaving surface treatment on the durability of the wood sand concretes. Therefore, in order to improve the durability of studied materials to freezing-thaw and to wet-dry cycling, the wood shavings are treated before incorporating them in sand concrete. The results demonstrated that the durability of these materials, concerning these two tests of durability, as well as other properties, is definitely better. In addition, the profit of lightening and heat insulation, obtained without treatment, is only slightly influenced.

1 INTRODUCTION

By definition, a sand concrete either does not comprise any gravels at all or only contains a small enough proportion whose the mass ratio (Sand/Gravel) remains higher than 1. If it contains gravels, the material would be called 'a loaded sand concrete' (PENPC 1994, Chauvin et al. 1988). This material, which was the subject of several researches currently, enters within the framework of the valorisation of local materials. This latter became a necessary solution to the economic problems of developing countries (Soufo 1993). In previous works, it was shown that the sand concretes are able to replace the conventional concretes in certain structures, along with the conclusion that the use of fillers is essential (Bederina 2000). Moreover, and in order to reuse the industrial wastes, which constitute environmental problems, wood shavings, coming from woodwork activities, have been incorporated in sand concrete. The effects of the addition of the latter on physico-mechanical properties have been determined (Campbell 1980). The weight is lower, the insulator capacity is better and the compressive strength is acceptable. In flexure, and at certain contents of wood shavings, the strength is higher (Bederina et al. 2005).

Because building materials are subjected, during their lifetime, to changes in temperature and changes of statements, sometimes dry and sometimes wet dending on the geographical location, a durability study vis-à-vis freezing-thaw and wet-dry cycling is necessary. Let's note that the durability is defined as the ability of a component to remain its physic-chemical properties after a mechanical damage. It is one of the most important properties for a material used in the building industry.

In this paper, we are going to study the durability of this composite, according local climatic condition, and to try to improve the obtained results by treatment of wood shavings.

2 NOMENCLATURE

DS: dune sand
RS: river sand
S: sand content, (kg/m^3)
F: filler content, (kg/m^3)
C: cement content, (kg/m^3)
B: wood content, (kg/m^3)
W: water content, (l/m^3)
ρ: apparent density, (kg/m^3)
ρ_s: specific density, (kg/m^3)

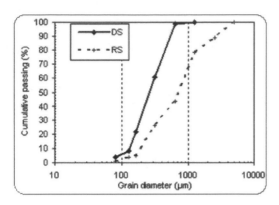

Figure 1. Particle size distribution of used sands.

Table 1. Densities of the used sands.

Characteristics	ρ kg/m^3	ρ_S kg/m^3
SD	1428	2596
SA	1482	2576

Table 2. Chemical analysis of the cement used.

SiO$_2$	Al$_2$O$_3$	Fe$_2$O$_3$	CaO	MgO	SO$_3$	PF
20.66	4.77	2.88	63.31	1.17	2.32	1.06

3 EXPERIMENTS

3.1 *Sand*

Two different sands were separately used for this study, a dune sand (DS) and a river sand (RS). The two sands are from the town of Laghouat. Results from the particle size distribution analysis of these sands, established according to standard test (NF P 18 560 1990), are presented in the Figure 1 and their densities have been listed in Table 1. The X-ray analysis of both dune and river sand demonstrates their essentially siliceous nature (Bederina et al. 2005).

3.2 *Cement*

The used cement is a Portland cement (type II) of class 45 whose denomination is "CPJ CEM II/A". The physical characteristics are the following: specific density 3078 kg/m^3 and specific surface 289 m^2/kg and the chemical analysis are shown in table 2.

3.3 *Fillers*

The fillers used in this work have been obtained by sifting (with a sieve opening of 0.08 mm) crushing waste from a quarry located in the region north of

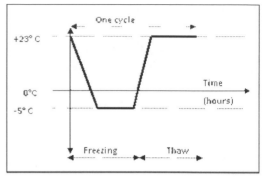

Figure 2. Freeze-thaw cycle used.

Laghouat. The chemical analysis conducted shows that these fillers are mainly composed of limestone (Bederina et al. 2005). The specific density (as measured by using the pycnometer) is of 2900 kg/m^3. The specific surface (as measured with Blaine's permeability meter according to the standard EN 196-6) is of 312 m^2/kg.

3.4 *Wood shaving*

The shavings consist of fir tree waste produced from woodworking activities; they display an irregular shape, with particle size distribution ranging from 0.1 mm to 8 mm (Fig. 2). These characteristics however are only given as an indication, since their significance has not been determined rigorously due to a lack of rigidity and the geometry of shavings. Specific density of the shavings was taken as the apparent density of a solid of wood block. The measured value amounted to approximately 512 kg/m^3. The apparent density of shavings is on the order of 160 kg/m^3. Water absorption, as measured after total immersion of wood block until weight stabilization and expressed by the "water/shaving" mass ratio, stands at approximate 36%.

3.5 *Admixture*

The admixture used is an Algerian superplasticiser of MEDAPLAST (SP40) type.

3.6 *Mixing and conservation*

The optimal compositions of the studied sand concretes, without the addition of any shavings, were given in a previous study (Bederina et al. 2005). These compositions (Table 3), were taken as the basic compositions in constituting the matrix of the studied composites. The material was lightened by incorporating wood shavings with proportions varying from 0 to 160 kg per cubic meter of sand concrete (from 0 to about 30% in volume of concrete). The used

Table 3. Composition of the two sands concretes (without shavings).

Material	C kg/m^3	S kg/m^3	F kg/m^3	W l/m^3
DS concrete	350	1305	200	245
RS concrete	350	1460	150	210

* The percentage of admixture is calculated in mass, compared to the cement mass: SP = 1.5%.

proportions are given in Table 3. The sand, the cement and filler, dried beforehand, were introduced into a mixer and mixed for 3 mn at slow speed. Once the mixture has become perfectly homogeneous, wet wood aggregates were added. Mixing then continued at slow speed for another 3 mn. The mixing water was added gradually. Material homogenisation was guaranteed by mixing at slow speed for 3 min, then at high speed for one more minute. Following setting of the (90% HR and 20°C), after 24 h, they were demoulded and kept in a dry environment (50% HR and 20°C) to remain close to local climatic conditions.

3.7 Wood treatment

Concerning the treatment of the shavings, different treatments have been used in former works and it was shown that the treatment by coating with cement gives good results, in particular in mechanical strength and shrinkage (Ledhem 1997, Gotteicha 2005, Bederina 2007). For this study, a milk prepared by mixing cement (CPJ-CEM II/A) with water according to a mass report/ratio "cement/wood" about 2.5, was used for coating the shavings.

3.8 Measurement tests

To study the durability of the studied material, we chose freezing – thaw test and wet-dry test.

3.8.1 Freezing – thaw test

There exists a certain number of protocols of tests to test the durability of concretes to freezing – thaw cycling. These procedures vary from a country to another according to the severity of the climate. The exact definition of the number of cycles has to translate the severity of the natural temperature. French standard AFNOR P18 205 considers three types of tests of freezing – thaw according to following conditions:

- Weak freezing: The temperature goes down to the lower part from −5°C only two days per annum at the maximum.
- Severe freezing: more than ten days per annum the temperature goes down than −10°C.
- moderate Gel: between weak freezing and severe freezing.

In our case, and according to the classification quoted above, the type of freezing which characterizes the local climatic conditions is "weak freezing". For these reasons, we inspired our tests from standard ASTM D 560. We chose 25 cycles of accelerated ageing where each one is characterized by four hours of freezing followed by a total immersion in water until the complete thaw (four hours approximately) (Figure 2). For this test we used an enclosure at adjustable temperature. Freezing was carried out in the air at a temperature of −5°C and the thaw in water at a temperature of 23°C. The test-samples used are cubes of $10 \times 10 \times 10 \, cm^3$.

3.8.2 Wet – dry test

This test, and for the same preceding reasons, was inspired from standard ASTM D 559. We chose 25 cycles of accelerated ageing each one of which consists in immersing the cubic test-cubes of $10 \times 10 \times 10 \, cm^3$ in water during five hours, followed by a drying in the drying oven during 42 hours at 70°C.

The principal parameter measured after freezing-thaw and wet-dry cycling is the compressive strength.

4 RESULTS AND DISCUSSION

In former studies, it was shown that it is completely possible to obtain acceptable resistances with certain defined proportions of wood shavings. In flexion, it was noticed that with a low content of wood the strength is slightly higher than that of the sand concrete without wood. This remark is much clearer in the case of SD-concrete (Bederina 2007).

It should be noted that the losses in mechanical strength caused by the addition of wood shavings were replaced by other interesting properties:

A considerable lightness is obtained by increasing the proportion of wood shavings in the concrete (Bederina 2007). This lightness makes it possible to appreciably deaden the energy of the installation of this type of concrete.

Another more important property, which makes us, perhaps, forget the losses in strength, is improved by introducing shavings into these sand concretes. It is about the good insulating capacity which these materials present. Thermal conductivity clearly decreased with the addition of shavings; the higher the proportion of the shavings is, the more thermal conductivity is improved (Bederina et al. 2007).

While combining the two behaviors, mechanical and thermal, we can say, that with well defined proportions of wood, it is possible to obtain good concretes which can be used as insulating- carriers materials in the local construction industry, where the climate is very hot in summer and very cold in winter and the constructions are generally little staged. With very high contents of wood, the mechanical strength is lower, but the thermal conductivity is very good, which also allows the use of this material as filler blocks ensuring a very good thermal comfort.

In order to improve the properties of these materials, an attempt at treatment of the shavings was carried out. The wood shavings were coated before incorporating them in the sand concretes, using cement milk. The results obtained demonstrated that, different properties are considerably improved and several problems are solved, while the profits obtained without treatment, are not or only slightly influenced. (Bederina 2007).

Concerning the effect of this treatment on the durability of these materials, objective of the present work, we can conclude the following:

4.1 Freezing-thaw cycling

The principal cause of the damage due to freezing in the concrete is the effect of bursting caused by water freezing in the capillary pores. While freezing, the volume of water increases and can create strong pressures and high tensions into the concrete. According to the concrete strength, these forces are the subject of an elastic absorption or can give place to weakening in the structure even a complete destruction of the concrete. More the wet concrete quickly and frequently freezes, more the detrimental effects are important. The study of the behavior of building materials with respect to freezing-thaw is thus necessary.

In our case, the studied material has been exposed to series of freezing – thaw cycles according to the procedure described above.

We must announce that the incorporation of wood shavings in the sand concretes improves its resistance to freezing-thaw cyclings and their treatment improves it more. The treatment of the shavings improves considerably the durability to freezing-thaw. More the quantity of wood in the sand concrete is high, more the improvement is better. This is normal owing to the fact that the treatment is applied to the wood shavings only. We note that improvements going up to 55% were recorded. For highlighting that well, we tried to trace the losses in strength according to wood content for the two studied concretes (Fig. 3).

In addition, the curves thus presented underline the composition least sensitive to freezing-thaw: for the two sand concretes and in both cases, shavings treated or not, the composition containing approximately 80 kg/m³ seems the best to resist to this phenomenon.

Lastly, it should be noted that the two studied sand concretes, dune and river, similarly behave with respect to freezing-thaw cycling.

4.2 Wet-dry cycling

Generally, and for all the considered compositions, we notice a light reduction in the compressive strength following the wet-dry cycling test. The higher wood

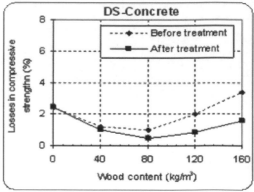

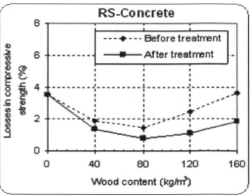

Figure 3. Loss in compressive strength due to freezing-thaw cycling before and after wood treatment.

content is, the more the effect of this test is important. But, it should, however, be noted that the losses in the recorded strengths, remain in the margins authorized for building materials ASTM D 560.

We must announce that, contrary to freezing-thaw, and since wood is sensitive to water, the incorporation of wood shavings in the sand concretes decreases the resistance to wet-dry cycling. But, the treatment of the shavings considerably improves it. As in the case of freezing-thaw test, more the quantity of wood in the sand concrete is high, more the improvement is clearer. This is also due to the fact that the treatment is applied to the wood shavings only. Let us note that improvements going up to 35% were recorded. For highlighting that well, we tried to trace the losses in strength according to wood content for the two studied concretes (Fig. 4).

Moreover, the curves thus presented underline the composition least sensitive to wet-dry cycling test, for the two sand concretes and in both cases, shavings treated or not, the optimal composition which seems the best to resist to this phenomenon is that which does not contain wood shavings.

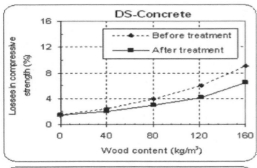

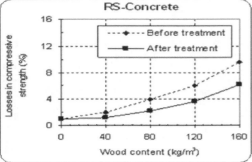

Figure 4. Loss in compressive strength due to wet-dry cycling before and after wood treatment.

Lastly, it should be noted that the two studied sand concretes, dune and river, similarly behave with respect to wet-dry cycling.

4.3 Comparison between the two tests of durability

In this part of study, we try to make a comparison between the effects of the two tests of durability on the studied sand concretes. Let us note that this comparison is made on the results obtained on sand concretes containing untreated shavings. It is advisable to note that although, at small scales, the structure is almost similar (Fig. 5), the two studied concretes differ in their behaviour with respect to freezing – thaw and wet–dry cycling according to the proportion of wood incorporated. Figure 6 shows that, without wood or with a small quantity of wood content, the effect of freezing-thaw cycling is slightly higher than the effect of wet-dry cycling. Around 20 kg/m³, in the case of dune sand and 40 kg/m³ in the case of river sand, the effect of the two tests is similar. On the other hand, beyond these contents, the effect of wet-dry is more marked. The difference considerably increases with important content of wood.

From this we can conclude, that it is the wood which leads to the degradation of the composite when the latter is subjected to wet-dry cycling. On the other hand, in freezing-thaw and at relatively average wood

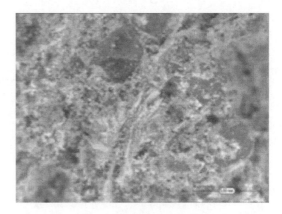

Figure 5. General aspect of the studied wood sand concrete G = 75.

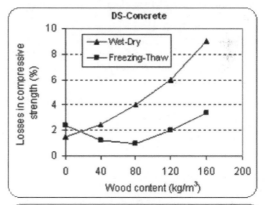

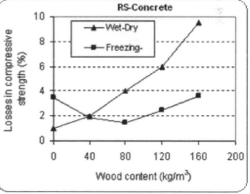

Figure 6. Comparison between the two tests of durability: freezing-thawet wet-dry cycling.

contents, wood is opposed to the degradation of the composite. The effect of the Wet-drying, which increases with wood content is perhaps due to the sensitivity of wood to this type of test (repeated swelling-deflation) and the effect of the freezing-thaw, which improves with the increase of the proportion of wood, is perhaps due to the porosity which becomes increasingly inter-connected by increasing the wood content.

5 CONCLUSION

This study underlined the effectiveness of the treatment of the wood shavings on the majority of the characteristics of the sand concretes. The treatment of the shavings by coating them with milk of cement before introducing them into the studied sand concretes, improved the majority of the characteristics of this composite without too much influencing the profits obtained by the addition of shavings. The mechanical strength, which constitutes the principal characteristic, is considerably improved. Concerning durability to freezing-thaw and wet-dry cycling, we notes:

- The resistance to freezing-thaw cycling is better;
- The resistance to wet-dry cycling, which constitutes a rather important problem for these materials, is clearly improved;

Finally, according to the wood proportion, we can formulate structural wood sand concretes, structural-insulator wood sand concretes or insulator wood sand concretes.

REFERENCES

Bederina, M. 2007. Caractérisation mécanique et physique des bétons de sable à base de déchets de bois, *Thèse de Doctorat, Ecole Nationale polytechnique, Alger, Algérie.*

Bederina, M., Khenfer, M.M., Dheilly, R.M. & Queneudec, M. (2005): Reuse of local sand: effect of lime stone filler proportion on the rheological and mechanical properties of different sand concrete. *Cement and Concrete Research*, 35(2005): 1170–1179.

Bederina, M., Marmoret, L., Mezreb, K. , Khenfer, M.M., Bali, A. & Quéneudec, M. 2006. Effect of the addition of wood shavings on the thermal conductivity of the sand concretes – experimental study and modelling. *Construction and Building Materials*, 21(2007):.662–668.

Campbell, M.D. 1980. Coutts RSP. Wood fiber reinforced composites. *J. Mater Sci*, 15(10):1962–70.

Chauvin, J.J. & Grimaldi, G. 1988. Les bétons de sable. *Bulletin de liaison des laboratoires des ponts et chaussées (LCPC)*, 157(1988): 9–15.

Gotteicha, M. 2005. Caractérisation des bétons de sable a base de copeaux de bois traités. *thèse de Magister soutenue en Jui 2005, Université Amar Télidji de Laghouat, Algérie.*

Ledhem, A. Contribution à l'étude d'un béton de bois. Mise au point d'un procédé de minimisation des variations dimensionnelles d'un composite Argile-Ciment-Bois, *Thèse de doctorat soutenue le 01 juillet 1997 à l'INSA de Lyon.*

NF P 18 560. 1990. Granulats – Analyse granulométrique par tamisage.

Presse d'Ecole Nationale des Ponts et Chaussées, 1994. Béton de sable, Caractéristiques et pratique d'utilisation. *Project SABLOCRETE.*

Soufo, Y.M. 1993. Matériaux locaux et construction de logements dans les 382 pays en voie de développement. *Ph.D. Université* de Montréal.

Excellence in Concrete Construction through Innovation – Limbachiya & Kew (eds)
© 2009 Taylor & Francis Group, London, ISBN 978-0-415-47592-1

Microstructure characteristics of cementitious composites at elevated temperatures

Y.F. Fu, W.H. Li & J.Q. Zhang
Research Institute of Highway, The Ministry of Communications, Beijing, P.R. China

J.J. Feng & Z.H. Chen
China University of Mining & Technology, Beijing, P.R. China

ABSTRACT: This paper presented an experimental study on the microstructural morphology of cementitious composites at high temperatures up to 500°C. A real-time SEM was used to observe the evolution of microstructure of neat hardened cement paste (HCP), PFA HCP, neat mortar and PFA mortar exposed to elevated temperatures. The SEM observation was focused on the changes of morphology, microvoids and microcracks with increasing temperatures. The thermal deterioration of HCPs were induced by the dehydration and decomposition of hydration products. The associated evolution of microstructure covered the shrinkage cracks, phase changes of calcium hydroxide and C-S-H gel. Stressed tests were carried out for comparing the HCPs' resistance against elevated temperatures with using a high temperature fatigue-testing machine (HTFTM). The whole process of thermal deterioration of HCPs was studied with considering both the changes in morphology and the external loads. The results showed PFA HCP has better resistance to high temperature than the neat HCP. The mechanisms of thermal deterioration for all HCPs were discussed with analyzing the changes in microstructural morphology.

1 INTRODUCTION

The compressive stress-strain relationships of concrete exposed to elevated temperatures were studied at a macro-level (Castillo & Durrani, 1990; Khoury *et al.*, 2002; Cheng *et al.*, 2004). However, these test results cannot identify the main contributor to the thermal degradation. It has been reported that using a SEM technique the thermal decomposition and the thermal cracks could be correlated to the degradation, but most of the previous observations were conducted under cool conditions so that they were suitable for understanding the residual properties of concrete after high temperatures. As for the thermal damage of concrete during a heating up process, it is of interest to observe the change of the microstructure of the cementitious materials in a real-time mode.

The decomposition of hydration products leads to an increase in porosity and the occurrence of microcracks, which dramatically degrade the concrete. Previously, attempts had been made to use mechanical tests together with micro-structural examinations of a heated hardened cement paste to quantify the relation between the decomposition of the hydrates and the strength deterioration of the paste material. In this respect, a number of semi-empirical equations were developed which related the compressive or tensile strengths of the paste with its porosity. Details can be found in the literature (Khoury, 1992). It was suggested that the increase in porosity and coarsening of the pore size distribution were expected to cause strength reduction, particularly above 300°C. Unfortunately, these experimental studies did not extend to the testing of mortar or concrete samples. Micro-structural examinations using SEM provide information on the thermal decomposition of constitutive materials and the crack formation of concrete. However, all the previous observations were carried out after the heated specimens were cooled down to ambient temperatures. The crack/damage patterns so observed might be different from those at high temperatures. Hence, there is still a need to observe the thermal cracking process during heating, especially to study its effects on mechanical properties.

The presented paper describes an experimental study of the tensile strength vs. temperature relation and the thermal cracking process of HCP and the companion mortar samples during heating up from room temperature to 500°C under a real-time SEM observation mode. The obtained results can be employed to explain the mechanism of the thermal damage of cement-based composites, such as HCP and mortar, at a meso-level. The results can also be extended to understand the thermal degradation of concrete at elevated

Table 1. Mix proportions.

No.	OPC, kg	PFA, kg	Water, kg	Sand, kg
H1	500	–	175	–
H2	475	25	175	–
H3	425	75	175	–
M1	500	–	175	1000
M2	475	25	175	750

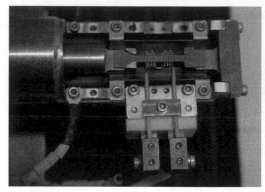

Figure 2. Test setup and heating chamber.

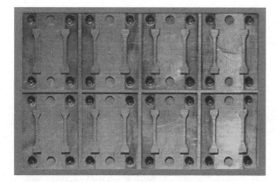

Figure 1. Specimens and steel molds.

temperatures, and used as input data for the numerical simulations.

2 EXPERIMENTAL METHODS

2.1 Materials and specimen preparation

One series of the HCP specimens and one series of the companion mortar specimens were prepared for testing. The mix proportions of the two test series of the specimens are shown in Table 1. All the specimens were cured in a water tank for 28 days.

The test samples were deliberately dried in an oven at 60°C for 6 hours before testing. The test samples in this study were then designed to be very small and thin (see Figure 1) so that a thermal steady state could be attained in a relatively short time, and the 'structural effect' or thermal gradient would be eliminated.

The specimens were then coated with gold prior to the heating and loading tests. To ensure consistency of the test results, at least three repetitive samples were examined for each case of study. The test series was denoted by '**H**' or '**M**' series for the HCP and the mortar specimens respectively.

2.2 Equipment

All specimens were tested using the high temperature fatigue-testing machine (HTFTM) of the Shimadzu Servopulser series, which is capable of applying thermal and external loads on a specimen simultaneously. The equipment has three components: a servo-controlled load testing machine, a temperature controlled heating device, and a SEM of Super scan SS-550, which is able to observe the change in morphology of material under elevated temperatures up to 800°C in a real-time mode. Figure 2 shows the schematic of the test setup and the heating chamber. The specimen is heated through a heating coil. The chamber is movable during the whole heating and loading process, so the thermal cracking process can be traced dynamically.

2.3 Experiment procedure

In this study, a steady thermal state test method was adopted to determine the mechanical properties of HCP and mortar specimens at elevated temperatures. Firstly, a test specimen was mounted in the heating chamber of the HTFTM with a constant clamping load (see Figure 2), and then was heated to a target temperature ranging from 100°C to 500°C at a temperature increment of 100°C at a rate of 5°C per minute. Once the target temperature was attained, a further uniaxial tensile load was applied to the specimen at the displacement rate of 0.001 mm/s till failure.

During the entire period of heating and loading, the SEM was used to observe the thermal damage and the cracking process in a real-time mode. The SEM analyses were conducted using the Super scan SS-550 with an acceleration voltage between 10 kV–25 kV. A schematic of heating process and loading process for stressed test was shown in Figure 3.

3 EXPERIMENTAL RESULTS AND DISCUSSIONS

3.1 Tensile strength vs. temperatures

In stressed tests, the evolution of tensile strength of HCPs is characterized by a gradual reduction with increasing temperatures (see Figure 4). For all HCPs, the residual strength at 500°C reached 40% or less

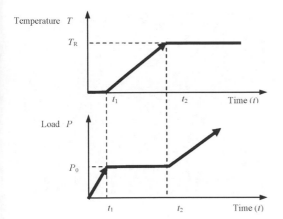

Figure 3. Heating regime of stressed test.

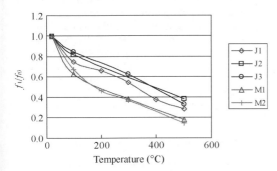

Figure 4. Tensile strength vs. temperature for HCPs and mortars.

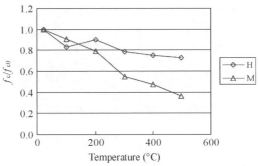

Figure 5. Compressive strength vs. temperature for HCP and mortar (Fu et al., 2005).

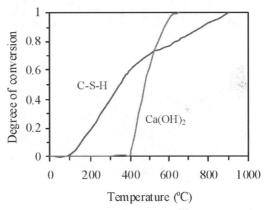

Figure 6. Degree of conversion of C-S-H gel and Ca(OH)$_2$ (Harmathy, 1970).

of their original values at room temperature. Generally, PFA HCP showed a better fire resistance than neat HCP during the whole heating up process. During 100°C to 500°C, all mortars suffered a more rapid reduction in tensile strength than HCPs. The residual strength of all the mortars reached about less 20% of their original values at room temperature.

Figure 5 shows the strength-temperature relationships of HCP and mortar in compressive test conditions. They demonstrate that the evolution of strength losses under compressive conditions is similar to that under tensile conditions during heating-up process. Sullivan and Poucher (1971) reported that the mortar beam has better performance in strength than concrete beam at elevated temperatures. The current experimental results show a good consistency in strength evolution with previous ones.

3.2 Morphology and microstructure of HCP and mortar

With increasing temperatures, calcium hydroxide formed in the process of cement hydration decomposes

into calcium oxide and water. The dehydration reaction occurs at about 400°C and is finished at about 600°C. Its decomposition reaches maximum rate at 500°C (see Figure 6). It can been found that 70% of the decomposition reaction has been finished at 500°C and completely dehydrated at 850°C. The reaction rate is most rapid at about 200°C.

A series of SEM micrographs illustrating various characteristics feature of HCPs and mortars at room temperature and high temperatures up to 500°C are shown in Figure 7 and Figure 8.

As shown in Figure 7a, HCP reveals a good framework of plate-like calcium hydroxide crystals, needle-or fiber-like CSH and crystals of ettringite and pores. Hydration products in HCP exposed to up to 200°C, did not show significantly distinct change in the morphology. Increasing temperature caused chemical reactions of dehydration of HCP and conversion of calcium hydroxide into calcium oxide, in which chemically bound water is gradually released to form evaporable water. A loose and discrete framework of

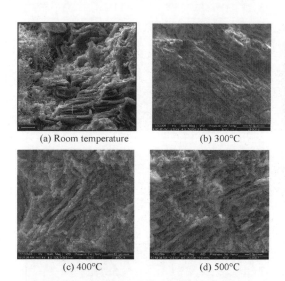

(a) Room temperature   (b) 300°C

(c) 400°C   (d) 500°C

Figure 7. Decomposition of calcium hydroxide.

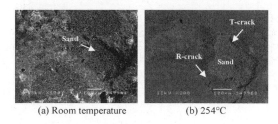

(a) Room temperature   (b) 254°C

Figure 8. Thermal cracks around sands.

decomposed hydration products is observed at 500°C in Figure 7d.

Therefore, the observed strength loss of HCP at increasing temperatures may be related to the loss of bound water, increased porosity, and consequently, increased permeability, which makes the concrete progressively more susceptible to further destruction.

Figure 8 demonstrated a real-time observation of thermally cracking propagation in mortar exposed to elevated temperatures. Tangential crack (T-crack) and radial cracks (R-crack) in the mortar specimen were formed due to the thermal expansion mismatch between the HCP and the sand aggregates with increasing temperatures. At room temperature, there are not cracks around the sand particle (see Figure 8a). Because of the unique thermal characteristics of HCP-expansion at lower temperatures and contraction at higher temperatures, T-cracks and R-cracks occurred with increasing temperatures (see Figure 8b).

It is further proved that T-cracks and R-cracks exist around sands in mortar at elevated temperatures through a SEM observation (see Figure 9). It is very clear that R-crack zones were formed around sands and T-cracks developed along the interfaces between

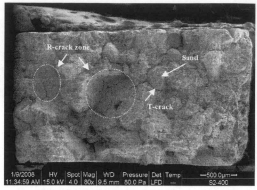

Figure 9. Thermal cracks on fractured surface of mortar.

the HCPs and the sand aggregates. Eventually, a crack network is formed in mortar specimen.

Certainly, the effects of the thermal decomposition of hydration products and the fine dehydration-induced cracks also existed in the HCP matrix of the mortar specimens. Hence, thermal induced-damage of the mortar specimens is expected to be more severe than that of the paste specimens at a given elevated temperature due to the formation of crack network.

3.3 Mechanism of thermal degradation

Apart from the non-stress-induced thermal decomposition of hydration products as reported in the literature (Chang et al., 1994; Bažant & Kaplan, 1996; Handoo et al., 2002; Fu et al., 2005), the dehydration-induced cracks, which were probably caused by the thermal mismatch of expansion/shrinkage among the different hydration products, were believed to be another factor affecting the mechanical properties of HCP under high temperatures (Fu et al., 2005). These micro-cracks were formed below 100°C and grew rapidly with increasing temperature. Studies (Felicetti & Gambarova, 1998) showed that 100°C is an important threshold temperature, beyond which dehydration speeds up. Above 200°C, an intense crack pattern developed (Fu, 2003). This also implies that the variations of the pore size and the porosity of HCP (Khoury, 1992; Chang et al., 1994) are not sufficient to qualify/quantify the thermal damage of the HCP.

A mortar was prepared by mixing cement with fine aggregates. When a mortar specimen was subjected to high temperatures, macro-thermal cracks appeared as a result of the increasing mismatch of expansion/shrinkage between the HCP and aggregates, the strength-weakening decomposition of hydrates, and dehydration-induced cracks. The rapid growth of the macro-crack intensity made the mortars more vulnerable to strength loss than the companion HCP (see Figures 4 and 5).

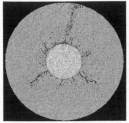 

(a) Numerical results   (b) Tested results

(Elbadry *et al.*, 2000)

Figure 10.   Thermal cracks.

Because of the unique thermal characteristics of a HCP-expansion at lower temperatures and contraction at higher temperatures, the T-cracks and R-cracks were formed in the HCP matrix around the sand aggregates at different temperatures (see Figures 8 and 9). It had been shown that the sand aggregates generally had a lower linear coefficient of thermal expansion than the HCP at a temperature below 100°C. Consequently, radial tensile stresses led to the formation of tangential cracks around the sand inclusions if the HCP-sand interface bond was weak. Further increase in temperatures made the HCP shrink. Consequently, the radial tensile stressed regions changed to be compressive and the tangential stresses were in tension, so that the bursting stresses around the sand inclusions eventually caused the development of radial cracks. The formation of tangential cracks and radial cracks due to thermal mismatch could lead to a dramatic degradation of the mechanical properties of the mortar/concrete in a macro-scale.

In order to perform a reliable numerical simulation of the thermal damage of a cement-based composite material, it is necessary to have realistic estimations of the temperature-dependent mechanical properties and more importantly expansion/shrinkage properties of each phase material (HCP and sand) in the formulation of the material models. The author has incorporated the current and previous findings to formulate a thermo-elastic damage model for the numerical simulation of a cement-based composite under high temperatures. Figure 10 gives a comparison in thermal cracks of numerical results with tested ones. The simulation is being verified using the experimentally obtained stress-strain relations of the heated companion mortar specimens, details of which will be reported in another publication.

4   CONCLUSIONS

The mechanical properties and changes in microstructure for HCP and mortar specimens were successfully measured/observed in a real-time mode under elevated temperatures up to 500°C. The main findings are summarized as follows:

1) The tensile strength of HCP and mortar suffered a gradual reduction with increasing temperatures up to 500°C. PFA HCP had a better resistance against high temperatures than neat HCP, while HCP had better performance in strength than mortar at elevated temperatures.

2) The real-time SEM examinations revealed that thermal decompositions of cement hydrates took place in the HCP specimens, as shown from the degeneration of the hydrates and the coarsening of the pore structures. The C-S-H gel and calcium hydroxide were dehydrated with increasing temperature. The decomposition and desiccation of hydration products reduced the HCP strength. At 500°C most of decomposition reaction was completed, so the residual strength reached a very low value.

3) The thermal damage of the mortar specimens was caused both by the chemical decomposition of the cement hydrates as same as to HCP, and by the thermal cracks (tangential, radial, and inclusion) originated from the thermal expansion mismatch between the HCP and the sand aggregates. The thermal cracks further weakened the mechanical properties of mortar, so the mortar had higher reduction in strength than HCP exposed elevated temperatures.

ACKNOWLEDGEMENTS

This research was supported by a grant from the National Science Foundation (NSF) of China (No. 50408029).

REFERENCES

Bažant, Z.P. & Kaplan, M.F. 1996. Concrete at high temperatures: Material properties and mathematical models. Longman House, England, UK.

Castillo, C. & Durrani, A.J. 1990. Effect of transient high temperature on high-strength concrete. *ACI Materials Journal* 87(1): 47–53.

Chang, W.T., Wang, C.T., Huang, C.W. 1994. Concrete at temperatures above 1000°C. *Fire Safety Journal* 23: 223–243.

Cheng, F.P., Kodur, V.K.R., Wang, T.C. 2004. Stress-strain curves for high strength concrete at elevated temperatures. *Journal of Materials in Civil Engineering* 16(1): 84–90.

Elbadry, M.M., Abdalla, H., Ghali, A. 2000. Effects of temperature on the behaviour of fiber reinforced polymer reinforced concrete members: experimental studies. *Canadian Journal of Civil Engineering* 27: 993–1004.

Felicetti, R., Gambarova, P.G. 1998. Effects of high temperature on the residual compressive strength of high strength siliceous concrete. *ACI Materials Journal* 95(4): 395–406.

Fu, Y.F. 2003. Thermal stresses and associated damage in concrete at elevated temperatures. *PhD thesis*, The Hong Kong Polytechnic University.

Fu, Y.F., Wong, Y.L., Poon, C.S., Tang, C.A., Lin, P. 2005. Experimental study of micro/macro crack development and stress-strain relations of cement-based composite materials at elevated temperatures. *Cement and Concrete Research* 34(5): 789–797.

Handoo, S.K., Agarwal, S., Agarwal, S.K., 2002. Physicochemical, mineralogical, and morphological characteristics of concrete exposed to elevated temperatures. *Cement and Concrete Research* 32(7): 1009–1018.

Harmathy, T.Z. 1970. Thermal properties of concrete at elevated temperatures. *ASTM Journal of Materials* 5(1): 47–74.

Khoury, G.A. 1992. Compressive strength of concrete at high temperatures: a reassessment. *Magazine of Concrete Research* 44(161): 291–309.

Khoury, G.A., Majorana, C.E., Pesavento, F., Schrefler, B.A. 2002. Modelling of heated concrete. *Magazine of Concrete Research* 54(2): 77–101.

Sullivan, P.J., Poucher, M.P. 1971. Influence of temperature on the physical properties of concrete and mortar in the range 20C to 400C. *Temperature and Concrete* SP-25, ACI, Detroit: 103–135.

Excellence in Concrete Construction through Innovation – Limbachiya & Kew (eds)
© 2009 Taylor & Francis Group, London, ISBN 978-0-415-47592-1

Durability of polyester polymer concrete under varying temperature and moisture

M. Robles
Universidad Politécnica de Madrid, Madrid and Prefabricados Uniblok (Grupo Ormazabal), Seseña, Spain

S. Galán & R. Aguilar
Universidad Politécnica de Madrid, Madrid, Spain

ABSTRACT: The application of polymer concrete to precast products in construction presents notorious advantages over traditional concrete: higher strength, lower permeability and shorter curing time. A better durability is usually predicated, but this is a research field where published data are scarce. In this paper, an experimental study about the influence of temperature and moisture on the durability of polymer concrete based in an unsaturated polyester resin (where the risk of hydrolysis with water exists) is described and the results are presented. The measured effects of the thermal degradation fatigue cycles are the mechanical properties (compressive and flexural strength) and the external variation of the specimens for various periods of time. The rate of degradation is also evaluated.

1 INTRODUCTION

The development of new composite materials possessing increased strength and durability when compared with conventional types is a major requirement of applications in repairs and in the improvement of infrastructure materials used in the civil construction industry. Polymer concrete (PC) is an example of a relatively new material with such high performance. Its excellent mechanical strength and durability reduce the need for maintenance and frequent repairs required by conventional concrete (Gorninski et al. 2004).

Polymer concrete is a composite material formed by polymerizing a monomer and an aggregate mixture. The polymerized monomer acts as the binder for the aggregates. Initiators and promoters are added to the resin prior to its mixing with the inorganic aggregates to initiate the curing reaction. Polymer concretes therefore consist of well-graded inorganic aggregates bonded together by a resin binder instead of the water and cement binder typically used in normal cement concretes (ACI 1986). The most common type of binder utilized in polymer concrete is unsaturated polyester because of its good properties and relatively low cost.

Durability is one of the most important properties of the materials used in the building industry. The relevant literature frequently reports the chemical durability in various aggressive environments. However, thermal durability, i.e., the ability of a component to retain its physical-mechanical properties during and after exposure to different, sometimes severe, thermal conditions is also very important. Building components, during their life-time, are often subject to changing temperatures. Therefore, sensitivity to thermal cycles is a relevant subject that must be taken into account in the evaluation of durability and service life of construction materials.

Most of the related literature is restricted to limited temperature range and a certain type of polymer. Traditional approaches are usually focused on temperatures above room temperature, which is understandable taking into account that common resins used in polymer concretes present glass transition temperature ranges above this temperature. However, resistance to very low temperatures, depending upon polymer concrete application, could also be a very important and even crucial subject (Ribeiro et al. 2004). (Ohama 1977) studied the resistance of polyester PC to hot water. Cylindrical PC specimens were immersed in boiling water for up to 1 year before being tested in compression and splitting tension. It was concluded that erosion depth in polyester PC increased with the immersion time and the compressive and tensile strengths decreased, with no appearance or weight change. The risk of hydrolysis with water was considered, since unsaturated polyester resin is an ester.

Accelerated test methods, especially with relatively new materials, can be used to understand and compare the degradation phenomena and to predict their service lifetimes. In this work, polymer concrete (PC) based on unsaturated polyester, whose polymer base matrix is said to degrade at elevated temperatures, was investigated for its behaviour under the influence of humidity and temperature in both static and cyclic conditions. Since the motivation for this work was to determine the durability of polymer concrete covers, it was considered important to simulate the types of aggressive environments that could conceivably be brought into contact with a cover. Therefore, the environments related to moisture and temperature were selected.

First, the effects of high temperatures, (always higher than the glass transition temperature of the unsaturated polyester resin used), with and without moisture are compared. Then, the effect of cyclic high and low temperatures on the material is studied. In this test, the samples were immersed in water for a number of cycles at elevated temperatures (90_C) followed by low temperature (−2_C) cycles. For all cases, moisture absorption and the weight loss of the samples are measured evaluating its influence on the mechanical properties (compressive and flexural strength) of the material. Also, the external variation of the specimens is observed.

2 MATERIALS AND METHODS

Unsaturated polyester was the type of resin selected as binder in this study.

2.1 Preparation of the specimens

The materials used for the elaboration of the specimens were the following:

- Resin: The resin used has been designed to be used in polyester concrete in conjunction with high filler content, a dicyclopentadiene (DCPD)/Orthophtalic polyester resin for casting with 1.5% of catalyst Methyl Ethyl Ketone Peroxide (MEKP), (solution at 50% styrene), and 0.1% of accelerator Cobalt Octoate, (solution at 6% styrene). This resin has good wetting properties, mechanical performance and fast curing. Its commercial name is Crystic U 1112 K from the Scott Bader company. The concentration of polymer used was 14% by weight of the dry materials.
- Aggregates: For concrete, clean sand with size between 0–4 mm was used for concrete. Also one type of gravel which gradation is 4–8 mm. The specific mass is 2.520 g/cm³ according to the UNE-EN 83133:1990 standard.
- Charge: It was decided to use calcium carbonate as the only charge to be added to the resin mixing. Calcium carbonate was used as a filler and composition with 12% by weight of aggregate. The specific mass of calcium carbonate is 2.70 g/cm³ according to the EN 1097-6 standard.

Automatic mixing achieves a better mixture and more homogeneous concrete. Therefore, the mixing procedure used to make the different polymer concretes was automatic mixing by means of a slow speed mechanical mixer fitted with a paddled stirrer, with 40 liters of capacity at 20°C and around 50% R.H. Samples were prepared by mixing required quantities of resin with accelerator, aggregates and catalyst. First, the resin was put into the mixer. Afterwards, the accelerator and the catalyst were added slowly while stirring. The UPE resin was mixed thoroughly with cobalt octoate, followed by the addition of MEKP. These mixing processes are highly exothermic in nature.

Polymer concretes of the same composition were prepared, cured and tested under identical conditions. PC with the binder formulations and mix proportions aforementioned was mixed and moulded to prismatic specimens 40_40_160 mm to determine flexural strength according to UNE 83305:1986 standard and to cast cubic specimens of 100_100_100 mm in order to determine the compressive strength according to UNE EN 12390-3:2003 "Tests for hardened concrete. Part 3. Method to obtain compressive strength of specimens" standard.

The composition of the PC used was 86% (weight) of sand, gravel and filler and 14% of resin. This composition was found to be the best in an optimization study conducted in a previous work. Since polymer concrete is a heterogeneous material, the properties of polymer concrete may be highly variable. In order to determine the performance of a composite in a particular environment, the composite is exposed to that environment in a precisely controlled manner. As mentioned above, polymerization was achieved through the addition of 1.5% MEKP by weight of resin used and 0.1% octoate cobalt. In all cases at least 3 days curing was allowed prior to any further treatments to ensure maximum cure of the specimens. No postcuring treatment was carried out on the samples. The composition of PC is summarized in table 1:

The aggregate materials were studied. These form the major component of the polymer concrete and have a great importance in the mixture. Local silica aggregates were chosen for all cases. These kind of aggregates have been fitted by the Fuller Method and Granular Bar Bones Optimum according to a previous

Table 1. Weight proportions of mixing.

Agent	Product	Dosage
Catalyst	MEKP (at 50%)	1:5% resin
Accelerator	OcCo (at 6%)	0:1% resin
Resin	Crystic U 1112K	14%
Aggregates	Sand/Gravel/Filler	86%

work. It is advisable the absent of moisture in aggregates. The humidity allowed is up to a maximum of 1%.

The number of posible ingredients and proportions that can be used in the material is limitless, and consequently the results of previous research are often applicable only to that material used. However, certain consistency do exist and it is to be further verified.

2.2 Experimental set-up and testing procedures

The influence of humidity and temperature in both static and cyclic conditions was investigated. The experiments designed were intended to determine:

1. The effects of high temperatures (always higher than the glass transition temperature of the unsaturated polyester resin used) with moisture. The objective of this study is to clarify the resistance of polyester resin concrete to hot water. Three cubic and three prismatic specimens were immersed in water at 85°C for 17 cycles and each cycle lasted four hours and then were left at 20°C, 50% R.H. for 3 days before the mechanical tests.

2. The effects of high temperatures (again, always higher than the glass transition temperature of the unsaturated polyester resin used) without moisture. In this case the cycle is of four hours for day. The temperature of cycle is 120–130°C. Four samples are exposed to high temperatures: two cubic and two prismatic in the oven. The total cycles were 70, but when 17 cycles have been completed, the mechanical properties were also tested with half of specimens. In figure 1 it is shown the warning curve in oven.

3. The effect of cyclic high and low temperatures on the material. The thermal fatigue cycles is used to evaluate the flexural and compressive deterioration by repeated thermal actions. In this test, the samples were immersed in water for a number of cycles at elevated temperatures (90°C) followed by low temperature (−2°C) cycles.

4. The effect of low temperatures with moisture. The specimens were immersed in water for 29 cycles with a temperature of −2°C. In figure 2 it can be seen the cooling curve in the thermostatic bath.

First, the effects of high temperatures with and without moisture are compared. Each day, for each immersion period, the appearance of test specimens was visually checked and their weight was measured. For all cases, moisture absorption and the weight loss of the samples are determined evaluating its influence on the mechanical properties of the material. After that, the specimens were tested for flexural and compressive strength in accordance with UNE 83305:1986 and UNE 83304:1984.

Cubic specimens were used for the axial compressive strength tests and prismatic specimens (beams) for the flexural tensile strength. As a consequence of the brittle nature of the material, flexural testing is the most practical method to use to determine the mechanical properties of polymer concrete (Griffiths and Ball 2000).

In all tests realized the specimens experience thermal shock because for four hours they are influenced by the test temperature and the rest of the time the temperature rises or descends to reach the room temperature. Therefore, all the specimens were allowed to cure for one day at room temperature and two days later. In all experiments the mixing temperature was measured in time by a thermocouple placed inside of the specimens.

In all cases of exposure to constant or changing temperatures (thermal fatigue cycles) the specimens are taken back to the initial environmental conditions (before the tests). The specimens are not tested at aging temperature but at room temperature after tempering. (Some authors (Ribeiro et al. 2004) have studied the mechanical properties at aging temperature as well as after specimen tempering). The weight of the

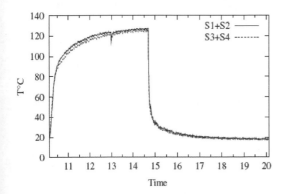

Figure 1.   Calibrated curve of high temperature in oven.

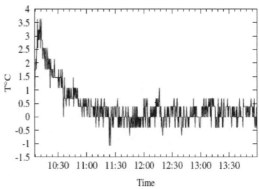

Figure 2. Calibrated curve of low temperature in thermostatic bath.

Table 2. Loss of weight.

Loss of weight	17 cycles	70 cycles
Cubic specimens	0.12%	0.20%
Prismatic specimens	0.16%	0.24%

Figure 3. Pathology detail at the end of hot cycle.

Figure 4. Comparison of specimens at the end of combined cycle test.

specimens, that were subjected to thermal actions, was recorded before and after exposure to thermal cycles.

3 RESULTS AND DISCUSSION

The data obtained reveals that the thermal fatigue cycles, to which the material is exposed during its lifetime, constitute severe environmental conditions. The test results: mass change, flexural and compressive strength change, for each test, as function of immersion period, are shown in figures.

A very useful means of noting the degradation of the resin is by visual inspection after exposure of specimens to environments for various periods of time.

3.1 Effects of high temperatures without moisture

In this case the cycle is four hours each day. The temperature of the cycle is 120–130_C. Four samples are exposed at high temperatures: two cubic and two prismatic inside the oven. When 17 cycles have been completed since the specimens were immersed half of specimens are tested. The specimens were found to develop yellowing.

The test ends when the loss of weight does not change. It happened with 70 cycles. But the most important loss (75%) occurs at 17 cycles. The total change in weight is not important as it is shown in the table 2. No appreciable reduction appeared in mechanical properties (compressive and flexural strength).

3.2 Effects of cyclic high and low temperatures

In this case, combinations of exposures are used to assess the synergistic effects of exposure types. The samples were immersed in water for 7 cycles at elevated temperatures (90°C) followed by 6 low temperature (−2°C) cycles. This test is called "combined cycle". The weight of the specimens, that were subjected to both cycles, was recorded before and after exposure to thermal cycles. The changes in the pH of the water where the specimens are immersed are controlled.

In the first part of the test, it was necessary to fill the container because of evaporation of water. When the material is immersed in hot water, in the specimens appear flakes, blisters and the colour becomes darker. In figure 3 it can be seen the pathology on the specimen after hot cycle. Also, erosion in the rim of the specimens develops. Therefore, when the second part

of the test starts the specimens have already suffered a prior degradation. When the test is finished appreciable discoloration and loss of gloss was observed for specimens as it is shown in figure 4.

Thermal damage caused by combined cycle had relevant influence on both mechanical properties: flexural and compressive strength. After 13 thermal fatigue cycles, a percentage drop of 12% occurs on compressive strength and 32% on flexural strength.

3.3 Effects of low temperatures with moisture

The specimens were immersed in cold water (−2°C) for 4 hours each cycle, with a number of cycles ranging from 14 to 47. Exposure to low temperature when the specimens are immersed in water to −2°C had no relevant influence on compressive strength. However, the behavior for the flexural strength is not clear. It will be necessary to repeat the test and to increase the immersion time of specimens or a larger number of cycles. The results may be explained by the low degree of water absorption and high watertightness, already know from previous studies (Ribeiro, Tavares, and Ferreira 2002).

Table 3. Weight gain at low temperatures with moisture.

Weight gain	16 cycles	29 cycles
Cubic specimens	0.029%	0.185%
Prismatic specimens	0.011%	0.29%

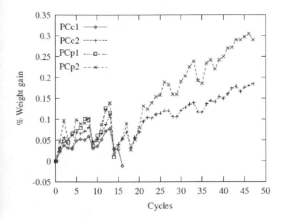

Figure 5. Weight gain at low temperatures with moisture.

Figure 6. Cubic and prismatic specimens after 17 cycles.

Neither scaling surface was observed nor any sediment appeared in the water containers after the 29 cycles. Therefore, specimen weight change was totally due to water absorption. This weight change is presented in table 3 and figure 5.

3.4 Effects of high temperatures with moisture

The specimens were immersed in hot water (85°C) for 4 hours and the number of cycles was 17. The more relevant changes can be seen in the picture of figure 6. The changes in the weight are shown in figure 7 and table 4.

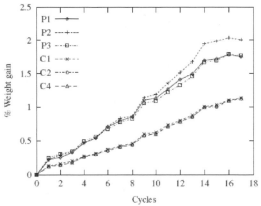

Figure 7. Weight gain at high temperatures with moisture.

Table 4. Weight gain at high temperatures with moisture.

Weight gain	17 cycles
Cubic specimens	1.13%
Prismatic specimens	1.76%

Changes in mechanical properties can be observed. The compressive strength suffered a decrease of 18.2%. Also the flexural strength suffered a great decrease. On the other hand the visual aspect of specimens was different. The specimens lost gloss and appeared yellowing.

4 SUMMARY AND CONCLUSIONS

Based on experimental results, the following conclusions are proposed:

- The continued cyclic exposure of unsaturated polyester concrete to high temperatures has not a negative effect on its flexural strength. Also no significant influence is observed in the compressive strength. The maximum cycling temperature is 15% higher than the polyester Tg, so it constitutes a much less severe condition to the polyester polymer network.
- In all test solutions, the percentage of weight change of the polyester PC specimens was very small, usually asymptotic. If the increment of weight is compared between the test at low and high temperatures it can be seen that the upper value occurs in the second case because the effect of temperature.
- In the case of low temperatures, additional immersion periods would be required because there were not conclusive results.
- In overall test the specimens developed defects, big or small, depending on the test.

- It is believed that the degradation observed is mostly confined to surface regions of the material.
- Exposures of samples to combined cycle and hot water showed the most important reduction in mechanical properties.

In conclusion, the data obtained reveals that the thermal fatigue cycles, to which the material is exposed during its lifetime, constitute severe environmental conditions.

ACKNOWLEDGEMENTS

This study has been partially supported by Prefabricados Uniblok (Grupo Ormazabal).

REFERENCES

ACI, C. (1986). Guide for use of polymers in concrete (ACI 548-1R-86). In American Concrete Institute, Detroit, pp. 32.

Gorninski, J., D. Dal Molin, and C. Kazmierczak (2004). Study of the modulus of elasticity or polymer concrete compounds and comparative assessment of polymer concrete and Portland cement concrete. Cement and Concrete Research 34, p. 2091–2095.

Griffiths, R. and A. Ball (2000). An assessment of the properties and degradation behaviour of glass fibre reinforced polyester polymer concrete. Composites Science and Technology 60, p. 2747–2753.

Ohama, Y. (1977). Hot water resistance of polyester resin concrete. In Proceedings of 20th Japan Congress on Materials Resistance, Tokyo, Japan, pp. 176–178.

Ribeiro, M., P. Nóvoa, A. Ferreira, and A. Marques (2004). Flexural performance of polyester and epoxy polymer mortars under severe thermal conditions. Cement and Concrete Composites 26, p. 803–809.

Ribeiro, M., C. Tavares, and A. Ferreira (2002). Chemical resistance of epoxy and polyester polymer concrete to acids and salts. Journal of Polymer Engineering 22(1), p. 27–44.

Excellence in Concrete Construction through Innovation – Limbachiya & Kew (eds)
© 2009 Taylor & Francis Group, London, ISBN 978-0-415-47592-1

Microstructure degradation of concrete in extreme conditions of dry and high temperature difference

A.M. She
School of Materials Science and Engineering, Wuhan University of Technology, Wuhan, P.R. China
School of Materials Science and Engineering, Tongji University, Shanghai, P.R. China

Z.H. Shui
School of Materials Science and Engineering, Wuhan University of Technology, Wuhan, P.R. China

S.H. Wang
School of Materials Science and Engineering, Wuhan University of Technology, Wuhan, P.R. China
Chifeng Transportation Bureau, Chifeng, P.R. China

ABSTRACT: Conditions of dry and high temperature difference are detrimental to the durability and service life of concrete structures. The microstructure features of concrete exposed in dry and high temperature difference environment is studied in this paper by using mercury intrusion porosimeter (MIP), micro-hardness testing and scanning electron microscopic (SEM). In the dry and large temperature difference environment, the total porosity and mean pore diameters of concrete have increased as well as the total pore volume. The pore size distribution curve shifts to the big pore diameter portion. Besides the changes of pore structure, the fraction of large pores volume increases remarkably. The extreme conditions also effected the formation of the interfacial transition zone (ITZ). The width of the ITZ increases, while the microhardness values decreases obviously compared to those samples cured in standard condition. The hydration products in the ITZ are loose and porous. The bond between the aggregate and cement paste is weakened, and some microcracks in the vicinity of the ITZ are found. It is revealed that the microstructure of concrete exposed in dry and high temperature difference conditions has been deteriorated.

1 INSTRUCTIONS

Climate with the typical characteristics of dry and big temperature differences widely exists in the northern part of China. For example, in Tongliao and Chifeng of Inner Mongolia, the annual precipitation is only 58.6% of the average level of the country and the maximal temperature difference in one year is nearly 70°C. In these years, as the improvement of the transportation infrastructure in China's northern regions, more and more concrete constructions are exposed to the harmful environment of dry and big temperature difference, where the extreme conditions are potentially detrimental to the long-term durability of concrete.

The macro-properties of concrete material is closely related to its microstructure. Therefore, studying the microstructure of concrete after exposure in extreme environment helps understanding the mechanisms of concrete deterioration. However, very little work has been carried out on the study of microstructure changes of concrete exposed in the extreme conditions of dry and big temperature difference.

In this paper, the microstructure deterioration of concrete, including the pore structure and the interfacial transition zone ITZ, has been studied by means of mercury intrusion porosimeter (MIP), microhardness test and scanning electronic microscopy (SEM).

2 EXPERIMENTS

2.1 Materials and mix proportions

Ordinary Portland cement (Blaine fineness 325 m^2/kg, density 3.1 kg/m^3) was used as binder in this investigation. The fine aggregate was river sand (fineness modulus 2.8). In addition, a crushed basalt rock was used as coarse aggregate. Their particle sizes were in the range of 5–20 mm (55%) and 20–40 mm (45%), respectively. Ordinary tap water was used in the

Table 1. Mix proportions of concrete.

Concrete grade	Water kg/m^3	Cement kg/m^3	Sand kg/m^3	Aggregate kg/m^3	w/c
C30	198	388	635	1179	0.51

Table 2. Compressive strength of concrete.

Group	Compressive strength (MPa)						Mean
A	39.76	40.05	39.44	38.63	39.41	41.2	39.75
B	32.98	33.42	33.01	33.43	32.75	33.18	33.13

concrete mixes. The mixture proportion of concrete is shown in Table 1. The water-cement ratio is 0.51.

2.2 Specimens preparation and exposure conditions

Fourteen cube specimens ($150 \times 150 \times 150$ mm in size) were prepared for experiments, and were divided into 2 groups averagely, namely group A and B, respectively. All specimens were demoulded 24 h after casting and then the A group were cured in standard conditions (temperature $20 \pm 2°C$, and relative humidity 95%) while the B group were exposed in outside with extreme conditions of dry and high temperature difference (during the exposure period, temperature between day and night: 9–27°C; average temperature difference in one day: 16°C; relative humidity 28%–37%, average relative humidity 32%).

After 28 days, 6 specimens of every group were used for compressive strength testing and the other one was used for microstructure investigation.

2.3 Methods of testing

2.3.1 Compressive strength test
Compressive strength test was performed on 150-mm concrete cubes according to the China Standard JTJ-053-94.

2.3.2 Mercury intrusion porosimeter (MIP)
The porosity and pore size distribution were measured by using the mercury intrusion porosimeter (MIP) with a maximum pressure of 207 MPa. The contact angle was 140° and the measurable pore size ranged from about 6 nm to 360 µm. Values of 485.0 dyn/cm were used for mercury surface tension. The Washburn equation was used to calculate the pore radii.

The samples in the shape of pellets of about 5 mm in size without coarse aggregate for pore structure testing were separated from the crushed specimens used. The samples were immerged in ethanol to stop the hydration immediately after being crushed and dried at about 105°C for 24 h before MIP test.

2.3.3 Measurement of microhardness
Microhardness test was used to investigate the interface between aggregates and cement matrix. The specimens for ITZ microhardness test were cut into slices of the size $100 \times 100 \times 10$ mm. Slices cut from

Table 3. Basic parameters of pore structure measured by MIP.

Group	Porosity %	Total pore volume mL/g	Total pore area m^2/g	Average pore diameter µm
A	11.42	0.0551	7.42	0.0997
B	13.81	0.0679	6.012	0.4645

the middle of the cube sample, containing the ITZ between aggregate and mortar, were polished with 600# paper then 1,500# paper to obtain an adequately smooth surface with a minimum of damage. Slices were then carefully sealed to avoid carbonation, which might lead to larger measured hardness values because the carbonation product ($CaCO_3$) is much harder than the hydration products.

2.3.4 SEM test
The micrographs of ITZ between aggregate and cement matrix were obtained by means of JSM-5610LV scanning electronic microscopy (SEM). The specimens for ITZ morphology observation were cut into prisms of $8 \times 8 \times 6$ mm.

3 RESULTS AND DISCUSSION

3.1 Compressive strength

As shown in table 2, samples from group A cured in standard conditions (SC) and B cured in extreme conditions (EC) of dry and high temperature difference have different compressive strength. Strength values of B only account 83.3% for that of the A and the rate of strength loss is 16.7%. The results showed that the mechanical properties of the concrete with the same components and materials but different cured conditions were changed. The EC environment has resulted in the reduction of the compressive strength.

3.2 Pore structure measured by MIP

3.2.1 Basic parameters
The measured pore structure by using MIP is shown in Table 3. As the values in Table 3 indicate, the pore structure variation of the group B cured in EC

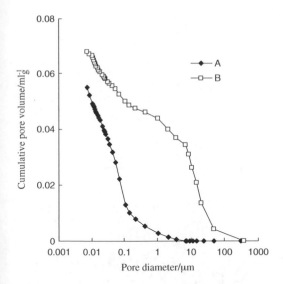

Figure 1. Cumulative pore volume of concrete samples.

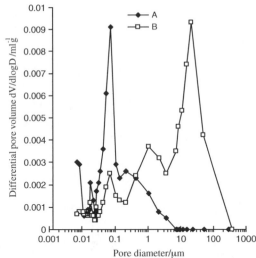

Figure 2. Pore size distribution of concrete samples.

environment is more remarkable than that of the group A cured in SC environment. The samples from group B cured in EC environment have higher porosity, total pore volume and average pore diameter, increasing by 20.9%, 23.2% and 36.9% respectively, than those of group A cured in SC environment. However, the total pore area of the group B becomes lower than that of the group A, decreasing by 19.0%. This effect can be related to the coarsening of the pore structure of the concrete from group B, and also can be used to explain the compressive strength loss as shown in section 3.1.

The variation of the porosity of the concrete can also be seen in Figure 1, where the cumulative pore volume curve of the group A subjected to SC environment is below that of the group B subjected to EC environment. Especially, the cumulative pore volume (from 0.1 to 100 μm) in the group B is significantly higher compared to group A, which can be identified the harmful effect of the EC environment to the pore structure.

3.2.2 Pore size distribution

The relationship of dV/dlogD and logD has been used to reflect the pore size distribution, in which V is mercury intrusion volume and D is pore diameter. It is accepted that the macro mechanical properties of material are always closely linked to microstructure. Thus, to a certain degree, the variation of pore structure reflects the deterioration of concrete subjected to EC environment.

From Figure 2, it can be seen that the dV/dlogD vs. logD curves for the two groups cured in different conditions are different as well. After been exposed in extreme conditions of dry and big temperature difference the peak value of dV/dlogD in the curve of group

B appears at 20 μm, while that of the group A appears at 0.08 μm, which is prominent smaller than that of group B. In addition, the highest peak of dV/dlogD of the group A occurs within narrow range from 0.05 to 0.1 μm while that of the group B ranges from 10 to 50 μm. This means, on the one hand, the curve of the group B has shifted to the big pore diameter portion compared to that of group A and a great increment of pore size occurred here; on the other hand, the micro-pores within the big pore diameter portion are most notably affected by the EC environment, and the micro-pores with big diameter have account for the most proportion of the total pore volume of group B.

3.3 Investigation of the ITZ

3.3.1 Microhardness test

Microhardness is a comprehensive parameter for various characteristics of ITZ. Microhardness includes the information of mean crystal size, crystal orientation index of CH and pore microstructure. The microhardness test results in ITZ of the group A and B are shown in Figures 3 and 4. The interface of aggregate-cement paste has been regarded as the original distance, where the left side of the 0 μm distance shows aggregate and the right presences cement paste. The trends of the microhardness profiles in the two figures are similar, in each of which there is a 'valley'. The 'valley' is contained in the ITZ. The ITZ is a weak zone, where the microhardness values are between those of the aggregate and the cement paste.

It is seen from Figure 4 that the extent of the weak zones (ITZ range) of the group B cured in EC environment increased to about 70 μm, while the width of the ITZ of the group A cured in SC environment is about

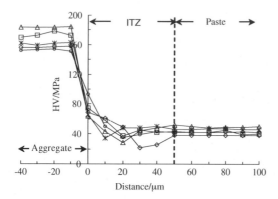

Figure 3. Microhardness in ITZ of group A.

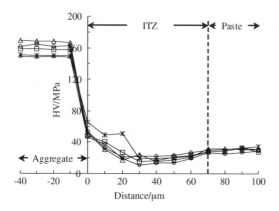

Figure 4. Microhardness in ITZ of group B.

50 μm. Due to the different cured conditions, the ITZ range of the group B increases 40% than that of the group A. In addition, the lowest microhardness values of the group A and group B, where each value is averaged from the results of five testing, are 31.44 MPa and 16.72 MPa, respectively. The lowest microhardness value of group B has decreased 46.82% compared to the result of the group A.

Higher microhardness values in the vicinity of the aggregate were observed in the specimens of group A than those of group B, and this may be due to the presence of stiff inclusions in the excited range around the indentation, which restrains the flow of material under the indentation. This investigation result indicates that EC environment with dry and big temperature difference is disadvantage to the hydration reaction of the cement and the formation of the dense cementitious matrix. This also explains why the specimens of the group B possess weak compressing strength.

3.3.2 SEM test of ITZ

The ITZ morphology of concrete examined in the SEM varies with the different cured conditions. Figure 5

(a) ×1000

(b) ×2000

Figure 5. SEM images in ITZ of group A.

and Figure 6 contain the SEM pictures (×1,000 and ×2,000 zoomed in) of aggregate–mortar ITZ at the curing age of 28 days. The study on SEM pattern of aggregation–mortar ITZ follows the regular results of the compressive strength test and microhardness test. It is summarized as follows. After cured in SC environment, the reaction products in the ITZ of the group A grow more compactly than those of the group B (compare Figure 5a to Figure 6a). The ITZ morphology shown in Figure 5 is much denser than the corresponding ITZ in Figure 6.

Furthermore, some microcracks present in the aggregate–mortar ITZ of group B, some of which are larger than 5 μm. This typical deterioration in ITZ may indicate the seperation of cement paste from the aggregate. It is apparent that extreme conditions are little contributed to the improvement of ITZ microstructure at most.

4 CONCLUSIONS

The results and conclusions are summarized as follows:

1. The compressive strength of the concrete exposed in extreme environment of dry and big temperature

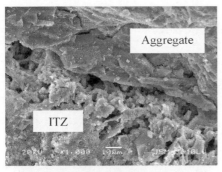

(a) ×1000

(b) ×2000

Figure 6. SEM images in ITZ of group B.

difference decreased compared to those cured in standard conditions.

2. Extreme environment has significant effect on the microstructure degradation of concrete. After exposed in extreme environment, the pore structure of the concrete become 'coarsening' as well as the aggregate-cement paste ITZ become weaker.

3. The results presented in this paper indicated that curing measures should be strengthened in extreme environment.

ACKNOWLEDGEMENTS

The authors are grateful to financial support from the Transportation Bureau of Inner Mongolia and to the staff of the Institute of Materials Recycling and Evaluation for their help in preparing and testing of the samples.

REFERENCES

Ahmen, H. Bushlaibi & Abdullah, M. et al 2002. Efficiency of curing on partially exposed high-strength concrete in hot climate. *Cement and Concrete Research* (32): 949–953.

Chan, Y.N. Luo, X., et al. 2000. Compressive strength and pore structure of high-performance concrete after exposure to high temperature up to 800°C. *Cement and Concrete Research* (30): 247–251.

Gao, J.M. Qian, C.X. 2005. ITZ microstructure of concrete containing GGBS. *Cement and Concrete Research* (35): 1299–1304.

Igarashi, S. Bentur, A. & Mindess, S. 1996. Microhardness testing of cementitious materials. *Adv Cem Based Mater* 4(2): 48–57.

Khandaker, M. & Anwar, Hossain 2006. High strength blended cement concrete incorporating volcanic ash: Performance at high temperatures. *Cement and Concrete Composites* (28): 535–545.

Laskr, A.L. 1997. Some aspects of evaluation of concrete through mercury intrusion porosimetry. *Cement and Concrete Research* 27(1): 93–105.

Olivier, J.P. & Maco, J.C 1995. Interfacial transition zone in concrete. *Advan Cem Bas Mat* 28(2): 30–38.

Raymond, A.C. & Kenneth, C.H. 1999. Mercury porosimetry of hardened cement pastes. *Cement and Concrete Research* 29(6): 933–943.

Shui, Z.H. & Wan, H.W. 2002. Distributions of chemical elements in aggregate-cement interfacial transition zone in old concrete (II). *Journal of Wuhan University of Technology* 24(5): 22–25.

Zhang, B. 1998. Relationship between pore structure and mechanical properties of ordinary concrete under bending fatigue. *Cem Concr Res* (28): 699–711.

Zhao, G.F. Peng, S.M. & Huang, C.K. 1999. Structure of concrete containing steel fiber. Beijing: *China Architecture and Building Press*, in Chinese.

Excellence in Concrete Construction through Innovation – Limbachiya & Kew (eds)
© 2009 Taylor & Francis Group, London, ISBN 978-0-415-47592-1

Intelligent exothermal Nano concrete with high thermal conductivity and designing and performing the automatic road temperature monitoring system

J. Poursharifi, S.A.H. Hashemi, H. Shirmohamadi & M. Feizi
Islamic Azad University of Qazvin, Iran

ABSTRACT: In the present project, effects of Nano materials, aggregates, age of the specimen, humidity, temperature, type of grading, amount of fine aggregates, steel fibers & water/cement ratio on increase of thermal conductivity and effects of using fly ash and slag as replacement for cement have been investigated and finally high exothermal & thermal conductive concrete specimen and an electronic system and the respective software, for measuring thermal conductivity coefficient has been produced. This controlling system can measure the average temperature of road and automatically provides and settles the temperature in a certain range to prevent the surface of roads, streets & runways from freezing by thermal sensors.

1 INTRODUCTION

One of the recent major challenges of transportation industry is safety of country road's surface and therefore prevention of accidents and easy transportation and traffic control.

Snow plowing and removing the blockages of roads and bridges' crown and airports in winter are major challenges of road and airport management in cold region of country including the very high costs to salt dispersing and maintenance of roads, harmful environmental effect of traditional road maintenance method, low traffic flow freezing whether and canceling the flights. This kind of road surface is a more efficient method than traditional ones due to intelligence, fully automatic monitoring high safety, long-life and better transferring and discharging the ice and snow from roads' surface.

This concrete has high thermal conductivity an may transfer generated heat inside the road easily into the surface and measures concrete temperature using thermal sensors which are placed near surface of concrete and then fixes it within a certain limits. We introduce the design and its laboratory results and finally present importance and application of this design in industry. SEM and TEM electron microscopy tests were also used to better recognition of functional mechanism of nano-materials. STA and RUL thermal analysis tests also demonstrate effects of nano-materials and metal fibers on thermal conductivity.

2 LABORATORY PREPARATION

There are two main procedures to test thermal conductivity:

1) Two parallel lines procedure (TLPP): this procedure was offered by "karsla" and is the most famous test on thermal conductivity. In this procedure two tubes are placed inside the specimen which one of them plays the role of thermal source and the other one as thermal sensor.

2) Temperature-flat plate-source procedure (PHS): basis of this procedure is similar to the first procedure but specimen should be cut very thin and a shaft is attached to appropriate positions using epoxy glue.

How to measure thermal conductivity in this research:

To measure thermal conductivity of concrete an instrument was designed in which available sensors may measure thermal conductivity of concrete per 30 seconds. Its error is about 1%. Since thermal conductivity of ordinary concrete is about 3 wkcal/m hr°c, we may say that instrument error is less than 0.03 kcal/m hr°c.

Instrument utilizes a thermal source inside the concrete and a thermal sensor is placed in a certain distance of it inside the concrete. When thermal source produces heat, sensor measures concrete temperature in every 30 seconds and transfer its result into the

Table 1. Comparison of thermal conductivity for different aggregates.

Type of ground	Thermal conductivity (kcal/m h°C)
Quartz	4.45
Granite	2.50–2.65
Limestrone	2.29–2.78
Marble	2.11
Basalt	2.47

Table 2. Physical characteristics of admixtures.

Classification	Specific gravity	Specific surface (m²/kg)	Place of production
FX	2.10	430	Boryung
BFS	2.90	450	Kwangyang

computer and then measures thermal conductivity in every temperature degree and at last plots thermal Figure of concrete.

3 MATERIALS

3.1 Cement

Cement in this study was Portland cement, type 2 which was produced by "Abeyek cement factory". Thermal conductivity of Portland cement totally is about 0.26 kcal/m hr°c.

3.2 Grading

Concrete in this study was self-compacting concrete (SCC) with metal fibers and nano- materials which are calculated as a percent of cementation agent volume. Quartz aggregates inside the rocky skeleton of structure is main characteristic of this concrete which includes 5%–8% porosity and provides an appropriate space for better cohesion of nano composite, cement and higher thermal conductivity.

On the other hand, since in this study we focused on high thermal conductivity of concrete quartz aggregate was used which contains higher thermal conductivity than other aggregates. Table (1) presents thermal conductivity characteristics of different aggregates comparing to quartz.

3.3 Admixtures

Fly ash and iron slag were used as cement alternative and BFS was used to minimize the water consumption. Table (2) presents the characteristics of admixtures.

3.4 Nano materials

Recent researches on thermal conductivity of nano materials have considerably extended thermal limits both in insulators and conductors. Therefore, if thermal limits are highly extended, it will be possible to considerable affect on thermal conductivity technology. While metal nano materials in concrete composites affect thermal conductivity of concrete so much, the main obstacle to achieve extra thermal conductivity is low thermal conductivity of interfacial zone between concrete and metal fibers & nano materials. This obstacle originates from low thermal conductivity of electron couples-fenon in interfacial zone between steel and non-steel materials. Further comprehensive researches are necessary to overcome this problem in interfacial zone of composite materials of concrete.

Three different nano materials were used to improve thermal conductivity characteristics and studying the effects of nano materials on concrete and correcting the infrastructures. Table (3) presents their characteristics.

3.5 Metal fibers

Studies conducted on fiber application in concrete confirm that metal fibers result in more pressure, bending and strain strengths of concrete than propylene fibers. Reinforced concrete with metal fibers therefore, illustrates higher stability and tolerance than concrete without fibers and reinforced concrete with propylene fibers. Concrete tolerance will improve against the temperature variations and thermal shocks of exothermal nanoconcrete in freeways bed and airports and prevents thermal cracks. Metal fibers are taken into account due to their high thermal conductivity owing to free electrons of metal and acceptable thermal maintenance & distribution. Amongst different metal fibers, steel fibers were selected due to their higher pressure & strain strength than other fibers and their anti corrosive characteristic. Poisson index of selected fibers also were close to Poisson index of hydrated cement gel to prevent lateral stresses.

4 EXPERIMENT VARIABLES

4 different nano materials were selected as main variables to study effect of nano materials on raising the thermal generation and thermal conductivity of concrete including AL_2O_3, $FE2O3$, TIO_2 and SIO_2. Metal fibers inside the concrete were used to homogenized thermal distribution and raising the thermal & strain strengths. Kind of fibers and their weight ratios sin the concrete were considered as secondary variables. Fly ash and slag were applied as cement alternative materials in order to increase concrete stability. Aggregate and grading and humid levels were considered as variables of design.

Table 3. Physical characteristics of nano materials.

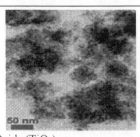

Titanium Oxide (TiO₂)
Purity: 99.99%
APS: 50 nm
SSA: 210 10 m²/g
Color: white
Morphology: spherical
Bulk density: 0.15 – 0.25 g/cm³
True density: 3.9 g/cm³

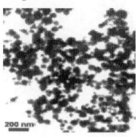

Iron Oxide (Fe₂O₃)
Purity: 99.9%
APS: 30 nm
SSA: > 50 m²/g
Color: red brown
Morphology: spherical
Bulk density: 1.20 g/cm³
True density: 5.24 g/cm³

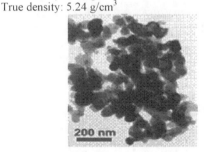

Aluminum Oxide (Al₂O₃, alpha)
(contains 5–10% theta)
Purity: 99.99%
APS: 30 nm (TEM)
SSA: ~25 m²/g
Color: white
Morphology: spherical
Bulk density: 0.52 g/cm³
True density: 3.97 g/cm³

5 MIXING DESIGN

A critical requirement to study quality control methods is precise study of specimen characteristics. It seems necessary to recognize various affecting parameters on thermal conductivity in order to improve behavioral mechanism of exothermal intelligent nanoconcrete with different kinds of nano materials and metal fibers.

Self-compacting concrete with over 20 years experience may be a correct selection for profiles which require reinforcement in all directions. This subject is more critical for road bed because their concrete should both tolerates static dynamic loads in all directions. We, therefore, selected self-compacting concrete with reinforced metal fibers in this study. some characteristics of self-compacting concrete include high efficiency, speed of execution, high strength high resistance to shock & bending and strain forces, and acceptable tolerance to dynamic and static loads in all directions.

Fresh self-compacting concrete (scc) tests in this study were conducted to select acceptable grading including the L-box, U-box, V-funnel, slump flow tests. 10*20 cm cylindrical moulds then were used.

To test effect of nano on SSC thermal conductivity, materials, at first, were test under different temperatures up to 1000°c through STA thermal test. This test also assessed functional procedure of nano materials in concrete structure, these materials as gel cell were added to concrete regarding to space weight of them. Bin fiber mixer, then, was used to better mixing these materials.

6 PREPARATION OF SPECIMENS

Cylindrical and plate specimens in this study were used for experiments. In the first place, cylindrical specimens were used to study different factors such as grading type, aggregate materials, and effect of nano materials and metal fibers on thermal conductivity of concrete. At last, 3 selected designs of first step were studied in the from of plate specimens with dimensions 30*30*7.5 cm in order to match with geometrical form of road and airport surface. These moulds were opened after 24 hours and soaked in water. 7, 28 and 90 days specimens were also studied by pressure, bending and strain strengths and thermal conductivity tests. Specimens of this study were as dry and saturated specimens in order to study applied behavior of exothermal intelligent nanoconcrete in road bed.

Specimens were also studied by water absorption and electrical resistance tests to assess their penetrative capacity and their functional procedure against the ions and their stability.

A thermal source thermal of element was placed in the centre of cylindrical moulds before concrete work

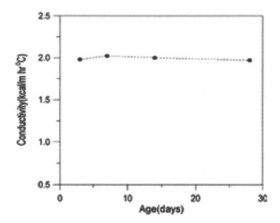

Figure 1. Effect of age on thermal conductivity.

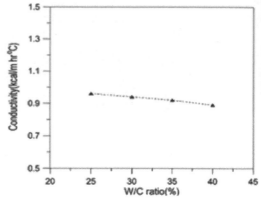

Figure 2. Effect of w/c ratio on thermal conductivity.

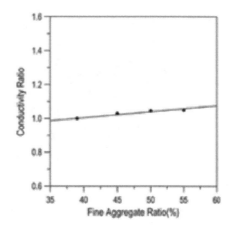

Figure 3. Effect of fine aggregates on thermal conductivity.

for thermal conductivity test. Mould was, then, filled by concrete and a thermal sensitive sensor was placed in distance 2cm of mould body.

7 EXPERIMENT RESULTS

7.1 Thermal conductivity doesn't depend on age

3,7,14 and 28 days specimens with temperature 20°c in saturation state were tested to determine effect of specimen age on thermal conductivity. As it can be seen from Figure (1), thermal conductivity of concrete doesn't depend on specimen age.

7.2 Thermal conductivity depends on water/ cement ratio

Some tests were conducted in concrete paste in different thermal and humid conditions to determine variations resulted from cement/water ratio on thermal conductivity of concrete. Since change of aggregate and its materials relates to its cement, concrete paste was used instead of hardened concrete in order to determine thermal dependency into cement/water ratio. Results in Figure (2) indicate that raise of cement and loss of cement/water ratio increase thermal conductivity and cement thermal conductivity may become more than water thermal conductivity.

7.3 Effect of admixtures on concrete thermal conductivity

Replacement of cement with fly ash & iron slag results in loss of thermal conductivity and raising the concrete stability. Optimized level of these materials may raise the stability.

7.4 Dependency of thermal conductivity into fine aggregate ratio

Thermal conductivity was measured after changing the fine aggregate ratio in terms of total aggregate in 4 different specimens and Figure (3) illustrates the results. There is a little increase of thermal conductivity in terms of fine aggregate ratio. This ascending process is because thermal conductivity of fine aggregates are higher than macro aggregates. This fact is also because fine aggregates result in homogenous distribution of aggregates inside the concrete mixture.

7.5 Effect of temperature on thermal conductivity

To test the effect of temperature concrete& thermal conductivity, they studied in temperatures 20, 40 and 60°c and results indicated that concrete thermal conductivity will lost through increasing the temperature of specimen. This is due to limitation of thermal conductivity in fine aggregates of electron couples-fanon

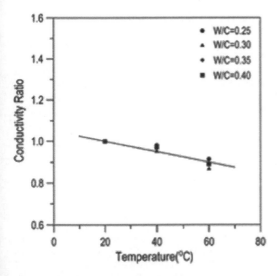

Figure 4. Effect of temperature on thermal conductivity.

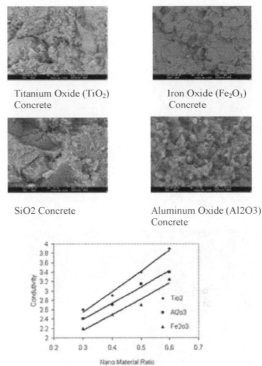

Titanium Oxide (TiO$_2$)   Iron Oxide (Fe$_2$O$_3$)
Concrete   Concrete

SiO2 Concrete   Aluminum Oxide (Al2O3)
                             Concrete

Figure 5. Effect of nano material percentage on thermal.

and fanon-fanon in interfacial zone matrix of cement particles together and with metal fibers and nano materials (Figure 4).

7.6 Effect of humidity on thermal conductivity

Humid conditions of specimen are a critical item amongst other parameters of material. When specimen condition is changed from completely dry into completely saturated condition, concrete thermal conductivity is raised because porosities of concrete is filled by water and as you know thermal conductivity of water is more than air.

7.7 Effect of nano materials on thermal conductivity

Maximum effect of raising the concrete thermal conductivity amongst the nano, materials of this study is on TiO$_2$. As Figure (5) presents, more percentage of nano materials will raise the thermal conductivity. Filling up the microscopic porosities amongst the aggregates by nano materials result in raising the concrete thermal conductivity. As following figures show, concretes with nano titanium and nano aluminum have regular crystal structures and have less porosity than concrete with nano ferrous and nano silicium. They, therefore, transfer the heat better.

7.8 Effect of metal fibers on concrete thermal conductivity

Application of metal fibers in concrete raises thermal conductivity and thermal distribution characteristic of

concrete. Metal fibers also improve mechanical characteristic of concrete including the raise of pressure & strain strengths and prevent the thermal cracks. Figure (6) illustrates application effect of fibers on raising the thermal conductivity.

7.9 Effect of fiber and nano interaction on raising the thermal conductivity

Regarding to positive effect of nano and fiber on raising the concrete thermal conductivity, Figure (7) illustrates these two parameters. Nano, in fact, raises thermal generation and thermal conductivity and metal fibers raise, the characteristic of concrete thermal distribution.

8 RESULTS

Amongst all affecting factors on electrical conductivity study results indicate that raising the nano materials improve concrete thermal conductivity. Adding the metal fibers into concrete mixture, on the other hand, may raise the thermal conductivity and characteristic of concrete thermal conductivity. Specimen age

315

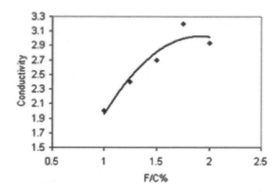

Figure 6. Effect of fiber/cement percentage ratio on thermal conductivity.

Concrete with fiber & nano- TiO_2.

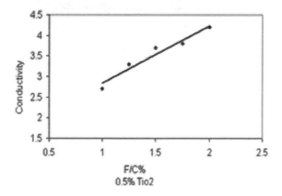

Figure 7. Effect of fiber/cement ratio on thermal conductivity of nano concrete.

parameter also has no significant effect on concrete thermal conductivity.

Application in industry:

Regarding to high cost of road maintenance in cold regions of country and harmful environmental effects of salt dispersion, it's necessary to replace traditional methods by novel and developed methods. Recent design may raise temperature of concrete using thermal source inside the concrete and prevent freezing the road surface. It's also possible to measure surface temperature using thermal sensitive sensors which are placed near the surface and automatically fix the surface temperature and receive information about freezing danger using an intelligent network, constantly. Application of recent design, that's replacement of thermal-generating concrete in hazardous are as may decrease many accidents and damages owing to cold weather and freezing the roads.

REFERENCES

D. Corr, S. P. Shah, "Concrete Materials Science at the Nanoscale", Proceeding of the International Conference (Application of Nanotechnology in Concrete Design), University of Dundee, Scotland, UK, ed. R. K. Dhir, M. D. Newlands, L. J. Csetenyi, London: Tomas Telford 2005, pp. 129–136.

H. Okamura, "Self-Compacting High-Performance Concrete", Concrete. International, 19 (1997), pp. 50–54.

Excellence in Concrete Construction through Innovation – Limbachiya & Kew (eds)
© *2009 Taylor & Francis Group, London, ISBN 978-0-415-47592-1*

Application of composite in offshore and marine structure

Q. Jafarpour, M. Balaei & B.B. Zadeh
Islamic Azad University-Tabriz Branch, Iran

ABSTRACT: High strength of concrete against corrosion is main reason for using it in building industry. Application of composite reinforcement and profile manufactured by pathogen in crease lifetime and reduce maintenance and manufacturing costs in marine and offshore corrosive environment. Composite application in corrosive condition of offshore saline water is considerable statistical data show that more than two milliards dollars spent on compensation on damages due to corrosion in offshore structures (military – unmilitary) every year. Need to reduce repairement and maintenance costs of different huge offshore and Meta offshore structures cause that engineers and specialists use new material with high relative profits regarding to similar materials (concrete, steel and wood). This paper aims to propose the advantages of composites and their application as a new solution in marine and offshore structures which concrete, steel and wood were employed as main material.

1 INTRODUCTION

The use of these types of complex material started since 1940 and like many other techniques and technologies, they had military application and were used in aerospace industries, such that the use of polymers and polymeric composites in aircraft and missile manufacturing industries increased up to 80% in U.S.A and West Europe. But they were used in other fields, such as construction, automobile industry and dock manufacturing industries (fast boats and ships and marine structures) because of their lightweight and high resistance.

2 STRENGTHENING OF STRUCTURES IN DIFFERENT INDUSTRIES USING COMPOSITES

Composites, synthetic and complex materials have recently found many different applications, such that the use of these materials is considered as one of the indexes of development in society. In Asia, countries like Japan, per capita consumption of composites is 4.5 kg, which is higher than the global standard (3.5 kg). Per capita consumption of these materials is 0.3 kg, which is one tenth of the global standard. This amount is even less than the amount used in our neighboring countries like Turkey (per capita consumption in Turkey: 1.5 kg).

2.1 *Oil & Gas Industries*

The use of composite structures causes savings up to 60% in oil industries, from oil platform constructions up to transmission lines, especially in places with high corrosion rate. For example, the use of composts is very economical and appropriate in marine environments. One of the interesting applications of composites is that we can strengthen the oil and gas pipes using these materials. Persons, who have activities in the field of gas and oil, know that removing the corrosions in these pipes with usual method is very costly and time-consuming for the experts. Using this method we can reinforce the corroded and worn pipes with a layer of composite without stopping the activity of the line. (2) Unfortunately, the directors and decision-makers have not good information about the applications of this technology and enormous amounts of country's budget are spent because of the losses resulting from corrosion each year. This problem can be solved using some strategies.

3 PALTROGEN TECHNOLOGIES IN PRODUCTION OF COMPOSITE PRODUCTS

Paltrogen process is one of the most important and fast methods for production of composite products. We can produce different types of composite profiles with fixed cross section and high speed using this method. Several Iranian companies have started activities to benefit from this technology in our country.

3.1 *Paltrogen Profiles*

Paltrogen profiles are among the most durable and strong engineering materials (FRP). Standard Fiberglas Reinforced Plastics are supplied in the market.

The quality of these profiles is determined with some factors like FRP, direction, situation, quality and type of the resins. These profiles are produced and supplied with a different limit of resistance and stability against corrosion and heat stability. The main production method of these profiles is Paltrogen process. (1). Paltrogen is one of the fast methods for production of composites reinforced with continuous fibers, which were first used in 1951. The products produced using these methods are of high resistance, lightweight and long life in chemical environments. In this process, the continuous fibers, as long fibers, woven material and material with short fibers are passes through a bath contain resin and are smeared with resin. After exiting the bath, the fibers are directed into a warm mold and the product gains its final shape inside the mold. The hardening and firing operations are carried out inside the mold and after the continuous product is taken out of the mold and cut to the intended length. So, the final product is ready to be supplied to the market without any need for other operations. The mold being used for this operation is made of steel and its length is 30–155 cm.

3.1.1 Advantages

The length of standard produced profiles is usually 25 mm to 5.3 m and the production speed in this process is 2–30 m/h, which depends on the shape of the product and resin used in the product. Usually, polyester resins, epoxy, fiberglass, carbon, aramid and polyethylene are used in his process. The resistance-weight ratio is high in the products produced using this method and they have good resistance against corrosion and good dimensional stability. The products produced using this method include: different types of timbers, drainpipes, frames, automobile springs, fishing poles, I-beams, hammer hands, tent poles, golf stick, ski poles, tennis rocket and other profiles. Production of parts with complex cross sections in continuous form is one of the unique characteristics of composites. We can only make the desired cross section only through changing the mold.

The percentage of fibers' weights to the total weight of the product in this process is higher in compare with other methods for production of complex materials.

Figure 1.   Paltrogen process.

So, paltrogen method is used for production of parts with high longitudinal resistance. Products produced using paltrogen method show more resistance against chemicals and one of the applications of these products is acidic and basic environments with high pH and because of the continuous process used in production of parts using paltrogen method, the produced parts don't have length limitation. High speed and using of simple and cheap equipments and less need for manpower decreases the expenses in production in compare with other methods for production of composites.

3.1.2 Disadvantages

However, using paltrogen method has some disadvantages, which cause limitation of its application range. As mentioned before, a mold with fixed output section is used for shaping the part in paltrogen method and it is not possible to produce parts with different section. The low shear resistance is among the other weaknesses of the products produced using this method. Also, these products are relatively weak against drilling operations.

Profiles resulting from paltrogen process are mostly used in marine structures. These profiles are cut in required sizes and are easily assembled in site using special fittings. Using these profiles can decrease the final weight of the structure up to 50% in compare with the steel structures. In corrosive conditions, the life of these structures is estimated several times of the steel structures and taking in to account the expenses of maintenance and repaid during the life of the building, we can say that the final expenses are deceases at least 10–15%.

Today, paltrogen profiles are mostly used in construction industries. For example, the paltrogen profiles are used in different types of roads, fences, shields, doors, windows and beams with I-shape cross section, angles and bridges. These profiles have attracted the attention of Europeans in construction of bridges and dock structures. These profiles are proposed as the best alternative for construction of marine structures in American countries.

The reasons are summarized in the four following factors:

1. Suitable primary price
2. Low repair and maintenance
3. High work life
4. Simplicity, fast installation and application

Now, more than tens of bridges in U.K are constructed using paltrogen profiles and in a country like Canada, big investments are carried out for production of profiles resistant against corrosion using paltrogen method. India, China, Malaysia and many other Asian countries have done extensive activities to produce and use these products in construction industry. The annual consumption of composite in the world is 6 million tons and as it was mentioned before, the industrial countries are pioneer in this field. Today, we witness the development of composite bridges all over the world. For example West Brock in Ohio (diagram 2) is the first bridge in which the composite structures are uses. This has done in the first phase of the 100 project. The aim of this project, which was an innovation by "Composite National Center", was to repair and replace 100 old bridges by composite (NCC).

In diagram 2, we can see the longest bet composite structure in Europe. Special resins are used in constructing the deck of this bridge. Besides the low maintenance expenses of this bridges, it should be mentioned that we didn't face any traffic problem while installing this bridge, because it tool only 24 hours.

According to the available statistics, the development of composite production process doesn't follow a unique pattern in different parts of the world. For example: Paltrogen process is of high attention in U.S.A because of special conditions of this country and the need for marine structure in most parts of this country (this has not been much welcomed in Europe). In West Europe, this market reached 3200 tons with 47% increase from 1995 to 2000.

Figure 2.

3.1.3 Role of producers in market expansion (case study: Paltrogen profiles)

By studying the statistics of composite-shaping technologies development, we can see that paltrogen method with a development rate of about 10% is expanding faster than other shaping methods, while paltrogen profiles are not popular in our country and only few companies have shown tendency for production of these profiles in the country during the several recent years. When we simultaneously need lightweight and corrosive, electrical and mechanical resistance, these profiles are the suitable and economical alternatives. Investments for establishing composite profiles production unit, taking into account the background for establishing its conversion industries, can directly create suitable jobs. In terms of the primary price and technical specification, these profiles can well compete with aluminum, iron and galvanized steel and in most of applications, they can be more suitable alternatives for metal profiles. Domestic production of these profiles is economical and has considerable added value. At present, the price of each kilo of this profile in world markets is 5.3 to 6 dollars. For each Iranian buyer, the transportation expenses, custom charges and warehousing expenses add 20% to this price, while these profiles are available in Iran with a price lower than the world price. Of course, the price of these profiles varies according to their application type.

3.1.4 Paltrogen profiles market

As it was said before, some electrical industries and oil companies are the consumers of paltrogen profiles. However, we can claim that the present rate of consumption is very low for the hidden potential of Iran market for consuming these products. There are two reasons for this hidden potential: The lack of information and distrust for new technologies. For example, one of the applications of these profiles is in guardrails in highways. Now, these guardrails are made of metal profiles or pipes, while paltrogen profiles are more secure and resistant against stroke in compare with metal and also are beautiful and of good appearance. If we study about the repair and maintenance expenses of metal and composite, we can sees that paltrogen profiles are more economical in long-term. But these

Table 1.

Other Cases	Byproducts	Products Resistant Against Corrosion	Transportation	Electrical Building	Industries	
17% 3500 Ton	7%	17%	15%	17%	32%	1995–2004

Figure 3.

Figure 4.

profiles have not entered the market because of distrust and lack of information.

Another example is the structures constructed in corrosive regions. Sometimes, these structures are exposed to destructive corrosion within less than five years. Composite profiles are the suitable and economical solution, but because of unawareness of the producers and lack of these products in the country, these materials have not been used in construction of structures. There are several other markets like the equipments of power transmission lines or agricultural applications, which are considered as good consumers of paltrogen profiles. We can open these markets by giving information and attracting the trust of the users and also create a very bright horizon for progress of composite industry in the country.

3.1.5 The role of producers

The producers have a very important role in removing the obstacles of the market and accepting these products by the users. The producers should spend time and energy to supply a new technology to the market. It is not possible to expect for users without investment for introducing and giving information. Even, it may be needed to wait for some time for the new product to be accepted by industries. Besides, the producers should be able to localize the imported technology to be compatible with the requirements of the market. In the 60th decade and beginning of the 70th decade, many persons, with non-production intention, took actions to establish production units in order to use the cheap foreign currency facilities. As some of these units were not familiar with the culture of production, they didn't pay the required expenses for suitable introduction of the goods to the market, improving the quality, innovation risk and localization of the imported technology and faced with production problems in the next years. Spending these expenses is one of the requirements of a successful production (especially in the modern fields) and the producer shouldn't avoid paying these expenses. Luckily, most of those who have activities in the field of composite in Iran are experts familiar with this industry and the culture of production. So, this capital should be used for developing the cooperation between these producers and consumption market.

3.2 Composite round bars

The construction industry has the biggest share of the composite materials market. Composite reinforcements have found good applications in construction, especially establishment of seaboard structures or other structures established in corrosive climatologic conditions. FRP composite round bars have been of attention of civil engineers in construction of big buildings because of their advantages. The main cause of using FRP round bars inside the concrete is prevention from corrosion and increasing of damping of vibrations created in the structure against vibration, very high tensile resistance (up to 7 times that of steel), acceptable module of elasticity, low specific weight (0.25 of steel), very high specific final resistance (10 times that of steel), good resistance against fatigue and creep, insulation against magnetic waves and good adhesion with concrete. Although using FRP round bars instead of metal ones will decrease the weight of the building, low weight is less important than the cases mentioned above when using these round bars. The reason of high damping coefficient of composites is their non-elastic properties, which damp the absorbed energy. While metal materials are elastic and don't damp the absorbed energy. So, composite materials will have better performance against earthquake vibrations and are the best alternative for resistance of structures against quakes. Applying FRP round bars instead of metal ones considerably decreases the losses resulting from corrosion. During destruction resulting from corrosion in concrete reinforced with metal round bar, first the metal bars inside the concrete are rusted and oxidized and then these oxides immigrate to the outer surface of concrete and are dissipated inside the concrete and destroy it.

So, after corrosion of the concrete and metal parts of the structure, the concrete structure will completely destroyed. Traditional methods like placing metal sheets on structure or increasing the thickness of concrete to face with corrosion not only will not solve the problem of metal corrosion but also will increase the weight of the structure and make it susceptible against earthquake. We can reinforce the outer surface of the concrete structure by composite materials and through using FRP round bars inside the concrete to solve this problem and prevent the performance of the structure from being failed against corrosion. This is the best method to fight against corrosion in a concrete structure. Technology of composite reinforcement's production is also paltrogen method. The paltrogen products have very long life and the production speed of a paltrogen product is also relatively high. Although the price of a paltrogen beam is higher that the metal beam, its good resistance against corrosion and earthquake can be a good justification for its higher price. In public applications, like construction of structures, if we need resistance against corrosion and earthquake, paltrogen beams can also have economic justification.

Besides, the good resistance of composite reinforcements against corrosion has introduced them as one of the best modern materials for construction of seaboard structures. These reinforcements are applied for reinforcing concrete columns of platforms, construction of bridges and other marine structures and their life is tens of years.

Many cases of damages resulting from corrosion are reported in coastal regions in southern parts of Iran, like docks and loading ports. In a dock in Imam Khomeini port, the signs of corrosion have been observed in some parts of the dock before completion of the final parts of dock, which shows the severity of damage resulting from corrosion. One of the important obstacles is general use of composite reinforcements (RP) by civil engineers is the lack of codes like the instructions for application of traditional construction materials. During the recent years, American Concrete Institute has presented a code (ACI-440) for design and constriction of reinforced concrete buildings with composite round bars (FRP), which is an important step towards developing the application of these reinforcements. Beside U.S.A, other countries like Japan and Canada have presented some codes regarding the using method of composite round bars in concrete structures. These codes explain the requirements for design and construction of concretes reinforced with composite reinforcements in construction of tall buildings. Considering the corrosive conditions of southern and central regions of Iran, national projects are being executed for producing composite reinforcements in composite department of polymer research institute and Iran petrochemical company. The related authorities should prepare codes for application of these reinforcements together with their production.

3.3 Paltrogen products in construction industry

Every day, the necessity of improving the quality of products and optimization in different industries creates a new application for these types of products, such that the necessity of using these products in industries that need high resistance and heat resistance or electrical properties is felt more than before because of the unique properties of composites. For example, the statistics show the increasing application of composites in construction industry. The application of composites in construction industry can be classified in three groups: supporting structures, non-supporting structures and structure resistant against corrosion. Paltrogen products have extensive application in the field of supporting structures and corrosion-resistant structures. Supporting structures have other applications including fences, rails, man ways (ways for one person around railroad and so on), components of primary and secondary structure (I-shaped beams, box beams and so on), bridges, bridge parts, supporting frames and reinforcements. Structures resistant against corrosion have many applications. These structures are highly used in docs, offshore structures, oil platforms and other marine installations and salty areas. Mostly, this types of composite application has been mostly discusses in this paper.

4 WEAKNESSES OF TRADITIONAL STRUCTURES (MARINE & OFFSHORE)

When a metal structure is exposed to salty waters, the surface of the metal starts rusting through an electrochemical interaction and gets corroded and is destroyed within few years. Reinforced concrete structures get rusted in corrosive marine environment very soon. On one hand, corrosion causes the loss of resistance and increases the volume of the metal skeleton inside the structure, which results in internal crack and fracture. On the other hand, the concrete loses its coherence because of contact with the humidity in the environment and starts cracking in outer side and fractures. Temperature changes in marine environment also fatigues the structures and destroys it as a result of contraction and expansion. The materials traditionally used in construction of different types of structures under climatologic conditions in South of Iran, especially under marine and seaboard conditions of Persian Gulf, are mainly made of steel and concrete. On the other side, the climatologic conditions of Persian Gulf are very rough and variable and severely attack the concrete, especially steel. In this regard, it has been necessary to fight against this attack and protect the materials from destructive factors from old times and the advanced countries in the world have carried out extensive researches in this regard and have developed a suitable technology. However, this scientific issue has not received much attention in Iran.

5 APPLICATION OF COMPOSITE IN MARINE & SEABOARD STRUCTURES

As you know, round bar corrosion in steel reinforced concrete is considered as an important problem in these environments. Many reinforced concrete structures have been seriously damaged because of contact with sulfates, chlorides and other corrosive factors. If the steel used in the concrete goes under higher tensions, the problem will be more critical. An ordinary reinforced concrete structure reinforced with steel round bars will lose some of its resistance in case of long term contact with corrosive factors like salts, acids and chlorides. Besides, when the steel is rusted inside the concrete, it will pressure the concrete around it and will crash and flakes the concrete. Some techniques have been developed and applied to prevent the corrosion of steel in reinforced concrete. Some of these techniques include covering the round bars with epoxy, injecting polymer to the surface of the concrete or cathodic protection. However, each of these methods has been successful to some extent and in some of the cases only. So, the experts and researchers of reinforced concrete have focused on complete removal of steel and replacing it with materials resistant against corrosion in order to completely remove the corrosion of round bars. In this regard, as FRP composites (Fiber Reinforced Plastics) are highly resistant against alkaline and salty environments, they were considered as a replacement for steel in reinforced concrete, especially in marine and seaboard structures. Unlike the traditional materials, composites show extraordinary resistance against highly corrosive environments like salty seawaters, chemical fluids, oil and gas. These materials have been proposed as a new method in construction of marine and seaboard structures, which have been made of steel, concrete and wood. Unlike concrete and steel, a marine structure made of composites (FRP) has high resistance against chemical and biologic corrosion resulting form marine microorganisms and don't go under structural contraction and expansion because of temperature changes in water. These structures don't need repair and inspection and live several times longer than the usual structures.

6 CONCLUSION & RECOMMENDATIONS

One of the properties of composite materials is their suitable performance against earthquake vibrations. As a result, these materials can be a good alternative for strengthening the structures against quakes. As our country has suffered from severe damages because of earthquakes and performance of structures in these natural disasters, it has the potential of using these materials. Taking into account what mentioned above, extensive researches should be carried out about the application of FRP composites in the concrete of marine and seaboard structures in south of Iran, especially in Persian Gulf Region. In this regard, it is proper to carry out suitable researches on different types of FRP composites (GFRP, CFRP, AFRP) and their suitable amount for marine structures established in Persian Gulf region. These researches include extensive theoretical researches about the behavior of reinforced concrete structures commonly used in marine regions (provided that they are reinforced with FRP composites). In this relation, it is necessary to carry out suitable experimental works on bending, tensile and compressive behavior of reinforced concrete parts reinforced with FRP composites. FRP composites are still unknown in Iran and their application in reinforced concrete in marine and seaboard structures have not been known for Iranian engineers and experts. Researches which will be carried out in this regard can result in preparation of an instruction or code for application of FRP in reinforced concrete as a material resistant against corrosion in marine and offshore structures in Iran. This movement can result in saving of billions of rials of capital, but unfortunately lots of capitals are wasted in different regions of the country like Urima Lake's bridges and southern parts of Iran (especially in marine and offshore regions) because of corrosion of round bars and destruction of concrete structures.

REFERENCES

Adaptable All-Composite Bridge Concept, John Uncer (Composite products Inc.)

Application of Advanced Composite Materials in Construction and Strengthening of Strategic Structures, 2004

Composite Bridge Deck, www.diabgroup.com

Composite Journal/Iran Composite Institute, issued 1,3,4,5

Composite, definition and Its World Niche/First Report, Industries Center Http://Kkhec.ac.ir

Iran Technology Analysis Network, Web site

Scientific Pole of Marine Technology & Sciences/Modern Marine Technologies Research Journal in The Field of Structure & Materials

Theme 4:   Protection against deterioration,
repair and strengthening

Excellence in Concrete Construction through Innovation – Limbachiya & Kew (eds)
© 2009 Taylor & Francis Group, London, ISBN 978-0-415-47592-1

Lateral strength evaluation of seismic retrofitted RC frame without adhesive anchors

T. Ohmura
Musashi Institute of Technology, Tokyo, Japan

S. Hayashi
Tokyo Institute of Technology, Kanagawa, Japan

K. Kanata
Taisei, Kanagawa, Japan

T. Fujimura
Taisei, Tokyo, Japan

ABSTRACT: The Japanese Building Standard Act was revised in 1981 considering limit states after yielding of structural members, on a basis of the lesson learned from Miyagi earthquake in 1978. It was found that poor seismic behavior of the buildings built before 1981 were exhibited due to inadequate horizontal retained forces and shortage of shear reinforcement of RC framed structures. Adhesive anchors are generally used to connect steel braces to the existing structure on the conventional way of seismic retrofitting. As the adhesive anchor requires drilling holes in the concrete structure, the retrofitting work causes nuisances such as vibration, noise and dust particles. These hazards make retrofitting work difficult and become one of the reasons which obstructs wide spread of seismic retrofitting. It is important to emphasize that seismic retrofitting without adhesive anchors would resolve those problems. We carried out cyclic loading tests and analysis of a series of framed specimens strengthened with steel braces to confirm the effect of the retrofit method without adhesive anchors, and proposed a evaluation method of lateral strength of those retrofitted frames.

1   INTRODUCTION

1.1   Background

A large earthquake occurred in Japan at around 05:14 p.m. on Monday, June 12, 1978. The magnitude 7.4 event was located off the coast of Miyagi Prefecture. The earthquake caused significant damage in the Tohoku region and especially Miyagi Prefecture.

This damage served as a lesson, and the Japanese Building Standard Act was revised in 1981. The new provisions of seismic design standard basically involved two limit states considering medium (i.e., relatively frequent) and large (i.e., feasible strongest) earthquakes. It is need to confirm that ultimate strength of buildings at a kinematically admissible plastic mechanism based on the capacity design.

Being insufficient seismic capacities of those buildings built before 1981, it is politically encouraged to improve structural properties which are specified stiffness, strength and ductility considering large earthquakes that are possibly to happen in the near future.

1.2   Research significance

Many methods of seismic retrofitting (Kanata & Kikuchi 2000, Kanata et al. 2003, Kei et al. 2000, Kuse 2006) have been developed considering structural types or deficiencies of those buildings. In the buildings with brittle columns, for example, the columns are wrapped with fiber-reinforced polymer to ensure column toughness. Shear walls or steel braces are used to enhance stiffness and strength for the buildings with insufficient lateral strength. Photo 1 shows a building retrofitted with a steel braces using anchorless system developed by Taisei. On a conventional way of seismic retrofit, steel braces are connected to an existing concrete frame by anchor bolts fixed in drilling holes with epoxy grouting. This method is commonly used to provide additional strength and lateral load-carrying capacity in existing substandard structures.

The method using adhesive anchors is shown in Figure 2. Many holes are drilled into the concrete structure and headed studs are inserted into the holes with epoxy adhesives. Then steel braces are set up and high strength mortar is grouted. This method causes nuisances such as vibration, noise and dust particles when the holes drilled. These nuisances obstruct prevalence of seismic retrofitting, and make quite difficult its execution while the buildings in service.

The outline of the anchorless (non-anchor) method developed by Taisei since 2000 is shown in Figure 3. Because of no anchor bolts, the flange of the perimeter frame surrounding two braces is arranged parallel to the bottom of beams, and the gap between the perimeter frame and the existing frame are grouted with high strength mortar. Although this anchorless method resolves the nuisances mentioned above, few prior studies have been conducted. It is necessary to confirm the responses of a RC frame applied the anchorless method under earthquake excitations.

Based on the study for anchorless retrofitting of RC framed structures by Kanata et al. (2000a,b), the seismic behavior of the retrofitted frames are clarified and design equations are proposed. And we performed FEM analyses of one of those specimens to simulate the load – deflection relationships (Ohmura 2006).

This paper presents the results of experimental and analytical studies with regard to a one-bay one story specimen applied anchorless method and confirm the analytical results are suited the experimental results. Then we propose a evaluation method of

lateral strength of the retrofitted frames applied the anchorless method.

2 EXPERIMENTAL PROGRAM

Cyclic loading tests were conducted on one-half scale models at the Taisei Technology Research Center in Japan.

The research program was undertaken in order to subject a frame specimen to reversed cyclic lateral displacements at the top girder under a constant axial load in columns.

2.1 Test specimens

Details of the frame specimen (S1 and S3) are shown in Figure 3. The material properties of three specimens are summarized in Table 1. The horizontal loading level was 1,600 mm from the base and the span was 3,000 mm. A 350 mm square cross-section was used for the columns. The section of the girder was 230 × 380 mm, having a slab of 100 mm thickness and 600 mm width on its top.

The column and girder reinforcements were arranged simulating an existing old building. The steel braces were made of SS400 steel, and their material properties were in accordance with this grade of steel; SS400 conforms to the Japanese Industrial Standards. The maximum aggregate size of the specimen was 20 mm.

The gap between the existing frame and steel braces was 30 mm. This gap was grouted with a high strength mortar without adhesive anchors. In other words, the steel braces were fixed into the frame with a high strength mortar by pressure grouted.

Specimens S1 and S3 had steel braces being kept elastic, and they were intended for punching shear

Photo 1. A building retrofitted with steel braces.

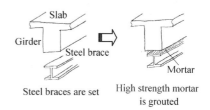

Steel braces are set   High strength mortar is grouted

Figure 2. The new method without anchors.

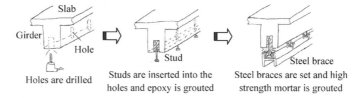

Holes are drilled   Studs are inserted into the holes and epoxy is grouted   Steel braces are set and high strength mortar is grouted

Figure 1. The method of using anchors.

failure of the columns. Specimen S41 had steel braces in relatively small sections, which were un-bonded braces being constrained by a steel pipe, intended for yielding of the braces before the occurrence of the punching shear failure.

2.2 Test setup, loading program, and instrumentation

The test setup is shown in Photo 2. The specimens were designed to be subjected seismic forces from the girder to simulate the behavior of a real structure under earthquake excitations.

The axial forces applied to each of the columns were 300 kN. The specimen was subjected to reversed cyclic lateral displacements while the normal axial stress in the columns was held constant. A constant axial load was applied and then a uniaxial lateral displacement history was applied at the top of the column.

The specimen consisted of one girder containing loading bars (4-D25) and two columns having a foundation block attached to the reaction floor.

A typical lateral displacement history consisting of cycles of "displacement control," increasing story drifts (i.e., 0.001, 0.002, 0.003, 0.005, and 0.01 %), was adopted.

3 ANALYSIS

Specimens S1, S3, and S41 have been modeled and are named *model S1, S3*, and *S41*.

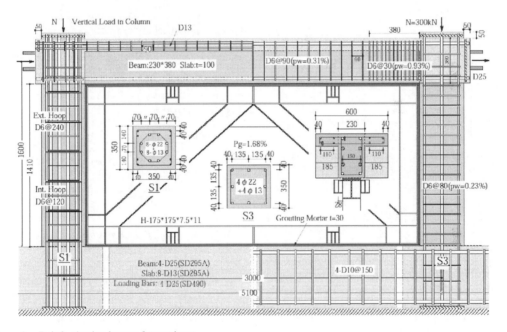

Figure 3. Reinforcing bar layouts for specimen.

Table 1. Properties of materials.

Specimen Name	σ_B (N/mm²)	Longitudinal bar of column	Hoop	Longitudinal bar of girder	Stirrup	Bar of slab	Steel brace
S1	19.4	φ13 SR235A (330)	D6 SD295A	D25 D490	D6 SD295A	D13 SD295A	H-175 × 175 ×
S3		φ22 SR235A (314)	(300)	(337)	(330)	(354)	7.5 × 11
S41	22.4	D16 SD345 (401) D19 SD345 (391)	D6 SD295A (300)	D19 SD345 (391)	D6 SD295A (300)	D13 SD295A (361)	H-125 × 125 × 6.5 × 9

Concrete: Light weight concrete with maximum aggregate size 15 mm.
Grout: Shrinkage-compensating grout over 45 N/mm².
Steel: SS400 (Japan Industrial Standard).
σ_B: Concrete compressive strength.
The number in the parenthesis indicates yield strength N/mm².

327

Photo 2.   Loading setup.

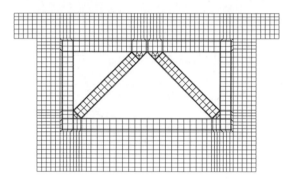

Figure 4.   Mesh (S1 and S3).

3.1 *Gridding*

The mesh of *model S1* and *S3* is shown in Figure 4. The high stiffness of the braces with the perimeter frame comparing to the bare frame caused large slippage at the horizontal joint. Because no shear transmission existed at the horizontal joint after cracking except the friction forces depending on the coefficient and the amount of the forces acting vertically to the joint. It is assumed that the shear capacity of the retrofitted frame used the anchorless method depends on the friction at the horizontal joint. To simulate the shear resisting mechanism of the specimen, a nonlinear finite element analysis applying monotonic loading is conducted under plane stress conditions. The essential information on the modeling is given below.

3.2 *Solution*

The column axial forces were applied at the first step and were kept constant afterwards. At the following steps, the lateral loads at the both ends of the girder were applied with the increment of 10 kN.

By applying the Newton method, the number of load steps required to minimize the work done by the unbalanced forces was determined. The number of iterations in each loading step was placed less than 100.

4 TEST AND ANALYSIS RESULTS

4.1 *Lateral load – deformation behavior*

The lateral load at the center of the girder was measured with two pressure gages provided with hydraulic jacks on the both ends of the girder. The story drift was measured with a displacement transducer at the center of the girder in mid span. Strain gages were attached on the reinforcement of the frame and the steel braces. Figure 5 shows the measured hysteresis curves of the lateral load – story drift.

For specimens S1 and S3, when the lateral load was about 400 kN at the first cycle, a bond slip was occurred at the connection between the perimeter frame's flange and column concrete. The stiffness of the specimen gradually decreased as the story drift increased, but the residual displacements was kept relatively small until a story drift was 1/200 rad. Then longitudinal bars at the end of the girder reached the yield stress and the plastic hinges were formed. Past a deformation angle of 1/160 rad, punching shear cracks were marked at the top portion of the column; even as loading progressed, strength did not increase.

Although the analytical initial stiffness was a bit lower than that in the experiment, good agreement was shown.

4.2 *Compression and shear stress of the grouted joint*

Figure 6 shows the story shear resistance mechanism. Figures 7 and 8 show the distribution of the compressive force and the shear force at the grouted joint of *model S3 respectively*.

The story shear force acting from the both ends of the beam is shared by the left and the right columns as well as the steel braces. Considering the equilibrium in the horizontal direction at the top of the column, it is assumed that the story shear force shared by steel braces is the summation of the compressive force reacting from the column and the shear friction force occurring at the bottom of the beam and the grouted mortar.

The compressive force is transmitted from the top portion of the left column to the steel braces, not more than the punching shear capacity of the column. And the shear friction force is transmitted depending on the coefficient of friction and the magnitude of the vertical shear acting the girder.

Distributions of these compressive forces and shear friction forces are shown in Figures 7 and 8.

328

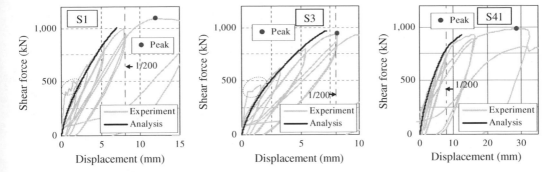

Figure 5. Force displacement hysteresis curve.

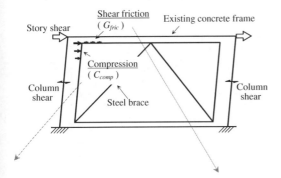

Figure 6. Story shear resistance mechanism.

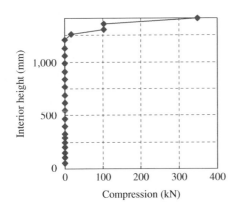

Figure 7. Compression of grout between the left column and steel brace on Model S3.

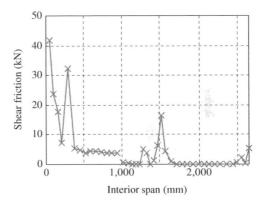

Figure 8. Shear friction of grout between the girder and steel brace on Model S3.

with grout stripped from the surface of the steel brace reaches zero.

The compressive force in the horizontal direction at the vertical joint of the left column is shown in Figure 8. The shear forces for portions with a compressive force of zero or less or with grout stripped from the surface of the steel brace reach zero in the same manner as portions with grout from the under side of the girder.

Some gaps occurred at the grouted joint, which meant the separation of the RC frame and the perimeter steel frame. The gaps cause the deterioration of the RC frame and make the deformation increase. These tendencies were observed in the experiments and analyses.

With regard to the right column, a lack of increase in story shear force borne by steel braces was observed in the test. This paper does not describe the compressive force and shear force (friction) for the portion with grout on the right–hand column.

Thus, compressive force transmitted by the top of the left column of the RC frame structure via portions with grout and shear force (friction) transmitted by the end of the under side of the RC frame structure's girder

Shear force carried by friction of grouted joint at the bottom of the girder are shown in Figure 9. The shear force is obtained by multiplying the compressive force in the vertical direction by a coefficient of friction of 0.8, so the distributions of the shear force and the compressive force are similar. The shear force for portions with a compressive force of zero or less or

329

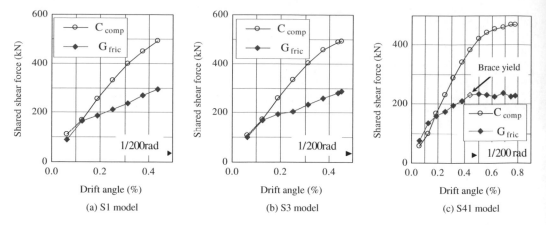

Figure 9.   Shared story shear force to compression and shear friction.

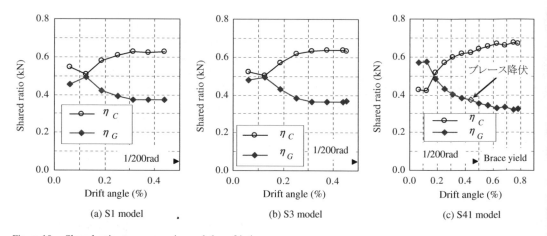

Figure 10.   Shared ratios to compression and shear friction.

via portions with grout were transmitted as the story shear force, which was borne by the steel braces.

4.3   Shared story shear force to compression and shear friction

Figures 9 (a) through (c) show shared story shear force to compression (C_{comp}) and shear friction (G_{fric}) for models *S1*, *S3*, and *S41*. In the analysis results, C_{comp} and G_{fric} increased as the story drift increased, but the increasing gradient of G_{fric} was less than that of C_{comp}.

4.4   Shared ratio to compression and shear friction

Figures 10 (a) through (c) show the comparison of the shared forces of C_{comp} to G_{fric} for models *S1*, *S3*, and *S41*. The shared story shear force ratios of C_{comp}

and G_{fric} to the story shear force shared of the brace (Q_{barace}) are shown below.

$$\eta_C = {C_{comp}} \big/ {Q_{brace}} \quad \dots\dots\dots \quad (1)$$

$$\eta_G = {G_{fric}} \big/ {Q_{brace}} \quad \dots\dots\dots \quad (2)$$

In the analysis results, η_C increased and η_G decreased, as the story drift increased.

4.5   Comparison of sheared story shared ratio to shear friction

Figure 11 shows a comparison of η_G of models *S1*, *S3*, and *S41*. In the analysis results, η_G for the three models showed a similar tendency of decreasing as the story drift increased.

330

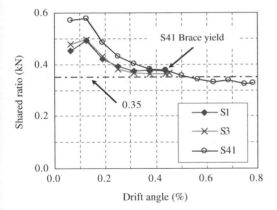

Figure 11. Shared ratios to shear friction.

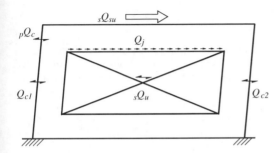

Figure 12. Story shear force resist mechanism in a conventional retrofitting method.

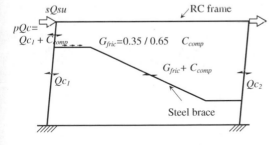

Figure 13. Story shear force resist mechanism in this paper.

In this analysis, η_G was expected to be 0.35 when the story drift angle was below about 0.5% before the brace yielded in compression.

4.6 Proposal of a lateral strength evaluation

Figure 12 shows the story shear force resistance mechanism in a conventional retrofitting method. In the Japanese seismic retrofitting guidelines[6], the lateral strength of a frame retrofitted with steel braces in a

conventional method using many anchor bolts is estimated considering story shear force resistance using the following equation.

$$_sQ_{su} = Min\left(_sQ_u + Q_{C1} + Q_{C2} \;,\;\; Q_j + _pQ_C + Q_{C2} \right) \cdots \cdots \quad (3)$$

where $_sQ_{su}$ = the lateral strength of the frame retrofitted with steel braces, Q_j = the shear strength of the joint at the bottom of the girder, $_sQ_u$ = the lateral strength of the steel frame, Q_{C1} = the ultimate strength of the left column, Q_{C2} = the ultimate strength of the right column, and $_pQ_C$ = the punching shear strength (Japan Building Disaster Prevention Association 2006) at the top portion of the left column.

The lateral strength of a frame retrofitted without adhesive anchor bolts cannot be evaluated using Eq. (3), because of the difference in the story shear force resistance mechanism.

Based on the above results, we propose the lateral strength evaluation method for the specimens below.

Figure 13 shows the story shear force resistance mechanism in the anchorless retrofitting method.

If η_G is 0.35, the lateral strength of a RC frame retrofitted with steel braces without adhesive anchor bolts at the grouted joint is shown below.

Q_j and C_{comp} are defined as

$$Q_j = G_{fric} = \frac{0.35}{0.65} \cdot C_{comp} \quad \cdots \cdots \cdots \quad (4)$$

$$C_{comp} = _pQ_C \quad \cdots \cdots \cdots \cdots \cdots \cdots \cdots \quad (5)$$

Using Eq. (3), (4) and (5), $_sQ_{SU}$ is given as below.

$$_sQ_{su} = 1.53 \cdot _pQ_C + Q_{C2} \quad \cdots \cdots \cdots \cdots \quad (6)$$

Q_{C2} was estimated using the equation below, because the right-hand columns of specimens S1, S3, and S41 failed due to shear force.

$$Q_{C2} = \left\{ \frac{0.053 \cdot p_t^{0.23} \cdot (18 + \sigma_B)}{M/(Q \cdot d) + 0.12} + 0.85\sqrt{p_w \cdot _s\sigma_{wy}} + 0.1 \cdot \sigma_0 \right\} \cdot b \cdot j \quad \cdots \cdots \quad (7)$$

where p_t is the ratio of the tensile longitudinal bar area, p_w is the ratio of the transverse reinforcement bar area, $_s\sigma_{wy}$ is the yield stress of the transverse reinforcement bar, σ_0 is the axial stress of the column, $M/(Q \cdot Ed)$ is the shear span ratio, d is the effective depth of the girder, σ_B is the concrete strength, b is the column width, and j is 0.875d.

Table 2 shows past experiment and calculation results (Kanata et al, 2003, Kanata et al. 2000, Kanata et at. 2002, Toshiharu 2006). These have been cited and compared. Figure 14 shows the comparison of experiment and calculation results for frame lateral strength. The ratios of the experimental results to the calculation results were from 1.17 to 1.73. The average was 1.33, showing that a safety allowance was present.

Table 2. Measured and calculated results.

Specimen	$pQ_{c \cdot cal}$ (kN)	Q_{c2} (kN)	bQ_j (kN)	G_{fric} (kN)	$sQ_{su \cdot cal}$ (kN)	$sQ_{su \cdot EXP}$ (kN)	$\dfrac{sQ_{su \cdot EXP}}{sQ_{su \cdot cal}}$	Ref.
S1	484	168		257	909	1,089	1.20	
S2	484	168	424		1,076	1,345	1.25	7
S3	392	158		208	758	965	1.27	
S10	510	164	166	270	1,110	1,566	1.41	8
S11	456	154	130	242	982	1,419	1.45	8
S12	470	157			627	1,086	1.73	8
S13	443	160	130	235	968	1,360	1.41	8
S20	439	151	332	233	1,155	1,480	1.28	2
S21	411	145	166	218	940	1,335	1.42	2
S22	364	124	166	193	847	1,193	1.41	2
S41	449	152		238	839	982	1.17	
S51	429	144	79	227	879	935	1.20*	9
S52	443	146		235	824	1,037	1.26	9

G_{fric} is $0.53 \times pQ_c$.
*: Calculated from Eq. (1).

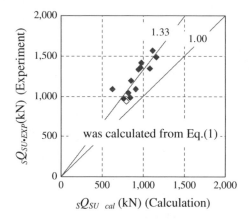

Figure 14. Lateral strength comparison of measured and calculated results.

5 SUMMARY AND CONCLUSIONS

RC framed specimens applying the anchorless method were tested and simulation was conducted using FEM analysis. The specimens represented typical substandard structures built before 1981.

This paper discussed the focus of the story shear resistance mechanism of a specimen retrofitted without adhesive anchor bolts. The following conclusions can be drawn.

1. Experimental and analytical results were compared for the force displacement hysteresis curve, and although the initial stiffness was lower in the analysis than in the experiment, for the most part good agreement was shown.

2. Seismic retrofitting with steel braces without adhesive anchors improved the seismic performance of the specimens, because story shear force was adequately conveyed to the steel braces by compression and shear friction via grout.

3. We have proposed a lateral strength evaluation method for a frame retrofitted with steel braces without adhesive anchors.

ACKNOWLEDGMENTS

The technical staff of Taisei is thanked for their support and many helpful comments in the course of this study. The support of JIP Techno Science staff is deeply appreciated. The analysis in this paper was sponsored by the Musashi Institute of Technology.

REFERENCES

Kazuhiro Kanata, Kenichi Kikuchi, "Strengthening Methods for R/C Frames with Infilled Panels by Using Friction at the Panel-Frame Joint", Proceedings of the Japan Concrete Institute, Vol. 22, No. 3, 2000, pp 1669–1674 (in Japanese).

Kazuhiro Kanata, Kenichi Kikuchi,, "Retrofitting methods for exsiting frames to enhance strength by using friction at steel-concrete joints", Proceedings of JCI symposium on evaluation of the effect of seismic retrofit of existing concrete structures, June 2000, pp 553–560 (in Japanese).

Kazuhiro Kanata, Maezawa Sumio, Kenichi Kikuchi, Tasirou Fujimura, "Evaluation for strengthened frames with Infilled Panels by Using Friction at the Panel-Frame Joint", Proceedings of the Japan Concrete Institute, Vol. 25, No. 2, 2003, pp 1483–1488. (in Japanese).

Kazuhiro Kanata, Maezawa Sumio, Kenichi Kikuchi, Tasirou Fujimura, "Strengthening Methods for R/C Frames with

Infilled Panels by using Friction at the Panel-Frame Joint (Part 4–6)", Summaries of technical papers of Annual Meeting Architectural Institute of Japan. C-2, Structures IV, 2002, pp 707–712 (in Japanese).

Nakamura Toshiharu, Kazuhiro Kanata, Nishikawa Yasuhiro, Suzuki Yumi, Tasirou Fujimura, "Strengthening Methods for R/C Frames with Infilled Panels by using Friction at the Panel-Frame Joint (Part 14-15)", Summaries of technical papers of Annual Meeting Architectural Institute of Japan. C-2, Structures IV, 2006, pp 593–596 (in Japanese).

Takahiro Kei, Toshihiro Kusunoki, Katsuyoshi Konami, Masaru Fujimura, "Experimental Study on Reinforced Concrete Infilled Shear Walls using Epoxy Resin", Summaries of technical papers of Annual Meeting Architectural Institute of Japan. C-2, Structures IV, pp 393–394, 31 July, 2000 (in Japanese).

Tetsuya Ohmura, Shizuo Hayashi, Kazuhiro Kanata, Tashiro Fujimura, "Seismic Retrofit of Reinforced Concrete Frames by Steel Braces Using No Anchors", 100th Anniversary Earthquake Conference, DRC-000041(CD-ROM), 18, April, 2006.

The Japan Building Disaster Prevention Association, "Seismic retrofit design guidelines and comment for existing reinforced concrete buildings", 2001 (in Japanese).

Youhei Kuse, Toshihiko Yamamoto, Kazuo Yamada, Takashi Kamiya, Yukio Ban, Youichi Ueda, "Experimental Study on No Anchor Seismic Retrofit of Fiber-Reinforced Concrete Wall (Part1 Experiment by 1-story 1-span Extension Walls Specimens The Outline and Results and Part2 Consideration of Adhesive Joint)", Summaries of technical papers of Annual Meeting Architectural Institute of Japan. C-2, Structures IV, pp 545–548, 31 July, 2006 (in Japanese)

Excellence in Concrete Construction through Innovation – Limbachiya & Kew (eds)
© 2009 Taylor & Francis Group, London, ISBN 978-0-415-47592-1

Assessment of deformation capacity of reinforced concrete columns

M. Barghi & S. Youssefi

K.N. Toosi University of Science and Technology, Tehran, Iran

ABSTRACT: The current seismic design philosophy allows nonlinear behavior of the structure to reduce strength requirements. This implies that the structure must be able to retain integrity under cycles of displacement in to nonlinear range. The recent move toward displacement-based design and assessment methods requires a model for the drift beyond which failure is expected. In this study by using the results of 253 cyclic loading tests on columns, a new model for estimation drift capacity has been presented. This proposed model can be used to estimate drift capacity of reinforced concrete columns with shear, shear flexural and flexural failure separately. Uncertainty of the results of this model is little because of the study has been concentrated on the three independent groups of the columns with one failure mode. The results of cyclic loading tests used in this study have been taken from PEER center database.

1 INTRODUCTION

It has been well established by experimental evidence that many existing reinforced concrete columns are vulnerable to shear failure after flexural yielding (Kokusho 1964, Ikeda 1968). Several models have been developed to represent the degradation of shear strength with increasing inelastic deformations (Priestley et al.1994, Sezen 2002). While these shear strength models are useful for estimating the column capacity for conventional strength-based designed assessment, the recent move toward displacement-based design and assessment methods (ATC 1996, ASCE 2000) requires a model for the drift beyond which shear failure is expected. Furthermore, after flexural yielding, the force demand on a column will be approximately constant, while the displacement demand will increase substantially, suggesting that a drift capacity model is more useful for columns experiencing flexural-shear and flexural failures. Such as these considered in this study, although the shear strength models relate the degradation of shear strength to displacement ductility, these models may not be appropriate for assessing the drift at shear failure. Most models for estimating the drift capacity of reinforced concrete columns are based on performance of columns with good seismic detailing. Such models assume the response is dominated by flexural deformations and use estimates of the ultimate concrete and steel strains to determine the ultimate curvatures that the section can withstand. These models are not applicable to older reinforced concrete columns with limited transverse reinforcements since the degradation of the shear strength begins before the flexural deformation capacity can be achieved. Furthermore, the calculation of ultimate strains assumes good crack control, provided by reasonably distributed reinforcement, such that deformation can be averaged over finite distances. Experimental studies and earthquake reconnaissance have shown, however, that the shear failure of older reinforced concrete columns often is associated with concentrated deformations along a limited number of primary cracks (Pantazopoulou, 2003). Hence, such models based on flexural performance can not be used.

In developing the necessary drift function, extensive test data were used. In the PEER database, 253 cyclic test data of reinforced concrete columns has been reported (PEER 1999). This database has been used in a systematic regression analysis. The pertinent data were all for square or rectangular sections.

2 DESCRIPTIONS OF CYCLIC LATERAL LOAD TESTS ON RECTANGULAR REINFORCED CONCRETE COLUMNS (PEER1999)

Three types of specimens were used in this study: double curvature specimens (Fig. 1-a), simple cantilever specimens (Fig. 1-b), and double-ended specimens (Fig. 1-c). To present the data from these three types of specimens in a uniform format, all data is reported

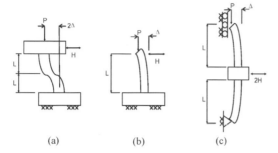

Figure 1. Different supports of columns.

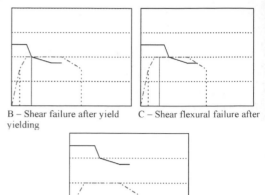

B – Shear failure after yield   C – Shear flexural failure after yielding

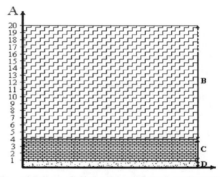

D – Flexural failure

Figure 2. Cyclic behavior of single reinforced concrete column.

in terms of an equivalent simple cantilever column. Thus, for cantilever specimens, the forces and deflections reported are simply the same as those reported by the original researchers. For double ended specimens, the lateral forces reported are one-half the forces applied to the central stub of the specimen, and the lateral deflections reported are the lateral deflections of the central stub relative to the two fixed end points. For double curvature specimens, the lateral forces reported are those applied to each end beam of the specimen, and the lateral deflections reported is one-half the relative lateral displacements of the two end beams. The reported length "L" of each specimen is shown in Fig. 1.

3 PROPOSED CLASSIFICATION OF FAILURE MODES OF SINGLE REINFORCED CONCRETE COLUMNS

At the beginning, a new method has been proposed for prediction of failure modes of single reinforced concrete columns (Barghi 2005).This method has been proposed based on observations from the extensive experimental data base. In this method, failure modes of columns are classified and predicted as flexural failure, shear failure after yielding of longitudinal reinforcement and shear failure based on the amount of "A" parameter. Differences between failure modes are presented by comparing envelope P-δ curve (dashed line) and the degrading capacity of shear strength along the deformational history (Fig. 2). Procedure to determine "A" parameter can be expressed as follows:

If initial shear strength before loading, V_n, flexural capacity, M_n, and shear span ratio, l/d of a column be determined, for prediction the failure mode, "A" parameter can be computed by equation 2 and the failure mode may be predicted based on the proposed limits that can be observed in Fig. 3.

$$A = \frac{V_n}{2\frac{M_n}{l}} \times \frac{l}{d} \qquad (2)$$

Figure 3. Proposed limits of a parameter for prediction failure mode of columns.

B-Flexural Failure   C-Shear-Flexural Failure   D-Shear Failure

4 PROPOSED EMPERICAL DRIFT CAPACITY MODELS

This paper introduces an empirical model based on observations from the extensive experimental data base. The goal of developing a new model is to reduce the coefficient of variation and provide a simple relationship that identifies the critical parameters influencing the drift at shear, shear flexural and flexural failure of columns. The coefficients in proposed equations were chosen on a least-squares fit to the data. The results from the database suggest that for columns with low levels of transverse reinforcement an increase in the axial load ratio tends results in a decrease in the drift ratio at failure. The influence of axial load on the drift ratio was incorporate in to the empirical model by including the variable $P/A_g f_c'$. These models were developed based on Elwood model for drift

336

ratio at shear failure (Elwood 2002). Based on these observations, the following empirical expressions are proposed to estimate the drift ratio of reinforced concrete columns.

Shear failure:

$$\rho'' \leq 0/004 \quad \frac{\delta}{l} = \frac{16}{500} + 4\rho'' - \frac{12}{500} \times \frac{v}{\sqrt{f'c}} - \frac{1}{40} \times \frac{P}{Agf'c} \geq \frac{1}{100} \quad (4)$$

$$\rho'' > 0/004 \quad \frac{\delta s}{l} = \frac{1}{30} + 5\rho'' - \frac{48}{1000} \times \frac{v}{\sqrt{f'c}} \geq \frac{1}{100} \quad (5)$$

Shear-Flexural failure:

$$\rho'' \leq 0/004 \quad \frac{\delta}{l} = \frac{45}{1000} + 4\rho'' - \frac{12}{500} \times \frac{v}{\sqrt{f'c}} - \frac{1}{40} \times \frac{P}{Agf'c} \geq \frac{2}{100} \quad (6)$$

$$\rho'' > 0/004 \quad \frac{\delta s}{l} = \frac{35}{1000} + 5\rho'' - \frac{48}{1000} \times \frac{v}{\sqrt{f'c}} \geq \frac{2}{100} \quad (7)$$

Flexural failure:

$$\rho'' \leq 0/004 \quad \frac{\delta}{l} = \frac{25}{1000} + 9\rho'' - \frac{1}{100} \times \frac{v}{\sqrt{f'c}} - \frac{1}{29} \times \frac{P}{Agf'_c} \geq \frac{2}{100} \quad (8)$$

$$\rho'' < 0/004 \quad \frac{\delta s}{l} = \frac{45}{1000} + 1.5\rho'' - \frac{15}{1000} \times \frac{v}{\sqrt{f'_c}} \geq \frac{2}{100} \quad (9)$$

In which $\rho'' =$ transverse reinforcement ratio; $= \frac{v}{\sqrt{f'_c}}$ maximum shear stress. $P/A_g f'_c =$ normalized axial load. These proposed equations for drift ratio are used in SI units.

5 EVALUATIONS OF PROPOSED DRIFT CAPACITY MODELS

Proposed model has been evaluated based on equation 3.

$$DI = \delta_m \Big/ \delta_s \quad (3)$$

which $DI =$ measured drift ratio divided by the calculated drift ratio $\delta_m =$ measured drift ratio at the failure of the column $\delta_s =$ drif capacity of the column calculated from proposed relationships. Drift of 253 columns under cyclic loading have been measured and calculated based on proposed equations. The mean of the measured drift ratio at failure divided by the drift capacity calculated according to proposed equations are 1.06, 1.05, 1.03 respectively and the coefficients of variation are 0.24, 0.31, and 0.30. The results are shown in Fig. 4. This model also can be used to predict failure mode of the columns.

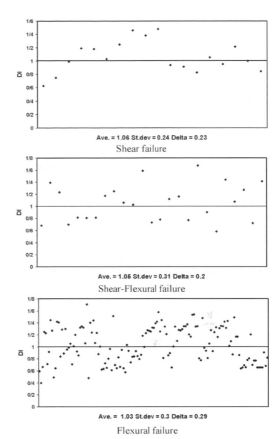

Ave. = 1.06 St.dev = 0.24 Delta = 0.23
Shear failure

Ave. = 1.05 St.dev = 0.31 Delta = 0.2
Shear-Flexural failure

Ave. = 1.03 St.dev = 0.3 Delta = 0.29
Flexural failure

Figure 4. Evaluation of proposed drift capacity models.

6 CONCLUSIONS

At the beginning, for prediction failure mode of the columns under cyclic loading, a new method has been proposed. Then all studies have been continued on three independent groups of the results of 253 cyclic loading tests of reinforced concrete columns. Then, the Elwood model for estimation maximum drift capacity of the columns has been developed. This proposed model can be used to estimate drift capacity of the reinforced concrete columns at shear, shear flexural and flexural failure separately and adequate accurately. Uncertainty of the results of this model is little because of the studies have been concentrated on the three independent groups of the columns with one failure mode and all critical parameters influencing the drift at failure were considered.

REFERENCES

Barghi M. 2005. Shear failure and damage index of reinforced concrete columns, *PhD dissertation, Department of Civil Engineering, Iran University of Science and Technology.*

Elwood 2002. *PhD dissertation, chapter 2.pdf, Berkeley Edu.*

FEMA273. 1997. NEHRP. Guidelines for the seismic rehabilitation of buildings, *Federal Emergency Management Agency, Washington DC.*

Ikeda, A.1968. Report of the Training institute for engineering teachers, *Yokohama National University, Japan.*

Kokusho, S. 1964. Report by Building Research Institute, *Building Research Institute, Tsukuba, Japan.*

Lynn, A. & Moehle, J. P 1996. Seismic evaluation of existing reinforced concrete, *Journal of Prestressed concrete Institute, V. 17.*

Maekawa, K. 1998. Shear resistance and ductility of RC columns after yield of main reinforcement. *Journal of materials, Concrete structures and pavements, JSCE, No.585/V- 38,233-247.*

Preistley et al. 1994. Seismic shear strength of reinforced concrete columns, *Journal of the Structural Division, ASCE, Vol. 120, No. ST8.*

Researches at the University of Washington, 1999. Column Data Base, *National, and Science Foundation Pacific Earthquake Engineering Research Center (PEER).*

Williams et al.1995. Seismic damage indices for concrete structures, a state-of-the-art review. *Earthquake Spectra, Earthquake Engineering Research Institute, Vol. 11, No. 2, pp. 319–349.*

Williams et al. 1997. Evaluation of seismic damage indices for concrete element loaded in combined shear and flexure *ACI Structural Journal, V.94.*

Young-Ji et al. 1985. Mechanistic seismic damage model for reinforced concrete, *Journal of Structural Engineering Vol. 111, No.4. ASCE.*

Excellence in Concrete Construction through Innovation – Limbachiya & Kew (eds)
© 2009 Taylor & Francis Group, London, ISBN 978-0-415-47592-1

Ductility of confined reinforced concrete columns with welded reinforcement grids

Tavio, P. Suprobo & B. Kusuma

Department of Civil Engineering, Sepuluh Nopember Institute of Technology (ITS), Surabaya, Indonesia

ABSTRACT: Concepts in ductile design have led to an increased interest in understanding the role of confinement in improving the seismic performance of reinforced concrete members. Confinement models have been developed by numerous researchers to describe the stress-strain behavior of concrete as a function of certain key parameters that are related to the amount and type of transverse reinforcement. Accurate constitutive models of confined concrete are necessary for use in a computer-based program developed for analysis of reinforced concrete columns subjected to a combination of flexural and axial loadings. This paper presents an analytical approach for computing the moment-curvature response of a section under various load levels by using an idealized stress-strain relationship for concrete confined by welded reinforcement grids. The predictions were then compared with the test results for the corresponding loading levels. Overall, it is found that the predictions for the moments were found to be underestimated, and vice versa for the curvature capacities.

1 INTRODUCTION

Prediction of the performance of a structure during its life-time and its behavior under various loading conditions is a key element in "performance-based design" as the future design methodology, especially for reinforced concrete structures. In performance-based seismic design, evaluation of the deformation capacity of reinforced concrete columns is of paramount importance. The deformation capacity of a column can be expressed in several different ways: (1) curvature ductility, (2) displacement ductility, or (3) drift.

Use of confining steel in the critical regions of columns designed for earthquake resistance is a common way of achieving ductile structural behavior. The lateral steel, in conjunction with the longitudinal steel, affects the concrete properties significantly depending on several factors, which includes distribution of steel including spacing of longitudinal and lateral steel, amount of lateral steel, and the type of anchorage of lateral steel.

The ductility of a reinforced concrete member is normally determined from the moment-curvature relationship. In the development of the analytical moment-curvature relationship, an appropriate constitutive model for confined concrete is of predominant importance. Hence, the analytical model of confined concrete has to be properly selected in order to provide an accurate prediction on the actual column response and ductility.

The goal of this study is limited to exploring possible analytical models of confined concrete which are capable of simulating the actual performance of a reinforced concrete column confined by welded reinforcement grid. The analytical work in the study is based on the moment-curvature analysis of a column section under various loading conditions based on the selected confined concrete models.

The research reported in this paper is a follow-up of the experimental work reported earlier by Saatcioglu & Grira (1996, 1999). The analytical models adopted in the study for confined concrete columns are available in the literature (Modified Kent & Park, 1982; Modified Sheikh & Uzumeri, 1986; Mander et al., 1988; Legeron & Paultre, 2003). The accuracy of the analytical models in predicting the actual moment-curvature response of a concrete column section confined by welded reinforcement grid under various loading stages was investigated.

The analytical models were implemented in a computer program developed for flexural analysis of a reinforced concrete column section. The results obtained from the selected analytical models were then compared to each other and validated against the experimental data of ten large-scale reinforced concrete square columns. Overall, the predictions on the moments in the moment-curvature responses of concrete columns confined by welded reinforcement grids were found to be underestimated, and vice versa for the curvature capacities.

2 DEFINITIONS OF DUCTILITY PARAMETERS

The required ductility of a structure, element or section can be expressed in terms of the maximum imposed deformation. Often it is convenient to express the maximum deformations in terms of ductility factors, where the ductility factor is defined as the maximum deformation divided by the corresponding deformation when yielding occurs. The use of ductility factors permits the maximum deformations to be expressed in non-dimensional terms as indices of post-elastic deformation for design and analysis. Ductility factors have been commonly expressed in terms of various parameters related to deformations, namely displacements, rotation, curvatures, and strains.

Several definitions for ductility have been proposed in the literature, since the actual behavior of confined reinforced concrete sections or members is not elastic-perfectly plastic. In the study, the ductility parameters are defined according to the definition introduced by Bae & Bayrak (2006). The ductility parameters are defined using the idealized backbone curves shown in Figure 1. Definition of the yield curvature (or displacement) can be defined as the straight line joining the origin and a point on the ascending branch (where $M = 0.75M_{max}$ or $V = 0.75V_{max}$). The line passes through the moment-curvature (or the load-displacement) curve at 75% of the maximum moment (or load) and then reaches the maximum moment (or load) to define the idealized yield curvature ϕ_1 (or yield displacement Δ_1), as shown in Figure 1. Failure of the column is conventionally defined when the post-peak curvature ϕ_2 (or post-peak displacement Δ_2)

reaches a point at which the remaining column strength has dropped to 80% of the maximum moment (or load). The curvature and displacement ductility factor are defined as follows:

$$\mu_\phi = \frac{\phi_2}{\phi_1} \tag{1}$$

$$\mu_\Delta = \frac{\Delta_2}{\Delta_1} \tag{2}$$

3 ANALYTICAL MODELS

Several models for the compressive stress-strain behavior of concrete confined by rectangular hoops have been proposed in recent years. These curves have generally been determined from uniformly loaded specimen, and give complete stress-strain curves up to very high concrete strains from which the concrete compressive stress block parameters at any extreme fiber concrete strain in a member may be calculated (Park and Paulay, 1975).

The stress-strain models suggested by Modified Kent & Park (MKP) 1982, Modified Sheikh & Uzumeri (MSU) 1986, Mander, Priestley & Park (MPP) 1988, Legeron & Paultre (L&P) 2003, were adopted in the study to calculate the theoretical flexural strength of the test columns. Details of these stress-strain models may be seen in the literature and will not be elaborated in detail here, but all take into account the increase in concrete ductility due to the confinement from lateral steels.

All of the confined concrete models studied here, were implanted in the program developed particularly for predicting the actual moment-curvature responses of the columns confined by welded reinforcement grids for each of the experimental loading cases.

4 APPLICATION OF ANALYTICAL MODELS

A computer program was developed to carry out the calculations for the theoretical moment-curvature relations of the test column specimens using the concrete stress-strain curves from the four selected analytical models. The required input data included cross-sectional dimensions of specimens, position, and amount of longitudinal steel including the location of laterally supported longitudinal bars, properties of longitudinal and transversal steels, stress in tie steel at the maximum moment, unconfined concrete strength (f_c'), and applied axial load. The section was divided into small elements, each one containing two kinds of elements, concrete core and cover.

The analysis procedure involved following steps:
(1) Assign a value to the extreme concrete compressive

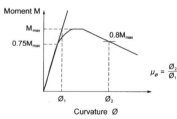

(a) Curvature ductility parameter

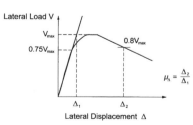

(b) Displacement ductility parameter

Figure 1.   Definitions of ductility parameters.

fiber strain (normally starting with a very small value); (2) assume a neutral axis depth; (3) calculate the strain and the corresponding stress at the centroid of each longitudinal reinforcement bar; (4) use appropriate stress-strain models for confined and unconfined concrete to determine stress values; (5) calculate axial force from equilibrium and compare with the applied force. If the difference lies within the specified tolerance, the assumed neutral axis depth is adopted. The moment capacity and the corresponding curvature of the section are then calculated. Otherwise, a new neutral axis depth is determined from the iteration (using secant method) and steps 3 to 5 are repeated until it converges; (6) set new concrete strain and return to step 2; (7) Repeat the whole procedure until the complete moment-curvature curve is obtained.

In the application of the analytical models, a few assumptions were needed to be made and these are explained here. In the first two models (Modified Kent & Park & Mander et al.), the lateral steel stress was suggested to be equal to the yield stress. The adoption of yield stress in these models has resulted in unconservative predictions of test results in general. To accommodate this observation, tie stress at the maximum moment was used in the next two models as suggested by Sheikh & Uzumeri (1982) and Legeron & Paultre (2003). In the application of the models, the unconfined concrete curve suggested by Kent and Park (1971) was adopted.

5 COMPARISON OF ANALYTICAL AND EXPERIMENTAL RESULTS

Figs. 2 to 11 show the comparisons between the moment-curvature relations for ten representative specimens (Saatcioglu and Grira, 1996, 1999), and those from the numerical analyses of the four selected models discussed previously. Ten large-scale columns of 350×350 mm in cross section and 1645 mm long were tested under simulated seismic loading. The columns were subjected to concentric compression of approximately 20 and 40 percent of their capacities, while subjected to incrementally increasing lateral deformation reversals. All models predict the moment capacities for ten specimens fairly well, whereas the predicted curvature capacities are not consistent.

The model by Modified Kent & Park (1982) consistently overestimated the moment capacities of the sections. The difference between test and analytical moment capacities was found to be about 13 percent, whereas curvature ductility was approximately

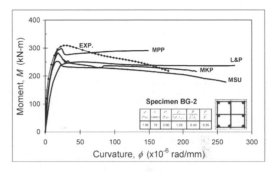

Figure 3. Comparison of Experimental and Analytical Curves for Specimen BG-2.

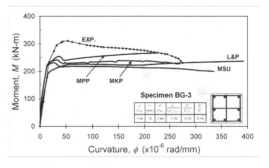

Figure 4. Comparison of Experimental and Analytical Curves for Specimen BG-3.

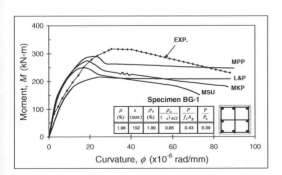

Figure 2. Comparison of Experimental and Analytical Curves for Specimen BG-1.

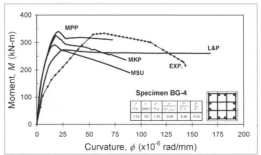

Figure 5. Comparison of Experimental and Analytical Curves for Specimen BG-4.

341

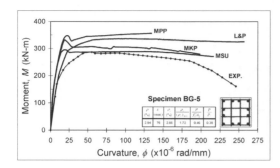

Figure 6. Comparison of Experimental and Analytical Curves for Specimen BG-5.

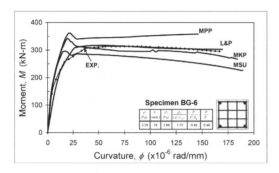

Figure 7. Comparison of Experimental and Analytical Curves for Specimen BG-6.

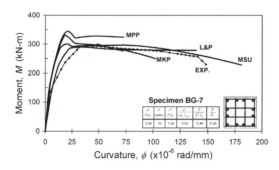

Figure 8. Comparison of Experimental and Analytical Curves for Specimen BG-7.

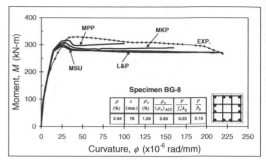

Figure 9. Comparison of Experimental and Analytical Curves for Specimen BG-8.

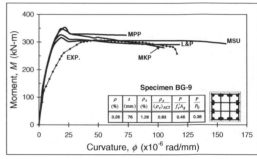

Figure 10. Comparison of Experimental and Analytical Curves for Specimen BG-9.

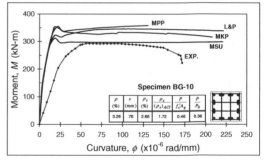

Figure 11. Comparison of Experimental and Analytical Curves for Specimen BG-10.

46 percent for most of the specimens. The Mander et al. (1988) also overestimated the flexural capacity in almost all the columns to the same extent as MKP model did, but the predicted curvature ductility overestimated of approximately 80 percent. This difference is even higher for specimens under moderate axial loading with high longitudinal reinforcement ratio of above 2 percent. For well-confined concrete columns tested under low axial loading, the predictions from these two models, although mostly unconservative, are reasonable (see Figs. 4 and 9).

Analytical results from Modified Sheikh and Uzumeri model have underestimated the sectional capacities for specimens under low to moderate axial load levels. The difference between the experimental and the computed moment and curvature capacities was found to be about 11 and 51 percent respectively, however, the moment capacities of the section for most columns are much smaller than those obtained from Modified Kent and Park, Mander et al., and Legeron and Paultre models. Legeron and Paultre model gives

Table 1. Comparison of Test Results with the Values Predicted for Strength and Ductility Capacities.

Specimen	$\frac{P}{P_0}$	ρ_s (%)	$\frac{\rho_s}{(\rho_s)_{ACI}}$	$(M_{exp})_{Max}$ kNm	Ductility factor		Theoretical moments, M_{th}				Theoretical curvature ductility			
					μ_ϕ^{**} $0.8M_{max}$	μ_Δ^{***} $0.8V_{Max}$	MKP M_{exp}/M_{th}	MSU M_{exp}/M_{th}	MPP M_{exp}/M_{th}	L&P M_{exp}/M_{th}	MKP μ_ϕ	MSU μ_ϕ	MPP μ_ϕ	L&P μ_ϕ
BG-1	0.39	1.00	0.65	317	3.6	3.5	1.16	1.27	1.10	1.47	4.5	3.3	9.4	4.4
BG-2	0.39	2.00	1.29	309	12.7	4.2	1.10	1.22	1.04	1.24	9.5	12.4	15.8	14.0
BG-3	0.18	2.00	1.29	310	11.1	5.0	1.29	1.35	1.16	1.32	5.1	24.2	15.6	27.4
BG-4	0.38	1.33	0.86	333	2.8	6.6	1.03	1.15	0.98	1.22	1.5	3.7	7.9	8.0
BG-5	0.38	2.66	1.72	287	9.6	5.0	0.86	0.97	0.81	0.85	3.3	9.7	14.5	12.4
BG-6	0.40	2.66	1.72	314	9.3	5.6	0.92	1.06	0.87	0.99	9.0	8.4	15.9	8.3
BG-7	0.38	1.26	0.83	293	6.9	4.9	0.89	0.98	0.85	1.01	3.1	14.7	7.7	6.9
BG-8	0.19	1.26	0.83	329	10.1	3.6	1.07	1.13	1.04	1.10	4.9	6.6	7.7	14.3
BG-9	0.38	1.26	0.83	295	5.3	4.9	0.85	0.94	0.84	0.91	1.0	8.4	7.7	6.6
BG-10	0.38	2.66	1.72	293	5.0	5.9	0.83	0.94	0.82	0.85	2.0	11.6	14.1	11.9

* Moments measured at peaks of moment-rotation relationship for columns tested by Saatcioglu and Grira (1996, 1999).
** Curvature ductility measured based Eq. (1) proposed by Bae and Bayrak (2006).
*** Displacement ductility measured based Eq. (2) proposed by Bae and Bayrak (2006).

the moment capacities slightly higher than those of the aforementioned three models by 18 percent.

Table 1 and Figs. 2 to 11 shows a comparison of the experimental data with four analytical recently developed confinement models, namely, Modified Kent & Park (1982), Modified Sheikh & Uzumeri (1986), Mander et al. (1988), and Legeron & Paultre (2003). It is clear that the all models considered in the study are not capable of predicting the curvature ductility with reasonable accuracy. This was expected because such models were developed for tied and/or spiral columns, which are not as ductile as columns that transversely reinforced with welded reinforcement grids.

Figs. 2 to 11 indicate that the MKP model can predict the experimental results more accurately than other models, particularly in terms of moment capacities of the moment-curvature curves. From the figures, it can also be seen that the prediction of MKP model is slightly better than the remaining models, especially in the curvature ductility.

In general, the analytical models, which were derived from concrete columns confined by traditional hoop or crossties resulted in inaccurate predictions on the actual moment-curvature relationship for columns confined by welded reinforcement grids. Therefore, a more proper analytical stress-strain relationship for concrete column confined by welded reinforcement grid is needed in order to predict the actual moment-curvature relationships under axial load and flexure quite well.

6 SUMMARY AND CONCLUSIONS

The analytical models proposed by several researchers for concrete columns confined by rectangular hoop and/or crossties were adopted in the analytical study to predict the actual moment-curvature relationships of column specimens confined by welded reinforcement grids (Saatcioglu & Grira, 1996, 1999). Most of the models resulted in inaccurate predictions because they did not consider all the key parameters in the reinforced concrete columns confined by welded reinforcement grids and such models were developed mainly for tied and/or spiral columns, which are not as ductile as the reinforced concrete columns that are transversely reinforced with welded reinforcement grids.

Although the original models predict the moment-curvature relationships of the confined concrete sections under a combination of axial and flexural loadings quite well, the modified models resulted in more accurate representations of the experimental results.

REFERENCES

Bae, S., and Bayrak, O., "Performance-Based Design of Confining Reinforcement: Research and Seismic Design

Provisions," *International Symposium on Confined Concrete*, ACI SP-238, China, 2006, pp. 43–62.

Grira, M., and Saatcioglu, M., "Concrete Columns Confined with Welded Reinforcement Grids," *Research Report* OCEERC 96-05, Ottawa Carleton Earthquake Engineering Research Center, Department of Civil Engineering, University of Ottawa, Sept. 1996, pp. 89.

Kent, D. C., and Park, R., "Flexural Members with Confined Concrete," *Journal of the Structural Division*, Proceedings, ASCE, V. 97, No. ST7, July 1971, pp. 1969–1990.

Legeron, F., and Paultre, P., "Uniaxial Confinement Model for Normal-and High-Strength Concrete Columns," *Journal of Structural Engineering*, Proceedings, ASCE, V. 129, No. 2, Feb. 2003, pp. 241–252.

Mander, J. B., Priestley, M. J. N., and Park, R., "Theoretical Stress-Strain Model for Confined Concrete," *Journal*

Structural Engineering, ASCE, V. 114, No. 8, Aug. 1988, pp. 1804–1826.

Park, R., and Paulay, T., "Reinforced Concrete Structures," John Wiley and Sons, Inc., New York, NY, 1975, 769 pp.

Park, R., Priestley, M. J. N., and Gill, W. D., "Ductility of Square-Confined Concrete Columns," *Journal of the Structural Division*, Proceedings, ASCE, V. 108, No. ST4, Apr. 1982, pp. 929–950.

Saatcioglu, M., and Grira, M., "Confinement of Reinforced Concrete Columns with Welded Reinforcement Grids," *ACI Structural Journal*, V. 96, No. 1, Jan.–Feb. 1999, pp. 29–39.

Sheikh, S. A., and Yeh, C. C., "Flexural Behavior of Confined Concrete Columns," *ACI Structural Journal*, V. 83, No. 3, May–June 1986, pp. 389–404.

Excellence in Concrete Construction through Innovation – Limbachiya & Kew (eds)
© 2009 Taylor & Francis Group, London, ISBN 978-0-415-47592-1

On the effect of FRP sheet composite anchorage to flexural behavior of reinforced concrete beams

C.B. Demakos & G. Dimitrakis

Department of Civil Engineering, Technological Education Institute of Piraeus, Greece

ABSTRACT: The effect of a composite material anchorage system upon the ductility and load capacity of reinforced concrete (RC) beams strengthened with carbon (CFRP) fabrics were investigated in this paper. Experimental results were obtained for three groups of RC beams, one with unstrengthened (control) RC beams and three other of RC beams strengthened with one CFRP bonded fabric, which are subject to four- point bending. A single ply of a CFRP fabric was bonded to the bottom of the beam with the strong direction of composite material being parallel to the longitudinal axis of the beam. In some beams, either double U-strip of GFRP fabric anchored CFRP reinforcement at its ends. or anchor tufts (one or two) of carbon fibers was embedded through the sheet in the tension zone of beam to anchor CFRP nearby its ends. Specifically, U-strips were bonded, in the form of external open stirrups at FRP ends. Unstrengthened beams were designed to fail in flexure mode by reinforcing them with dense stirrups at shear spans. Experimental evaluations in RC beams strengthened with one CFRP, which was anchored at its ends either with a double GFRP U-strip or with two CFRP anchor tufts, have shown that their loading capacity is about similar. In addition, the ductility ratio was slightly higher in beams with one carbon anchor tuft than that attained in beams with double glass U-strips. The stiffness of all beams strengthened with one CFRP sheet was about the same, independently of the anchorage applied. Specifically, RC beams strengthened by a CFRP fabric with two carbon tufts used as anchors of FRP, it seems that an increase of about 27% occured in the loading capacity, whereas the ductility ratio was inreassed by an amount of 67%, respectively, in comparison to the respective ones evaluated in RC beams strengthened with one unanchored CFRP. Such improved structural behavior has, indded, been experimentally verified by also evaluating the deflection at mid-span of beams as well as the deformations of concrete and FRPs developed at midpoint of upper beam face and surface of fabrics.

1 INTRODUCTION

One of the biggest problems troubling the international technical world is the need for strengthening RC beams. There are a lot of reasons that could force an engineer to increase the loading capacity of structure, such as increase of load requirements or damage induced to the structural members due to seismic or other action, or even cases of design and construction faults. There are two different solutions to overcome such a structural deficiency, resulting to a partial repair or complete replacement of the damaged part of the structure. Various techniques have been developed in the last years for strengthening RC beams concluding that the use of FRP materials for structural repair presents several advantages [Baaza & Missihoum (1996), Chajes & Thomson (1994), Crasto & Kim (1996), Nakamura & Sakai, Meier (1997)] as is the high strength to weight ratio of FRP, the corrosion resistance of composite materials,

the ease application and the little equipment needed and finally the increase of durability that make FRPs a very reliable solution. The flexural behavior of members strengthened with FRPs has been investigated [Hutchinson & Rahimi (1993), Triantafillou & Plevris. (1990), Wei & Saadatmanesh (1991)]. The results of all these studies lead to conclusion that the lack of an effective anchorage of the composite material didn't allow the RC member to reach its maximum flexural design capacity. Thus, researches focus their efforts to develop an easy application of FRP anchorage, in order to increase the load carrying capacity and ductility of strengthened beams. Further studies [Spadea & Bencardino (1998)] have shown that carefully designed external anchorages of the FRP sheets applied to RC beams can increase the load capacity of the beam and at the same time regain beam's ductility. Particularly, it has been shown that the absence of end sheet anchorages, at beams strengthen by FRP sheets can lead to a brittle and catastrophic brittle failure in the form of plate or sheet peeling [Hutchinson & Rahimi (1993) Ritchie & Thomas, 1991, Ahmed & Van Gemert, (1999)]. The ratio of the length of bonded plate within the shear span of the beam to the length of shear span has been considered as a parameter affecting the plate anchorage of strengthened RC beams with CFRP plates [Fanning & Kelly, (2000)]. The use of mechanical anchorage provided at the strip ends prevents a catastrophic brittle failure of the strengthened beam by strip detachment [Spadea & Bencardino (1988)]. It is considered that sheet end anchorages have a greater effect in shorter beams, with a high ratio of shear force to bending moment, than in longer beams. Usually anchorage is provided either by anchor bolts or cover plates or by FRP U-shaped external stirrups. The use of multiple small fasteners without any bonding in opposite to large diameter bolts, distributes the load more evenly over the strip and reduces stress concentrations at the holes in the strip, which can lead to premature failure [Lamana & Bank (2001)]. In a recent paper [Demakos (2008)], It was investigated the use of U-strips from FRPs combined with glass fiber bolts, as a composite anchorage system, that can be applied easily at FRP ends to strengthen RC beams. In this paper, beneficial effects of this anchorage system were presented on the loading capacity as well as on the ductility of strengthened beams.

The aim of the present paper is the evaluation of the most appropriate anchorage system applied in CFRP sheets used to strengthen RC beams and improving their structural response, which subject to four-point bending. The carbon sheet bonded at the bottom of the beam was anchored by two different ways; i.e. either by using double U-shaped external stirups from GFRP (Glass Fiber Reinforced Polymer) at the endings or by using one or two anchors from carbon fibers (Carbon Anchor) embedded in the lower face of the beam nearby every FRP end. To the best of our knowledge, the applied technique with the tuft anchors at the bottom face of the beam used as anchorage of FRPs has not been investigated and relevant bibliography does not exist. Among others, the results of the present paper will demonstrate the advantages and disadvantages of the each particular type of anchoring system applied.

2 EXPERIMENT

2.1 Specimens

The material used for the fabrication of RC beams were C16/20- concrete, and S500- steel class. For the need of experiments thirteen RC beams having dimensions 150 cm (length) × 20 cm (height) × 10 cm (width) were constructed. Their reinforcement consisted of two Φ8 rebars at upper and lower face. The shear reinforcement was formed by Ø8 mm steel bars at 110 mm centers at shear spans and

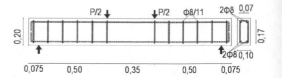

Figure 1. Details of RC control beams.

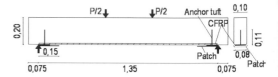

Figure 2. Details of RC beams strengthened with a CFRP sheet anchored by a CFRP tuft anchor. (B-CFRP-1AN-B).

Table 1. Properties of the composite materials.

Material	Elastic modulus (GPa)	Tensile strength (MPa)	Width (mm)	Thickness (mm)
CFRP	82.0	986	115	1.0
GFRP	20.9	575	250	1.3
Carbon Anchors	82.0	986	–	–

Φ8 mm steel bars at 350 mm centers at the midspan (Fig. 1). The material used as external reinforcement to strengthen the beam was a CFRP sheet of elastic modulus equal to 82 GPa. The epoxy resign used to bond the CFRP sheet on the beam had a low elastic modulus of 5 GPa. The anchors made of a tuft of carbon fibers 15 cm long characterised by an elastic modulus equal to 82 GPa. These anchors saturated in the epoxy resin mixture were inserted in drilled holes at the bottom face of the beam and the protruding tips of the fibers were then splayed circularly over the CFRP fabric (Fig. 2). Finally, the anchor was covered by a 150 × 80 mm patch of the same FRP (Fig. 2). The fibers of the patch were placed perpendicular to the fibers direction of CFRP shhet to improve better the effectiveness of anchorage. The U shape stirrups by GFRP had an elastic modulus equal to 20.9 GPa. The proprieties and dimensions of the composite material used are illustrated in Table 1. Five separate groups of RC beams were constructed as follows:

a. (BV) group were RC beams without external reinforcement referred as control beams.
b. (B-CFRP) group consisted of three beams with a CFRP sheet 11.5 cm wide and 1 mm thick, bonded with epoxy resign on the bottom face of the beams.
c. (B-CFRP-1AN-B) group consisted of three beams with a CFRP sheet 11.5 cm wide and 1 mm thick bonded with epoxy resign on the bottom face of

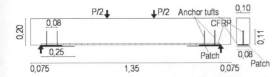

Figure 3. Details of RC beams strengthened with a CFRP sheet anchored by two CFRP tuft anchors. (B-CFRP-2AN-B).

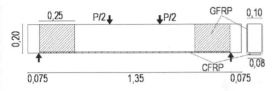

Figure 4. Details of RC beams strengthened with a CFRP sheet anchored by a double external U-shape GFRP jacket.(B-CFRP-2UG).

the beams. Their anchoring system included one carbon tuft anchor embedded in a depth of 11 cm through the fabric in the bottom of the beam. The anchor fibers were splayed circularly over the CFRP fabric at a 8 cm diameter. Over the anchor a patch of CFRP sheet 11.5 cm wide and 15 cm long was bonded with its fibers direction parallel to the axis of the beam to achieve a better anchorage of the fabric (Fig. 2).

d. (B-CFRP-2AN-B) group consisted of three beams with CFRP sheet 11.5 cm wide and 1 mm thick bonded with epoxy resign on the bottom face of the beams. Their anchoring system included two carbon tuft anchors at every end of the fabric being 8 cm far each other. Both anchors had 8 cm diameter and embedded at a depth of 11 cm inside the beam. A patch of CFRP 11.5 cm wide and 25 cm long was bonded over the anchors with direction of its fibers parallel to the axis of the beam to achieve a better anchorage effect (Fig. 3).

e. (B-CFRP-2UG) group consisted of three beams with a CFRP sheet 11.5 cm wide and 1 mm thick, bonded with epoxy resign on the bottom face of the beam and anchored by a double GFRP U-shape external stirrup on each CFRP end (Fig. 4).

The specimens were cured, after casting, in water for 28 days. Then, the edges of RC beams were rounded at their lower corners by a wire wheel of 1 cm radius, in order to avoid any injury of the fabric. Then, the specimens were cleaned from dust to achieve better contact and adhesion between concrete and fiber surface. At beams with carbon anchors, holes were drilled at the bottom face. Dust was removed by a air-pressure gun. The epoxy resign was two-component and spread by paintbrush over the bond area. FRPs were bonded at the beams as it was mentioned in previous paragraph.

Table 2. Experimental and theoretical ultimate load of strengthened RC beams with FRPs.

BEAM	Pmax (KN)	Pth. (KN)	Deflection (mm)	μ Ductility
BV1/2/3	44.1	45.4	13.61	2.8
B1/2/3-CFRP	76.2	74.0	09.90	1.5
B1/2/3-CFRP-2UG	95.3	99.5	16.91	1.9
B1/2/3-CFRP-1AN-B	91.3	99.5	15.10	2.7
B1/2/3-CFRP-2AN-B	96.8	99.5	14.58	2.5

2.2 Experiment setup

The beams were simply supported over an effective span of 1350 mm and subjected to a monotonic four-point bending loading. The load was applied stepwise through a 200 KN capacity, servo-hydraulic machine in force-controlled mode at the center of a stiffened spreader trapezoidal beam, which in turn distributed the load equally on a couple of identical bearing pads placed on the top of the beams. The load increment was controlled at a quasi-static rate of 0.2 KN/sec until collapse of beams occurred. Every 10 KN of load increment, crack developments on the beams were noted. The deflection of beam was measured at mid-span using a linear variable differential transducer (LVDT), and in the same time a load cell recorded the load automatically. The concrete deformations were measured by using strain gauges placed at mid-span, on the top of beams and at the middle of the CFRP sheets. An automatic data acquisition system was used to store the values of load, deflection and deformation.

2.3 Test results and discussion

Mean value results following the fracture of the beams appear on (Table 2). All the beams failed flexural. More specifically:

a. (BV1/2/3) group include beams failed in a ductile flexural mode. The mean value of the maximum load capacity of those beams of about 45 kN was greater than the expected theoretical.

b. (B1/2/3-CFRP) group include beams failed in a flexural mode without fracture of the upper face of the concrete. Both three beams failed because of the detachment of the CFRP sheet. This group of beams had a very similar behavior and the load of fracture was reached the theoretical one. This means that at a deformation of 7000 μs the CFRP sheet was detached, as it was calculated. Also it should be mentioned that those beams are more stiff behavior than the control beams. This happened because the CFRP sheet received the increase of the performed load and the RC beams became less ductile. From Table 2 it seems that this group of beams had the less deflection than all the others. But from Figure 5

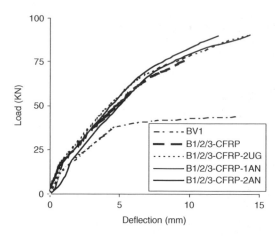

Figure 5. Load – deflection curves for all RC beams strengthened with fabrics.

Figure 6. CFRP cutting-off sheet nearly its tuft anchorage.

Figure 7. CFRP cutting-off at the middle span of beam.

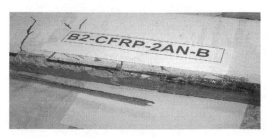

Figure 8. Complete CFRP cutting-off at the middle span of beam.

it is obvious that the stiffness slope in this group of beams is about similar with that obtained in the other strengthened beams.

c. (B1/2/3-CFRP-2UG) include three beams failed in flexural mode with a slight compression appeared on the upper flange of the concrete. In two beams of this group CFRP sheet was cut-off in the middle span (fig. 7) and led to a sudden failure of beams. At the third beam the sheet was cut-off at the left shear span of the beam. In three beams, a slight sliding of CFRP fabric observed at GFRP U-shape jacket. It should be mentioned that this slide (5–6 mm) did not affect the response of beam, as the sheet ripped.

d. (B1/2/3-CFRP-1AN-B) is a group of beams that failed in a flexural mode with a slight compression failure at the upper flange of the concrete. Beam 1 failure occured at a section 15 cm far from the end of beam (fig. 6). Beam 2 failed in a flexural-shear mode due to existing anchor at the right-hand end of the beam and detachment of CFRP fabric. Concrete in beam numbered 3 failed in flexural mode followed by a ctting off of the carbon sheet at a mid cross section of CFRP. (Fig. 8). This group of beams presented almost similar behavior in the load – deflection variation (Fig. 5) as did (B-CFRP-2UG) group of strengthened beams.

e. (B1/2/3-CFRP-2AN-B) group of beams failed in a flexural mode with a slight compression developed at the upper face of the concrete. In all beams ripping of CFRP sheet was observed. At first beam of this group, cutting-off of the sheet was observed at the left-hand side approximately 25 cm far from the end. At the remaining two beams the fabric cut off and led to failure at a mid cross section of the beam.

This group of beams presented the greater increase of stiffness than other groups of beams. This was probably caused to the fact that the distance between the internal anchors was smaller than the respective distance of anchors in (B-CFRP-1AN) group of beams. So the effective length of fabric in this case was smaller and those beams lost a part of their ductile behavior.

Generally speaking RC strengthened beams were stiffer and attained higher ultimate load than the control beams. In addition, their ductility was nearly similar to that attained in control beams. Most of the CFRP sheets applied were cut-off, which shows that a successful application of the anchors was made and the design of the experiment was sufficient. The CFRP sheets reached the fracture deformation of $11000\,\mu s$ and this is verified by the fact that the experimental ultimate load approached enough the theoretically evaluated ultimate load.

3 CONCLUSIONS

The aim of the present paper was to investigate the influence of a composite anchorage nearby FRP sheet ends to the optimization of flexural response of RC beams strengthened with composite laminates. A type of CFRP sheet was used as the main carbon fiber for external reinforcement. The fabrics were bonded with an epoxy mixture to the beam bottom. A composite material anchorage system consisting either of one or two carbon tuft anchors applied on the bottom face of RC strengthened beams or a double GFRP U-shape external jacket was used to anchor the fabric in order to avoid premature detachment of it from the concrete. Generally speaking, these anchors impose on the external reinforcement to be in contact with concrete, avoiding thus its premature debonding from the beam section. Thus, an eventual sliding of the external reinforcement from U-jackets is delayed and the strengthened beams can sustain higher loads at increased ductility levels. The conclusions derived from this study can be summarized as follows:

1. The use of anchorage of fabric at the ends of the beam with doubleU- form GFRP jackets provided higher stiffness of beams than in beams of other groups with one or two anchor tufts. On the other hand, the anchorage with one anchor tuft resulted in more flexible beams than those having two anchors.
2. The choice of anchorage technique of fabric depends on the project requirements and the engineer has to select which of the two methods will be applied. As concerns the number, i.e. one or two, of anchors to be used for the sheet anchorage, it seems that RC beams strengthened by a CFRP fabric with two anchor tufts fail in larger loads compared to the respective beams with one anchor (Table 2). In addition, the shearing failure mode of a single anchor tuft occured in RC strengthened beams shows that is preferable to use two anchors, instead of one, at each FRP end.
3. The anchorage of composite materials to the lower face of the beams using anchor tufts is a quite reliable method, giving solution in cases of beams, where the use of U-shape jackets is not advisable especially in case of constructionn difficulties. Due to their small size and their ease in application to repair of structural elements by only one worker, it is preferred to use anchor tufts instead of U-shape jackets.

REFERENCES

Ahmed, O. & Van Gemert, D. 1999. *Behavior of RC beams strengthened in bending by CFRP laminates*, Proc. 8th Int. Conf.On Struct. Faults and Repairs, Engineering Technics Press, Edinburgh, U.K.

Baaza, I.M., Missihoun, M. & Labossiere, P. 1996. *Strengthening of reinforced concrete beams with CFRP sheets*, Proc. 1st Int. Conf. on Compos. in Infrastruct. ICCI 96, Univ. of Arizona, Tucson, Ariz., 746–759.

Chajes, M.J., Thomson, T.A., Januszka, T.F. & Fin, W. 1994. *Flexural strengthening of concrete beams using externally bonded composite materials*, Constr. and Build. Mat. 8(3), 191–201.

Crasto, A.S., Kim, R.Y., Fowler, C. & Mistretta, J.P. 1996. *Rehabilitation of concrete bridge beams with externally bonded composite plates*. Part I, Proc., 1st Int. Conf. on Compos. in Infrastruct. ICCI ,Vol. 96, 857–869.

Demakos, C.B., 2008. *Investigating the Influence of FRP Sheet Anchorage to Structural Response of Reinforced Concrete Beams,* submitted for publication, Int. Scien. Conf. SynEnergy Forum, Spetses, Greece.

Fanning, P., & Kelly, O. 2000. *Smeared crack models of RC beams with externally bonded CFRP plates*, J. Computational Mech., 26(4), 325–332.

Hutchinson, A.R. & Rahimi, H. 1993. *Behavior of reinforced concrete beams with externally bonded fiber reinforced plastics*, Proc. 5th Int. Conf. on Struct. Faults and Repair, Vol. 3, Engineering Techniques Press, Edinburgh, U.K., 221–228.

Lamanna, A. J., Bank, L. C. & Scott, D.A. 2001. *Flexural strengthening of reinforced concrete beams using fasteners and fiber-reinforced polymer strips*, ACI Struct. J., 98(3), 368–376.

Meier, U. 1997. *Post strengthening by continuous fiber laminates in Europe*, Proc. 3rd Int. Symp. Non-Metallic (FRP) Reinforcement for Concrete Struct., Vol. 1, JCI, Tokyo, 42–56.

Nakamura, M., Sakai, H., Yagi, K. & Tanaka, T. 1996. *Experimental studies on the flexural reinforcing effect of carbon fiber sheet bonded to reinforced concrete beam,* Proc., 1st Int Conf. on Compos. in Infrastruct., ICCI, 96, 760–773.

Ritchie, P.A., Thomas, D.A., Lu, L.W., and Connelly, G.M. 1991. *External reinforcement of concrete beams using fiber reinforced plastics*, ACI Struct. J., 88(4), 490–500.

Spadea, G., Bencardino, F. & Swamy, R.N. 1998. *Structural behavior of composite RC beams with externally bonded CFRP*, J. of Comp. for Construcion, ASCE, 2(3), 132–137.

Triantafillou, T.C. & Plevris, N.1990. *Flexural behavior of concrete structures strengthened with epoxy-bonded fibre-reinforced plastics*, Proc., Int. Seminar on Plate Bonding Technique, University of Sheffield, Sheffield, U.K.

Wei, A., Saadatmanesh, H. & Eshani, M.R. 1991. *RC beams strengthened with FRP plates. II: Analysis and parametric study*, J. Struct. Engng., ASCE, 117(11), 3434–3455.

Excellence in Concrete Construction through Innovation – Limbachiya & Kew (eds)
© 2009 Taylor & Francis Group, London, ISBN 978-0-415-47592-1

FE modelling of the effect of elevated temperatures on the anchoring of CFRP laminates

D. Petkova, T. Donchev & J. Wen
Kingston University, London, UK

ABSTRACT: Carbon fibre reinforced polymers (CFRP) are type of relatively new composite material and its excellent structural performance determines its wide range of applications for various engineering solutions. In the last decades FRPs were found to be appropriate materials for strengthening of existing reinforced concrete structures subjected to different conditions- environmental or as a type of loading. The behaviour of the strengthened system has been the focus of investigation in different aspects of structural problems and challenges. One of these is the effect of elevated and high temperatures due to fire. A finite element model which represents the work of an anchorage of externally strengthened reinforced concrete beam with CFRP laminates is proposed in order to describe and explain the development of structural and temperature stresses and potential modes of failure during the heating and cooling process. Analysis of obtained results and comparison with similar research will be presented in order to assess the structural behaviour of the strengthened system.

1 INTRODUCTION

The excellent properties of fibre reinforced polymers (FRPs) have made them a popular solution for the increase of loadbearing capacity of timber, steel and concrete structural elements. The various types of polymer products define the wide range of shape and form of application – tendons, rods, fabrics, laminates both as external and internal reinforcement. The low self-weight, high strength and quick installation are some of the advantages of the FRP over the traditional strengthening materials.

Different research has been conducted in the last few decades on the behaviour of FRP strengthened structural elements. The effect and response to elevated temperatures and fire, however, is a relatively new area of investigation for the FRP materials. As the fire resistance of materials is an important requirement for fire safety design of structures, additional research is necessary in order to fully utilise the potential application of the polymers.

For the purpose of this paper, the behaviour of concrete specimens with externally attached CFRP laminates is analysed, reflecting the influence of different factors on the response of elevated temperatures. A finite element model of the anchoring zone of the laminate to the concrete surface is created to investigate the contribution of the different factors such as reduction of strength of the materials, coefficients of thermal expansion and the glass transition temperature of the adhesive to the interfacial stress between the laminate and the specimen.

2 LITERATURE REVIEW

Research on concrete elements strengthened with FRP at elevated temperatures is still in process of development. Green, Benichou, Kodur & Bisby (2007) presented a general overview of the North American and European approaches to the design of structures to address their performance in fire. In the North American method the structural strength is calculated based on the expected temperatures in concrete and reinforcement, FRP is recommended to be ignored in the analysis. The European approaches involve both thermal and structural analysis: the thermal analysis is conducted first and the temperature results are transferred to the structural analysis to calculate the strength of the structural element at a given time during the heat exposure. The authors concluded that more research was required to develop and experimentally validate the method, to better characterise the effects of elevated temperatures on the mechanical behaviour of FRP.

Camata, Pasquini & Spacone (2007) performed tests with different types of adhesives with glass transition temperature (T_g) higher than 85°C and 2 types of CFRP: pultruded laminates and unidirectional woven fabrics. It was observed that after the thermal load no cracks were visible, which was explained by the reduction of stiffness and the more uniform distribution of strain along the bond length. Failure occurred in concrete as a layer of the material remained attached to the laminate.

Klamer, Hordijk & De Boer (2007) observed two types of failure in their work. The specimens at the lower range(up to 40°C) failed due to concrete rupture leaving 1–3 mm of it on the adhesive layer, while at the upper temperature (50–75°C) range the failure occurred between the concrete surface and the adhesive. The results were explained as the stiffness of the adhesive was reduced as approaching to the glass transition temperature (T_g) of the material. From another set of experiments three temperature related effects were found to affect the failure load: the difference in the coefficient of thermal expansion of CFRP and concrete, which caused initial thermal stress distribution in the concrete at the interface with the adhesive; the reduction of Young's modulus with increase of temperature and the third effect was the reduced bond strength of the concrete-adhesive interface at elevated temperatures.

A study on the bond behaviour at elevated temperatures was conducted by Leone, Aiello & Matthys (2006) up to 80°C. In the case of bonded FRP laminate for all test temperatures, a mixed failure was obtained characterized by cohesion failure, adhesion failure at the epoxy- concrete interface and adhesion failure at the epoxy-FRP interface. Nevertheless, adhesion failures at the epoxy-FRP interface tended to be predominant at higher temperatures. A significant decrease of the maximum bond stress could be observed for elevated temperatures beyond the T_g (55°C). The bond interface was characterized by a less stiff behaviour.

The bond between the adhesive and concrete at elevated temperatures was investigated by Gamage, Al-Mahaidi & Wong, 2005 and 2006. It was found that exposure to temperatures higher than 80°C led to substantial decrease in the bond strength between the two materials.

Experiments with heating and loading of concrete prisms with attached CFRP were reported in literature. The temperature range was chosen by the glass transition temperature of the used adhesive (60–85°C). All authors discussed above have analysed the behaviour during the heating process. It is unclear, however, what the residual stresses are in the strengthened system after heating and cooling and their contribution to the structural stresses of the loaded samples which is the aim of this work.

3 FACTORS INFLUENCING THERMAL STRESSES BETWEEN LAMINATE AND CONCRETE

Due to their nature the polymers show the tendency to soften as their T_g is approached and at high temperatures they decompose. The coefficient of thermal expansion (CTE) is determined from the properties of the fibres and the resin. In the case of unidirectional laminates, the expansion of the CFRP in the two orthogonal directions is different. In longitudinal direction the CTE of the laminates is accepted to be zero due to the practically negligible expansion of the fibres and in the transverse direction its influence is ignored due to the limited distance of its application.

Information about the Young modulus of the CFRP is provided by the manufacturers and the corresponding values are included in the analysis. During the heating process the adhesive between the laminate and concrete exhibits significant reduction of its stiffness. The lap shear strength of the adhesive drops to 12.7 N/mm² at 45°C from 18.3 N/mm² at room temperature. The failure of specimens loaded at high temperatures has proved to be dependent on the decrease of strength of the adhesive which makes it the most critical element of the whole system. In the case of heating and cooling though the effect of glass transition should be taken into account as it could contribute to the redistribution of stresses in the constituents.

For the purpose of this research a 3D finite element model was used to estimate the effect of the elevated temperature on the strengthened system. Solid elements were used to represent the laminate, concrete prism with $75 \times 75 \times 300$ mm dimensions and the adhesive. The temperature loading was realised in two steps, analysing first the effect of heating and then the opposite process of cooling. The specific size of the specimens is accepted on the base of previous experimental investigations, see Donchev, Wen, & Papa (2007).

4 PRELIMINARY RESEARCH

In a previous research (Petkova, Donchev, Wen, Etebar & Hadavinia 2007) the behaviour of the anchoring zone of CFRP laminate externally attached to a concrete prism was investigated including considerations for elevated temperature conditions but ignoring the effects due to CTE. Both experimental investigation and theoretical FEA modelling were conducted. The thermal loading consisted of heating of the samples for at least 2 h at different constant temperatures up to 250°C to ensure evenly distributed temperature in the cross sections of the samples. The laminate was loaded in tension after the cooling of the specimens simulating longitudinal shear loading at this zone. The behaviour of the strengthened system did not exhibit substantial difference with specimens loaded up to 100°C. The decrease of strength of the system in the range 100–250°C was approximately linear which was mainly dictated by the reduction of the strength of the concrete. At 250°C the ultimate strength of

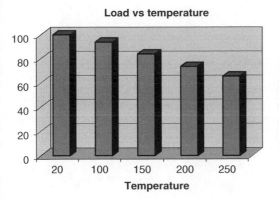

Figure 1. Decrease of strength of the strengthened system.

Table 1.

Material	Glass transition temperature	Coefficient of thermal expansion
Concrete C20	–	$10 \times 10^{-6}/°C$
CFRP laminate	100°C	$0 \times 10^{-6}/°C$
Adhesive	60°C	$33 \times 10^{-6}/°C$

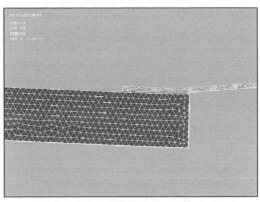

Figure 2. Deformed shape after heating to 60°C.

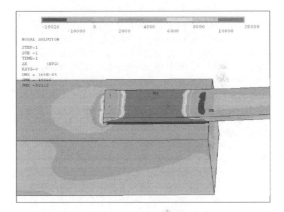

Figure 3. Normal stress distribution during heating up to 60°C.

the strengthened system was reduced to 65% of the strength at room temperature (fig. 1).

The predominant failure mode was through concrete cover separation which initiated with a vertical crack in the concrete at the laminate's edge. High concentration of stresses at the tip of the crack was observed and further development of stress concentration zone horizontally lead to the development of conditions for the separation of the concrete cover. Good agreement was found between the experimental and numerical results.

5 EFFECT OF THE DIFFERENT COEFFICIENTS OF THERMAL EXPANSION (CTE)

In reinforced concrete both the constituents – concrete and steel reinforcement – behave similarly in terms of thermal expansion when subjected to environmental conditions. In a strengthened system which includes adhesive and CFRP laminate, the change of temperature may cause significant stress development due to the difference of the CTEs (table 1). In the longitudinal direction the laminate has the lowest CTE while the adhesive could have 3 to 9 times higher coefficient compared to concrete.

5.1 Heating of the strengthened system

The free expansion of concrete at elevated temperatures does not lead to development of thermal stresses in terms of structural analysis. The attachment of the laminate, however, brings a limitation of the concrete to expand and is equivalent to a loading in compression for it along the bonding length. The laminate correspondingly is loaded in tension at the contacting zone with the concrete. Hence the deformed shape of the system during heating will be defined from compression in concrete close to the laminate and elongation in the far end of the prism (fig. 2).

From the three constituents, the adhesive is the one with highest CTE, which combined with characteristics of adjacent materials leads to the development of high compression stresses in the adhesive during the heating process (fig. 3).

The zone of the laminate's end is subjected to a local concentration of thermal stresses in the three constituents. The difference of the expansion rate of the laminate and the concrete results in conditions for the development of tension stresses in the concrete which reach high values of 6 MPa. The level of tensile stress and potential crack initiation at the zone close to the edge of the laminate of the strengthened system due to the elevated temperature are substantial.

Heating the materials above the T_g of the adhesive leads to a new process of stress relaxation at the interfaces of the constituents. Above 60°C the adhesive can be described as rubbery material with the tendency for free deformation. As the T_g of the adhesive is approached, its state is transformed from solid to viscous with the corresponding decrease of stiffness. The expansion of the concrete is no longer prevented by the limited expansion of the polymer and correspondingly the laminate is no longer subjected to tension. Therefore it could be stated that at temperatures above T_g the elements reach a state of relaxation of stresses and their deformation is free.

5.2 *Cooling of the materials and residual stress*

The next stage of the modelling is the cooling of the sample to room temperature. Once the relaxation of stress due to the softening of the adhesive is realized, the process of cooling is introduced at zero internal forces between the constituent materials. When the thermal condition decreases to temperatures under T_g, the adhesive is assumed to regain its strength and stiffness, causing interaction with the adjacent elements.

The performance of the strengthened system at cooling is the opposite of the analysis during heating. At room temperature the laminate is loaded in compression due to its lack of thermal contraction which prevents the concrete to return to its original state. The deformation of the strengthened system is again governed by the different behavior of concrete and laminate during temperature changes due to different CTE. The shape of the specimen is changed in accordance with the tendency for contraction of the concrete when subjected to decrease of the thermal load (fig. 4).

The process of cooling leads to a relaxation of the stress in the concrete at the zone of laminate's end.

The concrete is subjected to tension along the bonding area with values up to 4 MPa. The distribution of the tensile stresses in depth is similar to the formation of the horizontal crack which develops before the concrete cover separation mode of failure is observed.

The residual tensile stress at the adhesive on the other hand is significant and could lead to failure due to high concentration of stresses in this element of the strengthened system.

5.3 *Combination of residual thermal stress and stress due to external forces*

When specimens externally strengthened with CFRP laminates are heated, cooled and then subjected to tension, the remaining capacity and corresponding

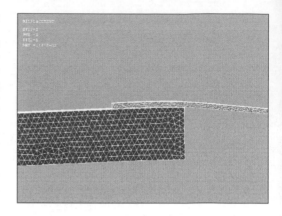

Figure 4. Deformed shape after cooling to room temperature.

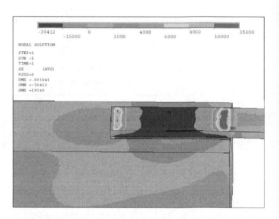

Figure 5. Normal stress distribution after cooling to room temperature.

failure mode is dictated by several different factors which make its prediction a complicated task. The thermal loading and its consequent effects on the materials, the development of microcracks in concrete, the phenomenon of glass transition temperature of strengthening materials and the condition of the interface between the adhesive and concrete are some of the problems that have to be considered. However, two main failure modes are observed from the experiments when heated up to 250°C and cooled – the concrete cover separation mode and interfacial failure between concrete and adhesive at the high value of the temperature interval.

The proposed theoretical model indicates residual tension stresses into the concrete and compression stresses into the laminate after heating above T_g and cooling. Such distribution of thermally induced stresses would have negative effect on the capacity and deformability of CFRP strengthened systems after subjected to elevated temperatures. However, it could be stated that the predominant failure of a strengthened system after heating and cooling is governed by the concrete as it is weaker in tension than any of the other materials in the strengthened system, and the development of the concrete cover separation is most likely to occur.

6 CONCLUSION AND RECOMMENDATIONS

As a result of the developed theoretical modelling, the following conclusions could be indicated:

- Temperature induced stresses in the concrete close to the edge of the laminate during heating are significant and close to the tensile strength of the concrete;
- During the heating process significant tensile stresses in the laminate and compression stresses in the concrete are generated. Those stresses are below the corresponding strength of the materials.
- When heated above T_g and cooled the system of concrete and CFRP laminate develops temperature induced stresses of compression in the laminate and tension in concrete. Such stresses could be reason for local cracking in the concrete and additional deformability at the strengthened system.

The obtained theoretical results will be subject of future experimental confirmation.

REFERENCES

BS EN 1992-1-1:2004. *Eurocode 2: Design of concrete structures- Part 1-1: General rules and rules for buildings.*

BS EN 1992-1-1:2004. *Eurocode 2: Design of concrete structures- Part 1-2: General rules – Structural fire design.*

Cadei, J. M. C., Stratford,T. J., Hollaway, L. C. & Duckett, W. G. 2004. *Strengthening metallic structures using externally bonded fibre reinforced polymers,* London: CIRIA.

Concrete Society, 1990. *Assessment and repair of fire-damaged concrete structures,* Technical report No. 33.

Camata, G., Pasquini, F. & Spacone, E. 2007. High temperature flexural strengthening with externally bonded FRP reinforcement. *Proceedings of the 8th international symposium on FRP reinforcement for concrete structures,* Patras, Greece.

Donchev, T., Wen, J. & Papa, E. 2007. Effect of elevated temperatures on CFRP strengthened structural elements, *11th International conference on Fire and materials, San Francisco, USA.*

Gamage, J. C. P. H., Al- Mahaidi, R.& Wong, M. B. 2006. Bond characteristics of CFRP plated concrete members under elevated temperatures, *Composite Structures,* Vol. 75, 2006, pp. 199–205.

Gamage, J. C. P. H., Wong, M. B. & Al- Mahaidi, R. 2005. Performance of CFRP strengthened concrete members under elevated temperatures, *Proceedings of the International Symposium on Bond behaviour of FRP in Structures,* pp. 113–118.

Green, M. F., Benichou, N., Kodur, V., Bisby, L. A., 2007. Design guidelines for fire resistance of FRP-strengthened concrete structures, *Proceedings of the 8th international symposium on FRP reinforcement for concrete structures,* Patras, Greece.

Karadeniz, Z. H. & Kumlutas, D. 2007. A numerical study on the coefficients of thermal expansion of fiber reinforced composite materials. *Composite Structures,* Vol. 78, pp 1–10.

Klamer, E., Hordijk, D. & De Boer, A. 2007. FE analyses to study the effect of temperature on debonding of externally bonded CFRP. *Proceedings of the 8th international symposium on FRP reinforcement for concrete structures,* Patras, Greece.

Klamer, E. L., Hordijk, D. A. & Janssen, H. J. M. 2005. The influence of temperature on debonding of externally bonded CFRP, *Proceedings of the 7th international symposium for FRP reinforcement for concrete structures,* Vol. 2, pp. 1551–1567.

Leone, M., Aiello, M.-A. & Matthys, S. 2006. The influence of service temperature on bond between FRP reinforcement and concrete, *Proceedings of the 2nd International Congress,* Naples, Italy.

Petkova, D., Donchev, T., Wen, J., Etebar, K. & Hadavinia, H. 2007. Effect of Elevated Temperatures on the Bond between FRP and Concrete, *Proceedings if the 11th International conference,* Vol. 1, pp. 783–790.

Rabinovitch, O. 2007. On thermal stresses in r.c. beams strengthened with externally bonded FRP strips. *Proceedings of the 8th international symposium on FRP reinforcement for concrete structures,* Patras, Greece.

SBD Weber Product catalogue. Bedford, UK: Dickens Hose.

Excellence in Concrete Construction through Innovation – Limbachiya & Kew (eds)
© *2009 Taylor & Francis Group, London, ISBN 978-0-415-47592-1*

Rehabilitation and strengthening of a hypar concrete shell by textile reinforced concrete

R. Ortlepp, S. Weiland & M. Curbach
Technische Universität Dresden, Germany

ABSTRACT: Textile reinforced concrete (TRC) is suitable for strengthening of already existing concrete structures. The ultimate load bearing behavior as well as the serviceability can be increased significantly. In this paper some results of an investigation on a Hyperbolic Paraboloid Shell will be presented. The reinforced concrete shell with a thickness of 8 cm covers an area of 38×39 m without any support. The actual state of the concrete shell was examined. Based on a FEM analysis a necessary strengthening method had to be planned. The strengthening layer is made of TRC, in which fine grained concrete is reinforced by several layers of textile reinforcement (e.g. carbon fabrics) with a polymer coating. This is the first practical application of TRC for strengthening. The paper covers information about architectural design, structural engineering design, constructive aspects, conception of reinforcement and manufacturing of the strengthening layer. Furthermore design engineering and experimental analysis including the issue of technical approval by the authorities are detailed.

1 INTRODUCTION

The first practical application of the innovative strengthening method using textile reinforced concrete (TRC) was carried out in October/November 2006 in the retrofit of a reinforced-concrete roof shell structure at the University of Applied Sciences in Schweinfurt, Germany. Since TRC had not yet been standardized as a construction material, a single "special-case" technical approval was sought from and granted by the appropriate authorities for this particular application of TRC.

One significant advantage of textile reinforcement is its deformability which allows it to adapt easily to complex, free-form geometries. Therefore, strengthening layers on profiles or shell-shaped components are possible. As such, fibers can be positioned in the desired direction so as to fully exploit their load-bearing capacities.

This simple reinforcement technology offers clear and definite advantages for the repair and strengthening of both flat and curved surface bearing structures. Simple and tested application procedures can be relied upon since the application is merely a multiply application of a fine-grained matrix – which is quite similar to that of modern plaster with embedded reinforcement and can be used and viewed similarly to standard plaster tissues. A further advantage beyond that of lamellar strengthening methods is the existence of two-dimensional load transfer which has the propensity to reduce any tendency for delamination.

2 INITIAL STATE OF THE SHELL

2.1 *Shell construction*

In 2002, the rehabilitation of a hyperbolic paraboloid, 1960's roof structure of a lecture hall at the University of Applied Sciences at Schweinfurt was set to be repaired and modernized with other buildings on the university's campus. The 80 mm thick rein-forced concrete hypar-shell spans the quadratic outline of the "Grosser Hoersaal" (Large Lecture Hall; see Figure 1) with a side length of 27.60 m and a maximum span of approximately 39 m. The shell is sloped in a north-south direction so that the southern most high point lies at 12 m and the northern most high point 7 m above the lowest point of the shell. The low points, lying in an east-west direction, are connected to massive pylons. The hull thickness measures 80 mm in the central area and increases in the direction of the edge girder to a maximum of 367 mm at its highest points and 772 mm at the pylon supports. The edge girders are supported by four pillars along the girder axis which are located at 6.50 m from the highest points. The external underside and frontside surfaces were completed in exposed concrete; at the time of measurement, the top surface had an extremely defective roof sealant layer.

Figure 1. Steel reinforced concrete hypar-shell of FH Schweinfurt.

On-site investigations, a technical survey, and equipment appraisal were conducted by Arge Schalenbau Rostock under the direction of Mr. U. Müther, Dr. B. Hauptenbuchner, and Prof. U. Diederichs (all experts on shell structures) in preparation for the repair of the structure. Decisive deviations were found, however, at the southern highest point, in particular, where a drop of up to 200 mm occurred. On-going surveying conducted between 2002 and 2006, as well as FEM modeling both in the deformed and un-deformed states, provided no indication concerning the critical structural state of increasing deformations.

According to Diederichs et al. (2006), the deformed shell form can be retained and shape reversal is not necessary. The re-calculation and subsequent specifications for the reinforced concrete structure, however, showed exceptional stresses in the cantilevered areas. This could be found in the upper reinforcement layer which also indicated that the load-carrying capacity was at its ultimate state. Necessary rehabilitation measures were to be done in the areas where the overstressed areas were found to exist (see Figure 2). Strengthening and rehabilitation of these particular areas were recommended in response to snowfall levels and wind factors.

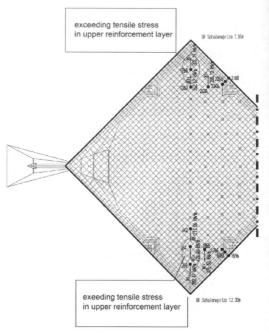

Figure 2. Overstressed areas in cantilevered portions of the shell (cut-out).

Figure 3. Textile adapted to curved hypar shell structure.

2.2 Choice of strengthening method

Classical, as well as established, procedures for the strengthening of reinforced concrete structures, such as shotcrete strengthening or lamellae made from steel or Fiber-Reinforced Polymers (FRP), were discussed as methods by which rehabilitation of this structure could be brought to fruition. TRC couples the advantages of a two-dimensional reinforcement layer like shotcrete with a low dead weight like FRP. The rehabilitation proposal compiled by Arge Schalenbau in

Rostock, in cooperation with the Institute of Concrete Structures at the Technische Universität Dresden, planned to safeguard the necessary bearing-capacity through TRC strengthening using three layers of textile carbon fabric.

Optimally, textile reinforcement adapts (i.e., takes the form of) to the curved surface of the shell (Fig. 3). Strengthening with TRC represents an excellent, new alternative to the strengthening of curved, thin-walled reinforced concrete surface structures.

Figure 4. Layers of textile reinforced concrete (TRC).

3 TEXTILE REINFORCED CONCRETE

3.1 *Characteristics*

Textile reinforced concrete (TRC) is a relatively new, high performance composite. It consists basically of textile reinforcements that are embedded in thin layers of concrete (Fig. 4). Textile reinforcements are produced from the processing of high performance fibers (i.e., fibers with high strength, high ductility and high durability) with flat or spatial reinforcing structures by utilizing state-of-the-art textile technology. Compared to steel reinforcement, textile reinforcing elements are extremely filigree. As a result, a composite material with very remarkable properties, as outlined below, is developed:

– There is no concrete cover necessary for corrosion protection of the reinforcement since the reinforcing materials used do not exhibit corrosion under normal environmental conditions. Therefore, very thin strengthening layers can be produced due to this "waiver" of concrete cover.
– Textile reinforcements have a much larger surface area than ordinary steel-bar reinforcement. Thus, very high bond forces can be introduced into the concrete which, in practice, is evidenced by the ability to use short anchoring lengths and the presence of very dense crack patterns.
– Fibers produced from either alkali resistant glass (AR-glass) or carbon possess a distinctively higher strength than that exhibited by standard steel-bar reinforcement. Current third generation textile reinforcements support strengths of clearly over $1000 \, N/mm^2$.

In addition to providing a pure increase in the ultimate loading capacity of reinforced concrete, TRC is also suitable for additional repair applications. TRC exhibits properties, which follow, that appear to be suitable for the strengthening of RC structures:

– The use of additional TRC strengthening layers in RC structures is proven to have a positive influence on subsequent concrete cracking. The number of cracks increases while crack spacing (between cracks) decreases and crack widths are simultaneously reduced.
– Members become stiffer. This effect is not only based on the application of additional material but also on the fact that members remain in an un-cracked state I, as demonstrated, for a longer time period.
– It is possible to produce more advanced repairs with the use of a mineral-based matrix. In addition, pure re-profiling, re-passivation and the provision of a new or additional dense concrete cover are valuable options.
– Surface design of the strengthening layer can vary within a wide range. A large range of design options is available today for plasters can also be utilized in TRC-application when suitable mixtures for matrices are used and appropriate treatment is present in the final layer.

3.2 *Bonding behavior*

An adequate bond between the textile fibers and the fine-grained concrete matrix covering is a precondition for the complete introduction of forces into the textile fabric and subsequent activation of all individual filaments within the reinforcing roving. In the same way, a functioning adhesive tensile bond is required between the TRC strengthening layer and the existing concrete substrate to establish and generate adequate interaction between the existing and new concrete of the strengthened shell structure. Experimental tests of the adhesive tensile bond and anchoring of the TRC strengthening layer were conducted by the Technische Universität Dresden during the course of the shell's rehabilitation.

3.2.1 *Textile pull-out tests*

The pull-out behavior of the textile reinforcing carbon fabric has been analyzed based upon specific textile pull-out tests. The goal of these experimental investigations was the determination of the filament yarn pull-out lengths of the textile reinforcing fabric. Contrary to simple pull-out tests on individual filament yarns, these tests analyzed the influence of cross-reinforcement on the pull-out forces. According to Xu et al. (2004), this is particularly important to impregnated (coated) textiles since the knot resistance is changed and the yarn cross-section is also strongly affected along its axis by the impregnation/coating. An adequate pull-out resistance is of special importance so as to eliminate the possibility of failure due to "pull-out" of the fabric from the matrix during the applied use.

The experimental set-up for the investigation of the filament yarn pull-out lengths was developed on the basis of the uni-axial tension tests of Jesse & Curbach (2003), as well as bond tests concerning end anchoring of TRC strengthening described by Ortlepp

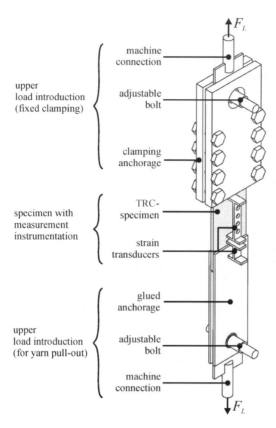

Figure 5. Test set-up for the textile pull-out tests.

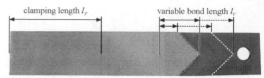

clamping length l_r variable bond length l_v

Figure 6. Specimen for the textile pull-out tests.

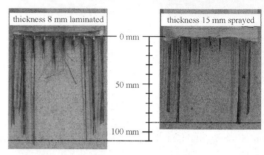

thickness 8 mm laminated   thickness 15 mm sprayed

0 mm

50 mm

100 mm

Figure 7. Evaluation of the filament yarn pull-out length; influence of fine-grained concrete matrix.

et al. (2006), Figure 5. The filament yarn pull-out tests include the failure mode filament yarn pull-out at the end anchoring of the TRC strengthening.

At the upper end of the specimen, a clamping anchorage was selected as the load introduction. A grip length of 200 mm was chosen for the clamping anchorage (Fig. 6, left). The lateral compressive stress applied and the large bond length are necessary to prevent a bond failure at the wrong end of the specimen. The adhesive bond length at the end of the specimen examined varies from 0 to 200 mm. The maximum bond length tested was limited to 200 mm. Larger anchoring lengths do not appear appropriate relative to the short anchoring lengths of TRC strengthening layers on an old concrete substrate as determined by Ortlepp et al. (2006).

The specimens are manufactured on steel formwork as plates measuring 1200 mm × 600 mm. The individual textile layers are produced by hand using a method of lamination. The plates are left in the formwork and covered with wet cloths for up to three days. The specimens are then stored in a climatically controlled chamber after removing the formwork. Several days prior to testing, the specimens are trimmed, with a stone saw, to fit a width of 100 mm and the corresponding appropriate length. In order to examine a bond length range of 50 mm (i.e., several bond lengths) by using only one, single specimen, the lower end of the specimen is cut at a 45-degree angle (Fig. 6). Several specimens with varying bond length ranges of up to a maximum of 200 mm bond can then be tested in one series. One to two days before the test, the lower load introduction plates are glued on the specimens.

On the basis of the tested specimens, the anchoring length of the filament yarns can be determined by measuring the protruding fibers after pull-out failure of the specimens tested. The specimens with short bond lengths fail by complete pull-out of all filament yarns. The filament yarns fail by tension break beginning with the desired filament yarn pull-out length. That is, specimens with very long bond lengths fail by tension break of all filament yarns. The desired filament yarn pull-out length for the examined textiles result from the specimen with the bond length range where a mixed failure occurs (Fig. 7).

The examination of the sprayed test specimen (Tab. 1, lines 4–5) resulted in a required anchorage length 20 mm, shorter than in the case of the manually laminated test specimen having a thinner layer thickness (Tab. 1, lines 6–9). With the reinforced strengthening layer of the hypar shell, a sufficiently thick layer was manufactured, so that an anchoring length of 90 mm could be attained in order to transmit forces of the textile reinforcement into the surrounding fine-grained concrete matrix.

Due to the changing moments in the shell, it can be assumed that the maximum tensile force existing in the TRC layer could no longer be attained at the shell's

Table 1. Pull-out lengths of textile reinforcement in fine grained concrete matrix.

Test No.	Matrix	Procedure*	Embedded length [mm]	Thickness [mm]	Failure type**	Pull-out length [mm]
1	T 110	S	70–120	25.6	P	120***
2	T 110	S	70–120	20.3	P	120***
3	T F	S	70–120	18.6	P	120***
4	SFB	S	70–120	15.4	P + T	90
5	SFB	S	70–120	17.9	P + T	90
6	SFB	L	40–100	7.8	P	100***
7	SFB	L	90–140	7.8	P + T	110
8	SFB	L	110–160	7.8	P + T	110
9	SFB	L	150–200	7.9	T	–

*S = sprayed, L = laminated.
**P = Pull-out, T = Tensile break.
***The maximum pull-out length is limited by the specimen length (bond range).

Table 2. Adhesive tensile strength of textile reinforcement in the fine concrete matrix.

Test No.	Adhesive tensile strength [N/mm²]	Failure type
1	3.18	Textile layer
2	4.02	Textile layer/fine-grained concrete
3	3.28	Textile layer
4	4.11	Textile layer
5	3.19	Textile layer
6	3.75	Textile layer
7	4.39	Textile layer/fine-grained concrete
8	3.78	Textile layer/fine-grained concrete
9	3.99	Textile layer
10	3.90	Textile layer
11	3.37	Textile layer
12	3.34	Textile layer
13	3.51	Textile layer
14	3.29	Textile layer
15	4.20	Textile layer/fine-grained concrete
16	4.12	Textile layer/fine-grained concrete

edge. Rather, the tensile force existing in the strengthening layer was gradually transmitted into the existing concrete substrate to the shell's edge via the entire available length. The anchoring force, at the edge of the shell construction, laid far below the ultimate load of the strengthening layer. Therefore, a very short bond length, circumvents any potential problems that could arise concerning the anchoring.

3.2.2 Adhesive tensile tests

The adhesive tensile load-carrying capacity of the TRC strengthening layer was examined on a separately manufactured specimen whose geometry corresponded to a free-drilled cylinder for the adhesive tensile testing of the coating systems according to the DAfStb-guideline (Deutscher Ausschuss für Stahlbeton 1991/1992). These cylindrical specimens of the TRC layer had a diameter of 50 mm and a height of approximately 10 mm. Adhesive strength measurements were ascertained by the use of the transportable testing set, DYNA Z 15. Sixteen individual tests were taken successively 28 days after applying the strengthening layer. The results of the individual tests are presented in Table 2.

On average, an adhesive tensile strength of 3.71 N/mm² resulted; this value is clearly larger than the adhesive tensile strength of the existing concrete. Since sufficient load-carrying capacity of the shell was already proven as a condition for the feasibility of the reinforcement measure, the load-carrying capacity of the adhesive bond is guaranteed based upon existing results.

4 DESIGN OF THE TRC STRENGTHENING

4.1 Old shell construction

The concrete of the old shell construction was dense and firm showing negligibly small carbonated depths

(Diederichs et al. 2006). The upper steel bar reinforcement was not sufficiently dimensioned for the eccentric static stress in the cantilevered areas under dead load, wind, and snow. Bending moments mainly arise due to wind and snow loads. The existing steel bar reinforcement could be substantiated with the appropriate corresponding level of safety for stresses only due to dead-load. Consequently, the TRC strengthening was to be designed for snow and wind loads.

4.2 TRC strengthening

TRC consists of two primary components – fine grained concrete and textile fabric. The mixture of the fine-grained concrete matrix used for Collaborative Research Center 528 is shown in Table 1. According to Jesse (2005) the matrix is specified as having the following variables:

- Density $\rho_{fc} = 2170 \, \text{kg/m}^3$;
- Young's Modulus $E_{fcm} = 28,500 \, \text{N/mm}^2$;
- Compressive Strength $f_{fck} = 76.3 \, \text{N/mm}^2$;
- Flexural Tensile Strength $f_{tk,fl} = 7.11 \, \text{N/mm}^2$

The textile reinforcement used, having a yarn fineness of 800 tex (cross-section of a roving $A_{f,Roving} = 0.45 \, \text{mm}^2$), was produced from carbon rovings produced by Toho Tenax Europe GmbH. The fabric was manufactured on a multi-axial warp-knitting machine, model MALIMO, at the Institute for Textile and Clothing Technology at the Technische Universität Dresden. A roving distance of 10.8 mm was chosen in the warp-direction and 18 mm in the weft direction (Fig. 8).

Table 3. Material composition of fine-grained concrete used in CRC 528 (Jesse 2005).

Materials	Amount [kg/m^3]
Cement CEM III/B 32,5 LH/HS /NA	628.0
Fly ash	265.6
Elkem Mikrosilica (Suspension)	100.5
Sand 0/1	942.0
Water	214.6
Plasticizer Woerment FM 30 (FM)	10.5

Figure 8. Carbon textile.

Strain-specimen-tests (displacement controlled direct tension tests) to measure strength according to Jesse & Curbach (2003), Jesse (2005) provided the following results for TRC containing the carbon roving textiles mentioned above:

– Fiber fatigue stress up to 2400 N/mm^2;
– Ultimate strain ≈ 0.9–1.2%;
– Ultimate tensile strength $f_{tu} = 1400$ N/mm^2;
– Ultimate limit strain $\varepsilon_u = 0.8\%$

The material safety factor was conservatively assumed to be $g_t = 2$ due to the innovativeness of the task at hand. Current and on-going investigations by the Collaborative Research Center 528 at the Technische Universität Dresden include the evolution of undeveloped security concepts and partial safety factors in the area of TRC.

The strengthening was designed for snow and wind loads. The existing reinforcement on the upper/outer side was not computationally derived. Using this conservative approach, the existing pre-deformation of the shell could remain dormant (i.e., neglected) while representing another load-carrying-reserve. A limit strain of 8‰ is defined as a result of the compatibility with the textile-strengthening layer. The ultimate strain of the steel bar reinforcement could only be used in a relative manner.

In order to simplify the situation and remain conservative in the approach, the upper/outer reinforcement

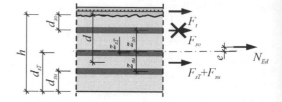

Figure 9. Forces within the strengthened shell cross-section.

situation was neglected (Fig. 9). The force of the TRC strengthening layer F_t could, thus, be, once again, conservatively divided according to the law of the lever.

$$F_t = N_{ed} \frac{z_{s,u+T} + e}{z_{to} + z_{s,u+T}} \tag{1}$$

$$F_{s,u+T} = N_{ed} \frac{z_{to} - e}{z_{to} + z_{s,u+T}} \tag{2}$$

The necessary amount of textile reinforcement required (a_{to}) could then be determined using the permissible ultimate strength f_{ttd} of the fiber within a mineral matrix using the following equations and designations.

$$a_{to} = \frac{F_{to}}{f_{ttd}} \quad \text{with:}$$

$$f_{ttd} = \frac{f_{tu}}{\gamma_t} = \frac{1400 \frac{N}{mm^2}}{2.1} = 666.7 \frac{N}{mm^2} \tag{3}$$

The necessary amount of steel reinforcement required ($a_{s,u+T}$) in the two lower layers could then be determined by

$$a_{s,u+T} = \frac{F_{s,u+T}}{f_{std}} \quad \text{with:}$$

$$f_{td} = \frac{f_{yk}}{\gamma_s} = \frac{420}{1.15} = 365.2 \frac{N}{mm^2} \tag{4}$$

The textile fabric consisted of a carbon roving with a fineness of 800 tex ($= 800$ g/1000 m) spaced at intervals of 10.8 mm in the warp direction. The roving fiber cross-section $A_{f,Roving}$, with a density of carbon of $\rho = 1{,}790$ kg/m^3, is:

$$A_{f,Roving} = \frac{\text{fineness}}{\text{density}} = \frac{803 \text{ tex}}{1790 \frac{kg}{m^3}} = 0.449 \text{ mm}^2 \tag{5}$$

Two layers of textile reinforcement were chosen; furthermore, an additional layer without regard for static requirements, was inserted to fulfill safety and limit state concerns.

5 SUMMARY

The first time the innovative textile reinforcement technology was used involved the rehabilitation of a reinforced concrete, hypar shell structure located on the campus of the University of Applied Science in Schweinfurt, Germany. The paper shows that TRC is an outstanding alternative and addition to the current strengthening and repair methods of rehabilitation used to date. Contrary to the use of well-established, traditional steel-reinforcement methods and despite the use of substantially thinner concrete layers, higher load-carrying capacity, as well as improved serviceability, can clearly be obtained from the use of textile reinforcement. The total thickness of reinforcement, which only amounted to 15–18 mm, not only extended the useful life of the structure, but it also made this structure safe for future use.

ACKNOWLEDGEMENTS

The authors would like to thank everyone who contributed to this project. with special thanks to the Arge Schalenbau Müther-Hauptenbuchner-Diederichs, as well as Schweinfurt State Building Authorities for their support and interest in this innovative project.

Furthermore, the authors express their gratitude to the management and team of specialists from TORKRET AG for their commitment and constructive co-operation during the entire course of the project. Special thanks also go to the Highest Building Authority in the Bavarian Internal Ministry of State in Munich for their competent support and special-case technical approval. Thanks also go to the Institute for Textile and Clothing Technology at the Technische Universität Dresden and the GWT-TUD mbH, whose substantial support contributed to the success of this ambitious project. Finally, the authors wish to thank all members of the Collaborative Research Centre 528, as well as the German Research Foundation (DFG), for their support.

REFERENCES

Deutscher Ausschuss für Stahlbeton (ed.) 1991/1992. *Richtlinie für Schutz und Instandsetzung von Betonbauteilen, Teil I bis IV*. Berlin: Beuth.

Diederichs, U., Hauptenbuchner, B. & Müther, U. (Arge Schalenbau) 2006. *Ermittlung und Dokumentation des Erhaltungszustandes der Stahlbeton-Hyparschale über dem großen Hörsaal der Fachhochschule Schweinfurt und Ausarbeitung von Vorschlägen zur Sanierung*. Rostock, 26.07.2006.

Jesse, F. & Curbach, M. 2003. Strength of Continuous AR-Glass Fibre Reinforcement for Cementitious Composites. In Naaman, A.E. & Reinhardt, H.-W. (eds), *High Performance Fibre Reinforced Cementitious Composites HPFRCC-4. Proceedings of the Fourth International RILEM Workshop*: 337–348. Bagneux: RILEM

Jesse, F. 2005. *Tragverhalten von Filamentgarnen in zementgebundener Matrix*. Dissertation, Technische Universität Dresden.

Ortlepp, R., Hampel, U., Curbach, M. 2006. A new Approach for Evaluating Bond Capacity of TRC Strengthening. *Cement and Concrete Composites* 28(7): 589–597.

Xu, S., Krüger, M., Reinhardt, H.W. & Ozbolt, J. 2004. Bond characteristics of carbon, alkali resistant glass, and aramid textiles in mortar. *Journal of Materials in Civil Engineering* 16(4): 356–364.

Excellence in Concrete Construction through Innovation – Limbachiya & Kew (eds)
© 2009 Taylor & Francis Group, London, ISBN 978-0-415-47592-1

Corrosion mitigation in concrete beams using electrokinetic nanoparticle treatment

K. Kupwade-Patil, K. Gordon, K. Xu, O. Moral, H. Cardenas & L. Lee
Louisiana Tech University, Ruston, Louisiana, USA

ABSTRACT: Chloride induced reinforcement corrosion is a major cause of civil infrastructure deterioration. This study evaluated the corrosion behavior of reinforced concrete beams when subjected to electrokinetic nanoparticle (EN) treatment. This treatment utilized a weak electric field to transport nanoscale pozzolans directly to the steel reinforcement. Each beam was subjected to saltwater exposure followed by electrochemical chloride extraction (ECE) and EN treatment. The specimens were re-exposed to saltwater following treatment. Results from this test indicated significantly lower corrosion rates among the EN treated specimens (0.006 mils per year) as compared to the untreated controls (0.87 mils per year). Scanning Electron Microscopy (SEM) was used to examine the microstructural impact of the treatment process. A reduction in porosity (adjacent to the steel) of 40% was observed due to EN treatment. During treatment application, the electric field also caused chlorides to be drawn away from the reinforcement and extracted from the concrete beam. After the chloride had been extracted, the nanoparticles apparently formed a physical barrier against chloride re-penetration.

1 INTRODUCTION

1.1 *Reinforcement corrosion in concrete*

Reinforcement corrosion in concrete is a major durability problem. A prime cause for corrosion initiation is due to the ingress of chlorides, derived either from the use of de-icing salts or exposure to marine environments (Bentur *et al.* 1997). Chloride induced corrosion leads to the reduction in reinforcement cross sectional area via formation of corrosion products. The increased volume of corrosion products results in tensile cracking and leads to structural failure. Electrochemical chloride extraction (ECE) is commonly used re-habilitation technique for corroded reinforced concrete. This study examined the impact of driving nanoscale pozzolans to the reinforcement while chlorides were being extracted.

1.2 *Techniques to mitigate reinforcement corrosion in concrete*

The ECE process involves removal of chloride ions from concrete cover by applying a constant current for a period of six to eight weeks (Marcotte *et al.* 1999). The positive terminal of the power source is connected to the external anode while the negative terminal is tied to the embedded reinforcement (cathode). This initiates movement of current between the anode and cathode resulting in the extraction of chloride ions that

are driven out of the concrete toward the anode. While this approach is effective it suffers from drawbacks. The chloride extraction effectiveness has been reported at only 40% after seven weeks of ECE (Orellan *et al.* 2004). While only half of the chlorides were removed adjacent to the steel reinforcement, at the same time, a significant amount of alkali ions were drawn to the reinforcement. Though the associated rise in pH is passivating to the steel the high alkali content could cause localized alkali silica reactions. Another study also reported that a significant amount of bond strength loss (19–33%) between the concrete and the reinforcement was observed when a relatively high current density was applied (Rasheeduzzafar *et al.* 1993).

A relatively new approach to mitigate reinforcement corrosion in concrete is the electrokinetic injection of pozzolanic nanoparticles (Cardenas and Kupwade-Patil 2007). It is called electrokinetic nanoparticle (EN) treatment. Previous work analyzed the impact on reinforcement corrosion when 24-nm pozzolanic nanoparticles were driven into the pores of relatively immature concrete (Kupwade-Patil 2007). Cylindrical specimens of reinforced concrete were subjected to initial saltwater exposure followed by ECE and EN treatment. The nanoparticles acted as pore-blocking agents inhibiting the re-entry of chlorides. The treatment had reduced the calcium hydroxide content ~8%, indicating the probable formation of C-S-H. A 25% increase in tensile strength was also observed.

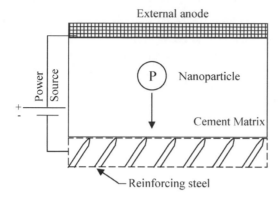

Figure 1. Electrokinetic Nanoparticle (EN) treatment in cement matrix.

Table 1. Materials set-up for reinforced concrete beam.

Beam specifications

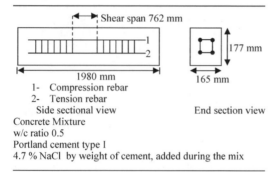

1- Compression rebar
2- Tension rebar
Side sectional view   End section view

Concrete Mixture
w/c ratio 0.5
Portland cement type I
4.7 % NaCl by weight of cement, added during the mix

Table 2. Chemical composition of cement.

Chemical	Al_2O_3	Fe_2O_3	SO_3	C_3A	CO_2	$CaCO_3$	CaO	SiO_2
Wt %	4.5	3.7	2.7	6.0	1.2	2.9	64	24

2 EXPERIMENTAL

2.1 Beam specifications

Ten reinforced concrete beams (length: 1.98 m, width: 0.17 m, length: 0.18 m) were cast using Portland cement Type I. All beams had two No. 4 (12.7 mm diameter) rebars and two No. 6 (19 mm diameter) rebars as longitudinal reinforcement. No. 4 bars acted in compression while the No. 6 bars were in tension. Commercially available stirrups (9.5 mm diameter) were used to form the rebar cages of each beam. The shear reinforcement stirrups were placed 102 mm apart throughout each beam. Specimen set-up, curing conditions and concrete mixture details are summarized in Table 1. Cement composition and properties of the concrete are shown in Table 2. Sodium chloride

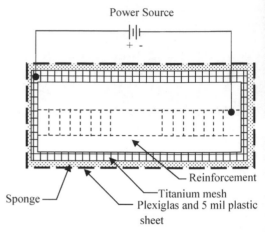

Figure 2. Circuit diagram for ECE and EN treatment (Top view).

(NaCl) was added to the mix in order to simulate the use of beach sand.

2.2 ECE and EN treatments

Initial saltwater exposure was conducted for twelve weeks on eight specimens by placing the beams in NaCl (4.7 wt %) immersion. Two beams were not subjected to initial saltwater exposure and were lime water cured as untreated controls. The other six specimens were subjected to ECE and EN treatments at the end of the initial saltwater exposure. The treatment circuit for both the ECE and EN treatments are shown in Figure 2. Mixed metal oxide titanium mesh was wrapped along the base and sides serving as the external electrode. The mesh was surrounded by sponge material. This was covered by a plexiglass outershell. Treatment liquid was circulated along the sponge between the plexiglass and the beam. A constant current density of 1 A/m² was maintained. The positive pole of the power source was connected to the titanium mesh and the negative terminal was connected directly to the concrete reinforcement. Six specimens were subjected to ECE for two weeks followed by six weeks of EN treatment. Two of the specimens were subjected to ECE for all eight weeks at selected dosages. During EN treatment the electric field drew nanoparticles from the fluid flowing through the sponge and drove them into the pores of the concrete. At the end of six weeks all treatments were discontinued and all the specimens were subjected to 24 weeks of post saltwater exposure. Corrosion potential and corrosion rate measurements were recorded periodically during the study.

2.3 Electrochemical measurements

After the EN treatment all the specimens were exposed to saltwater exposure for 24 weeks. The corrosion

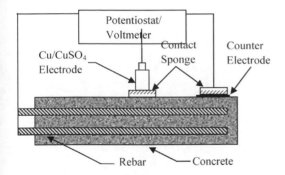

Figure 3. Corrosion rate measurement setup.

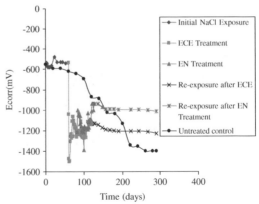

Figure 4. Corrosion potentials of concrete reinforcement.

potential was determined by measuring the voltage difference between a reference electrode and the embedded steel. A $Cu/CuSO_4$ reference electrode was used in this study. The potentials were recorded using a high input impedance ($100\,M\Omega$) voltmeter with a $1\,mV$ resolution. The corrosion rate of the reinforcement was determined using a linear polarization resistance (LPR) technique with a Solartron potentiostat (model no. 1287, manufactured by Roxboro Group Company, UK). A copper/copper sulfate reference electrode was used as a reference electrode while mixed-metal oxide coated titanium wire mesh was used as a counter electrode (See Figure 3). LPR measurements were performed at a scan rate of $0.1\,mv/s$ from $-20\,mV$ to $+20\,mV$. The duration of each scan was between 7 to 8 minutes. A plot of current density versus potential was generated (Ahmad 2006). Polarization resistance (R_p) of the corroding metal is the slope of the potential (E) versus current density (log i) curve at the corrosion potential (E_{corr}) point. The IR drop was corrected using ($R_p = R - R_\Omega$) where R is the total resistance measured by polarization resistance procedure and R_Ω is the concrete resistance. The surface area, equivalent weight and density of steel (ρ) were $60.70\,cm^2$, $25.50\,gm$ and $8.0\,gm/cm^3$ respectively. I_{corr} was calculated by (Jones 1996),

$$I_{corr} = \frac{B_a . B_c}{2.3\,(B_a + B_c)\,R_p} \qquad (1)$$

where B_a (mv/decade) and B_c (mv/decade) are the anodic and cathodic tafel curve slopes and R_p ($k\Omega cm^2$) is the polarization resistance. The resulting value of I_{corr} was used to calculate the corrosion rate (CR) which was derived from the Faraday's law as (Tayyib and Khan, 1988)

$$CR = \frac{K_1 \, x \, I_{corr} \, x \, EW}{d} \qquad (2)$$

where CR = corrosion rate (mills per year), K_1 = Faraday's constant as used in the corrosion rate

equation (mpy g/µAcm), EW = Equivalent Weight, d = density ($8.02\,g/cm^3$).

2.4 Microstructure and pore-structural analysis

Porosity measurements were conducted on samples taken from locations adjacent to the reinforcement. A small sample (5 to 7g) of each specimen was ground to pass a No. 30 sieve. The particles that passed were collected, and the initial weight of the sample was taken as W_1. The powdered sample was dried at $105°C$ until the weight stabilized at the value W_2. The porosity of the sample was calculated by,

$$Porosity = \frac{W_1 - W_2}{W_1} \times 100\,(\%) \qquad (3)$$

Microscopic imaging samples were extracted from the broken concrete cylinders and mounted in epoxy. Polishing was conducted using 60, 120, 150, 320 and 600 grit size papers with a Vector beta-grinder manufactured by Buehler, Lake Bluff, IL. Micropolishing was conducted with 5, 3 and 0.25 Buehler micro-cloths using a non-aqueous lubricant (propylene glycol). After each stage of polishing the specimens were rinsed in ethyl alcohol to remove loose material. Microstructural analysis was conducted using a Hitachi S-4800 field emission scanning electron microscope (FE SEM). Quantitative elemental analysis was done using Genesis Microanalysis software manufactured by Ametek Inc.

3 RESULTS AND DISCUSSION

3.1 Corrosion measurements

Figure 4 contains a plot of the corrosion potential as observed during this study. The corrosion potential

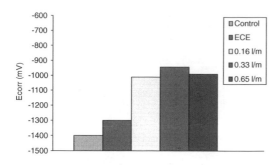

Figure 5. Comparison of half-cell potential (Ecorr) values measured on treated and untreated specimens.

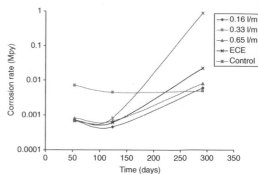

Figure 6. Corrosion rates using LPR technique.

measurements reported represent the average values of two specimens for the control and ECE cases and the average of six specimens for the EN treated case. During the initial saltwater exposure the corrosion potential dropped to -536 mV. ECE was carried out for 14 days on all the specimens causing the average corrosion potential to decrease to -1207 mV. During the ECE process the chlorides moved away from the reinforcement while sodium, potassium and calcium ions residing in the concrete pore fluid were attracted to the rebar, causing a negative trend in corrosion potentials. ECE was continued for an additional six weeks on two of the beam specimens while the other six were subjected to EN treatment over the same period.

In case of the EN treatments, significant increase in corrosion potential was observed from -1207 mV to -940 mV. This increase in corrosion potential may have been influenced by the alumina content of the treatment.

The average corrosion potential values measured on EN treated specimen dosage (0.16, 0.33 and 0.65 liters of particle per meter of beam) and untreated specimens (ECE and controls) at the end of post saltwater exposure are shown in Figure 5. The figure shows that corrosion potentials of untreated specimens were more negative than the treated specimens. The average difference in corrosion potential between untreated and treated cases, was 340 mV. Studies conducted by Suryavanshi and others observed that elevated C-A-H content was associated with relatively positive corrosion potentials (Suryavanshi et al. 1998). As a point of speculation the alumina content of the EN treatment may have produced C-A-H that could have induced the relatively positive corrosion potentials. Future work will need to examine this further.

The results obtained during the corrosion rate measurements of the beam specimens at different time intervals are shown in Figure 6. Corrosion rates of 0.006 Mpy, 0.004 Mpy and 0.008 Mpy was observed at the end of 292 days for the 0.16, 0.33 and 0.65 l/m particle dosages respectively. The ECE and

Table 3. Corrosion current density values at the end of post saltwater-exposure Error: $\pm 0.0001\%$.

Treatment type	I_{corr} ($\mu A/cm^2$)	CR (mpy)	Cl (%)	Severity of corrosion
Control	2.120	0.87	3.64	Moderate to high
ECE	0.050	0.02	0.45	Passive
Treated (0.65)	0.019	0.008	0.06	Passive
Treated (0.33)	0.011	0.004	0.00	Passive
Treated (0.16)	0.014	0.006	0.06	Passive

Table 4. Guidelines for evaluating rate of corrosion.

Corrosion current density I_{corr} ($\mu A/cm^2$)	Severity of corrosion
$I_{corr} < 0.1$–0.2	Passive state
$0.2 < I_{corr} < 0.5$	Low rate
$0.5 < I_{corr} < 5.5$	Moderate to high rate
$5.5 < I_{corr} < 100$	High rate

control specimens exhibited 0.022 and 0.87 Mpy at the end of 292 days. An average corrosion rate of 0.006 Mpy for the EN treated specimens was observed as compared to the untreated controls (0.87 Mpy). The untreated controls exhibited a higher average corrosion rate by a factor of 145 as compared to the EN treated specimens. The corrosion rates, corrosion current densities and chloride contents at the end of post saltwater exposure are shown in Table 3. The corrosion current density (I_{corr}) values can be used directly to examine the corrosion severity.

Guide lines for evaluating the severity of corrosion are shown in Table 4. The average I_{corr} values of 2.12, 0.05 and 0.01 were observed for the control, EN treated and ECE specimens respectively. The controls

Table 5. Porosity values.

Treatment type	Particle dosage (l/m)*	Porosity (%)
EN	0.16	4.7
EN	0.33	5.6
EN	0.65	7.3
ECE	Nill	9.8
Control	Nill	17.1

Error: ±13.
* Liters of particle delivered per meter of beam length.

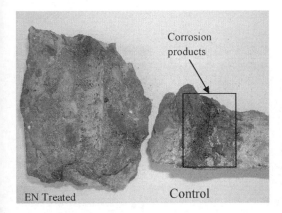

Figure 7. Imprints of corrosion products on EN treated and control specimens.

exhibited a moderate to high corrosion rate, while EN treated and ECE specimens demonstrated a passive state using the guidelines shown in Table 4.

3.2 Porosity and SEM observations

Results obtained from porosity testing are shown in Table 5. The EN treated specimens exhibited a 65% porosity reduction when compared to the control specimens and 40% when compared to the ECE cases. The reduction in porosity of the EN treated specimens appeared to protect the embedded reinforcement from chloride re-entry. The nanoparticles appeared to occupy the pores, causing an increase in tortousity, while also reducing the volume porosity. The lower chloride values associated with the lower porosities are not surprising (Refer to Tables 3 and 5). These observations appear to confirm that the nano-scaled pozzolans blocked the pores to some extent and slowed down the re-entry of the chlorides.

Imprints of corrosion products were observed on the reinforcement/concrete interface of the control specimens (Refer to Figure 7). SEM imaging combined with EDAX analysis was used to characterize the amount

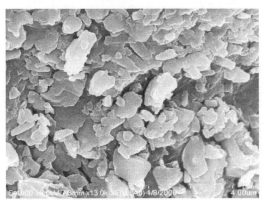

Figure 8. Cement paste in the corroding specimen (non-polished section) revealing the accumulation of free NaCl crystals.

Figure 9. Friedel's salt formation (2000×) on the control specimen in the vicinity of steel/concrete interface.

and morphologies of various corrosion phases. Following chloride ingress, some chlorides react with the cement matrix and others remain free in the cement pore solution, where they can cause localized corrosion on the steel surface. Free chlorides in the form of NaCl crystals were observed in the vicinity of the steel bars of the control specimens as shown in Figure 8. Formation of Friedel's salt was also observed in the control specimens in the vicinity of the steel/concrete interface (Figure 9). Friedel's salt ($3CaO \cdot Al_2O_3 \cdot CaCl_2 \cdot 10H_2O$) formation depends on the amount of tricalcium aluminate content in the cement (Koleva *et al.* 2006, Koleva *et al.* 2008). Friedels's salt formations can later be subject to dissolution of the chloride complex phases and future release of chloride ions. Specifically carbonation of cement paste with bound chloride ions increases the risk for later releases initiating corrosion attack. This potential suggests a residual vulnerability from ECE that could possibly be mitigated by the internal

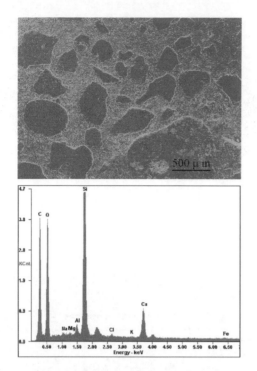

Figure 10.   BSE image and EDAX analysis (0.65 l/m).

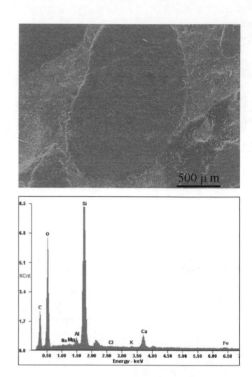

Figure 12.   BSE image of 0.16 (l/m) and EDAX analysis.

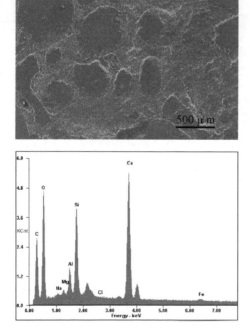

Figure 11.   BSE and EDAX analysis (0.33 l/m).

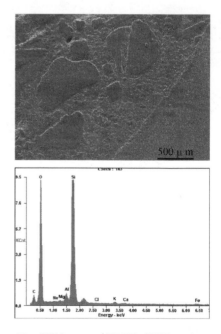

Figure 13.   BSE image and EDAX of ECE specimen.

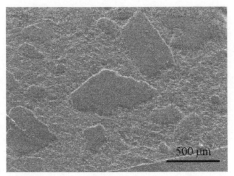

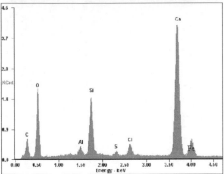

Figure 14. BSE image and EDAX analysis of control specimen.

structural barrier to chloride penetration that appears to be provided by EN treatment.

Back-scattered electron (BSE) images were taken on the polished control and EN treated specimens. Figures 10–14 show that EN treated specimens exhibited lower average chloride content by weight of cement of 0.04% as compared to the chloride extracted (0.45%) and control specimens (3.64%). Compare this value to the ACI 22R-01 code which allows the threshold chloride content of 0.08% for non-prestressed concrete. These results show that EN treatment was effective in keeping the chloride contents below the ACI allowable values.

4 CONCLUSIONS

The EN treatment was successful in mitigating moderate to high reinforcement corrosion. The corrosion potentials of ECE treated specimens were more negative than the EN treated specimens. The untreated controls exhibited a higher corrosion rate by a factor of 145 as compared to the EN treated specimens. The EN treated specimens exhibited a porosity reduction of 40% as compared to the ECE treated specimens and 65% as compared to the untreated controls. This porosity reduction adjacent to the rebar combined with the lowered corrosion rates and chloride contents suggests that the nanoparticle treatments yielded an effective barrier to chloride penetration.

REFERENCES

ACI 222R-01, 2006, Protection of metals in concrete against corrosion, MI, American concrete institute.

Ahmad Z, 2006, Principles of corrosion engineering and corrosion control, Netherlands, Butterworth-Heinemann.

Bentur A, Diamond S and Berke S, 1997, Steel corrosion in concrete, London, E & FN Spon.

Cairns J and Melville C 2003. The effect of concrete surface treatments on electrical measurements of corrosion activity, *Construction and Building Materials*, Vol 17, 301–309.

Cardenas H and Kupwade-Patil K, Corrosion Remediation using chloride extraction concurrent with electrokinetic pozzolan deposition in concrete. New solutions for environmental pollution, Proc. Intern. Symp on electrokinetic remediation, 12–15 June 2007, Vigo.

Jones D, 1996, Principles and prevention of corrosion, New Jersey, Prentice Hall.

Koleva D, Hu J, Fraaij A, Stroeven P, Boshkov N and Wit J 2006. Quantitative characterization of steel/cement paste interface microstructure and corrosion phenomena in mortars suffering from chloride attack, *Corrosion Science*, Vol. 48, 4001–4019.

Koleva D, Breugel K, Wit J, Westing E, Copuroglu O, Veleva L and Fraaij A 2008, Correlation of microstructure, electrical properties and electrochemical phenomena in reinforced mortar. Breakdown to multi-phase interface structure. Part I: Microstructural observations and electrical properties, *Material Characterization*, Vol 59, 290–300.

Kupwade-Patil K, 2007, A new corrosion mitigation strategy using nanoscale pozzolan deposition, Masters thesis, Louisiana Tech University.

Marcotte T, Hansson C and Hope B, 1999, The effect of electrochemical chloride extraction treatment on steel-reinforced mortar Part II: Microstructural characterization, *Cement and Concrete Research*, Vol 29, 1561–1568.

Orellan J, Ecsadeillas G and Arliguie G, 2004, Electrochemical chloride extraction: efficiency and side effects, *Cement and Concrete Research*, Vol 34, pp.227–234.

Rasheeduzzafar A, Ali G and Al-Sulaimani G 1993. Degradation of bond between reinforcing steel and concrete due to cathodic protection current, *ACI Materials Journal*, Vol. 90, pp.8–15.

Suryavanshi A, Scantlebury J and Lyon S 1998, Corrosion of reinforcement steel embedded in high water – cement ratio concrete contaminated with chloride, *Cement and Concrete Composites*, Vol. 20, 263–381.

Tayyib and Khan, 1988, Corrosion rate measurements of reinforcing steel in concrete by electrochemical techniques, *ACI Materials Journal*, Vol. 85, 172–177.

Excellence in Concrete Construction through Innovation – Limbachiya & Kew (eds)
© 2009 Taylor & Francis Group, London, ISBN 978-0-415-47592-1

Long-term durability of reinforced concrete rehabilitated via electrokinetic nanoparticle treatment

K. Gordon, K. Kupwade-Patil, L. Lee, H. Cardenas & O. Moral
Louisiana Tech University, Ruston, Louisiana, USA

ABSTRACT: The deterioration of reinforced-concrete (RC) infrastructure is a growing problem. Marine environments play a significant role in structural deterioration since conditions are conducive to corrosion. The crippling nature of corrosion makes it one of the leading contributors of uncertainty when trying to determine the remaining service life of a structure. A technology under development to mitigate corrosion is electrokinetic nanoparticle (EN) treatment. This paper uses a time-dependent reliability-based procedure to assess the effectiveness of EN treatment on the service life of RC beams. The rehabilitation of corrosion-damaged RC beams involved two weeks of chloride extraction (ECE) followed by six weeks of nanoparticle treatments. The nanoparticle treatments varied by concentration of silica-alumina (SiAl) nanoparticles applied. Corrosion potentials and current densities were obtained from measurements on beams exposed to a saltwater solution of 5% sodium chloride (NaCl) by weight. The beams were evaluated following 0, 90, and 180 days of saltwater exposure. The effects of chlorides on the steel reinforcement, as well as the contributions of the SiAl nanoparticles to long-term durability and load-bearing capacity were evaluated before and after exposure to the saltwater solution.

1 INTRODUCTION

1.1 Corrosion in reinforced concrete

The deterioration of reinforced concrete (RC) structures is due largely in part to the corrosion of the reinforcing steel. Much of the corrosion is precipitated by the presence of chlorides. Chlorides can be introduced to concrete structures in two ways. In regions prone to icy weather, chlorides are often introduced by the use of deicing salts, which are applied to concrete surfaces. In marine environments, chlorides are introduced by naturally occurring salts in both the air and water. When the chlorides penetrate the concrete, reactions take place, thereby making the reinforcing steel susceptible to corrosion (Page and Sergi 2000).

As the embedded steel corrodes, the formation of products induces stresses on the surrounding concrete, causing internal cracking. The reduction in cross-sectional area of the reinforcement adversely affects the capacity of the structure. The combination of cracking and decreased capacity shortens the service life and affects the long-term durability of the structure.

Electrochemical chloride extraction (ECE) is a method often used to halt corrosion in RC structures. Electrokinetic nanoparticle (EN) treatment is a developing technology used to both rehabilitate damaged concrete and prevent continued ingress of chlorides.

1.2 Rehabilitation techniques

ECE is used to remove chlorides from concrete. It is carried out by the use of the embedded steel, an electrolyte, a conductive mesh, and a power source. The power source is used to apply a constant current, with the embedded steel serving as the cathode and the mesh serving as the anode. The electrolyte aids in the transport of ions. With the application of a current, negatively charged chloride ions are drawn out of the concrete. The process can last, typically, for a period of four to eight weeks (Velivasakis et al. 1998).

EN treatments are conducted in a similar fashion to ECE. However, the primary objective of an EN treatment is to drive silica-alumina nanoparticles, or pozzolans, into the concrete. When inside the concrete, the nanoparticles undergo a reaction with calcium hydroxide, which is already present in the concrete. This reaction results in the formation of calcium silicate hydrate, which reinforces the strength of the concrete. This also leads to a decrease in concrete permeability, as shown in previous research (Cardenas and Struble 2006). The decrease in permeability is what helps prevent the further ingress of chlorides.

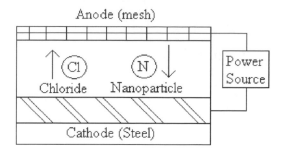

Figure 1. ECE and EN treatment configuration.

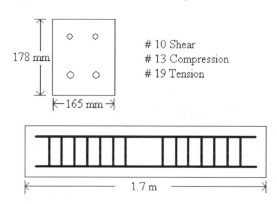

10 Shear
13 Compression
19 Tension

Figure 2. Beam design.

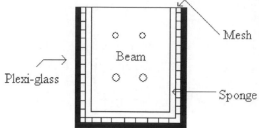

Figure 3. Treatment configuration for beams.

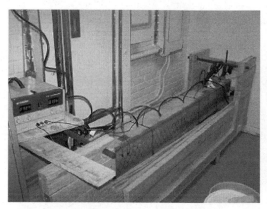

Figure 4. Actual treatment setup.

Figure 1 shows the general setup of both ECE and EN treatments.

2 EXPERIMENTAL DESIGN AND METHODOLOGY

2.1 Beam design

A total of ten beam specimens were used for the experiment. Each was designed with both longitudinal and transverse reinforcement. Longitudinal reinforcement consisted of two compression bars and two tension bars. Rebar sizes for the compression reinforcement and tension reinforcement were #13 (No. 4) and #19 (No. 6), respectively. Transverse reinforcement was constructed using #10 (No. 3) bars. Spacing for the transverse reinforcement was 101.6 mm (4 in). The beams were each constructed to lengths of 1.7 m (5.5 ft). Beam details are shown in Figure 2.

The concrete used to construct the beams was designed with a water/cement ratio of 0.5 and Portland Type 1 cement. The strength of the concrete was 2.8 N/m^2 (4000 lb/in^2). To expedite corrosion of the reinforcing steel, sodium chloride was mixed into the center portion of eight of the beams. The two remaining beams served as controls.

2.2 Saltwater exposure

After curing for twenty-eight days, the eight salted beams were exposed to a 5% NaCl solution in order to simulate saltwater exposure. The remaining controls were not exposed to saltwater. Initial exposure occurred for a period of twelve weeks. Following ECE and EN treatments, the beams underwent an extended saltwater exposure for a period of twenty-four weeks.

2.3 ECE and EN treatment

Following the initial saltwater exposure, the beams were subjected to ECE for a period of two weeks. A six-week period of EN treatment followed. Only six of the beams underwent EN treatments. The remaining two beams continued with chloride extraction. Figure 3 shows the setup used for both ECE and EN treatments.

Each of the treated beams was wrapped with a sponge material along the sides and bottom. The beams were then wrapped with titanium mesh.

Plexi-glass was used to hold the sponge and mesh along the surface of the beam. Figure 4 shows the actual treatment setup. During EN treatments, the treatment solution was constantly circulated. Particle concentrations of 0.16, 0.33, and 0.65 liters of

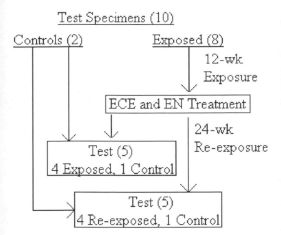

Figure 5. Breakdown of treatments and testing.

Figure 6. Specimen during testing.

particles delivered per meter of beam length (l/m) were maintained. Both the ECE and EN treatments were carried out with a constant current density of 1 A/m². Figure 5 shows a breakdown of the beam exposures and treatments.

2.4 Data collection

Prior to constructing the beam specimens, strain gages were placed along the length of the longitudinal steel. Strain gages were placed along one compression bar and one tension bar, as shown in the following figure. Wire was attached to each strain gage and extended beyond the length of the beam. The wires were attached to data collection equipment during testing.

2.5 Beam testing

The beams were tested in two sets of five specimens. The first set consisted of four beams that were exposed to saltwater only for the initial period of twelve weeks, then subjected to ECE and EN treatments. A control specimen was also included in the first set. The second set of beams consisted of a control beam, in addition to four treated beams that spent an additional twenty-four weeks in a saltwater bath. The spacing of the supports and loading point were configured to induce shear failure in the beam in order to study the durability of FRP shear rehabilitation in the second phase of the research. Here the results of EN treatment on the load bearing capacity are reported.

Each of the beams was tested using a three point loading configuration, with a spacing of 1.5 m (5 ft) between supports. Load was applied in increments of 8.9 N (2000 lb) so that progression of cracking could be noted. Each beam was loaded to failure. Figure 6 shows a beam during loading.

2.6 Shear analysis and prediction

The theoretical shear capacity of the beam was calculated using Equation 1 (ACI 318-05).

$$V_n = 2 \cdot \sqrt{f_c'} \cdot b_w \cdot d + \frac{A_v \cdot f_y \cdot d}{s} \qquad (1)$$

where, f_c' = concrete compressive strength; b_w = base width of beam section; d = effective depth of reinforcing steel; A_v = area of shear reinforcement; and s = spacing of shear reinforcement. The theoretical shear capacity of the constructed RC beam is 104.24 kN. To account for the effects of corrosion on shear capacity, the reduction in steel reinforcement area due to corrosion was introduced to equation 1 (Lee et al 2004).

$$A_s(t) = \frac{n \cdot \pi}{4} \cdot (D_0 - 2 \cdot C_r \cdot t)^2 \qquad (2)$$

where $A_s(t)$ = time-dependent cross-sectional area of steel; n = number of steel rebar; D_0 = initial rebar diameter; C_r = yearly rate of corrosion in mm/yr; and t = time in years. The theoretical corrosion rate value used in Equation 2 is 0.0115 mm/yr (Andrade and Alonso 2001). Applying Equation 2 to the shear reinforcement, the following time-dependent equation was developed for predicting shear capacity.

$$V_u(t) = \phi_v \cdot \left(2 \cdot \sqrt{f_c'} \cdot b_w \cdot d + \frac{A_v(t) \cdot f_y \cdot d}{s} \right) \qquad (3)$$

where $V_u(t)$ = the time-dependent shear capacity of the beam; ϕ_v = shear reduction factor; and $A_v(t)$ = the time-dependent cross-sectional area of the shear reinforcement. The change in shear capacity was also evaluated considering rehabilitation with a carbon fiber-reinforced polymer (CFRP) composite. For the purposes of the current investigation durability of FRP rehabilitation is integrated into the time-dependent analysis by the results of accelerated aging experiments. Predictions for FRP composite tensile modulus and tensile strength for both CFRP composites are modeled with an Arrhenius rate relationship (ARR) for wet lay-up CFRP composites immersed in deionized water at 23°C for approximately two years (Abanilla 2004). The percent retention of tensile modulus for CFRP composite are shown in the following equations.

$$y_m(t) = -0.4182 \cdot \ln\left(t \cdot 365 \frac{day}{yr}\right) + 100\% \tag{4}$$

$$E(t) = E_{frp} \cdot y_m(t) \tag{5}$$

$$f_{fe}(t) = \varepsilon_{fe} \cdot E(t) \tag{6}$$

where $y_m(t)$ = percent retention of the modulus of elasticity; $E(t)$ = time-dependent modulus of elasticity of FRP; E_{frp} = modulus of elasticity of FRP; $f_{fe}(t)$ = effective strength of the FRP; and ε_{fe} = rupture strain of the FRP. Equation 7, modified for time-dependency, was used to calculate shear capacity of the composite.

$$\Psi_f V_f(t) = \frac{A_{fv} \cdot f_{fe}(t) \cdot (\sin\alpha + \cos\alpha) \cdot d_f}{s_f} \tag{7}$$

where A_{fv} = area of FRP shear reinforcement; f_{fe} = effective stress in the FRP; α = angle of inclination of the FRP fibers; d_f = depth of FRP shear reinforcement; s_f = spacing of FRP shear reinforcement; and Ψ_f = FRP strength reduction factor. Combining Equation 3 and Equation 7 results in the following equation used to calculated theoretical time-dependent shear capacities with FRP rehabilitation.

$$V_u(t) = \phi_v\left(2\sqrt{f_c'}b_wd + \frac{A_v(t)f_yd}{s} + \Psi_f V_f(t)\right) \tag{8}$$

3 RESULTS AND DISCUSSION

3.1 Twelve-week exposure

Figure 7 shows the load-displacement curves for the specimens exposed to only twelve weeks of saltwater exposure, and subjected to ECE and EN treatments. Maximum values for each beam specimen are shown in Table 1.

As shown in Table 1, the control specimen carried a maximum load of 97.5 kN (21.9 kip), which was approximately 6% less than the theoretical prediction. Beams with values exceeding the control

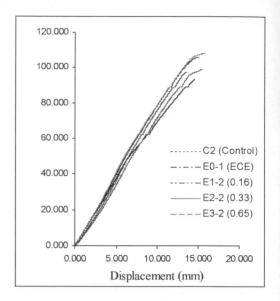

Figure 7. Load-displacement curves for 12-week exposure.

Table 1. Maximum test values for twelve-week exposure.

Specimen	Particle conc. (l/m)	Load (kN)	Displacement (mm)
C2 (control)	–	97.5	13.6
E0-1 (ECE)	–	93.1	14.6
E1-2 (EN)	0.16	105.7	15.1
E2-2 (EN)	0.33	108.2	15.9
E3-2 (EN)	0.65	98.9	15.5

value were the EN treated beams E1-2, E2-2, and E3-2. The beams carried loads of 105.7 kN (23.7 kip), 108.2 kN (24.3 kip), and 98.9 kN (22.2 kip), respectively. The ECE treated specimen carried a load of 93.1 kN (20.9 kip), which is less than the control value.

With respect to the capacities of the beams specimens, it appears that the twelve-week saltwater exposure did have a slight effect on the strength of the beams, as did the EN treatments. Specimen E0-1, subjected to ECE only, showed a slight decrease in strength. This may be due to the more porous state of the concrete following ECE. The porous nature may have led to a small decrease in the concrete strength. Each of the EN treated specimens showed an increase in load capacity. There was a gradual increase in capacity with increase in particle concentration, with exception of specimen E3-2. Although this specimen was treated with a more concentrated solution, the capacity was proven to be only slightly higher than that of the control. The greatest increase in strength was obtained with the 0.33 particle concentration.

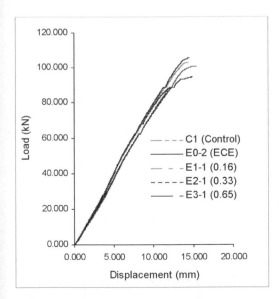

Figure 8. Load-displacement curves for twenty-four-week re-exposure.

Table 2. Maximum test values for twenty-four-week re-exposure.

Specimen	Particle conc. (l/m)	Load (kN)	Displacement (mm)
C1 (control)	–	100.7	15.4
E0-2 (ECE)	–	88.4	12.0
E1-1 (EN)	0.16	102.9	14.4
E2-1 (EN)	0.33	105.6	14.4
E3-1 (EN)	0.65	94.4	14.9

3.2 Twenty-four week exposure

Figure 8 shows the load-displacement curves for the specimens re-exposed to saltwater following ECE and EN treatments. Maximum values for each specimen are shown in Table 2.

As shown in Table 2, the control specimen carried a maximum load of 100.7 kN (22.6 kip) which is approximately 3% less than the theoretical prediction. Like the initial five specimens, the beams with load capacities exceeding the control value were E1-1, E2-1, and E3-1. The beams carried loads of 102.9 kN (23.1 kip), 105.6 kN (23.7 kip), and 94.4 kN (21.2 kip), respectively. The ECE treated specimen carried a load of 88.4 kN (19.88 kip), which is less than the control value.

Subjected to ECE without any EN treatment, specimen E0-2 was shown to have a greater decrease in load capacity. This may be due in part to the chlorides being able to penetrate the concrete more easily. The

Table 3. Corrosion rates following re-exposure.

Specimen	Particle conc. (l/m)	Corrosion rate (mm/yr)
C1 (control)	–	0.022
E0-2 (ECE)	–	0.000508
E1-1 (EN)	0.16	0.0001524
E2-1 (EN)	0.33	0.0001016
E3-1 (EN)	0.65	0.0002032

highest capacities were obtained with the EN-treated specimens. Specimen E3-1, despite receiving the highest dosage of particles, had a capacity that was actually lower than the control capacity.

3.3 Post re-exposure corrosion rates

Corrosion rates were determined for specimens re-exposed to the saltwater bath. Table 3 shows each specimen with its corresponding corrosion rate. The EN-treated specimens were shown to have the lowest corrosion rates than those of the ECE-treated and control specimens. Considering EN-treated specimens alone, the corrosion rate tends to decrease as particle concentration increases. However, the corrosion rate was slightly higher for specimen E3-1, which received the 0.65 particle treatment. Based on the experimental load capacities and corrosion rates, it appears that the highest particle treatment does not provide as much of a structural advantage as the lower concentrations. Further examination into the cost of such treatments may show that treatments with lower concentrations of particles may lead to savings in maintenance and rehabilitation costs.

3.4 Theoretical and experimental shear comparison without FRP

Using Equation 3, shear values were calculated with both theoretical and experimental corrosion rates. Figure 9 shows the time-dependent shear capacities over a span of fifty years, without the contribution of FRP. As expected, there was a greater decline in the theoretical shear and control shear values. This can be attributed to the higher corrosion rates that were used in calculations.

Figure 10 shows only the time-dependent shear capacities of the ECE and EN-treated specimens. Capacities varied with treatment concentration. With the experimental corrosion rates, it is shown that the specimens subjected to EN treatments are able to sustain their respective shear capacities at a slightly better rate than the specimen subjected to ECE only. This shows that over the life of beam, it may be more

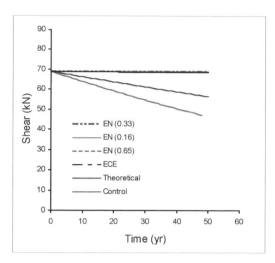

Figure 9. Time-dependent shear capacities without FRP.

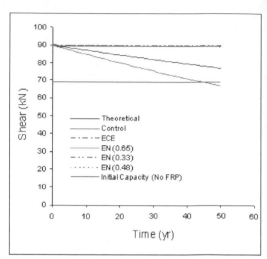

Figure 11. Time-dependent shear capacities with FRP.

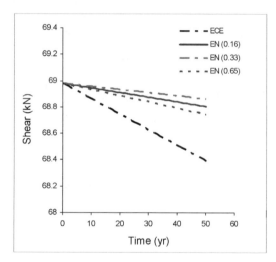

Figure 10. Time-dependent shear capacities for treated specimens without FRP.

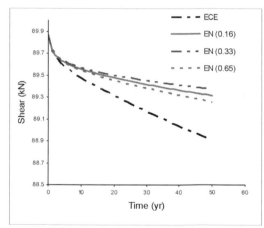

Figure 12. Time-dependent shear capacities for treated specimens with FRP.

beneficial to reinforce the contributions of ECE by continuing with some form EN treatments.

3.5 Theoretical and experimental shear comparison with FRP

The service-life contribution of FRP rehabilitation is shown in the Figure 11. The capacities are plotted along with the design shear capacity for the beam section shown in Figure 2. The FRP was applied as one layer using a U-wrap configuration. It is shown in Figure 11 that the application of FRP causes an instantaneous increase in load capacity. Over the span of fifty years, the specimens subjected to ECE and EN

treatments will fail to even reach the original shear capacity that each was designed to carry. It appears that from the FRP alone, the life of the specimens can be extended far beyond the span of fifty years. As with specimens that received only ECE and EN treatments, Figure 12 shows that when FRP is applied, the specimen that the received the 0.33 EN treatment has the most effective life span. Similarly the ECE treated specimen showed the greatest decrease in capacity.

3.6 Long-term maintenance with FRP

With the ability of FRP to greatly contribute to the capacity of an RC structure, it would be more

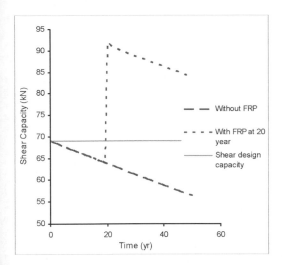

Figure 13. Theoretical FRP rehabilitation at twenty years CO.

able to maintain capacities well above the design shear value. In addition to experimental load capacity results, the long-term evaluation of capacities using experimental corrosion rates showed that the greatest maintenance of strength was achieved when ECE was continued with EN treatment, particularly with a concentration of 0.33 l/m. It was also shown that the use of FRP in addition to ECE and EN treatment has the potential to greatly increase the service-life of the reinforced concrete specimens. Although the experimental capacity values show that the treatments are advantageous, further study needs to be conducted in order to adequately characterize how and why the ECE and particle treatment have such an effects on the concrete specimens.

appropriate to use FRP rehabilitation as a form of maintenance.

Figure 13 shows how FRP rehabilitation can be used to extend the life of an RC structure. The theoretical capacities and corrosion rate were used to generate the curves. As shown in the plot, the FRP does not have to be applied immediately to the specimens. It can be applied at regular intervals in a maintenance schedule, if needed, to restore the capacity of a structure. In the figure, FRP was applied at twenty years.

4 CONLUSIONS

It can be concluded from the experimental data that ECE and EN treatments are moderately successful at increasing the capacities of reinforced concrete specimens. After both the twelve-week and twenty-four-week exposure conditions, the specimens were

REFERENCES

ACI, American Concrete Institute 2005. Building code requirements for structural concrete, *ACI 318-05*, Reported by ACI Committee 318.

ACI, American Concrete Institute 2002. Guide for the design and construction of externally bonded FRP systems for strengthening concrete structures, *ACI 440.2R-02, Emerging Technology Series*, Reported by ACI Committee 440, October, First Printing.

Andrade C and Alonso C 2001. On-site measurements of corrosion rates of reinforcements, *Construction and Building Materials*, Vol 15(2–3), 141–145.

Cardenas H and Struble L 2006. Electrokinetic nanoparticle treatment of hardened cement paste for reduction of permeability, *ASCE Journal of Materials in Civil Engineering*, Vol 18 (4), 554–560.

Lee L et al 2004. Investigation of integrity and effectiveness of RC bridge deck rehabilitation with CFRP composites,La Jolla: Dept. of Structural Engineering, Univ. of California, San Diego.

Page C and Sergi G 2000. Developments in cathodic protection applied to reinforced concrete, *ASCE Journal of Materials in Civil Engineering*, Vol 12 (1), 8–15.

Velivasakis E et al. 1998. Chloride extraction and realkalization of reinforced concrete to stop steel corrosion, *Journal of Performance of Constructed Facilities*, Vol 12 (2), 77–84.

Excellence in Concrete Construction through Innovation – Limbachiya & Kew (eds)
© 2009 Taylor & Francis Group, London, ISBN 978-0-415-47592-1

The use of glass-ceramic bonding enamel to improve the bond between concrete and steel reinforcement

C.L. Hackler
Porcelain Enamel Institute, Alpharetta, GA, USA

C.A. Weiss, Jr.*, J.G. Tom*, S.W. Morefield**, M.C. Sykes* & P.G. Malone*
*US Army Engineer Research and Development Center, Vicksburg, MS, USA
**US Army Engineer Research and Development Center, Champaign, IL, USA

ABSTRACT: Typically concrete does not form a strong bond to reinforcing steel. The paste interface around the steel is a boundary where interpore water accumulates during the setting process. The water-rich paste and the deposition of crystals of portlandlite, produces a weak, permeable or interfacial zone. The weak interfacial zone makes it relatively easy to separate the steel reinforcement from the surrounding concrete. Coating the reinforcing steel with a glass-ceramic layer containing hydraulically reactive silicates such as calcium silicate and aluminoferrites reduces the water in the interfacial zone, creates a glass-ceramic coupling layer resulting in a concrete-to-steel bond that is over four times stronger than the typical bond. The glass enamel forms a protective cover that minimizes corrosion of the reinforcing steel when the chemistry of the surrounding paste changes due to carbonation and chloride infiltration and puts the steel at risk to oxidation. Enameling technology can be used on all forms of steel reinforcement including rebar, steel fiber, or steel decking.

1 INTRODUCTION

Reinforced concrete is the most common construction material in the global infrastructure with uses ranging from drainage pipes, bridges, and overpasses. One of the fundamental problems related to reinforced concrete construction is the interface between the steel and the surrounding concrete. There is no strong chemical bond between the metal and the surrounding hydrated calcium silicate paste. As reinforced concrete cures and gains strength, the water in the concrete is distributed unevenly and the boundary between an impervious surface (steel) and paste develops a water-rich layer that results in a weak, pervious zone adjacent to the surface. The hydration reaction produces an interpore fluid in this zone that is a saturated calcium hydroxide solution. Crystals of soft calcium hydroxide form in the interfacial transition zone and further weaken the metal-paste junction.

A similar interfacial bonding problem has been encountered in other composite materials (e.g., fiberglass) and the solution has been to develop a coupling layer, such as the organosilanes, that form strong bonds with each of the dissimilar materials. The problem of joining concrete and steel can be addressed through a similar technique by fusing a reactive glass-ceramic to the surface of the steel before it is imbedded in fresh concrete. The technology for making and using glass and glass-ceramic coatings for engineering applications is well-established (Majumdar & Jana 2001). A coupling layer that can bond steel to concrete can be produced by using low-melting point, alkali-resistant enameling glasses to fuse the calcium silicate gel-producing ceramics and glasses in portland cement to the surface of the reinforcing steel.

An additional advantage of the described coupling layer is a significant increase in corrosion resistance of the steel. As conditions in the concrete change, due to carbonation and chloride-migration, the steel becomes vulnerable to corrosion (Funahashi et al. 1992; Qian 2004; Cheng et al. 2005). Without an intervening barrier to prevent corrosion, the build-up of iron oxide on the reinforcement steel places the surrounding concrete in tension and produces cracking and delamination of the concrete (Li et al. 2007; Zhou et al. 2005). As corrosion progresses and the concrete cracks, the contact pressure normal to the steel-concrete interface is reduced. The corrosion results in a decrease in the cross-sectional area of the steel and a decrease of the friction coefficient between the steel and the concrete (Amleh & Ghosh 2006). The net effect is a reduction of the reinforcement in the concrete.

Table 1. Composition range of a typical alkali-resistant groundcoat enamel for steel.

Constituent	Amount (%)	Range (%)
Silicon dioxide, SiO_2	42.02	40–45
Boron oxide, B_2O_3	18.41	16–20
Sodium oxide, Na_2O	15.05	15–18
Potassium oxide, K_2O	2.71	2–4
Lithium oxide, Li_2O	1.06	1–2
Calcium oxide, CaO	4.47	3–5
Aluminum oxide, Al_2O_3	4.38	3–5
Zirconium dioxide, ZrO_2	5.04	4–6
Copper oxide, CuO	0.07	nil
Manganese dioxide, MnO_2	1.39	1–2
Nickel oxide, NiO	1.04	1–2
Cobalt oxide, Co_3O_4	0.93	0.5–1.5
Phosphorus pentoxide, P_2O_5	0.68	0.5–1
Fluorine, F_2	2.75	2–3.5

Figure 1. Examples of test rods prepared with various samples of glass frits.

A project was conducted to examine the effects that can be produced by fusing a glass-ceramic coating that contains water-reactive silicate and ferrite compounds (those found in portland cement) to a steel surface prior to embedding the steel in concrete. This vitreous enamel is different from previous polymer or metal coatings in that it firmly bonds a reactive silicate phase to the steel and develops an outer layer of cement grains capable of bonding to the surrounding hydrating paste. Additionally, the glassy coating isolates the surface of the steel to minimize corrosion by protecting the steel from changes in chemical conditions as concrete ages.

2 TEST METHODS

2.1 Preparation of coated steel specimens

Mild steel (ASTM C 1018) test specimens with a smooth, glass bead-blasted surface; 6.3-mm in diameter by 75-mm in length were used in the investigation. One end of the rod was threaded to allow it to be attached to the test apparatus. The length of the rod permitted it to be embedded in mortar to a depth of 63-mm.

The test rod surfaces were prepared for ground coat enameling using a grit scrubbing process and an alkaline washing process. The composition of the glass frit applied to the test rods varied depending on the manufacturer with the exact composition of most formulations being proprietary. All manufacturers were asked to furnish an alkali-resistant formulation that would be a suitable ground coat for a two-firing application. The composition for a typical glass frit prepared for this application is given in Table 1.

The porcelain enamel coating was fired onto the steel rods at temperatures from 745 to 850°C. Firing times were typically from 2 to 10 minutes depending on the mass of metal to be heated and the size of the furnace. No attempt was made to obtain an even or smooth coating as would normally be the case for porcelain enamels for appliances, bathtubs, etc. (Figure 1). The enamel coatings had an average thickness of 0.8-mm.

2.2 Preparation and testing of bond strength test samples

The enameled test rods were embedded in a mortar prepared in accordance with ASTM C 109, "Standard Test Method for Determining Compressive Strength of Hydraulic Cement Mortars." The proportion of the standard mortar was one part cement (Type I-II) to 2.75 parts of standard graded sand (ASTM C 778). The water-to-cement ratio was maintained at 0.485. Test cylinders were prepared for each mortar batch and tested to determine the unconfined compressive strength at seven days. All test results were within the limits recognized as standard for this ASTM mixture proportion.

Each enameled test rod was inserted into a 50-mm diameter, 100-mm long cylinder mold filled with fresh mortar. The rod was clamped at the top so that a 63-mm length of the coated portion of the rod was in contact with the mortar. Each cylinder was tapped and vibrated to remove entrapped air and consolidate the mortar. The samples were placed in a 100% humidity cabinet at 25°C and cured for seven days. After seven days, the test cylinders were demolded and mounted in the test apparatus. The force required to pull the rod out of the mortar was measured using a servo-hydraulic material test system. Each series of test rods was prepared in triplicate.

2.3 Preparation of samples for examination of corrosion

A second series of tests were conducted to examin corrosion resistance. The tests on the bare rods, enameled

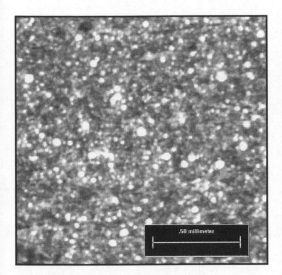

Figure 2. Photograph of polished section of vitreous enamel.

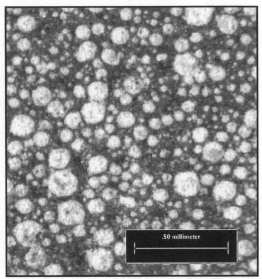

Figure 3. Photograph of polished section of vitreous enamel with embedded cement grains.

rods and enameled rods with portland-cement were performed by exposing three sets of identically prepared rods to a 3% sodium chloride solution in partly saturated sand. The goal of the testing was to provide corrosion inducing conditions similar to those found in carbonated (non-alkaline) portland-cement concrete that was contaminated with chloride. The pH of the wet, drained sand ranged form 6.0 to 6.5 and the temperature was maintained at 25°C. Corrosion will only occur on the enamel coated rods if there is a defect in the coating. This is because of the typically high electrical resistance of the enamel. Therefore, a defect was intentionally made in the enamel coating to exposed the metal rod. The defect was made by drilling through the enamel thereby exposing the bare metal of the rod. Vitreous enamel typically has a volume resistivity of 1×10^{14} ohm-cm, so a perfect enamel surface is an insulator. Defects were introduced in each of the coated rods. All of the enameled rods were tested using the procedure outlined in ASTM C 876 and showed potentials more negative than -0.35 CSE (copper sulfate electrode) indicating corrosion was occurring. The test rods were examined and photographed after 72 hours of salt water exposure.

2.4 Preparation of samples for hydration testing

Enameled test rods were examined to determine if the enamel was reacting with contacting water to hydrate the cement grains embedded in the glass (Figures 2 and 3). The presence of the hydration reaction was determined by measuring the pH of the surface of the rods enameled with and without the embedded cement

Table 2. Comparison of average bond strengths.

Treatment	Average peak force (N)	Std. deviation (N)	Average bond strength (MPa)
Steel rods, uncoated embedded in mortar	2,618	466	2.06
Enameled rods without portland cement embedded in mortar	3,498	541	2.70
Rods with enamel containing portland cement embedded in mortar	11,125	235	8.79

grains. Test rods with hydrating cement develop an elevated surface pH. The surface of each test rod was abraded with corundum paper and the surface was flooded with distilled water. The pH of the surface water film was determined by using pH test strips. Solutions with known pH were used to verify the colorimetric response.

3 RESULTS

3.1 Pull-out bond tests

The results of the pull-out testing for coated and uncoated rods are presented in Table 2. Each series of

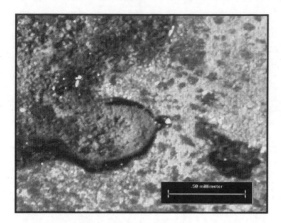

Figure 4. Photograph of surface of test rod after exposure to a 3% salt solution for 72 hours.

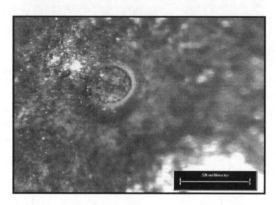

Figure 5. Photograph of surface of test rod after 72 hour of exposure to 3% salt solution.

test rods was prepared in triplicate. The results are presented as the average value and the standard deviation. The average bond strength is calculated as the force for pullout distributed across the area of the metal-enamel interface.

3.2 Corrosion test results

An example of the effect on the surface of the bare steel after 72-hours exposure is shown in Figure 4. Active corrosion was noted where steel was exposed in defects created on the enameled rods (Figure 5), both with and without the portland cement addition. Since the enamel did not debond, corrosion should remain a local occurrence and not cause excessive expansion.

3.3 Hydration test results

The abraded surface of rods containing only glass had a neutral to mildly acidic pH. Similarly treated samples containing both glass and portland cement and glass showed a surface pH that was strongly alkaline ranging from 10.5 to 11.5. The elevated pH indicates the portland cement grains were reacting with the distilled water to form calcium silicate hydrate gel and calcium hydroxide (Taylor 1997).

4 DISCUSSION AND CONCLUSIONS

The glass-ceramic coating containing the layer of portland cement can significantly increase the bond strength between the metal and the surrounding mortar. The increased bond strength (up to four times that of bare steel) suggests that the hydration of the cement on the surface takes up the excess water that would normally accumulate on the surface of the reinforcement.

The glass layer on the surface of the steel protects the underlying steel and only portions of the steel that were purposely exposed by removing the glass layer showed any corrosion. The enamel layer provided protection with or without the addition of the portland cement.

The increase in the pH observed when the surface of the glass-ceramic coating was wetted indicates that the calcium silicates and aluminoferrite compounds in the embedded ceramic would still hydrate resulting in an increase in bond strength.

Data collected in this investigation indicate that a glass-ceramic coating on reinforcing steel used in concrete can improve the bond strength between the steel and the surrounding concrete. The glass bonded to the surface of the steel effectively protects the steel from corrosion even in an aggressive (high chloride) environment.

The hydration of the embedded portland cement produces a saturated calcium hydroxide solution. The elevated pH on the surface of the steel can help to restore the passivity of the steel surface. The calcium silicate hydrate gel produced by the reaction with water can provide additional protection for the steel.

ACKNOWLEDGMENT

The tests described and the resulting data presented herein, unless otherwise noted, were obtained from research conducted under the AT22 Basic Research Military Engineering Program of the US Army Corps of Engineers by the Department of the Army. Permission was granted by the Director, Geotechnical and Structures Laboratory to publish this information.

REFERENCES

Amleh, L. & Ghosh, A. 2006. Modeling the effect of corrosion on bond strength at the steel-concrete interface with finite-element analysis. *Canadian J. of Civil Eng.* 33(6): 673–682.

Cheng, A., Huang, R. Wu, J. & Chen, C. 2005. Effect of rebar coating on corrosion resistance and bond strength of reinforced concrete. *Construction and Building Mater.* 19(5): 404–412.

Funahashi, M. Fong, K-F. & Burke, N. 1992. Investigation of rebar corrosion in partially submerged concrete. STP 1137-EB. ASTM International, West Conshohocken, PA.

Li, C.Q., Zheng, J.J. Lawanwisut, W. & Melchers, R.E. 2007. Concrete delamination caused by steel reinforcement corrosion. *J. Mat. In Civ. Engrg.* 19(7): 591–600.

Majumdar, A. & Jana, S. 2001. Glass and glass-ceramic coatings, versatile materials for industrial and engineering applications. *Bull. Mater. Sci.* 24(1): 69–77.

Qian, S. 2004. *Preventing rebar corrosion in concrete structures.* Publication No. NRCC-47625, National Research Council Canada, Institute for Research in Construction, Ottawa, Canada.

Taylor, H. 1997. *Cement Chemistry.* Thomas Telford Publ., London. 459 p.

Zhou, K., Martin-Pérez, B. & Lounis, Z. 2005. *Finite element analysis of corrosion-induced cracking, spalling and delamination of RC bridge decks.* Publication No. NRCC-48147, National Research Council Canada, Institute for Research in Construction, Ottawa, Canada.

Excellence in Concrete Construction through Innovation – Limbachiya & Kew (eds)
© 2009 Taylor & Francis Group, London, ISBN 978-0-415-47592-1

A new design approach for plate-reinforced composite coupling beams

R.K.L. Su & W.H. Siu

Department of Civil Engineering, The University of Hong Kong, Hong Kong, PRC

ABSTRACT: Plate-reinforced composite (PRC) coupling beam is fabricated by embedding a vertical steel plate into a conventional reinforced concrete coupling beam to enhance its strength and deformability. Shear studs are weld on the surfaces of the steel plate to transfer forces between the concrete and steel plate. Based on extensive experimental studies and numerical simulation of PRC coupling beams, an original and comprehensive design procedure is proposed to aid engineers in designing this new type of beams. The proposed design procedure consists of four main parts, which are (1) estimation of ultimate shear capacity of beam, (2) design of RC component and steel plate, (3) shear stud arrangement in beam span, and (4) design of plate anchorage in wall piers. The design procedure developed was validated by comparing the predicted beam capacities with those from non-linear finite element analyses.

1 INTRODUCTION

Coupling beams in coupled shear walls are often the most critical members in tall buildings subject to earthquake or wind loads. To ensure the survival of shear walls under high-intensity cyclic loading, these beams, which normally have limited dimensions, should possess high deformability and good energy absorption while being able to resist large shear forces. Conventional reinforced concrete (RC) coupling beams with longitudinal reinforcement and vertical stirrups for taking flexure and shear forces respectively have been proven unsatisfactory to fulfil the requirements. Brittle failures in the form of diagonal or sliding cracking are often encountered in these beams under seismic loading.

Various alternative forms of coupling beams, such as RC coupling beams with diagonal reinforcement (Paulay & Binney 1974), steel coupling beams (Gong & Shahrooz 2001), rectangular steel tube coupling beams with concrete infill (Teng et al., 1999), were proposed and investigated. Although these alternatives perform much better than the conventional RC coupling beams, none of them concurrently satisfies demands on high deformability, good energy absorption, low stiffness degradations, easy construction and minimum disturbance to slab or wall detailing. With the aims of fulfilling all these requirements, Lam et al. (2003) proposed a new alternative which is fabricated by embedding a steel plate in a conventional RC beam and using shear studs to couple the steel plate and the concrete (herein referred to as plate reinforced composite (PRC) coupling beams).

Eight half-scale PRC coupling beam specimens with span/depth ratios of 2.5 and 1.17, respectively, had been tested (Lam et al., 2003; Lam et al., 2005; Su et al., 2006; Su et al., in press a) to quantify the strength, deformability, energy dissipation, shear stud force distributions and internal plate forces. The excellent shear capacity, very good energy absorption and deformability of PRC coupling beams were demonstrated experimentally. The performance of deep PRC coupling beams was found to be comparable to that of diagonally reinforced coupling beams, while the former can be designed for large capacities by simply providing a thicker plate without facing the problem of steel congestion (Su et al., in press a). The shear studs in the wall regions contributed considerably in improving inelastic performance while those in the beam span could only slightly increase the beam capacity of PRC coupling beams (Su et al., 2006). To ensure desirable inelastic performance of PRC coupling beams, the embedded steel plate has to be effectively anchored in the wall piers by providing shear studs on the plate surfaces in the wall regions. Design models of simplified bearing stress distribution in the plate anchorage were proposed (Lam et al., 2005; Su et al., 2006). Nearly one hundred non-linear finite element models of hypothetical PRC coupling beams with different ratios of clear span length to overall depth $1 \le l/h \le 4$ and steel plate depth to overall beam depth $0.95 > h_p/h > 0.8$ were analyzed (Su et al., in press b; Lam et al., submitted.) to provide a better understanding of the behaviours of shear studs in the embedded beam region and in wall regions. The effects of the variations in span/depth ratio, plate anchorage

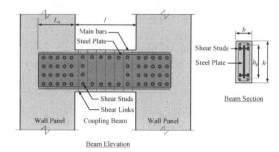

Figure 1. Typical arrangement of PRC coupling beams.

length, plate thickness and longitudinal reinforcement ratio were investigated. Equations for quantifying the shear stud forces were established and a set of non-dimensional design charts for determining the internal forces of the embedded steel plate were constructed. Based on the comprehensive experimental and numerical results, this paper attempts to establish a unified design procedure to aid engineers in designing PRC coupling beams.

2 FEATURES OF PRC COUPLING BEAMS

Fig. 1 shows a typical arrangement and geometry of a PRC coupling beam. In this design, a steel plate is vertically embedded into the conventional RC beam section across the whole span. Throughout the span, shear studs are welded on both vertical faces of the plate along the top and the bottom longitudinal reinforcement to enhance the plate/ RC composite action. The plate is anchored in the wall piers and shear studs are provided in these regions to increase the plate bearing strength. With the embedded steel plate of a PRC coupling beam framing into the wall piers, a continuous shear transfer medium far less affected by concrete cracking at the beam-wall interfaces during inelastic stage is provided, thus preventing brittle failure and increasing the rotational deformability of the beam. The experimental study (Su et al., in press a) further indicated that the steel plate is effective in taking both shear and bending forces for deep coupling beams.

PRC coupling beams are flexible in design and easy to construct. By using different amounts of longitudinal reinforcement and steel plate, the flexural capacity of the beam can be easily adjusted to suit different magnitudes of design moment. Unlike other approaches, such as embedding steel sections in coupling beams, the insertion of steel plate has the least disturbance to reinforcement details, so that vertical, lateral and longitudinal reinforcement from walls, slabs and coupling beams respectively can be harmoniously integrated together. The vertical arrangement of steel plate allows concrete to be placed and compacted easily, so

honeycomb type of defects can be avoided. Furthermore, the cast-in steel plate can naturally be protected by the surrounding concrete against fire and lateral buckling. Small holes through the plate to accommodate pipes and conduits are also possible. As shear studs are welded on the plate to couple the concrete element and the steel plate, it is much simpler, faster and economical than fabricating compound steel sections.

3 PROPOSED UNIFIED DESIGN PROCEDURE

The design procedure of PRC coupling beams described in this section is applicable to normal practical ranges of span/depth ratios ($1 \le l/h \le 4$) and plate depth/ beam depth ratios ($0.95 > h_p/h > 0.8$).

3.1 Ultimate shear capacity of PRC coupling beams

PRC coupling beams were recommended to be designed for shear stresses not exceeding 12 MPa for grade 60 concrete. For other concrete grades lower than grade 60 concrete, the following ultimate shear stress capacity of PRC coupling beam may be adopted,

$$v_u = V_u /(bd) \le 1.5\sqrt{f_{cu}} \le 12\text{MPa} \qquad (1)$$

where V_u is the ultimate shear force resisted by the beam, b is the beam width, d is the effective depth measured from top fiber to centre of longitudinal tensile steel and f_{cu} is the characteristic concrete cube strength. It should be noted that the material partial safety factor (1.25) of shear has been incorporated in Equation (1).

3.2 Shear resistance design of steel plate

In the design of PRC coupling beams, a steel plate is cast in a conventional RC coupling beam to supplement the RC component for resisting shear. The steel plate is required to take up the additional shear when the design ultimate shear V_u exceeds the maximum allowable shear in the RC component $V_{RC,allow}$, which varies depending on the concrete grade. Numerical investigation (Su et al., in press b) revealed that the load shared by the steel plate should not be more than $0.45V_u$ even for beams with a small span/depth ratio and embedded with a thick plate. Thus the plate shear demand $V_{p,req}$ is expressed as

$$V_{p,req} = V_u - V_{RC,allow} \le 0.45V_u \qquad (2)$$

According to BS5950 (BSI, 1990), there are two possible cases for the design shear in a steel plate, low shear load and high shear load. Low shear load is defined as

$$V_{p,req} \leq 0.6\left(0.6 f_{up} / \gamma_{mp}\right)\left(0.9 t_p h_p\right) = 0.324 f_{up} t_p h_p / \gamma_{mp} \tag{3}$$

where f_{up} is the ultimate strength of the steel plate, γ_{mp} is the partial safety factor of the plate and t_p is the thickness of the steel plate. This load case usually corresponds to medium-length and long coupling beams with span-to-depth ratios, $l/h \geq 1.5$.

Alternatively, high shear load is defined as

$$V_{p,req} > 0.324 f_{up} t_p h_p / \gamma_{mp} \tag{4}$$

This loading case is more often associated with short coupling beams ($l/h < 1.5$). In this case, bending and ultimate tensile strength (f_{up}) of the steel plate have to be reduced by a factor $1-\rho_1$, where ρ_1 has been given in BS5950 and is expressed as,

$$\rho_1 = 2.5 V_{p,req} / (0.54 f_{up} t_p h_p / \gamma_{mp}) - 1.5 \tag{5}$$

It is noted that in the cases of low shear load, $\rho_1 = 0$.

Based on BS5950, the shear capacity V_p of steel plates can be obtained as:
for low shear load condition

$$V_p = 0.324\left(f_{up} / \gamma_{mp}\right) t_p h_p \geq V_{p.req} \tag{6a}$$

for high shear load condition

$$V_p = 0.6\left(f_{up} / \gamma_{mp}\right) \times 0.9 t_p h_p = 0.54 f_{yp} t_p h_p \geq V_{p.req} \tag{6b}$$

At the beam-wall joints, the plate is subject to combined bending, axial and shear forces. In order to reserve sufficient load-carrying capacity for the steel plate to resist the bending and axial forces, for the cases of high shear load, the design shear load is advised to be

$$V_{p,req} \leq 0.8 V_p \tag{7}$$

such that the stress reduction factor ρ_1 is less than 0.5.

For sizing of plate, the depth of plate h_p can be determined geometrically. Based on Equations (2), (6) and (7), a suitable plate thickness t_p can be selected for short coupling beams ($l/h \leq 1.5$), of which the design is controlled by shear force only. For the cases of long coupling beams where the design is controlled by bending, sizing of steel plate and reinforcement will be described in the next section.

3.3 Bending resistance design of steel plate and RC section

In typical RC coupling beam designs, the same amount of top and bottom reinforcements is often provided as both reinforcements are required for taking tension under reversing cyclic loads. Also, because of the plate/RC interaction in a PRC coupling beam, the RC component will be under an axial compression and the standard design procedure for RC beams in BS8110 (BSI, 1997) cannot be applied to determine the required longitudinal reinforcement. A new design procedure for design PRC coupling beams under bending is proposed herein.

Under partial plate/RC composite action, the flexural strains of the concrete and the steel plate will not be the same. Horizontal forces F_x would be exerted on the RC part and the steel plate respectively in equal magnitudes but at opposite directions. The previous numerical and experimental investigations found that beam-wall joints are the most critical location for the plate design and yielding often occurs at the ultimate loading stage. The simplified stress blocks of member forces at beam-wall joints are shown in Fig. 2. The force of longitudinal compression steel can be expressed as

$$C_s = \varepsilon_{sc} E A_s = \frac{x - (h - d)}{x} \cdot \varepsilon_c E A_s \leq \frac{f_y A_s}{\gamma_{ms}} \tag{8}$$

where ε_{sc} is the strain of longitudinal compression steel, ε_c is the ultimate compressive strain of concrete, E is the Young's modulus of steel bars, x is the neutral axis depth, f_y is the yield strength of reinforcement, A_s is the area of longitudinal tensile or compressive reinforcement and γ_{ms} is the partial safety factor of reinforcement. Using the simplified rectangular stress block, the compression of concrete can be obtained and expressed in Equation (9).

$$C_c = \frac{0.67 f_{cu} \times 0.9(b - t_p)x}{\gamma_{mc}} = \frac{0.603 f_{cu}(b - t_p)x}{\gamma_{mc}} \tag{9}$$

where γ_{mc} is the partial safety factor of concrete. When the concrete compressive strain has reached its ultimate value ($\varepsilon_c = 0.0035$), the deformation of the longitudinal tensile steel has usually exceeded the yield limit and the tensile force of the reinforcement can be calculated by Equation (10).

$$T_s = \frac{A_s f_y}{\gamma_{ms}} \tag{10}$$

Due to the horizontal force interaction between the RC part and the steel plate, a net compressive force F_x is exerted on the RC section at the beam-wall joints. In the parametric study (Su et al., in press b), the tensile

389

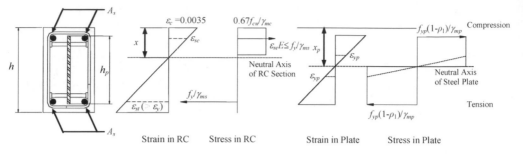

Figure 2. Strain and simplified stress diagrams of beam section under ultimate sagging moment.

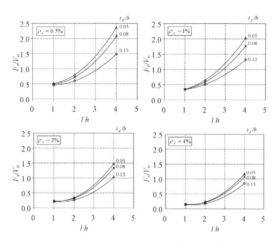

Figure 3. Axial force of steel plate at beam-wall joint.

·

force F_x was found to be dependent on the steel ratio of the longitudinal bars, the plate thickness to beam width ratio, and the span/depth ratio of the beam. The plots of F_x/V_u against l/h with various steel ratios ρ_s are reproduced in Fig. 3 for reference.

When $\varepsilon_{sc} < \varepsilon_y$, where ε_y is the yield strain of reinforcement, the neutral axis depth of RC section is derived as,

$$x = \frac{F_x + \left(\dfrac{f_y}{\gamma_{ms}} - \varepsilon_c E\right)A_s + \sqrt{\left[F_x + \left(\dfrac{f_y}{\gamma_{ms}} - \varepsilon_c E\right)A_s\right]^2 + 2.412\dfrac{f_{cu}}{\gamma_{mc}}(b - t_p)(h - d)\varepsilon_c EA_s}}{1.206\dfrac{f_{cu}}{\gamma_{mc}}(b - t_p)}$$

(11a)

Alternatively, when $\varepsilon_{sc} \geq \varepsilon_y$, the neutral axis depth can be simplified to

$$x = \frac{\gamma_{mc}F_x}{0.603 f_{cu}(b - t_p)}$$

(11b)

From the equilibrium, the neutral axis depth of the steel plate can be expressed as

$$x_p = \frac{h}{2} - \frac{\gamma_{mp}F_x}{2 f_{up}t_p(1 - \rho_1)}$$

(12)

The compression of steel plate is equal to

$$C_p = \frac{f_{up}(1 - \rho_1)h_p t_p}{\gamma_{mp}} \cdot \frac{1}{2} - \frac{F_x}{2}$$

(13)

And the tension of steel plate is

$$T_p = \frac{f_{up}(1 - \rho_1)h_p t_p}{\gamma_{mp}} \cdot \frac{1}{2} + \frac{F_x}{2}$$

(14)

Taking moment at the neutral axis of the RC section, the bending moment capacities of RC section and steel plate are expressed, respectively, as

$$M_{RC} = T_s(d - x) + C_s(x - (h - d)) + 0.55C_c x$$

(15)

$$M_p = T_p\left(x_p - x + \frac{h_p}{4} + \frac{\gamma_{mp}F_x}{4t_p f_{up}(1 - \rho_1)}\right) + C_p\left(x - x_p + \frac{h_p}{4} - \frac{\gamma_{mp}F_x}{4t_p f_{up}(1 - \rho_1)}\right)$$

(16)

By choosing a suitable steel ratio of longitudinal reinforcement A_s and plate thickness t_p, the total bending capacity of the section can be designed to be greater than the ultimate design moment M_u, i.e.

$$M_{RC} + M_p \geq M_u$$

(17)

3.4 Shear resistance design of RC section

After determining the steel area of reinforcement A_s, the shear reinforcement can be provided to resist the shear force $V_u - V_{u,req}$ and the corresponding shear stress, i.e.

$$v_{RC,req} = \frac{V_u - V_{p,req}}{(b - t_p)d} \quad (18)$$

where $v_{RC,req}$ is the required shear stress in RC part. The design concrete shear strength according to BS8110 (BSI, 1997) is

$$v_c = \frac{0.79}{\gamma_{mv}} \cdot \left(\frac{100A_s}{(b-t_p)d}\right)^{\frac{1}{3}} \left(\frac{400}{d}\right)^{\frac{1}{4}} \left(\frac{f_{cu}}{25}\right)^{\frac{1}{3}} \quad (19)$$

where $\dfrac{100A_s}{(b-t_p)d} \le 3$ and $\dfrac{400}{d} \ge 1$.

For span/depth ratio $l/h \le 4$, shear enhancement described in the clause 3.4.5.8, BS8110 may be used. The design shear stress may be increased by $2d/a_v$, where a_v is the distance measured from the concentrated load to the support and it may be taken as half of the span length of the coupling beam. In any case, the design shear stress should not be higher than $0.8\sqrt{f_{cu}}$ or 5 MPa, according to BS8110 (or 7 MPa, according to the Hong Kong code of practice for structural use of concrete 2004 (Buildings Department, 2004).

$$\frac{A_{sv}}{s_v} \ge \frac{(b-t_p) \cdot (v_{RC,req} - v_c)}{f_{yv}/\gamma_{mv}} \quad (20)$$

where A_{sv}, s_v and f_{yv} are the area, the spacing and the yield strength of transverse reinforcement respectively. It should be noted that Equation (20) has only considered the shear area from the shear links. The shear area contributed from the vertical steel plate is ignored conservatively.

3.5 Shear stud arrangement in beam span

Shear studs are required in the beam span to serve four functions as illustrated in Fig. 4. Based on the observations from the parametric study (Su et al., in press b), design equations for estimating the required shear connection strengths, and the numbers of shear studs required in turn, for these functions were proposed and listed below.

Function 1: Vertical stud forces for inducing shear on steel plate
Shear studs should be provided within a width of $0.3h_p$ away from the beam-wall joint at each beam end for transferring the plate shear force. The required transverse shear connection force for inducing shear on the steel plate $P_{t1,req}$ is

$$P_{t1,req} = 0.065 \frac{\rho_p^{0.75} V_u}{\rho_s^{0.1}} \quad (21)$$

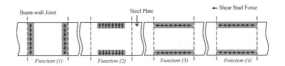

Figure 4. Functions of shear studs in beam span.

where $\rho_p = 100 h_p t_p/(hb)[\%]$ is the ratio of plate sectional area to the beam sectional area and $\rho_s = 100 A_s/(hb)[\%]$ is the steel ratio.

Function 2: Vertical stud forces for maintaining tension tie effect of steel plate
Such forces are provided within the central $(l - 0.6h_p)$ region near the top and the bottom plate fibres, and the required transverse shear connection strength for providing tension tie effect $P_{t2,req}$ was found to be

$$P_{t2,req} = 0.3(l/h)^{0.65} \rho_s^{0.1} V_u \quad (22)$$

Function 3: Horizontal stud forces for inducing moment on steel plate
It was proposed that the required longitudinal shear connection force within the beam span for transferring moment $P_{l1,req}$ would be

$$P_{l1,req} = 0.165\left(\frac{l}{h}\right)^{0.3} \rho_s^{0.45} \rho_p^{0.4} V_u \quad (23)$$

Function 4: Horizontal Stud Forces for Inducing Axial Force on Steel Plate
The required longitudinal shear connection forces within the beam span for inducing axial force on steel plate $P_{l2,req}$ was expressed as

$$P_{l2,req} = \max\begin{cases} 0.0217\left(\dfrac{l}{h}\right)^{2.2} \dfrac{M_u}{\rho_s^{0.75} \rho_p^{0.15} h_p} \\[3mm] 0.106\left(\dfrac{h}{l}\right)^{0.4} \dfrac{V_u}{\rho_s^{0.07} \rho_p^{0.2}} \end{cases} \quad (24)$$

It should be noted that the contribution of natural plate/RC bonding to transfer shear forces is ignored in the design.

The numerical results showed that shear studs provided in the central beam region could not be effectively mobilized. By arranging all shear studs near the beam-wall joints and near the top and the bottom fibers of the steel plate, where shear studs could be effectively mobilized, a high degree of shear stud mobilization could be assumed.

According to BS5950, the shear stud force Q is taken as $0.8Q_k$ and $0.6Q_k$ under positive and negative moments respectively when designing conventional composite beams with RC slabs and structural

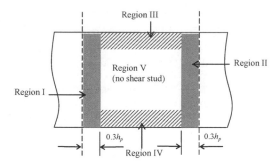

Figure 5. Regions of steel plate in beam span for shear stud arrangement.

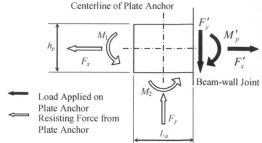

Figure 6. Simplified plate anchorage design model.

steel beams interconnected by shear studs. As the plate/RC interface slips in a PRC coupling beam are unlikely to be as large as in the case of conventional composite beams under positive moments, $Q = 0.6Q_k$ is considered when calculating the numbers of shear studs required in different regions in the beam span.

The steel plate of a PRC coupling beam can be divided into five regions in the beam span according to the different shear stud arrangements (Fig. 5). Based on the above proposals, and assuming the width of Regions I and II to be $0.3h_p$, the required numbers of shear studs (n_{req}) in different regions are expressed as follows:

Region I or Region II:

$$n_{req} = \frac{\sqrt{\left(0.5P_{t1,req}\right)^2 + \left[0.3\left(h_p/l\right)\left(P_{l1,req} + P_{l2,req}\right)\right]^2}}{0.6Q_k} \quad (25)$$

Region III or Region IV:

$$n_{req} = \frac{\sqrt{\left(0.5P_{t2,req}\right)^2 + \left[\left(0.5 - 0.3h_p/l\right)\left(P_{l1,req} + P_{l2,req}\right)\right]^2}}{0.6Q_k}$$
$$(26)$$

3.6 Design of plate anchorage in wall piers

The simplified plate anchorage design model adopted in this paper is depicted in Fig. 6.

3.6.1 Design plate anchorage loads near the beam-wall joints

The plate anchors of a PRC coupling beam are designed to take up the ultimate plate moment M_p' and part of the ultimate plate shear F_y' (as part of the shear transfer will take place in the beam span). They also need to resist an axial force F_x' jointly induced by

the plate/RC interaction under bending and the beam elongation upon cracking of concrete.

The load-carrying capacity of PRC coupling beams and the plate anchorage design loads near the beam-wall joints are controlled by the shear and flexural capacities of the steel plate. The shear and flexural capacities of PRC coupling beams are expressed, respectively, in Equations (27) and (28).

Shear capacity

$$V_{PRC} = V_{RC} + V_p \geq V_u \quad (27)$$

Flexural capacity

$$M_{PRC} = M_{RC} + M_p \geq M_u \quad (28)$$

where V_{RC}, V_p, M_{RC} and M_p can be calculated from Equations (6), (15), (16) and (18) to (20).

The PRC coupling beam is flexural-controlled when $V_{PRC} > 2M_{PRC}/l$ or shear-controlled when $V_{PRC} < 2M_{PRC}/l$. When the beam is flexural-controlled, the plate has reserved shear but not flexural capacity. In such cases, the capacities of the steel plate, the plate anchorage and the PRC coupling beam are all governed by the yield moment of the steel plate. The design applied moment of the plate anchorage (equal to the yield moment) can be obtained by taking moment about the centroid of the plate, and is expressed in Equation (29).

$$M_p' = C_p\left[\frac{\left(h_p + h\right)}{2} - x_p\right] \qu(29)$$

Assuming the plate has fully yielded with the stress distribution as shown in Fig. 2, the compression of steel plate can be determined and into Equation (29), and then

$$M_p' = \left[\left(\frac{h_p}{2}\right)^2 - \left(\frac{h}{2} - x_p\right)^2\right]\frac{f_{up}t_p}{\gamma_{mp}}\left(1 - \rho_1\right) \quad (30)$$

The shear load taken by the steel plate near the beam-wall joint, which is less than or equal to the shear capacity of the steel plate, is estimated by Equation (31).

$$V_p' = V_{PRC}\left(\frac{M_p'}{M_{PRC}}\right) \le V_p \tag{31}$$

Conversely, when the beam is shear-controlled, the shear load taken by the steel plate near the beam-wall joint may be obtained from Equation (6), such that, $V_p' = V_p$. The design plate anchorage moment of the steel plate may then be estimated as,

$$M_p' = M_{PRC}\left(\frac{V_p'}{V_{PRC}}\right) \le M_p \tag{32}$$

It was observed in the parametric study (Su et al., in press b) that the plate anchor of a PRC coupling beam would take up about 50 to 75% of the design plate shear V_p'. As the shear studs in beam span near each beam-wall joint have been designed (in Equation 21) to transfer $0.5P_{t1,req}$ to the plate, the remaining vertical force required to be taken by the plate anchor is

$$F_y' = V_p' - 0.5P_{t1,req} \tag{33}$$

The axial force F_x' induced on a plate anchor is equal to F_x acting on the plate and can be determined from Fig. 3. The design plate anchorage loads obtained (M_p', F_x' and F_x') will be used for calculating the bearing stress distribution and designing the shear stud arrangement in plate anchors.

3.6.2 Bearing stress distributions and shear stud arrangements in plate anchors

Taking moment about the centroid of beam section at the beam-wall joint, the required resisting moments of the plate anchor can be expressed as:

$$M_1 + M_2 = 0.5F_y'L_a + M_p' \tag{34}$$

The required moment resistance has to be determined in conjunction with the plate anchorage length L_a of which the minimum length is given in Figure 7. Note that slightly longer plate anchorage length than the recommended minimum value may be assumed first, as the recommendation is based on the arrangement of shear studs at minimum allowable spacing throughout the whole anchor, which is not necessarily the case in the design. The distributions of resisting moments, which depend on the geometry of the plate anchor, were investigated in the parametric study (Lam et al., submitted) and are plotted in Fig. 8.

Assuming high degrees of shear stud mobilizations (i.e. $Q = 0.6Q_k$), the required numbers of shear studs

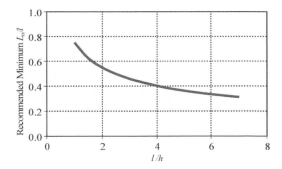

Figure 7. Recommended minimum plate anchorage length.

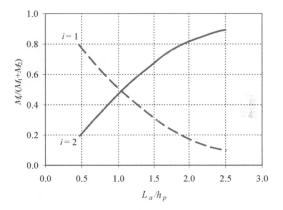

Figure 8. Distributions of resisting moments in plate anchor.

n_{req} in different regions are calculated from Equations (35) and (36). The design envelopes for the bearing stress distributions in the vertical and the horizontal directions are shown in Fig. 9 for arranging shear studs in Regions I and II in the plate anchors.

Region I (width $L_I = F_y'L_a^2/6M_2$):

$$n_{req} = \frac{\sqrt{\left(\frac{L_I}{L_a}\right)^2\left(\frac{w_x h_p}{2}\right)^2 + \left(w_{y1}L_I\right)^2}}{0.6Q_k} \tag{35}$$

Region II (width $L_{II} = (1 - F_y'L_a/6M_2)L_a$):

$$n_{req} = \frac{\sqrt{\left(\frac{L_{II}w_x h_p}{L_a}\right)^2 + \left(w_{y2}L_{II}\right)^2}}{0.6Q_k} \tag{36}$$

where

$$w_x = \frac{6M_1 + 2F_x'h_p}{h_p^2} \tag{37}$$

393

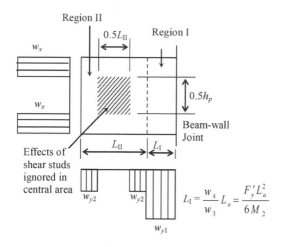

Figure 9. Regions of plate anchor with different shear stud arrangements and simplified design bearing stress blocks.

$$w_{y1} = \frac{6M_2 + L_a F_y'}{L_a^2} \qquad (38)$$

$$w_{y2} = \frac{6M_2 - L_a F_y'}{L_a^2} \qquad (39)$$

Note that although the effects of shear studs in the shaded area in Fig. 9 are ignored, evenly distributed shear studs are provided in Region II for simplicity. Furthermore, BS5950 states that the minimum allowable shear stud spacing is five times and four times the nominal shank diameter in the directions along and perpendicular to the major shear stud action respectively. As the major shear stud actions can either be in the horizontal or in the vertical directions, it is recommended that a minimum shear stud spacing of five times the nominal shank diameter be provided in all cases.

4 VALIDATION OF THE PROPOSED DESIGN PROCEDURE

An extensive non-linear finite element analysis (Su et al., in press b) was conducted to investigate the load-carrying capacity of the PRC beams and the behaviour of shear studs under different combinations of beam geometries, plate geometries, and reinforcement details. Nearly 100 models of hypothetical PRC coupling beams which were set within a normal practical range to simulate real coupling beams (i.e. 1 to 4 for span/depth ratios, 250 to 1000 mm for beam depths, and 250 mm for wall thicknesses and beam widths) were analyzed in the study. The material parameters assumed in the study were as follows: $f_{cu} = 60$ MPa, $f_y = 460$ MPa, $E_s = 200$ GPa,

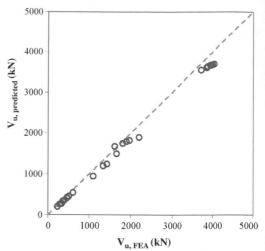

Figure 10. Comparing the strengths of PRC coupling beams obtained from the proposed design procedure and finite element analysis (Su et al., in press b).

$f_{yp} = 355$ MPa ($t_p \leq 16$ mm) or 345 MPa (16 mm $< t_p \leq 40$ mm), $E_p = 205$ GPa. Adopting the geometric and material properties in the parametric study and assuming all the material partial safety factor to be unity in Section 3, the load-carrying capacity of the PRC coupling beams was determined in accordance with the proposed design procedure and the results together with those obtained from the finite element analysis are plotted in Figure 10. The predicted design capacity in general underestimated that from the numerical analysis by around 10%. Consistent results from the proposed design procedure and finite element analysis are observed over a wide range of span/depth ratios, plate thicknesses and steel ratios of PRC coupling beams. The reliability and accuracy of the proposed design procedure are demonstrated.

5 CONCLUSIONS

With the aim of providing the construction industry with a practical, effective and economical coupling beam to resist high shear force and large rotational demand from large wind or seismic loading, plate-reinforced composite (PRC) coupling beam was developed. The effectiveness and efficiency of this new form of beams were demonstrated by extensive experimental studies and numerical simulations. In this paper, an original and comprehensive design procedure is proposed to ensure proper beam detailing for desirable performances of PRC coupling beams. The proposed design procedure consists of four main parts, which are (1) estimation of ultimate shear capacity of beam, (2) design of RC component and steel plate, (3) shear stud arrangement in beam span, and (4) design of plate anchorage in wall piers. The design procedure developed was validated by comparing the predicted beam capacities with those from non-linear finite element analyses. Lastly, the practicality of PRC coupling beams in terms of integration of steel plate together with neighbouring reinforcement, easy of concreting and no special requirement for protecting steel plate against fire and lateral buckling was highlighted.

ACKNOWLEDGEMENTS

The work described in this paper has been fully supported by the Research Grants Council of Hong Kong SAR (Project No. HKU7168/06).

REFERENCES

BSI. 1990. BS5950 Structural use of steelwork in building, Part 3: Design in composite construction, Section 3.1: Code of Practice for Design of Simple and Continuous Composite Beams. London: British Standards Institution, London.

BSI. 1997. BS8110, Part 1, Code of Practice for Design and Construction, Structural Use of Concrete, British Standards Institution, London.

Buildings Department. 2004. Code of Practice for Structural Use of Concrete 2004, The Government of the Hong Kong Special Administrative Region.

Gong, B. & Shahrooz, B.M. 2001. Concrete-steel composite coupling beams I: Component testing. Journal of Structural Engineering, ASCE 127(6): 258–271.

Paulay, T. & Binney, JR. 1974. Diagonally reinforced coupling beams of shear walls. ACI Special Publication SP-42: 579–598.

Lam, W.Y., Su, R.K.L. & Pam, H.J. 2003. Strength and ductility of embedded steel composite coupling beams. Advances in Structural Engineering 6(1): 23–35.

Lam, W.Y., Su, R.K.L. & Pam, H.J. 2005. Experimental study on embedded steel plate composite coupling beams. Journal of Structural Engineering, ASCE 131: 1294–302.

Lam, W.Y., Su, R.K.L. & Pam, H.J. submitted. Behavior of plate anchorage in plate-reinforced composite beams, Steel and Composite Structures.

Su, R.K.L., Pam, H.J. & Lam, W.Y. 2006. Effects of shear connectors on plate-reinforced composite coupling beams of short and medium-length spans. Journal of Constructional Steel Research 62: 178–188.

Su, R.K.L., Lam, W.Y. & Pam, H.J. in press a. Experimental study of plate-reinforced composite deep coupling beams, The Structural Design of Tall and Special Buildings.

Su, R.K.L., Lam, W.Y. & Pam, H.J. in press b. Behaviour of embedded steel plate in composite coupling beams, Journal of Constructional Steel Research.

Teng, J.G., Chen, J.F. & Lee, Y.C. 1999. Concrete-filled steel tubes as coupling beams for RC shear walls. Proceedings of the 2nd International Conference in Advanced Steel Structures, Hong Kong: 391–399.

Analytical modeling of FRP-strengthened RC exterior beam-column joints

T.H. Almusallam, Y.A. Al-Salloum, S.H. Alsayed & N.A. Siddiqui
Department of Civil Engineering, King Saud University, Riyadh, Saudi Arabia

ABSTRACT: In this paper a model has been presented for the prediction of shear strength of FRP strengthened exterior beam-column joint using stress equilibrium and strain compatibility equations. The model predictions are compared with the experimental observations. The predictions show good agreement with the test results. The model is further extended to predict variation of shear stress in the joint at different stages of loading. Here again predictions are fairly good. Effect of column axial load and effectiveness of FRP quantity on joint shear strength have also been investigated through parametric studies.

1 INTRODUCTION

Shear failure of exterior RC beam-column joints is identified as the principal cause of collapse of many moment-resisting frame buildings during recent earthquakes. It is, therefore, imperative to develop an effective and economic strengthening technique for upgrading the vulnerable beam-column joints in existing structures. Out of many conventional and relatively old strengthening techniques, joint strengthening using fiber-reinforced polymers (FRP) has the advantages of simplicity of application and less need for skilled labor. The use of FRP composites for strengthening is relatively a modern way, and generally, most effective due to advantages like fast and easy application; high strength/weight ratio and corrosion resistance.

FRP strengthened reinforced concrete (RC) beam-column joints have been the subject of extensive research over the past decade. The reviews bring the general conclusion that externally bonded composite materials (e.g. Glass FRP or Carbon FRP), attached to the faces of the joint by using epoxy resin, significantly improve the joint shear strength and ductility. The review shows that only a very limited work is available on analytical modeling of FRP-strengthened RC beam-column joints. Present work is an effort in the same direction.

FRP strengthened reinforced concrete (RC) beam-column joints have been the subject of extensive research over the past decade. The reviews bring the general conclusion that externally bonded composite materials, attached to the faces of the joint by using epoxy resin, may significantly improve the joint shear strength and ductility. The review shows that a substantial work is available on experimental studies of FRP upgraded joints. However, a limited work is available on analytical modeling of FRP-strengthened joints. Using the analogy to steel stirrups Gergely et al. (1998) computed the FRP contribution to the shear capacity of the RC joint. Gergely et al. (2001) repeated this analogy and, based on limited test results, fixed the FRP strain to a certain value for prepared concrete surface (0.0021 for concrete surfaces prepared with a wire brush, and to 0.0033, for water-jetted concrete surfaces). In addition to detailed experimental study on FRP strengthened RC beam-column joint, Ghobarah and Said (2001) also proposed a design methodology for fibre jacketing to upgrade the shear capacity of existing beam-column joints in reinforced concrete moment resisting frames. El-Amoury and Ghobarah (2002) proposed a simple design methodology for upgrading reinforced concrete beam-column joints using GFRP sheets. Antonopoulos and Triantafillou (2002) proposed a powerful model which uses stress equilibrium and strain compatibility to yield shear strength of the beam-column joint with known geometry, reinforcement quantities, and externally bonded FRP. They programmed their proposed model to predict shear strength of FRP strengthened joints. In the present study, authors have applied aforementioned model on present FRP strengthened exterior joint and also extended this model to predict some other governing parameters such as variation of shear stress in the joint at different stages of loading. To implement available and extended formulations for various predictions of FRP-strengthened exterior joints, a comprehensive computer program was developed and its results were

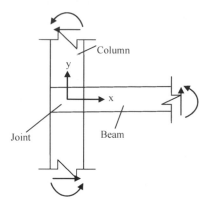

Figure 1. Moment and shear forces acting at the exterior joint.

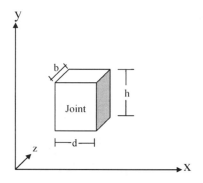

Figure 2. Dimensions of joint.

compared with the experimental observations. Effects of axial load and FRP quantity on joint shear capacity are also investigated through parametric studies.

2 PROBLEM FORMULATION

To predict the shear capacity and some other governing parameters for FRP-strengthened RC beam-column joint, joint is idealized as a 3D element (Figs. 1 and 2) with dimensions d (width of column), b (width of beam), and h (height of beam). Having idealized the joint as a 3D element following assumptions has been made to derive the governing equations:

- Shear stress τ distribution is uniform over the boundaries of the joint.
- To have generalization in the formulation, the joint is assumed to be loaded at the time of strengthening, and therefore, a set of initial normal strains, ε_{0x} and ε_{0y} exists in the transverse (beam) and longitudinal (column) directions respectively. An initial shear strain, γ_0, has also developed due to loading at the time of strengthening.

- Strengthening of the joint is carried out through the use of unidirectional sheets placed in two orthogonal directions (vertically and horizontally).

2.1 Governing states

Under the action of seismic forces, joints are assumed to pass through various states of deterioration. In the present study following basic states of joint behaviour have been analyzed:

i. Before yielding of horizontal and vertical steel reinforcement.
ii. After yielding of effective horizontal reinforcement but before yielding of effective vertical reinforcement.
iii. After yielding of effective vertical reinforcement but before yielding of effective horizontal reinforcement.
iv. After yielding of both horizontal and vertical reinforcement.
v. Compressive crushing of concrete.
vi. Failure of FRP.

These states are same as considered by Antonopoulos and Triantafillou [5] in their analysis of exterior joints. Detailed derivation and formulation for theses states can be found in Alsayed et al.(2008).

3 SOLUTION PROCEDURE

To incorporate the above states in the analysis of RC joints strengthened with FRP sheets, a comprehensive computer program was developed. This program requires user to input various material and geometric properties in order to predict shear capacity and other governing parameters. Program major input consists of:

i. The material properties of concrete, steel and FRP.
ii. The variables related with geometry of the joint;
iii. The variables which determine bond conditions;
iv. The horizontal and vertical axial forces in normalized form; and
v. At the time of strengthening, values of initial strains.

4 DISCUSSION OF RESULTS

4.1 Comparison with test results

In order to have confidence in present formulation and developed program, shear capacity of control (Figs. 3 and 4) and strengthened specimen (Figs. 5 and 7) are predicted, and predicted results are compared with the test results. A summary of various data employed for the predictions are shown in Table 1. The joints of these specimens were inadequately detailed (with no stirrups in the joint core, Fig. 4), and the strengthening system was designed such that failure would occur due

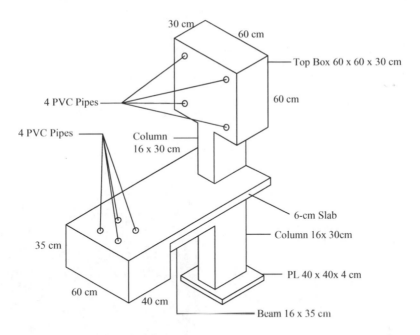

Figure 3. Schematic diagram of RC exterior joint specimen.

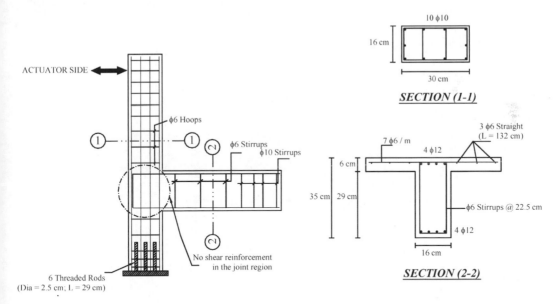

Figure 4. Reinforcement details of the specimen.

to shear. Earthquake loads were simulated by applying an alternating force (in a quasi-static cyclic pattern) at the end of the beam (Fig. 6). The displacement controlled loading sequence for each specimen consisted of three cycles at series of progressively increasing (by 1 mm) displacement amplitudes in each direction (push and pull) until a displacement corresponding to failure of the specimen was reached. A detailed description of the experiment and its test set up are presented in Al-Salloum and Almusallam (2007).

Last three columns of Table 2 shows the experimentally observed shear strength, analytically predicted shear capacity and ratio of these two values, respectively. The ratios obtained in the last column clearly indicate that predicted shear capacities are in a good agreement with experimental values. This comparison, thus, add a sufficient degree of confidence on present formulation, and developed program in order to use them for shear capacity prediction of FRP strengthened beam-column joints.

4.2 *Prediction of joint shear stress*

Figures 8 and 9 illustrate the comparison of experimentally observed and predicted joint shear stress variation with transverse strain in the beam bars. Due

to malfunctioning of some of the strain gauges beyond yielding of bars, comparison is restricted up to the yield. These curves again show that the prediction is in good agreement with experimental shear stress almost for all stages of loading.

4.3 *Effect of FRP reinforcement*

Using the present formulation [6], effect of quantity of FRP reinforcement was studied on parametric basis. This study is significant, as to study the effect of such parameter experimentally is very expensive

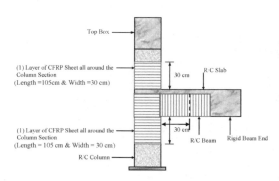

Figure 5.   Schematic diagram showing FRP strengthening scheme.

Figure 7.   Picture showing FRP strengthening scheme.

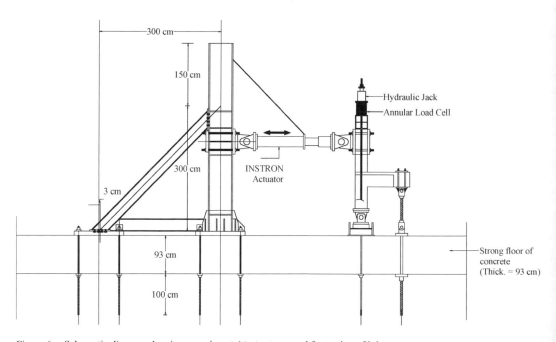

Figure 6.   Schematic diagram showing experimental test set-up used for testing of joints.

Table 1. Data for the prediction of shear capacity of control and FRP upgraded exterior joints.

Parameters	Value
Size of beam	160×350 mm
Size of column	160×300 mm
Longitudinal steel ratio, ϕ_y	0.016
Beam top or bottom steel ratio, ϕ_b	0.009
Modulus of elasticity of steel E_s	2.0×10^5 MPa
Yield strength of longitudinal steel, p_{uy}	420 MPa
Yield strength of transverse steel, p_{ux}	420 MPa
Concrete strength, f_c'	30 MPa
Average tensile strength of concrete p_{ctm}	3.0 MPa
Column reinforcement ratio inside joint core	0.016
Type of FRP	Unidirectional CFRP sheet
Ultimate tensile strength in primary fiber direction	745 MPa
Elastic modulus of FRP sheets in primary fibers direction, E_x	61.5×10^3 MPa
Fracture strain of FRP sheets	1.2%
Thickness of the sheet, t_f	1.0 mm
FRP reinforcement ratio ϕ_f	0.0125
Shear modulus of FRP sheets, G_{yx}	2.51×10^3 MPa
Poisson's ratio of FRP sheets, $v_{xy}(=v_{yx})$	0.25
Mechanical Anchorage	No
Applied vertical axial force, F_v	20% of column axial load capacity

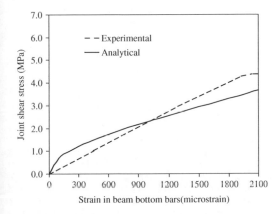

Figure 8. Joint shear stress in control specimen.

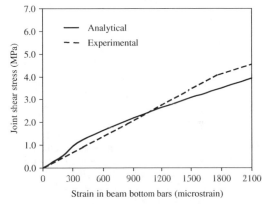

Figure 9. Joint shear stress in strengthened specimen.

and time consuming. The baseline data for this parametric study is same as given in Table 1. Figure 10 shows the variation of joint shear strength with amount of FRP, keeping all other parameters same as shown in Table 1. The amount of FRP is presented as FRP reinforcement ratio, a dimensionless form, defined as $\phi_f = nt_f/b$, where b is the width of beam, n is the number of joint sides covered by the FRP, and t_f is thickness of the sheet. As we can see from this figure that a little increase in the amount of FRP, considerably increases the shear strength of the joint. This can be attributed to the confinement of joint core. As the quantity of FRP increases, confinement increases, which in turn increases the shear strength of the joint. It is to be

noted that this curve is drawn with the assumption that joint is the weakest link. However, if beams are significantly weaker than the joint, the sub assemblage can fail before any significant failure in the joint.

4.4 Effect of axial load

Figure 11 shows the variation of shear capacity of the joint with axial load. It is obvious from the figure that as axial load increases joint shear strength also increases. This is due to the fact that with the increase of axial load confinement of joint core increases which consequently increases the shear capacity of the joint.

401

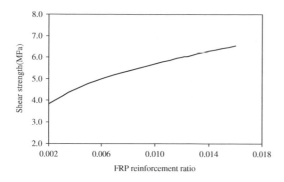

Figure 10. Effect of FRP reinforcement on joint shear strength.

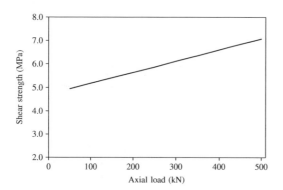

Figure 11. Effect of axial load on shear strength of the joint.

5 CONCLUSIONS

In this paper a procedure for analytical prediction of joint shear strength of exterior beam-column joints, strengthened with externally bonded CFRP sheets, was presented. The predicted shear capacities and joint shear stress variations for control and FRP-strengthened beam-column joints were compared with experimental observations and they were found in good agreement with the experimental results. The effect of FRP quantity on shear strength and on various strains was studied. It was observed that as the quantity of FRP increases, confinement increases, which in turn increase the shear strength of the joint. The effect of axial load was also investigated and it was observed that with the increase of axial load confinement of joint core increases which consequently increases the shear capacity of the joint.

ACKNOWLEDGEMENTS

Authors acknowledge the financial support provided by King Abdualaziz City for Science and Technology (KACST) under grant Number AR-21-40.

REFERENCES

Alayed, S.H., Almusallam, T.H. And Al-Salloum, Y.A. 2008. "Seismic Upgrade of Beam-Column Connections in Existing RC Buildings using FRP Composite Laminates," Final Report, King Abdulaziz City For Science And Technology (KACST), Project Ar-21-40, Riyadh, Saudi Arabia.

Almusallam, T.H. And Al-Salloum, Y.A. 2007. "Seismic Response of Interior Beam-Column Joints Upgraded with FRP Sheets. II: Analysis and Parametric Study", Journal of Composites for Construction, ASCE, Vol. 11(6): 590–600.

Al-Salloum, Y.A. And Almusallam, T.H. 2007. "Seismic Response of Interior Beam-Column Joints upgraded with FRP Sheets. I: Experimental Study", Journal of Composites for Construction, ASCE, 11(6): 575–589.

Antonopoulos, C. And Triantafillou, T.C. 2002. "Analysis of FRP-Strengthened Beam-Column Joints", Journal of Composites for Construction, ASCE, 6(1): 41–51.

El-Amoury, T. And Ghobarah, A. 2002. "Seismic Rehabilitation of Beam-Column Joint using GFRP Sheets", Engineering Structures, 24: 1397–1407.

Gergely, J., Pantelides, C.P. And Reaveley, L.D. 2000. "Shear Strengthening of RC-Joints using CFRP Composites", Journal of Composites for Construction, 4(2): 56–64.

Gergely, J., Pantelides, C.P., Nuismer, R.J. And Reaveley, L.D. 1998. "Bridge Pier Retrofit Using Fiber-Reinforced Plastic Composites", Journal of Composites for Construction, 2(4): 165–174.

Ghobarah, A., And Said, A. 2001. "Seismic Rehabilitation of Beam-Column Joints using FRP Laminates", Journal of Earthquake Engineering, 5(1): 113–129.

Excellence in Concrete Construction through Innovation – Limbachiya & Kew (eds)
© 2009 Taylor & Francis Group, London, ISBN 978-0-415-47592-1

Environmentally-friendly self-compacting concrete for rehabilitation of concrete structures

V. Corinaldesi & G. Moriconi

Dept. of Materials and Environment Engineering and Physics, Università Politecnica delle Marche, Italy

ABSTRACT: Self-compacting concrete technology can be particularly useful for rehabilitation of concrete structures due to its high fluidity, allowing to fill wide areas independently of their complicated shape and narrow section. Moreover, its cohesiveness guarantees adequate adhesion to the existing substrate as well. For this purpose, several self-compacting concretes were prepared by using different recycled aggregates coming from a suitable treatment of either rubble or concrete from building demolition as well as concrete from precasting plants. Self-compacting concretes were prepared without any filler addition. Other very-fine material required to achieve cohesiveness was supplied by the recycled aggregates themselves. Self-compacting concretes were characterized in the fresh state by means of slump-flow, V-funnel, and L-box with horizontal steel bars tests. They were then characterized in the hardened state by means of compression and splitting tension tests, in addition to drying shrinkage measurements. However, the shrinkage of the self-compacting concrete, expected to be higher because of the lower inert/cement, could cause differential strain with respect to the substantially stable old concrete substrate. This risk can be avoided provided that right adjustments of the self-compacting concrete mixture are adopted, as shrinkage-reducing admixture and/or fibres are added.

1 INTRODUCTION

Deterioration of concrete structures is a major problem in civil engineering, which is mainly associated with contamination, cracking and spalling of the cover concrete. As a consequence, the serviceability of the deteriorated structures becomes an important issue and therefore the most cost-effective solution is often to use patch repair, which involves the removal of deteriorated parts and reinstatement with a fresh repair mortar (see Fig. 1).

Compatibility in a repair system is the combination of properties between the repair material and the existing concrete substrate which ensures that the combined system withstands the applied stresses and maintains its structural integrity and protective properties in a certain exposure environment over a designed service life (Hassan et al. 2001, Emmons et al. 1993, Emmons & Vaysburd 1996).

Among other factors, the dimensional stability is probably the most important controlling the volume changes due to shrinkage, thermal expansion, and the effects of creep and modulus of elasticity (Emberson & Mays 1990a, Emberson & Mays 1990b, Morgan 1996).

Instead of a traditional repair mortar, a self-compacting concrete can be advantageously used for rehabilitation of concrete structures due to its high

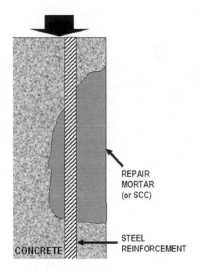

Figure 1. Schematic diagram of patch repair.

fluidity, allowing to fill wide areas independently of their complicated shape and narrow section. Moreover, its cohesiveness can guarantee adequate adhesion to the existing substrate as well.

In general, for self-compacting concrete production a high volume of very fine materials is necessary in order to make the concrete more fluid and cohesive.

For this purpose, in previous works (Corinaldesi & Moriconi 2003, Corinaldesi & Moriconi 2004a, Corinaldesi & Moriconi 2004b) rubble powder (that is a powder obtained from a suitable treatment of rubble from building demolition) was used as mineral addition in order to ensure adequate rheological properties of the self-compacting concrete in the absence of any viscosity modifying admixture, which is usually employed to guarantee sufficient cohesiveness of the self-compacting concrete (Khayat & Guizani 1997, Khayat 1997). In fact, in the absence of any viscosity-enhancing admixture (Khayat 1999), a lower water to cementitious material ratio is necessary to provide stability and, consequently, some fillers (passing the sieve ASTM n°100 of 150 μm) must be employed.

For further simplification of the self-compacting concrete (SCC) mixture proportion, in this work an attempt was made to use some recycled aggregate fractions rich of very fine material (substantially the same previously called 'rubble powder') as 100% replacement of both virgin aggregate and filler.

Three different self-compacting concretes were prepared by using different recycled aggregates coming from a suitable treatment of either rubble or concrete from building demolition as well as concrete from precasting plants. In this way an environmental benefit can also be achieved by reducing not renewable resources consumption and rubble disposal in landfill.

2 EXPERIMENTAL PART

2.1 Materials

A commercial portland-limestone (20% maximum limestone content) blended cement type CEM II/A-L 42.5R according to EN-197/1 was used. The Blaine fineness of cement was $410 \, m^2/kg$, and its specific gravity was $3.05 \, kg/m^3$.

A fine recycled rubble aggregate (FRecRub) with maximum particle size of 8 mm was used coming from a recycling plant in which demolition waste is suitably treated. Its mean composition is 72% concrete, 25% masonry and 3% bitumen.

In alternative, a fine recycled concrete aggregate (FRecCon) with maximum particle size of 8 mm was used coming from the same recycling plant in which, in this case, only waste made of concrete was suitably treated. Its composition is 100% concrete. No further processing of these two aggregate fractions was made in laboratory.

Finally, in combination with the fine recycled rubble fraction, a further recycled concrete aggregate was used, indicated as 'CRecCon', based on concrete scraps obtained as rejection from precast concrete production (50-55 MPa strength class), which were refined in laboratory from a thirty-minute dry milling in order to allow the gradation reported in Figure 2 with a maximum grain size equal to 12 mm.

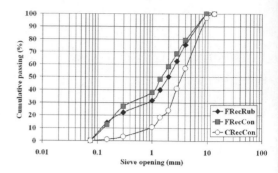

Figure 2. Grain size distribution curves of the recycled aggregate fractions.

Table 1. Physical properties of aggregate fractions.

Fraction	Water absorption (%)	Specific gravity in SSD* condition (kg/m³)	Passing 150 μm sieve (%)
Fine recycled rubble (FRecRub)	12	2206	14.3
Fine recycled concrete (FRecCon)	9	2310	12.7
Coarse recycled concrete (CRecCon)	7	2460	1.0

(*)SSD = saturated surface-dried.

The main physical properties of aggregate fractions, determined according to EN 1097-6 and EN 933-1, are reported in Table 1.

A common water-reducing admixture based on polycarboxylate polymers (30% aqueous solution) was also used for preparing concrete mixtures.

2.2 Self-compacting concrete mixture proportions

The self-compacting concrete mixture proportions are reported in Table 2. All the concrete mixtures were prepared with a water to cement ratio of 0.54 without any filler addition, by adding a water reducing admixture at a dosage of 1.15% by weight of cement.

The dosage of the recycled aggregate fractions was suitably chosen in order to achieve a volume of very fine materials (including the cement too) in the range $190–200 \, l/m^3$, which is the recommended value in order to obtain concrete self-compactability. In addition, it was verified that, in this way, the volume of the coarse material was under $340 \, l/m^3$ and the ratio between sand (i.e. the fraction between 4 and 0.15 mm) and mortar fell within the recommended range 0.47–0.53.

Table 2. Mixture proportions of self-compacting concretes.

Mixtures	RecRub	RecRub +Con	RecCon
Water/cement	0.54	0.54	0.54
Water (kg/m³)	190	190	190
Cement (kg/m³)	350	350	350
Fine recycled rubble (FRecRub) aggregate (kg/m³)	1500	1125	–
Fine recycled concrete (FRecCon) aggregate (kg/m³)	–	–	1500
Coarse recycled concrete (CRecCon) aggregate (kg/m³)	–	375	–
Superplasticizing admixture (kg/m³)	4	4	4
Volume of very fine (<0.150 mm) materials (l/m³)	209	187	199
Volume of coarse (>4 mm) materials (l/m³)	162	193	136
Volume of sand/ volume of mortar	0.50	0.51	0.53

Table 3. Results of tests on fresh concrete.

Tests		RecRub	RecRub+ Con	RecCon
Slump test	Φ_{fin} (mm)	665	720	670
	t_{fin} (s)	12	7	1
V-funnel	t_{fin} (s)	10	8	3
L-box	$\Delta H_{fin}^{(1)}$ (mm)	140	100	70
	$t_{edge}^{(2)}$ (s)	12	8	7
	$t_{stop}^{(3)}$ (s)	18	10	9

(1) difference in the concrete level between the beginning and the end of the box;
(2) elapsed time to reach the opposite edge of the box;
(3) elapsed time to establish the final configuration.

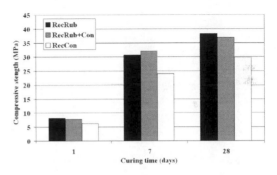

Figure 3. Compressive strengths vs. curing time.

Since the first mixture (RecRub), prepared by using the 'FRecRub' fraction showed a quite high volume of very fine materials (209 l/m³, see Table 2), the second mixture (RecRub+Con) was prepared by suitably combining the 'FRecRub' with the coarser 'CRecCon' fraction (at 75% and 25% respectively), in order to optimize the SCC mixture proportions. On the other hand, the last mixture (RecCon) was prepared by using the only 'FRecCon' fraction.

3 RESULTS AND DISCUSSION

3.1 Fresh concrete behaviour

Fresh concrete performance was evaluated trough slump flow, V-funnel and L-box with horizontal steel bars tests in order to verify the likely self-compactability. Results obtained are reported in Table 3.

All the concrete mixtures fulfil SCC requirements and in particular the 'RecRub+Con' mixture perform very well in terms of fluidity and the 'RecCon' mixture perform very well in terms of mobility through narrow sections.

Any segregation phenomena were never noticed.

3.2 Mechanical performance of hardened concrete

Compressive strength of SCCs was evaluated after 1, 7 and 28 days of wet curing since ingredients' mixing on three cubic (100 × 100 × 100 mm) specimens for each curing time and each concrete mixture. Results obtained are reported in Figure 3.

The concrete mixtures 'RecRub' and 'RecRub+Con' achieved a concrete strength class equal to 30 MPa while the concrete mixture 'RecCon' achieved a concrete strength class equal to 25 MPa.

In terms of tensile strengths the mean values of the results obtained were 2.29, 2.64 and 2.24 MPa for the 'RecRub', 'RecRub+Con' and 'RecCon' mixtures respectively. The positive effect on tensile strength of the coarse recycled concrete fraction (characterized by rough surface and better physical properties, see Table 1) is quite evident in the mixture 'RecRub+Con'. These values were obtained by means of splitting tension tests carried out after 28 days of wet curing since concrete preparation on five cubic specimens (100 × 100 × 100 mm) for each concrete mixture.

The tangent elastic modulus of SCC was also evaluated according to BS 1881 part 5. The results obtained are reported in Figure 4.

Elastic modulus values are about 20% less than the expected values for ordinary concretes belonging to strength class of 30 and 25 MPa. The reason probably lies on the kind of aggregate used and it confirmed

405

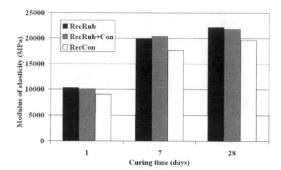

Figure 4.   Elastic moduli vs. curing time.

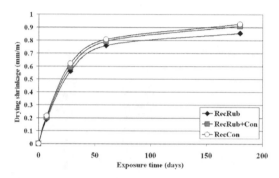

Figure 5.   Drying shrinkage vs. time of exposure to 50% R.H. at 20°C.

previous results obtained on recycled-aggregate concretes by the authors (Corinaldesi & Moriconi 2001, Corinaldesi & Moriconi 2006).

Mismatch in the modulus of elasticity becomes of great concern in repairs when the applied load is parallel to the bond line in a combined system. The material with the lower modulus deforms more and, therefore, transfers the load through the interface to the higher modulus material (Mailvaganam 1997). For this reason, for the design of an efficient repair system the modulus of elasticity of both the repair concrete and the concrete substrate should be similar. As a consequence, these SCC can be used together with low strength (15–20 MPa) concrete substrate, which indeed is the most common concrete strength class for those old structures requiring rehabilitation.

3.3 Drying shrinkage

Three prismatic ($100 \times 100 \times 500$ mm) concrete specimens for each concrete mixture were exposed to 50% relative humidity and a temperature of 20°C for 6 months in order to evaluate their drying shrinkages. Results obtained are reported in Figure 5.

The results obtained showed too high shrinkage strains, affecting SCC potential compatibility with concrete substrate.

In fact, as the fresh repair material tends to shrink, the concrete substrate (relatively old) restrains it. The differential movements cause tensile stresses within the concrete (creep in such a situation is an advantage, as it releases part of these stresses). As shrinkage proceeds, the stresses accumulate and could cause cracks and failure if exceed the tensile capacity of the repair material or the bond strength at the interface.

In the case of these SCC mixtures, tensile strengths are relatively low while drying shrinkages are quite high. For this reason, some expedients are needed in order to improve SCC performance under drying. An idea can be the addition of shrinkage-reducing admixtures to the SCC mixtures, which allow higher concrete stability in hot-dry environment (Collepardi et al. 2005, Berke et al. 2003, Nakanishi et al. 2003). The use of steel fibres proved also to be very effective in reducing self-compacting concrete shrinkage (Corinaldesi & Moriconi 2004c).

4   CONCLUSIONS

All the concrete mixtures, prepared by using recycled aggregate fractions rich of very fine materials as 100% replacement of both virgin aggregate and filler, fulfil SCC requirements. For this reason, these mixtures can also be called 'environmentally-friendly self-compacting concretes'. As a matter of fact, the use of recycled materials produces an environmental benefit, achieved by reducing norenewable resources consumption and rubble disposal in landfill.

In terms of mechanical performances in general, and modulus of elasticity in particular, these SCC can be used together with low strength (15–20 MPa) concrete substrate, which indeed is the most common concrete strength class for those old structures requiring rehabilitation.

On the basis of the data collected, the drying shrinkages measured for all the SCC mixtures seems to be very high and cracking is likely to occur if they are used as patch repair of old deteriorated concrete structures without any proper adjustment.

Future development of the research program concerns the study of self-compacting concrete mixtures prepared with shrinkage-reducing admixtures and/or different fibre types in order to verify their effectiveness in reducing excessive deformability under drying.

REFERENCES

Berke, N.S., Li, L., Hicks, M.C., Bal, J., 2003. Improving concrete performance with shrinkage-reducing admixtures. In V.M. Malhotra (ed.), Superplasticizers and Other Chemical Admixtures in Concrete: Proc.s of the Seventh CANMET/ACI Int. Conf., Berlin, Germany, 2003 (SP-217), Farmington Hills, Michigan (U.S.A.): American Concrete Institute, 37–50.

Collepardi, M., Borsoi, A., Collepardi, S., Ogoumah Olagot, J.J., Troli, R., 2005. Effect of shrinkage reducing admixture in shrinkage compensating concrete under non-wet curing conditions, *Cement & Concrete Composites* 27, 704–708.

Corinaldesi, V., Moriconi, G., 2001. Role of Chemical and Mineral Admixtures on Performance and Economics of Recycled Aggregate Concrete. In V.M. Malhotra (ed.), *Fly Ash, Silica Fume, Slag and Natural Pozzolans in Concrete, Proc.s of the Seventh CANMET/ACI Int. Conf., Chennai (Madras), India, 2001 (SP-199)*, Farmington Hills, Michigan (U.S.A.): American Concrete Institute, Vol. 2, 869–884.

Corinaldesi, V., Moriconi, G., 2003. The Use of Recycled Aggregates from Building Demolition in Self-Compacting Concrete. In O. Wallevik and I. Nielsson (eds.), *RILEM Proceedings PRO 33, Proc.s of the 3rd Int. RILEM Symp. on "Self-Compacting Concrete", Reykjavik, Iceland, 2003*, Bagneux (France): RILEM Publications S.A.R.L., 251–260.

Corinaldesi, V., Moriconi, G., 2004a. The Role of Recycled Aggregates in Self-Compacting Concrete. In V.M. Malhotra (ed.), *Fly Ash, Silica Fume, Slag and Natural Pozzolans in Concrete, Proc.s of the Eighth CANMET/ACI Int. Conf., Las Vegas, U.S.A., 2004 (SP-221)*, Farmington Hills, Michigan (U.S.A.): American Concrete Institute, 941–955.

Corinaldesi, V., Moriconi, G., 2004b. Self-Compacting Concrete: A Great Opportunity for Recycling Materials. In E. Vázquez, C.F. Hendricks and G.M.T. Janssen (eds.), *RILEM Proceedings PRO 40, Procs. of the Int. RILEM Conf. on the "Use of the Recycled Materials in Building and Structures", Barcelona, Spain, 2004*, Bagneux (France): RILEM Publications S.A.R.L., 600–609.

Corinaldesi, V., Moriconi, G., 2004c. Durable fiber reinforced self-compacting concrete. *Cement and Concrete Research* 34: 249–254.

Corinaldesi, V., Moriconi, G. 2006. Behavior of Beam-Column Joints Made of Sustainable Concrete under Cyclic Loading. *Journal of Materials in Civil Engineering* 18(5): 650–658.

Emberson, N.K., Mays, G.C., 1990a. Significance of properties mismatch in the repair of structural concrete, Part I: Properties of repair systems. *Magazine of Concrete Research 42(152)*, 147–160.

Emberson, N.K., Mays, G.C., 1990b. Significance of properties mismatch in the repair of structural concrete, Part II: Axially loaded reinforced concrete members. *Magazine of Concrete Research 42(152)*, 161–170.

Emmons, P.H., Vaysburd, A.M., McDonald, J.E., 1993. A rational approach to durable concrete repairs. *Concrete International 15(9)*, 40–45.

Emmons, P.H., Vaysburd, A.M., 1996. System concept in design and construction of durable concrete repairs. *Construction and Building Materials 10(1)*, 69–75.

Hassan, K.E., Brooks, J.J., Al-Alawi, L., 2001. Compatibility of repair mortars with concrete in a hot-dry environment. *Cement and Concrete Composite 23*, 93–101.

Khayat, K.H., 1997. Use of Viscosity-Modifying Admixture to Reduce Top-Bar Effect of Anchored Bars Cast with Fluid Concrete. *ACI Materials Journal* 94(4), 332–340.

Khayat, K.H., Guizani, Z., 1997. Use of Viscosity-Modifying Admixture to Enhance Stability of Fluid Concrete. *ACI Materials Journal 94(4)*, 332–340.

Khayat, K.H., 1999. Workability, Testing and Performance of Self-Consolidating Concrete. *ACI Materials Journal 96(3)*, 346–353.

Mailvaganam, N.P., 1997. Repair and rehabilitation of concrete structures: Current practice and emerging technologies. In: *Evaluation, Repair and Retrofit of Structures using Advanced Methods and Materials. First Int. Civil Engineering "Egypt-China-Canada" Symp., Cairo, 1997*, 137–167.

Morgan, D.R., 1996. Compatibility of concrete repair materials and systems. *Construction & Building Materials 10(1)*, 57–67.

Nakanishi, H., Tamaki, S., Yaguchi, M., Yamada, K., Kinoshita, M., Ishimori, M., Okazawa, S., 2003. Performance of a Multifunctional and Multipurpose Superplasticizer for Concrete. In V.M. Malhotra (ed.), *Superplasticizers and Other Chemical Admixtures in Concrete: Procs. of the Seventh CANMET/ACI Int. Conf., Berlin, Germany, 2003 (SP-217)*, Farmington Hills, Michigan (U.S.A.): American Concrete Institute, 327–342.

Excellence in Concrete Construction through Innovation – Limbachiya & Kew (eds)
© 2009 Taylor & Francis Group, London, ISBN 978-0-415-47592-1

Shrinkage-free fiber-reinforced mortars

S. Monosi & O. Favoni

Dept. of Materials and Environment Engineering and Physics, Università Politecnica delle Marche, Italy

ABSTRACT: Shrinkage compensating mortars are in general cementitious materials containing expansive agents that produce a compressive stress due to the restraint exerted by a metallic reinforcement. This compressive stress reduces or disappears at later ages. To be successful, the mortar must be wet-cured because the expansion can occur only in the presence of water. The present work is devoted to studying the behaviour of expansive mortars fibre reinforced with fibres, containing CaO as expansive agent and a special superplasticizer based on a polyacrylate with a special chemical group acting as shrinkage reducing admixture (SRA). The aim was to verify if it is possible to maintain expansion or at least to avoid shrinkage at later ages even in the absence of wet curing. The expansion effectiveness on the bending behaviour of fibre reinforced conglomerates was also investigated. The presence of fibres different in shape and/or material did not modify nor restraint expansion and shrinkage trend. Not even the flexural strength was modified by the simultaneous presence of fibres and expansive agent, despite the fact that this combined addition should have produced a prestressed state localized around the fibres. In addition it seems that the shrinkage free superplasticizer does not negatively influence the compressive strength of cement mortars.

1 INTRODUCTION

Several researchers have studied the performances of expansive cement mortars (Collepardi et al. 2005, Collepardi et al. 2006a, Maltese et al. 2005, Monosi & Fazio 2005) and several papers have reported that good results can be obtained by the combined use of expansive agents based on CaO and SRA (Shrinkage Reduction Admixture). In particular, the desired expansion can be achieved and can last for long periods by the simultaneous presence of both admixtures, even without wet curing. Moreover some authors noticed that the first expansion appears greater than in the presence of the expansive agent alone.

Concrete industrial floors (Collepardi et al. 2006b) and restoration mortars (Coppola 2000) are two typical structures that can benefit from the above findings. Pavements are not covered by formworks during the first curing days, therefore they are suddenly exposed to the open air and generally show very poor expansion. Restoration mortars are employed to repair damaged structures and in order to do so, they have to be sound and strongly adhere to the previous concrete. The old concrete is very stable because it has almost completed its hygrometric shrinkage; on the contrary the new mortar still has to begin its contraction. Tensions localized on the interface between old and new concrete can cause crack formation and material detachment that abolish the benefit of restoration.

The aim of the present work was to study the dimensional stability of expansive fibre-reinforced mortars designed for structural restoration purposes. In addition, compressive and tensile strength were tested and evaluated.

2 MATERIALS AND METHODS

The actual experimental work was carried out by studying five different types of cement mortars manufactured with an ordinary blended Portland cement (CEM II/A-L type) and natural sand with maximum diameter of 4 mm. An expansive agent based on CaO was used. The prefixed workability, measured by vibrating table, corresponded to a slump flow of 140 mm in the absence of fibres; such consistency was reached by adding an acrylic superplasticizer or a special superplasticizer modified by the presence of a special chemical group acting as shrinkage reducing admixture (SRA) (Collepardi et al. 2006b). In previous works an acrylic superplasticizer and SRA were used in the same mixes, but added separately (Collepardi et al. 2005, Collpardi et al. 2006a). By selecting this new admixture, the aim was to check whether operability is actually simplified (only one admixture to add and to dose) and whether the efficacy is the same or different.

Table 1. Composition of the reference mixes (without fibres).

Composition (kg/m^3)	M0	M1	M2
Cement	450	400	400
CaO	—	45	45
Sand	1450	1450	1450
Water	225	220	215
Superplasticizer	2.70	3.60	1
Ginius	—	—	8.5

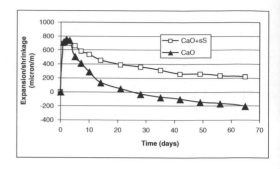

Figure 1. Expansion/shrinkage of unreinforced mortars with and without special Superplasticizer (sS).

Moreover, different types of fibres were used: a) flexible steel fibres, usually utilized for restoration mortars, b) stiff steel fibres c) structural polymeric fibres, recently introduced as alternative to the type b) ones. The mixes composition is shown in Table 1.

Three types of specimens were produced: cubic specimens ($5 \times 5 \times 5$ cm) in order to test compressive strength, prismatic specimens ($7 \times 7 \times 28$ cm) to test flexural strength and prismatic specimens ($5 \times 5 \times 24$ cm) with a main bar to measure the restrained expansion and/or shrinkage. The specimens used to test compressive strength and flexural strength were demoulded at 1 day and immediately exposed to an open environment at 60–65% of R.H. Reinforced specimens, 6 hours after casting, were demoulded and measured; the recorded value represented datum point or zero point for the subsequent measurements. After demoulding, these specimens, were also exposed to an open environment at 60–65% of R.H.

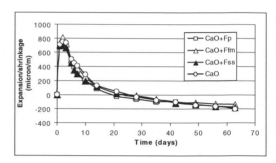

Figure 2. Expansion/shrinkage of expansive reinforced mortars and of one without fibres (Fp: polymeric fibres; Ffm: flexible metal fibres; Fss: stiff steel fibres).

3 RESULTS

The results are presented beginning from the restrained expansion-shrinkage trend. In Figure 1 the performances of expansive mortars without fibres and with or without modified superplasticizer can be observed.

In the presence of the special superplasticizer, expansion continues for a long time, in fact it equals 180–200 μm/m after two months in the environment at 60–65% of R.H. but the first expansion value remains unchanged, contrary to that reported by other authors. Since the cause of the higher first expansion is not clear, it is impossible to find an explanation for this result. Without the special superplasticizer, expansion is already zero at 25 curing days, after which the shrinkage phase begins.

Figure 2 shows curves of dimensional change vs time of the expansive fibre-reinforced mortars (with flexible metal, stiff steel and polymeric fibres) and for comparison, of the only expansive mortar without reinforcement, in order to evaluate which fibres could influence the dimensional variations.

One can observe an almost complete overlap of the four curves, that is, full coincidence of the extent of

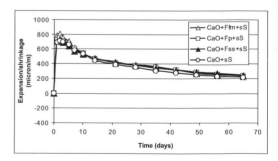

Figure 3. Expansion/shrinkage of expansive reinforced mortars and of one without fibres with special superplasticizer.

expansion-shrinkage and of the period during which these dimensional variations occur.

In the same way, all mortars with special superplasticizer with or without fibres show a similar trend (Fig. 3); after a curing time of about two months they preserve an expansion of 180–200 μm/m. Full coincidence of the dimensional change that was observed in previous mixes without SRA, is here confirmed.

Table 2 shows the compressive strength values of only three mortars, without fibres to check the special superplasticizer effect on increase in strength.

Table 2. Compressive strength of mortars without fibres, with and/or without expansive agent.

Time (days)	Compressive strength (MPa)		
	CEM	CEM + CaO	CEM + CaO + G
1	45	37	35
3	55	52	47
7	63	60	56
14	68	62	63
28	65	65	64

Table 3. Flexural strength of all fibre-reinforced mortars (Fss: stiff steel fibres; Ffm: flexible metallic fibres; Fp: polymeric fibres).

	Flexural strength (MPa)
M0 (60 g/l)	12
M0 (30 g/l)	12
M0 (10 g/l)	11
M1 + Fss(60 g/l)	14
M1 + Fss(30 g/l)	14
M1 + Ffm (30 g/l)	12.5
M1 + Fp (10 g/l)	12.5
M2 + Fss (60 g/l)	13
M2 + Fss (30 g/l)	12.5
M2 + Ffm (30 g/l)	12.5
M2 + Fp (10 g/l)	12

The results obtained during the whole curing time, point out a moderate negative effect when CaO replaces Portland cement. If the superplasticizer is also present, the decrease in strength is more evident up to seven days.

Table 3 shows the flexural strength of the mortars cured for twenty eight days. The total number of mortars tested was higher because other mixes were added; in particular, two different dosages of stiff steel fibres (the highest and the lowest of the recommended percentages) were used, to further investigate which typology in terms of shape, stiffness and aspect ratio would give the best results. Moreover one non-expansive fibre-reinforced mortar was added to evaluate whether a prestress state due to the simultaneous presence of expansive agent, special superplasticizer and fibres, could positively influence the results of flexural strength.

The recorded values (Table 3) all ranging between 12 MPa and 14 MPa do not seem to differ from each other beyond the experimental error. The higher values achieved in the presence of stiff metallic fibres are slightly significant because they are independent of the

dosages; moreover, these values were obtained from the expansive mix without special superplasticizer which has totally lost its expansion and consequently, the hypothetic prestress state at that time (Fig. 2).

4 CONCLUSIONS

The combined use of expansive agent based on CaO and a special superplasticizer with SRA, allows to retain expansion for a long time (over two months). On the other hand the increase in the first expansion found by other authors is not confirmed.

The presence of metallic and/or polymeric fibres added at different dosages does not change mortar performances in terms of restrained expansion-shrinkage with time. This can probably be ascribed to the fibres small size which is not able to confine expansions and for the same reason cannot modify the extent of shrinkage.

The combined use of expansive agent and special superplasticizer do as not significantly influence the bending behaviour of the fibre-reinforced mortars thus confirming the above hypothesis. The fibres are not able to confine cement matrix expansion, and consequently, they do not generate a prestress state, effective on first crack formation due to bending stress.

REFERENCES

Collepardi, M., Borsoi, A., Collepardi, S., Troli, R., Ogoumah Olagot, J.J., 2005. Effects of Shrinkage Reducing Admixtures in Shrinkage-Compensating Concretes Under Non-Wet Curing Conditions, *Cement Concrete Composites*, Volume 27, Issue 6, 704–708.

Collepardi, S., Corazza, G., Fazio, G., Monosi, S., Troli, R., 2006a. Shrinkage-Free Mortars for Repairing Damaged Concrete Structures, Proc. Eighth CANMET/ACI International Conference on "Superplasticizers and Other Chemical Admixtures in Concrete", SP-239, 375–387.

Collepardi, M., Troli, R., Bressan, M., Liberatore, F., Sforza, G., 2006b. Crack-Free Concrete for Outside Industrial Floors in the Absence of Wet Curing and Contraction Joints, Proc. Eighth CANMET/ACI International Conference on "Superplasticizers and Other Chemical Admixtures in Concrete", Supplementary Papers, 103–115.

Coppola, L., 2000. Concrete Durability and Repair Technology, Proc. Fifth CANMET/ACI International Conference on "Durability of Concrete", 1209–1220.

Maltese, C., Pistolesi, C., Lolli, A., Bravo, A., Cerulli, T., Salvioni D., 2005. Combined Effect of Expansive and Shrinkage Reducing Admixtures to Obtain Stable and Durable Mortars, *Cement and Concrete Research*, **35**(12), 2244–2251.

Monosi, S., Fazio, G., 2005. Malte da restauro senza ritiro (Shrinkage-Free Restoration Mortars, in Italian), *L'Edilizia*, 141, 40–43.

Excellence in Concrete Construction through Innovation – Limbachiya & Kew (eds)
© *2009 Taylor & Francis Group, London, ISBN 978-0-415-47592-1*

Comparison between design codes and procedures for concrete beams with internal FRP reinforcement in balanced case failure

M. Kadhim & T. Donchev
Kingston University, London, UK

S. Al-Mishhdani
University of Technology, Baghdad, Iraq

I. Al-Shaarbaf
Al-Nahrain University, Baghdad, Iraq

ABSTRACT: Fibre reinforced polymers (FRP) have emerged as a promising alternative material for reinforced concrete structures. FRP reinforcing bars are non-corrosive and can thus be used to eliminate the corrosion problem invariably encountered with conventional steel reinforcement. In addition FRP materials are extremely light, versatile and possess high tensile strength which is making them almost ideal material for reinforcement. This paper is proposing a method for design for concrete with FRP Reinforcement based on BS 8110 stress–strain model for concrete in compression for balanced case failures. It presents an overview and discussion on the design codes and guidelines for application of internal FRP reinforcement. The analysis of the available sources includes a comparison of main design approaches and basic principles for design of FRP reinforcement for reinforced concrete elements loaded on bending.

1 INTRODUCTION

Many engineering societies such as; ACI (American Concrete Institute), CSA (Canadian Standards Association), ISIS Canada Research Network (Intelligent Sensing for Innovative Structures), JSEC (Japan society of civil engineering) and other establishment have increased their interest in FRP as an internal reinforcement and have subsequently published several codes and guidelines of design concrete with FRP as internal reinforcement. In Japan, JSEC were first to publish and establish recommendations for design with FRP internal reinforcement. In America, ACI published several guides that dealt with FRP reinforcing (ACI 440.1R-01, ACI 440.1R-03, ACI 440.1R-06 and ACI 440.1R-07). In Canada, ISIS and CSA, published several of modules and standards of FRP reinforced concrete. In Europe, the FIB Task Group 9.3 published several reports for reinforced concrete with FRP. And in Italy, the Italian National Research Council presented the Guide for the Design and Construction of Concrete Structures Reinforced with Fiber-Reinforced Polymer Bars.

2 PROPERTIES OF FRP REINFORCED CONCRETE

One of the most important properties of FRP is its resistance to environmental conditions. FRP are non-corrosive, non-conductive, high tensile strength and lightweight which make it perfect as an internal reinforcement in extreme situations. In comparison with steel reinforcement, FRP bars have higher strength, lower modulus of elasticity and more brittle as material.

Stress-strain curves and properties of FRP reinforcement are shown in Table 1 and Fig. 1 respectively.

3 SECTION ANALYSIS IN BENDING

A section analysis in bending for concrete reinforced with FRP can be carried out with regard to ultimate strength theory. This philosophy of section analysis is similar as to that used for steel reinforcement, whilst taking into consideration difference in properties for FRP.

Table 1. Usual tensile properties of reinforcing bars.

	Steel	GFRP	CFRP	AFRP
Nominal yield stress (MPa)	276–517	N/A	N/A	N/A
Tensile strength (MPa)	483–690	483–1600	600–3690	1720–2540
Elastic modulus, (GPa)	200	35–51	120–580	41–125
Yield strain, %	1.4–2.5	N/A	N/A	N/A
Rupture strain, %	6–12	1.2–3.1	0.5–1.7	1.9–4.4

GFRP (Glass FRP), CFRP (Carbon FRP), AFRP (Aramid FRP).

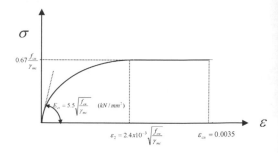

Figure 2. Stress–strain of concrete in compression.

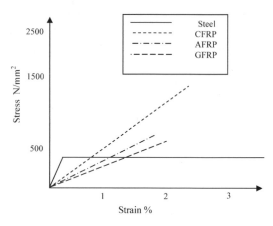

Figure 1. Stress-strain curves for reinforcing materials.

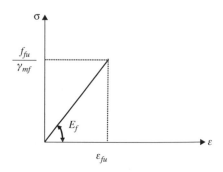

Figure 3. Stress–strain of FRP in tension.

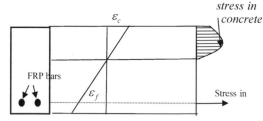

Figure 4. Typical strain and stress distribution in concrete reinforced with FRP.

3.1 Assumptions

The codes and design guidelines establish many procedures and recommendations to determine flexure capacity. Most of the procedures and the recommendation based on the following common assumptions:

- The plane section remain plane at any stage of loading
- A perfect bond exist between the FRP and concrete
- strength of concrete and FRP in tension zone is not considered and neither the strength of FRP in the compression zones
- FRP are linear elastic until failure
- Ultimate strain of concrete is either 0.003 or 0.0035

In addition to the above assumptions, the proposed method uses BS8110 stress-strain model of concrete in compression and is a modification the behaviour of steel for FRP in tension as is shown in Fig. 2 and Fig. 3

3.2 Flexural capacity

Flexural capacity depends on the properties of concrete (compressive strength and ultimate strain), the properties FRP (tensile strength and the modulus of elasticity) and the shape of section.

According to the assumptions that are made using the codes and design guidelines, linear distribution for the strain of concrete along with strain of FRP are adopted. The stress distribution for the concrete is non linear as shown in Fig. 4.

4 FAILURE MODES

The modes of flexural failure that occur in concrete reinforced by FRP can be classified as follows:

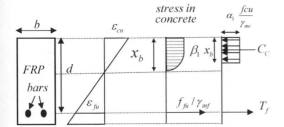

Figure 5a. Strain and stress distributions in balanced failure.

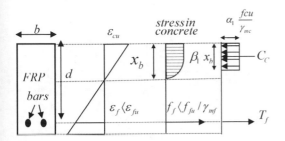

Figure 5b. Strain and stress distributions in compression failure.

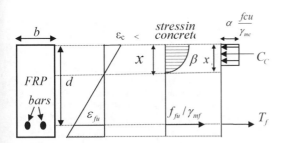

Figure 5c. Strain and stress distributions in tension failure.

- Balanced failure (the simultaneous crushing of concrete and the FRP Rupture). See Fig 5a
- Compression failure (concrete crushing). See Fig. 5b
- Tension failure (FRP rupture). See Fig 5c

All the above failures are brittle but from structural viewpoint, compression failure is preferred in compression with tension failure because compression failures arte more ductile. Most codes and design guidelines allow for all the above failures, although some codes give a greater a safety factors for tension failure (for example, ACI increased safety factor about 18% depending on ratio of FRP reinforcement). This paper will discuss only balance failure case.

4.1 Balanced failure

The criteria of balanced failure are that concrete is crushed when it reaches maximum strain at the same time as FRP rupture. The calculations for balanced failure are as follow:

Basic equilibrium is,

$$C_C = T_f \tag{1}$$

where

$$Cc = \alpha_1 \frac{f_{cu}}{\gamma_{mc}} \beta_1 x_b b \tag{2}$$

and

$$T_f = A_f \frac{f_{fu}}{\gamma_{mf}} \tag{3}$$

From compatibility

$$\frac{\varepsilon_{cu}}{x_b} = \frac{\varepsilon_{cu} + \varepsilon_{fu}}{d} \Rightarrow x_b = \frac{\varepsilon_{cu}}{\varepsilon_{cu} + \varepsilon_{fu}} d \tag{4}$$

Substituting equations 2, 3 and 4 in 1 will result in following:

$$\alpha_1 \frac{f_{cu}}{\gamma_{mc}} \beta_1 \frac{\varepsilon_{cu}}{\varepsilon_{cu} + \varepsilon_{fu}} d\, b = A_f \frac{f_{fu}}{\gamma_{mf}} \tag{5}$$

$$\Rightarrow$$

$$M_u = A_f \frac{f_{fu}}{\gamma_{mf}} (d - \beta_1 x_b / 2) \tag{6}$$

The ultimate moment capacity is

$$\rho_{fb} = \frac{A_f}{b\,d} = \alpha_1 \beta_1 \frac{\gamma_{mf}}{\gamma_{mc}} \frac{f_{cu}}{f_{fk}} \frac{\varepsilon_{cu}}{\varepsilon_{cu} + \varepsilon_{fu}} \tag{7}$$

Expressed via strength of the concrete:

$$M_u = \alpha_1 \frac{f_{cu}}{\gamma_{mc}} \beta_1 x_b (d - \beta_1 x_b / 2) \tag{8}$$

To get the dimensionless relation, substituted equation 4 in equation 7

$$M_u = \alpha_1 \frac{f_{cu}}{\gamma_{mc}} \beta_1 \frac{\varepsilon_{cu}}{\varepsilon_{cu} + \varepsilon_{fu}} d \left(d - \frac{\beta_1 d \varepsilon_{cu}}{2(\varepsilon_{cu} + \varepsilon_{fu})} \right) \tag{9}$$

$$\Rightarrow$$

$$\frac{M_u}{b\,d^2 f_{cu}} = \frac{\alpha_1 \beta_1 \varepsilon_{cu}}{\gamma_{mc}(\varepsilon_{cu} + \varepsilon_{fu})} \left(1 - \frac{\beta_1 \varepsilon_{cu}}{2(\varepsilon_{cu} + \varepsilon_{fu})} \right) \tag{10}$$

Table 2. Details of proposed method.

ε_{cu}	α_1	β_1	γ_{mc}	γ_{mf}		
0.0035	0.67	0.9	1.5	CFRP	AFRP	GFRP
				1.3	1.7	2.7

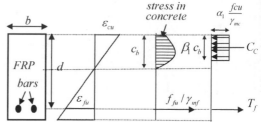

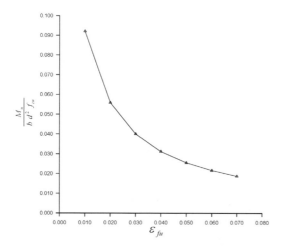

Figure 6. Relation between the ultimate strain in FRP and the dimensionless parameter (M_u/bd^2f_{cu}).

The basic data is as indicated in Table 2. The dimensionless parameter M_u/bd^2f_{cu} decreases with the increase of ultimate strain of FRP as shown in Fig 6.

The balance ratio from other design methods depend on the philosophy of design each code include the different maximum strain and stress distribution for concrete ε_{cu}, α_1, β_1 and maximum stress allowable in FRP (Fig. 7). For instance, ACI 440.1R-06[3] adopted Equation 11, whilst ISIS[4] used Equation 12 and Pilakoutas et al[6]. proposed equation 13 derived from EC2 (Eurocode 2)

$$\rho_{fb} = 0.85 \; \beta_1 \; \frac{fc'}{f_{fu}} \; \frac{E_f \; \varepsilon_{cu}}{E_f \; \varepsilon_{cu} + f_{fu}} \tag{11}$$

$$\rho_{frpb} = \alpha_1 \; \beta_1 \; \frac{\phi_c}{\phi_{frp}} \; \frac{f_c'}{f_{frpu}} \left(\frac{\varepsilon_{cu}}{\varepsilon_{cu} + \varepsilon_{frpu}} \right) \tag{12}$$

$$\rho_{fb} = \frac{0.81(f_{ck}+8) \; \varepsilon_{cu}}{f_{fk}\left(\dfrac{f_{fk}}{E_{fk}} + \varepsilon_{cu} \right)} \tag{13}$$

As seen in Fig. 8, the balance ratio increases approximately in liner fashion along with increased concrete strength. The proposed method give good results compared with other design methodologies.

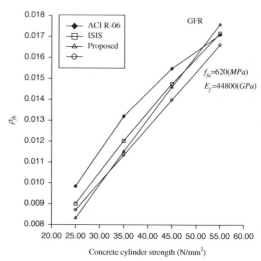

Figure 7. Strains and stress distribution model of ACI and ISIS for a section in balance case.

Figure 8. Comparison for balance ratio of different design methods for CFRP. with assume that the concrete exposed to earth conditions and cube strength = 1.25 times of cylinder strength.

5 CONCLUSION

1. The British standards do not include procedures for concrete reinforced with FRP, therefore this paper presents a proposed method for design concrete with FRP reinforcement in balance case failure based on the philosophy of the British Standard
2. Good correspondence between the proposed method and other design methods is established
3. The balance ratio of FRP reinforcement is increased with an increase of the strength of concrete

6 NOTATION

ε_c strain of concrete
ε_{cu} ultimate strain of concrete
ε_f strain of FRP
ε_{fu} ultimate strain of FRP

f_{fu}	tensile stress of FRP reinforcement
f_{fu}	ultimate tensile stress of FRP reinforcement
f_{fk}	characteristic value of tensile strength of FRP reinforcement
f_{cu}	ultimate concrete cube compressive strength
f_{ck}	characteristic value of concrete compressive strength
f_c'	concrete cylinder compressive strength
γ_{mc}	Partial safety factor for concrete
γ_{mf}	Partial safety factor for FRP
ϕ_c	Reduction factor for concrete
ϕ_f	Reduction factor FRP
α_1 & β_1	block stress parameters
E_f	modulus of elasticity of FRP reinforcement
ρ_f	FRP reinforcement ratio
ρ_{fb}	balanced FRP reinforcement ratio

REFERENCES

"FRP Reinforcement in Reinforced Concrete Structures". FIP Task Group 9.3 (2007).

"Reinforcing Concrete Structures with Fiber Reinforced Polymers," Design Manual No. 3, The Canadian Network of Centers of Excellence on Intelligent Sensing for Innovative Structures, ISIS Canada Corporation, Winnipeg, Manitoba, Canada.

ACI 440.1R-06 (2006) "Guide for the Design and Construction of Structural Concrete Reinforced with FRP Bars," ACI Committee 440, American Concrete Institute.

ACI 440R-07 (2007) "Report on Fiber-Reinforced Polymer (FRP) Reinforcement for Concrete Structures," ACI Committee 440, American Concrete Institute.

CAN/CSA-S806-02, "Design and Construction of Building Components with Fibre-Reinforced Polymers", Canadian Standards Association.

CNR-DT 203/(2006) – "Guide for the Design and Construction of Concrete Structures Reinforced with Fiber-Reinforced Polymer Bars."

Japan Society of Civil Engineers (JSCE) 1997 "Recommendation for Design and Construction of Concrete Structures Using Continuous Fiber Reinforced Materials," Concrete Engineering Series 23, ed. by A. Machida.

Excellence in Concrete Construction through Innovation – Limbachiya & Kew (eds)
© 2009 Taylor & Francis Group, London, ISBN 978-0-415-47592-1

Experimental behavior of repaired and strengthened self-compacted RC continuous beams

K.M. Heiza

Department of Civil Engineering, Menoufiya University, Shbeen El-Koom, Egypt

ABSTRACT: The utilization of the advanced composites materials under a name of fiber reinforced polymers has received a special attention in Egypt during the last decade. These materials have been applied in many fields of repair and strengthening the different structural elements. The technique of adhesively bonding traditional and advanced materials to the surfaces of reinforced concrete beams is being adopted in this research to repair and strengthen self-compacting reinforced concrete beams. An experimental program was carried out to investigate the behavior of two-span continuous reinforced concrete beams repaired and strengthened using both traditional and advanced materials by different techniques. The experimental program consists of twenty-two of self-compacting reinforced concrete beam models that have a constant cross section of 10×25 cm and of a total length 330 cm. These models were classified into four groups according to the method of repair and strengthening. The first group contains three control beams. The second group contains five beams strengthened by steel plates *(SP)* without preloading and three beams repaired by steel plates after preloading level 50% of the failure load. The third group consists of five beams strengthened by E-glass fiber reinforced polymers *(GFRP)* laminates and three beams are repaired after preloading level 50% of the ultimate load. The latest group consists of three beams strengthened by carbon fiber reinforced polymers *(CFRP)* laminates. The behaviors of the tested beams are compared at different loading levels. The failure loads, deflections, curvature and strain values at different loading stages as well as the crack propagation patterns for the tested beams were recorded. The main conclusions and recommendations for practical applications were introduced.

1 INTRODUCTION

Externally bonded advanced composite materials, are currently being studied and applied around the world for the repair and strengthening of structural concrete members (El-Shiekh 1996). FRP composite materials are of great interest to the civil engineering community because of their superior properties such as high stiffness and strength as well as ease of installation when compared to other traditional repair materials. Researches reveal that strengthening using FRP provides a substantial increase in post-cracking stiffness and ultimate load carrying capacity of the members subjected to flexure, torsion and shear (Mahmut et al 2006, Panchacharam & Belarbi 2002, Damian & David 2002).The technique of adhesively bonding steel or fiber reinforced plastics as additional laminates to the surfaces of reinforced concrete beams and slabs is being adopted worldwide to repair and strengthen the RC buildings and bridges (Tjandra & Tan 2003, Neale & Labossiere 1997) because it is inexpensive, easy to apply, causes minimal disruption to moving traffic and negligible losses in headroom. International building codes such as American Code

(ACI Committee 440) (2001) and Egyptian Code of Practice (ECOP-2005) (2005) have been introduced for the design, construction and strengthening of RC structures using FRP.

The main object of the research work is to compare the efficiency of the different materials as well as the location of the additional layers for repair and strengthening the high performance self-compacting reinforced concrete beams. Steel plates are chosen to represent the traditional materials the both glass and carbon fibers reinforced polymers are chosen to represent the advanced composite materials.

2 EXPERIMENTAL TEST PROGRAM

2.1 Test beam details

An experimental program was carried out to investigate the behavior of two-span continuous self-compacting reinforced concrete beams repaired and strengthened using both traditional and advanced materials by different techniques. The experimental program consists of twenty-two of reinforced concrete

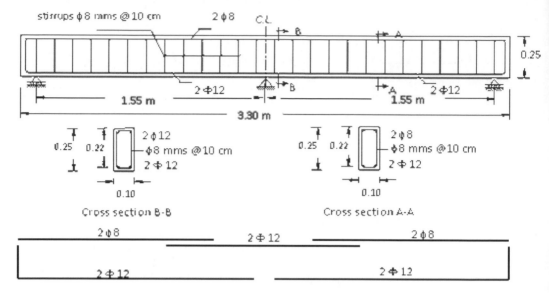

Figure 1. Typical dimensions and reinforcement details of the control beam Bo.

Table 1. Mechanical properties of steel.

Steel type	Yield stress (kg/cm^2)	Tensile strength (kg/cm^2)	Elongation %	Modulus of elasticity (t/cm^2)
High tensile bars	3600	5200	12	2100
Mild steel bars	2400	3500	20	2050
Steel plates	2350	3450	25	2100

beam models that have a constant cross section of 10×25 cm, and a total length of 330 cm. Each beam has a lower reinforcement 2 Φ 12 mm and upper main reinforcement at the negative moment zone 2 Φ 12 mm of high tensile steel. Stirrup hangers 2 φ 8 mm were used to hang stirrups 8 mm in diameter @ 10 cm of mild steel. Figure (1) shows the dimensions and reinforcement of the test models. Table 1 shows the mechanical properties of the used steel.

2.2 Material properties for self-compacting concrete

The properties of the materials such as cement, aggregates, additives and reinforcing steel used for preparing self-compacting concrete beams tested in this study were as follows:

Cement: A locally produced ordinary Portland cement complied with E.S.S.373/91 requirements was used.

Aggregate: The fine aggregate was siliceous natural sand. The coarse aggregate was crushed dolomite No.1, 2 with maximum nominal size 12.5, 25.4 respectively.

Steel reinforcement: The main bottom and the top reinforcement over the intermediate support used for all beams was high tensile strength steel of 12 mm diameter. Smooth mild steel bars of diameter 8 mm were used as stirrup hangers and transverse stirrups.

Viscosity enhancing agent (VEA) (super-plasticizer): The super-plasticizer used in this experimental program under a commercial name of Sika-Viscocrete 5-400) from Sika Egypt. VEA matches with the American specifications [ASTM C-494-type F, G].

Fly ash: The mineral admixture used in this experimental program is fly ash under a commercial name of SUPPER POZZ-5.

Self-compacting concrete mixes: Ten trail mixes are designed according to CBI method and prepared for casting the self-compacting concrete with some variation of mixing proportions. Table 2 summarizes the proportions for the different material as well as the characteristic compressive strength after 3, 7 and 28 days. It is noticed that concrete mix, M6 gives the best results, and it is chosen to be used in the experimental program. Testes were carried out according to ECCS-203[9] and recommendations of Housing and

Table 2. Self-compacting concrete mix proportions.

	Mix proportions						f_{cu} after 3-days kg/cm^2	f_{cu} after 7-days kg/cm^2	f_{cu} after 28-days kg/cm^2
	C Kg/m^3	W Kg/m^3	F.A Kg/m^3	C.A Kg/m^3	M.A Kg/m^3	VEA Kg/m^3			
M1	450	180	853	698	45	4.50	415	522	634
M2	450	180	853	698	45	6.75	351	469	658
M3	450	180	853	698	45	9.00	401	474	700
M4	450	180	853	698	67	4.50	421	577	685
M5	450	180	853	698	67	6.75	348	588	640
M6	450	180	853	698	67	8.10	442	471	705
M7	450	180	853	698	67	9.00	320	471	612
M8	450	180	853	698	90	4.50	273	497	530
M9	450	180	853	698	90	6.75	345	410	552
M10	450	180	853	698	90	9.00	204	308	411

Where
C = Cement, F.A. = Fine aggregate (sand), C.A. = Coarse aggregate (crushed dolomite)
W = Water, M.A. = Mineral Admixture (fly ash), VEA = Viscosity enhancing agent.

Table 3. The experimental program.

Group	Beam code	Material	Description	Applied position of repair and strengthening layers	Pre-Loading level
A	Bo$_1$	–	Control	–	–
	Bo$_2$	–	Control	–	–
	Bo$_3$	–	Control	–	–
B	BSST	SP	Strengthening	top surface at −ve moment	–
	BSSB	SP	Strengthening	bottom surface at +ve moments	–
	BSSTB	SP	Strengthening	both top and bottom surfaces	–
	BSS−ve||	SP	Strengthening	both vertical sides at −ve moment	–
	BSS+ve||	SP	Strengthening	both vertical sides at +ve moments	–
	BRS∩	SP	Repair	∩-shape at −ve moment	0.5 P$_f$
	BRSU	SP	Repair	U-shape at +ve moments	0.5 P$_f$
	BRS-U	SP	Repair	U-shape at both −ve & +ve moments	0.5 P$_f$
C	BSGT	GFRP	Strengthening	top surface at −ve moment	–
	BSGB	GFRP	Strengthening	bottom surface at +ve moments	–
	BSGTB	GFRP	Strengthening	both top and bottom surfaces	–
	BSG−ve||	GFRP	Strengthening	both vertical sides at −ve moment	–
	BSG+ve||	GFRP	Strengthening	both vertical sides at +ve moments	–
	BRG∩	GFRP	Repair	∩-shape at −ve moment	0.5 P$_f$
	BRGU	GFRP	Repair	U-shape at +ve moments	0.5 P$_f$
	BRGU	GFRP	Repair	U-shape at both −ve & +ve moments	0.5 P$_f$
D	BSCT	CFRP	Strengthening	top surface at −ve moment	–
	BSCB	CFRP	Strengthening	bottom surface at +ve moments	–
	BSCTB	CFRP	Strengthening	both top and bottom surfaces	–

Building Research Center (2002) to define the general properties for both concrete and steel.

2.3 Repair and strengthening schemes

Nineteen models of self-compacting RC beams were repaired and strengthened by different methods and were classified into thee groups according to the material of the repair and strengthening as follows:

Group A: contains three control beams.

Group B: contains five beams strengthened by steel plates *SP* without pre-loading and three beams were repaired by steel plates after pre-loading level 50% of the failure load. Mild steel plates were used as a tradional method for both repair and strengthening beams.

Table 4. Physical and mechanical properties of Glass and Carbon Fiber Laminates.

	Length L/ roll m	Width mm	Thickness mm	Weight gm/m^2	Tensile strength N/mm^2	Tensile modulus, E N/mm^2	Strain at failure, ε %	Fiber orientation
GFRP	50	1000	0.17	430	2250	70000	3.1	two dim.
CFRP	45.7	610	0.13	230	3500	230000	1.5	one dim.

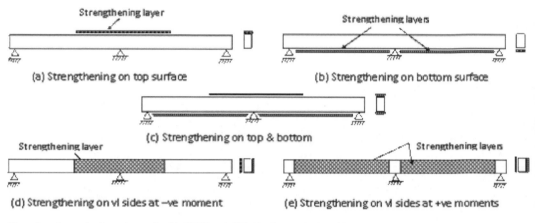

(a) Strengthening on top surface

(b) Strengthening on bottom surface

(c) Strengthening on top & bottom

(d) Strengthening on vI sides at −ve moment

(e) Strengthening on vI sides at +ve moments

Figure 2. Strengthening schemes by SP, GFRP and CFRP in Groups B, C and D.

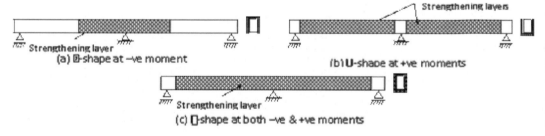

(a) ▯-shape at −ve moment

(b) U-shape at +ve moments

(c) ▯-shape at both −ve & +ve moments

Figure 3. Repair by SP and GRFP in Groups B and C.

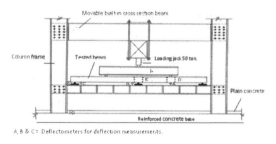

A, B & C= Deflectometers for deflection measurements.

Figure 4. Testing.

Characteristic properties of those steel plates are given in Table 2. External plates, 0.1 mm thickness, are fixed to beam sides using epoxy-resin mortar (Exuit-222 from Egyptian Swiss chemical industries Co.).

Group C: consists of five beams strengthened by E-glass fiber reinforced polymers, *GFRP* laminates and three beams are repaired after preloading level 50% of the failure load. Sika wrap Hex-430G is used with epoxy impregnating resin matrix Exuit-50.

Group D: consists of three beams strengthened by carbon fiber reinforced polymers, *CFRP* laminates Sika Wrap Hex 230C is used with epoxy impregnating resin matrix Exuit-50. Table 3. Summarizes the tested beams, and show the method of repair or strengthening.

2.4 Test set-up and instrumentation

Beams were tested under two concentrated loads up to failure. A steel frame of 200 ton capacity was used for testing beams in Menoufiya lab. Loads were applied in

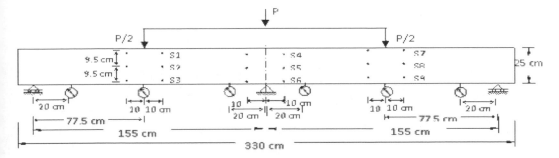

Figure 5.   External instrumentation for the loaded beam.

increments of 1.0 ton using a hydraulic jack of 50-ton maximum capacity. Figure 4 shows the loading rig.

Three dial gauges of 0.01 mm accuracy and a total capacity of 25 mm were fixed to measure the deflection at mid-spans and the rotation of the end and the intermediate support. Demec points were arranged and fixed on the painted side of each tested beam in three rows at the center of each span under the concentrated loads and at the intermediate support. Concrete strains were measured by mechanical strain gauges of 200 mm gauge length and 0.001 mm accuracy. A magnifying lens was used to observe the crack propagation clearly. Cracks were traced and marked at each load increment. Figure 5 shows the arrangement of dial gauges and demec points.

Control beams are tested up to failure. Repaired beams are loaded first up to 50% of the failure loads and then loads are released. Beams are then repaired or strengthened by the specified methods and then loaded up to failure.

3   ANALYSIS AND DISCUSSION OF TEST RESULTS

Beams detailed in Table 3 were tested, and the results were analyzed and compared. The load-deflection curves were plotted, the first cracking and failure loads were recorded. Finally the crack propagation were marked and photographed at failure.

3.1   *Deflection*

Figures 6a, b, c compares the load deflection curves at mid-span for strengthened self-compacting RC beams in different groups using the same strengthening material with the results of the control beam Bo, while Figs. 6d,e compares the repaired beams.

At the ultimate failure load of the control beam, deflections of beams strengthened by SP in Group B, BSST, BSSB, BSSTB, BSS−ve|| and BSS+ve|| were about 85%, 63%, 45%, 37% and 30% of Bo respectively (Fig. 6a.) Beams strengthened by GFRP

in Group C, BSGT, BSGB, BSGTB, BSG−ve|| and BSG+ve|| were about 51%, 40%, 22%, 18% and 16% of that recorded for Bo respectively (Fig. 6b.), while beams strengthened by CFRP in Group D, BSCT, BSCB, BSCTB, were about 25%, 21% and 11% of that recorded for Bo respectively (Fig. 6c.). It is shown from these curves that strengthening of the self-compacting RC continuous beams at both vertical sides at +ve moments gives best results.

Repaired beams by SP in Group B, BRS∩, BRSU and BRS∩ were about 81%, 51% and 37% (Fig. 6d), whereas repaired beams by GFRP in Group C, BRG∩, BRGU and BRG∩ were about 31%, 20% and 14% (Figure 6e). It is noticed that the U-shapes and ∩-shapes exhibited significant decrease in deflections more than the ∩-shapes. It is recommended to use U-shapes because it is easy to be applied in common cases.

Figure 7 shows a comparison between load-deflection curves at mid-spans for strengthened beams by different materials for different schemes and Figure 8 compares load-deflection curves for repaired beams. It is noticed that using advanced composite materials GFRP and CFRP improve the behavior and decrease greatly values of deflection at mid-spans more than using traditional materials SP.

3.2   *Cracking and ultimate loads*

First cracking and ultimate failure loads were recorded for all tested beams. Crack patterns are marked at different loading stages.

Figure 9 compares the both cracking and failure loads for all tested beams in the different groups. With respect to the control beam B_O, the increases in the initial cracking load recorded for strengthened beams in Group B, BSST, BSSB, BSSTB, BSS−ve|| and BSS+ve|| were about 10%, 20%, 35%, 40% and 50% respectively. The increases for beams in Group C, BSGT, BSGB, BSGTB, BSG−ve|| and BSG+ve|| were about 25%, 40%, 60%, 70% and 80%, while for beams in Group D, BSCT, BSCB, BSCTB, were about 35%, 55% and 80% of that recorded for Bo respectively.

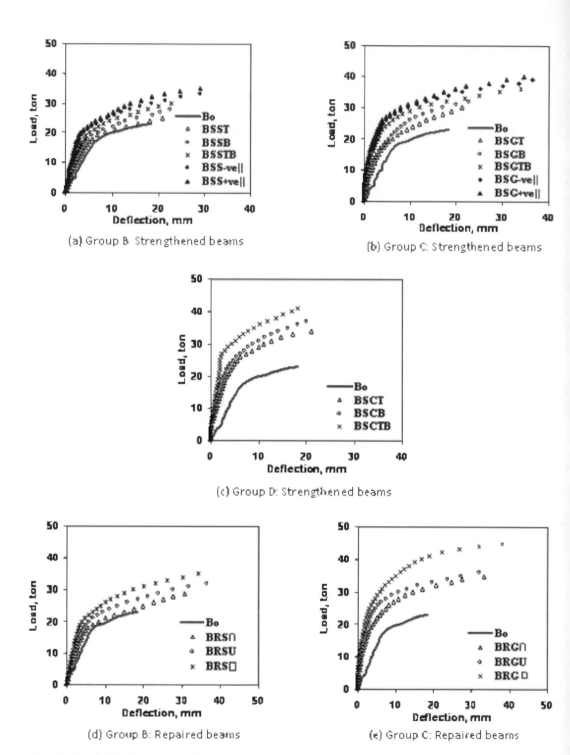

Figure 6.   Load-deflection curves at mid-spans for strengthened and repaired beams in different groups.

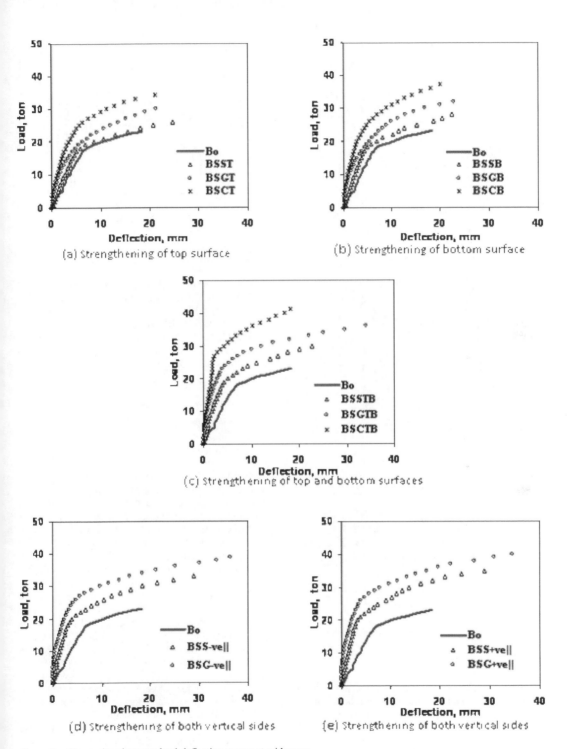

Figure 7.   Comparison between load-deflection curves at mid-spans.

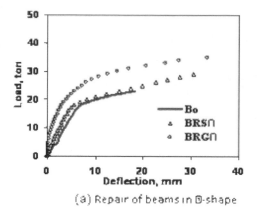
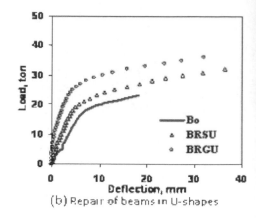

(a) Repair of beams in ∩-shape

(b) Repair of beams in U-shapes

Figure 8. Comparison between load-deflection curves for repaired beams.

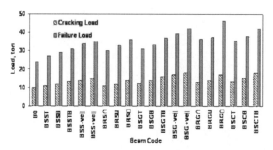

Figure 9. Cracking and failure loads for tested beams.

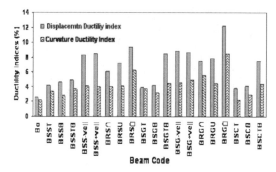

Figure 10. Displacement and curvature ductility indices for strengthened and repaired beams.

The increases in ultimate capacity with respect to the control beam for beams strengthened by SP in Group B, BSST, BSSB, BSSTB, BSS–ve|| and BSS+ve|| were about 12%, 21%, 29%, 42% and 50% respectively. The increases for beams in Group C, BSGT, BSGB, BSGTB, BSG–ve|| and BSG+ve|| were about 29%, 40%, 54%, 64% and 75%, while beams in Group D, BSCT, BSCB, BSCTB, were about 46%, 58% and 75% of that recorded for Bo respectively.

Failure loads of repaired beams by SP in Group B, BRS∩, BRSU and BRSU were increased by about 25%, 37% and 50%, whereas repaired beams by GFRP in Group C, BRG∩, BRGU and BRGU were increased by about 52%, 56% and 92%. It is noticed that the U-shapes exhibited significant increase in capacity more than the ∩ and U shapes.

3.3 Displacement and curvature ductility

In this investigation, the displacement and curvature ductility of the tested beams are compared. Displacement ductility index, μd, is defined as the ratio between the maximum deflection due to the ultimate load and the maximum deflection at the first yielding of the tension reinforcement. Curvature ductility index, μc, is defined as the ratio between the maximum curvature at failure due to the ultimate load and the maximum curvature at the first yielding of the tension reinforcement. Figure 10 compares the displacement ductility index μd at the mid-spans, and the curvature ductility index, μc at the end supports, for different specimens.

With respect to the control beam B_O, the increases in the displacement ductility index for strengthened beams on top surfaces, BSST, BSGT, BSCT, were about 62%, 50%, 46%, while for strengthened beams on bottom surfaces, BSSB, BSGB, BSCB, were about 77%, 62%, 58%. For beams strengthened on both top and bottom surfaces, BSSTB, BSGTB, BSCTB, the increases were about 92%, 227% and 188% of that recorded for Bo respectively. The increases in the displacement ductility index for beam strengthened at both vertical sides in –ve moments, BSS–ve||, BSG–ve||, were about 220%, 242%, while for strengthening in +ve moment zones were about 227%, 235% respectively. Displacement ductility index of repaired beams in ∩-shapes, BRS∩, BRG∩, were increased by about 135%, 188% and repaired beams in U-shapes, BRSU, BRGU, were increased by about 181%, 204%, whereas

426

repaired beams in ∪-shapes, BRS∪, BRG∪ were increased by about 265%, 369 respectively.

The increases in the curvature ductility index, with respect to the control beam B_O for strengthened beams on top surfaces, BSST, BSGT, BSCT, were about 55%, 68%, 0%, while for strengthened beams on bottom surfaces, BSSB, BSGB, BSCB, were about 27%, 45%, 32%. For beams strengthened on top and bottom surfaces, BSSTB, BSGTB, BSCTB, the increases were about 68%, 100% and 95% of that recorded for Bo respectively. The increases in the displacement ductility index for beam strengthened at both vertical sides in negative moment zones, BSS–ve∥, BSG–ve∥, were about 86%, 105%, while for strengthening in positive moment zones were 82%, 127% respectively. Curvature ductility indices of repaired beams in ∩-shapes, BRS∩, BRG∩, were increased by about 82%, 155% and repaired beams in U-shapes, BRSU, BRGU, were increased by about 86%, 100%, whereas repaired beams in ∪-shapes, BRS∪, BRG∪ were increased by about 186%, 288 respectively. Both displacement and curvature ductility indices were increased wherever repair or strengthening on vertical beam sides especially in positive moment zones.

Table 4 compares the properties of GFRP and CFRP wraps used in repair and strengthening.

Figure 2, and 3 show the strengthening and repair techniques applied in this research.

4 CONCLUSIONS

Out of this study, the following conclusions are drawn:

(1) All methods used in this research for repair and strengthening the self-compacting RC continuous beams were effective in restoring and improving the overall behavior in terms of flexural rigidity, initial cracking and ultimate load carrying capacities. The use of the advanced composite materials, GFRP, CFRP exhibited significant increases more than the traditional materials, SP.

(2) Results of strengthening schemes at the bottom faces of the continuous beams in positive moment zones by both traditional and advanced materials used in this research provided better results than strengthening the top face in negative moment zone.

(3) Applying the strengthening layers at vertical sides of tested beams at the positive moment zones give the most efficient method used in this research for strengthening.

(4) Although, ∪-wrapped shapes give the best retrofitting results, it is not always practical to use the composite material from all sides of the beams. It is recommended to use the repair and strengthening layers in U-shapes in positive moment zones.

(5) Substantial increases in cracking strength and ultimate capacity were observed when RC beams were strengthened with CFRP wraps more than the other materials.

(6) Using GFRP as repair and strengthening material increases both displacement and curvature ductility indices with respect to other materials.

REFERENCES

ACI Committee 440. "Guide for the Design and Construction of Externally Bonded FRP Systems for Strengthening Concrete Structures." *American Concrete Institute*, Detroit, MI, 2001.

Damian I. K., and David D. M. "Testing of Full-Size Reinforced Concrete Beams Strengthened with FRP Composites", *Oregon State University, Department of Civil, Construction and Environmental Engineering*, Report SPR 387, June 2002.

Egyptian Code of Practice ECCS-203, "Design and Construction of Reinforced Concrete Structures", Ministry of Building Construction, 2002.

Egyptian Code of Practice, ECOP, for Design and Construction of Reinforced Concrete Structures Using Fiber Reinforced Polymers, 2005.

Housing and Building Research Center, Strength of Material and Quality Control Department, "*State-of-the Art-Report on Self-Compacting Concrete*" October 2002.

Mahmut E, Andrea R., John J. and Antonio N. "Flexural Fatigue Behavior of Reinforced Concrete Beams Strengthened with FRP Fabric and recurred Laminate Systems", *Journal of Composites for Construction* © ASCE, pp 433, 442 September/October 2006.

Neale, K.W., and Labossiere, P., "State-of-the-Art Report on Retrofitting and Strengthening by Continuous Fiber in Canada", *Non-Metallic (FRP) Reinforcement for Concrete Structures*, Japan Concrete Institute, Tokyo, Japan, V. 1, pp. 25–39, 1997.

Panchacharam, S., and Belarbi, A. "Torsional Behavior of Reinforced Concrete Beams Strengthened with FRP Composites," First FIB Congress, Osaka, Japan, October 13–19, 2002.

The First Middle East Workshop on Structural Composites Advanced Composite Materials: *State-of-the-Art Report – Sponsored by The Egyptian Society of Engineers* – Sharm El-Shiekh, Egypt-June 14–15, 1996.

Tjandra, R.A. and Tan, K.H., "Strengthening of Reinforced Concrete Continuous Beams with External Tendons," *6th International Symposium on FRP Reinforcement for Concrete Structures (FRPRCS-6)*, Singapore, July 8–10, 2003, Vol. 1, pp. 723–732.

Excellence in Concrete Construction through Innovation – Limbachiya & Kew (eds)
© 2009 Taylor & Francis Group, London, ISBN 978-0-415-47592-1

Performance-based durability testing, design and specification in South Africa: latest developments

M.G. Alexander & H. Beushausen
University of Cape Town, Cape Town, South Africa

ABSTRACT: Over the last decade, an approach to improving the durability of reinforced concrete construction has been developed in South Africa. The philosophy involves the understanding that durability will be improved only when unambiguous measurements of appropriate cover concrete properties can be made. Such measurements must reflect the in situ properties of concrete, influenced by the dual aspects of material potential and construction quality. Key stages in formulating this approach were developing suitable test methods, characterising a range of concretes using these tests, studying in-situ concrete performance, and applying the results to practical construction. The paper discusses the latest developments in durability specification practice in South Africa and attempts to show a sensible way forward for practical application of the DI approach. The approach is an integrated one in that it links durability index parameters, service life prediction models, and performance specifications. As improved service life models become available, they can be implemented directly into the specifications. Concrete quality is characterised in-situ and/or on laboratory specimens by use of durability index tests, covering oxygen permeation, water absorption, and chloride conduction. The service life models in turn are based on the relevant DI parameter, depending on whether the design accounts for carbonation-induced or chloride-induced corrosion. Designers and constructors can use the approach to optimise the balance between required concrete quality and cover thickness for a given environment and binder system.

1 INTRODUCTION

Deterioration of reinforced concrete is often associated with ingress of aggressive agents from the exterior such that the near-surface concrete quality largely controls durability. The bulk of durability problems concern the corrosion of reinforcing steel rather than deterioration of the concrete fabric itself (Figure 1). The problem is then cast in terms of the adequacy of the protection to steel offered by the concrete cover layer, which is subjected to the action of aggressive agents such as chloride ions or carbon dioxide from the surrounding environment.

For concrete structures, durability is generally defined as the capability of maintaining the serviceability over a specified period of time without significant deterioration. In general, design concepts for durability can be divided into prescriptive concepts and performance concepts. Prescriptive concepts are based on material specification from given parameters such as exposure classes and life span of the structure. However, durability is a material performance concept for a structure in a given environment and as such it cannot easily be assessed through intrinsic material properties. Performance concepts, on the other hand, are based on quantitative predictions for durability from exposure conditions and measured material parameters.

As in other countries, durability problems in South Africa derive mainly from inadequate attention to durability with regard to both design and construction. This has resulted in extensive deterioration of concrete, which is mainly related to reinforcement corrosion. In response to this situation, 3 durability index tests, namely oxygen permeability, water sorptivity and chloride conductivity were developed

Figure 1. The bulk of durability problems concern the corrosion of reinforcing steel.

(Alexander et al, 2001, Mackechnie & Alexander 2002, Beushausen et al, 2003, Streicher & Alexander 1995, Mackechnie 2002). The concrete surface layer is most affected by curing initially, and subsequently by external deterioration processes. These processes are linked with transport mechanisms, such as gaseous and ionic diffusion and water absorption. Each index test therefore is linked to a transport mechanism relevant to a particular deterioration process.

Material indexing of concrete requires quantifiable physical or engineering parameters to characterise the concrete at early ages. The 3 index tests have been shown to be sensitive to important material, constructional, and environmental factors that influence durability. Thus, the tests provide reproducible engineering measures of the microstructure of concrete. The tests characterise the quality of concrete as affected by choice of material and mix proportions, placing, compaction and curing techniques, and environment.

In the South African approach, 'durability indexes' are quantifiable physical or engineering parameters which characterise lab or in-situ concrete and are sensitive to material, processing, and environmental factors such as cement type, water: binder ratio, type and degree of curing, etc. Increasingly, design specifications for structures for which durability is of special concern include limiting values for chloride conductivity (marine environment) and oxygen permeability (risk of carbonation-induced corrosion). The paper discusses the latest developments in durability specification in South Africa, using the Durability Index test methods linked to oxygen permeability and chloride conductivity.

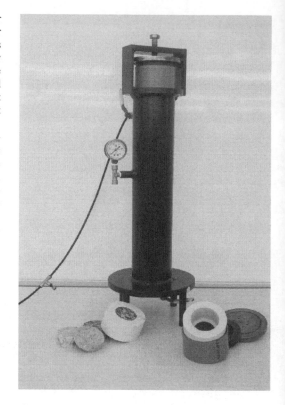

Figure 2. Test set-up for the Oxygen Permeability Index test (OPI).

2 DURABILITY INDEX TEST METHODS

The Durability Index test methods comprise oxygen permeability, chloride conductivity and water sorptivity. As mentioned above, this publication concerns the application of the former two test methods for design specifications. Test equipment and test procedures are described in detail in the literature developed (Alexander et al, 2001, Mackechnie & Alexander 2002, Beushausen et al, 2003, Streicher & Alexander 1995, Mackechnie 2002) and basic principles are discussed in the following.

The Oxygen Permeability Index (OPI) test method consists of measuring the pressure decay of oxygen passed through a 25 mm thick slice of 68 mm diameter core of concrete placed in a falling head permeameter (Figure 2). The oxygen permeability index is defined as the negative log of the coefficient of permeability. Common OPI values for South African concretes range from 8.5 to 10.5, a higher value indicating a higher impermeability and thus a concrete of potentially higher quality. Note that oxygen permeability

Figure 3. Test set-up for the chloride conductivity test.

index is measured on a log scale, therefore the difference between 8.5 and 10.5 is quite substantial. An empirical prediction model for carbonation was formulated using the oxygen permeability test. Using this approach, 50 year carbonation depths may be predicted for different environments.

The chloride conductivity test apparatus (Figure 3) consists of a two cell conduction rig in which concrete core samples are exposed on either side to a 5 M NaCl chloride solution. The core samples are preconditioned before testing to standardize the pore water solution (oven-dried at 50°C followed by 24 hours vacuum saturation in a 5 M NaCl chloride solution). The movement of chloride ions occurs due to the application of a 10 V potential difference. The chloride conductivity is determined by measuring the current flowing through the concrete specimen. The apparatus allows for rapid testing under controlled laboratory conditions and gives instantaneous readings.

Chloride conductivity decreases with the addition of fly ash, slag, and silica fume in concrete, extended moist curing and increasing grade of concrete. Portland cement concrete for instance generally has high conductivity values with only high-grade material achieving values below 1.0 mS/cm. Slag or fly ash concrete in contrast has significantly lower chloride conductivity values. While the test is sensitive to construction and material effects that are known to influence durability, results are specifically related to chloride ingress into concrete. Correlations between 28-day chloride conductivity results and diffusion coefficients after several years marine exposure have shown to be good over a wide range of concretes (Mackechnie & Alexander 2002).

3 APPLICATION OF THE DURABILITY INDEX APPROACH

3.1 General

The sensitivity of the South African index tests to material and constructional effects makes them suitable tools for site quality control. Since the different tests measure distinct transport mechanisms, their suitability depends on the property being considered. Durability index testing may be used to optimise materials and construction processes where specific performance criteria are required. At the design stage the influence of a range of parameters such as materials and construction systems may be evaluated in terms of their impact on concrete durability. In this way, a cost-effective solution to ensuring durability may be assessed using a rational testing strategy (Ronnè et al, 2002).

The durability indexes, obtained with the above test methods, have been related to service life prediction models. Index values can be used as the input parameters of service life models, together with other variables such as steel cover and environmental class, in order to determine rational design life. Limiting index values can be used in construction specifications to provide the necessary concrete quality for a required life and environment. Thus, a framework has been put in place for a performance-based approach to both design and specification.

Table 1. Environmental Classes (Natural environments only) (after EN206).

Carbonation-Induced Corrosion	
Designation	Description
XC1	Permanently dry or permanently wet
XC2	Wet, rarely dry
XC3	Moderate humidity (60–80%) (Ext. concrete sheltered from rain)
XC4	Cyclic wet and dry

Corrosion Induced by Chlorides from Seawater	
Designation	Description
XS1	Exposed to airborne salt but not in direct contact with seawater
XS2a*	Permanently submerged
XS2b*	XS2a + exposed to abrasion
XS3a*	Tidal, splash and spray zones Buried elements in desert areas exposed to salt spray
XS3b*	XS3a + exposed to abrasion

*These sub clauses have been added for South African coastal conditions

3.2 Service life prediction models

Two corrosion initiation models have been developed, related to carbonation – and chloride – induced corrosion. The models derive from measurements and correlations of short-term durability index values, aggressiveness of the environment and actual deterioration rates monitored over periods of up to 10 years. The models allow for the expected life of a structure to be determined based on considerations of the environmental conditions, cover thickness and concrete quality (Mackechnie & Alexander 2002, Mackechnie 2001). The environmental classes are related to the EN 206 classes as modified for South African conditions (Table 1), while concrete quality is represented by the appropriate durability index parameter. The oxygen permeability index is used in the carbonation prediction model, while the chloride model utilises chloride conductivity. The service life models can also be used to determine the required value of the durability parameter based on predetermined values for cover thickness, environment, and expected design life. Alternatively, if concrete quality is known from the appropriate DI, a corrosion-free life can be estimated for a given environment.

3.3 Specifying durability index values

Two possible approaches to specifying durability index values are a deemed-to-satisfy approach and a rigorous approach. The former is considered adequate for the majority of reinforced concrete construction and represents the simpler method in which limiting DI values are obtained from a design table, based on binder type and exposure class, for a given cover depth (50 mm for marine exposure and 30 mm for carbonating conditions).

The rigorous approach will be necessary for durability–critical structures, or when the design parameters assumed in the first approach are not applicable to the structure in question. Using this approach, the specifying authority would use the relevant service life models developed in the concrete durability research programme in South Africa. The designer can use the models directly and input the appropriate conditions (cover depth, environmental classification, desired life, and material). The advantage of this approach is its flexibility as it allows the designer to use values appropriate for the given situation rather than a limited number of pre-selected conditions.

3.3.1 Examples for the deemed-to-satisfy approach

This approach mimics structural design codes: the designer recommends limiting values which, if met by the structure, result in the structure being 'deemed-to-satisfy' the durability requirements.

The carbonation resistance of concrete appears to be sufficiently related to the early age (28 d) Oxygen Permeability Index (OPI) value, so that OPI can be used in a service life model. The environments that require OPI values to be specified in the South African context are XC3 and XC4 (Table 1), with XC4 considered the more critical because steel corrosion can occur under these conditions. Two design scenarios with standard conditions and required minimum OPI values are shown in Table 2.

Chloride resistance of concrete is related to its chloride conductivity, and therefore this index can be used to specify concrete performance in seawater environments. Table 3 presents chloride conductivity limits for common structures (50 years service life). Different values are given for different binder types, since chloride conductivity depends strongly upon binder type. The horizontal rows give approximately equal performance (i.e. chloride resistance) in seawater conditions for the different binders. Binder types are restricted to

blended cements for seawater exposure, since CEM I on its own has been shown to be insufficiently resistant to chloride ingress.

3.3.2 Example for the rigorous approach

As an example of practical implementation of the rigorous approach, consider the case of specifying a marine structure for a 50-year design life, subject to the environmental conditions given in Table 1. Combining the relevant durability index of chloride conductivity with the appropriate service life model yields the data given in Table 4. It should be noted that the DI values are presented here for purposes of illustration only. The relative values are more important than the absolute values as these will vary in response to regional and environmental variations.

Table 3. Maximum Chloride Conductivity Values (mS/cm) for Different Classes and Binder Types: Deemed to Satisfy Approach – Common Structures (Cover = 50 mm).

EN206 Class	Binder combination		
	70:30	50:50	90:10
	CEMI:FA	CEMI:GGBS	CEMI:CSF
XS1	3.0	3.5	1.2
XS2a	2.45	2.6	0.85
XS2b, XS3a	1.35	1.6	0.45
XS3b	1.1	1.25	0.35

Table 4. Limiting DI values based on rational prediction model: maximum chloride conductivity (mS/cm) (50 year life).

Expo-sure class	Cover (mm)	Max. chloride conductivity (mS/cm) for various binder types		
		100% CEM I	30% FA	50% GGCS
XS3b	40	0.45	0.75	1.05
	60	0.95	1.35	1.95
	80	1.3	1.8	2.6
XS0b	40	1	1.85	2.5
	60	1.85	2.95	3.9
	80	2.5	3.75	4.8

[shaded]	Impractical mixes; concrete grade > 60 Mpa
[dark shaded]	Not recommended: grades < 30 MPa, and/or w/b > 0.55
[white]	Acceptable mixes. Grades 30 to 60 Mpa

Table 2. Deemed to Satisfy OPI values (log scale) for carbonating conditions.

	Common Structures	Monumental Structures	
Service Life	50 years	100 years	100 years
Minimum Cover	30 mm	30 mm	40 mm
Minimum OPI	9.7	9.9	9.7

The table shows the trade-off between material quality (i.e. chloride conductivity) and concrete cover, with lower quality (represented by a higher conductivity value) allowable when cover is greater. The dependence of the conductivity on binder type is also illustrated, with higher values permissible for blended binders at any given cover, based on their superior chloride ingress resistance. These higher values translate into less stringent w/b ratios. Therefore, a conservative approach is recommended at present, with mixes for which the concrete grade may be less than 30 MPa, and/or the w/b may be greater than 0.55, not being recommended. However, in these cases, the particular cover and binder can be used, but the conductivity value will be over-specified.

3.4 Establishing limiting values for concrete mixtures

To establish limiting DI values for concrete mixtures and evaluate compliance with durability requirements, the following two aspects need to be considered:

– Statistical variability of test results (hence selection of appropriate characteristic values for Durability Indexes)
– Differences between as-built quality (in-situ concrete) and laboratory-cured concrete
– The consideration of the two above aspects is discussed below and illustrated by an example in Table 5.

3.4.1 Characteristic values versus target values

The values determined in Tables 2, 3 and 4 are characteristic values to be achieved in the as-built structure,

Table 5. As-built chloride conductivity values (mS/cm) vs. potential target values (hypothetical case).

Design level	Binder combination		
	70:30	50:50	90:10
	CEMI:FA	CEMI:GGBS	CEMI:CSF
In-situ char. (XS3a)	1.35	1.60	0.45
Potential characteristic (as delivered to site)	1.22	1.44	0.41
Potential target	1.09	1.30	0.37

A: (x 0.90): target vs. characteristic value
B: (x 0.90 or x 0.82+0.20): as-delievered vs. as-built quality

not target (average) values. The material supplier must aim at target values that will achieve the required characteristic values with adequate probability. The variability inherent in concrete performance needs to be considered when evaluating the test results, similar to the approach that is adopted with strength cubes. Since durability is a serviceability criterion, the limitations may not need to be as stringent as for strength. It is proposed that a 1 in 10 chance be adopted at this stage for the Durability Index tests with a margin of 0.3 below for the OPI, and 0.2 mS/cm above for the chloride conductivity test.

3.4.2 As-delivered concrete quality versus as-built concrete quality

A clear distinction must be drawn between material potential and in-situ construction quality. Although specifications are usually only concerned with as-built quality, the processes by which such quality is achieved cannot be ignored. There are two distinct stages and responsibilities in achieving concrete of a desired quality. The first is material production and supply, which could be from an independent party such as a ready-mix supplier. A scheme for acceptance of the as-supplied material must be established so that the concrete supplier can have confidence in the potential quality of the material. The second stage is the responsibility of the constructor in ensuring that the concrete is placed and subsequently finished and cured in an appropriate manner. It is ultimately the as-built quality that determines durability and the constructor has to take the necessary steps and precautions in the construction process to ensure that the required quality is produced. If the as-built quality is found to be deficient, the specification framework must have an internal acceptance scheme that is able to distinguish whether the deficiency arises from the as-supplied material or the manner in which it was processed by the constructor. To enable this, a two-level quality control system has been proposed in South Africa, with testing of both material potential and as-built quality. Material potential is represented by as-supplied concrete specimens with a laboratory-controlled wet curing period (5 days), while as-built quality is determined using in-situ sampling of concrete members.

As a general rule, concrete in the as-built structure may be of lower quality compared with the same concrete cured under controlled laboratory conditions described above. To account for the improved performance of laboratory concrete over site concrete, the characteristic values for the durability indexes of the laboratory concrete should be:

– For OPI: a margin of at least 0.10 greater than the value determined in Sect. 3.3.
– For chloride conductivity: a factor no greater than of 0.90 times the value determined in Sect. 3.3.

433

4 CLOSING REMARKS

The paper describes the development of the Durability Index approach to addressing problems of reinforced concrete durability in the South African context. The approach is an integrated one in that it links durability index parameters, service life prediction models, and performance specifications. As improved service life models become available, they can be implemented directly into the specifications. Concrete quality is characterised in-situ and/or on laboratory specimens by use of durability index tests, covering oxygen permeation, water absorption, and chloride conduction. The service life models in turn are based on the relevant DI parameter, depending on whether the design accounts for carbonation-induced or chloride-induced corrosion. Designers and constructors can use the approach to optimise the balance between required concrete quality and cover thickness for a given environment and binder system. More work remains to be done, in particular generating correlations between indexes and actual structural performance. Only in this way will the usefulness of the approach be assessed.

REFERENCES

Alexander, M.G., Mackechnie, J.R. and Ballim, Y. (2001), 'Use of durability indexes to achieve durable cover concrete in reinforced concrete structures', Materials Science of Concrete, Vol. VI, Ed. J. P. Skalny and S. Mindess, American Ceramic Society, 483–511.

Beushausen, H., Alexander, M.G., and Mackechnie, J. (2003), 'Concrete durability aspects in an international context', Concrete Plant and Precast Technology BFT, vol. 7, 2003, Germany, pp. 22–32.

Mackechnie, J.R. and Alexander, M.G. (2002), 'Durability predictions using early age durability index testing', Proceedings, 9th Durability and Building Materials Conference, 2002. Australian Corrosion Association, Brisbane, 11p.

Mackechnie, J.R., (2001), Predictions of Reinforced Concrete Durability in the Marine Environment – Research Monograph No. 1, Department of Civil Engineering, University of Cape Town, 28 pp.

Mackechnie, J.R., and Alexander, M.G., (2002), 'Durability Predictions Using Early-Age Durability Index Testing,' Proceedings of the Ninth Durability and Building Materials Conference, Australian Corrosion Association, Brisbane, Australia, 11 pp.

Ronnè, P.D., Alexander, M.G. and Mackechnie, J.R. (2002), 'Achieving quality in precast concrete construction using the durability index approach', Proceedings: Concrete for the 21st Century, Modern Concrete Progress through Innovation, Midrand, South Africa, March 2002.

Streicher, P.E. and Alexander, M.G. (1995), 'A chloride conduction test for concrete', Cement and Concrete Research, 25(6), 1995, pp. 1284–1294.

Excellence in Concrete Construction through Innovation – Limbachiya & Kew (eds)
© 2009 Taylor & Francis Group, London, ISBN 978-0-415-47592-1

Mechanical properties and durability of FRP rods

J.R. Nasiri, A. Bahari & O.J. Farzaneh
Islamic Azad University-Tabriz Branch, Iran

ABSTRACT: Another application of FRP rods in construction was developed to retrofit and repair reinforced concrete (RC) and masonry structures, using a recently developed technology known as near surface mounted (NSM) rods. The application of FRP rods in new or damaged structures requires the development of design equations that must take into account the mechanical properties and the durability properties of FRP products. Several concerns are still related to the structural behavior under severe environmental and load conditions for long-time exposures. In this paper an effective tensile test method is described for a mechanical characterization of FRP rods. An effort has also been made to develop an experimental protocol to study the effects of accelerated ageing on FRP rods. The studies showed the effectiveness of the proposed tensile test method and the Influence of aggressive agents on durability of the tested FRP rods.

1 INTRODUCTION

1.1 Use of FRP rods in civil engineering structures

Several technologies were developed for manufacturing FRP materials as laminates, rods, filament wound tanks and many others. FRP materials have a number of advantages when compared to traditional construction materials such as steel and concrete. FRP materials have been utilized in small quantities in the building and construction companies for decades. At the moment, numerous successful applications using FRP composites for repair, strengthening and reinforcement of concrete and masonry structures such as bridges, piers, columns, beams, walls, walkways, pipelines etc. have been reported. Nearly 600 Articles concerning FRP use in the construction industry published between 1972 and 2000 are available. This expresses a measure of the significant potential that FRP materials showed in the last decades. Japan, Canada and U.S.A. developed design codes for use of FRP in civil engineering applications, but further reviews and experimental data are still needed. FRP reinforcement in concrete structures should be used as a substitute of steel rebars for that cases in which aggressive environment produce high steel corrosion, or lightweight is an important design factor, or transportation cost increase significantly with the weight of the materials. In the last year a new technology has also been developed using FRP rods in structural rehabilitation of deficient RC structures. In fact, the use of Near Surface Mounted (NSM) FRP rods is a promising technology for increasing flexural and shear strength of deficient RC and PC members. Another application

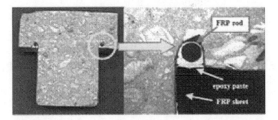

Figure 1. Anchorage system using FRP rods.

of FRP rods is the anchorage of FRP externally bonded sheets in RC members, used for shear strengthening, as it can be seen in Figure 1.

FRP rods were also used for structural repointing of masonry structures, (Tinazzi et al., 2000; De Lorenzis, 2000 (C)). An example of FRP installation procedure is illustrated in Figure 2. Recent installations of FRP rods were also conducted in Europe for the repair of historical buildings such us churches and ancient monuments (La Tegola, 2000 A & B). In Figure 3 it can be seen how Aramid FRP (AFRP) rods were installed for the structural rehabilitation of a cracked masonry column.

2 DURABILITY AND MECHANICAL TESTS OF FRP REINFORCEMENT

2.1 Durability of FRP used in construction

In order to expand the use of FRP in civil structures, relevant durability data must be available in the building codes and standards. In general durability of

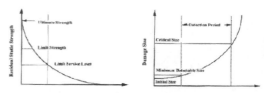

FRP rods

Figure 2. FRP installation for structural repointing in masonry walls.

Figure 3. Column confinement with AFRP rods.

a structure and of a material can be defined as the ability to resist cracking, oxidation, chemical degradation, delamination, wear, and/or the effects of foreign object damage for a specific period of time, under the appropriate load conditions and specified environmental conditions. This concept is realized in design through the application of sound design principles and the principles of damage tolerance whereby levels of performance are guaranteed through Relationships between performance levels and damage/degradation accrued over specified periods of time. In this sense, damage tolerance is defined as the ability of a material or a structure to resist failure and continue performing prescribed levels of performance in the presence of flaws, cracks, or other forms of damage for a specified period of time under specified environmental conditions. The overall concept is illustrated schematically in Figure 4.

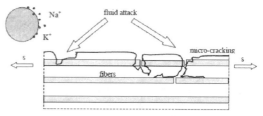

Figure 4. Concepts of durability and damage tolerance to design.

Figure 5. Fluid attack in FRP rods.

The following different damage mechanisms are distinguished in order to classify the potential problems related to the long-term behavior of FRP rods used in civil engineering:

- Effects of solutions on mechanical properties
- Creep and stress relaxation
- Fatigue and environmental fatigue damages
- Weathering

All these mechanisms can be considered as a consequence of the attack by external agents including:

- Moisture and aqueous solutions
- Alkaline environment
- Thermal effects (freeze-thaw cycling, high temperatures)
- Fatigue loads
- Ultra-violet (UV) radiation
- Fire

Hence, in the stressed FRP reinforcement micro cracks in thin outer matrix skin may arise, thus leading to the loss of the ingress of aggressive fluids as shown in Figure 5.

3 CONCLUSIONS

A set of conclusions is drawn herein, in order to furnish information and recommendations that could help in the characterization and development of FRP reinforcement for concrete structures. An improvement of long term behavior of FRP reinforcement may result from efforts by researchers and manufacturers in selecting appropriate materials.

3.1 Test protocol for characterization and durability investigation of FRP rods

An experimental method was used for characterization of FRP rods and for investigations. Of durability effects due to environmental exposure and alkaline exposure. A combination of physical and mechanical tests is proposed. Electronic microscopy was also used to observe the effects of the degradation phenomena. The following conclusions are reported in order to help future researchers and engineers for conducting characterization and durability studies:

– Grouted anchors with alignment devices allowed performing tensile tests that showed fiber rupture for different cross section geometry and surface conditions.

Expansive grout may substitute epoxy resin to develop a gripping force for tensile test. Particular benefits of the proposed protocol can be summarized:

1. No damages due to gripping force
2. Perfect tensile stress developed during the test
3. Easy preparation of the specimens
   – Ratio between test length and diameter of rebars did not affect tensile test results.
   – Environmental combined agents were used to simulate external conditions in an environmental chamber and alkaline accelerated exposure was used to simulate cementations environment in which the rods are embedded during the service life. A pH of 12.6 was chosen and K+ and Na+ were introduced because their chemical attack generates glass fiber damages.
   – Short shear span test according to ASTM D4475 is recommended in order to study resin properties.
   – Gravimetric measures are recommended after any solution immersion, since the weight increase furnishes, without any other information, a measure of potential degradation of the system.

3.2 Durability and structural safety: Design recommendations

Although design guidelines were drawn in different countries, including USA, Japan, Canada and UK, recommendations and coefficients that could take into account the long-term behavior of FRP reinforcement were not well defined. Several studies were conduced and provisions for mechanical and durability characterization were furnished. Provisional values can be provided using also the results of experimental studies. In particular it was observed that GFRP presents higher sensitivity to external agents, including alkaline cementitious environment, while CFRP can be used with less concerns as was also demonstrated in previous researches. With reference to ACI 440H (ACI provisions), an environmental knock-down factor Ce can be used to compute the FRP design strength from experimental results, and recommendations should be furnished also regarding resin degradation:

– Ce = 0.90 can be used for CFRP reinforcement
– Ce = 0.70 can be used for GFRP reinforcement
– Residual tensile strength should not be less than 75% after experimental accelerated aging according to the proposed protocol – Residual transverse properties should not be less than 65% after experimental accelerated aging according to the proposed protocol – Weight increase should not be more than 2.5% for CFRP rods and not more than 2% for GFRP rods after accelerated fluid immersion – Extreme environmental conditions or specific environments should be investigated using a coefficient of reduction of 0.8 for all the acceptance criteria mentioned above.

4 RECOMMENDATIONS FOR FUTURE WORKS

The first limitation of the experimental work is the absence stress during the accelerated aging of the rods. Other aspects should be investigated, and further recommendations should be provided in order to establish quality specifications that will help to draw common design guidelines. Therefore the following recommendations for future work are provided:

– Tensile stress comparable to service loads should be applied during further durability test to see the effect of the applied load.
– Combined effects of fluid penetration (alkali, acid etc.) and environmental agents could provide more information on durability in aggressive environments.
– Resin properties should be investigated for all the products that are candidate to substitute steel reinforcement in construction, since a degradation of the resin accelerate fiber damages.
– Creep experimental investigations are needed, especially for prestressing tendons, in order to establish coefficients for prestressed FRP rods.
– Further tests are needed in order to validate this method for rectangular CFRP rods with smooth surface.
– The same conditioning regimen should be provided using water or other solution (marine water, acid solutions etc.) in order to study the effects of accelerate diffusion mechanisms that cause fluid penetration. This is essential for marine structures or other members immersed in a solution or subjected to aggressive vapors during their service life.

REFERENCES

ACI, (2000), "Guide for the Design and Construction of Concrete Reinforced with FRP bars", American Concrete Institute Committee 440.

Aiello M.A. and Ombres L., (2000), "Load-Deflection Analysis of FRP Reinforced.

Alkhrdaji, T.; Nanni, A.; Chen, G.; and Barker, M. (1999), "Upgrading the transportation Infrastructure: Solid RC Decks Strengthened with FRP", *Concrete International*, ACI, Vol. 21, No. 10, October, pp. 37–41.

Al-Zahrani, M.M., Nanni, A., Al-Dulaijan, S.U., and Bakis, C.E., (1996), "Bond of Fiber Reinforced Plastic (FRP) Rods to Concrete," *Proc. 51st Ann. Conf. of the Composites Institute, Soc. Plastics Engineers*, New York, 1996, pp. 3A.1–3A8.

Ashour, S.A. and Wafa, F.F, (1993), "Flexural Behavior Of High-Strength Fiber Reinforced Concrete Beams" *ACI Struct. J.*, Vol. 90, No. 3, pp 279–287.

ASTM (2000), American Society for Testing and Materials, ASTM A 312/A 312M "Standard specification for seamless and welded austenitic stainless steel pipes", October 2000.

ASTM, (1982), American Society for Testing Materials E 632 "Standard Practice for Developing Accelerated Tests to Aid Prediction of the Service Life of Building Components and Materials".

"Concrete Flexural Members", *J. Comp.Constr.*, ASCE, Vol. 4, No. 4, pp. 164–171.

Homam S.M., and Sheikh S.A., "Durability of Fiber Reinforced Polymers Used in Concrete Structures", *Proc. 3rd International Conference on Advanced Materials in Bridges and Structures*, Ottawa, Canada, August 15–18th 2000, pp. 751–758.

La Tegola A., La Tegola A., De Lorenzis L., and Micelli F., (2000), "Applications of FRP materials for repair of masonry structures", *Proceedings of the Technology Transfer Seminar Advanced FRP Materials for Civil Structures*, October 19th, 2000 Bologna, Italy.

Noritake, K., Mukae, K., Kumagai, S, and Mizutani, J., (1993), "Practical Applications of Aramid FRP Rods to Prestressed Concrete Structures" *Proceedings of the International Symposium on Fiber-Reinforced-Plastic Reinforcement for Concrete Structures*, Ed. by Antonio Nanni and Charles W. Dolan, Vancouver, Canada, Mar 28–31, 1993, American Concrete Institute, Detroit, MI, 1993, pp. 853-873. (ACI SP-138) No. 1, pp. 12–19.

Tannous F.E., and Saddatmanesh H., (1999), "Durability of AR Glass Fiber Reinforced Plastic Bars", J. Comp. Constr., Vol. 3.

Excellence in Concrete Construction through Innovation – Limbachiya & Kew (eds)
© 2009 Taylor & Francis Group, London, ISBN 978-0-415-47592-1

FRP composites in fabrication, rehabilitation and strengthening of structure

Z. Aghighi & M. Babaha
Islamic Azad University, Iran

ABSTRACT: Application of FRP composites in civil engineering has been gained attention of authors. Light weight, extradinary resistance against corrosion is main characteristics of these composites in multiple forms in building engineering. Since these materials are used in different purposes by different properties, diverse items are Fabricated by hose materials in allover the world. Recognition and multiple applications of composites are important in our country. In this paper, types and different capabilities of FRP composites are investigated.

1 INTRODUCTION

Mentioned composites consist of two parts; microscopic elements and insoluble particles firm early periods, different types of composites have been used by civil engineers. Wood is a natural composite, multilayer boards have been used as laminated natural composite; in general, concrete has been used as a composite by distinguishable elements from other composed material in fabrication of composed materials by wide spectrum application. These materials consist of elements by different mechanical behavior, but resulted composite can have entirely different properties relative to own conformed particles. Fibers are primary and loading part of composites the second part is called resin or adds beside material which connect fibers it each other. Fiber is prepared by elastic, brittle and resistant. Material thickness of these fibers is 5–25 micron. Depend on their types. Today, fibers are manufactured in different shapes size and types. Glass, carbon, aramid or vinyl on fibers are available. The name of FRP manufactured by this material begins by first letter of fiber composite. These materials consist of CFRP. GFRP, AFRP, VFRP as a FRP manufactured by carbon, glass. Aramid and vinylon fibers. They have own mechanical and physical specifications.

2 TYPES OF FIBERS

Fiber glass which divided into four groups: A- glass, Z- glass, E- glass and S- glass. Resistance in alkaline environment is main characteristics of these fibers; their modulus of elasticity is lower than 100 GPA. Carbon fiber: carbon fibers can be divided into two categories:

1- Artificial fiber by chemical name of poly acrylionitril (PNN)
2- Fiber by bituminous origin obtains from coal, these fibers cheaper than PAN but their resistance and modulus of elasticity is low ($E_N = 2.17...$)

2.1 Aramid fiber

Aramid is one the simplest form of aromatics poly aramid. artificial fibers were manufactured by German deponent under the title of coolar for first time. Today, there are four types of coolar fibers, generally their tensile strength is 55 percent and their shcar resistance is 180% more than fiber glass, laboratory research has been shown that the practical lily of coolar fiber is more than glass and carbon fibers, in addition, tensile strength of coolar fiber is 10 percent lower than tensile strength of carbon fiber, the price of carbon fiber is tow fold than coolar fiber.

2.2 Resins

Resin- or matrix used in FRP production act as an adhesive culture for contact of fibers, so resin by low resistance do not play considerable role in mechanical properties of composite. In addition to adhesive property, resin in FRP composite acts as a protection factor against environmental factors and distribution factor of stress on composite laminate. Adhesive used in composite could be thermoplastic or tempest compound. Thermoses adhesives are hard material against high temperature and it doesn't loose by increasing temperature. Thermoplastic resins are melted by high temperature and are harden by low temperature. Resins mode up by unsecured polyesters, vinglaster and epoxy, amino, phenol, meta cry late and outran are thermoses resins. PVC, polyethylene and polypropylene (PP) are thermoplastic resins. Different shapes of composite in civil engineering currently. Two kinds of FRP composites are used in building and rehabilitation of structures in civil engineering. Composite bars and coating composites, in addition to these types, there are composites in shape, angle and etc in different oart of world. Many researches have been conducted on FRP composites by special technique.

3   FRP BARS

One of the FRP applications in using as a bar. Most of the reinforced concrete structures are damaged in corrosive environment by attack of sulfates, chlorides and other corrosive factors. Common damages are erosion and corrosion in reinforcing bars. Replacement and repairmen of damaged concrete structures are costly practice in as over the world. For elimination of this problem, several techniques have been developed. Their success was relative. Replacement of reinforced bar in reinforced concrete structures by a resistant material against corrosion is optimal method. Since FRP is high resistive against corrosion, the idea of using this material has been rooted. Coolarm glass (E-glass) and carbon fibers are used in manufacturing of composite bars. Yielding stress of these bars are minimum 660 Mpa and maximum 300 Mpa and their modulus of elasticity vary from 5/41 GPA and 147 GPA. Maximum tolerable strain is between 1/3–3/6 percent. Resistance against corrosion, high strength, special weight, electrical conductivity, good behavior in fortune and approximately zero creep are characteristics of FRP. Bending of bars is main disadvantage.

4   FRP COAT (SHEATH)

FRP coats are used mainly for improving recent structures behaviors or repairmen of damages in current structures. These coats (ream basement) stick on external surface of reinforced concrete. There are three types of coats sued in repairmen and rehabilitation of concretes.

4.1   Hand- made coats- (sheath)

In this kind, at first, the external surface of concrete is prepared and ane layer adhesive is applied then applied fibers are stuck by hand in one or different directions. After drying, one layer composite created

stick on lateral. It should be noted that, applied fiber is woven as a bag (one or multiple direction) in factory.

4.2   Pre-cast panels or sheets

In this case, the composites are fabricated as a double sheets or belts. In flat parts like slabs, beams, FRP prefabricated nets are stuck on clean surface. Width of these belts vary from 50–15 mm. in this method, the surface of concrete is cleaned by sand and pressure then inspected in order to complete adhesion between concrete and FRP. These panels are used in pillar with pre-cast sheets, in ing-shape. After preparation of concrete surface, one layer. Adhesive is stuck mentioned panel is stuck in specified direction. By pulling wires on stuck sheet, extra adhesive come out between sheet and concrete.

4.3   Machined sheets

In this system fibers are used dry or pre-soaked in adhesive material. A thermal chamber. Or preservation oven is employed for caring of stuck layers on concrete rolled sections. From other composites, it can be dressed to rolled section. These sections are manufactured in different shapes like angled, square, and etc. several studies have been carried out in Fabric citing structures by FRP composite frame instead of steel frame or reinforced concrete. Structures strengthen systems by FRP composites these is two main goals in reinforced concrete structure strengthening:

4.3.1   *Rehabilitation of instance structures*
4.3.2   *Repairmen of damaged structures there are wide spectrum structures donor meet earthquake regulations criterion*

Or are made according traditional regulations. FRP composites are used in order to their rehabilitation and strengthening. When repairmen of a building cause weakness in some elements, such a system could be useful. It can be refered to structure strengthening in order to enhancement of alive loads, increasing traffic loads (in bridges) and necessity of placing heavy machineries in some parts of present buildings that did not predicated in primary designing. In this direction, help in employability of system, control of factors like creep and reducing stress in reinforced bars, controlling crack wide are other factors of using ERP sheaths. Also, when increasing depth or reinforced bar is necessary in bending element, FRP sheets provide this requirement. Using FRP coats is could rehabilitation. Preventing fractures in structures because of fire. Earthquake, passing heavy vehicles, corrosion of steel reinforced bars and expiration of material useful life by local and apparent damages is other reason for using composite external coats this kind usage of FRP sheets is soled repairmen. In recent years, many researches have been conducted in possibility of using polymeric composites in sliding strength inland prevention of corrosion in wood, steel and reinforced concrete structures. According to these studies, using composites is ineffective technique in repairmen and rehabilitation of reinforced concrete structures even in wood steel and masonry structures.

Enhancing loading capacity, reinforced concrete pillars plasticity by eternal coats stuck on pillar is one of the successful applications of FRP composites. In this process, multiple layers of composite are attack on prepared surface in linear or spiral shapes in order to strengthening of shaping and loading capacity of around of pillar. In 1990s, investigations showed that application of FRP composites enhance plasticity of pillars especially in seismic focus lateral extension; consequently, by providing spiral stress, plasticity in pillar is increased. In addition, composites at as an effective system for prevention of corrosion in hard condition (like Persian golf environmental conditions) Application of FRP composites are extended to other structural elements like slabs and beams. For strengthening these elements like slabs and beams, for strengthening these elements, FRP sheets and straps are stuck under, upper poor lateral of reinforced concrete slabs or beams. In beams and slobs bending, shearing and strain strengthening can be expected. In addition, improving exploitability of slabs and beams by using FRP coats has been gained attention. Using FRP composite sheets in rehabilitation of connects proper ting is new slope in studies of structures strengthening. In addition to mentioned application, Foundation and deck of bridges are structures which their strengthening by FRP coats have been studied several times under presumes water pipes by large thickness are corrosive FRP coats are wed in their strengthening.

5 CAPABILITIES OF FRP COMPOSITES IN STRUCTURAL ELEMENTS STRENGTHENING

Based on past studies, it is observable that FRP composites are applicable in strengthening structural elements and increasing their capacity because of special properties. Also, most of current structure problems are solved by correct application of FRP composites. These problems are seen in normal buildings. Bridges and foundations, in general, followings are addressed as potential capabilities' of FRP in structures repairmen.

– increasing loading capacity and enacting pillar, beam, slab and reinforced concrete connects plasticity
– strengthening of steel and concrete tanks
– strengthening of marine and shore structures.

– strengthening of explosion resistance structures
– strengthening of wood pillars and beams
– strengthening of reinforced chimney or masonry material
– strengthening of rein forced concrete walls and masonry materials reinforced and UN reinforced
– strengthening of channels
– strengthening of steel or concrete pipes
– strengthening of hospital and ancient structures In addition to mentioned specifications, floorings can be addressed as advantages of composite application specially FRP sheets
– Low weights
– Free length (without crack)
– Low thickness (about millimeter)
– Simple and easy transportation
– Simple performances of sheets
– Economical applications (in large project, without need to complex equipment)
– High tensile resistances (even pressure)
– Diversity in modulus of elasticity value
– Considerable fatigue strength
– High resistances in alkali environment zero corrosion
– Clean finished surface after performance (this specification is special for pre- fabricated strop)

6 EXISTENCE REGULATIONS

Until 1990s, a few specialist organizations conducted research in development of standards and designing recommendations in using of composites in building application. ASCE published two designing guidebooks by collaboration of SCAP committee in 1980. Currently, SCAP has published new guidelines in designing of composite connections. ACI has been carried out different research in present decade by ACI 440 committee. The aim of this committee is publishing complete review collection and handbook of required receipts for special designing means FRP application beside concrete. In addition to mentioned collections, ASTM institute and AASHTO publish reports about building application of FRP composites. International building conference [CBO] has published two valuable reports about strengthening of concrete and masonry structures by using FRP composites (AC125, AC 178) in all over the world; several regulation reports have been published. These reports are based on carried out research (by SCI japans concrete institute), case, Euro code Swedish bridge code, Australian industry code, structures strengthening by FRP code committee which are active in Egypt and Saudi Arabia.

7 CONCLUSIONS

Using FRP composites in different forms has special advantages in civil engineering. FRP reinforced bars by high resistance and zero corrosion in corrosive environment is proper real averment for steed. FRP rolled sections propose optimal perspective in manufacturing building Skelton because of having high resistance and low weight. One of the important applications of DRP composites is external coat. These coats and sheaths are used in rehabilitation and strengthening of building against earthquake and in applications which requires structure strengthening in order to increasing alive load. Employing these coats when present structures do not meet special criterion of new sods is proper solution in world and Iran, different reasons for using FRP material in different projects are presented. Rough and corrosive environment of presented. Rough and corrosive environment of south of Iran especially the shores of Persian golf is proportionate clarification for replacing reinforced bars and steel sections by FRP sectional bars. Alsip, the weakness of old structure frames in country and inefficiency of structures against earthquake even in large city like Tehran- using FRP strengthening panels in structural element specially in connection parts is necessary.

REFERENCES

FRP composites for structural rehabilitation by Dr D. Mostofi Nejiad.

Holloway, L.C. and M.B., (2001) strengthening of Rein forced Coerce structures, Boston New York.

Homam S.M., and Sheikh S.A., "Durability of Fiber Reinforced Polymers Used in Concrete Structures", *Proc. 3rd International Conference on Advanced Materials in Bridges and Structures*, Ottawa, Canada, August 15–18th 2000, pp. 751–758.

La Tegola A., La Tegola A., De Lorenzis L., and Micelli F., (2000), "Applications of FRP materials for repair of masonry structures", *Proceedings of the Technology Transfer Seminar Advanced FRP Materials for Civil Structures*, October 19th, 2000 Bologna, Italy.

Mosallam, A; chakrabarti, P., Lau, E. and Elsanadely, E., "Application of polymer composites in seismic Repair and Rehabilitation of Reinforced coruscate Connections," internet file.

*Theme 5: Environmental, social and economic
sustainability credentials*

Excellence in Concrete Construction through Innovation – Limbachiya & Kew (eds)
© 2009 Taylor & Francis Group, London, ISBN 978-0-415-47592-1

Effect of GGBFS and GSS on the properties of mortar

İ. Yüksel
Department of Civil Engineering, Zonguldak Karaelmas University, Turkey

R. Siddique
Department of Civil Engineering, Thapar University, Patiala (Punjab), India

Ö. Özkan
Alaplı Vocational School, Zonguldak Karaelmas Univ., Zonguldak, Turkey

J.M. Khatib
School of Enggineering and the Built Env., Univ. of Wolverhampton, West Midlands, UK

ABSTRACT: This study investigated the strength of mortars made with cements incorporating Ground Granulated Blast-Furnace Slag (GGBFS) and Ground Steel Slag (GSS) as partial replacement of Portland cement clinker in different ratios of replacement and fineness. Combination of 70% GGBFS and 30% GSS was partially replaced from 20% to 80% of Portland cement clinker. Tests were performed for sulphate resistance, chloride attack, and high temperature resistance tests. Test results showed that as fineness and replacement ratio of slag was affected volume expansions. Fine ground slag combination has increased sulphate attack durability. Increases in compressive strength were observed as replacement ratio was smaller than 40%. Behavior of mortar specimens of high temperature resistance test are similar to that of the mortars made with ordinary Portland cement. It was concluded that 40% replacement is suitable for 3800 g/cm^2 Blaine fineness of slag combination. The replacement ratio can be raised to 80% if the Blaine fineness of the combination increased up to 4600 g/cm^2.

1 INTRODUCTION

Industrial wastes are generally considered as a major source of environmental problems in the world. Land disposal that is a partial solution for this problem causes secondary pollution problems and extra costs. For this reason, more efficient solutions such as alternative recovery options need to be investigated. In this respect, cement and concrete industry could be an important consumer of industrial by-products or solid wastes. Fly ash (FA), blast-furnace slag (BFS), and silica fume (SF) are currently used in cement and concrete industry. In Europe, every year, nearly 12 million tons of steel slags (SS) get generated. Today about 65% of the produced SS is used on qualified fields of application. But the remaining 35% of the slag are still dumped (Motz & Geiseler 2001). On the other hand, it is generally agreed that Portland cement clinker production is expensive and ecologically harmful. For this reason, various studies have investigated about usage of slags in cement production. However most of these studies concerns mostly one type of slag that is mainly GGBFS. Especially, a combination of

GGBFS and GSS as Portland clinker substitution has not yet been reported in the literature.

Slag cement can be defined as a product of a ground mixture of slag, Portland cement clinker, and gypsum. Some admixtures also can be added to it. BFS, SS are mostly used in slag cement. Steel slag cement can be used in common civil and industrial buildings as well as many special structures, such as road, underground, seaport, mass concrete construction, etc. (Dongxue et al. 1997). The principal constituents of GGBFS and GSS are silica, alumina, calcia, and magnesia, which make up almost 95% of the composition. Minor elements include manganese, iron, and sulphur compounds, as well as trace amounts of several others. Depending on the cooling process, three types of BFS are produced: air-cooled, expanded, and granulated. BFS and SS are also used in cement manufacturing, concrete aggregates, agricultural fill, and glass manufacturing. They are also used as a mineral supplement and liming agent in soil amendment (Kalyoncu 1999). According to Shi & Qian (2000) free calcium oxide content increases the basicity of the SS. The reactivity of SS increases with its basicity. The C$_3$S content in SS

445

is much lower than that in Portland cement. Therefore, SS can be regarded as a weak Portland cement clinker. However it has the tendency of volume expansion in case of its free CaO content is high (Shi & Day 1999).

BFS has been known for its cementitious properties if it is rapidly cooled and subsequently grounded to fineness equivalent to that of the cement. The strength development during the early days is slower for blast-furnace cement under normal curing temperatures. Also, late setting is the disadvantage of BFS cements. In spite of these disadvantages, the incorporation of glassy BFS in the cement improves strength and durability properties of concrete at later age (Wu et al. 1990). Pal et al. (2003) reported that the fineness of slag, activity index, and slag/cement ratio are the parameters that affect the strength of slag concrete. The benefits derived from the use of supplementary cementing materials in the cement and concrete industries can be divided into three categories: engineering, economic, and ecological. Reducing water requirement at a given consistency, enhancement of ultimate strength, impermeability, durability to chemical attack, and improved resistance to thermal cracking are the benefits in the engineering category (Bouzoubaâ & Fournier, 2005). Samet & Chaabouni (2004) suggested a slag cement composition that contained 61% clinker, 35% GGBFS, 3% gypsum, and 1% limestone. They were reported that the replacement of a part of clinker by slag led to an acceptable extension of the setting time, an improvement of the rheological behavior, a very good stability to expansion, and an improvement of the compressive strength at the age of 28 days.

This study investigated strength properties of mortars made with cements incorporating GGBFS and GSS as partial replacement of Portland cement clinker in different ratios of replacement and fineness. High temperature resistance, sulphate and chloride resistance were investigated. Also, specific weight, initial and final setting times, and expansion values of composite cements were investigated.

2 RAW MATERIALS

Granulated blast-furnace slag and steel slag were obtained from Ereğli Iron&Steel Factories (ERDEMIR) in Turkey. The chemical compositions of raw materials were shown in Table 1.

CEN standard sand and CEM I 42.5 N type cement (OPC) were used in preparation of mortar specimens. Loss on ignition values of OPC, GGBFS, and GSS were determined as 1.97%, 1.08%, and 1.11%, respectively.

3 METHODS

Granulated blast-furnace slag and steel slag were grinded separately in laboratory for three different

Table 1. Chemical compositions of raw materials.

Constituent (%)	OPC	GGBFS	GSS
CaO	63.62	37.80	58.53
SiO_2	20.57	35.14	10.72
Fe_2O_3	2.84	0.73	15.30
Al_2O_3	5.11	17.54	1.71
MgO	1.59	5.50	4.27
SO_3	3.00	0.70	0.04

Table 2. Mixture proportions of raw materials in cements.

Cement code	Blaine fineness of slag (cm^2/g)	Materials in the cement (%)		
		OPC	GGBFS	GSS
S0	3400	100	0	0
S1a	3800	80	14	6
S1b		60	28	12
S1c		40	42	18
S1d		20	56	24
S2a	4200	80	14	6
S2b		60	28	12
S2c		40	42	18
S2d		20	56	24
S3a	4600	80	14	6
S3b		60	28	12
S3c		40	42	18
S3d		20	56	24

Blaine fineness values, which were 3800, 4200, and 4600 cm^2/g. Then 30% GSS and 70% GGBFS were blended in a blender to homogenize every mixture for various fineness values. Three basic composite cement series (S1, S2, and S3) were made. Each of these series was divided into four groups designated with letters a, b, c, and d respectively. Blended slag combinations replaced Portland cement clinker from 20 to 80% with an increment of 20% by weight for all series. Details of mix proportions of cements are shown in Table 2. Specific weight, initial and final setting times, and expansion values of these cements were determined according to the standards EN 196-3, and EN 196-6.

Mortar specimens $(40 \times 40 \times 160 \, mm \times mm \times mm)$ were manufactured for each group as shown in Table 2 complying with the procedure given in EN 196-1. In addition, one reference mortar group (S0) was produced with OPC having 3400 cm^2/g Blaine fineness value. Three specimens were produced for each group. Water/Cement/Sand ratio (0.5/1/3) was constant for all groups. All specimens were spooled in laboratory in their moulds for the first 24 hours. After that, they were demoulded and put in a water tank for wet curing (temperature was $20 \pm 2°C$) for 7 days.

Resistance to sulphates and chloride was evaluated by comparing residual compressive strength of specimens of each group exposed to chloride or sulphate solutions with respect to compressive strength of reference group for the same age. For this purpose, the first one of the three specimens was put in 4% $NaSO_4$ solution at the end of 7th day. The second specimen was put in 4% $MgSO_4$ solution, and the third one was put in 4% $NaCl$ solution. They were exposed to these solutions till the 90th day. On the other hand, the reference group was continued to cure in water till the end of the 90th day. Each solution was stirred every day twice and their concentrations were checked once a week. The specimens taken from corroding solutions and reference specimens were cut into three equal parts. Uniaxial compression test was conducted with these parts ($40 \times 40 \times 40$ mm) at the 90th day. The compressive load was applied to lateral surfaces regarding casting direction of specimens. 1 kN/s loading rate was applied until the specimen was crushed.

Mortar specimens produced for high temperature resistance tests were taken from water tank at the 28th day and then spooled in laboratory. Tests were conducted at the age of 90th day. Specimens were exposed to predetermined temperatures in a muffle furnace. Temperature ranged from 200, 400, 600, and 800°C. Five sub-groups of specimens were used for the four different temperatures and for reference. The inner volume of the muffle furnace was 9 dm^3 and the rate of heating was 6°C/min. When the inner temperature of the furnace reached the target temperature, the heating was stopped and the specimens were taken out. They were placed in a room for cooling where the relative humidity ranged 60 to 70 percent for 24 hours. The compressive strength of specimens was measured after cooling.

4 DISCUSSION OF RESULTS

4.1 Physical properties of cements

Specific weight, initial and final setting times, and expansion values of cements are shown in Table 3. It is evident that replacement of GGBFS and GSS combination reduced specific weight of cement according to the results of this study. Since the ratio of GGBFS/GSS was equal to 7/3 the specific weight of GGBFS was governed the specific weight of the new composite cements. Specific weight of OPC, GSS and GGBFS used in this study were equal to 3.14, 3.3 and 2.9 respectively. It was known that specific weight of GSS is varying with iron content significantly. From Table 1, it can be seen that Fe_2O_3 content equals 15.3%. The specific weight of combination with GGBFS and GSS can be calculated as 3.02. As replacement ratio

Table 3. Physical properties of composite cement groups.

Cement code	Specific weight	Vicat setting time (hr:min)		Volume expansion (mm)
		Initial	Final	
S0	3.12	03:15	04:10	0.5
S1a	3.07	04:20	05:40	0.5
S1b	3.05	04:20	05:10	1.0
S1c	3.04	04:20	05:30	1.0
S1d	3.02	04:30	05:40	1.5
S2a	3.06	04:10	05:30	1.5
S2b	3.05	04:10	05:20	1.5
S2c	3.03	04:30	05:30	1.5
S2d	3.03	04:20	05:20	2.0
S3a	3.08	04:00	05:30	1.5
S3b	3.05	04:00	05:10	2.0
S3c	3.04	04:10	05:20	2.0
S3d	3.02	03:50	05:00	2.0
EN 197-1 limit		Min. 1:00	–	Max. 10

of GGBFS and GSS combination increases, specific weight of cement groups decreased to the value of 3.02 (Table 3).

Setting times were retarded with the increase in replacement ratio of GGBFS and GSS combination. In addition, as fineness of GGBFS and GSS increased setting time was slowly decreased. These results are similar to the results reported in the literature. Altun&Yılmaz (2002), Özkan (2006) were concluded that replacement of GGBFS and GSS as single or as combined form retarded the setting time. Chemical properties of the additives (MgO and Al_2O_3 content) may be the reason for this slow reaction. Moreover fineness of replacement materials and replacement ratio increased volume expansions. The increases in volume expansion were explained with amounts of free CaO and MgO in steel slag by Altun & Yılmaz (2002). The upper and lower limit values given in EN 197-1 were not passed in spite of increases observed in setting time and volume expansions.

4.2 Sulphate effect

Compressive strength is the main feature that allows an appreciation of cement quality (Binici & Aksoğan 2006). That is why the effect of exposure of mortars to sulphates was determined in terms of compressive strength. Table 4 shows the variation of compressive strength of the mortar specimens exposed to sulphate attack. It can be seen from Table 4 that as the replacement ratio increased residual compressive strength increased at first in all series. However, this trend was not continued till the 80% replacement ratio for both of sodium sulphate and magnesium sulphate resistance.

Figure 1 that shows the compressive strength index of specimens. The strength index is the ratio of f_s/f_c

Table 4. Compressive strength of reference and mortar specimens exposed to sulphate attack.

Cement type	Compressive strength (MPa)		
	In water	Na$_2$SO$_4$	MgSO$_4$
S0	64.49	66.46	62.49
S1a	68.84	70.87	66.87
S1b	65.33	67.99	64.34
S1c	54.87	64.13	61.44
S1d	40.45	43.76	39.67
S2a	70.45	72.24	68.09
S2b	69.98	73.24	68.55
S2c	55.61	58.23	54.86
S2d	54.14	57.98	52.13
S3a	72.34	75.87	70.67
S3b	66.26	70.23	65.01
S3c	65.53	71.23	63.98
S3d	61.20	69.82	61.24

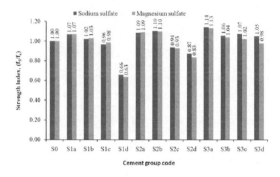

Figure 1. Relative residual compressive strength of specimens exposed to sulphate attack.

where f_s is denoted as residual strength of specimens exposed to sulphate attack, and f_c represents strength of control specimens exposed to sulphate attack. An obvious increase in relative residual compressive strength was observed for the replacement ratios 20 and 40% in S1 and S2 series. The maximum increase (7%) was observed for S1a group (20% replacement) exposed to Na$_2$SO$_4$ or MgSO$_4$ solutions. Although the residual strength shows a decrease trend, it still remains over the 1.00 for the S1b group (with 40% replacement ratio) for both of solutions. About 40% decreases were observed for high (80%) replacement ratios in S1 series. The decrease rate of residual strength in S1d group was 36.5% for specimens cured in MgSO$_4$ solution and this decrease was regarded as conspicuous. There wasn't considerable performance difference between Na$_2$SO$_4$ and MgSO$_4$ resistance in S1 series specimens.

The performance of S2 series was similar to S1 series. The relative residual strength was increased for low replacement ratios (20 to 40%) and decreased for high replacement ratios (60 to 80%). The only difference for this series with respect to S1 series was the fineness of the slag combination which was higher than S1 series. After all, high relative residual strength gain for low replacement ratio was observed and strength loss for high replacement ratio was less when compared with that of S1 series. For instance, this difference was reduced to 16.6% for S2d group, while it was 36.5% for S1d group. Resistance to sodium sulphate and magnesium sulphate effect of mortar specimens was also similar in S2 series.

The behavior of S3 series was quite different from the other two series, and that this difference was in positive direction. It can be seen that there was a linear strength change between groups. The greatest relative residual strength was observed for S3a group and the residual strength of all groups of S3 series was higher than the strength of S0. The only exception was for S3d group with the value of 98 near to 100 for MgSO$_4$ resistance. It was believed that fineness of slag composition results in these good relative residual strength values. Resistance to sodium sulphate was better than resistance to magnesium sulphate in this series.

Expected results were obtained from all three series. Fineness of slag combination and replacement level were parameters having effect on durability of mortars that were produced with partial replaced OPC. Finely ground slag combination has increased sulphate attack durability. Generally, it was sufficient for low level replacement if the fineness value of slag combination was equal to fineness of OPC. However the slag combination should be increased for high level replacement ratio such as 60 or 80%. It was known from the past researches that C-S-H gels results in chemical reaction between pozzolans and calciumhydroxide that is a hydration product of Portland cement clinker. GGBFS reduces water permeability of mortar specimens by filling pore spaces with C-S-H. On the other hand, as fineness of GGBFS or GSS was increased the rate of the reaction was also increased. GGBFS and GSS was improved microstructure of cement paste. At the same time, they helped to decrease calcium hydroxide content of the paste. As a result, durability of mortars made with cement incorporating GGBFS and GSS was improved. Torii and Kawamura (1994), Dongxue et al. (1997), Wu et al. (1999), Temiz et al. (2006), and Binici at al. (2006) have also stated that GGBFS and GSS replacement was increased sulphate resistance of mortars (Wu et al. 1990, Binici & Aksoğan 2006, Torii & Kawamura 1994, Temiz et al. 2007). That is why the best results were attained in S3 series in this study.

Compressive strength of mortars exposed to sulphate attack was compared with strength of mortars

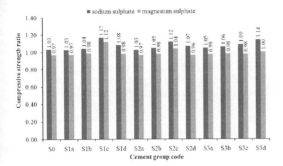

Figure 2. Compressive strength ratio of specimens exposed to sulphate attack.

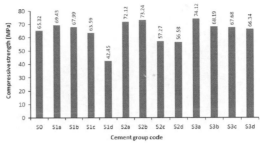

Figure 3. Resistance of series exposed to sodium chloride attack.

cured in water that all of these specimens were produced with the same cement group (Figure 2). The compressive strength ratio shown in Figure 2 was the ratio of P_S/P_W where P_s was denoted compressive strength of specimens exposed to sulphate attack, and P_w represents strength of reference specimens of the same group cured in water.

It is evident from Figure 2 that compressive strength ratio of specimens exposed to sodium sulphate was equal to 1.0 or greater than 1.0. Contrary, compressive strength ratio of specimens exposed to magnesium sulphate was equal to 1.0 or less than 1.0. This shows that strength of specimens exposed to sodium sulphate attack was higher than the strength of specimens cured in water in the same test group. It also indicates that magnesium sulphate can be more aggressive because of reactions due to the presence of magnesium ions. These magnesium ions can decompose C-S-H gels and sulfoaluminates.

The best behavior was indicated by series S3 specimens if comparisons between series were considered. On the other hand, increase in slag replacement ratio caused increase in compressive strength of specimens exposed for sulphate resistance. These increases reached up to 17% in S1c, S2c, and S3d groups for sodium sulphate attack. The worst behavior was observed in S2d group for MgSO$_4$ attack. Only 4% decrease was recorded in compressive strength ratio. Fineness of GGBFS and GSS combination was 4200 cm^2/g which was higher than fineness of OPC clinker. As a result, Figure 2 shows that replacement ratio or fineness of GGBFS and GSS has not a considerable effect on compressive strength ratio.

4.3 Chloride attack

Figure 3 shows compressive strength of cement groups' specimens exposed to chloride attack. Behavior of cement groups was found to be similar to that of the sulphate attack. An increase in

compressive strength was observed when replacement ratio was smaller than 40%. Contrary, the strength was decreased when replacement ratio was in between 60 and 80% replacement levels. The decrease rate in S1 series was observed higher than S2 series.

Compressive strength of all groups in S3 series was observed to be higher than the strength of control group (S0). Nominal increase such as 3.61% and 1.56% were obtained for S3c and S3d groups (high level replacement). This result showed durability of mortar specimens to sodium chloride attack was increased in S3 series. The reason of this result was caused by fineness of slag combination. Sodium chloride improves the strength of plain concrete in most case and can actives the reactivity of GGBFS or GSS within blend cement. However, the most serious problem of sodium chloride for durability was that chloride ions intrudes inner of hardened mortar and causes rust of reinforcing steel bar.

Effect of GGBFS and GSS on chloride attack was similar to the effect as in the case of sulphate attack. Increases in permeability of mortar specimens facilitate chloride ions into the mortar. Therefore cements used for concrete subjected chloride attack must be minimized chloride diffusion. These cements should have high C$_3$A content. On the other hand, high content of C$_3$A was not preferred for durability to sulphate attacks. GGBFS and GSS combination improves microstructure by filling pore spaces with C-S-H gels.

4.4 High temperature

Comparison of residual compressive strength with respect to S1, S2, and S3 series were shown in Fig. 4, 5, and 6, respectively. Residual compressive strengths were discussed considering temperature, replacement ratio, and fineness of slag combination.

Behavior of mortar specimens produced with cement groups were similar to that of the mortars made with Portland cement. As expected, residual compressive strength was decreased with the increase in temperature. Decreases in strength have also been

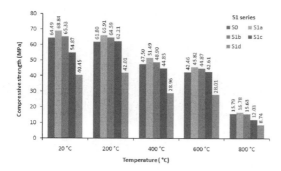

Figure 4. Comparison of residual compressive strength of specimens exposed to high temperature with respect to reference specimens in S1 series.

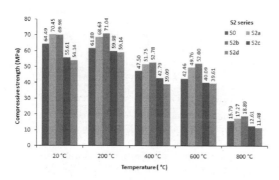

Figure 5. Comparison of residual compressive strength of specimens exposed to high temperature with respect to reference specimens in S2 series.

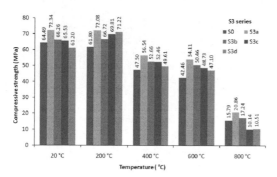

Figure 6. Comparison of residual compressive strength of specimens exposed to high temperature with respect to reference specimens in S3 series.

reported in literature due to high temperature (Özkan 2006, Cülfik & Özturan 2002, Savva et al. 2005). It was recognized that decomposition of $Ca(OH)_2$ begins at temperatures above 400°C. At about 600°C, all the mixtures were almost completely dehydrated. At temperatures over 600°C, as applied in this study, the C–S–H gels, used as the binders undergo dehydration

and lose cementing ability. $Ca(OH)_2$, one of the most important compounds in cement paste, turns into CaO at 530°C. An almost 33% of shrinking is reported to be inevitable during this process (Yüzer et al. 2004). Residual compressive strengths at 200°C were similar to the strengths of reference mortars in this study. However considerable strength losses were observed over 400°C. The percentage of decrease in strength was of the order of 80% for the temperature of 800°C.

It can be seen from Figures 4, 5, and 6 that when the replacement ratio was below 40% the residual strength was higher than that of reference for all the series at all temperatures. This implies that partial replacement of Portland cement with slag combination positively affect high temperature response of mortars. The maximum increase in residual strength was 6.75% for S1a group at 200°C. The maximum increase in residual strength with respect to reference was 22.47% for S2b at 600°C. The maximum increase in residual strength with respect to reference was 32.11% for S3a group at 800°C. As fineness of slag combination increased, the rate of increase in residual strength was observed for 20–40% replacement ratios. Fineness of slag combination has an important effect on residual strength. When the replacement ratio was 60% some residual strength values were more than reference value for S1 and S3 series. When replacement ratio was increased to 80%, the corresponding residual strength values at 200, 400, and 600°C in S3 was still higher than reference value. This shows that even high replacement ratios such as 80% the high temperature durability was similar to normal Portland cement mortars. Fineness of slag combination affects positively high temperature response of mortars produced with the cement groups in this study. As a result, finely ground BGFS-SS mixture can be used as substitution material of Portland cement clinker. The replacement ratio can be selected regarding fineness of slag combination. Higher fineness results higher residual compressive strength.

5 CONCLUSIONS

Based on the tests performed in this study the following conclusions and suggestions could be drawn:

a) GGBFS-GSS combination could be used as partial replacement material for OPC to produce composite slag cement. An optimum replacement ratio depends on fineness of slag combination.
b) Replacement of GGBFS-GSS combination decreased specific weight, and retarded setting times of the composite cement produced in this study.
c) Increase in specific surface area of GGBFS-GSS combination results in expansions. However measured expansion values were still suitable as per the standard EN 197-1.

d) The investigated properties of Portland slag cement in this study to chloride and sulphate attacks were better than durability of OPC. Resistance to sodium sulphate was better than resistance to magnesium sulphate in general.

e) Fineness of slag composition increases high temperature durability of Portland slag cement mortar. Cement produced in this study shows better high temperature resistance than OPC up to 400°C. After 400°C, it was similar to the high temperature resistance of OPC.

f) According to the results of this study, 40% replacement ratio was proposed as an optimum replacement ratio for 3800 g/cm^2 Blaine fineness of slag composition.

g) 4600 g/cm^2 Blaine fineness was needed to increase the replacement ratio for higher replacement ratios such as 80%.

h) More researches on the other durability characteristics and micro structural investigations should be performed in the future.

i) Experiments such as chloride and sulphate resistance should be carried out for longer time periods.

REFERENCES

Altun, İ.A. & Yılmaz, İ. 2002. Study on steel furnace slags with MgO as additive in Portland cement. *Cem. Concr. Res.* 32(7):1247–1249.

Binici, H. & Aksoğan, O. 2006. Sulfate resistance of plain and blended cement. *Cem. Concr. Comp.* 28(1):39–46.

Bouzoubaâ, N. & Fournier, B. 2005. Current situation with the production and use of supplementary cementitious materials (SCMs) in concrete construction in Canada. *Can. J. Civ. Eng.* 32(1):129–143.

Cülfik, M.S. & Özturan, T. 2002. Effect of elevated temperatures on the residual mechanical properties of high-performance mortar. *Cem. Concr. Res.* 32(5):809–816.

Kalyoncu, R.S. Coal combustion products. 1999. *U.S. Geological Survey Minerals Yearbook Vol. 1.* Washington DC: USGS.

Li, D.X., Fu, X.H., Wu, X.Q., Tang, M.S. 1997. Durability study of steel slag cement. *Cem. Concr. Res.* 27(7): 983–987.

Motz, H & Geiseler, J. 2001. Products of steel slags an opportunity to save natural resources. *Waste Management* 21(3):285–293.

Özkan, Ö. 2006. Heat effects on cements produced with GBSF and SS additives, *Journal of Materials Science.* 41(21):7130–7140.

Pal, S.C., Mukherjee, A., Pathak, S.R. 2003. Investigation of hydraulic activity of ground granulated blast furnace slag in concrete. *Cem. Concr. Res.* 33(9):1481–1486.

Samet B., Chaabouni, M. 2004. Characterization of the Tunisian blast-furnace slag and its application in the formulation of a cement. *Cem. Concr. Res.* 34(7): 1153–1159.

Savva, A., Manita, P., Sideris, K.K. 2005. Influence of elevated temperatures on the mechanical properties of blended cement concretes prepared with limestone and siliceous aggregates, *Cem. Concr. Comp.* 27(2):239–248.

Shi, C. & Day, R.L. 1999. Early strength development and hydration of alkali-activated blast furnace slag/fly ash blends. *Adv. Cem. Res.* 11(4):189–196.

Shi, C.J. & Qian, J.S. 2000. High performance cementing materials from industrial slag – a review. *Resour. Conserv. Recycl.* 29(3):195–207.

Temiz, H., Köse, M.M., Köksal, S. 2007. Effects of portland composite and composite cements on durability of mortar and permeability of concrete. *Con. and Build. Mat.,* 21(6):1170–1176.

Torii, K. & Kawamura, M. 1994. Effects of fly ash and silica fume on the resistance of mortar to sulfuric acid and sulfate attack. *Cem. Concr. Res.* 24(2):361–370.

Wu, X.Q., Jiang, W.M., Roy, D.M. 1990. Early activation and properties of slag cement. *Cem. Concr. Res.,* 20(6):961–974.

Yüzer, N., Aköz, F., Öztürk, L.D. 2004. Compressive strength–color change relation in mortars at high temperature, *Cem. Concr. Res.* 34(10):1803–1807.

Excellence in Concrete Construction through Innovation – Limbachiya & Kew (eds)
© *2009 Taylor & Francis Group, London, ISBN 978-0-415-47592-1*

Comparative study on behaviour of concrete-filled steel tubular columns using recycled aggregates

R. Malathy & E.K. Mohanraj
Department of Civil Engineering, Kongu Engg. College, Perundurai, Tamil Nadu, India

S. Kandasamy
Department of Civil Engineering, Government College of Engineering, Tamil Nadu, India

ABSTRACT: In this paper, an attempt was made with steel tubular columns in-filled with recycled aggregate concrete instead of normal conventional concrete so as to utilize the construction and demolition debris for effective recycling in construction works. The behaviour of circular and square concrete-filled steel tubular sections (CFSTs) with partial replacement of coarse aggregate by recycled aggregates under axial load is presented. The effects of steel tube dimensions, shapes and the confinement of concrete are examined. 12 specimens were tested with strength of concrete as 20 MPa and a D/t ratio 22.3, 25.3 & 36.0. The columns were 76 & 89 mm in diameter and 72 & 91 mm in square are 350 & 900 mm in length. From the test results it was noted that square column saving 30 % of steel when compare to circular column. Also it was observed that the load carrying capacity of steel tubular columns in-filled with recycled aggregate concrete is higher than that of conventional concrete and it saves 10 % cost of concrete. Hence this research would give a solution for effective solid waste management as well as cost effective.

1 INTRODUCTION

Steel members have the advantages of high tensile strength and ductility, while concrete members have the advantages of high compressive strength and stiffness. Composite members combine steel and concrete, resulting in a member that has the beneficial qualities of both materials. The two main types of composite column are the steel-reinforcement concrete column, which consists of a steel section encased in reinforced or unreinforced concrete, and the concrete-filled steel tubular (CFST) columns, which consists of a steel tube filled with concrete.

CFST columns have many advantages over steel-reinforcement concrete columns. The major benefits of concrete filled columns are (i) Steel column acts as permanent and integral formwork (ii) The steel column provides external reinforcement (iii) The steel column support several levels of construction prior to concrete being pumped.

Although CFST columns are suitable for all tall buildings in high seismic regions, their use has been limited due to a lack of information about the true strength and the inelastic behaviour of CFST members. Due to the traditional separation between structural steel and reinforced concrete design, the procedure for the designing CFST column using the American Concrete Institute's (ACI) code is quite different from the Load and Resistance Factor Design (LRFD) method suggested by the American Institute of Steel Construction's (AISC).

2 NOTATION

D   outside dimension of column
t   wall thickness of steel tube
L   length of the column
P_{ue} measured ultimate load of the column
f_y yield strength of steel
f_{cc} characteristics cube compressive strength of concrete
f_{cr} flexural strength of concrete
f_{ct} split tensile strength of concrete

3 CONCRETE FILLED STEEL TUBULAR SECTIONS

Circular tubular columns have an advantage over sections when used in compression members, for a given cross-sectional area, they have a large uniform flexural stiffness in all directions. Filling the tube with concrete will increase the ultimate strength of the member without significant increase in cost. The main effect of concrete is that it delays the local buckling of the tube wall and the concrete itself, in the restrained state, is able to sustain higher stresses and strains that when is unrestrained.

The use of CFSTs provides large saving in cost by increasing the floor area by a reduction in the required cross-section size. This is very important in the design of tall buildings in cities where the cost of letting spaces are extremely high. These are particularly significant in the lower storey of tall buildings where short columns usually exist. CFST can provide an excellent monotonic and seismic resistance in two orthogonal directions. Using multiple bays of composite CFST framing in each primary direction of a low to medium-rise building provides seismic redundancy while taking full advantages of the two way framing capabilities of CFSTs (Hajjar, 2002).

3.1 Past research

Experimental research on CFT columns has been ongoing worldwide for many decades, with significant contribution having been made particularly by researchers in Australia, Europe and Asia. The vast majority of these experiments have been on moderate scale specimens (less than 200 mm in diameter) using normal and high-strength concrete.

Neogi et al., (1969) investigated numerically the elasto-plastic behaviour of pin-ended, CFST columns loaded either concentrically or eccentrically about one axis. It was assumed complete interaction between the steel and concrete, triaxial and biaxial effects were not considered. Eighteen eccentric loaded columns were tested, in order to compare the experimental results with the numerical solutions. The conclusions were that there was a good agreement between the experimental and theoretical behaviour of columns with L/D ratios greater than 15, inferred that triaxial effects were small for such columns. Where for columns with smaller L/D ratios, it showed some gain in strength due to triaxial effect.

A series of tests had been carried out by O'Shea & Bridge (1996) on the behaviour of circular thin-walled steel tubes. The tubes had diameter to thickness D/t ranging between 55 and 200. The tests included; bare steel tubes, tubes with un-bonded concrete with only the steel section loaded, tubes with concrete in filled with the steel and concrete loaded simultaneously and tubes with the concrete infill loaded alone. The test strengths were compared to strength models in design standards and specification. The results from the tests showed that the concrete infill for the thin-walled circular steel tubes has little effects on the local buckling strength of the steel tubes. However, O'Shea &

Bridge (1997) found that concrete infill can improve the local buckling strength for rectangular and square sections. Increased strength due to confinement of high-strength concrete can be obtained if only the concrete is loaded and the steel is not bonded to the concrete. For steel tubes with a D/t ratio greater than 55 and filled with 110–120 MPa high-strength concrete, the steel tubes provide insignificant confinement to the concrete when both the steel and concrete are loaded simultaneously. Therefore, they considered that the strength of these sections can be estimated using Eurocode 4 with confinement ignored.

The influence of local buckling on behaviour of short circular thin-walled CFSTs has been examined by O'Shea & Bridge (1997). Two possible failure modes of the steel tube had been identified, local buckling and yield failure. These were found to be independent of the diameter to wall thickness ratio. Instead, bond between the steel and concrete infill determined the failure mode. A proposed design method has been suggested based upon the recommendations in Eurocode 4 (1994).

Kilpatrick et al., (1997a, b) examined the applicability of the Eurocode 4 for design of CFSTs which use high-strength concrete and compare 146 columns from six different investigations with Eurocode 4. The concrete strength of columns ranged from 23 to 103 MPa. The mean ratio of measured/predicted column strength was 1.10 with a standard deviation of 0.13. The Eurocode safely predicted the failure load in 73% of the column analyzed.

Brauns (1998) stated that the effect of confinement exists at high stress level when structural steel acts in tension and concrete in compression and that the ultimate limit state material strength was not attained for all parts simultaneously. In his study, the basis of constitutive relationships for material components, the stress state in composite columns was determined taking into account the dependence of the modulus of elasticity and Poisson's ratio on the stress level in concrete.

O'Shea & Bridge (2000) tried to estimate the strength of CFSTs under different loading condition with small eccentricities. All the specimens were short with a length-to-diameter ratio of 3.5 and a diameter thickness ratio between 60 and 220. The internal concrete had a compressive strength of 50, 80 and 120 MPa. From those experiments O'Shea and Bridge concluded that the degree of confinement offered by a thin-walled circular steel tube to the internal concrete is dependent upon the loading condition. The greatest concrete confinement occurs for axially loaded thin-walled steel with only the concrete loaded and the steel tube used as pure circumferential restraints. Eurocode 4 has been shown to provide the best method for estimating the strength of circular CFSTs with the concrete and steel loaded simultaneously.

Table 1. Specimen properties and measured ultimate load.

Reference Columns	D (mm)	t (mm)	D/t (mm)	L	L/D	Weight of f_y (Mpa)	f_{cu} (Mpa)	P_{ue}	Steel (kG)
C1-HS	76.0	3.0	25.3	900	11.8	260	NA	148.55	4.86
C2-PC	76.0	3.0	25.3	900	11.8	260	25.03	264.40	
C3-RAC	76.0	3.0	25.3	900	11.8	260	28.14	265.50	
C4-HS	89.0	4.0	22.3	350	3.93	260	NA	283.20	2.93
C5-PC	89.0	4.0	22.3	350	3.93	260	25.03	599.20	
C6-RAC	89.0	4.0	22.3	350	3.93	260	28.14	625.80	
S7-HS	72.0	2.0	36.0	900	12.5	260	NA	170.10	3.95
S8-PC	72.0	2.0	36.0	900	12.5	260	25.03	270.80	
S9-RAC	72.0	2.0	36.0	900	12.5	260	28.14	283.10	
S10-HS	91.0	3.6	25.3	350	3.85	260	NA	376.10	3.45
S11-PC	91.0	3.6	25.3	350	3.85	260	25.03	650.45	
S12-RAC	91.0	3.6	25.3	350	3.85	260	28.14	689.80	

For axially loaded thin-walled steel tubes, local buckling of the steel tube does not occur if there is sufficient bond between the steel and concrete. For concrete strength up to 80 MPa, Eurocode 4 can be used with no reduction for local buckling. For concrete strength in excess 80 MPa, Eurocode 4 can still be used but with no enhancement of the internal concrete confinement and no reduction in the steel strength from local buckling and biaxial effects from confinement. Thin-walled circular axial compression and moment can be designed using the Eurocode 4 with no reduction for local buckling.

Mandal, et al., (2002) reported that the quality of RAC is found to be improved considerably with the addition of fly ash. This, in turn, improves the durability of RAC against sulphate and acid attack. Therefore, the results of this study provide a strong support for the feasibility of using recycled aggregates instead of natural aggregates for the production of concrete.

Ramamurthy & Gumaste (1998) reported that the compressive strength of recycled aggregate concrete is relatively lower and the variation depends on the strength of original (demolished) concrete from which the aggregates have been obtained. This reduction is mainly caused by the bond characteristics of recycled aggregate and the fresh mortar of the recycled concrete.

4 EXPERIMENTS

A total of twelve specimens of Circular (designated C) and Square (designated S) sections were tested for this study. The column specimens were classified into three different groups. Each group consists of four specimens filled with plain concrete (designated PC), partial replacement of coarse aggregates by recycled

Table 2. Concrete properties.

Type of Concrete	fck (MPa)*	fcr (MPa)*	fct (MPa)*
Plain Concrete	25.03	3.06	2.26
Partial replacement of coarse aggregate by C&D debris 25% (Recycled Aggregate Concrete)	28.14	3.07	3.01

*average of three cubes, prisms and cylinders respectively

aggregate concrete (designated RAC) and the rest of the column specimens were tested as hollow sections for comparison (designated HS).

All the specimen properties and measured test results are given in Table 1. All the specimens were fabricated from circular and square hollow steel tube and filled with two types of concrete. The average values of yield strength and ultimate tensile strength for the steel tube were found to be 260 and 320 MPa respectively. The modulus of elasticity was calculated to be 2.0×10^5 MPa. In the present experimental work, the parameters of the test specimens are the shape of specimen, size of specimen, strength of concrete and D/t ratio of columns. All the selected parameters are within the ranges of practical limits.

The concrete mix was obtained using the following dosages: 3.75 kN/m^3 of Portland cement, 5.23 kN/m^3 of sand, 11.62 kN/m^3 of coarse aggregate with maximum size 12 mm, and 0.192 m^3 of water. Construction & demolition debris by weight basis are taken. In order to characterize the mechanical behaviour of concrete, three cubic, three prismatic and three cylindrical specimens were prepared from each concrete and

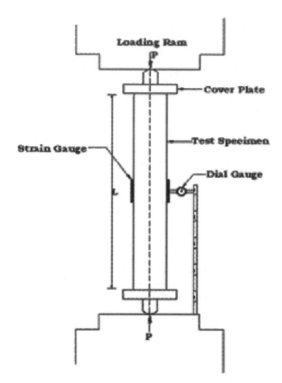

Figure 1. Test Set up of Concrete Filled Steel Tubular Column in Electronic UTM (1000 kN).

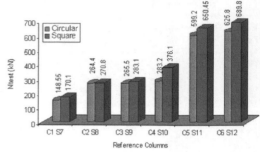

Figure 2. Comparison of load carrying capacity of all columns.

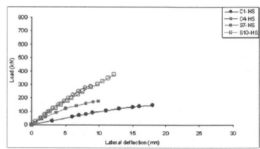

Figure 3. Load – Lateral deflection for hollow column.

tested. The mean values of the strength related properties of concrete at an age of 28 days are summarized in Table-2. During preparation of the test specimens, concrete was cast in layers and light tamping of the steel tube using wooden hammer was performed for better compaction. The specimens were cured for 28 days in a humidity-controlled room.

4.1 Test setup and procedures

All the tests were carried out in an Electronic Universal Testing Machine of a capacity 1000 kN. The columns were hinged at both ends and axial compressive load applied. Test set up of columns as shown in Figure 1. A pre-load of about 5 kN was applied to hold the specimen upright. Dial gauges were used to measure the lateral and longitudinal deformations of the columns. The load was applied in small increments of 20 kN. At each load increment, the deformations were recorded. All specimens were loaded to failure.

5 TEST RESULTS AND DISCUSSIONS

The use of recycled aggregate as a filling material in concrete increases the load carrying capacity to a

greater extent compared with that of plain concrete filled columns and reduces the lateral displacements. Figure 2 shows the load carrying capacity of square and circular column of varying cross section. From the Figure 2 it was noted that load carrying capacity of square column of all sizes, hollow as well as in-filled with plain concrete and recycled aggregate concrete is more than that of circular columns. Fig. 3 shows the load – lateral deflection pattern of hollow tubular circular (76 & 89 mm) and square (72 & 91 mm) columns.

From Figure 3 it was observed that hollow square column perform well than the hollow circular column. When compare to circular column, the strength of the square column is 15–33% higher. Also from Table 1, when comparing strength to weight ratio of 72 mm hollow square column is about 41% more than that of 76 mm hollow circular column and 91 mm hollow square column is about 13% more than 89 mm hollow circular column. Hence about 30% of steel can be saved when square columns are preferred for the same load capacity. For the same load carrying capacity instead of 76 mm dia hollow circular column 60 mm side square column and 89 mm dia hollow circular column 75 mm side square column are suggested to increase the working space area.

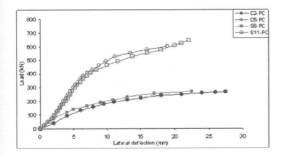

Figure 4. Load – Lateral deflection for plain concrete column.

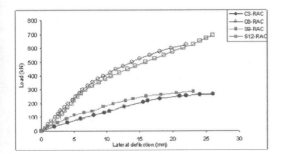

Figure 5. Load – Lateral deflection for recycled aggregate concrete column.

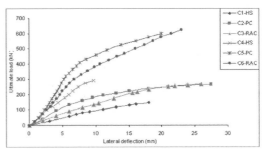

Figure 6. Load – Lateral deflection for circular column.

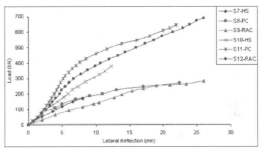

Figure 7. Load – Lateral deflection for square column.

Figure 4 shows the load – lateral deflection pattern of circular and square columns in-filled with plain concrete. From fig. 4 it is also noted that tubular columns in-filled with plain concrete of M_{20} grade is taking more load than hollow column. Comparing fig. 3 and fig. 4, the strength of in-filled column with normal concrete is increased about 60 to 112%. When comparing square and circular column, square column is taking about 3 to 9% more load than that of circular column.

Figure 5 shows the load – lateral deflection pattern of circular and square columns in-filled with recycled aggregate concrete. From fig. 5 it is noted that tubular column in-filled with recycled aggregate concrete taking 5–10% more load than that of normal concrete. When compare to hollow column, recycled aggregate concrete in-filled column taking 66–121% more load. When compare to circular recycled aggregate concrete in-filled column, square recycled aggregate concrete in-filled column taking 6–10% more load. Hence if square recycled aggregate concrete in-filled columns are adopted, there will be a saving in cost for steel about 30% and saving in concrete of 10% and hence totally 40% cost saving can be achieved. Figures 6 & 7 compares the load – lateral deflection pattern for circular and square columns of hollow section, in-filled with normal concrete and recycled aggregate concrete.

6 CONCLUSIONS

The results obtained from the tests on composite columns presented in this paper allow the following conclusions to be drawn.

- Square column taking 15–33% more load than circular column. Hence size reduction for square column is possible.
- For square column 30% cost saving can be achieved. When compare to circular column of same cross section.
- Concrete in-filled columns are taking 60–112% more load than hollow columns.
- Recycled aggregate concrete filled column is taking 66–121% more load than hollow column and 6–10% more than that plain concrete.
- When recycled aggregate concrete is used 10% cost saving in concrete can be achieved. When compare to normal concrete.
- In recycled aggregate concrete in-filled square column 40% cost saving can be achieved. When compare to circular column in-filled with plain concrete.
- Usage of recycled aggregate concrete in steel tubular column, not only a waste minimizing technique, also it saves cost and reduction of size of columns increases the working space.

REFERENCES

Brauns J, 1998, Analysis of stress state in concrete-filled steel column. J. Constructional Steel Research, 49(2): 189–196.

Eurocode 4 DD ENV 1994-1-1, 1994, Design of Composite Steel and Concrete Structures. Part 1.1, General Rules for Buildings (with UK National Application Document) London, British Standards Institution.

Hajjar J.F., 2002, Composite steel and concrete structural systems for seismic engineering, J. Constructional Steel Research, 59 (58); 703–723.

Ramamurthy, K., Gumaste, K.S. 1998, Properties of recycled aggregate concrete, The Indian Concrete Journal, 49–53.

Kilpatrick A., Rangan B.V., 1997, Behaviour of high-strength composite column. In: Composite construction – conventional and innovate, Innsbruck, Austria; 789–794.

Kilpatrick A, Taylor T., 1997, Application of EC4 design provisions to high strength composite columns.

In: Composite construction-conventional and innovate, Innsbruck, Austria; 561–566.

O'Shea, M.D., Bridge, R.Q. 2000, Design of circular thin-walled concrete filled steel tubes. Journal of Structural Engineering, ASCE, Proc. 126, 1295–1303.

Neogi PK, Sen HK, Chapman JC, 1969, Concrete-filled tubular steel columns under eccentric loading. Structural Engineering, 47(5); 195–197.

O'Shea M, Bridge R. 1997, The design for local buckling of concrete filled steel tubes. In: Composite Construction – Conventional and Innovate, Innsbruck, Austria; 319–324.

O'Shea M, Bridge R., 1996, Circular thin-walled tubes with high strength concrete infill. Composite construction in steel and concrete II. Irsee (Germany); ASCE; 780–793.

S. Mandal, et al., 2002, Some studies on durability of recycled aggregate concrete, The Indian Concrete Journal, 385–388.

Excellence in Concrete Construction through Innovation – Limbachiya & Kew (eds)
© 2009 Taylor & Francis Group, London, ISBN 978-0-415-47592-1

A comparative study of using river sand, crushed fine stone, furnace bottom ash and fine recycled aggregate as fine aggregates for concrete production

S.C. Kou & C.S. Poon

Department of Civil and Structural Engineering, The Hong Kong Polytechnic University, Hong Kong

ABSTRACT: This paper compares the properties of concretes that are prepared with the use river sand, crushed fine stone (CFS), furnace bottom ash (FBA), and fine recycled aggregate (FRA) as fine aggregates. The investigation included testing of compressive strength, drying shrinkage and resistance to chloride-ion penetration of the concretes. The test results showed that when designing the concrete mixes with a similar slump value, at all the test ages, when FBA was used as the fine aggregates to replace natural aggregates, the concrete had higher compressive strength, lower drying shrinkage and higher resistance to the chloride-ion penetration. But the use of RFA led to a reduction in compressive strength but increase in shrinkage values. The results suggest that both FBA and FRA can be used as fine aggregates for concrete production.

1   INTRODUCTION

Natural materials such as river sand and crushed fine stone are generally used in concrete as fine aggregates. However, with the booming in urban infrastructure development and the increasing demand on protecting the natural environment, especially in build-up areas such as Hong Kong and some southern Chinese cities in the Pearl River Delta, the availability of the natural resources is diminishing rapidly. Other sources of fine aggregates are urgently needed.

Furnace bottom ash (FBA) is a waste material generated from coal-fired thermal power plants. Unlike its companion – pulverised fuel ash (PFA), it usually has much lower pozzolanic property which makes it unsuitable to be used as a cement replacement material in concrete. However, as its particle distribution is similar to that of sand which makes it attractive to be used as a sand replacement material especially in concrete masonry block production. But few studies have been done on exploring the feasibility of using FBA for making concrete.

Previous studies carried out by Bai et al on using FBA as a natural sand replacement material in concrete indicated that, although FBA has no adverse effect on the strength of concrete, beyond 30% replacement level, the permeation properties of the concrete would be detrimentally affected (Bai & Basheer 2003a,b). The porous structure of the FBA particles has been considered to have caused the increase in the permeation properties. However, the porous nature of the aggregate is believed to be a benefit for reducing the shrinkage of concrete (Collins & Sanjayan 1999, Kohno et al, 1999), which is considered to be due to its "internal curing effect" through slow release of moisture from the saturated porous particles (Weber & Reinhardt 1997, Bentz & Snyder 1999).

Recycled aggregates are produced from the re-processing of mineral waste materials, with the largest source being construction and demolition (C&D) waste. The coarse portion of the recycled aggregates has been used as a replacement of the natural aggregates for concrete production. The potential benefits and drawbacks of using recycled aggregates in concrete are well understood and extensively documented (Dhir et al. 1999, Abou-Zeid et al. 2005, Poon et al. 2002, Eguchi et al. 2007, Evangelista & De Brito 2007, Etxeberria et al. 2007, Gomez-Soberon 2007. In general, the quality of recycled aggregates is inferior to those of natural aggregates. The density of the recycled aggregates is lower than the natural aggregates and the recycled aggregates have a greater water absorption value compared to the natural aggregates. As a result, a proper mix design is required for obtaining the desired qualities for concrete made with recycled aggregates (Lin et al. 2004, Bairagi et al. 1990).

In addition to the coarse recycled aggregates, fine recycled aggregates (<5 mm) can also be used to replace natural fine aggregates in the production of concrete. Khatib (2005) reported that when natural fine aggregates in concrete were replaced by 0%, 25%,

50%, 75% and 100% fine recycled aggregates and the free water/cement ratio was kept constant for all the mixes, the 28-day strength of the concrete developed at a slower rate. Furthermore, the concrete mixtures containing fine recycled aggregates had higher shrinkage than the natural aggregates concrete. Evangelista et al. (2007) indicated that the use of fine recycled concrete aggregates up to 30% replacement ratios would not jeopardize the mechanical properties of concrete.

This paper compares the properties of concretes that are prepared with the use river sand, crushed rock fine, furnace bottom ash, and recycled fine aggregates as fine aggregates. The investigation included testing of compressive strength, drying shrinkage and resistance to chloride-ion penetration of the concretes.

2 EXPERIMENTAL DETAILS

2.1 Materials

The cement used was the ASTM Type I Portland cement complying with BS EN 197–1:2000.

The coarse aggregate used was 10 and 20 mm crushed natural granite. The natural fine aggregates used were river sand sourced from the Pearl river and crushed fine stone (CFS, granite) obtained from a local quarry. Both materials comply with BS EN 12620:2002. The FBA used was obtained from a local coal-fired power plant. Before the FBA was used, it underwent a process of sieving so that all materials used in the experiment were <5 mm. The fine recycled aggregate (FRA) was obtained from a local C&D waste recycling plant.

The physical and chemical properties of the materials used are shown in Tables 1 and 2. Figure 1 shows the particle size distributions of the FBA, the CFS, the FRA and the river sand used in this study.

2.2 Mixture proportions

Three series of concrete mixes were prepared. In the concrete mixes (Series A, B and C), natural river sand was replaced by the FBA, CFS and FRA at replacement levels of 0%, 25%, 50%, 75% and 100% by mass, respectively. In the experimental programme, the concrete mixes were designed to have a constant slump. All concrete mixes were designed with a slump value of 60-80 mm. The cement content was fixed at 386 kg/m³. As such, the free water content (and hence the water-cement ratio) varied and the amount of water added was adjusted to achieve the targeted slump range. Table 3 shows the detailed mix proportions.

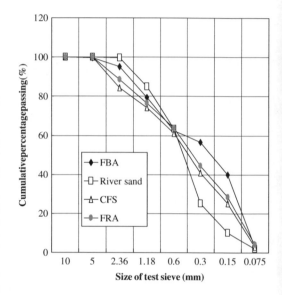

Figure 1. Comparison of particle size distributions of FBA, river sand, crushed fine stone and fine recycled aggregate.

Table 1. Chemical composition (% by mass) of cement and FBA.

	SiO_2	Al_2O_3	Fe_2O_3	MgO	CaO	Na_2O	K_2O	TiO_2	SO_3	Others	LOI
Cement	19.6	7.33	3.32	2.54	63.15	–	–	–	2.13	–	2.97
FBA	60.7	18.3	6.56	1.28	3.25	0.89	2.12	0.95	0.82	1.00	4.13

(LOI: loss on ignition)

Table 2. Properties of aggregates.

	Granite					
Property	10 mm	20 mm	CFS	River sand	FBA	FRA
Density (SSD) (kg/m³)	2620	2620	2610	2620	2190	2310
1-h water absorption (%)	0.48	0.47	0.89	0.38	28.9	2.38

2.3 Details of specimen

For each concrete mix, twelve 100 mm size cubes were cast to determine the compressive strength. Three $75 \times 75 \times 285$ mm prisms with an indentation at the centre of the two ends were cast to determine the drying shrinkage. Two 100×200 cylindrical specimens were cast to determine the resistance to chloride penetration.

All concrete specimens were prepared in accordance with BS 1881: Part 125:1986 using a 0.1 m³ capacity laboratory pan mixer. All specimens were cast in two layers and compacted on a vibrating table until no more air bubbles appeared. They were covered with a plastic sheet and left in the mould in the laboratory at 22($\pm$1)°C for 24 hrs. After that, different curing regimes were used as described below: (1) The 100 mm cubes were cured in water 27($\pm$1)°C until they were tested at 3, 7, 28 and 90 days to determine the compressive strength, (2) the concrete prisms were covered with a damp Hessian cloth and a plastic sheet. After 1 day, the covers were removed and the specimens were wiped clean, and then the initial length was measured. The prisms were then stored in an environmental chamber at 23($\pm$1)°C and 50($\pm$1)% RH until the drying shrinkage was tested at the ages of 1, 4, 7, 14, 28, 90 and 112 days. (3) The concrete cylinders were cured in water at 27($\pm$1)°C until the curing ages of 28 days and 90 days. The cylinders were then cut by a diamond saw to obtain 100 mm diameter × 50 mm thick concrete discs for the chloride-ion penetration test.

2.4 Test procedures

The workability of fresh concrete was measured by the slump test, in accordance with BS 1881: Part 102: 1983. The compressive strength was measured by crushing three 100 mm cubes in accordance with BS

1881: Part 116: 1983. The drying shrinkage values were determined following ASTM C490-07. The resistance to chloride penetrability of concrete was determined in accordance with ASTM C1202-94. The resistance of concrete against chloride ion penetration is represented by the total charge passed in coulombs during a test period of 6 hour.

3 RESULTS AND DISCUSSION

3.1 Property of fresh concrete

The property of fresh concretes is shown in Figure 2. It can be seen that the workability was maintained at approximately the same value by reducing the free

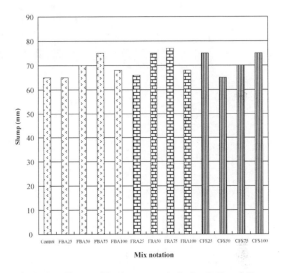

Figure 2. Slump of fresh concrete in Series A, B and C.

Table 3. Mix proportion of concrete mixes in Series A, B and C.

Series No	Mix notation	SRL* (%)	Cement	Water	W/C	Sand	FBA	FRA	CFS	Coarse Aggregate
Control		0	386	205	0.53	652	–	–	–	1110
A	FBA25	25	386	190	0.49	494	167	–	–	1127
	FBA50	50	386	170	0.44	318	343	–	–	1126
	FBA75	75	386	150	0.39	138	529	–	–	1135
	FBA100	100	386	130	0.34	–	725	–	–	1184
B	FRA25	25	386	200	0.52	490	–	164	–	1114
	FRA50	50	386	195	0.51	307	–	331	–	1086
	FRA75	75	386	190	0.49	130	–	500	–	1073
	FRA100	100	386	185	0.48	–	–	671	–	1086
C	CFS25	25	386	208	0.54	487	–	–	162	1105
	CFS50	50	386	211	0.55	322	–	–	323	1099
	CFS75	75	386	214	0.56	161	–	–	482	1093
	CFS100	100	386	217	0.57	–	–	–	640	1087

*SRL: Sand replacement level

461

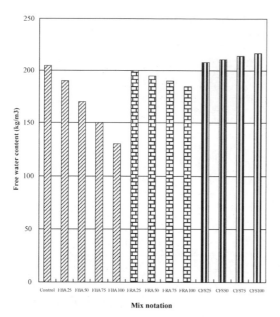

Figure 3. Free water content of concrete at fixed slump range (60–80 mm).

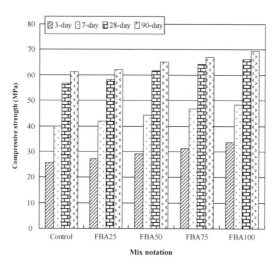

Figure 4. Compressive strength of concrete mixes in Series A (FBA).

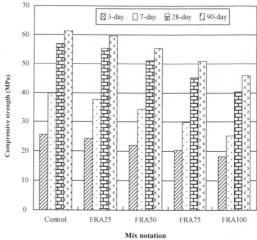

Figure 5. Compressive strength of concrete mixes in Series B (FRA).

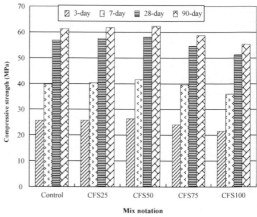

Figure 6. Compressive strength of concrete mixes in Series C (CFS).

water in Series A and B (Figure 3) when FBA and FRA were used to replace river sand. This was due to the FBA and FRA had higher water absorption values than that of the river sand. However, for the case of CFS (Series C), more free water was needed to produce the same workability due to the angular shape of the CFS when compared to river sand.

3.2 Compressive strength

Figures 4, 5 and 6 present the compressive strength results of the concrete mixes in Series A, B and C at the ages of 3, 7, 28 and 90 days, respectively. It can be seen from Figure 4 that the compressive strength of the FBA concrete was higher when compared with that of the control concrete at all the test ages. The improved in compressive strength should be attributed to the decrease in free W/C. This is due to the fact that for a given slump of concrete, the high water absorption properties of FBA would lead to a reduction of free water required, and hence an overall reduction in W/C. This would result in higher compressive strength as shown in Figure 4.

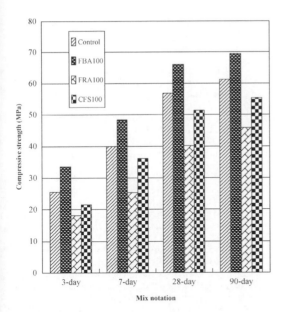

Figure 7. Comparison of compressive strength of concrete mixes prepared with 100% FBA, FRA and CFS as fine aggregate.

Figure 5 shows the compressive strength of the concrete decreased with an increase in the FRA content at all the test ages. This is because similar to the case of FBA, the free water required for the fixed slump for the case of FRA was also decreased. But due to the water absorption value of FRA was a lot lower than that of FBA (see Table 2), the water reduction effect on FRA concrete was not as significant as that on the FBA concrete. Under such a circumstance, the effect of the relative weaker FRA on concrete strength would lead to an overall reduction of compressive strength.

Moreover, it can also be seen from Figure 6 that at replacement levels of 75% and 100%, the compressive strength of the CFS concrete decreased when compared with the control. This was due to the increase in free W/C ratio to compensate for the decrease in slump when the angular CFS was used to replace sand.

Figure 7 shows the comparison of the compressive strength of the concrete made with 100% FBA, 100% FRA and 100% CFS. The results show at all the test ages the concrete made with FBA had the highest compressive strength while the FRA concrete had the lowest compressive strength.

3.3 Drying shrinkage

The drying shrinkage results of the concrete mixes are presented in Figure 8. It shows that the drying shrinkage values of all the FBA concretes are lower than that of the control concrete. This was due to the fact the concrete mixes were prepared with a fixed slump range, and with the increase in FBA content, the free water

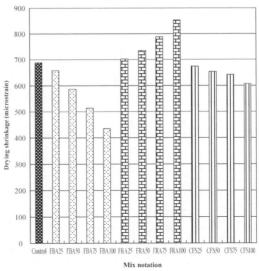

Figure 8. Drying shrinkage of concrete mixes at 112 days.

decreased (Table 3). However, Figure 8 also shows the drying shrinkage of the FRA concrete increased with an increase in the FRA content probably due to the instability of the adhered mortar in the FRA. Moreover, the drying shrinkage values of all the CFS concretes were lower than that of the control concrete due to the CFS having larger particle sizes than that of the river sand.

3.4 Chloride ion penetration

The test results of chloride ion penetration of the concrete mixes are shown in Figure 9. The results indicate that the resistance to chloride penetration of all FBA, FRA and CFS concrete mixes was higher than that of control concrete and was in the order of FBA > CFS > FRA > sand.

4 CONCLUSIONS

Based on the present investigation, the following conclusions can be drawn:

1. At a fixed slump value, the use of FBA and FRA was able to reduce to free water requirement of the concrete mixes.
2. With the use of a lower free W/C ratio, the FBA concrete had the highest compressive strength values.
3. But the use of FRA led to a reduction in compressive strength despite the use of a lower free W/C. This might be due to the inherent weaker mechanical properties of FRA.
4. The drying shrinkage decreased with the increase of the FBA content. FRA increased the drying shrinkage of the concrete.

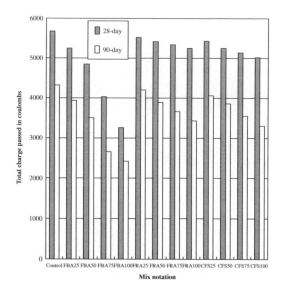

Figure 9. Total charge passed in coulombs of concrete mixes at 28 days and 90 days.

5. The resistance to chloride penetration of all FBA, FRA and CFS concretes was higher than that of the control concrete.
6. It is feasible to use FBA and FRA as fine aggregate in preparing concrete mixes.

ACKNOWLEDGEMENTS

The authors would like to thank the Hong Kong Polytechnic University for funding support. The FBA for this research was provided by the China Light and Power Co. Ltd and the FRA was provided by the Civil Engineering Development Department of the Hong Kong SAR Government.

REFERENCES

ABOU-ZEID, M.N., SHENOUDA, M. N., MCCABE, S., AND EL-TAWIL, F. A. Reincarnation of concrete, Concrete International, 2005, 27, 2, 53–59.

ASTM, C. 1202, Standard test method for electrical indication of concrete's ability to resist chloride ion penetration, American Society of Testing Materials, 1997.

ASTM, C490-07. Standard Practice for Use of Apparatus for the Determination of Length Change of Hardened Cement Paste, Mortar, and Concrete, American Society of Testing Materials, 1997.

BAI Y, BASHEER P.A.M. Properties of concrete containing furnace bottom ash as a sand replacement material. In: Proceedings of structural faults and repair (CD-ROM), London, July 1–3; 2003.

BAI Y, BASHEER PAM. Influence of furnace bottom ash on properties of concrete. Proc 1 Civil Eng Str Build 2003;156(1):85–92.

BAIRAGI, N.K., VIDYADHARA, H. S., AND RAVANDE, K. Mix design procedure for recycled aggregate concrete," Construction and Building Materials, 1990, 4(4), pp. 188–193.

BENTZ, D.P., SNYDER KA. Protected paste volume in concrete – extension to internal curing using saturated lightweight fine aggregate. Cement Concrete Res 1999; 29:1863–7.

BRITISH STANDARDS INSTITUTION. Aggregates for concrete. BSI: London, BS EN 12620:2002.

BRITISH STANDARDS INSTITUTION. Cement – part 1: composition, specifications and conformity criteria for common cements. BSI: London, BS EN 197 – 1:2000.

BRITISH STANDARDS INSTITUTION. Method for determination of compressive strength of concrete cubes. BSI: London, B.S. 1881: Part 116; 1983.

BRITISH STANDARDS INSTITUTION. Method for determination of slump. BSI: London, BS 1881: Part 102; 1983.

BRITISH STANDARDS INSTITUTION. Method of mixing and sampling fresh concrete in the laboratory. BSI: London, B.S. 1881: Part 125; 1986.

COLLINS F, SANJAYAN JG. Strength and shrinkage properties of alkali-activated slag concrete containing porous coarse aggregate. Cement Concrete Res 1999; 29:607–10.

DHIR, R. K., LIMBACHIYA, M. C., AND LEELAWAT, T. Suitability of recycled concrete aggregate for use in BS 5328 designated mixes, Proceedings of ICE – Structures and Buildings, 1999, 134, 257–274.

EGUCHI, K. TERANISHI, K. NAKAGOME, A. KISHIMOTO, H. SHINOZAKI K AND NARIKAWA. Application of recycled coarse aggregate by mixture to concrete construction. Construction and Building Materials, 21 (2007) 1542–1551.

ETXEBERRIA, M. VÁZQUEZ, E. A. AND BARRA M. Influence of amount of recycled coarse aggregates and production process on properties of recycled aggregate concrete. Cement and Concrete Research, 37 (2007) 735–742.

EVANGELISTA, L. AND DE BRITO J. Mechanical behaviour of concrete made with fine recycled concrete aggregates. Cement and Concrete Composites, 29 (2007) 397–401.

GOMEZ-SOBERON, J.M.V. Relationship between gas absorption and the shrinkage and creep of recycled aggregate concrete. Cement, Concrete and Aggregates, 25(2), 42–48.

KHATIB, J.M. Properties of concrete incorporating fine recycled aggregate, Cement and Concrete Research 35 (2005) 763–769.

KOHNO K, OKAMOTO T, ISIKAWA Y, SIBATA T, MORI H. Effects of artificial lightweight aggregate on autogenous shrinkage of concrete. Cement Concrete Res 1999; 29:611–4.

LIN, Y.H., TYAN, Y.Y., CHANG, T.P., AND CHANG, C.Y. An assessment of optimal mixture for concrete made with recycled concrete aggregates", Cement and Concrete Research 2004, 34, pp. 1373–1380.

POON, C. S., KOU S. C. AND LAM L. Use of recycled aggregates in moulded concrete bricks and blocks, Construction and Building Materials, 2002, 16, pp. 281–289.

WEBER S, REINHARDT HW. A new generation of high performance concrete: concrete with autogenous curing. Adv Cem Based Mater 1997; 6:59–68.

Excellence in Concrete Construction through Innovation – Limbachiya & Kew (eds)
© *2009 Taylor & Francis Group, London, ISBN 978-0-415-47592-1*

Feasibility of using low grade recycled aggregates for concrete block production

C.S. Poon, S.C. Kou & H.W. Wan
Department of Civil and Structural Engineering, The Hong Kong Polytechnic University, Hong Kong

ABSTRACT: The Hong Kong Government has set up sorting facilities to separate the inert portion of the construction waste (rock, rubble concrete, asphalts, sand, brick, tile, soil etc) from the non-inert portion (i.e. paper, timber, bamboo, plastic, metals) to facilitate better construction waste management. The intention is that the inert portion can be "reused" as a fill material for land reclamation. The non-inert portion would be disposed of landfills. It should be noticed that the characteristics of the inert construction waste are significantly different from that of crushed concrete rubbles that are mostly derived from demolition waste streams. This is due to the presence of higher percentages of non-concrete components (e.g. >10% soil, brick, tiles etc) in the sorted construction waste. This paper presents the results of a laboratory study to explore the feasibility of using the inert portion of the sorted construction waste (low grade recycled aggregates) for concrete block production. Three Series of concrete block mixtures were prepared by using the low grade recycled aggregates to replace natural coarse granite and 0, 25, 50, 75 and 100% replacement levels of crushed stone fine (crushed granite <5 mm) in the concrete blocks. Test results on properties such as density, compressive strength, transverse strength and drying shrinkage are presented. The results gathered would form a part of useful information for recycling the low grade recycled aggregates.

1 INTRODUCTION

In Hong Kong, a huge quantity of construction wastes is produced every day representing a large fraction of the total solid waste stream. Construction wastes are normally composed of concrete rubble, brick, tile, sand, dust, timber, plastic, cardboard, paper, and metals. The disposal of the wastes has become a severe social and environmental problem in the territory. Government sources have indicated that there is an acute shortage of landfill space in Hong Kong as Hong Kong's landfills are expected to be full within 6–10 years' time (CEDD 2007). To tackle this problem, the Government has introduced a construction waste charging scheme to encourage waste producers to minimize wastage and recycle reusable materials from the waste stream. Under the scheme, Government has set up sorting facilities to separate the inert portion of the construction waste (rock, rubble concrete, asphalts, sand, brick, tile etc) from the non-inert portion (i.e. paper, timber, bamboo, plastic, metals). It should be noticed that the characteristics of the sorted construction waste is significantly different from that of crushed concrete rubbles that is mostly derived from demolition waste streams. The latter, after appropriate crushing and sieving, can be reused

as recycled aggregates in various forms of construction. The sorted construction waste, however, can at best be regarded as a low-grade recycled aggregate. This is due to the presence of higher percentages of non-concrete components (e.g. soil, brick, tiles etc) in the sorted construction waste. In the past, this "inert" portion of the construction waste was reused for reclamation projects. But the recent public objections to public filling have greatly reduced this disposal outlet. The possibility of recycling these wastes in the construction industry is thus of increasing importance. In addition to the environmental benefits in reducing the demand on land for disposing the waste, the recycling of construction wastes can also help to conserve natural materials.

The feasible use of C&D waste as aggregates for precast concrete masonry blocks production has received much research interest in recent years (Poon & Kou 2004, Poon et al. 2004, Poon & Chan 2006, Poon et al. 2006, Chini et al. 2001). Poon et al. (2006) reported that small percentages of substitution of coarse and fine natural aggregates by recycled aggregates had minor effects on the compressive strength of the blocks produced but the compressive strength would decreased at high levels of replacement. However, the flexural strength increased with the use of recycled aggregates. It was also reported that the use of recycled aggregate which contained a significant percentage of crushed clay bricks reduced the density, compressive strength and tensile strength of the blocks produced due to the high water absorption capacity of the crushed clay bricks (Poon et al. 2002).

The Hong Kong Government has recently set up sorting facilities to separate the inert portion of the construction waste (rock, rubble concrete, asphalts, sand, brick, tile, soil etc) from the non-inert portion (i.e. paper, timber, bamboo, plastic, metals) to facilitate better construction waste management. The intention is that the inert portion can be "reused" as a fill material for land reclamation. The non-inert portion would be disposed of at landfills. Besides the non-inert components, the sorting facility separates the inert part of the C&D waste into four main fractions according to sizes: (a) bigger than 250 mm, (b) between 250 mm and 150 mm, (c) between 150 mm and 50 mm and (d) smaller than 50 mm. (Fig. 1)

This paper presents the results of a laboratory study to explore the feasibility of using the inert portion of the sorted construction waste (low grade recycled aggregates) for concrete block production.

2 EXPERIMENTAL DETAILS

2.1 Materials

Cement
In this study, ASTM Type I Portland cement was used and the corresponding properties are shown in Table 1.

Recycled aggregate (RA)
The recycled aggregates used in this study were sorted C&D wastes sourced from a sorting facility in Hong Kong. The two main products of the sorting facility, namely size fractions between 150–50 mm, and <50 mm were used as recycled coarse and fine aggregates. As shown in Table 2, the <50 mm fraction mainly contained soft soil and old concrete rubbles. It also contained a small amount of natural stone, clay bricks, and other impurities such as small pieces of

(a) Size fraction A, ≥250mm   (b) Size fraction B,150~250mm

(c) Size fraction C, 50~150mm   (d) Size fraction D, ≤50mm

Figure 1(a~d). Photos of different particle size fractions of construction wastes.

Table 1. Chemical composition of cement.

SiO$_2$ (%)	Fe$_2$O$_3$ (%)	Al$_2$O$_3$ (%)	CaO (%)	MgO (%)	SO$_3$ (%)	LoI (%)	Specific mass (g/cm^3)	Specific surface area (cm^2/g)
19.61	3.32	7.33	63.15	2.54	2.13	2.97	3.16	3520

wood, paper, tiles and metals. The fraction 150–50 mm contained mainly old concrete rubbles and a small amount of natural stones and other impurities similar to that of the <50 mm fraction although the percentage of the impurities were smaller in the 150–50 mm fraction. The impurities were not removed before the experiment.

The <50 mm fraction underwent a further process of mechanized sieving to produce recycled coarse aggregates, 10/5 mm, and recycled fine aggregate, 5/0 mm, according to the particle size requirements of British Standard BS 812. The recycled aggregates were referred to as RCAI and RFAI, respectively. The 150–50 mm fraction underwent a further process of mechanized crushing and sieving to produce recycled coarse aggregate (RCAII) and recycled fine aggregate (RFAII).

The properties of recycled aggregates were tested according to British Standard methods (BS 812: Part 2: 1975) and the results are given in Table 3. The soil content was determined according to Chinese Standard GB/T14685-2001. Fig. 2 shows the grading curves of the recycled fine aggregates (RFI and RFII).

As shown in Table 3, the water absorption capacity and the soil content were the major differences between the aggregate obtained from the two different fractions of the sorting plant. The soil content of recycled fine aggregates, RFAI and RFAII, were 23.2%

and 7.7%, respectively. And the soil content of recycled coarse aggregates, RCAI and RCAII, were 8.0% and 0.7%, respectively.

Crushed fine stone
Crushed fine stone (CFS) obtained from a local quarry with a fineness modulus (FM) of 3.3 was used the natural fine aggregate. Its grading curve is also given in Fig. 2.

2.2 Mix proportioning

A total of three series of concrete block mixes were prepared. In Series A, the blocks were prepared using RCAI the coarse aggregates. An aggregate-to-cement ratio of 10:1 was used. A0, the control, was produced with the use of CFS as the fine aggregate. In the other mixes (A1 to A4), RFAI was used at levels of 0, 25, 50, 75 and 100% to replace CFS.

In Series B, 100% RCAII and 100% RFAII were used as the coarse and fine aggregates respectively.

Table 2. Constituents of recycled aggregates.

Material	Constituent (% by weight)	
	50~150 mm	<50 mm
Old concrete	74.4	13.4
Natural stones	10.7	2.3
Clay bricks	13.0	4.5
Gravel soil	0.5	78.6
Other impurities (glass, metals, wood, pitch, plastic, paper, etc.)	1.4	1.2

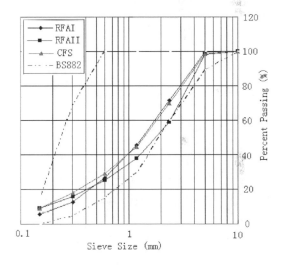

Figure 2. Particle size distribution of the recycled fine aggregates and crushed fine stone.

Table 3. Properties of recycled aggregates.

Properties	Coarse aggregates (5~10 mm)		Fine aggregates (<5 mm)		
	RCAI	RCAII	RFAI	RFAII	CFS
Density-SSD (kg/m³)	2263	2315	1988	2005	2098
Density-oven-dry (kg/m³)	2133	2159	1671	1837	1960
Water adsorption (%)	6.25	5.81	18.92	13.48	7.06
Water-soluble sulphate content (g/L)	0.02	0.01	/	/	/
Ten percent fine value (kN)	72	88	/	/	/
Soil content (%)	8.0	0.7	23.2	7.7	/

CFS-Crushed fine stone

Table 4. Mix ratio (by mass) of block mixtures in Series A (A/C = 10).

| No. | Coarse aggregate | Fine aggregate | | | Total aggregate | Cement | Soil content (%) |
	RCAI	RFAI	CFS	RT			
A0	3.5	0	6.5	0	10	1	2.8
A1	3.5	1.625	4.875	25	10	1	6.6
A2	3.5	3.25	3.25	50	10	1	10.3
A3	3.5	4.875	1.625	75	10	1	14.1
A4	3.5	6.5	0	100	10	1	17.9

A0-Control sample, CFS was used as fine aggregate.
RT-replacement ratio of CFS by RFAI.

Table 5. Mix ratio (by mass) of block mixtures in Series B (100% RCAII and RFAII, varying A/C).

| No. | Coarse aggregate | Fine aggregate | Total aggregate | Cement | Soil content (%) |
	RCAII	RFAII			
B1	2.8	5.2	8	1	5.3
B2	3.5	6.5	10	1	5.3
B3	4.2	7.8	12	1	5.3

Table 6. Mix ratio (by mass) of block mixtures in Series C (A/C = 12).

| No. | Coarse aggregate | Fine aggregate | | | Total aggregate | Cement | Soil content (%) |
	RCAI	RFAI	CFS	RT			
C0	4.2	0	7.8	0	12	1	2.8
C1	4.2	3.9	3.9	50	12	1	10.3
C2	4.2	7.8	0	100	12	1	17.9

C0-Control sample.
RT-replacement ratio of CFS replaced by RFAI.

The aggregate-to-cement (A/C) ratios were 8:1, 10:1 and 12:1 for B1, B2 and B3 mixes respectively.

In Series C, an aggregate-to-cement ratio of 12:1 was used. C1 was prepared with 100% RCAI as the coarse aggregate and 50% RFAI and 50% CFS as fine aggregates. C2 mixture was produced with 100% RCAI as the coarse aggregate and 100% RFAI as the fine aggregate. The mix proportions of the blocks in Series A, B and C are shown in Tables 4, 5 and 6, respectively.

2.3 Fabrication of blocks

The blocks were fabricated in steel moulds with internal dimensions of 200 mm in length, 100 mm in width, and 60 mm in depth. The mixed materials used were approximately 2.8 kg for each block. The materials were put into the mould in three layers of about equal depth. After each of the first two layers was filled, compaction was applied manually using a hammer and a wooden stem. After the third layer was filled, a compressive force at a rate of 600 kN/min was applied for about 50s to mechanically compact the mix within the mould. Excess materials were then removed with a trowel. The fabricated blocks, in the steel moulds, were covered by a plastic sheet and left at room temperature and relative humidity of about 50%. The blocks were then demoulded one day after casting and were cured (covered by a hemp bag to maintain a RH of over 90%) at room temperature (21°C) until testing.

2.4 Testing

2.4.1 Density
The density of partition blocks was determined using a water displacement method as per BS 1881 Part 114 for hardened concrete.

468

2.4.2 Compressive strength

The compressive strength was determined using a compressive testing machine with a maximum capacity of 3000 kN. The load, increased at a rate of 450 kN/min, was applied to the nominal area of the blocks. Prior to the loading test, the blocks were soft capped with two pieces of plywood.

2.4.3 Transverse strength

The transverse strength of the block specimens was determined in accordance with BS 6073 (1981). The test was carried out by a three-point bending test with a supporting span of 180 mm and a height of 60 mm, using a material testing machine with a maximum load capacity of 30 kN.

2.4.4 Drying shrinkage

The drying shrinkage of the specimens was determined in accordance with BS 6073 (1981). After 28 days of curing, the specimens were immersed in water at room temperature for 24 hours, and then initial length of the specimens was measured. After the initial reading, the specimens were conveyed to a drying-chamber with a temperature of 23°C and a relative humidity of 55%. Length measurements were made again 1, 3, 7, 14 days after the initial measurement.

3 RESULTS AND DISCUSSIONS

3.1 Density

The density values of the blocks specimens are shown in Fig. 3. The presented values are the average of three measurements. The results indicated that the density of the blocks decreased with an increase in the RFA content. This is due to the fine recycled aggregates had lower densities when compared to the CFS. Moreover, the aggregate-to-cement ratio had only a minor influence on the density of the blocks produced.

3.2 Compressive strength

The 7-day and 28-day compressive strengths of the blocks in Series A, B and C are given in Fig. 4. The presented values are the average of three measurements. It can be seen that the compressive strength of the blocks in all three Series decreased with an increase in the RFA content except for A1 and A2 specimens. For the case of RFA1 in Series A, a replacement level of 50% or lower resulted in essentially no or only small reduction in compressive strength. Moreover, the compressive strength of the blocks increased when either RFA II was used or with a decrease in the aggregate-to-cement ratio.

Specimens A1 and A2 (25 and 50% replacement of CFS by low grade recycled aggregates) had higher compressive strength compared to the corresponding

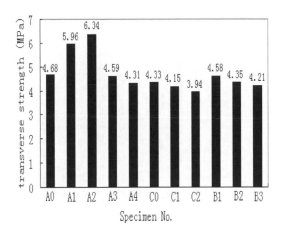

Figure 3. Density of concrete blocks specimens.

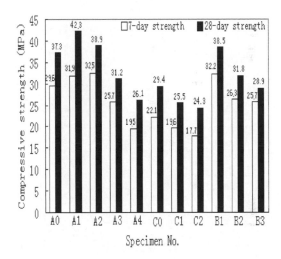

Figure 4. Compressive strengths of concrete blocks at 7-day and 28-day.

blocks made with 100% natural fine aggregates. This may be due to the RFAI contained a higher percentage of fine soil which was capable of filling up the voids more effectively that that of CFS in the blocks produced. Comparing the blocks prepared with the same A/C and fine aggregates replacement level (A4 and B2), B2, which was produced made with RCAII had a higher compressive strength that of A4.

3.3 Transverse strength

The results of the transverse strength for the blocks are given in Fig. 5. The presented values are the average of two measurements. A similar trend to that of compressive strength was observed. The transverse strengths of all block specimens were much higher than that of the requirements of BS6073 ($\geq$0.65 MPa).

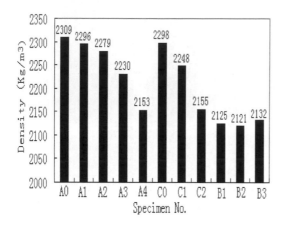

Figure 5. Transverse strength of concrete blocks at 28-day.

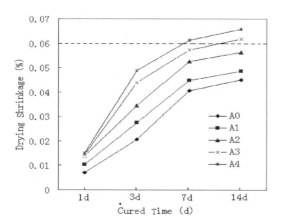

Figure 6. Drying shrinkage values of block specimens in Series A.

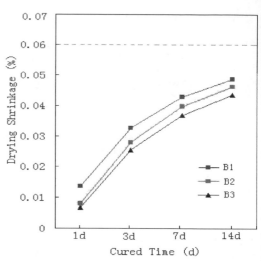

Figure 7. Drying shrinkage values of block specimens in Series B.

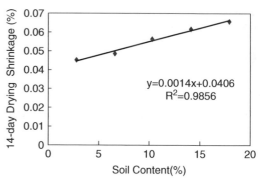

Figure 8. Relationship between drying shrinkage and soil content of blocks in Series A.

3.4 Drying shrinkage

The results of drying shrinkage of the blocks in Series A and B are shown in Figs.6 and 7, respectively. The present data are the average of three measurements. The results show the drying shrinkage of the blocks increased with an increase in RFA content, particular for RFAI. Fig. 8 shows the drying shrinkage of the blocks increased with an increase in soil content. For the case of RFAI and with an A/C ratio of 10, the 14-day drying shrinkage values of the blocks specimens prepared with <50% fine aggregate replacement level (A3 and A4) were within the prescribed limit of BS6073 (0.06%). Also, the drying shrinkage of the blocks decreased either when RFAII was used (compare A4 and B2) or with an increase in the aggregate-to-cement ratio.

4 CONCLUSIONS

Based on the results of this study, the following conclusions can be drawn:

1. The density of blocks decreased with the increase in RFA content.
2. The compressive strength of the blocks decreased with an increase in the RFA content.
3. The drying shrinkage of the blocks increased with the increase in the RFA content.
4. The soil content in the recycled fine aggregate was an important factor in affecting the properties of the blocks produced.
5. The compressive strength, transverse strength of the blocks increased but the drying shrinkage of the

blocks decreased with a decrease in the aggregate-to-cement ratio.

6. With the same A/C ratio and fine aggregate replacement level, blocks prepared with RFAII performed better than that of RFAI.

7. The results show that the low grade recycled aggregates obtained from the construction waste sorting facility has potential to be used as aggregates for making non-structural pre-cast concrete blocks.

ACKNOWLEDGEMENTS

The authors wish to acknowledge the Research Grants Council (PolyU 5259/06E) and the Hong Kong Polytechnic University for funding support.

REFERENCES

BS 1881 PART 114. Testing concrete: methods for determinations of density of harden concrete. British Standards Institution; 1983.

BS 6073: PART 1: Precast concrete masonry units, Specification for precast concrete masonry units. British Standards Institution, 1981.

BS 812. PART 2: Methods for determination of physical properties. British Standard Institution, 1975.

CHINI, A.R., KOU, S.S, ARMAGHANI, J.M., DUXBURY J.P. Test of recycled concrete aggregate in accelerated test track. J Transp Eng. 127(2001) 486–492.

Civil Engineering and Development Department (CEDD), HKSAR, 2007. Statistic on generation of C&D materials.

POON, C.S., CHAN, DIXON. Feasible use of recycled concrete aggregates and crushed clay brick as unbound road sub-base [J]. Construction and Building Materials, 20(2006) 578–585.

POON, C.S., CHAN, DIXON. Paving blocks made with recycled concrete aggregate and crushed clay brick. Construction and Building Materials, 20(2006) 569–577.

POON, C.S., KOU, S.C., LAM, L. Use of recycled aggregate in molded concrete bricks and blocks [J]. Construction and Building Materials, 16 (2002) 281–289.

POON, C.S., KOU, S.C. Properties of Steam Cured Recycled Aggregate Concrete [A].Proceedings of the International Conference [C]. Kingston: the University of Kingston, 2004, 14–16.

POON, C.S., SHUI, Z.H., LAM, L., Influence of Moisture States of Natural and Recycled Aggregates on the Slump and Compressive Strength of Hardened Concrete [J]. Cement and Concrete Research, 2004, 34(1) 31–36.

POON, C.S. CHAN DIXON. Effects of contaminants on the properties of concrete paving blocks prepared with recycled concrete aggregates. Construction and Building Materials, 21 (2007):164–175.

POON, C.S., QIAO, X.C. CHAN, DIXON. The cause and influence of self-cementing properties of fine recycled concrete aggregates on the properties of unbound sub-base. Waste management, 26(2006):1166–1172.

Excellence in Concrete Construction through Innovation – Limbachiya & Kew (eds)
© *2009 Taylor & Francis Group, London, ISBN 978-0-415-47592-1*

Utilization of glass cullet for the manufacture of binding material used in the production of concrete

A.S. Belokopytova
Penosytal Ltd., Russia

P.A. Ketov, V.S. Korzanov & A.I. Puzanov
Perm State Technical University, Russia

ABSTRACT: Utilization of secondary raw materials and waste materials in construction allows us to reduce building costs and to improve ecological situation. One of the most abundant wastes is glass cullet, which cannot be totally utilized nowadays by means of the existing technologies. In the given research work the methods of manufacture a binding material on the basis of glass cullet are discussed. The material can be used for the manufacture of monolithic structures, similar to concrete ones. It have been discovered that amorphous silicon dioxide can interact with Na+ ions, which results in formation of new phases in the system and strengthening of the composition. Mixtures of dispersed glass with amorphous silicon dioxide were tested by various methods. The conditions for obtaining strong binders on the basis of dispersed glass have been developed. As a result, glass concrete with the compressive strength up to 50 MPa has been obtained.

1 INTRODUCTION

Glass cullet as a raw material possesses a number of valuable properties: high strength, chemical stability, accessibility and relatively low cost. At present glass cullet can not be completely utilized, and it is one of the most abundant wastes. According to the United States Environmental Protection Agency (2005), the annual production of glass cullet since the beginning of 90-s has stabilized and is equal to 12.7–13.3 million tons. The amount of utilized glass cullet is equal to 2.7–2.8 million tons. That is, 9.9–10.7 million tons of glass cullet cannot be utilized and is accumulated on disposal damps, which is harmful for the environment. It should be noted that the tendency is not going to change. Similar situation is observed in other countries. Utilization of glass cullet is a serious problem for municipalities all over the world.

The quantity of utilized glass cullet depends mainly on the technology of its collection, for the main method of utilization of glass cullet is addition to melt glass on glass-manufacturing plants. The unsorted glass cullet cannot be used for this purpose, so new technologies of its utilization are required.

Earlier (Ketov 2003) it was reported that unsorted glass cullet can be recycled to produce heat insulation material – foamed glass. But foamed glass has low density, so in respect to the problem of glass cullet utilization the production of heavy product, like glass concrete, is more prospective. Besides, the technology of foamed glass production, mentioned above, includes the step of formation of semi-finished granules – glass-based heavy and strong material.

That is, the problem of obtaining glass-based binding materials is of interest in relation both to glass cullet utilization and making high-quality semiproduct in the technology of foamed glass.

2 RESULTS AND DISCUSSION

It is known that strong binding material can be produced by adding alkali to glass powder. For example, Jones et al. (2003) produced concrete-like material from the dope, containing glass powder and calcium hydroxide. Besides, glass powder can partially or totally replace cement in poured-in-place concrete structures, as it was suggested by Dyer & Dhir (2001). But in this case there is a possibility of alkali-silica reaction, which may cause deterioration and cracking of the concrete (Dhir et al, 2003).

In our judgment, fine glass powder should be considered as independent binding material. To obtain strong glass concrete it is not necessary to mix glass powder with other binders, for example, with liquid glass, as we did earlier (Ketov 2007). We suggest that

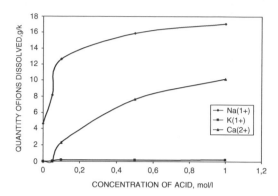

Figure 1. The influence of the concentration of acid in the initial solution (mol/l HCl) on the quantity of Na$^+$, K$^+$ and Ca^{2+} ions, eliminated from the glass (grams per kilogram of glass powder). Green bottle glass, average particle size – 35 μm.

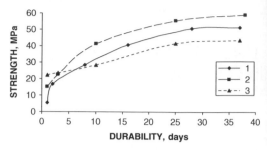

Figure 2. Growth of compression strength of compositions, containing glass powder and alkali solutions with time. The samples contained the following components per 1000 g of glass powder: 1. 50 g of silica gel, 100 g of NaOH, 320 g of water; 2. 50 g of silica gel, 230 g of liquid glass, 200 g of water; 50 g of acid-treated glass, 230 g of liquid glass, 200 g of water.

the main thing is to understand the mechanism of hardening of fine glass powder in alkali media.

It has been experimentally shown that in aqueous medium ion-exchange elimination of Na$^+$, K$^+$ and Ca^{2+} ions from the glass surface takes place. Several grades of glass were chosen for the experiment. The samples were ground in a ball mill to the powder with average particle size of 40–50 μm. The first task was to investigate ion-exchange ability of glass. For this purpose samples were mixed with a solution of hydrochloric acid of certain concentration, mass ratio solution/powder was equal to 1. The obtained suspensions where mixed at 25°C for 30 hours. The concentration of ions in the aqueous phase was measured by the method of flame-ionization photometry. It was found out that the concentration of ions in the solution increases with the concentration of acid. Concentration of K$^+$ ions was negligible, as well as the concentration of K$_2$O in the glass. The specific feature of this ion-exchange process is that the quantity of Ca^{2+} ions eliminating from the glass surface was very small in case of zero and low concentration of acid. As a rule, Ca^{2+} ions were discovered in the solution when the initial concentration of acid exceeded 0.1 M. In difference to other components, the concentration of Na$^+$ ions in the solution has always been rather high. Typical dependence for green bottle glass is shown on Figure 1.

Evidently, silicate glass, under the action of water and acid, undergoes hydrolysis, with the elimination of the cations into the aqueous phase. As the result of the process, a film of hydrated silicon dioxide is formed on the surface of glass particles.

So, we suggest, that hardening of glass-based compositions in alkalescent medium may result from the interaction of hydrated silicon dioxide and Na$^+$

ions, followed by the formation of hydrated sodium polysilicates.

In practice this suggestion can be proved by obtaining glass based binding materials, with addition of amorphous silicon dioxide and Na$^+$ ions.

So, samples of glass powder containing amorphous silicon dioxide have been prepared. The following methods of introduction of silicon dioxide into the mixture have been used: 1. intergrinding of silica gel with glass cullet; 2. treatment of glass powder with 0.2 M solution of hydrochloric acid followed by rinsing with water and drying. For tempering of the samples pure water, liquid glass or sodium hydroxide solutions were used.

Hardening of the composition did not happen in all cases when pure water was used or amorphous silicon dioxide was not present in the mixture. Otherwise, the concrete strength increased and was measured on standard press machine. The growth of compression strength of some compositions is shown on Figure 2.

It is obvious that hardening of the compositions takes place in all cases, when amorphous silicon dioxide and Na$^+$ ions in alkalescent solution are present in the system. Besides, the process goes more intensive after the addition of soluble silicates, such as liquid glass. Probably it can be explained by the fact that the formation of polysilicates is promoted in the presence of anionic form of silicic acids. Otherwise, when the initial solution does not contain soluble silicon compounds, certain time is required for their dissolving and reaching a definite concentration. The process of hardening can be accelerated also by preliminary treatment of glass with hydrochloric acid, which results in formation of active hydrated silicon dioxide directly on the surface of glass particles.

The hypothesis about the formation of hydrated sodium polysilicates of variable composition, resulting in hardening of the mixture is validated by the results

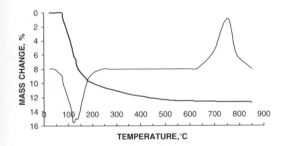

Figure 3. The results of thermogravimetric analysis of glass based concrete samples. 1 – graphic chart of mass loss (in % mass.) with temperature (0C). 2-graphic chart of differential temperature.

of thermogravimetric analysis of the obtained concrete samples. The corresponding data is given on the Figure 3. The mass loss at the temperature higher than 500°C is more than 0.4%, which cannot be explained by some kinetic reasons or experimental error. The mass loss is observed for all samples. The absence of definite step, corresponding to thermal decomposition of the hydrated polysilicate formed can be explained by its variable composition. Besides, for all the samples the elimination of heat in the temperature range from 630 to 840°C was observed. Probably it can be explained by crystallization of glass.

3 CONCLUSION

It is discovered that fine powder of common glass can be used for the production of poured-in-place concrete structures. The process of glass concrete hardening is based on the interaction of amorphous silicon dioxide and free Na^+ ion contained in alcalescent solution. For the production of binding materials various sources of the dispersed amorphous silicon dioxide and free Na^+ ions can be used. The obtained compositions show satisfactory strength and rate of hardening.

ACKNOWLEDGEMENTS

The authors would like to acknowledge the financial help provided by JCS Perm Production of Foamed Silicates for this project.

REFERENCES

A. Ketov. Peculiar Chemical and Technological Properties of Glass Cullet as the Raw Material for Foamed Insulation // Recycle and Reuse of Waste Materials: International Symposium / Dundee. United Kingdom, 2003, P. 695–704.

Dhir R.K., Dyer T.D., Tang M.C. Expansion due to alkali-silica reaction (ASR) of glass cullet used in concrete // Recycling and Reuse of Waste Materials: Proceedings of the International Symposium 9–11 September 2003, Dundee UK. P. 751–761.

Dyer T.D., Dhir R.K. Use of glass cullet as a cement component in concrete // Recycling and Reuse of glass Cullet: Proceedings of International Symposium 19–20 March 2001, Dundee UK. P. 157–166.

Jones T.R., Pascoe R.D., Hegarty P.D. A novel ceramic (casamic) made from unwashed glass of mixed colour // Recycling and Reuse of Waste Materials: Proceedings of the International Symposium 9–11 September 2003, Dundee UK. P. 577–585.

Municipal solid waste generation, recycling, and disposal in the United States: Facts and Figures for 2003 // United States Environmental Protection Agency, Rep. No. 05–18, Washington, April. 2005.

Кетов П.А., Пузанов С.И., Корзанов В.С. Использование вяжущих свойств дисперсных силикатных стекол при утилизации стеклобоя // Строительные материалы.-№5.- 2007.- С.66 67.

Excellence in Concrete Construction through Innovation – Limbachiya & Kew (eds)
© *2009 Taylor & Francis Group, London, ISBN 978-0-415-47592-1*

Methodology for the prediction of concrete with recycled aggregates properties

J. de Brito & R. Robles

Instituto Superior Técnico, Technical University of Lisbon, Portugal

ABSTRACT: The definition of expedient procedures in order to estimate the properties of concrete with recycled aggregates is the main objective of this study. The experimental results used for this research were gathered from international campaigns developed on this subject. With these values, a relationship was established between some of the properties of hardened concrete and the density and water absorption of the aggregates used in the mixture and also the compressive strength of concrete at the age of 7 days. The properties of hardened concrete with recycled aggregates under analysis were compressive strength, splitting and flexural strength, modulus of elasticity, chloride penetration, shrinkage, creep, carbonation penetration and water absorption. The workability and density were the properties analysed for fresh concrete. In order to compare all the campaigns, the graphic analyses of each property were not made with absolute values, but instead with the relationship between those for the concrete with recycled aggregates and the one with natural aggregates only. The density and water absorption of all the aggregates in the mixture, for each substitution rate, were calculated in order to represent the exact proportion of each type of aggregates (natural and recycled). This new method will allow the estimation of the variation of the properties of concrete with recycled aggregates by obtaining the results of the three parameters mentioned above.

1 INTRODUCTION

1.1 *General information*

The construction industry is one of the economic sectors with greater responsibility and contribution to natural resources depletion and production of solid waste. Within this sector, the activities related with the use of concrete, from production to demolition, have a preponderant role.

According to the organization Strategic Development Council (2002), each year around 6 billion tons of concrete are produced, equivalent to 1 ton per human being in the planet. The amount of natural, and finite, resources needed to maintain this level of production of concrete is a great problem to be solved in a short-term future.

On the other hand, the demolition of old structures also produces a great environmental impact. As referred by Masood et al (2001), concrete demolition waste in the European Union and United States of America is up to 100 million tons each year. The high prices for transportation of waste and the lack of authorized landfill places are some of the obstacles to this activity.

Therefore, the two aspects (the need of resources to produce new concrete and the high economic and environmental cost of demolition waste) lead to the need of developing technologies for using recycled aggregates (RA) in the production of concrete.

So far, the major use of recycled aggregates has been as backfill and base course pavement construction. Although it may be considered as a re-use of this material, it is actually a "down-cycling" process in terms of its properties, because the potentiality of this resource is not being fully used. The production of structural concrete with RA is the best way of inverting this tendency and contributing to an effective sustainability of the process.

This study is centred on the research for experimental campaigns on concrete with recycled aggregates (RA) done by investigators worldwide and the graphic analysis of those results in order to relate the properties of hardened concrete with the properties of the aggregates (natural and recycled) used. Along with this study, another is being developed, with the same subject, dedicated to the analysis of similar experimental campaigns done in Portugal in the last years, namely at Instituto Superior Técnico.

1.2 *Scope and methodology of the investigation*

A research for international experimental campaigns was the first step of this investigation. With the collected information, a database was created referring the most important properties of the aggregates and the experimental test procedures of each campaign.

A common conclusion to all the investigations done about this subject is a generalized reduction of the mechanical and durability properties of the concrete with RA, with the increase of the substitution rate of natural aggregates (NA) with RA, when compared with concrete with NA only. The main objective of this study is the definition of procedures that allow the estimation of the properties of the concrete with recycled aggregates by knowing the density and the water absorption of the aggregates, natural and recycled, used in the production of the concrete. Another parameter for this estimation can also be the results of the compressive strength of the concrete at the age of 7 days. The influence of the aggregates properties in the behaviour of the concrete properties is commonly recognized by many authors. Limbachiya et al (2000), in their study about high-strength concrete with RA, refer, in the conclusion chapter, the relationship between the density and water absorption of the recycled and natural aggregates.

To establish the correlation between the properties of the concrete with RA and the three parameters mentioned, the following methodology was adopted:

- analysis and organization of the data available from each experimental campaign, including the information about the test results for the properties of the aggregates used in the production of the concrete;
- calculation of the exact value of the density and water absorption of the aggregates used in the mixture, through the mix proportions of the concretes (with NA only and with RA) and the individual density and water absorption of the aggregates (natural and recycled);
- graphical analysis of the relationship between the substitution rate of NA by RA and each property of concrete;
- graphical analysis of the variation of the ratio between the properties of concrete with RA and the one with NA only (reference conventional concrete) and the substitution rate of NA by RA;
- graphical analysis of the variation of the ratio between the properties of concrete with RA and the reference concrete and the ratio between the weighed value of density of aggregates in the mixture of concrete with RA and the reference concrete;
- graphical analysis of the variation of the ratio between the properties of concrete with RA and the reference concrete and the ratio between the weighed value of water absorption of aggregates in the mixture of concrete with RA and the reference concrete;
- graphical analysis of the variation of the ratio between the properties of concrete with RA and the reference concrete and the ratio between the

compressive strength at the age of 7 days of concrete with RA and the reference concrete.

After obtaining the graphical results for each experimental campaign, and to establish a relationship between the properties of concrete with RA and those of reference concrete for the largest amount of tests possible, the results of each property in the different campaigns were plotted in the same graphic.

A statistical analysis is made of each graphic through a regression line and the correspondent correlation coefficient. In order to simulate the real physical behaviour of the properties analysed with the regression line, this line was forced to go through the correspondent value of the reference concrete. However, this "correction" on the positioning of the regression line contributes to the reduction of the correlation coefficient.

2 DATABASE

Taking into account the international perspective of this investigation, the research for the database was mainly done on the Internet. The search for information was also done through articles in scientific magazines, compilations of conferences or seminars and degree, master and doctoral thesis. Due to the lack of information on some of the articles, direct contact with the authors was also tried through electronic mail, but in the majority of the cases no response was obtained. Among all the sources analysed it is fair to point out the Internet site of the São Paulo University considering the amount and quality of information available (technical articles and master and doctoral theses) on this subject. Unfortunately, the majority of the investigation centres does not follow the same procedure by allowing other investigators to access the information. A significant number of scientific articles are only available on a commercially basis. The criteria adopted for the selection of information on the articles analysed was:

- availability of the experimental results on the properties of the recycled aggregates, particularly the water absorption and mass density;
- availability of the experimental results concerning the largest amount of hardened and fresh concrete properties (mechanical and durability), particularly the compressive strength at the age of 7 days;
- the largest number of substitution rates of NA by RA;
- the largest number of unchanged parameters (w/c ratio, aggregate dimension composition, workability, type of curing, and others) in the experimental procedure of concrete with RA production;
- data information about the concrete with natural aggregates only.

From all the campaigns analysed only a small number was considered useful for this investigation, since the majority of the campaigns did not fulfil the criteria mentioned above. Actually, a general tendency was noted concerning the experimental procedures. Most of the investigators choose to keep constant the compressive strength level in all the concrete with RA (with different substitution rates) and the NA concrete, by increasing the amount of cement in the mixture or adjusting the water/cement ratio. This kind of experimental procedure leads to unfeasible and incomparable results when the focus of the investigation is the influence of the RA properties. It becomes impossible to measure the real effect of each percentage of RA replacing NA. Another handicap detected on the research was the lack of information related to the tests specifications or the procedures adopted. In some of the documents observed the test results were only presented graphically which, in most of the cases, makes the analysis for this study impossible.

Most of the campaigns were centred on coarse recycled aggregates. Only Khatib (2004) approaches the exclusive influence of fine recycled aggregates on the properties of the concrete with RA. It was also noticed that the majority of the investigations were about recycled concrete aggregates or a mixture of recycled ceramic and concrete aggregates. The number of campaigns only oriented for the study of recycled ceramic aggregates is very reduced, and for this reason this subject should be more developed (specially because of the particular properties of the ceramic aggregates, such as the high water absorption).

The information presented in the database was organized in order to be used by other investigators in an easy and fast way. The objective is to clearly identify the main properties analysed in each campaign. The following criteria were adopted for the database presentation:

- RA origin: concrete, ceramic, both or a mixture of debris;
- size of the substituted aggregates: coarse, fine or both;
- parameters kept constant (unchanged criteria for the production of all the concretes: w/c ratio, size distribution, compressive strength, mix proportions, amount of cement, and others);
- varying parameters, that define the objective of the investigation (substitution rate of NA by RA, w/c ratio, amount of fly ashes or synthetic fibres added, age of the testing, and others);
- tests to the aggregates (density, water absorption, size distribution, bulk density, compressive strength, and others);
- tests on fresh concrete (workability, density, bleeding, air content);
- tests on hardened concrete (compressive strength, chloride and carbonation penetration, shrinkage, creep, water absorption, density, porosity, permeability, flexural and splitting strength).

After the collection of all the information, six campaigns were considered within the criteria defined for proceeding with this study. The reference of the campaigns are: Carrijo (2005), Kou et al (2004), Leite (2001), Soberón (2002), Cervantes et al (2007) and Katz (2003).

3 GRAPHICAL ANALYSIS

The graphical analysis started with the relationship between the concrete properties and the substitution rate of NA with RA. After this representation, and to proceed with the analysis of the concrete properties behaviour as a function of the density and water absorption of the aggregates and the 7-day compressive strength of the concrete, the calculation of the weighed value for the density and water absorption of all the aggregates in the mixture was done.

3.1 Calculation of the exact density and water absorption of the aggregates in the mixture

The weighed value for the density of the aggregates in the mixture depends on 2 factors: the density of the individual aggregates (depending on their origin) and the proportion of each type of aggregates used in the mixture to produce the concrete. To calculate the mentioned value equation 3.1 was used, where the percentage of each type of aggregates is multiplied by the correspondent density.

$$D_{mix} = \frac{FA}{100} \times \left[\frac{subst_{FRA} \times D_{FRA} + (100 - subst_{FRA}) \times D_{FNA}}{100} \right] +$$

$$\frac{(100 - FA)}{100} \times \left[\frac{subst_{CRA} \times D_{CRA} + (100 - subst_{CRA}) \times D_{CNA}}{100} \right]$$

(1)

where: D_{mix} = weighed density of the aggregates in the mixture of concrete; FA = percentage of fine aggregates used in the mixture; $subst_{FRA}$ = substitution rate of fine recycled aggregates by fine natural aggregates; $subst_{CRA}$ = substitution rate of coarse recycled aggregates by coarse natural aggregates; D_{FRA} = density of the fine recycled aggregates; D_{FNA} = density of the fine natural aggregates; D_{CRA} = density of the coarse recycled aggregates; D_{CNA} = density of the coarse natural aggregates.

For the calculation of the water absorption of the aggregates used in the mixture, a similar equation was adopted where the density values were replaced by the water absorption values for each type of different aggregates used. For each substitution rate a different value of density and water absorption of the aggregates in the mixture was obtained.

Table 1. Qualitative classification for the correlation coefficient.

Classification	Correlation coefficient
very good	$R^2 \geq 0,95$
good	$0,80 \leq R^2 < 0,95$
acceptable	$0,65 \leq R^2 < 0,80$
not acceptable	$R^2 < 0,65$

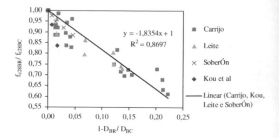

Figure 1. Variation of the ratio between the 28 day compressive strength of concrete and the ratio between the density of the aggregates in the mixture for the campaigns of Carrijo, Leite, Soberón and Kou.

3.2 Relationship between the properties of the concrete and the three parameters

The information collected on the search for international campaigns allowed to establish correlations between nine properties of hardened concrete with recycled aggregates (compressive strength, splitting and flexural strength, modulus of elasticity, chloride penetration, shrinkage, creep, carbonation penetration and water absorption) and the density and water absorption of the aggregates in the mixture and the compressive strength at the age of 7 days.

In order to compare the different campaigns, the absolute values were converted into relative values by dividing the results of the concrete with recycled aggregates (BR) by the results for the concrete with natural aggregates only, the reference conventional concrete (BC).

Table 1 shows the qualitative criteria adopted to evaluate the correlation coefficient obtained by the regression lines in each graphic.

3.2.1 Compressive strength

Compressive strength is the most common tested property of hardened concrete, and for this reason, it was possible to obtain results in 4 campaigns: Carrijo (2005), Leite (2001), Kou et al (2004) and Soberón (2002). The general trend identified for this property indicates a reduction of strength with the increase of the substitution rate of NA with RA.

Figure 1 shows the variation of the ratio between the 28 day compressive strength of concrete (f_c) and the ratio between the densities (D) of the aggregates in the mixture for the campaigns of Carrijo, Leite, Soberón and Kou. The correlation coefficient is considered good and a linear relation between the parameters can be identified. The reduction of the density of RA comparing to NA, due to the higher percentage of attached mortar of RA, contributes to the reduction of the ratio between the compressive strength of concrete.

The same analysis was performed with the variation of the ratio between the 28 day compressive strength of concrete and the ratio between the water absorption of the aggregates in the mixture for the campaigns of Carrijo, Leite, Soberón and Kou and is presented in Figure 2. The correlation coefficient

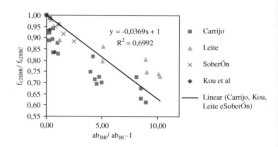

Figure 2. Variation of the ratio between the 28-day compressive strengths of concrete and the ratio between the water absorption of the aggregates in the mixture for the campaigns of Carrijo, Leite, Soberón and Kou.

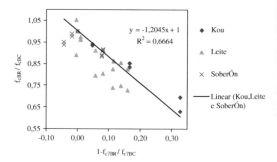

Figure 3. Variation of the ratio between the 28 and 90-day compressive strength of concrete and the ratio between the 7-day compressive strength of concrete for the campaigns of Leite, Soberón and Kou.

is considered acceptable, pointing towards a tendency for a linear behaviour between the ratios.

Figure 3 shows the variation of the ratio between the 28 and 90-day compressive strength of concrete and the ratio between the 7-day compressive strength of concrete for the campaigns of Leite, Soberón and

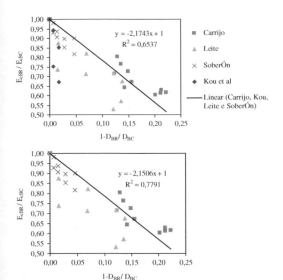

Figure 4. Variation of the ratio between the 28 and 90-day modulus of elasticity of concrete and the ratio between the density of the aggregates in the mixture for the campaigns of Carrijo, Leite, Soberón and Kou (top) and without Kou (bottom).

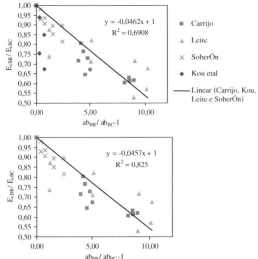

Figure 5. Variation of the ratio between the 28 and 90-day modulus of elasticity of concrete and the ratio between the water absorption of the aggregates in the mixture for the campaigns of Carrijo, Leite, Soberón and Kou (top) and without Kou (bottom).

Kou. The inexistence of data about the 7-day compressive strength of the concrete in the campaign of Carrijo (2005) excluded the author in this particular analysis. The negative values in the abscissa axis mean that in the campaign of Soberón (2002), some of the results for the 7-day compressive strength of the concrete with RA were higher than the conventional concrete. This particular behaviour is not to be expected and contradicts the majority of the investigations; nevertheless, the values were included in the analysis contributing to the reduction of the correlation coefficient, considered as acceptable.

3.2.2 Modulus of elasticity

Modulus of elasticity results were obtained from the campaigns of Carrijo (2005), Leite (2001), Kou et al (2004) & Soberón (2002). In the majority of the investigations, the modulus of elasticity decreases when the substitution rate of RA for NA increases. This behaviour is mostly attributed to the lower stiffness of RA compared to NA. The higher porosity of RA is responsible for the higher deformation of these aggregates when compared to NA, and this effect is also reflected in the concrete with RA when compared to conventional concrete.

Figure 4 shows the variation of the ratio between the 28 and 90-day modulus of elasticity of concrete and the ratio between the densities of the aggregates in the mixture. On the right side of the figure the

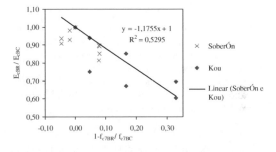

Figure 6. Variation of the ratio between the 28 and 90-day modulus of elasticity of concrete and the ratio between 7-day compressive strength of concrete for the campaigns of Soberón and Kou.

same ratio is presented but without the results of Kou et al (2004), because the ratio between the densities of the aggregates in the mixture is very low, and for this reason, it has a negative contribution to the correlation coefficient. The correlation coefficients are considered acceptable in both cases.

The same variation for the ratio between the water absorption of the aggregates in the mixture is presented in Figure 5. The correlation coefficients are considered acceptable and, in the case of the analysis without the results of Kou et al (2004), good. It is possible to identify a linear tendency in the variation between the relations.

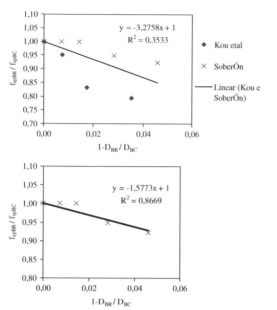

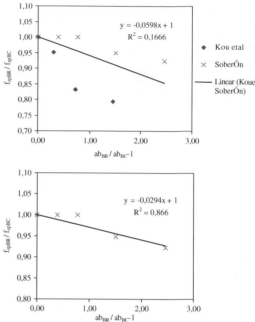

Figure 7. Variation of the ratio between the 90-day tensile strength of concrete and the ratio between the density of the aggregates in the mixture for the campaigns of Soberón and Kou (top) and without Kou (bottom).

Figure 8. Variation of the ratio between the 90-day tensile strength of concrete and the ratio between the water absorption of the aggregates in the mixture for the campaigns of Soberón and Kou (top) and without Kou (bottom).

Figure 6 shows the variation of the ratio between the 28 and 90-day modulus of elasticity of concrete and the ratio between the 7-day compressive strength of concrete for the campaigns of Kou and Soberón. The correlation coefficient is considered not acceptable and, for this reason, it is not possible to identify a linear relationship in the variation.

3.2.3 Tensile strength

The campaigns of Kou et al (2004), Soberón (2002) & Leite (2001) tested concrete tensile strength at different ages (28 and 90 days). The results of Leite (2001) were only at the age of 28 days and very inconstant. Generally, the results of tensile strength graphics obtained confirm the scatter of test results for this property between different campaigns. Figure 7 shows the variation of the ratio between the 90-day tensile strength of concrete and the ratio between the density of the aggregates in the mixture for the campaigns of Soberón and Kou (left) and without Kou (right). The correlation coefficients are considered not acceptable for the two campaigns and good for the results of Soberón (2002) only.

Figure 8 shows the same correlation but for the water absorption of the aggregates in the mixture. The correlation coefficients obtained are also considered not acceptable and good.

The variation of the ratio between the 90-day tensile strength of concrete and the ratio between the 7-day

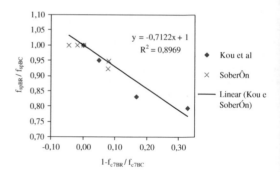

Figure 9. Variation of the ratio between the 90-day tensile strength of concrete and the ratio between the 7-day compressive strengths for the campaigns of Soberón and Kou.

compressive strength for the campaigns of Soberón and Kou is shown in Figure 9. In this case the correlation factor is considered good.

3.2.4 Flexural strength

For the flexural strength graphic analysis, the test results of Leite (2001) at the age of 28 and 90 days are used. From Figure 10, the variation of the ratio between

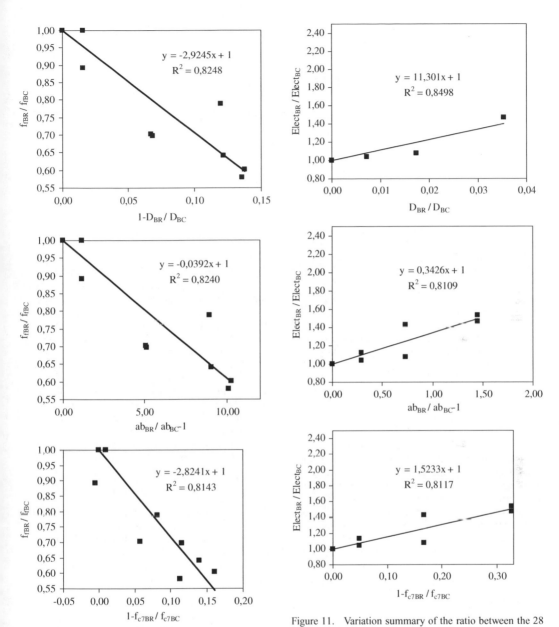

Figure 10. Variation summary of the ratio between the 28 and 90-day flexural strength of concrete and the ratio between the density (top), the water absorption (centre) of the aggregates in the mixture and the 7-day compressive strength of concrete (bottom) for the campaign of Leite.

Figure 11. Variation summary of the ratio between the 28 and 90-day electric charge measured and the ratio of the density (top), water absorption (centre) of the aggregates in the mixture and the 7-day compressive strength of concrete (bottom) for the campaign of Kou.

the 28 and 90-day flexural strength of concrete and the ratio between the three parameters allow to conclude the existence of a linear relationship of the variation. In the three graphics the correlation coefficients are considered good.

3.2.5 *Chloride penetration*

The chloride penetration results were obtained by Kou et al (2004) through the test defined in the ASTM C1202-94. This norm establishes the relationship between the electric charge across concrete during

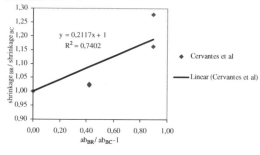

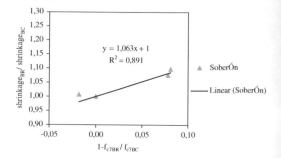

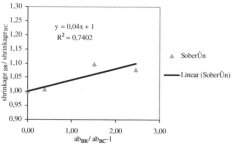

Figure 12. Variation of the ratio between the 28 (top) and 90-day (bottom) shrinkage of concrete and the ratio between the water absorptions of the aggregates in the mixture for the campaigns of Cervantes and Soberón.

Figure 13. Variation of the ratio between the 90-day shrinkage of concrete and the ratio between the compressive strength of concrete for the campaign of Soberón.

3.2.6 Water absorption

The water absorption of the concrete was tested by Soberón (2002) according to the UNE 83-310-90 norm. It is expected that an increase of the substitution rate of NA with RA increases the water absorption of concrete, mostly because of the higher water absorption of RA compared to NA (due to the mortar attached to the first ones). Figure 14 summarises the variation of the ratio between the 28-day water absorption of concrete and the ratio of the density and water absorption of the aggregates in the mixture and the 7-day compressive strength of concrete for the campaign of Soberón (2002). The correlation coefficients are considered very good for the variation with the ratio between the properties of the aggregates in the mixture and not acceptable for the variation with the ratio between the 7-day compressive strength of concrete.

3.2.7 Creep

Soberón (2002) tested the creep resistance of concrete at the age of 90 days. The reduction of the stiffness of RA compared with NA contributes to a higher creep with the increase of the substitution rate of the RA for the NA. Also, a hypothetical increment of the w/c ratio, to balance the higher water absorption of RA compared with NA, can contribute to higher values of creep in the concrete with RA. Figure 15 shows the variation summary of the ratio between the 90-day creep of concrete and the ratio of the density and water absorption of the aggregates used in the mixture and the 7-day compressive strength of concrete. The correlation coefficients obtained are considered very good for the variation with the ratio between the properties of the aggregates in the mixture and acceptable for the variation with the ratio between the 7-day compressive strength of concrete.

a certain period of time and the chloride penetration resistance of concrete. The higher values of electric charge correspond to a lower resistance against chloride penetration. It is expected that the chloride penetration resistance decreases with the increase of the substitution rate of NA with RA. Figure 11 summarises the variation of the ratio between the 28 and 90-day electric charge measured and the ratio between the density and water absorption of the aggregates used in the mixture and the 7-day compressive strength of concrete for the campaign of Kou et al (2004). The correlation coefficients obtained are all considered good, expressing the trend for a linear relationship of the variations.

Figure 12 shows the same correlation but for the water absorption of the aggregates in the mixture. Coincidently, the correlation coefficients are equal and both considered acceptable.

Since the 7-day compressive strength results for the Cervantes et al (2007) campaign are not consistent, the analysis of the correlation with this parameter is not presented. Figure 13 shows the variation of the ratio between the 90-day shrinkage of concrete and the ratio between the compressive strength of concrete for the campaign of Soberón. The correlation coefficient obtained is considered good.

3.2.8 Carbonation

The experimental campaign of Katz (2003) analyzed the carbonation effect in concrete for a conventional

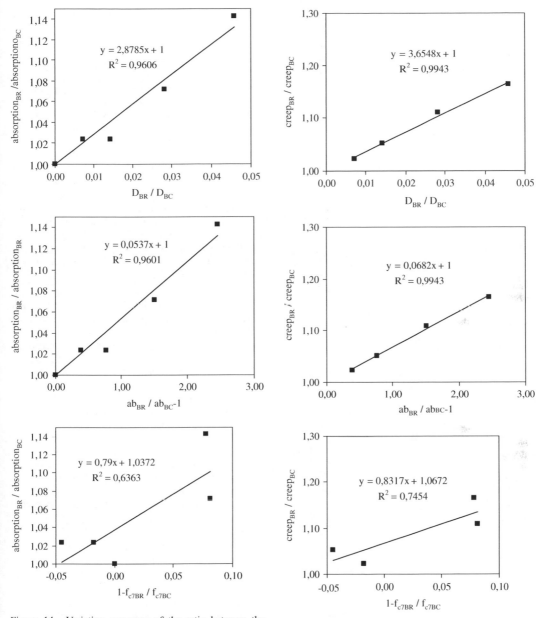

Figure 14. Variation summary of the ratio between the 28-day water absorption of concrete and the ratio of the density (top), water absorption (centre) of the aggregates in the mixture and the 7-day compressive strength of concrete (bottom) for the campaign of Soberón.

Figure 15. Summary of the ratio between the 90-day creep of concrete and the ratio of the density (top), water absorption (centre) of the aggregates in the mixture and the 7-day compressive strength of concrete (bottom) for the campaign of Soberón.

concrete and a concrete with recycled aggregates only. For this study, and in order to collect the largest amount of results, values obtained in the three areas of the concrete specimen tested (top, bottom and sides) were used. Like the resistance against chloride penetration,

the carbonation penetration resistance is reduced with an increase of the substitution rate of RA for NA. This behaviour is mainly justified by the higher porosity of RA compared to NA.

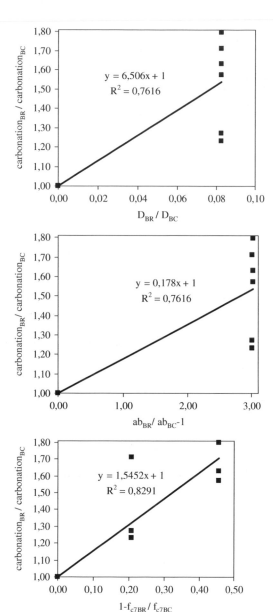

Figure 16. Variation summary of the ratio between the 7-day carbonation penetration of concrete and the ratio of the density (top), water absorption (centre) of the aggregates in the mixture and the 7-day compressive strength of concrete (bottom) for the campaign of Katz.

Figure 16 shows the variation of the ratio between the 7-day carbonation penetration of concrete and the ratio of the density and water absorption of the aggregates used in the mixture and the 7-day compressive strength of concrete for the campaign of Katz (2003).

The correlation coefficients are considered acceptable and good.

4 CONCLUSIONS

The search for international experimental results for this study revealed great differences in the test procedures and organization of the published information. Most of the campaigns accepted the variation of more than one parameter, including the w/c ratio, making the analysis of the effect of the substitution rate impracticable. These obstacles excluded a great number of campaigns to be used in the graphical analysis developed.

Based on the selection of six campaigns, it was possible to analyse nine properties of the hardened concrete. The relationship between these properties and the density and water absorption of the aggregates used in the mixture and the 7-day compressive strength of concrete allowed the following conclusions:

- with very few exceptions, it is possible to establish a linear relationship for the variation of the ratio between the concrete properties and the ratio of the three parameters mentioned;
- generally, the density of the aggregates used in the mixture showed higher correlation coefficients in the graphical analysis with the hardened concrete properties;
- the 7-day compressive strength seems to be the most inadequate parameter to estimate the long-term concrete properties, since the lower correlation coefficients were, in general, obtained with this property; this behaviour can be justified by the influence of the variation of mixture procedures from one campaign to the other and by the higher scatter of results for young concrete;
- the lower results were obtained with the tensile strength and can be justified with the greater variability of this property compared with the compressive strength for example (a trend common to conventional concrete).

Notwithstanding the variability of factors introduced by each investigator in the experimental procedures, it is possible to validate a methodology of estimation of the properties of the concrete with recycled aggregates. The major advantage of this procedure is related with the low cost and short time needed to obtain the results to estimate the long-term properties of hardened concrete. This innovative methodology was registered as a Portuguese patent (No. PT103756). The generalization of this procedure can, in the future, allow construction promoters to decide, in an economic and fast way, about the use of RA in the construction of new concrete structures.

Table 2 shows all the correlation obtained with this study (most were not mentioned in this summary). The

Table 2. Summary of the correlation between the different concrete properties and the density and water absorption of the aggregates in the mixture and the 7-day compressive strength of concrete.

Property	Campaigns	Density		Water absorption		f_{c7d}*	
		R^2	declivity	R^2	declivity	R^2	declivity
f_{c28}	Carrijo/ Kou/ Leite /Soberón	0,8697	1,8354	0,6692	-0,0369	0,6339	1,3551
	Carrijo/ Leite /Soberón	0,8927	1,8284	0,7271	-0,0368	-	-
f_{c90}	Kou/ Leite/ Soberón	0,6514	1,4486	0,5778	-0,0192	0,7616	1,0539
	Leite/ Soberón	0,8152	1,4169	0,7716	-0,0190	-	-
f_c	Carrijo/ Kou/ Leite/ Soberón	0,8370	1,7693	0,5905	-0,0310	0,6664	1,2045
	Carrijo/ Leite/ Soberón	0,8734	1,7583	0,6250	-0,0308	-	-
E_{c28}	Carrijo/ Kou/ Leite/ Soberón	0,6636	2,1841	0,7003	-0,0456	0,3047	1,7111
	Carrijo/ Kou/ Soberón	0,7591	1,9224	0,7300	-0,0506	0,6356	1,3738
E_{c90}	Kou/ Soberón	0,7565	4,9946	0,6593	-0,0934	0,4215	0,9771
E_c	Carrijo/ Kou/ Leite/ Soberón	0,6537	2,1743	0,6908	-0,0462	0,2538	1,4166
	Carrijo/ Leite/ Soberón	0,7791	2,1506	0,8250	-0,0457	-	-
	Carrijo/ Kou/ Soberón	0,7211	1,9579	0,7107	-0,0515	0,5295	1,1755
f_{sp28}	Leite/ Soberón /Kou	0,3331	1,5305	0,2508	-0,0188	0,3019	0,7110
	Leite/ Soberón	0,4906	1,3441	0,4693	-0,0180	-	-
f_{sp90}	Soberón /Kou	-	-	-	-	0,6356	0,6321
	Soberón/ Kou	0,3533	1,4282	0,1666	-0,0598	0,8969	0,7122
	Soberón	0,8669	3,2758	0,8660	0,0294	-	-
f_{sp}	Leite/ Soberón /Kou	0,2652	3,2758	0,1372	-0,0198	0,4904	0,7115
	Leite/ Soberón	0,5358	1,3530	0,5094	-0,0183	-	-
	Soberón /Kou	-	-	-	-	0,7858	0,6721
f_f	Leite	0,8248	2,9245	0,8240	-0,0392	0,8143	2,8241
Chloride$_{28}$	Kou	0,8101	-14,0860	0,8888	0,4108	0,8903	-1,8266
Chloride $_{90}$	Kou	0,8821	-16,8710	0,8455	0,2744	0,8460	-1,2199
Chloride	Kou	0,8498	-11,3010	0,8109	0,3426	0,8117	-1,5233
Shrinkage$_{28}$	Cervantes	0,6280	-3,7310	0,7402	0,2117	-	-
Shrinkage $_{90}$	Soberón	0,7397	-2,1448	0,7402	0,0400	0,8910	-1,0630
Shrinkage	Soberón and Cervantes	0,5890	-3,1945	0,3626	0,0525	-	-
Absorption$_{28}$	Soberón	0,9606	-2,8785	0,9601	0,0537	0,6363	-0,7900
Creep$_{90}$	Soberón	0,9943	-3,6548	0,9943	0,0682	0,7454	-1,0672
Carbonation$_7$	Katz	0,7616	-6,506	0,7616	0,1780	0,8291	-1,5452

correlation coefficient acceptable $(0,65 \leq R^2 < 0,80)$
correlation coefficient good $(0,80 \leq R^2 < 0,95)$
correlation coefficient very good $(R^2 \geq 0,95)$
* excluded the campaign of Carrijo (2005) due to lack of data
- in the campaign of Kou et al (2004), only the results for the concrete with 0% of fly ash and with the cure by immersion were used.
- in the campaign of Carrijo (2005), the results for the concrete with RA classified as "ash" and "red" and with density under 2,2 g/cm³ were used.
- in the campaign of Leite (2001), the test results and the values obtained by statistical method by the author were used.
- the regression lines were obtained automatically by the commercial software used and were conditioned to go through the point correspondent to the conventional concrete.

correlation coefficient classification was identified by different shades.

REFERENCES

Buttler, A.; "Concrete with coarse recycled concrete aggregates – Influence of the age of recycling on the properties of the recycled aggregates and concrete" (in Portuguese), Masters Dissertation, University of São Paulo, 2003

Carrijo, P.; "Analysis of the influence of the density of coarse aggregates from construction and demolition waste on the mechanical performance of concrete" (in Portuguese), Master Dissertation, University of São Paulo, São Paulo, 2005

Cervantes, V.; Roesler, J.; Bordelon, A.; "Fracture and drying shrinkage properties of concrete containing recycled concrete aggregate", Technical note CEAT, University of Illinois, 2007

Etxeberria, M.; Vázquez; E., Marí; "Recycled concrete aggregate as a structural material", Materials and Structures DOI 10.1617/s11527-006-9161-5, 2006

Katz, A.; "Properties of concrete made with recycled aggregate from partially hydrated old concrete", Cement and Concrete Research, V. 33 (5), pp 703–711, 2003

Khatib, J. M.; "Properties of concrete incorporating fine recycled aggregate", Cement and Concrete Research, V. 35 (4), p 763–769, 2005

Kou, S.C.; Poon, C.S.; Chan, D.; "Properties of steam cured recycled aggregate fly ash concrete", International RILEM Conference on the Use of Recycled Materials in Buildings and Structures, Barcelona, pp 590–599, 2004

Larrañaga, M.E.; "Experimental study on microstructure and structural behaviour of recycled aggregate concrete", PhD Thesis, Polytechnic University of Catalonia, Barcelona, 2004

Latterza, L.; Machado, E.; "Concrete with coarse recycled aggregate: properties in the fresh and hardened states and application in light precast elements" *(in Portuguese)*, Engineering and Structures Journal, No. 21, pp 27–58, 2003

Leite, M.; "Evaluation of the mechanical properties of concrete made with recycled aggregates from construction and demolition waste" *(in Portuguese)*, PhD Thesis in Civil Engineering I, Federal University of Rio Grande do Sul, Porto Alegre, 2001

Levy, S.; "Contribution to the study of the durability of concrete made with concrete and masonry waste" *(in Portuguese)*, PhD Thesis in Civil Engineering, Polytechnic School of São Paulo, São Paulo, 2001

Limbachiya, M.C.; Dhir, R.K.; Leelawat, T.; "Use of recycled concrete aggregate in high-strength concrete", Materials and Structures, V. 33, pp 574–580, 2000

Masood, A.; Ahmad, T.; Arif, M.; Mahdi, F.; "Waste management strategies for concrete", Environ Eng Policy, V. 3, pp 15–18, 2002

Soberón, J.G.; "Porosity of recycled concrete with substitution of recycled concrete aggregate, an experimental study", Cement and Concrete Research, V. 32 (8), pp 1301–1311, 2002

Strategic Development Council; "Roadmap 2030: The U.S. Concrete Industry Technology Roadmap", United States, 2002

Tsujino, M.; Noguchi, T.; Tamura, M.; Kanematsu, M.; Maruyama, I.; "Application of conventionally recycled coarse aggregate to concrete structure by surface modification treatment", Journal of Advanced Concrete Technology, V. 5 (1), pp 13–25. 2007

Excellence in Concrete Construction through Innovation – Limbachiya & Kew (eds)
© 2009 Taylor & Francis Group, London, ISBN 978-0-415-47592-1

Geopolymeric concrete based on industrial wastes

F. Colangelo & R. Cioffi
University Parthenope of Naples, Italy

L. Santoro
University Federico II of Naples, Italy

ABSTRACT: In this paper the results of the synthesis of geopolymer-based building materials made with a kaolinitic residue are reported. The preparation of hardened materials takes place through a calcination step followed by a polycondensation step. The calcination step was carried out at temperatures ranging from 500 to 750°C and times ranging between two and six hours. Optimum calcination conditions were found by evaluating the reactivity of the calcined products by differential scanning calorimetry. The polycondensation step was carried out at temperatures ranging from 25 to 85°C by reaction with sodium or potassium silicate. The compositions of polycondensation systems stoichiometric for the synthesis of polysialate, polysialatesiloxo and polysialatedisiloxo geopolymers were tested, but only the first two were successful. An original quantitative analytical method was employed to determine the amounts of reacted silicate and water after polycondensation. The mortars obtained were tested for unconfined compressive strength, apparent density, porosity, surface area and pore size distribution. The results show that it is possible to produce good quality building materials starting from a kaolinitic residue.

1 INTRODUCTION

The term "geopolymers" was first introduced by Davidovits (1979) to designate a new class of three dimensional silico-aluminate materials. Since then, interest in these new materials has steadily grown as both the ceramic and binder industries can draw great advantage from the implementation of new technologies for the manufacture of geopolymer-based products.

Geopolymers comprise three classes of inorganic polymers that, depending on the silica/alumina ratio, are based on the following three different monomeric units: (-Si-O-Al-O-), polysialate (PS), $SiO_2/Al_2O_3 = 2$; (-Si-O-Al-O-Si-O-), polysialatesiloxo (PSS), $SiO_2/Al_2O_3 = 4$; (-Si-O-Al-O-Si-O-Si-O-), polysialatedisiloxo (PSDS), $SiO_2/Al_2O_3 = 6$.

The synthesis of geopolymers takes place by polycondensation and can start from a variety of silico-aluminates. Davidovits (1989, 1991, 1993), Palomo et al. (1992) and Barbosa et al. (2000) used metakaolinite to obtain geopolymers by reaction with alkali metal (Na or K) silicate. Alternatively, Ikeda et al. (1998), van Jaarsveld and van Deventer (1999), Phair et al. (2000), Xu and van Deventer (2000), Xu et al. (2001), Swanepoel and Strydom (2002), Lee and van Deventer (2002, 2004), van Jaarsveld et al. (2002,

2003) and Bakharev (2005) and Andini at al. (2008) have shown that geopolymers can be obtained starting from many raw silico-aluminates; polycondensation takes place by reaction with alkali metal silicate.

The synthesis of geopolymers relies on the following reactions (Xu and van Deventer, 2000).

The structure of geopolymers can be either amorphous or crystalline, depending on the condensation temperature. Amorphous polymers are obtained at temperatures ranging from 20 to 90°C, while crystalline polymers are obtained at 150–200°C. The structure of crystalline geopolymers resembles that of zeolite A (Davidovits, 1991).Geopolymeric materials are attractive because they have excellent mechanical properties, fire resistance and durability (Davidovits and Davidovics, 1988; Palomo et al., 1992). Furthermore, the reaction pathway requires either metakaolinite, obtained by calcining kaolinite at temperatures of the order of 600–700°C, or raw silicoaluminates. There is thus great interest in the reduced energy requirement for the manufacture of new materials based on geopolymers.

The applications of geopolymer-based materials cover many fields. New ceramics, cements, matrices for hazardous waste stabilization, fire-resistant materials, asbestos-free materials and high-tech materials are some of the potential uses of geopolymers (Davidovits, 1991; van Jaarsveld et al., 1997, van Jaarsveld et al., 1999, Kriven et al., 2004 and Bell et al., 2005).

All aspects of research on geopolymers were dealt with in the International Conference "GEOPOLY-MERE '99" held in Saint-Quentine, France on June 30–July 2 1999, which claimed to be the first conference in twenty years to focus on the science behind new materials for the third millennium.

The literature pertinent to geopolymers deals mainly with product characterization, while less effort has been made to understand the qualitative and quantitative features of the polycondensation process.

In this work, a calcined kaolinitic residue has been used for the synthesis of geopolymers. Both the calcination step and the polycondensation step (reaction of metakaolinite with sodium or potassium silicate) have been studied from a quantitative point of view. The reactivity of calcination products obtained under different experimental conditions has been ranked by measuring the heat evolved during the initial polycondensation phase, thus optimizing the calcination process. Following this, an original quantitative method has been developed for the measurement of the amount of silicate and water reacted during the polycondensation reaction. This has been carried out under different experimental conditions: temperature, polycondensation time and composition of the starting mixture.

The kaolinitic residue used in this work comes from a process of silica sand extraction for glass manufacture. Its use enables optimal exploitation of natural resources, especially because its aim is to produce good quality materials starting from a residue. Furthermore, this process can potentially be extended to other silico-aluminate natural materials and industrial wastes such as clays, pozzolana, fly ash and so on (Andini et al., 2008).

2 MATERIALS AND METHODS

The kaolinitic residue used for geopolymer synthesis was characterized by means of chemical analysis, X-ray diffraction, Fourier transform infrared spectroscopy (FTIR) and thermal analysis (Chianese, 1998). The chemical composition was: SiO_2, 39.41%; Al_2O_3, 35.38%; K_2O, 5.69%; CaO, 5.55%; Fe_2O_3, 1.15%; Na_2O, 1.14%; MgO, 0.11%; loss on ignition at 1050°C, 11.35%. X-ray diffraction analysis showed that the material is mainly composed of kaolinite with impurities of α-quartz, calcite and K-feldspar. The FTIR analysis was essentially employed to check the Al coordination number, which proved to be VI. Thermal analysis (heating rate 10°C/min.) showed that the loss of crystallization water takes place within the temperature range 450–650°C. This causes the crystalline structure to collapse and changes the Al coordination number from VI to IV (checked by FTIR analysis).

Table 1. Grams of alkali metal hydroxide, silica and water per gram of calcined residue.

Geopolymer	NaOH	KOH	SiO_2	H_2O
K-PS	–	0.5	0.023	0.4
K-PSS	–	0.5	0.44	0.5
K-PSDS	–	0.5	0.86	1.0
Na-PS	0.36	–	0.023	0.5
Na-PSS	0.36	–	0.44	1.0
Na-PSDS	0.36	–	0.86	2.0

The calcination step was carried out at 500, 550, 650 and 750°C for 2, 4, and 6 hours.

The polycondensation step was carried out at 25, 40, 60 and 85°C for times ranging from 15 min. to 28 days, depending on the reaction temperature. The polycondensation reaction was monitored by measuring the heat evolved by differential scanning calorimetry (DSC). This technique proved to be a useful tool for evaluating the reactivity of the calcined kaolinitic residue as a function of calcination temperature and time.

The polycondensation reaction was carried out with three different mixtures of the calcined residue with alkali metal silicate solution to get PS, PSS and PSDS geopolymers. The alkali metal silicate solution was prepared by dissolving appropriate quantities of reagent grade silica (cristobalite) in water with sodium or potassium hydroxide. The amount of water used for the preparation of the alkali metal silicate solution was such as to ensure constant workability of the mixes. Taking into account that the molar composition of the kaolinitic residue is $0.23(K_2O + Na_2O) \cdot Al_2O_3 \cdot 1.89SiO_2$, the stoichiometric amounts of alkali metal hydroxide and silica to be added to get the three different types of geopolymers were calculated as shown in Table 1.

The mixtures formed into small cylindrical samples $(d \times h = 2 \times 3\ cm^2)$ which were cured at the four different temperatures for various times. These samples were used for the quantitative determination of water and silica consumed during the polycondensation reaction. The amounts of reacted alkali metal silicate and water at any polycondensation condition (time and temperature) were determined as follows. Each sample was ground and washed with a 0.1 M KOH solution to leach all the unreacted alkali metal silicate. This procedure is similar to the one usually used to stop hydration in cements and was carried out at room temperature and for the time required to completely disgregate the sample. Checks were made to ensure that the low alkalinity of the washing solution avoided unreacted metakaolinite dissolution (Maffucci, 1999). Then, the leached solid was rinsed with acetone and with diethyl ether and oven dried at 40°C. The cumulative amount of reacted alkali metal silicate and water was obtained by weight difference between the solid

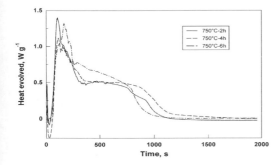

Figure 1. DSC traces of polycondensation systems with kaolinitic residue calcined at 750°C for different times.

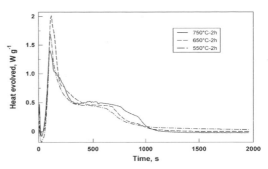

Figure 2. DSC traces of polycondensation systems with kaolinitic residue calcined 2 hours at different temperatures.

recovered after the above treatments and the residue initially employed. The amount of reacted water was determined by weight loss of the recovered solid on ignition at the same temperature as that of the residue calcination step.

Specific surface area, porosity and pore size distribution were measured by means of a Carlo Erba 1990 Sorptomatic apparatus employing the BET method of analysis.

Normalised mortars samples 4x4x16 cm³ in size (UNI EN 196-1) were also prepared for the measurement of unconfined compressive strength, density, specific surface area, porosity and pore size distribution.

3 RESULTS AND DISCUSSION

Of the six systems listed in Table 1, those labelled K-PSDS and Na-PSDS (SiO_2/Al_2O_3 ratio equal to six) did not harden at any of the temperatures tested. Thus, only the other four systems were tested.

3.1 Optimization of the calcination step

The calcination step was optimized in terms of time and temperature by evaluating the reactivity of the products obtained in the different conditions tested through the measurement of the heat evolved during the polycondensation step.

Figure 1 shows the DSC traces obtained with the system of type K-PSS at 85°C reaction temperature. The mixes were made with the kaolinitic residue calcined 2, 4 and 6 hours at 750°C. It can be seen that the curves are very similar to each other, indicating that the calcination time has little effect on the reactivity of the product obtained at 750°C. Heat is evolved very rapidly up to about 300 s; then, the rate of heat evolution lowers without very much variation up to about 800 s. After this time, the rate decreases rapidly approaching zero.

From the quantitative point of view and for the sake of comparison, the reactivity of the products obtained

in the different calcination conditions can be evaluated taking into account the heat of reaction as given by the area below each curve. These values are 532, 572 and 574 J g⁻¹ for 2, 4 and 6 hours calcination time, respectively. It is clear that at 85°C, for residue calcined at 750°C, the maximum reactivity is reached just after 4 hours calcination but that an almost equally high reactivity is reached at 2 hours. In fact, the heat evolved in this condition is about 93% of that evolved when the calcination time is longer.

Figure 2 shows the effect of calcination temperature for a uniform two hours calcination. Again, the DSC traces are very similar, giving values of heat evolved of 528 J g⁻¹ in the case of the residue calcined 2 hours at 650°C and 487 J g⁻¹ when the calcination is carried out at 550°C for the same time. From these data it can be inferred that the reactivity of the product of a two hour calcination at 650°C is almost the same as that of the residue calcined at 750°C for the same time and very close to the maximum reactivity attainable at the higher temperature for longer calcination times. In fact, the heat evolved after reaction of the residue calcined 2 hours at 650°C is about 92% of that evolved when the residue undergoes the calcination step at 750°C for 4–6 hours.

On the other hand, the reactivity of the residue calcined 2 hours at 550°C is significantly lower, as the heat evolved is only about 85% of the maximum. This is in agreement with the results of thermal analysis carried out on the raw kaolinitic residue which showed that calcination is not complete at 550°C (Chianese, 1998). DSC measurements carried out at the polycondensation temperatures of 60 and 40°C confirmed that the reactivity of the calcined residue increases as both temperature and time of calcination increase. No DSC measurement was possible at 25°C due to the extremely low rate of heat evolution. Then, it was concluded that the optimum calcination time and temperature are respectively 2 hours and 650°C. These conditions were ensured when calcining the residue for the remaining part of the work.

491

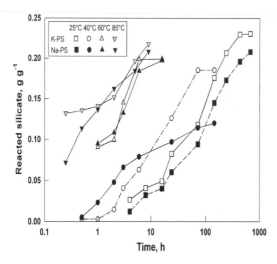

Figure 3.   Amount of reacted silicate of systems of type PS.

3.2   *Chemical characterization of the polycondensation step*

The quantitative data of the amounts of reacted water and reacted water plus alkali metal silicate were used to obtain the amount of reacted alkali metal silicate at the different polycondensation conditions. These results are reported in Figures 3 and 4 for the systems (Na,K)-PS and (Na,K)-PSS, respectively. While Figures 1 and 2 show no measurable heat evolution after about 1000 s (0.27 hours), Figures 3 and 4 show that polycondensation continues beyond that time. This implies that the reaction has an initial heat producing phase followed by a different, scarcely heat producing phase.

Figure 3 refers to PS type geopolymers and shows that in all cases the amount of reacted alkali metal silicate increases with reaction time and that ultimate values greater than 0.2 grams per gram of calcined residue are attained at any temperature (with the exception of Na-PS at 40°C). Also, no great difference is observed at a given temperature between Na-PS and K-PS geopolymers.

The values of reacted alkali metal silicate can be easily converted in terms of percentages of the initial quantity added (see Table 1). The maximum percentages of reacted alkali metal silicate are about 57% for Na-PS geopolymers and about 44% for K-PS geopolymers, indicating that the fractional conversion is significantly higher when NaOH is used for the preparation of the initial mixtures.

Figure 4 shows that even for PSS type geopolymers no significant effect of the alkali metal is observed on the absolute amount of reacted alkali metal silicate. In this case, the values are higher than those found for PS type systems, proving that a geopolymer of different composition is really formed. It is also seen that the amount of reacted alkali metal silicate beyond

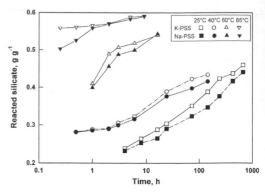

Figure 4.   Amount of reacted silicate of systems of type PSS.

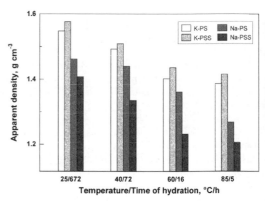

Figure 5.   Apparent density of geopolimeric mortars of types PS and PSS polycondensed at different temperatures/times.

the shortest times investigated increases slightly with reaction time. The maximum values are observed at 85°C (about 0.6 g/g for both Na-PSS and K-PSS). As the temperature decreases, the values of reacted alkali metal silicate decrease steadily. However, even at 25°C they are higher than those found for PS type geopolymers. In the case of PSS geopolymers, the maximum percentages of reacted alkali metal silicate (at 85°C) are about 75% for Na-PSS and about 64% for K-PSS, confirming the higher alkali metal silicate conversion degree for the systems with added NaOH.

3.3   *Physico-mechanical characterization of geopolymers*

Figure 5 shows the apparent density of the mortars obtained at the longest reaction times at each temperature, that is 28 days at 25°C, 3 days at 40°C, 16 hours at 60°C and 5 hours at 85°C. It is seen that the apparent density decreases with polycondensation temperature and is higher for K-PS and K-PSS on one side in comparison to Na-PS and Na-PSS on the other side. A higher density of KOH-based products would

Table 2. Porosity and specific surface area of geopolymeric mortars obtained at 25°C.

Geopolymer	Porosity, $m^3 g^{-1}$	Specific surface area, $m^2 g^{-1}$
K-PS	0.0139	38.23
K-PSS	0.0825	55.45
Na-PS	0.0176	29.19
Na-PSS	0.0811	44.32

be expected as potassium is heavier than sodium. No evidence is found of the effect of the type of polymer, as

K-PSS products have a higher density than that of K-PS products, while Na-PS products show higher density that Na-PSS products.

All density values lie within the range 1200–1600 kg m^{-3}, so that lightweight products could be manufactured starting from these inorganic polymers; for normal cementitious binders the density ranges between 1900 and 2300 kg m^{-3}.

The mortars obtained after hardening at 25°C were also submitted to porosity characterization. The data regarding specific surface area and specific pore volume are reported in Table 2. These results show that a clear difference exists between systems of types PS and PSS. Both specific pore volume and surface area are greater for PSS type systems.

The effect of added alkali metal is of lesser significance. Figure 6 shows the pore distribution with reference to three classes of size which, according to Anderson and Pratt (1985), are as follows: macropores (above 50 nm), mesopores (between 50 and 2 nm) and micropores (below 2 nm). Again, it is seen that a clear-cut difference exists between PS and PSS type systems. In PS type systems micropores prevail, while mesopores prevail in the case of PSS type systems. As before, the effect of added alkali metal is not significant.

Figure 7 shows the values of unconfined compressive strength measured on the same samples in Figure 5. It is seen that the effect of the type of polymer and alkali metal are correlated to apparent density; the products obtained with KOH show better strength and higher density. The effect of polymer type is not significant except for the products obtained at 60 and 85°C with added NaOH. It is very remarkable that at 25 and 40°C, K-PS and K-PSS products give higher values than that required for cementitious pastes (32.5 MPa). Furthermore, the values of mechanical strength obtained at 25 and 40°C with added NaOH are equally interesting for the manufacture of pre-formed building blocks.

The remarkable decrease in strength observed at 60 and 85°C (as well as the decrease in density) may be due to the heating in consequence of the exothermic polycondensation which can cause partial water evaporation with formation of micro cavities.

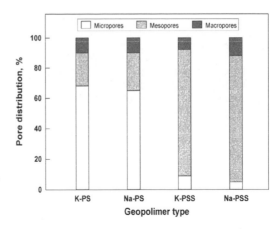

Figure 6. Distribution of pores of geopolimeric mortars polycondensed 28 days at 25°C.

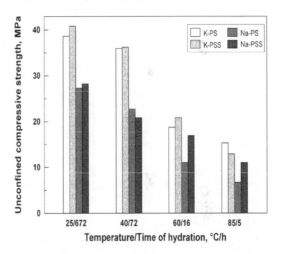

Figure 7. Mechanical strength of geopolimeric mortars of types PS and PSS polycondensed at different temperatures/times.

The observed favourable effect of KOH on strength is in agreement with literature results (van Jaarsveld et al., 1999; van Jaarsveld and van Deventer, 1999; Xu and van Deventer, 2000 and Xu et al., 2001). However, this effect can be only partially explained on a physico-structural basis. In fact, while the density data are in agreement with the mechanical strength data, porosity and pore size distribution data are not. These are the same for Na-PS and K-PS products on one side and for Na-PSS and K-PSS products on the other. Equally, differences in microstructural properties cannot be invoked, as Palomo et al. (1992) proved that geopolymer-based materials are structurally featureless and substantially amorphous.

A possible explanation of the favourable effect of potassium on mechanical strength could stem from an observation by McCornick and Bell (1989). These authors consider that, due to its larger size, potassium may favour a greater degree of polycondensation. However, our quantitative data on the amounts of reacted alkali metal silicate and water are not in agreement with this consideration. Therefore, it is more likely that the larger size of potassium has a favourable effect on crosslinking even if the polycondensation degree is left unchanged.

4 CONCLUSIONS

The experiments have proved that it is possible to produce geopolymer-based good quality building materials through calcination and polycondensation of a kaolinitic residue.

As far as the calcination step is concerned, the use of differential scanning calorimetry proved to be very useful for optimization purposes, due to the possibility to evaluate the reactivity of the products obtained in the different calcination conditions tested.

The polycondensation step was carried out with potassium and sodium silicate and addressed towards the formation of polysialate, polysialatesiloxo and polysialatedisiloxo products, but it was successful only for the first two geopolymers. An original quantitative method was developed for the determination of the quantities of metal alkali silicate and water reacted during the polycondensation step.

The results of physico-mechanical characterization are highly appealing for the development of processes for the manufacture of both hydraulic binders and pre-formed building blocks. In this regard, the already known favourable effect of potassium was confirmed and further understanding was obtained from the polycondensation quantitative results.

The work is a contribution to the development of feasible pathways for the optimum exploitation of natural resources.

ACKNOWLEDGMENT

MUR (Italian Ministry of University and Research) – PRIN-2006 (Research Project of National Interest) is gratefully acknowledged for financial support to this research.

REFERENCES

Anderson, J.R. and Pratt, K.C., 1985. Introduction to characterization and testing of catalysts. Academic Press Australia, North Ryde, N.S.W., Australia.

S. Andini, R. Cioffi, F. Colangelo, T. Grieco, F. Montagnaro and L. Santoro, 2008. Coal Fly Ash as Raw Material for Manufacture of Geopolymer-Based Product. Waste Management. 28, 416–423.

Barbosa, V.F.F., MacKenzie, K.J.D. and Thaumaturgo, C., 2000. Synthesis and characterization of materials based on inorganic polymers of alumina and silica: sodium polysialate polymers. Int. J. Inorg. Mater., 2(4): 309–317.

Chianese, B., 1998. Thesis for Master Degree in Chemistry. University of Naples "Federico II", Naples, Italy.

Davidovits, J., 1979. SPE PATEC '79, Society of Plastic Engineering, Brookfield Center, USA.

Davidovits, J., 1989. Geopolymers and geopolymeric materials. J. Therm. Anal., 35(2): 429–441.

Davidovits, J., 1991. Geopolymers: inorganic polymeric new materials. J. Therm. Anal., 37(8): 1633–1656.

Davidovits, J., 1993. Geopolymers cement to minimize carbon dioxide greenhouse-warming. Ceram. Trans., 37: 165–182.

Davidovits, J. and Davidovics, M., 1988. Geopolymer room temperature ceramic matrix for composites. Ceram. Eng. Sci. Proc., 9: 835–842.

Ikeda, K., Nunohiro, T. and Iikuza, N., 1998. Consolidation of silica sand slime with a geopolymer binder at room temperature and the strength of the monolith. Chem. Pap., 52(4): 214–217.

Lee, W.K.W. and van Deventer, J.S.J., 2002. The Effect of Inorganic Salt Contamination on the Strength and Durability of Geopolymers. Colloids and Surfaces A: Physicochem. Eng. Aspects 211, 115–126.

Lee, W.K.W. and van Deventer, J.S.J., 2004. The Interface between Natural Siliceous Aggregates and Geopolymers. Cement and Concrete Research 34, 195–206.

Maffucci, L., 1999. Ph.D. Thesis. University of Sassari, Sassari, Italy.

McCornick, A.V., and Bell, T., 1989. The solution chemistry of zeolite precursors. Catal. Rev. Sci. Eng., 31: 97–127.

Palomo, A., Macías, A., Blanco, M.T. and Puertas, F., 1992. Physical, chemical and mechanical characterization of geopolymers. In: Proc 9th Int. Cong. Chemistry of Cement, 23–28 November 1992, New Delhi, India, vol. 5, pp. 505–511.

Phair, J.W., van Deventer, J.S.J. and Smith, J.D., 2000. Mechanism of polysialation in the incorporation of zirconia into fly ash-based polymers. Ind. Eng. Chem. Res., 39: 2925–2934.

Swanepoel, J.C. and Strydom, C.A., 2002. Utilisation of Fly Ash in a Geopolymeric Material. Applied Geochemistry 17, 1143–1148.

van Jaarsveld, J.G.S., van Deventer, J.S.J. and Lorenzen, L., 1997. The potential use of geopolymeric materials to immobilise toxic metals: Part I. Theory and applications. Miner. Eng., 10(7): 659–669.

van Jaarsveld, J.G.S., van Deventer, J.S.J. and Schwartzman, A., 1999. The potential use of geopolymeric materials to immobilise toxic metals: Part II. Material and leaching characteristics. Miner. Eng., 12(1): 75–91.

van Jaarsveld, J.G.S. and van Deventer, J.S.J., 1999. Effect of the metal alkali activator on the properties of fly ash-based polymers. Ind. Eng. Chem. Res., 38: 3932–3941.

Xu, H. and van Deventer, J.S.J., 2000. The geopolymerization of alumino-silicate minerals. Int. J. Miner. Process., 59: 247–266.

Xu, H., van Deventer, J.S.J. and Lukey, G.C., 2001. Effect of alkali metal on the preferential geopolymerization of stilbite/kaolinite mixtures. Ind. Eng. Chem. Res., 40: 3749–3756.

Excellence in Concrete Construction through Innovation – Limbachiya & Kew (eds)
© 2009 Taylor & Francis Group, London, ISBN 978-0-415-47592-1

Restoration mortars made by using rubble from building demolition

F. Colangelo, R. Sommonte & R. Cioffi
University Parthenope of Naples, Italy

ABSTRACT: In this paper valorisation and recycling of construction and demolition debris have been analyzed. Specifically, aspects concerning the realization of selective demolition and employment of mobile plants have been investigated regarding quality of artificial aggregates. A demolition work in a building yard situated in Naples area has been carried out. Criteria requested by the selective demolition practice have been applied and a mobile crusher has been used in order to select three different types of artificial aggregates. In particular, aggregates have been produced from bricks, mortars and concrete. 50% and 100% of each of the three above aggregates have been employed in order to prepare six restoration mortars. The mortars have been characterized and afterwards employed and tested. The results have been used for a technical-economical comparison with commercial ones.

1 INTRODUCTION

The construction industry is one of the most important industries contributing to development worldwide, but it is using natural resources and disposing of construction and demolition wastes (C&DW) in very large quantities. The waste legislation aims at reducing use of natural resources and at increasing waste recycling. In the field of demolition industry, Germany is the country that makes the most demolitions with about 59 million tons of C&DW, followed by the UK with 30 million tons, France with 24 million tons and Italy with 20 million tons. Regarding the recycling percentage of construction and demolition wastes, Italy is placed behind many European countries. In fact Germany recycles 17%, UK the 45%, France 15%, while Italy only 9% (CEPMC 2004).

Bressi G. (2006) reports that in Italy about 46 million tons of C&DW (European Waste Catalogue code: 17.XX.XX) are produced with a recycling rate of about 12%.

The C&DW are composed of heterogeneous materials, due to locally different raw and building materials. This waste, after screening and separating of lightweight material unsought and grain size analysis, becomes a recycled aggregate, according to EN 12620.

The natural aggregates employed in building structural applications are about 60% of the total ones. Then, a huge amount of aggregate is used in fields in which lower physical-mechanical properties are required. Therefore, the latter fraction can be more easily replaced with artificial aggregates.

In Italy, most part of civil demolition is carried out by mechanical means as machine with telescopic arms equipped with pincers, hydraulic excavator and shears which allow a partial separation in three main types of materials mainly containing: wood, iron and a mixture of concrete, bricks and other impurities. This separation is certainly cursory and insufficient to ensure the valorisation of mineral components, such concrete and bricks.

Selective demolition is a process able to obtain more homogenous residual materials and separate hazardous wastes.

Both in new and restoration construction activities, mortars play a primary role. Aggregates employed in mortars preparation are finer than the ones used in concrete.

Marques et al. (2005) believe that in a correct restoration work it is necessary to guarantee the physical-chemical and mechanical compatibility between rehabilitation mortars and old ones. These properties can be successfully obtained by different materials combinations. Furthermore, the authors sustain the difficulty in defining rules and methodologies for the preparation of suitable rehabilitation mortars.

Several studies have been successfully carried out to evaluate fresh and hardened properties and durability of concrete made by substituting coarse natural aggregates with artificial ones, named: coarse recycled concrete aggregates (CRCA) coming from concrete demolition (Colangelo et al. 2004 a and b).

Other works have shown that the use of fine recycled concrete aggregates (FRCA) can have detrimental effects on the fresh and hardened properties of concrete

(Corinaldesi & Moriconi 2007). Poon et al. (2006) have proved that FRCA can be used in manufacture of mould concrete bricks and blocks where workability is less important. In all cases the density of natural aggregates (NA) was higher than the recycled concrete aggregates (RCA) and, therefore, the latter have a higher water absorption value compared to those NA. The results of this study have shown that the FRCA self-cementing capacity have a significant influence on the permeability concrete specimens. In this work, the authors established that the optimum RCA content and maximum dry density in the road sub-base concrete containing RCA were higher and lower than to those of prepared with NA, respectively. Concrete strength decreased when recycled aggregate was used and the rate of such decrease varied according to the nature of aggregates.

Sagawa et al. (2007) have reported the relationship between microstructure of neo-formed cementitious phases and performance of mortar containing recycled fine aggregates. In particular, mortars compressive strength was lower in respect to one incorporating natural aggregates. While, carbonation speed of mortar incorporating recycled fine aggregate was higher than the ones made with NA. The authors have also proven that microstructure of cement paste-artificial aggregates interfaces was porous and loose when high water absorption RCA were used.

Chen et al. (2002) have found that with high water/cement ratios the compressive strength of concrete made with RCA was similar to that of conventional concrete. While at lower water/cement ratios the compressive strength of RCA-concrete is much lower. High-strength RCA-concrete might be achieved increasing the amount of cement or cementitious materials.

In this research restoration mortars prepared by replacement of natural aggregate with rubble coming from concrete, bricks and mortars have been studied. The mortars have been tested before and after sulphate attack to evaluate their durability. The study also includes an environmental-economic analysis.

2 EXPERIMENTALS AND DISCUSSION

2.1 Mixtures

CEM II/A-LL 42.5 R, according to UNI EN 197-1 was employed in same quantities for all specimens. The mixtures were composed by CEM II A-LL 42.5 R, natural aggregate (N), recycled aggregates (RA), superplasticizer (SP) and viscosity modifying agent (VMA). The mixtures are indicated with a code composed by a letter that means the type of artificial aggregate used: "C" for concrete, "B" for bricks aggregate, "M" for mortar. The letter is followed by a number

Table 1. Mix proportion of mortars [g].

Mixture	CEM	N	B	C	M	SP	VMA
N 100	500	1350				11	0.3
B 50	500	675	675			16	
B 100	500		1350			18	
C 50	500	675		675		12	0.3
C 100	500			1350		13	0.3
M 50	500	675			675	15	0.2
M100	500				1350	17	

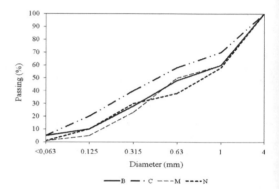

Figure 1. Grain size distribution, wt %.

equal to the percentage of replacement of natural aggregate. "N" is the letter used for reference mixture. Table 1 reports the mortars composition.

2.2 Aggregates

Selective demolition has been carried out in a building yard operating on a school in the province of Naples. Afterwards the rubbles have been treated in a mobile plant to obtain different fractions with specific grain size distribution and physical and mechanical characteristics. After crushing and sieving three selected aggregates, from bricks, mortars and concrete, have been obtained.

The grading of these aggregates must satisfy the limits set by the standard EN 12620. According to UNI 2332 regulation, in the case of RCA particles having size higher than 0.075 mm, mechanical sieving analysis was carried out. Before testing, the aggregates have been oven dried to constant weight, according to EN 933-1 and UNI 8520 standards. Due to superficial adhesion, in presence of flat or elongated particles with size ranging from 0 to 5 mm, grain size distribution of the three different rubbles employed could be very different.

Figure 1 shows results of sieving analysis carried out on the different types of aggregates used.

Table 2. Water absorption (%) and loose bulk density (kg/m^3) of aggregates.

	Aggregates			
	N	B	C	M
Water absorption	1.5	4.0	3.8	4.5
Loose bulk density	1610	1270	1530	1320

Table 3. Water vapour permeability coefficient of mortars, μ $[gr\,cm^{-2}\,h^{-1}]$.

Mortars						
N 100	B 50	B 100	C 50	C 100	M 50	M 100
μ 11.3	12.7	12.9	12.0	12.7	12.5	13.0

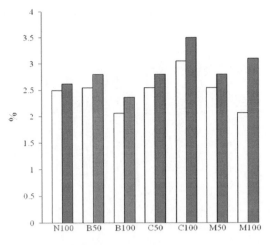

Figure 2. Water absorption of mortars (wt. %).

The water absorption and loose bulk density of the aggregates were determined according to UNI EN 1015-18 and EN 1097-6 standards. The results are reported in Table 2.

The high water absorption of aggregates influences mode and times of mixing because of the water and additive steady reducing during mixing phase. These results are in agreement to those showed by Desai & Limbachiya (2007). This produces a quick decrease of initial workability.

The phenomenon increases during mixing by "crushing-rubbing" of the particles, increasing fine fraction and therefore the specific surface of aggregates. It produces a higher water demand and decreases the efficiency of the additives. The addition of a suitable superplasticizer allows to obtain mortars with strength comparable to traditional ones.

According to EN 197-1 standard, the Fratini-pozzolanicity test was carried out on aggregates obtained from bricks demolition. The amount of CaO reacted was equal to 60.0 mg/g.

2.3 Mortars

The water vapour permeability coefficient of mortars (μ) was measured according to UNI EN 1015-19. As well known, lower μ values indicates lower wall-transpiration. The results are reported in Table 3.

Ravindrajah and Tam (1987) reported that the mechanical resistance of mortars containing RCA decreases as the aggregate size decreases. It is due to the fact that a bigger amount of soft cement paste remains attached to the natural aggregates when the size of RCA particles decreases. Chen et al. (2002) also found that if the amount of finer aggregates increases mortar strength decreases.

Compressive strength of mortars was determined in accordance with UNI EN 1015-11. The tests were carried out on prismatic specimen ($40 \times 40 \times 160\,mm^3$), at 20°C and R.H. $\geq$90%, in according to EN 196.

The dynamic modulus of elasticity was determined according to Rilem NDT1.

2.4 Tests in situ

All the mortars were applied to two different surface present in a building yard. Well cleaned concrete and brickwork walls were utilised for in situ experimentations. The test areas were of about $60 \times 50\,cm^2$ (UNI EN 1015-12). After 28-day curing, it was possible to see that even the mortars B100, C100, M100 didn't show cracks due to shrinkage.

Furthermore, the costs and the performances, such as mortars amount (kg/m^2) and applying times (hours/m^2), have been evaluated during wall-applications.

2.5 Durability

28-day curing all mortars were exposed to $MgSO_4$ fog room for 360 days at 25 ± 2°C and 100% R.H. After exposition, all specimens were tested to determine water absorption, compressive strength and modulus of elasticity. The tests results are illustrated in Figures 2, 3, & 4, respectively.

The results reported in the above figures show that the chemical resistant of RA-mortar seems to be comparable to that of the NA-mortars. In particular, the mechanical performances decrease observed for RA-mortars is equal or not much lower than NA-mortars, while the water permeability measured on RA-mortars

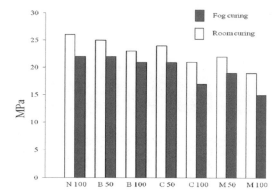

Figure 3. Compressive strength of mortars [MPa].

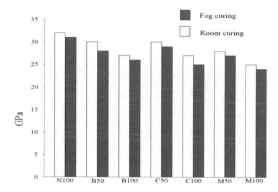

Figure 4. Modulus of elasticity of mortars [GPa].

is higher than NA-mortars. More specifically, the specimens C100 and M100 showed major permeability increases.

Table 4 shows the comparison between the properties of the NA- and RA-mortars and two commercial ones, here named A and B, before ad after fog curing. It can be seen that comparing with B-mortar, the compressive strengths of RA-mortars are slightly better with the exception of the systems C100 and M100 after fog curing. While A-mortars, is characterised by very similar strengths, before and after fog curing. The mortar A shows highest vapour permeability values, while the behaviour of the mortar B is very similar to the others.

3 ECONOMICAL AND ENVIRONMENTAL ANALYSIS

Table 5 reports cost and performance for all the mortars evaluated during all assessing phases. The RA-mortars are cost-effective, while, regarding to the performances, the quantities of RA-mortars needed

Table 4. Physical and mechanical properties of the mortars.

Mortars	Compressive strength [MPa]		Vapour Permeability, μ [gr cm^{-2} h^{-1}]		Specific weight [kg/m^3]
	RC	FC	RC	FC	
N 100	26	22	11.3	12.8	2306
B 50	25	22	12.9	13.8	2117
B 100	23	21	12.7	13.6	1919
C 50	24	21	12.7	14.4	2193
C 100	21	17	12.0	14.2	2081
M 50	22	19	13.0	14.8	2163
M 100	19	15	12.5	14.5	2013
A	25	23	20.0	22.1	1900
B	20	18	10.0	12.5	1700

Table 5. Cost and performance of the mortars.

Mortars	Costs		Performance	
	Production [€/m^2]	Application [€/kg]	Quantity [kg/m^2]	Time [h/m^2]
N100	8.28	0.180	43.0	0.27
M50	6.67	0.175	41.5	0.27
M100	5.85	0.170	40.3	0.28
C50	6.67	0.175	41.5	0.27
C100	7.07	0.170	38.3	0.28
B50	6.67	0.175	41.5	0.27
B100	5.51	0.170	41.6	0.28
A	6.30	0.210	34.0	0.25
B	5.58		26.6	

Table 6. Environmental analysis relating to 1 kg of aggregate.

	Processes			
	Demolition	Crashing	Landfill	Recycling in mortars
Eco indicator (mPt)	4.28	4.85	5.12	4.37

are higher than the commercial ones. Finally, the application times are very similar for all the mortars.

Environmental analysis, shown in Table 6, evaluates the energy demand during partial demolition of the studied building and crashing, landfilling and recycling processes of wastes. The analysis has been carried out using the Eco-indicator'99 method, created for the Dutch Environmental Ministry on 2000. It is based on an impact assessment methodology that transforms the data of an inventory table into damage scores that can be aggregated to one single score/point that expresses the total environmental load of a product or a process. This index is simple to understand and represents a fair compromise between effort and time, it is expressed in millipoints (mPt) (Neri, 2008).

4 CONCLUSIONS

The replacement of recycled materials to natural aggregates has been studied in this work and the following conclusion can be drawn:

- The mixtures containing brick-wastes, in total or partial replacement of natural aggregates, show lower mechanical performances. In this case, a double amount of superplasticizer was added.
- The mortars prepared by using concrete-wastes in replacement of 50% of natural aggregate show a slight decrease in mechanical properties.
- The RA-mortars can be properly utilized as restoration- mortars.
- The economical and environmental analysis show that RA mortars are economically sound and environmental sustainable.

ACKNOWLEDGMENT

MUR (Italian Ministry of University and Research) – PRIN -2006 (Research Project of National Interest) is gratefully acknowledged for financial support to this research.

REFERENCES

Bressi G. 2006. Annual Report, Edited by National Association Producers of Recycled Aggregates (ANPAR), Milan, Italy.

Council of European Producers of Materials for Construction (CEPMC) 2004, Environment Working Group Report.

Chen H.J., Yen T., Chen K.H. 2002. Use of building rubbles as recycled aggregates. Cement & Concrete research 3:125–132.

Colangelo F., Marroccoli M., Cioffi R. 2004 a, Properties of self – levelling concrete made with industrial waste, International RILEM Conference on the Use of Recycled Materials in Building and Structures, Barcelona, Spain.

Colangelo F., Marroccoli M., Cioffi R. 2004 b, Exact use of construction and demolition wastes: Regional Guide Lines, Edited by Camera di Commercio di Potenza, Italy.

Corinaldesi V., Giuggiolini M. and. Moriconi G. 2002, Use of rubble from building demolition inmortars, Waste Management, 22(8), 893–899.

Corinaldesi V., Morioni G. 2007 Behaviour of cementitious mortars containing different kinds of recycled aggregate, Construction and Building Materials: 123–131.

Corinaldesi V., Morriconi G. 2005, Recycling of demolition wastes in concrete manufacture: Part I- General

principles and yard applications, EncoJournal, 5: 45–54.

Desai S.B., Limbachiya M.C. 2006 Coarse recycled aggregate- A sustainable concrete solution, The Indian Concrete Journal.

Dutch Ministry of Housing, 2000 Eco-indicator 99 – Manual for designers.

Gonzalez-Fonteboa B., Martınez-Abella F. 2008. Concretes with aggregates from demolition waste and silica fume. Materials and mechanical properties, Building and Environment 43:429–437.

Hwang E.H., Ko Y.S., Jeon J.K., 2008. Effect of polymer cement modifiers on mechanical and physical properties of polymer-modified mortar using recycled artificial marble waste fine aggregate. Journal of Industrial and Engineering Chemistry.

Industrial Consult, AICARR Conferences: Life Cycle of plants in engineering process- Bologna 2006-Torino 2006-Napoli 2007.

Koulouris A. Limbachiya M.C., Fried A.N., Roberts J.J., 2007 Use of recycled aggregate in concrete application: case studies. International Conference Sustainable Waste Management and Recycling: Construction Demolition Waste.

Limbachiya M.C., Leelawat T., Dhir R.K. 2000, Use of recycled concrete aggregate in high-strength concrete, Materals and structures, vol. 33 pp 574–580.

Marques S.F., Ribeiro R.A., Silva L.M, Ferreira V.M., Labrincha J.A. 2005. Study of rehabilitation mortars: Construction of a knowledge correlation matrix, Cement and Concrete Research 36 1894–1902.

Miranda L.F.R., Selmo S.M.S 2005. CDW recycled aggregate renderings, Construction and Building Materials 20:615–624 625–633.

Neri P. 2008, Toward Environmental analysis of building, Alinea Editor.

O'Farrell M., Sabir B.B., Wild S. 2006. Strength and chemical resistance of mortars containing brick manufacturing clays subjected to different treatments, Cement & Concrete Composites 28: 790–799.

Poon C.S., Lam C.S. 2007. The effect of aggregates to cement (A/C) ratio and types of aggregates on the properties of pre-cast concrete blocks. Cement & Concrete Composites.

Poon C.S., Chan D.2007. The use of recycled aggregate in concrete in Hong Kong. Resources, Conservation and Recycling 50: 293–305.

Poon C.S., Qiao X.C. and Chan D. 2006.The cause and influence of self-cementing properties of fine recycled concrete aggregates on the properties of unbound subbase. Waste Management, Vol 26, Issue 10, 2006, pages 1166–1172.

Ravindrarajah R.S. and Tam C.T. 1987. Recycling concrete as fine aggregate in concrete. International Journal of Cement Composites and Lightweight Concrete, Vol 9: 235–241.

Sagawa Y., Matsushita H. and Kawabata Y. 2007. Microstructure of new cement matrix phase and performance of mortar incorporating recycled fine aggregate, International Seminar on Durability and Lifecycle Evaluation of Concrete Structures.

Shui Z., Xuan D., Wan H., Cao B. 2007, Rehydration reactivity of recycled mortar from concrete waste experienced to thermal treatment, Construction and Building Materials.

UNI EN 1015. Methods of test for mortar for masonry. Determination of flexural and compressive strength of hardened mortar; 2001.

Vivian W.Y. Tam, K. Wang, C.M. Tam 2007. Assessing relationships among properties of demolished concrete, recycled aggregate and recycled aggregate concrete using regression analysis, Journal of Hazardous Materials.

Vivian W.Y. Tam, C.M. Tam, Y. Wang 2007. Optimization on proportion for recycled aggregate in concrete using two-stage mixing approach, Construction and Building Materials 21:1928–1939.

Vivian W.Y. Tam, X.F. Gao, C.M. Tam, C.H. Chan 2008. New approach in measuring water absorption of recycled aggregates, Construction and Building Materials 22: 364–369.

Excellence in Concrete Construction through Innovation – Limbachiya & Kew (eds)
© 2009 Taylor & Francis Group, London, ISBN 978-0-415-47592-1

High strength products made with high volume of rejected fly ash

X.C. Qiao, B. Zhou & J.G. Yu
School of Resource and Env. Enggr, East China University of Science and Technology, Shanghai, China

ABSTRACT: The rejected coal fly ash (RFA) with a particle size larger than 45 μm has remained relatively unused due to a high carbon content and large particle size. It is hence necessary to develop appropriated activation and preparation methods in order to reuse RFA. The mineralogy and chemical properties of RFA have been characterized. RFA has been blended with ordinary Portland cement and chemical additives. The blended materials were compressed to form specimens which were cured at 60°C for 7 days. Physical properties and micro-structures of the specimens prepared have been investigated. The results show that main minerals in RFA are mullite and quartz. The addition of Na_2SO_4 benefits to the strength development of specimen with RFA. Na_2SiO_3 and $Ca(OH)_2$ significantly elevate strength development of cement-free RFA sample. The samples with 85 wt. % of ground RFA, 10 wt. % of $Ca(OH)_2$ and 5 wt. % of Na_2SiO_3 had compressive strength of 48.1 MPa at 7 days. Adding polypropylene emulsion into RFA specimen has little effect on the strength development.

1 INTRODUCTION

Coal fly ash (CFA) is a by-product generated from burning coal during the generation of electricity. The finer fraction of CFA (mostly <45 μm) produced by passing the raw ash through a classifying process is generally used in building construction and road sub-base et al. (Conner 1990, Baykal et al. 2004). The low-grade coal fly ash, as well as rejected fly ash (RFA), constitutes about 70% of the CFA and has been rejected from the ash classifying process due to a high carbon content and large particle size (>45 μm). In China, more than 2,600 million tons of this material is currently being dumped in ash-bin and remained unused.

The activation of fine CFA has been studied widely and the addition of chemicals Na_2SO_4 (Bakharev 2005, Qian et al. 2001) and water glass (Swanepoel & Strydom 2002, Bakharev 2005b) et al., elevating the curing temperature (Bakharev 2005b, Katz 1998) and mechanical grind (Payá et al. 1996, 2000) have been proved to be successful methods. However, it was found in previous research that these methods have less effect on activating RFA (Poon et al. 2003, Qiao et al. 2006). It is therefore necessary to extend more research study on the further application of RFA.

This study investigated the feasibility of producing high strength products made with high volume of RFA. The products prepared have been characterized by compressive strength, X-ray fraction analysis and scanning electron microscopy observation.

2 EXPERIMENTAL DETAILS

2.1 Materials

The rejected fly ash (RFA) was obtained from a local coal fired power plant and it corresponded to low-calcium Class F. Only 9.04 wt. % of this RFA passed through a 45 μm sieve. Figure 1 shows its X-ray diffraction (XRD) pattern. A commercially available ordinary Portland cement was used as a blending agent. Reagent grade $Ca(OH)_2$, Na_2SO_4, Na_2SiO_3 and commercial polypropylene emulsion (water content <55%) were used as additives. The mechanically treated RFA was prepared by grinding as-received RFA (density = 2.1 (kg/m³)) in a laboratory ball mill to pass

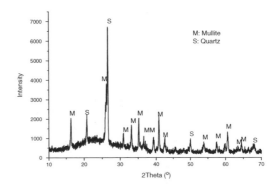

Figure 1. X-ray diffraction pattern of RFA.

through a 125 μm sieve. The chemical compositions of the RFA were measured using X-ray fluorescence and the results are given in Table 1.

2.2 Preparation of specimens

Six mixes were prepared at a water to total solid ratio (w/s) of 0.2 using a compression testing machine to form specimens 20 mm × 40 mm with bulk densities of 3.4 g/cm³. The specimens produced were cured at 60°C for 7 days in a fog tank with a relative humidity of 98%.

2.3 Compressive strength test

Three specimens per mix were subject to compressive strength test after curing for 7 days using a compression testing machine (TYE-2000B). The strength results reported were the average of three specimens with variations of not more than 10%. The fractured pieces of the specimens after the compression testing were preserved for other tests. To stop the hydration reactions, the fractured pieces of the specimens were soaked in acetone for 7 days and were then dried at 50°C for one week in an oven.

2.4 X-ray diffraction

The oven dried fractions collected from 7 day compressive strength test were ground to pass through a 150 μm sieve and were then used to determine the crystalline phases by X-ray diffraction (XRD). A RINT2000 vertical goniometry using a Cu target was employed and run at a 2 theta of 0.04° per step.

2.5 Scanning electron microscopy (SEM)

A JSM-6360 scanning electron microscopy was used to study the morphology of oven dried fracture surfaces of each sample.

3 RESULTS AND DISCUSSION

3.1 Characterization of RFA

The X-ray diffraction results in Figure 1 show that main minerals in RFA are mullite and quartz. Mullite is a kind of refractory material which is resistant to most chemical attack. This must be related with the low chemical reactivity of RFA besides high carbon content and large particle size. The CaO/SiO_2 weight ratio (S) in pozzolanic materials is usually related with their chemical reactivity and the S of blast furnace slag is usually between 0.8 and 1.2. Thus, the low CaO/SiO_2 weight ratio of 0.05 in RFA suggests its low reactivity.

3.2 Compressive strength

The compressive strength results of specimens cured for 7 days are shown in Figure 2. The strength of specimen with 85 wt. % of RFA and 15 wt. % ordinary Portland cement (M1) is 22.5 MPa and is only a little higher that of specimen M2 which substitute Na_2SO_4 for 1/3 of cement. It means that the addition of Na_2SO_4 benefits to the strength development of specimen with RFA. When the $Ca(OH)_2/Na_2SiO_3$ weight ratio is 0.5 the strength of specimen M3 is 16.3 which is lower than that of specimen M1. However, the strength of specimen M4 reaches 48.1 MPa when the $Ca(OH)_2/Na_2SiO_3$ weight ratio increases to 2. The addition of polypropylene emulsion shows little effect on elevating the compressive strength of specimen M5 and M6. However, the advantageous effect of Na_2SO_4 is proved again in specimen M5. The results in previous study (Poon et al. 2003, Qiao et al. 2006) show that the compressive strength of ground RFA specimen is generally low when the water to solid ratio is 0.3 or higher, even with the addition of cement or

Table 1. Chemical compositions of RFA.

SiO₂	Fe₂O₃	Al₂O₃	TiO₂	CaO
55.22	3.48	31.47	1.25	2.75
MgO	SO₃	P₂O5	Na₂O	I.L
0.65	0.79	0.26	0.26	1.64

Table 2. Mix proportions (wt. %).

	M1	M2	M3	M4	M5	M6
RFA	85	85	85	85	85	85
Ca(OH)₂			5	10		
Na₂SO₄		5			5	
Na₂SiO₃			10	5		
Cement	15	10			10	15
Polypropylene					15	15

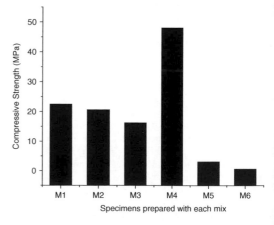

Figure 2. Compressive strength of specimens prepared with each mix.

chemical activators. However, the results in Figure 2 show that the strengths of 7 day cured specimens prepared under appropriate methods are high enough to be used as building materials. The cement-free specimen activated by $Ca(OH)_2$ and Na_2SiO_3 (M4) has even higher strength than specimen M1 contained cement and ground RFA. It shows that compression method coupled with chemical activator is useful to increase the strength development of RFA contained specimen.

3.3 X-ray diffraction

Figure 3 shows the X-ray diffraction pattern of every cured specimen. New mineral calcite was formed after 7 days of curing. It is related with the reaction between carbon dioxide from atmosphere and calcium hydroxide from the addition of chemical $Ca(OH)_2$ or the hydration of cement. The mullite and quartz in RFA can be still detected in every specimen. However, the peak intensities decreased and changed with different specimen. The detailed changes in the first main peak intensity of mullite and quartz of each specimen are shown in Table 3. The addition of Na_2SO_4 decreased the peak intensity of quartz as well as mullite. It is consistent with the compressive strengths results. The peak intensities of both mullite and quartz in specimen M4 are lower than those in specimen M3. It shows that $Ca(OH)_2/Na_2SiO_3$ weight ratio of 2 increases the reaction between RFA and chemicals $Ca(OH)_2$ and Na_2SiO_3. The activation effect of Na_2SiO_3 in blast furnace slag/fine coal fly ash mixtures has been

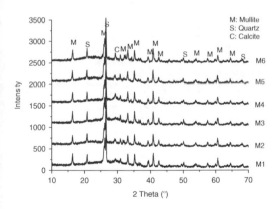

Figure 3. X-ray diffraction of each specimen.

Table 3. Intensities of the first main peaks of quartz and mullite measured in each specimen.

	M1	M2	M3	M4	M5	M6
Quartz	1229	1022	1270	1086	1091	1104
Mullite	755	615	719	639	619	682

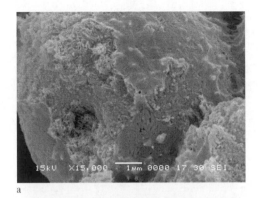

a

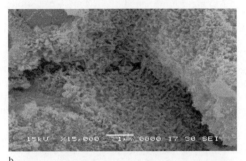

b

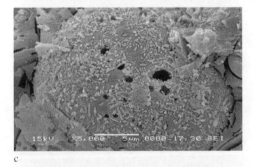

c

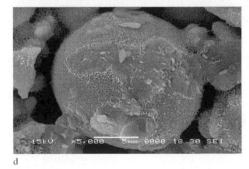

d

Figure 4. SEM observation of specimen (a: M1, b: M2, c: M3, d: M4).

503

generally thought to be able to form calcium silicate hydrated with calcium source (Richardson 2004). However, the results in Table 3 show that minerals in rejected coal fly ash (RFA) can be induced to join the hydration reaction after the addition of $Ca(OH)_2$ and Na_2SiO_3. This means that RFA presents some different reaction properties from blast furnace slag and fine coal fly ash. The methods used for activating fine coal fly ash can not be applied directly to RFA.

3.4 *Scanning electron microscopy (SEM)*

The scanning electron microscopy (SEM) observation results are shown in Figure 4. Specimens M5 and M6 are not measured due to their weak strength. Figure 4a shows that short stick like hydration products appeared in specimen M1 contained RFA and cement. This hydration product is though to be calcium silicate hydrated with high Ca/Si ratio (Taylor 1990). After blending Na_2SO_4 with RFA and cement, foil like calcium silicate hydrated with lower Ca/Si ratio are formed in specimen M2 (Fig. 4b). This is possible to be related with the formation of aluminate sulfate covered on the surface of reactants in the specimen, which retards the diffusion of calcium ions to join reactions (Taylor 1990). Similar results also occurred in specimens M3 and M4 (Fig. 4c, d).

4 CONCLUSIONS

Application of rejected coal fly ash (RFA) is paid more and more attention in China, even in some other countries that coal fire is a key energy source. This study applied chemical activators, mechanical grind, elevating curing temperature and compression method to the activation of RFA. Potentially high strength products were prepared at 7 curing days. However, the properties of long time cured specimens are necessary to be studied in the future research. Some conclusions obtained from this study are listed as follows.

1. High strength cement-free RFA products can be prepared through adding $Ca(OH)_2$ and Na_2SiO_3.
2. Activation methods used for fine coal fly ash do not have the same activation effect on the application of RFA. The strength of RFA specimen can be increased through compressing in the preparation process of specimen.
3. The addition of Na_2SO_4 has advantageous effect on the strength development of specimen.

ACKNOWLEDGEMENT

This research was supported by Shanghai Leading Academic Discipline Project (Project Number: B506).

REFERENCES

Bakharev, T. 2005a. Durability of geopolymer materials in sodium and magnesium sulfate solutions, Cement and Concrete Research, 35(6): 1233–1246.
Bakharev, T. 2005b. Geopolymeric materials prepared using Class F fly ash and elevated temperature curing, Cement and Concrete Research 35(6): 1224–1232.
Baykal, G., Edinçliler, A. & Saygılı, A. 2004. Highway embankment construction using fly ash in cold regions. Resources, Conservation and Recycling 42(3): 209–222.
Conner, J.R. 1990. Chemical Fixation and Solidification of Hazardous Wastes. New York: Van Nostrand Reinhold.
Katz, A. 1998. Microscopic study of alkali-activated fly ash, Cement and Concrete Research 28(2): 197–208.
Payá, J., Monzó, J, Borrachero, M.Y. & Peris-Mora, E. 1996. Mechanical treatment of fly ashes. Part II: particle morphologies in ground fly ashes (GFA) and workability of GFA-cement mortars. Cement and Concrete Research 26(2): 225–235.
Payá, J., Monzó, J, Borrachero, M.Y., Peris-Mora, E. & Amahjour, F. 2000. Mechanical treatment of fly ashes: Part IV. Strength development of ground fly ash-cement mortars cured at different temperatures. Cement and Concrete Research 30(4): 543–551.
Poon, C.S., Qiao, X.C. & Lin, Z.S. 2003. Pozzolanic properties of reject fly ash in blended cement pastes. Cement and Concrete Research 33(11): 1857–1865.
Qian, J.S., Shi, C.J. & Wang, Z. 2001. Activation of blended cements containing fly ash. Cement and Concrete Research 31(8): 1121–1127.
Qiao, X.C., Poon, C.S. & Chung, E. 2006. Comparative studies of three methods for activating rejected fly ash. Advances in Cement Research, 18(4): 165–170.
Richardson, I.G. 2004. Tobermorite/jennite- and tobermorite/calcium hydroxide-based models for the structure of C-S-H: applicability to hardened pastes of tricalcium silicate, h-dicalcium silicate, Portland cement, and blends of Portland cement with blast-furnace slag, metakaolin, or silica fume. Cement and Concrete Research 34(9): 1733–1777.
Swanepoel, J.C. & Strydom, C.A.2002. Utilisation of fly ash in a geopolymeric material,Applied. Geochemistry 17(8): 1143–1148.
Taylor, H.F.W. 1990. Cement Chemistry. Academic Press: New York.

Excellence in Concrete Construction through Innovation – Limbachiya & Kew (eds)
© 2009 Taylor & Francis Group, London, ISBN 978-0-415-47592-1

Industrial symbiosis – effective resource recovery within the UK's construction industry

R. Kirton
Entec, UK

D.H. Owen
International Synergies Limited, UK

E.J. Probert
Swansea University, UK

ABSTRACT: The UK National Industrial Symbiosis Programme (NISP) is the first national industrial symbiosis (IS) programme of its kind in the world. Delivered as a holistic programme, this approach is impacting across the entire resource hierarchy delivering significant resource efficiency and consumption reductions, accruing true bottom line environmental, economic and social benefits. IS works to bring together companies and organisations from all sectors regardless of size or turnover. The approach generates environmental benefits such as reduced greenhouse gases and landfill diversion, as well as significant economic benefits including cost reductions and new sales. These benefits in turn encourage greater economic activity, which stimulate further social benefits with the creation of new businesses and jobs, whilst also encouraging process improvements through innovation and technology development. Through case study examples, this paper will illustrate the leading UK industrial symbiosis best practice being developed across the construction sector. Highlighting successful projects which have worked closely to identify sustainable resource recovery and management solutions for all involved, this paper will demonstrate how the IS approach provides more than simple bi-lateral solutions to the movement of materials, water and energy.

1 INTRODUCTION

1.1 *Introduction to resource efficiency in the construction industry*

The concept of industrial ecology (IE) is not new to academia (Playfair 1892, Frosch & Gallopoulos 1989, Graedel & Allenby 1995) but is rapidly becoming the chosen method of engagement for businesses to improve their resource efficiency, delivering bottom line environmental, economic and social benefits. The approach of IE is for businesses to move from a linear system to a closed loop system, whereby the wastes from one process become the raw materials for another.

A number of industrial ecology methodologies have been tested (Ayres & Ayres 2002, Chertow 2000). Applied industrial symbiosis has repeatedly emerged at the forefront with technology and innovation advances in the UK impacting on the ability of businessesto reuse its waste resources in new processes and applications.

The construction industry in the UK has significantly advanced in its approach towards the reuse of materials and the consumption of resources (Department of Trade and Industry, 2006). This advance is being spurred on by the requirements of the EU Landfill Directive (European Parliament and Council Directive 1999/31/EC on the landfill of waste) to pre-treat all non-hazardous waste, including commercial and industrial waste, in order to reduce its volume and enhance its recovery. New drivers are also being introduced in England in the form of the Site Waste Management Plans Regulations 2008. These will mean that companies will need to prepare plans showing how waste will be managed before starting work on-site. In addition, a forthcoming strategy for sustainable construction (Department for Business, Enterprises and Regulatory Reform, 2007) is likely to suggest a zero waste to landfill approach. Industrial symbiotic relationships have played an important role in recent years to help the industry to think outside of the box, finding a diverse range of uses for resources that would otherwise have been sent to landfill, and there will be pressure to increase these activities in the future.

The industry has received considerable assistance from the resource efficiency programmes supported by the UK Department for Environment Food and Rural Affairs (Defra), with funding being placed to develop new markets for recyclate and to stimulate market interest in the reuse of materials. Further programmes have focused on the top tiers of the waste hierarchy; eliminating and minimizing waste. However none have been potentially as successful in targeting the entire resource hierarchy as the National Industrial Symbiosis Programme.

1.2 About the National Industrial Symbiosis Programme

The National Industrial Symbiosis Programme (NISP) is now a well established UK-wide business opportunity programme which has grown rapidly since its National launch in 2005. The programme now has in excess of 10,000 industry members from across the UK delivering significant business opportunity through resource efficiency and waste minimisation. NISP is part funded by Defra's Business Resource Efficiency and Waste (BREW) programme, the Welsh Assembly Government, the Scottish Government and Invest Northern Ireland.

As a holistic programme, NISP impacts across the entire resource hierarchy, delivering greater resource efficiency and consumption reductions by encouraging the adoption of industrial symbiosis (IS) by business. NISP works to influence the upstream production side of the sustainable production and consumption paradigm, bringing together companies and organisations of all sectors and sizes, generating cost reductions and new sales, as well as creating significant environmental benefits such as reduced greenhouse gases. This encourages economic activity which generates further social benefits with the creation of new businesses and jobs, and stimulates process improvements by industry through innovation and new technologies. Such innovation in turn also creates opportunities for businesses to penetrate global environmental technology markets (Laybourn & Lombardi 2007).

As a business opportunity programme, NISP facilitates the identification of sustainable resource management solutions for both companies and organisations. By identifying and creating productive synergies between companies and organisations NISP provides more than simple bi-lateral movement of materials, water and energy (Laybourn 2008). NISP's holistic approach therefore enables it to actively deal with all resources including water, energy, materials, logistics, assets, expertise etc. By working successfully across the entire resource hierarchy NISP successfully markets actual business opportunity as the real mechanism in encouraging resource efficiency.

NISP remains the first and only industrial symbiosis (IS) initiative in the world to be operated on a national scale and its novel yet highly successful approach for effective synergy facilitation has attracted considerable international attention. Praised across the world, NISP has already been cited as the European Commission's Environmental Technologies Action Plan (ETAP) exemplar programme with real potential for replication across Europe whilst also being twice ranked 1st by UK Government in its recent league table of *Government* funded programmes (in 2005/06 and 2006/07).

Industrial symbiosis support is already being delivered internationally through DfID's Sustainable Development Dialogue programmes (SDD) in both China and Mexico (with further potential deployment in Brazil, South Africa and India before end of 2008.) A further pilot programme has also been successfully deployed and established in Illinois, USA supported by the US EPA, City of Chicago and Chicago Manufacturing Centre. Stimulated by ETAP, European interests in NISP's IS approach is also increasing with substantive discussions for the establishment of IS pilots having now begun in France, Hungary, Eire, Portugal, Estonia, Germany and Romania.

Consequently, due to the hugely successful results, the programme, its approach and terminology are therefore increasingly being emulated by other programmes in the market, both in the UK and internationally.

Through its common sense approach to better management and sustainable use of natural resources NISP has, in only 33 months, already delivered engagement with over 9,900 industry members, generated more than £119 million in additional industry sales, saved over 5.2 million tonnes of virgin raw materials whilst reducing industrial water use by over 2.5 million tonnes and diverting over 2.95 million tonnes of waste from landfill. The programme has also delivered actual costs saving to industry of over £97 million, secured over £75 million private capital investment in reprocessing and recycling facilities whilst reducing over 2.88 million tonnes of CO_2 (equivalent to over 0.78 million tonnes of Carbon).

Notwithstanding that the programme has helped create over 1,400 jobs across the country, NISP has delivered a Total Economic Value Added (TEVA) of £117m (NISP 2007).

2 RESOURCE RECOVERY IN THE UK CONSTRUCTION INDUSTRY

2.1 Benefits of NISP in the construction industry

The National Industrial Symbiosis Programme has been working with the construction industry since the programme's inception. In that time, significant bi-lateral exchanges of resources have taken place, impacting greatly on the amount of waste being sent to landfill. In addition to this however are the social and economic benefits that the programme has generated for the industry.

Currently the programme is working with some of the key construction businesses, working in all areas from new development, redevelopment and demolition projects. In total, the programme is currently working on 286 construction related projects throughout the UK. These projects typically being at various stages of intervention ranging from idea stage through to discussion, implementation and completion.

The NISP engagement methodology is such that work is progressed through a staged approach beginning with the initial idea between the Practitioner and the Company. From here, the idea is progressed to a discussion between all parties concerned, looking at the practicalities of the reuse of materials, the legislative requirements for transfer and any trials of the materials that may be required. The 'Synergy' is then progressed further through negotiation and implementation where the financial and economic considerations including transportation are determined and agreed. At this point the project is ready for completion, with either one off movements of material taking place, or, more often than not for long term projects dealing with regular supply of resources, timetables of transfer established and initiated.

Almost 100 synergy projects have been completed in the construction sector since the NISP programme began. Of these, several projects are considered exemplars in the level of intervention and the environmental, social and economic benefits that have been attributed and verified as originating from the NISP programme.

The following sections elaborate on some of these achievements in delivering bottom line economic benefits in association with real environmental and social improvements, using selected case study examples.

2.2 Environmental benefits

The issue of increasing landfill costs, combined with businesses striving towards more sustainable site practices and corporate social responsibility is driving the construction industry's movement and desire for 'Zero Waste to Landfill'. Notably, businesses operating in London are demonstrating a desire to reduce their carbon footprint to play their own role in achieving London's bid to become the greenest city in the world.

Tube Lines is one such company that NISP has been working with to help the business meet its milestone targets. The programme has been working with the company to implement a number of pilot projects demonstrating the environmental and financial benefits of industrial symbiosis in action. This has included the identification of material going into and out of the projects, developing site waste management

plans and identifying sustainable alternatives to landfill, reusing the material directly where possible. The benefits of this synergy resulted in the diversion of 99% of material from landfill with associated carbon savings.

It is important to note that the savings from the project have been quantified and externally verified, offering further credibility that IS is delivering real and sustainable resource efficiency and consumption reductions. The total environmental benefit from this project alone resulted in 4,400 tonnes of waste diversion from landfill, 2,853 tonnes of virgin material being saved from use due to the reuse of existing 'waste' materials and a saving of approximately 46 tonnes of CO_2 emissions from landfill diversion and transportation savings. By connecting Tube Lines with a number of key solution providers and sustainable suppliers, NISP London is continuing to help to develop a number of strategic pilot synergies in support of Tube Lines' delivery of their zero waste to landfill strategy.

Another London project which has generated impressive savings is that between Metronet and Foster Yeoman. The concept of industrial symbiosis is focused on closing the loop of the resource cycle and reusing materials in other applications rather than sending them to landfill. However, as previously stated, the NISP programme offers more than simple bilateral movement of waste and resources. NISP also works to imbed long term culture change within organisations to look holistically at their processes and to identify better methods of working to maximizing the potential for reuse of materials. This example between Metronet and Foster Yeoman typifies how the programme has achieved such result.

Metronet were already sending non contaminated track ballast to Foster Yeoman for conversion into a type 1 aggregate. A large proportion however was still being sent to landfill after being consigned as hazardous. The intervention of the NISP London team helped the company identify that a large proportion of this material could be reclassified and subsequently reused. This resulted in the diversion of 67,149 tonnes of waste from landfill and a carbon saving of 725 tonnes.

The concept of industrial symbiosis is not new to many businesses; however it is often the application and implementation of innovation and technology that results in the diversion of materials from landfill. An example of such activity has been a project involving Jack Moody Holdings, Lafarge Roofing and Akristos Ltd.

NISP has been actively engaged with Akristos since the programme began. The micro sized business has been working with a number of construction companies across the UK, looking at how innovation and technology can be used to find new applications for waste streams.

The NISP West Midlands team were instrumental in introducing Jack Moody Holdings, another member of the programme and Akristos, who were at the time representing the interests of Lafarge Roofing. Through this meeting, Akristos identified an opportunity to help Lafarge Roofing divert tile waste generated by the production process by working with Jack Moody Holdings to manufacture the waste into a secondary aggregate. Without NISP's intervention in introducing these parties the resulting symbiotic relationship would not have been identified. This synergy resulted in the diversion of 1,800 tonnes of waste tiles from landfill and the saving of 1,800 tonnes of virgin materials.

As well as demonstrating the ability of the programme to encourage new ideas, creating new products from waste resources this example highlights the networking capacity of NISP leading to previously un-thought of business opportunities. One of the strengths of the NISP programme, and where it has been particularly successful in delivering IS methodologies, has been in its ability to network effectively a range of businesses across traditional sector boundaries that would otherwise not have come together. This approach is used throughout the UK, often via networking meetings and one to one member introductions. It is through this unique approach that the IS programme in the UK has become so successful. This approach was used when introducing Scott Wilson to Weybrook Park Golf Club. The South-East team were instrumental in bringing these companies together to assist in the planning applications and re-development of the site.

This example typifies where IS in the UK has resulted in more than a bi-lateral movement of waste resources. The scope of the programme in working across the waste hierarchy and aiding businesses to navigate the legislative process has been key in generating environmental benefits. From this project alone the programme was successful in identifying sustainable supplies of inert materials for use on the Weybrook Park golf course with a diversion of 300,000 tonnes of construction materials from landfill, 300,000 tonnes of virgin materials saved and a carbon saving of over 800 tonnes.

Such examples, however, do not fully depict the entire achievements of the completed construction related IS projects in the UK with only a handful of these projects being discussed as exemplars of best practice in this paper which outline the variety of engagement methods employed by the programme.

From the 95 or so completed construction related synergies, over a million tonnes of waste has been saved from going to landfill, being diverted into new innovative projects that would otherwise have used virgin materials. The additional carbon savings has been quantified to approximately 200,000 tonnes;

having been fully verified and externally audited by independent sources.

2.3   Economic benefits

The simultaneous cost savings to the construction industry generated from the intervention of the NISP has amounted to over £18.5 million with additional sales reaching over £11.5 million. These figures are the combined effect of savings generated from landfill gate fees, landfill taxes, transportation cost savings and general efficiency improvements. However, the level of private investment realised was a little over £2.5 million. It is important to note that although in comparison this figure is relatively low, the risk involved in these projects can inevitably be high. This investment was made on the basis of sound business plans to reuse materials opposed to landfill, and in an effort to stimulate the UK market for reuse of recyclate and redundant resource streams rather than transporting material overseas for reuse.

In the instance of the Tube Lines project, through working with the NISP programme and adopting the principles of industrial symbiosis into its resource efficiency activities, the company reported a cost saving of over £22,000 at the end of the project with a further reported outcome of £8,000 additional sales for local recycling solutions. These achievements were, at a time of increasing costs and financial obligations, a welcomed outcome to the project offering further incentive to adopt IS into the culture of the business.

In the case of Metronet the cost savings realised have been much more substantial. With cost savings of over £80,500 and additional sales in excess of £1.7 million the benefits realised by this project required considerable effort and participation from the company, engaging actively with other NISP members to explore and share resource efficiency measures and best practice.

Adopting industrial symbiosis into the day to day culture of an organization can lead to many significant and quantifiable benefits. When businesses are facing increasing financial pressures in addition to the desire to be viewed as more corporately responsible, the efficient adoption of the IS approach is becoming more of a necessity. The approach enables businesses to prove compliance with legislation, demonstrates a commitment to continual improvement and resource efficiency, whilst at the same time facilitate considerable economic benefits to all participating companies.

By far the most impressive of the examples identified however are the results reported by Weybrook Park Golf Club. Through adopting IS into their business culture, the company has reported a saving of over £6.6million in landfill diversion savings in addition to a further £900,000 in reduced virgin aggregate for the site. By being able to reuse its own material on site much of these savings come directly off the bottom line.

The intention of these examples is to demonstrate that IS can and does offer far more to the economy than bi-lateral solutions to the movement of waste materials, water and energy. The Programme is a business efficiency driver which also delivers tangible resource efficiency benefits. Importantly, these projects are reproducible within the construction industry, opening up the potential of IS to a wider audience.

2.4 *Social benefits*

Finally, the social benefits of the programme are equally beneficial. When we discuss social benefits such as job creation, jobs safeguarded and businesses created or sustained, the expectations can often be disproportionate to the amount of work required to sustain these changes.

To put this into perspective, a business would be required to significantly increase its productivity and output in order to successfully create new jobs, and in addition to this a business is equally required to prove sustainability above and beyond process improvement to sustain existing jobs. Often when we consider efficiencies in business, we reflect on the process improvements of a business, reducing the inefficiencies of the companies and adopting more lean processes to become more competitive in the market place.

The social benefits that the construction companies who have engaged with NISP have reported are therefore even more significant. With respect to Weybrook Park Golf Club, the project created at least one additional job in the management of material supplies, and it sustained a number of other positions within the company.

It is important to note however that the biggest social benefits are not always aligned to the amount of waste diverted from landfill and reused, or the amount of cost savings generated. In the instance of a demolition company, the McGrath Group based in the east of London, the company invested systematically in advancing its processes to produce a range of secondary products, which have been recovered from its demolition projects.

The company has worked extensively with NISP and adopted the principles of IS into the heart of the business, investing significantly in the right equipment to maximize the potential to recover as much reusable material as possible. Through the networking approach adopted by NISP, the McGrath Group has proactively engaged with other NISP members to diversify its existing construction activities entering into the market of recycling tyres for equestrian surfaces. This business diversification from the demolition and construction sector into the recycling sector has created over 150 jobs within the company.

Examples such as these, although not common place in terms of significance, do promote the activities of IS to be far more than a discussion regarding waste reuse and exchange. It is equally as important to note that the principles of IS can be adopted by any business regardless of size, as illustrated by Akristos, and most importantly can be adapted to become viable based on economies of scale.

3 CONCLUSIONS

The National Industrial Symbiosis Programme is the world's leading programme to deliver the methodologies of industrial symbiosis on a national level. The programme has been in operation within the UK since April 2005 and in that time has generated significant environmental benefits to the Construction Industry amounting to some 1.2 million tonnes of waste diverted from landfill, some 1 million tonnes of virgin materials saved from use and a carbon equivalent reduction of over 220,437 tonnes.

These savings are further supported by the economic benefits that the programme generates for both the participating businesses, and HM Treasury. From the 95 completed construction related projects, NISP has supported a total cost saving of over £18.7 million, supported by additional sales in excess of £11.5 million and private investment of approximately £2.6 million. These outcomes are all externally and independently verified and audited, and prove the additionality that IS is delivering within the Construction Industry.

Many of these results are achievable and replicable throughout the UK and overseas, offering long term, sustainable solutions to resource efficiency and consumption. The case study examples highlighted within this paper demonstrate the real life experiences of the National Industrial Symbiosis Programme has affected the culture and operations of businesses within the sector, affording more than a bilateral solution to waste and resource issues, but the additional sustainable benefits of cost savings and business generation.

Finally, some of the greatest impacts of the the programme are realised in the social benefits reported with a total of 43 jobs being created within the construction industry alone. A further 48 jobs being safeguarded and a total of 8 new businesses having been created as a result of the programme's intervention. These results support the statement that IS stimulates the economy creating more than simple resource efficiency solutions, offering long term options for the sector which can be sustained and built upon through the continued use of technology, innovation and culture change.

REFERENCES

Ayres, R.U & Ayres, L.W. 2002. *A handbook of industrial ecology*. Cheltenham: Edward Edgar.

Chertow, M. 2000. Industrial symbiosis: literature and taxonomy. *Annual Review of Energy and the Environment* 25: 313–337.

Department of Trade and Industry 2006. *Review of sustainable construction*. http://www.berr.gov.uk/files/file34979.pdf.

Department for Business, Enterprises and Regulatory Reform 2007. *Draft strategy for sustainable construction*. http://www.berr.gov.uk/files/file40641.pdf.

Frosch, R.A. & Gallopoulos, N.E. 1989. Strategies for manufacturing. *Scientific American* 261(3): 144–152.

Graedl, T. & Allenby, B. 1995. *Industrial Ecology*. Englewood Cliffs, NJ, USA: Prentice Hall.

Laybourn, P. 2008. A fruitful collaboration, *The Environmentalist* 54: 27.

Laybourn, P. & Lombardi, R. 2007. The Role of audited benefits in Industrial Symbiosis: The UK National Industrial Symbiosis Programme. *Measurement and Control* 40(8): 244–247.

NISP. 2007. *Response by NISP to the House of Lords Select Committee for Science and Technology Inquiry into Waste Reduction*. http://www.parliament.uk/documents/upload/st1NISP.pdf.

Playfair, L. 1892. Waste products made useful. *North American Review* 155(432): 560–569.

Effects of different glasses composition on ecosustainable blended cements

M.C. Bignozzi & F. Sandrolini

Dept. of Applied Chemistry and Materials Science, Bologna University, Italy

ABSTRACT: The use of different ground glass waste as 25 wt% of ordinary Portland cement replacement is reported and discussed with the aim to establish if chemical glass composition plays an effective role in developing pozzolan activity. Physical and mechanical properties of the new binders are reported and compared to those of 100% CEM I 52.5 R. The presence of ions with different mobility in glass waste appears to have a correlation with mortar mechanical strength development for the investigated curing time.

1 INTRODUCTION

In the last years, cement industry made several efforts to reach a high level of energy efficiency: Portland clinker production requires elevate temperature process with consequent large fuel consumption as well as CO_2 emissions. Different cement constituents (such as natural pozzolana, fly ash, silica fume, granulated blast-furnace, etc.) are added to clinker not only to limit its production, but also to promote cement durability. Moreover, the use of industrial waste for blended cements production has been already assessed, as reported in European standard EN 197-1 where 27 different cement types are exploited.

Recently several kinds of waste have been selected and studied as new possible constituents for cement: for example, slag derived from pulverized municipal solid waste incinerator bottom ash (Lin et al. 2008), ferroalloy industry waste such as SiMn slag and Mn oxide filter cakes (Frías et al. 2008), polishing and glazing ceramic sludges (Andreola et al., 2007), etc. Many researches were carried out on glass waste: its high amorphous nature and elevate silica content made it particularly attractive for new blended cement designing, although fineness and grain size distribution are very important parameters in view of cement strength development. High pozzolan activity was observed for ground lime-glass having particles $\leq 38\,\mu m$ (Shao et al. 2000) and for glass waste coming from glass beads manufacturers with an average size almost the same or smaller than the cement one (Shi et al. 2005).

Another feature that can play an important role in pozzolan reaction development between glass waste and cement products is glasses chemical composition.

The effect of 10–40 wt% of cement replacement by white, green and amber powder cullet from beverages containers was studied by Dhir et al. (2001) and their pozzolan reaction was confirmed. The investigated glasses were mainly soda-lime-silica glass type (SiO_2: 70–72%; Na_2O: 14–17%; CaO: 6–11 wt%).

The aim of this paper is to go on investigating the effects of glass chemical composition on mortar mechanical strength development. Accordingly, four different type of ground glass waste, characterized by different origin and oxide composition, were used as 25 wt% of ordinary Portland cement (OPC, CEM I 52.5R) replacement: setting time, soundness, mechanical strength at different curing time (2, 7, 28 and 90 days), pozzolan activity index were determined. The use of ground glass waste was compared to OPC behavior: the obtained results are here reported and discussed.

2 EXPERIMENTAL DETAILS

2.1 Materials

CEM I 52.5 R (EN 197-1, Italcementi, Calusco d'Adda (BG) Italy) was selected for paste and mortar samples preparation to evaluate the effects of ground glass waste addition to OPC. Five different glass waste were used: soda-lime glass (named SLG) coming from food containers cullet (kindly supplied by Bormioli Rocco, Fidenza (PR) Italy), borosilicate glass (named BG) and amber borosilicate glass (named ABG) coming from pharmaceutical containers cullet (kindly supplied by Bormioli Rocco, Fidenza (PR) Italy) and two kinds of crystal glass waste (named CGB and CGC) coming from production of tableware, giftware and home

Table 1. Chemical composition of glass waste.

	SLG wt %	BG wt %	ABG wt %	CGB wt %	CGC wt %
SiO_2	70.40	68.60	65.90	61.60	61.70
Al_2O_3	2.06	5.64	5.88	0.13	<0.05
TiO_2	<0.05	<0.05	2.89	<0.05	<0.05
Fe_2O_3	<0.05	<0.05	0.78	<0.05	0.16
CaO	11.20	1.53	1.42	<0.05	<0.05
MgO	1.47	<0.05	<0.05	<0.05	<0.05
K_2O	1.21	1.24	1.16	7.05	6.96
Na_2O	12.69	8.06	7.49	3.86	3.96
ZrO_2	<0.05	<0.05	0.17	<0.05	<0.05
B_2O_3	0.69	11.56	10.34	0.81	0.72
ZnO	<0.05	0.82	0.78	<0.05	<0.05
BaO	0.12	2.95	2.86	<0.05	<0.05
PbO	<0.05	<0.05	<0.05	26.38	26.35
Sb_2O_5	<0.05	<0.05	<0.05	0.27	0.26

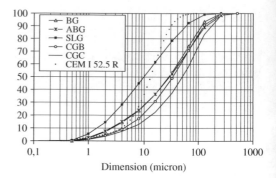

Figure 1. Grain size distribution of glass waste and CEM I 52.5 R.

décor items in crystal (kindly supplied by CALP, Colle di Val d'Elsa (SI) Italy). A representative batch of each glass waste was used and their average chemical composition by atomic spectroscopy is reported in Table 1. All the glass waste was used as received, except for grinding operations carried out on a laboratory ball mill. Grinding time was properly adjusted considering dimensions of starting glass and its nature with the aim to reach grain size distributions comparable with that of cement.

Grain size distributions of the ground glasses and CEM I 52.5 R, determined by laser granulometer (Mastersizer 2000, Malvern Instruments), are reported in Figure 1. Particle size distributions of BG, ABG and CGB are very similar, but coarser than that of CEM I 52.5 R and SLG. Their average size is about 30 μm. CGC exhibits the coarsest particles, with only about 20 vol% smaller than 20 μm. SLG has a larger grain size distribution for particles with size ≥20 μm compared to CEM I 52.5 R, whereas for particles with size <15 μm the reverse situation occurs.

Silica sand with normalized grain size distribution according to EN 196-1 was used for mortar preparation.

2.2 Experimental tests and samples preparation

Initial cement setting time and soundness were determined on pastes containing 75 wt% of cement and 25 wt% of ground glass waste by Vicat method and Le Chatelier apparatus respectively, according to EN 196-3. For comparison, initial setting time and soundness of pastes prepared with 100% CEM I 52.5 R were also determined.

Minislump test (Zappia et al. 1990) was carried out to assess paste workability using a water/binder (W/B) ratio equal to 0.50.

Mortar samples were prepared by a Hobart mixer, according to the normalized formulation and procedure reported in EN 196-1. As binder, 100% of CEM I 52.5 R was used in reference sample (named REF) and 75% of CEM I 52.5 R + 25% of ground glass waste was used for mortar samples respectively named SLG-M, BG-M, ABG-M, CGB-M, CGC-M.

Mortar workability was determined by using a flow table according to UNI 7044.

Prismatic samples (40 × 40 × 160 mm) were prepared and cured for 2, 7, 28, 90 days at 20°C and R.H. >95%. When necessary, the curing period of mortar samples is indicated with the relevant number after the name of the mortar (e.g. SLG-M_2 for two days of curing).

The activity index, defined by EN 450-1 for concrete fly ash as the compressive strength ratio between samples containing 75% of CEM I 52.5 R + 25% of waste and reference mortar with 100% of CEM I 52.5 R, was determined after 28 and 90 days of curing.

2.3 Characterization

Flexural and compressive strength of mortar samples was determined by an Amsler-Wolpert machine (100 kN) at constant displacement rate of 50 mm/min. Pore size distribution measurements were carried out by mercury porosimeter (Carlo Erba 2000) equipped with a macropore unit (Model 120, Fison Instruments).

3 RESULTS AND DISCUSSION

Results of initial setting time and soundness are collectively reported in Table 2 for all the investigated pastes. The presence of ground glass waste, compared to CEM I 52.5 R, increases setting time slightly for SLG, BG and ABG, more appreciably for CGB and CGC. This behaviour is probably due to organic impurities in the ground glass waste (boiling water soluble

512

Table 2. Setting time and soundness for the investigated mixes (average of two measurements).

CEM wt %	Glass waste wt %	Glass waste type	Setting time h:min	Soundness mm
100	0	–	2:58	0
75	25	SLG	3:35	0
75	25	BG	3:33	0
75	25	ABG	3:33	0
75	25	CGB	4:01	0.2
75	25	CGC	4:16	0.3

Table 3. Minislump and flow table test results.

Binder	W/B*	Minislump cm	Flow table test %
CEM I 52.5 R	0.5	17	70
CEM I + SLG	0.5	16	58
CEM I + BG	0.5	17	74
CEM I + ABG	0.5	18	82
CEM I + CGB	0.5	18	80
CEM I + CGC	0.5	17	71

* W/B: water/binder ratio

fraction is about 3–5%) and to their different average grain size.

The expansion registered for all the samples falls within EN 197-1 requirements (≤ 10 mm).

Data collected by workability measurements are reported in Table 3: both tests, carried out on paste and mortar, show that ground glass waste does not meaningfully modify workability in comparison with CEM I 52.5 R. The lowest values were observed when SLG is used, according to its highest content of very fine fraction (50 vol% $\leq 10 \mu$m).

In Table 4 compressive strength at 2 and 28 days of the investigated mortar is reported and compared to the requirements set by EN 197-1. For both early and standard curing times mortar samples with ground glass waste exhibit compressive strength lower than REF and than the limits set by EN standard for 52.5 R strength class. However, regardless of glass waste type, the determined compressive strength results higher than the limits for 42.5 R and 32.5 R cement strength classes. SLG, BG and ABG in the relevant mortar lead to similar compressive strength values, whereas mortar samples containing ground crystal waste show the lowest strength for both curing times.

The mechanical strength development as function of curing time is reported for all the investigated samples in Figures 2–3. Mortar containing ground glass waste lead to a general increase in both flexural and compressive strength moving from 2 to 90 days of curing, however such increase is more evident for

Table 4. Compressive strength at 2 and 28 days of curing and relevant limits by EN 197-1.

		Compressive strength	
		MPa 2 days	MPa 28 days
REF		37.9	58.9
SLG-M		26.5	45.6
BG-M		28.0	49.3
ABG-M		27.6	46.5
CGB-M		24.8	43.4
CGC-M		23.6	43.3
Limits by EN	32.5 R	≥ 10.0	≥ 32.5
197/1	42.5 R	≥ 20.0	≥ 42.5
for CEM	52.5 R	≥ 30.0	≥ 52.5

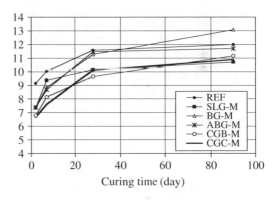

Figure 2. Flexural strength as function of curing time for the investigated mortar.

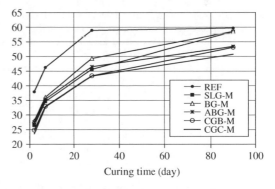

Figure 3. Compressive strength as function of curing time for the investigated mortar.

compressive strength and for samples where SLG and BG were used.

SLG-M_90 and BG-M_90 almost reach the same compressive strength exhibited by REF_90. The mechanical properties of REF increase till 28 days of

Table 5. Activity index and limits by EN 450-1.

	Activity Index	
	28 days	90 days
SLG-M	77.4	98.0
BG-M	83.7	98.1
ABG-M	78.9	89.5
CGB-M	73.7	89.0
CGC-M	73.5	85.0
Limits by EN 450/1	≥75	≥85

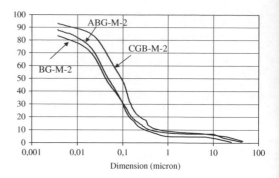

Figure 4. Porosity distribution of mortar samples containing BG, ABG and CGB after 2 days of curing.

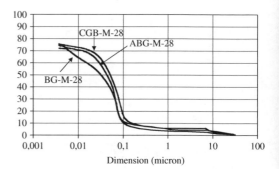

Figure 5. Porosity distribution of mortar samples containing BG, ABG and CGB after 28 days of curing.

curing and remain constant moving to 90, accordingly with OPC behavior. With the aim to quantify pozzolan activity of the used ground glass waste, the activity index was calculated at 28 and 90 days of curing: Table 5 reports the obtained values and limits set by EN 450-1. All the ground glass waste show an activity index at 90 days overcoming the standard requirement, however the glass activity is almost equal to 100% only for SLG and BG, about 90% for ABG and CGB and only 85% for CGC.

At 28 days, BG exhibits the highest activity, whereas for CGB and CGC it is slightly lower than the required limit.

All the investigated ground glass waste appears to undergo pozzolan reaction, at least in terms of strength development. Concerning pozzolan reaction products, when soda-lime glass waste is involved as Portland cement replacement, C-S-H gel formation with a Ca/Si ratio of about 1.5–1.7 and including sodium has been reported (Bignozzi et al. in press). For the other ground glass waste, investigations are currently running,

As pozzolan reaction is strongly influenced by grain size distribution and average dimension, the calculated activity index must be evaluated also considering this aspect. If 40 μm is considered as superior limit for lime glass particles size to exhibit pozzolan behavior (Shao et al. 2000), the investigated glass waste SLG, BG, ABG, CGB and CGC show respectively 83, 60, 58, 55 and 45 vol% with dimension ≤40 μm. Accordingly, SLG and CGC respectively exhibit the highest and lowest activity index at 90 days of curing for the relevant mortar samples. However, BG, ABG and CGB show very similar grain size distributions, but BG activity index is higher than those of ABG and CGB, at both curing time.

BG behavior can be explained by its chemical composition (Table 1): it has high amount of sodium that is easily mobilized at pH about 9–12 and low content of transition elements (such as iron, titanium, zirconium) that usually strongly reduce glass dissolution rate (Trocellier et al. 2005). ABG has almost the same sodium content of BG, but higher amount of iron, titanium, zirconium (Table 1) and this explains its lower pozzolan activity.

CGB and CGC show typical chemical composition for crystal glasses, with high content of lead (about 26%). The lowest activity index determined for this glass waste can be explained by their low content of sodium (about 4%). Moreover, high amount of lead seems to favor the formation of high swelling gel thus leading to the detrimental alkali-silica-reaction (ASR). This negative reaction, recently observed (Saccani et al., 2008), is evident for the investigated ground crystal glass waste, but not for SLG, BG and ABG.

Microstructure studies were carried out by means of porosimeter measurements. As general trend total open porosity decreases with curing time for reference and ground glass waste based samples. No meaningful differences were detected between the investigated samples, then only pore size distribution of BG-M, ABG-M and CGB-M after 2 and 28 days of curing are reported in Figures 4–5. The effect of BG, ABG and CGB on microstructure development can be more easily compared as these ground glass waste has a very similar grain size distribution.

At both curing time, porosity increases in the following order: CGB-M > ABG-M > BG-M. The difference between size porosity is more evident for dimension <1 μm at 2 days of curing and for dimension <0.1 μm at 28 days.

The development of very compact microstructure for BG-M-28 agrees with mechanical data and the high value of activity index also determined at 28 days.

4 CONCLUSIONS

The use of different ground glass waste as 25 wt% of ordinary Portland cement replacement highlights that chemical glass composition plays an important role in developing pozzolan activity. As glass durability depends on its chemical composition and environmental conditions (T, pH, etc.), in a similar way ground glass waste should be influenced by the same factors in developing pozzolan action with consequent increase in mortar mechanical strength.

From the obtained results, ground glass waste coming from soda-lime and clear borosilicate glasses appears to be the most promising for new ecosustainable blended cements production. Moreover, should this recycling route be considered attractive for other kinds of glass waste, the amount of transition metals must be evaluated and high lead content avoided.

REFERENCES

Andreola, F., Barbieri, L., Lancellotti, I., Piccagliani, V., Rabitti, D., Bignozzi, M. C. & Sandrolini, F. 2007 Valorization of polishing and glazing ceramic sludges in new blended cement. *In Proceedings of VI International Congress: "Valorisation and recycling of industrial waste", L'Aquila, Italy, 27–29 June, 2007.*

Bignozzi, M. C., Saccani, A. & Sandrolini, F. 2008. Matt waste from glass separated collection: a reactive addition to cement. *Submitted to Cement and Concrete Research.*

Dyer T.D. & Dhir R.K. 2001. Chemical reactions of glass cullet used as cement component. J Mater Civ Eng. 13: 412–7.

Frías, M. & Rodríguez, C. 2008. Effect of incorporating ferroalloy industry wastes as complementary cementing materials on the properties of blended cement matrices. *Cement & Concrete Composites 30: 212–219.*

Lin K.L., Chang, W.C. & Lin D.F. 2008 Pozzolanic characteristics of pulverized incinerator bottom ash slag. *Construction and Building Materials 22: 324–329.*

Shao, Y., Lefort T., Moras, S. & Rodriguez, D., 2000. Studies on concrete containing ground waste glass. *Cem. Concr. Res. 30: 91–100.*

Shi C, Wu Y, Riefler C. & Wang H. 2005. Characteristics and pozzolanic reactivity of glass powders. *Cem Concr Res. 35: 987–93.*

Saccani, A. & Bignozzi, M. C. 2008. An insight in the behavior of glasses with different chemical composition to be used in cementitious composites as aggregates or reactive fillers. *In Proceedings of the 13th International Conference on Alkali-Aggregate Reactions in concrete, Trondheim, Norway, 16–20 June 2008.*

Trocellier, P., Djanarthany, S., Chêne, J., Haddi, A., Brass, A.M., Poissonnet S.& F. Farges. 2005. Chemical durability of alkali-borosilicate glasses studied by analytical SEM, IBA, isotopic-tracing and SIMS. *Nuclear Instruments and Methods in Physics Research Section B. 240 (1–2): 337–344.*

Zappia, G., Sandrolini, F. & Motori, A. 1990. Premix PCC materials: mechanical properties as function of tecnological parameters. *Materials and Structures 23: 436–441.*

Excellence in Concrete Construction through Innovation – Limbachiya & Kew (eds)
© 2009 Taylor & Francis Group, London, ISBN 978-0-415-47592-1

Possible utilization of wheat husk ash waste in the production of precast concrete elements

J. Zhang
Inner Mongolia University of Science and Technology, Baotou, Inner Mongolia, China

J.M. Khatib & C. Booth
School of Engineering and the Built Environment, University of Wolverhampton, UK

R. Siddique
Department of Civil Engineering, Thapar University, Patiala (Punjab), India

ABSTRACT: Large quantities of wheat husk is produced when wheat grains is extracted for human consumption in the northern part of China. The wheat husk is normally burnt and hence wheat husk ash (WHA) is generated. WHA is mainly destined to landfill. In this paper the possible utilization of WHA in the production of pre-cast concrete elements is investigated. For this reason, three series of concrete mixes containing different amounts of WHA were prepared in order to determine their workability and compressive strength. In the first series (series 1) WHA is used to increase the workability while maintaining the same compressive strength and no increase in cement content. In the second series (series 2) mixes containing WHA were used to increase the compressive strength while maintaining the same workability and without increasing the cement content. In series 3 concrete mixes are prepared to have similar compressive strength. There is reduction in cement content in this series as the WHA increases. The compressive strength testing was conducted after a total steam curing period of 8 hours. The results show that by carefully selecting the mix proportion, WHA can be used in concrete as % addition by mass of cement to increase workability without reducing the compressive strength. It can also lead to an increase in strength while maintaining the same workability. Furthermore, using WHA can lead to a reduction in the cement content in the concrete mix without reducing the compressive strength.

1 INTRODUCTION

Wheat is the main agricultural crop in the country side of the northern part of China. After the extraction of wheat grain, wheat husk is generated in large quantities. Wheat husk is considered to be a kind of waste material. Farmers use only a small volume of wheat husk by mixing it with the soil as fertilizer. The rest of the husk is usually burnt in the open air. Wheat husk ash (WHA) is produced as a result of the burning process. The burning in open air causes environmental problems. This is exacerbated further when the wind blows and thus transports the ash to nearby cities causing immediate health problems. Also, the farmers have to deal with this large quantity of WHA. There is a need, therefore, to properly use the WHA in order to reduce its environmental impact including land and air pollution.

When burnt, the wheat husk has a calorific value of about 3500 kcal/kg. Thus, burning wheat husk as a fuel in boilers could be one of the suitable ways to reuse the husk in an efficient and controlled manner. Taking into account that big amount of WHA, attention should be paid to its disposal and possible utilization. A strong incentive to the utilization of WHA is represented not only by the large quantities generated, but also by the people's awareness towards its right disposal, in order to reduce its direct impact on human health.

There has been a great deal on the use of pozzolanic material such as pulverized fuel ash in the production of concrete (Malhotra 1999, Zhang & Ushakov 1996, Zhang & Yan 2003, Zhang & Song 2006, Zhang & Khatib 2006, Zhang & Zhang 2002, Zhang & Yan 2005). The chemical composition of WHA is similar to other pozzolanic materials such as pulverized fuel ash, as will be reported in the following section. Therefore, this paper reports the results of a preliminary investigation on the workability and strength properties of concrete incorporating WHA. Concretes were steam cured to resemble the conditions in a precast concrete factory.

2 EXPERIMENTAL

2.1 *Materials*

The sample of wheat husk was taken from a paper making factory in Inner Mongolia. The wheat husk was burnt in a small experimental furnace with a temperature between 450°C and 500°C. The WHA produced was analysed using X-ray diffraction for its chemical composition. Also the physical properties were determined. Table 1 shows the chemical compositions and physical properties WHA.

Portland Cement (CEM1-42.5) from a cement factory in Inner Mongolia was used. The coarse aggregate was crushed gravel and had a maximum size of 10 mm and the sand used had a fineness modulus of 2.2.

2.2 *Details of concrete mixes*

Three series of concrete mixes were used in order to assess the usage of WHA in concrete production. The first series (series 1) consists of the control mix (A-0) and two other mixes (A-1) and (A-2). Mix A-0 had

Table 1. Compositions and properties of wheat husk ash (WHA).

SiO_2	%	53.94
Al_2O_3	%	13.23
CaO	%	4.40
Fe_2O_3	%	4.10
MgO	%	1.58
Loss on ignition	%	1.5
Loose density	kg/m^3	650
Average particle size range	μm	6–10
Specific surface area	m^2/kg	550

a proportion of 1 (cement) : 2 (sand) : 4 (aggregate) without any WHA. In mixes A-1 and A-2, $50 \, kg/m^3$ and $80 \, kg/m^3$ of WHA were respectively added to the mixes. The mixes in series 1 were chosen so that the workability is increased with increasing WHA content while keeping the same cement content and similar 28-day compressive strength. Details of mixes A-0, A-1 and A-2 are given in Table 2.

In the second series (series 2) the WHA was used in such a way that the workability was kept the same for all mixes while increasing the WHA content in order to increase the compressive strength. The cement content of the mixes was kept the same. The series contained in addition to the control mix (A-0), two other mixes; mix B-1 and mix B-2 where $60 \, kg/m^3$ and $90 \, kg/m^3$ of WHA was used in the mix. Further details of mixes B-1 and B-2 are presented in Table 3.

In the third series (series 3) the WHA was used in such a way to decrease the quantity of cement in the concrete without reducing the compressive strength or workability. This series consists of the control mix (A-0) and mixes (C-1) and (C-2) with cement content of $230 \, kg/m^3$ and $210 \, kg/m^3$ respectively and the WHA content was $55 \, kg/m^3$ (C-1) and $90 \, kg/m^3$ (C-2). Table 4 presents details of mixes C-1 and C-2.

It is worth noting that no admixtures were used in any of the above mixes in order to keep the cost of concrete production to a minimum.

2.3 *Testing and curing*

For each mix 3 cubes of 70 mm in size were cast. Soon after casting, specimens were covered and placed in a room at 20°C until demoulding. The specimens were demoulded after 24 hours and placed in a steam

Table 2. Details of mixes for series 1.

Mix code	Quantities (kg/m^3)					W/C^2	W/B^3	$\%WHA^4$
	Cement	WHA^1	Water	Sand	Gravel			
A-0	265	0	208	540	1180	0.79	0.79	0
A-1	265	50	235	500	1180	0.89	0.75	19 (16)
A-2	265	80	240	480	1180	0.91	0.70	30 (23)

[1] Wheat husk ash [2] Water to cement ratio [3] water to binder ratio (binder consists of cement and WHA) [4] % addition by mass of cement (% by mass of binder)

Table 3. Details of mixes for series 2.

Mix code	Quantities (kg/m^3)					W/C2	W/B3	%WHA4
	Cement	WHA1	Water	Sand	Gravel			
A-0	265	0	208	540	1180	0.79	0.79	0
B-1	265	60	215	480	1180	0.81	0.66	23 (19)
B-2	265	90	220	440	1180	0.83	0.62	34 (25)

chamber (~100°C) for 8 hours, in order to resemble concrete curing in the production of precast elements. After this, specimens were removed from the chamber and tested for compressive strength.

3 RESULTS AND DISCUSSION

The slump values of series 1 mixes are presented in Figure 1. There is noticeable increase in workability (i.e. high slump values) with the increase in WHA content. The slump value for mix containing 19% WHA (A-1) is nearly twice that of the control mix (A-0) and that containing 30% WHA is more than twice of the control.

Figure 2 shows the compressive strength results after 8 hours of steam curing for series 1 mixes. Despite of the increased water content in mixes A-1 and A-2 compared to the control while keeping the cement content the same, all concretes exhibit slightly higher strength than the control. Although the water content in series 1 was increased with the increase in WHA content, no reduction in compressive strength was observed.

Also the concretes containing WHA in this series were more workable than the control without increasing the cement content and without including any chemical admixtures and hence keeping the cost to a minimum.

Figure 3 shows the slump values of series 2 mixes. In series 2, the WHA content was higher than those of series 1. The cement content was kept the same for all mixes. The workability of all mixes is similar in that the slump for all mixes was between 20 and 25 mm.

The compressive strength values after 8 hours of steam curing for series 2 mixes are shown in Figure 4. Despite of the increase in water content, the increase in WHA content has led to an increase in compressive strength without the need to increase the cement content. This suggests that the WHA contribute to strength enhancement.

The slump values for series 3 mixes are shown in Figure 5. The cement content in this series decreased with the increase in WHA content. The slump values were similar for all mixes in this series.

Figure 6 presents the compressive strengths at 28 days of curing for series 3 mixes. Despite of the

decrease in cement content and increase in water content, mixes incorporating 24% WHA content exhibit slightly higher strength (3%) than the control mix whereas the mix containing 43% WHA show 6% lower strength than the control. In other words, replacing 35 kg of cement with 55 kg of WHA, in one cubic meter of concrete, results in a similar strength to the control mix. This is achieved despite of the higher water requirement in the presence of increasing WHA content in order to keep the workability of the concretes almost the same.

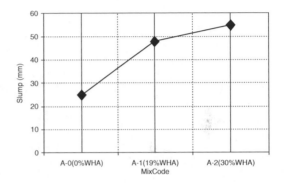

Figure 1. Slump values for Series 1 mixes with cement content of 265 kg/m^3 and varying water to cement ratio.

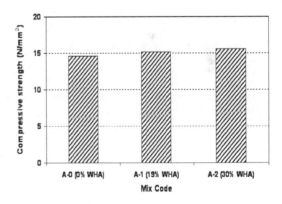

Figure 2. Compressive strength at 28 days of curing for Series 1 mixes with cement content of 265 kg/m^3 and varying water to cement ratio.

Table 4. Details of mixes for series 3.

Mix code	Quantities (kg/m^3)					W/C2	W/B3	% WHA[4]
	Cement	WHA1	Water	Sand	Gravel			
A-0	265	0	208	540	1180	0.79	0.79	0
B-1	230	55	210	520	1180	0.91	0.74	24 (19)
B-2	210	90	214	415	1180	1.02	0.71	43 (30)

519

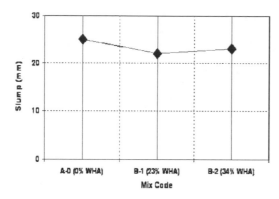

Figure 3. Slump values for Series 2 mixes with cement content of 265 kg/m^3 and varying water to cement ratio.

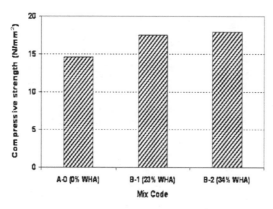

Figure 4. Compressive strength at 28 days of curing for Series 2 mixes with cement content of 265 kg/m^3 and varying water to cement ratio.

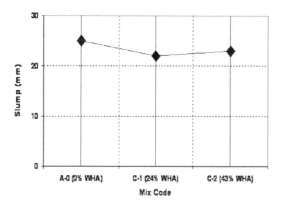

Figure 5. Slump values for Series 3 mixes with varying cement content and water to cement ratio.

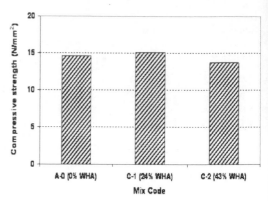

Figure 6. Compressive strength at 28 days of curing for Series 3 mixes with cement content of 265 kg/m^3 and varying water to cement ratio.

The WHA is a waste material. The above results suggest that WHA has the potential to be used in concrete production. The incorporation of WHA in concrete mixes contributes to strength increase. According to the chemical composition and physical properties, WHA consists of silica (50%) and alumina (13%) which indicate, that it is a pozzolanic material. More over WHA has substantially higher surface area (550 m^2/g) than cement and an average particle size between 8 and 10. Therefore, WHA can be classified as a pozzolanic material, which is a material that is able to react with calcium hydroxide (product of cement hydration) to produce further gel that causes an increase in compressive strength. A pozzolanic material is a material, which itself has little or no cementitious value (i.e. an ability to bond particles together) but by reacting with calcium hydroxide and water forms cementitious compounds. The high silica content, large specific area and fine particle size are some of the reasons why WHA could play a great role as a mineral admixture in concrete.

In order to produce highly pozzolanic wheat husk ash, it is necessary to control the burning process. The research in this paper has shown that controlled incineration of WHA at about 500°C will produce amorphous wheat husk ash, which is highly pozzolanic. This paper reports a preliminary work studying the possible use of WHA in construction. A deeper investigation is required in order to assess the proper use of WHA in the production of concrete. Other concrete properties including durability need to be performed, so that better understanding of using this material (i.e. WHA) can be obtained.

4 CONCLUSIONS

It is clear from the results of the experiment that, wheat husk ash can be used as a mineral admixture in concrete. If it is added to concrete with optimum quantity, some properties of the concrete can be improved. All concretes in this work were subjected to steam curing for 8 hours. They show that wheat husk ash can be incorporated in concrete to improve workability and have a beneficial effect on compressive strength compared to the control concrete. In general, the results of the experiment are encouraging and would allow the use of WHA for practical applications.

REFERENCES

Malhotra, V.M., Making concrete greener with fly ash, Concrete International, May 1999, Vol. 21, No. 5, pp 61–66.

ZHANG J.S., Ushakov, V.V. The effect of fly ash on the strength of concrete, Journal of Baotou university of iron/steel technology, 1996, Vol. 13, No. 4, pp 118–122.

ZHANG J.S., YAN C.J., The study of fly ash aggregate with higher compressive strength and fly ash aggregate concrete, New building materials, August 2003, No. 269, pp12–14.

ZHANG J.S., SONG L.P., The study and utilization of the bottom ash of Baotou power plant-3, Multipurpose utilization of mineral resources, April 2006, No. 2, pp40–43.

ZHANG J.S., KHATIB, J.M., Experimental study on mechanical properties and durability of bearing wall brick of fly ash from different electric fields, New building materials, June 2006, No.303, pp 17–19.

ZHANG J.S., ZHANG F., Experimental study of utilization of fly ash to construct ditch in farming field, Water saving Irrigation, October 2002, No.115, pp22–26.

ZHANG J.S., YAN C.J., Expert system for fly ash comprehensive utilization, Fly ash comprehensive utilization, October 2005, No. 93, pp17–18.

Excellence in Concrete Construction through Innovation – Limbachiya & Kew (eds)
© 2009 Taylor & Francis Group, London, ISBN 978-0-415-47592-1

Developing viable products using recycled rubber tyres in concrete

H.Y. Kew
Kingston University, London, UK

M.J. Kenny
University of Strathclyde, Glasgow, UK

ABSTRACT: The growing problem of waste tyre disposal in the UK can be alleviated if new recycling routes can be found for the anticipated surplus of tyres. One of the largest potential routes is in construction, but usage of waste tyres in civil engineering is currently very low. This study investigates the potential of incorporating recycled rubber tyres into Cement CEM 1 concrete and concrete blocks. It was found that rubberised concrete exhibited very low workability and a marked reduction in strength which inhibits its use for general structural applications. However, the potential was found for producing low strength products such as rubberised concrete block with beneficial properties. The production method replicated that used in industry was produced successfully. As part of the effort in developing new construction materials, it is essential to establish its economic viability as well as its technical viability.

1 INTRODUCTION

The disposal of waste tyres is becoming a major waste management problem in the UK. It is estimated that 40 million car and truck tyres are being discarded annually and this figure is expected to increase over the next 20 years, in line with the increase in traffic (Wallingford, 2005).

Landfill has been one of the most convenient ways of disposing of waste tyres. However, landfill is no longer a viable option due to the implementation of European Union legislation, which currently bans the disposal of whole tyres and shredded tyres in landfill sites. There is, therefore, an urgent need to identify alternative solutions in line with the UK Government's waste management hierarchy. One of the largest potential recycling routes is in construction, but currently only about 5% of tyres are recycled in civil engineering applications. However, the potential market in civil engineering application is enormous. For example, about £2 bn is spent annually in the UK on concrete products.

Since concrete is a low cost, versatile composite materials there is potential to use rubber tyres in crumb or chip form as replacement for natural aggregates in concrete (Pierce & Williams, 2004, Labbani et al., 2004, Siddique & Naik, 2004, Khatib & Bayomy, 1999, Li et al., 1998, Fedroff et al., 1996, Toutanji,

1996, Eldin a& Senouci, 1993). The resulting composite material is generally known as rubberised concrete or rubcrete. However, incorporating rubber tyre chips in Cement CEM 1 produces a decrease in concrete strength which can be substantial if the rubber content is high. For example, Eldin & Senouci (1993) discovered an 85% reduction in compressive strength and a 50% reduction in tensile strength when the coarse aggregate in the concrete was fully replaced by rubber chips. This strength reduction greatly inhibits the possible uses and market potential of rubberised concrete.

This paper considers the effectiveness and potential of using rubber aggregate chips as a replacement for natural aggregate in Cement CEM1 concrete and concrete blocks. The use of recycled rubber in concrete blocks was considered to have greater potential due to their much lower strength requirements and method of manufacture which obviates some of the preparation difficulties of ordinary rubberised concrete. Due to the considerable strength reductions reports by previous authors for high rubber contents, smaller rubber contents were used in the present study. In addition, the effect of coating the rubber particles with cement paste was investigated as a possible means of improving the distribution of rubber aggregate particles through the mix. The economics of tyre recycling is also discussed in order to assess the potential marker for rubberised concrete products such as blocks.

(a) 20 mm plain rubber aggregate

(b) 20 mm rubber aggregate coated
with cement paste

Figure 1.   20 mm rubber aggregate particles.

Table 1.   Mix proportions of the control mixes.

| Materials | Mix Proportions (kg/m^3) | | | |
	Mix A (0.55)*	Mix B (0.48)*	Mix C (0.55)*	Mix D (0.48)*
Cement CEM1	382	438	382	438
Water	210	210	210	210
Fine aggregate	543	526	543	526
Coarse aggregate (10 mm)	–	–	422	409
Coarse aggregate (20 mm)	1266	1227	844	818

* water/cement ratio.

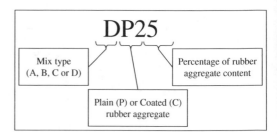

Figure 2.   Coding for rubberised concrete sample.

2   EXPERIMENTATION

2.1   Mix design and materials

The first part of the study was to produce rubberised concrete with sufficient strength to be used in a variety of structural application. Coarse rubber aggregate of 20 mm maximum size ($G_s = 1.14$) and angular shape was used to replace the 20 mm natural aggregate. Figure 1(a) and (b) show plain rubber aggregate and rubber aggregate coated with cement paste, respectively.

Four mixes were designed with a targeted compressive strength of 40 MPa and the design was carried out according to BS EN 206-1:2000 and Design of Normal Concrete Mixes (1975). The mix proportions vary in each case, specifically the water/cement ratio and the proportions of 10 mm and 20 mm aggregate. Table 1 shows the quantities of the constituents of the four control mix designs: A, B, C and D, for one cubic meter of concrete. For rubberised concrete; in Group P, the coarse aggregate of the control mix was replaced by plain rubber aggregate whilst in Group C, the coarse aggregate of the control mix was replaced by rubber aggregate coated with cement paste. For each group, three batches were made in which the 20 mm coarse aggregate was replaced at 10, 25 and 50% by volume of 20 mm rubber aggregate. The sample coding is shown in Figure 2. No mineral or chemical admixtures were added. All mixes were prepared and cured using standard methods.

To evaluate the fresh concrete properties, workability (slump) was measured in accordance with BS

EN 12350-2:2000. For hardened concrete properties, all mixes were tested for compressive and flexural strength at 28 days, whereas the splitting tensile strength test was carried out at 14 days. The compressive, splitting tensile and flexural strength tests were carried out using standard cube, cylinder and beam samples in accordance with BS EN 12390: Parts 3, 6 and 5, respectively.

3   TEST RESULTS

3.1   Workability

Figure 3 shows the workability test results for all mixes for control and rubberised concrete. The results show that increasing the percentage of rubber aggregate to up 50% produced zero slump value. Similar trend was observed for all mixes. The reduction in the workability of the concrete can be attributed to a combination of lower unit weight and the higher friction between the long angular shape of rubber aggregate and the mixture.

3.2   Ease of preparation and finishing

It was found that rubberised concrete samples can be prepared and finished to the same standard as the normal concrete. However, mixes containing higher rubber aggregate content required more effort and

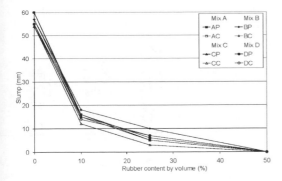

Figure 3. Effect of rubber aggregate on the workability of concrete (slump values).

work to smooth the finish surface. No special care was necessary for rubberised concrete mixes and a similar duration of compaction times between samples was maintained. No difficulties were encountered due to rubber aggregate except mixes containing higher rubber aggregate content.

3.3 Strength

The strength results are illustrated for compressive strength in Figure 4 and splitting tensile strength in Figure 5.

The results show that the addition of rubber aggregate resulted in a significant reduction in concrete strength. Of all the mixes, Mix D was the best performing, in which the smallest reductions in both compressive and tensile strengths were observed. Mix D has a lower water/cement ratio of 0.48 as compared to Mix A and C. Also, the use of smaller sized coarse aggregate of 10 mm in place of a proportion of the 20 mm aggregate will tend to produce concrete with lower voids content, both for plain and rubberised concrete. The compressive strength was reduced by up to 60% for DP samples and 28% for DC samples. The splitting tensile strength was reduced similarly by up to 48% for DP samples and 40% for DC samples.

The reduction in compressive and splitting tensile strength can be expected since the relatively compressible rubber aggregate particles tend to produce weak inclusions in the concrete. However, coating the rubber aggregate with cement paste does produce a smaller strength reduction. This is probably due to an enhanced adhesion between the rubber chips and the cement paste. In addition, the cement paste increases the weight of the chips, reducing their tendency to float during mixing and producing a more uniform mix.

Mix D was then chosen for further strength testing due to its better performance as rubberised concrete in the preceding strengths tests compared to the other mixes. The flexural strength test results for Mix D,

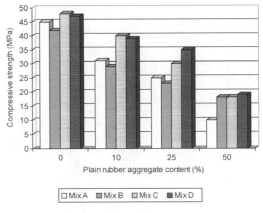

(a)

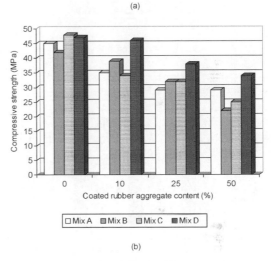

(b)

Figure 4. Compressive strength of rubberised concrete: (a) plain rubber aggregate and (h) coated rubber aggregate.

however, show than an increase in strength for rubber aggregate contents up to 20% was observed, as shown in Figure 6. This indicates a possible improvement in flexural strength if a smaller of rubber aggregate content is used.

It was also observed that rubberised concrete did not exhibit typical compression failure behaviour (Figure 7). The presence of rubber aggregate tends to hold the sample fragments together at failure. Likewise, the rubberised concrete samples did not split into two halves under split tension loading or break into two pieces under flexural loading as for conventional concrete.

3.4 Summary

It is evident from the test results that adding rubber aggregate into Cement CEM 1 concrete produced

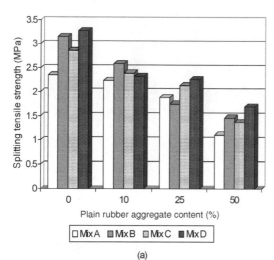

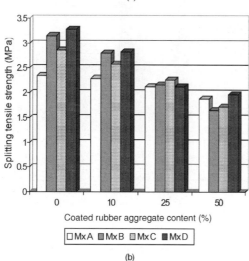

Figure 5. Splitting tensile strength of rubberised concrete: (a) plain rubber aggregate and (b) coated rubber aggregate.

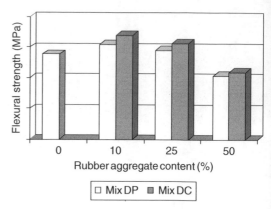

Figure 6. Flexural strength of rubberised concrete of Mix D.

Figure 7. Cube samples of rubberised concrete under compression loading.

a marked reduction in concrete strength and lower workability. This findings has greatly inhibits the development of rubberised concrete for general structural applications. It was, therefore to accept the inherent low workability and strength properties of rubberised concrete and investigate more viable low strength applications such as concrete blocks. This was done in the second part of this study.

4 CONCRETE BLOCKS

The second part of the study was to investigate the potential of low strength rubberised concrete products for use in dwelling construction. Concrete blocks

seemed worthy of investigation since the strength requirements are low and the lack of workability of rubberised concrete is less of a problem due to the mechanised method of production. Incorporating recycled rubber also gives the potential for enhanced properties such as thermal and acoustic insulation.

The standard concrete blocks in masonry construction are a high-density, general purpose, load-bearing block widely used as traditional construction techniques for dwelling construction. The blocks are available in solid, hollow and cellular format with compressive strengths ranging from 2.8 to 20 N/mm^2. The most commonly available strengths are 3.5 N/mm^2 and 7.3 N/mm^2. With a compressive strength of 7.3 N/mm^2, these aggregate concrete blocks can be used in all levels of dwellings of up to three storeys, giving no concentrated load design problems.

The standard blocks available commercially are manufactured in a different way to normal in-situ concrete and use a different mix design. The concrete block mix is discharged into the moulds and compacted using pressure and vibration to ensure that uniform concrete is produced. The compacted blocks are pressed out of the mould onto a moving conveyor belt and are then loaded onto a rack for curing.

For the concrete block laboratory production, a concrete block mix was developed, which is similar to that used in industry. The mix proportions (Mix E) are shown in Table 2. The mix differs from those used

Table 2. Mix proportions of control blocks (Mix E).

Materials	Mix Proportions (kg/m^3)
Cement CEM 1	150
Water	130
Fine aggregate	900
Coarse aggregate (6 mm)	350
Coarse aggregate (10 mm)	600

previously in this study that it is relatively lean with a low cement content and high water/cement ratio of 0.87. This type of mix is appropriate since the compressive strength requirements for concrete block are quite low and a lean mix also produces a lighter block than in-situ concrete. The aggregate size range is also smaller with a much higher proportion of sand used in the mix. The preparation and testing of the blocks replicated the normal commercial manufacturing process as closely as possible.

For rubberised concrete blocks mixtures, the 10 mm rubber aggregate either plain (Group P) or coated with cement paste (Group C) were used as a replacement of an equal part of the 10 mm coarse aggregate at 10, 25, 50 and 100% by volume, which form the four batches made from each group. All mix parameters were kept constant i.e. the cement content, water/cement ratio and the aggregate content. The sample coding used for this mix is similar to that used for rubberised concrete, as described previously in Section 2.1.

Smaller blocks with dimensions of $290 \times 215 \times 140$ mm were cast and tested. The smaller size blocks have recently been introduced to meet the requirement of the manual handling regulations. The compressive strength was determined after 28 days using standard method and procedures.

4.1 Test results

A rubberised concrete block containing 50% rubber aggregate is shown in Figure 8. Aesthetically, the surface texture of the block has not markedly changed and the finish and appearance is similar to the standard concrete blocks and it is only on very close inspection that rubber can be seen in the block. Other than this, the rubberised concrete blocks can be finished to the same standard as control concrete blocks.

Figure 9 shows the average compressive strength for the control and rubberised concrete blocks. It can be seen that both rubberised concrete blocks containing 10% of plain rubber aggregate and coated rubber aggregate produced higher strength than the control block. The strength of these rubberised concrete blocks was expected to be lower than the control block and the reason for the variation is unclear.

Figure 8. Rubberised concrete blocks with 50% rubber aggregate.

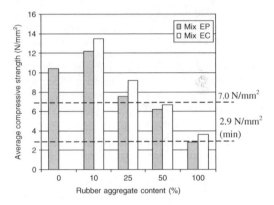

Figure 9. Average compressive strength for control and rubberised concrete blocks (6 blocks were tested for each sample).

Nevertheless, the compressive strength exceeds the required value of 7.3 N/mm^2 as specified in BS EN 771-3:2003 and BS 8103-2:2005 except for blocks containing 50 and 100% of rubber aggregate. However, the latter blocks exceed the minimum required strength value of 2.9 N/mm^2 for load bearing walls for low rise housing. The majority of housing in the UK is not individually designed for structural performance.

Rubberised concrete blocks containing 50 and 100% rubber aggregate could also potentially be used in cellular and hollow blocks construction, since the strength requirement is much lower, which is 3.5 N/mm^2. Cellular and hollow blocks are particularly beneficial for producing walls for which dense solid concrete blocks might be too heavy to lay rapidly, for producing reinforced masonry and to incorporate services within the thickness of the wall.

Rubberised concrete blocks can also be incorporated into floors, for example as infill blocks to beam and block flooring, which effectively results in the construction team handling one masonry materials for the entire structure. It also offers the advantage of an immediate working platform after erection.

4.2 *Summary*

Laboratory trials of rubberised concrete blocks were generally successful in producing a block which meets the required strength criteria. The production method replicated that used in industry and was found to overcome many of the difficulties of rubberised concrete preparation such as the lack of workability of the mix since the requirement for maintaining a workable mix is not so essential. This reduces the difficulties of controlling the water/cement ratio and workability, as long as the mix can be fed into the mould without any difficulties.

The mix design and production process can be further optimised to allow the use of larger quantities of rubber chips. However, the cost of incorporating rubber aggregate in concrete blocks should not be onerous provided that sufficient added value can be obtained. This can be achieved through producing blocks with enhanced properties such as reduced self-weight and improved thermal insulation.

5 COST ANALYSIS AND MARKET POTENTIAL

As previously mentioned, the number of used tyres which are recycled in civil engineering applications is very low and there is a pressing need markets for products incorporating recycled tyres. However, any rubberised concrete products developed for the market need to be feasible in terms of cost, including material costs and production processes. This section considers the market potentials of rubberised concrete products.

An important principle in terms of promoting recycling is that incorporating recycled materials in new products and processes should be cost-neutral, particularly for industry where the UK Government is promoting an approach of producer responsibility on a voluntary basis in order to improve used tyre recover rates. The pricing strategy for used tyre products such as granulated and crumb rubber should reflect this basis approach, otherwise the potential for the development of rubberised concrete products will remain low. A viable economic model for tyre recycling would produce benefits for both suppliers and users of products. The most important factors which affect the economics of tyre recycling in rubberised concrete are outline below.

5.1 *Tyre reprocessing economics*

The tyre recycling industry in the UK is dependent on a relatively small number of reprocesses to deliver growth in the market and sustainable reprocessing capacity. The products they produce vary widely in terms of quality and cost. For example, rubber aggregates produced are highly variable in terms of aggregate size, composition and price, with each supplier producing grades to meet a particular niche market. This leads to customer dependency on a single supplier for a specific material and tends to restrict large scale market development.

Tyre recycling processes involve the reduction of used tyres into smaller pieces such as chip and crumbs sizes for reuse or further processing. For most current uses of recycled tyres, the production processes attempt to add value to the basic material. This can be achieved by, for example, reducing the size, or by separating out the various components such as rubber, steel and fibre to produce a purer material. In addition, more values can be added by treating the crumb rubber in some way to improve its characteristics. However, as the amount of processing increases, the production costs and hence the price of the material also increase. This strategy is beneficial for producers and customers alike where the added value improves the profit margin for the producer and the cost of the new material is less than the material it replaces. This is the case for crumb rubber used in new tyre components as a replacement for virgin materials and also for crumb and chipped rubber used a surfacing for sports grounds and equestrian areas. In contrast, the current cost of recycled rubber presents a difficulty for use in rubberised concrete as it is much more expansive than the mineral aggregates it replaces. However, the costs tend to reflect the value of competitor products and the size of the market. Therefore, the economics of using recycling rubber in concrete can be expected to change, including the production costs, as the market potential of new products develops.

5.2 *Statutory measures*

Current UK Government policy is to reduce demand for virgin materials and encourage the use of recycled materials by promoting a market solution through a mixture of statutory regulation and economic measures. The Aggregate Levy, which was introduced in April 2002 with the aim to reduce demand for virgin aggregates, encourage the use of recycled materials and address the environmental costs associated with quarrying. The tax applies to sand, gravel and crushed rock and is charged at £1.60 per tonne. These materials are all used in standard concrete mixes and should be included in the material replacement costs for virgin materials. However, the current costs of these virgin materials will include the Aggregate Levy.

5.3 *Cost comparison for rubberised concrete*

There are several approaches, which can be taken to analyse the costs to rubberised concrete, bearing in

528

Table 3. Replacement value of rubber aggregate by weight.

a) Replacement value of rubber aggregate by weigh

	Material cost range (£/tonne)	
	Minimum	Maximum
10–20 mm aggregate	15.00	30.00
Process change cost	2.25	4.50
Acceptable price for rubber aggregate	12.75	25.50
Actual price of rubber aggregate	80.00	160.00

b) Replacement value of rubber aggregate by volume

	Material cost range (£/m³)	
	Minimum	Maximum
10–20 mm aggregate	40.35	80.70
Process change cost	6.05	12.11
Acceptable price for rubber aggregate	34.29	68.60
Actual price of rubber aggregate	91.20	182.40

Note: The Aggregate Levy will increase to £1.95/tonne effective 1 April 2008.

mind the instability of the current market situation following the implementation of the landfill ban.

5.3.1 *Replacement value of virgin materials*

The first approach adopted to analyse the costs of rubberised concrete is to consider the replacement value of virgin materials used in current products. This calculates the acceptable price for rubber aggregate based upon the current price of virgin materials less an allowance for the cost of process change. In this approach, the principle is that the use of rubber aggregate should be cost-neutral. The acceptable price for rubber aggregate can then be compared with the actual price. The process change costs are dependent on the particular application and are therefore difficult to estimate at present. However, in the case of the production of precast concrete units, additional costs are likely due to the increased difficulty in preparing concrete mixes, such as segregation during mixing and surface finishing. A process change allowance of 15% has been assumed in this analysis, based on previous studies (Owen, 1998).

The cost data has been calculated by weight of materials and also by volume, and is given in Table 3. Since rubber aggregate would be used to replace a specific volume of mineral aggregate, cost analysis by volume

is more appropriate. It should be noted that the cost analysis by volume is tentative, as the density of the materials will vary depending on the specific application. It can be seen that the use of rubber aggregate in concrete mixes cannot be sustained on the basis of the replacement value of virgin materials, although the analysis is much more favourable when based on volume. The Aggregate Levy makes very little positive difference to the economics of using rubber aggregate as a replacement material. The cost of aggregates is highly dependent on geography and haulage costs and while the British Aggregates Association (BAA) estimates that the imposition of the tax increased the cost of mineral aggregates by between 12–50%, cost of mineral aggregates remain much lower than current rubber aggregate prices.

The cost of rubber aggregates also varies widely depending on the source of the rubber and the amount of processing during production. The supplier used in the present investigation produces rubber chip from truck tyres which undergoes a high level of processing to remove steel and fibre components. The cost of these chips is therefore at the higher end of the range given in Table 3. However, the processing requirements for rubber aggregate used in concrete are likely to be less stringent; raising the possibility that production costs could be reduced is there is sufficient demand for the material.

5.3.2 *Cost of incorporating rubber aggregate*

The second approach is to consider the value added to concrete products as a results of using rubber aggregate. In general, the value added in production and processing will determine the viability of any type of recycling. This is the case for basic rubber crumb and chips products as well as any products incorporating these materials. The first stage is to determine the additional production costs for potential rubberised concrete products, which can be then be set against the benefits of using these products. The products considered are concrete blocks and in-situ concrete used in applications such as floor slabs.

The additional material costs are given in Table 4 for various potential rubberised concrete products. The table shows the cost analysis for the various mixes (Mix A to D) used in the present study as well as for a concrete blocks mix of Mix E developed in laboratory. The material substitution costs for rubber aggregate per cubic metre of concrete are given for various mixes as are the additional costs for the smaller blocks. It should be noted that the cost analysis is tentative at this stage and further detailed analysis of material costs and process change costs should be carried out in conjunction with block manufacturers. It can be seen that the additional costs per block vary between 4% and 34% depending on the concrete mix design and cost of the rubber aggregates.

Table 4. Additional material costs for rubberised concrete products.

		Concrete mix				
		Mix A	Mix B	Mix C	Mix D	Mix E
Volume of 20 mm aggregate per m³ of concrete (m³)		0.471	0.456	0.314	0.304	0.230†
Volume of rubber aggregate for 50% rubber content (m³)		0.236	0.228	0.157	0.152	0.115
Mass of rubber aggregate per m³ of concrete (kg)		269.04	259.9	179.0	173.3	131.1
Cost of rubber aggregate per m³ of concrete (£)	Min.	21.53	20.79	14.32	13.86	10.49
	Max.	43.05	41.59	28.64	27.73	20.98
Replacement cost of rubber aggregate per m3 of concrete (£)*	Min.	5.70	5.44	3.77	3.63	2.76
	Max.	27.22	26.24	18.09	17.50	13.25
290 × 215 × 140 mm laboratory concrete blocks						
Additional retail cost per block (£)	Min.	0.05	0.05	0.03	0.03	0.02
	Max.	0.24	0.23	0.16	0.15	0.12
Increase in retail cost per block (%)	Min.	6.7	6.7	4.0	4.0	2.7
	Max.	32.0	30.1	21.3	20.0	16.0

Note: Price is based on October 2005

* Based on rubber aggregate costs of £80 – £160/tonne less mineral aggregate costs of £25/tonne

† Based on volume of 10 mm aggregate per m³ of concrete (m³).

For the commercially available blocks, the estimated increase in retail costs range between about 4% and 17%, depending on the cost of the rubber aggregates. Provided that the cost of rubber aggregate can be kept to the lower end of the range, it can be seen that the cost increase should not be onerous for manufacturers. As mentioned previously, the less stringent processing requirements for rubber aggregate used in concrete are likely to further reduce the cost of rubber aggregate in this application, giving improved prospects for rubberised concrete blocks production.

5.3.3 Summary

It can be concluded that the economics of using recycling rubber in concrete would have to change considerably for it to be viable as a cost-neutral replacement for virgin materials. This is possible as markets for new products develop due to the imposition of the landfill ban in 2006. It follows that the viability of rubberised concrete depends on the ability to produce products with enhanced properties and characteristics for which there is a market demand.

6 CONCLUSIONS

The results of the present study show clearly that the use of rubber aggregate in Cement CEM 1 mixes produces a marked reduction in compressive. This inhibits its use as in-situ concrete for general structural applications. However, there is potential for producing materials and products with enhanced properties, such as improved flexural strength and reduced weight.

A rubberised concrete product with greater market potential is concrete blocks. The present study was successful in producing a rubberised concrete block which meets the required strength criteria. The standard production method used was found to overcome many of the difficulties of rubberised concrete preparation such as the lack of workability of the mix. The mix design and production process can be further optimised to allow the use of larger quantities of rubber chips. The cost of incorporating rubber aggregate in concrete blocks should not be onerous provided that sufficient added value can be obtained. This can be achieved through producing blocks with enhanced properties such as reduced weight and improved acoustics and thermal insulation. For example, the new UK building regulations, which have more stringent requirement for thermal insulation, should enhance the viability of relevant rubberised concrete products due to the low thermal conductivity of rubber.

The essential first step in developing a new material such as rubberised concrete is to investigate its technical viability in suitable applications, and a number of studies have already been undertaken. However, attention must be given to the economic viability and market potential of rubberised concrete products. Currently the economics of using recycled rubber in concrete are highly unfavourable to the development of new rubberised concrete products with the possible exception of some small niche markets. Current rubber crumb and chip production is geared to high value applications such as spots and playground surfacing, using mainly highly processed rubber from truck tyres. A low cost rubber chip for use in rubberised concrete applications has not yet been developed and tested. This situation may now improve following the imposition of the landfill ban in the UK in 2006 as the cost of alternatives to recycling increases.

ACKNOWLEDGEMENTS

The authors would like to acknowledge the Onyx Environmental Trust for funding this study. The authors also thank Charles Lawrence Recycling for supplying the rubber chips used in this study.

REFERENCES

British Standard Institution 2000. BS EN 12350-2: Testing fresh concrete: Slump test, UK: BSi.

British Standard Institution 2002. BS EN 12390-3: Testing hardened concrete: Compressive strength of test specimens, UK: BSi.

British Standard Institution 2000. BS EN 12390-6: Testing hardened concrete: Splitting tensile strength of test specimens, UK: BSi.

British Standard Institution 2000. BS EN 12390-5: Testing hardened concrete: Flexural strength of test specimens, UK: BSi.

British Standard Institution 2000. BS EN 206-1: Concrete. Specification, performance, production and conformity, UK: BSi.

British Standard Institution 2003. BS EN 771: Part 3: Specification for masonry units: Aggregate concrete masonry units (dense and light-weight aggregates), UK: BSi.

British Standard Institution 2005. BS 8103: Part 2: Structural design of low rise buildings: Code of practise for masonry walls for housing, UK: BSi.

Design of Normal Concrete Mixes 1975. Department of Environment, Building Research Establishment (BRE), Transport and Road Research Lab.

Eldin, N.N. & Senouci, A.B. 1993. Rubber-tyre particles as concrete aggregate, *Journal of Materials in Civil Engineering*, 5(2): 478–496.

Fedroff, D., Ahmad, S. & Savas, B.Z. 1996. Mechanical properties of concrete with ground waste tyre rubber, *Transportation Research Record*, No. 1532: 66-72.

Khatib, Z.K. & Bayomy, F.M. 1999. Rubberised Portland cement concrete, *Journal of Materials in Civil Engineering*, 11(3): 206–213.

Labbani, F., Benazzouk, A., Douzane, O. & Queneudec, M. 2004. Effect of temperature on the physico-mechanical properties of cement-rubber composites. In M.C. Limbachiya and J.J. Roberts (eds), *Sustainable Waste Management & Recycling: Used/Post-Consumer Tyres*; Proc. intern. conf., Kingston University, 14–15 September 2004. UK: Thomas Telford.

Li, Z., Li, F. & Li, J.S.L. 1998. Properties of concrete incorporating rubber tyre particles, *Magazine of Concrete Research*, 50 (4): 297–304. UK: Thomas Telford.

Owen, K.C. 1998. Scrap tyres: A pricing strategy for a recycling industry, *Corporate Environmental Strategy*. 15 (2): 42–50.

Pierce, C.E. and Williams R.J. 2004. Scrap tyre rubber modified concrete: Past, present and future. In M.C. Limbachiya and J. J. Roberts (eds), *Sustainable Waste Management & Recycling: Used/Post-Consumer Tyres*; Proc. intern. conf., Kingston University, 14-15 September 2004. UK: Thomas Telford.

Siddique, R. & Naik, T.R. 2004. Properties of concrete containing scrap-tyre rubber – An overview, *Waste Management*, Vol. 24: 563–569.

Toutanji, H. A. 1996. The use of rubber tyre particles in concrete to replace mineral aggregate, *Cement & Concrete Composite*, Vol. 18: 135–139.

Wallingford, H.R. 2005. Sustainable re-use of tyres in port, coastal and river engineering, Report SR669: Guidance for planning, implementation and maintenance, Release 1.0.

Excellence in Concrete Construction through Innovation – Limbachiya & Kew (eds)
© 2009 Taylor & Francis Group, London, ISBN 978-0-415-47592-1

Investigation into the potential of rubberised concrete products

H.Y. Kew, K. Etebar & M.C. Limbachiya
Kingston University, London, UK

M.J. Kenny
University of Strathclyde, Glasgow, UK

ABSTRACT: Currently, the implementation of research in rubberised concrete has been poor with few examples of successful use in industry or product development. However, the recycling of waste tyres in civil engineering applications is currently very low, so that the demand for viable new products remains pressing. Rubberised concrete exhibits lower workability and substantially reduced compressive strength in which these characteristics have greatly inhibited the development of viable rubberised concrete products. The approach adopted in this study is to accept the inherent low strength and workability properties of rubberised concrete and develop viable low strength applications such as concrete blocks with have beneficial characteristics and good economic viability. This study seeks to take advantage of the low thermal conductivity of rubber to develop thermally efficient rubberised concrete products which can be used in dwelling construction. Improving the energy efficiency of buildings is an important part of the UK Government's drive to conserve energy and reduce national CO_2 emissions.

1   INTRODUCTION

The need to develop infrastructure which is more sustainable is an immediate and growing challenge for the civil engineering profession. The requirement for sustainable development, coupled with the need to recycle waste and improve energy efficiency, has led to a requirement to develop new materials for use in construction which minimise the use of energy, preserve virgin materials and which are economically viable.

In general, tyre recycling rates have been static over the past few years, which is jeopardising the UK's ability to recycle the anticipated surplus of waste tyres as the results of the imposition of the landfill ban in 2006. There are several reasons for this lack of implementation. Firstly, rubberised concrete exhibits lower workability, reduced abrasion resistance and substantially a marked reduction in strength (Kew et al., 2004). These findings have greatly inhibited the development of viable rubberised concrete products. However, Moroney (2003) has been suggested that granulated rubber could be used to improve the properties of concrete in three particular aspects – freeze/thaw resistance, impact resistance and thermal efficiency.

Much effort has been made by previous researchers to improve the strength of rubberised concrete by using additives and chemically surface treatments of the rubber aggregate (Rostami et al., 1993, Tantala et al., 1996, Li et al., 1998, Segre & Jokes, 2000, Li et al., 2004, Moroney, 2003). However, the use of additives and surface treatments will further increase the production costs and reduce the viability of rubberised concrete. Furthermore, the previous studies have used highly processed and therefore expansive crumb and chipped rubber grades, such as truck tyres chips at £180 per tonne, which cost considerably more than the natural aggregate they are replacing (£15–£25 per tonne). Currently, recycled truck tyre rubber has established markets in the UK in areas such as sports surfacing and general flooring and there is no unused production. However, sustainable markets for car tyres have yet to be established and a large proportion of these have no other options of disposing after the imposition of the landfill ban.

The approach which adopted in this study is to accept the inherent low strength and workability properties of rubberised concrete with the aim of developing viable low strength applications such as concrete blocks which have beneficial characteristics and good economic viability. The study seeks to take advantage of the low thermal conductivity of rubber to develop thermally efficient rubberised concrete products which can be used in dwelling construction. The use of rubber aggregate, which has a much lower thermal conductivity than natural aggregate, should have a beneficial effect on the thermal conductivity of concrete product, which could help meet the more stringent UK Building Regulations, in line with the UK Government's drive to conserve energy and address global warming. The development of viable products with enhanced properties such as improved thermal insulation will give industry greater confidence in the use of rubberised concrete, so that it becomes much more than just a disposal option for a waste material.

2 EXPERIMENTATION

The main focus of this study is to determine the thermal insulating properties of two rubberised concrete products, namely rubberised concrete and rubberised concrete blocks at different rubber aggregate content using the thermal probe method.

2.1 Mix design and test materials

Table 1 shows the quantities of the constituent of Mix A and B. The test materials used for rubberised concrete mix (Mix A) were cement, sand and coarse aggregate of 10 and 20 mm maximum sizes and 20 mm rubber aggregate ($G_s = 1.14$). The mix was designed with a targeted compressive strength of 40 MPa and the design was carried out according to BS EN 206-1:2000 and Design of Normal Concrete Mixes (1975).

Whilst, for rubberised concrete blocks (Mix B) the test material used were cement, sand and coarse aggregate of 6 and 10 mm maximum sizes and 10 mm rubber aggregate, used from the same source as above. The mix developed is similar to that used in the industry and it differs from the above that it is relatively lean with a low cement content and high water/cement ratio of 0.87. This type of mix is appropriate since the compressive strength requirements for concrete block are quite low.

Table 2 summaries the rubber contents of Mix A and Mix B, respectively. Rubber aggregate of 20 mm and 10 mm were used as a replacement for an equal part of the 20 mm coarse aggregate of Mix A and 10 mm coarse aggregate of Mix B, respectively at 10, 25, 50 and 100% by volume.

2.2 Sample preparation

Three standard 100 mm cubes from batches of Mix A and B were prepared for determining the thermal conductivity of rubberised concrete and rubberised concrete blocks at different percentage of rubber contents, respectively. The dimension of the test samples were in accordance with BS 874:1986.

The mixing, casting and sample preparation of Mix A were conducted in accordance to BS 1881-125:1986. Whilst, the casting and preparation of Mix B

Table 1. Mix proportions of control concrete (Mix A) and control concrete block (Mix B).

Materials	Mix Proportions (kg/m^3)	
	Mix A (0.48)*	Mix B (0.87)*
Cement CEM1	438	150
Water	210	130
Fine aggregate	526	900
Coarse aggregate (6 mm)	–	350
Coarse aggregate (10 mm)	409	600
Coarse aggregate (20 mm)	818	–

*water/cement ratio.

Table 2. Summary of rubber contents for rubberised concrete and rubberised concrete blocks mixes.

Mixes	Sample coding	Rubber Content (%)
A (rubberised concrete)	A	0
	A10	10
	A25	25
	A50	50
	A100	100
B (rubberised concrete blocks)	B	0
	B10	10
	B25	25
	B50	50
	B100	100

replicated the normal commercial manufacturing process as closely as possible. It should be noted that in both preparation, some modifications were made in order to accommodate the addition of rubber aggregate into the mix.

After a 28-day of curing, three holes corresponding to the length of the probe were drilled into each of the cube samples. The size of the holes was such that the probe fits vertically and tightly into the hole ensuring that the probe was in good thermal contact with the sample. The samples were oven-dried at 70°C until reaching a constant weight. It is to be noted that the temperature was maintained at 70°C because too high temperatures can induce cracks due to the low tensile strength of the concrete and the elastic of rubber aggregate may also expand, hence affecting the characteristic of the concrete.

2.3 Test equipment and test procedures

2.3.1 Thermal probe method: theory
The thermal probe method (line source of heat), transient method is based on the fact that the ultimate temperature rise and the rate of temperature rise of an embedded heated body depend on the thermal conductivity of the surrounding medium. This method was

chosen in this study to determine the thermal conductivity of rubberised concrete due to it's relatively ease of use, rapid test time and involves minimal cost.

There is a linear relationship between the temperature of the probe and the natural algorithm of the time (ln t) for a given input of energy. The slope of this line ($Q/4\pi\lambda$) is related to the apparent thermal conductivity of the material (Gibbon & Ballim, 1998, Manohar et al., 2000). The relationship that describes the operation of the thermal probe is based on an infinitely long, thin heat source embedded in an infinite homogenous medium. The temperature rise T as a function of t and an analytical representation of this relationship is given by Carslaw & Jaeger (1959) and is expressed as:

$$T(t) = -\frac{Q}{4\pi\lambda} Ei\left(-\frac{r^2}{4Dt}\right) \tag{1}$$

where T = temperature increase (K); Q = power input per unit probe length (W/m); λ = thermal conductivity (W/mK); Ei = exponential integral; r = radius distance from probe (m); D = thermal diffusivity of the medium (m²/s); and t = time (s).

By expanding the exponential integral, Ei of Equation 1, the expression is:

$$Ei(-x) = -\gamma - \ln\left(\frac{1}{x}\right) + x - \frac{1}{4}x^2 + \frac{1}{9}x^3 + \ldots \tag{2}$$

where γ = Euler's constant (= 0.5772156); and $x = r^2/4Dt$

As $r^2/4Dt$ is small, the higher order terms can be ignored, so combining Equation 1 and 2, gives:

$$T(t) = \frac{Q}{4\pi\lambda}\left[\ln t + \ln\left(\frac{4D}{r^2}\right) - \gamma\right] \tag{3}$$

For a given interval $t_2 - t_1$, the rise in temperature ΔT is given by:

$$\lambda = \frac{Q}{4\pi(T_2 - T_1)}\ln\left(\frac{t_2}{t_1}\right) \tag{4}$$

From Equation 4, ΔT and ln t are linearly related and a plot of temperature rise versus the logarithm of time will give a straight line, as shown in Figure 1 with a slope of $Q/4\pi\lambda$. If the value of the slope and the power, Q are known, the thermal conductivity, λ of the material can be determined.

The general shape of the $T - \ln t$ plot includes three distinct portions in which the first portion is an initial transient portion (non-linear), which is a response to the probe heating. The second portion of the curve is linear and represents a quasi-steady-state condition for heat transfer from the probe (Nicolas et al., 1993).

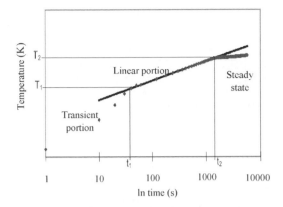

Figure 1. Thermal probe temperature against time graph (ASTM D 5334, 1995).

The slope of this linear portion of the curve represents the heat conduction through the medium under investigation and is used in Equation 4 to determine the thermal conductivity, λ. The time taken to achieve this condition depends on the sample size. There is no recommended samples size but Hanson et al. (2004) reported that in general, the sample can be cylindrical, square or rectangular. Kim et al. (2003) recommended a minimum sample size of $80 \times 80 \times 20$ mm which is smaller than the sample size used in this study. Bouguerra (1999) used standard 100 mm cubes whilst ASTM D 553 (1995) recommends cylindrical samples with a minimum diameter of 50 mm. None of the studies, however, specify the position of the probe, as long as the diameter of the hole drilled is equal to the diameter of the probe so that probe fits tightly into the hole. However, provided that the linear quasi-steady-state condition can be identified from the graph, the thermal conductivity of the sample can be obtained.

With the passage of time, the probe temperature will level off to a steady-state value because of an isothermal specimen boundary.

2.3.2 Test equipment

The thermal probes used in this study were constructed as per the instruction in the BS 874:1986 test method. The probe as shown in Figure 2 consists of an electrical heating wire, representing a perfect linear heat source and incorporates a thermocouple, capable of measuring the variation of temperature of this line of source.

Figure 3 illustrates the test set-up. A constant current source was used to produce a five-volt of constant current supply to power the heater wire of the thermal probe. A data acquisition system was used to collect and produce a digital readout of temperature variation, which is connected to a computer system. Lab View software was used to monitor and record the temperature-time variation.

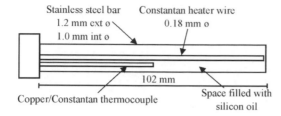

Figure 3. Schematic diagram of the thermal probe used.

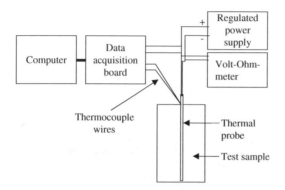

Figure 4. Schematic diagram of test equipment.

2.3.3 Test procedures

After the test samples achieved constant weight through oven-dried, they were allowed to come into equilibrium with the temperature of the laboratory. The room was maintained at constant room temperature of 20°C with 65% relative humidity throughout the duration of the test. Before inserting the probe into the hole, the hole was filled with a tiny amount of thermal grease such as epoxy resin in order to provide better thermal contact between the sample and the probe. The inserted probe and the test sample were then allowed to attain equilibrium with the room temperature. This is to make certain that isothermal conditions were attained before the heater was powered due to the friction associated with inserting the probe, which might cause a measurable temperature increase. The duration of the probe to attain this equilibrium was about 10 to 15 minutes.

Then, the data acquisition system and heater power were switched on together. The temperature change was measured at every 1s and collected using the data acquisition system and processed using Lab View software. The duration of the experiment was approximately 1500s, which will be sufficient to produce reasonable curve showing the three distinct portions. The power to the probe heater was remained constant.

At the end of the test, the data acquisition and heater power were switched off together. The heated probe will then be allowed to cool to ambient temperature before using it again for the next test. A minimum of 10 measurements from each sample were carried out.

Table 3. Verification results of thermal probe using reference materials.

Material	Average $\lambda_{measured}$ (W/mK)	$\lambda_{published}$ (W/mK) (Kubicar and Bohac, 2000)
Perspex 7740	0.25	0.26 ± 0.03
Plywood	0.19	0.18 ± 0.06
Polyvinyl Chloride (PVC)	0.19	0.16 ± 0.03
Leighton Buzzard sand	0.30	0.28 ± 0.03

2.4 Verification of thermal probe method

In order to verify and provide reasonable acceptance in using thermal probe method, a number of reference materials with reported thermal conductivity were tested. Among the materials used were Perspex 7740, plywood, Polyvinyl Chloride (PVC) and Leighton Buzzard sand.

Three 100 mm cubes from each material were prepared and three holes corresponding to the length of the probe were drilled into each of the cube samples (except for Leighton Buzzard sand). At least 10 measurements from each material were carried out using the same test procedures as described above.

The results of the verification tests are presented in Table 3. The measurements of each material produced reasonable curves with easily recognisable linear portion on the semi-logarithm graph. It should be noted that the method of interpreting the results and the calculation to determine the thermal conductivity values were the same as for rubberised concrete samples.

As can be seen from the Table 3, the thermal conductivities measured from the reference materials were within the range of values given with the reported values. The overall agreement between the measured and the published thermal conductivity values indicated that thermal probe method could be used to determine the thermal conductivity of concrete with acceptable accuracy.

3 TEST RESULTS AND DISCUSSIONS

In general, the tests conducted on both rubberised concrete of Mix A and rubberised concrete blocks of Mix B samples resulted in reasonable curves with recognisable linear portions on the semi-logarithm graph. As the thermal conductivity results reported herein are for oven-dried samples, the measured thermal conductivity values were corrected by multiplying each of the values by a moisture correction factor of 1.22 (for unprotected type of exposure – 3% moisture content) as listed by ACI (2005). Applying the correction factor will give a more realistic value for

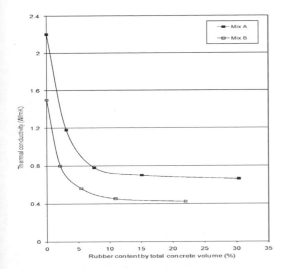

Figure 5. Apparent thermal conductivity of Mix A and Mix B.

in-service conditions. With the corrected thermal conductivity values, direct comparison can be made with the published data and also data from other studies.

The apparent thermal conductivity test results of both Mix A and B at different rubber contents are shown in Figure 5. It should be noted that because the concrete mixes are different and different sizes of rubber aggregate were used, comparison can only be reasonably made by determining the rubber content by total concrete volume.

As can be seen from Figure 5, the thermal conductivity of concrete is higher than that of concrete blocks. This is expected because of the higher density of concrete as compared to concrete blocks. The concrete sample, which has a density of 2500 kg/m^3, produced a thermal conductivity value of 2.20 W/mK whilst the concrete blocks sample, which has a lower density of 2170 kg/m^3 produced a lower thermal conductivity value of 1.50 W/mK. The thermal conductivity of concrete is directly related to the dry density, in which the thermal conductivity decreases as the density decreases because more air or voids is entrapped in the material. The denser the normal concrete, which has a lower estimated air content of 2% produced higher thermal conductivity as compared to the lighter concrete blocks, which has an estimated 12% of air content.

It was also observed that for each sample, the percentage of rubber content by total concrete volume of rubberised concrete blocks is lower than that of rubberised concrete and naturally, the thermal conductivity of rubberised concrete blocks should be higher than that of rubberised concrete. This is due to the higher air content of rubberised concrete blocks compared with rubberised concrete. The concrete block

mix, which is relatively lean with a low cement content has less filler in the form of cement paste to fill up the pores within the concrete mixture, hence the higher air content.

Nevertheless, both mixes exhibited lower thermal conductivity value when rubber aggregate was incorporated. This was expected because of the presence of rubber aggregate of low conductance produced lower heat conduction than natural aggregate. In both mixes, it was observed that the apparent thermal conductivity decreased with the increasing percentage of rubber aggregate. The decrease of apparent thermal conductivity of both mixes was more pronounced at low rubber contents of up to 25%, while there was little further decrease in thermal conductivity at rubber contents greater than about 50%.

3.1 Comparison with other research findings

Currently, no studies have been reported where the determination of thermal conductivity of rubberised concrete was made using a similar method of measurement used in this study. However, it is useful to compare the results of the present study with other studies where the thermal conductivity of rubberised concrete was measured. The only published studies are Moroney (2003) in which steady-state guarded hot box was used and Laidoudi et al. (2004) in which the transient plane source method was used.

It should be noted that both studies did not indicate any information about measured thermal conductivity being corrected to a standard 3% moisture content by volume although both studies over-dried their test samples. Therefore, only the uncorrected results of the present study were compared with the results of these previous studies, assuming that these reported results were not corrected with a correction factor. Figure 6 shows the comparison of the results of the present study with findings obtained from these studies. It should be noted that only the rubberised concrete results can be compared as no tests were carried out by these authors on rubberised concrete blocks.

The results of the present study show a reasonably similar trend to previous studies. It was also observed that all studies show a more pronounced decrease in measure thermal conductivity at small rubber contents of up to about 5%, while there was little further decrease in thermal conductivity at rubber contents greater than about 15%. However, the trends of the present study and Laidoudi et al. (2004) are more consistent than Moroney (2003), which shows a very high variability in the measured thermal conductivity. Nevertheless, in general, the findings of previous studies are consistent with the present study up to rubber contents of about 20–30%. However, there is a lack of consistent experimental data reported by previous studies at higher rubber contents.

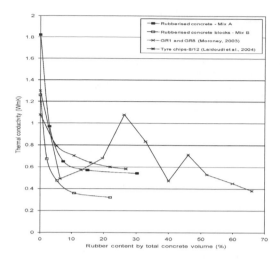

Figure 6. Comparison of apparent thermal conductivity of Mix A and Mix B with other research findings.

4 VERIFICATION ANALYSIS TO VALIDATE TEST RESULTS

Three analytical models have been selected for this study to help validate the thermal conductivity results of rubberised concrete (Mix A) and rubberised concrete blocks (Mix B) at different percentages of rubber content. These models were developed to predict the thermal conductivity of normal concrete and are therefore relevant to the present study: they are Campbell's model, Brailsford and Major's model and Valore's model. These models, which have been improved and developed, are based on the earliest model by Maxwell (1892), in which the author proposed that the thermal conductivity of concrete can be estimated by using a two-phase system composed of coarse aggregate in the dispersed phase enveloped in a solid continuous phase of mortar.

Campbell-Allen & Thorne (1963) considered the structure of concrete as being a set of cubes of aggregate of uniform size arranged systematically in the mixture of cement paste. The authors argued that the concrete was regarded as a suspension of coarse aggregate in a continuous matrix of mortar. It is therefore possible to calculate the conductivity of the concrete as a function of the conductivity of the aggregate λ_a and the conductivity of the mortar λ_m. The expression for the conductivity of concrete on this basis is:

$$\lambda_c = \lambda_m \left(2M - M^2\right) + \frac{\lambda_m \lambda_a (1 - M)^2}{\lambda_a M + \lambda_m (1 - M)} \quad (5)$$

where M $= 1 - (1 - P)^{1/3}$; P $=$ volume of mortar per unit volume of concrete; $\lambda_m =$ thermal conductivity of

mortar (W/mK); $\lambda_a =$ thermal conductivity of aggregate (W/mK); and $\lambda_c =$ thermal conductivity of concrete (W/mK).

Brailsford & Major (1964) extended Maxwell's original equation by considering a random two-phase system as having regions of both single phase in the correct proportion, embedded in a random mixture of the same two phases having a conductivity equal to the average value of the conductivity of the two-phase system, which is being calculated i.e. it represents the average conductivity of a random distribution of spheres of conductivity λ_1, in a continuous phase of conductivity λ_0. The authors' model of the thermal conductivity of a two-phase system, in which the continuous phase was assumed to be mortar, is expressed as:

$$\frac{\lambda_c}{\lambda_0} = \left[1 - 2f_0\left(\frac{1 - \lambda_1/\lambda_0}{2 + \lambda_1\lambda_0}\right)\right]\left[1 + f_1\left(\frac{1 - \lambda_1/\lambda_0}{2 + \lambda_1\lambda_0}\right)\right]^{-1} \quad (6)$$

where $\lambda_c =$ thermal conductivity of concrete (W/mK); $\lambda_0 =$ thermal conductivity of the continuous phase (W/mK); $\lambda_1 =$ thermal conductivity of the dispersed phase 1 (W/mK); $f_0 =$ fractional volume of the continuous phase; and $f_1 =$ fractional volume of dispersed phase 1.

Brailsford & Major (1964) then extended the simple limited two-phase model to enable evaluation of three or more phase systems and is expressed as:

$$\lambda_c = \frac{\left[\lambda_0 f_0 + \frac{\lambda_1 f_1 (3\lambda_0)}{(2\lambda_0 + \lambda_1)} + \frac{\lambda_2 f_2 (3\lambda_0)}{(2\lambda_0 + \lambda_2)}\right]}{\left[f_0 + \frac{f_1 (3\lambda_0)}{(2\lambda_0 + \lambda_1)} + \frac{f_2 (3\lambda_0)}{(2\lambda_0 + \lambda_2)}\right]} \quad (7)$$

$$f_0 + f_1 + f_2 = 1$$

where $\lambda_2 =$ thermal conductivity of the dispersed phase 2 (W/mK); and $f_2 =$ fractional volume of the dispersed phase 2.

Valore (1980) proposed that the thermal conductivity of a discrete two-phase system, such concrete can also be calculated by knowing the volume fraction and the thermal conductivity values of cement paste (mortar) and aggregates. The author described that in the case of concrete, highly conductive aggregates are the thermal bridge and they are surrounded by the lower conductive cement paste and/or fine aggregate matrix (mortar). The thermal conductivity is given by the expression:

$$\lambda_c = \lambda_m \left[V_a^{2/3} \left/ \left(V_a^{2/3} - V_a + \frac{V_a}{\frac{\lambda_a V_a^{2/3}}{\lambda_m} + 1 - V_a^{2/3}}\right)\right.\right] \quad (8)$$

where λ_c = calculate thermal conductivity of concrete (W/mK); λ_m = thermal conductivity of mortar W/mK); λ_a = thermal conductivity of aggregate (W/mK); and V_a = volume friction of aggregate.

Tinker (1984) reported that a major problem associated with the application of these models in practise is being able to obtain values for all parameters used such as the thermal conductivity of the solid continuous phase and the dispersed phase. However, Ganjian (1990) argued that most of the parameters can be easily determined with the exception of the thermal conductivity of the solid continuous phase, which is technically very difficult to obtain by measurement; hence the author suggested that the thermal conductivity of the solid continuous phase can be calculated by substituting the value of the measured thermal conductivity of concrete into the model.

This approach was adopted in this study to estimate the thermal conductivity of the continuous phase, which is the mortar (λ_m) by substituting the value of the measured thermal conductivity of control concrete of Mix A and Mix E, respectively.

4.1 Rubberised concrete (Mix A)

Equation 5, 6 and 8 were used to predict the thermal conductivity of control concrete. Whilst, the rubberised concrete, which contains rubber aggregate, was assumed to be a three-phase system, hence Equation 7 was used apart from Equation 5 and 8 to predict the thermal conductivity of rubberised concrete at different percentage of rubber content.

The results are presented and compared with experimental results in Table 4 and illustrated in Figure 7. The table shows that the experimental results show a similar trend to the models. However, for rubberised concrete samples of A25 and A50, the models over-predicted the thermal conductivity by approximately 30 to 50%. This could be due that it is difficult to predict the thermal conductivity of concrete and furthermore the presence of rubber aggregate in the concrete mix could have altered the characteristic of the mix.

However, it is interesting to note that among all the models, Brailsford and Major's model predicted values that are closest to the experimental results. This could be due to the fact that the parameters were substituted directly with correct proportion into the equation and using volumetric ratio, hence reducing possible errors.

4.2 Rubberised concrete blocks (Mix B)

Similar procedures were used to predict the thermal conductivity of concrete blocks using Equation 5, 6 and 8. However, it should be noted that unlike the conventional concrete mix, the concrete block is designed to achieve a semi-dry mix that contains relatively high air content. As such, the thermal conductivity of the

Table 4. Comparison of experimental results (Mix A) with predicted models.

Samples	Thermal conductivity values (W/mK)			
	Experiment results	Campbell-Allen and Thorne	Brailsford and Major	Valore
A	2.20	–	2.10 (−5%)	1.95 (−11%)
A10	1.18	1.17 (−0.8%)	1.20 (+2%)	1.25 (+6%)
A25	0.77	1.08 (+40%)	1.10 (+41%)	1.13 (−47%)
A50	0.75	0.98 (+30%)	0.95 (+25%)	1.10 (+45%)
A100	0.66	0.76 (+15%)	0.67 (+2%)	0.76 (+15%)

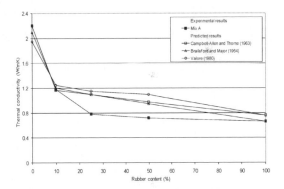

Figure 7. Comparison of experimental results (Mix A) with predicted models.

dispersed phase includes the coarse aggregate and air and the value was determined using the volumetric ratio. The thermal conductivity of rubberised concrete blocks at different percentage of rubber contents were predicted using Equation 7.

The results are presented and compared with experimental results in Table 5 and illustrated in Figure 8. The data shows that the experimental results are in very close agreement with the models. The models underestimate the thermal conductivity of sample B and B25 by approximately 3 to 25%, whilst the models over-predicted the thermal conductivity of sample B50 and B100 by approximately 5 to 15%. It is also interesting to note that among all the models, Brailsford and Major's model predicted values that are closest to the experimental results by 1 to 5%.

4.3 Summary

The verification analysis shows a similar trend with the present experimental results, in which the decrease in

539

Table 5. Comparison of experimental results (Mix B) with predicted models.

| Samples | Thermal conductivity value (W/mK) | | | |
	Experiment results	Campbell-Allen and Thorne	Brailsford and Major	Valore
B	1.50	–	1.48 (−1%)	1.30 (−13%)
B10	0.80	0.75 (−6%)	0.75 (−6%)	0.70 (−12%)
B25	0.65	0.69 (+6%)	0.65 (0%)	0.63 (−3%)
B50	0.45	0.52 (+15%)	0.50 (+5%)	0.48 (+7%)
B100	0.42	0.48 (+14)	0.40 (−5%)	0.44 (+5%)

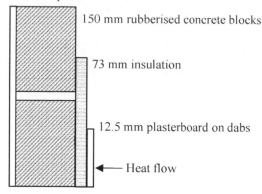

12.5 mm plasterboard on dabs

150 mm rubberised concrete blocks

73 mm insulation

12.5 mm plasterboard on dabs

Heat flow

Figure 9. Solid masonry wall using rubberised concrete blocks.

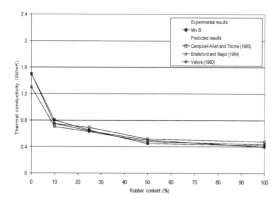

Figure 8. Comparison of experimental results (Mix B) with predicted models.

thermal conductivity was more pronounced at low rubber content. This supports the findings of the present study. These models are sufficiently to validate the experimental results and can be applied with reasonable confidence to predict the thermal conductivity of rubberised concrete products.

5 POTENTIAL APPLICATIONS OF RUBBERISED CONCRETE PRODUCT

The maximisation of energy conservation in cold weather and reduction of the environmental impacts associated with heating of buildings are important. Indeed, heating of buildings in the UK account for 27% of the total CO_2 emissions (Concrete Society, 2003) and for this reason, in April 2002 the UK Building Regulations becoming more stringent in terms of thermal transmittance requirements. This section describes examples in which the thermally efficient achieved by rubberised concrete products could be

beneficially used in dwelling construction in helping to meet the more stringent UK Building Regulations (Approved Document Part L).

There are a number of potential applications where the thermal efficiency achieved by rubberised concrete and rubberised concrete blocks could be beneficial used. In all cases, consideration must be given to the strength requirement as incorporating rubber aggregate reduces the strength. However, recent study conducted by Kew (2007) discovered that rubberised concrete blocks at higher rubber contents exceed the minimum required strength as specified in BS EN 771-1:2003.

It should be noted that it is not within the scope of this study to go deeply into the energy performance of dwellings/buildings. The focus of this study is to investigate a range of potential applications for thermally efficient rubberised concrete and rubberised concrete blocks to be used in dwelling construction.

5.1 Masonry external walling

A range of examples of external walling including full cavity, partially filled cavity, cavity clear and solid masonry wall constructed using rubberised concrete blocks containing different percentages of rubber content were considered and studied. It was found that that the use of rubberised concrete blocks in solid masonry wall produced a significant reduction in U-values as compared to other types of external walling.

Figure 9 shows an example of a solid masonry wall construction in which rubberised concrete blocks could be potentially used. The wall thickness is 250 mm consists of plasterboard on the outer leaf, internal insulation and plasterboard on the inner leaf. Table 6 shows the U-values achieved when using rubberised concrete blocks containing different percentage of rubber content.

Table 6. U-values calculation for solid masonry walls.

Components	Thermal resistance, R (m²K/W)				
	B	B10	B25	B50	B100
External surface resistance	0.06	0.06	0.06	0.06	0.06
12.5 mm plaster boards on daps	0.08	0.08	0.08	0.08	0.08
150 mm concrete blocks	0.10	–	–	–	–
150 mm rubberised concrete blocks	–	0.20	0.23	0.33	0.37
73 mm cavity batt insulant	2.92	2.92	2.92	2.92	2.92
12.5 mm plasterboard on daps	0.08	0.08	0.08	0.08	0.08
Internal surface resistance	0.12	0.12	0.12	0.12	0.12
Total thermal resistance, R	3.36	3.46	3.50	3.60	3.62
U-value ($1/\Sigma$ R) (W/m²K)	0.30	0.29	0.28	0.27	0.26

As shown in the table, the calculated U-value using concrete blocks is 0.30 W/m²K, whilst the use of rubberised concrete blocks produced 0.26–0.29 W/m²K. The results show that the use of rubberised concrete blocks containing different percentage of rubber content achieve the requisite U-values of 0.30 W/m²K for wall.

5.2 Floor construction

Two examples of floor construction, which are beam and block flooring and solid ground floors in which rubberised concrete and rubberised concrete blocks could be beneficially used were studied. It was found that the beam and block flooring shows the best potential applications. This popular form of flooring system, which is a suspended flooring using the same concrete blocks used for the walls as infill blocks in conjunction with the inverted 'T' concrete beams. Cement/sand screeds is applied to provide a level surface finish. This type of flooring has a number of advantages such as simplicity in both design and site control, in which exactly the same blocks can be used for both walls and floors, cost saving, in which long spans are readily achieved without intermediate support, reduces build time due to its dry construction process and thermal and sound requirements are easily achieved.

Figure 10 shows an example in which rubberised concrete beam and infill rubberised concrete blocks could be potentially applied. Table 7 shows the U-values achieved when using rubberised concrete and rubberised concrete blocks containing different percentage of rubber content, respectively.

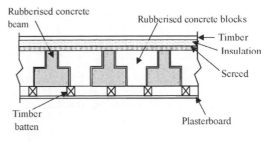

Figure 10. Beam and block flooring.

Table 7. U-value calculation for beam and block flooring.

Components	Thermal resistance, R (m²K/W) Rubber content (%)				
	0	10	25	50	100
External surface resistance	0.14	0.14	0.14	0.14	0.14
15 mm timber	0.083	0.083	0.083	0.083	0.083
71 mm cavity batt insulation	2.84	2.84	2.84	2.84	2.84
50 mm screed	0.121	0.121	0.121	0.121	0.121
515 mm concrete beam (Mix A)	0.23	0.44	0.67	0.72	0.78
275 mm infill concrete blocks (Mix B)	0.18	0.34	0.42	0.60	0.69
Airspace	0.18	0.18	0.18	0.18	0.18
12.5 mm timber batten	0.07	0.07	0.07	0.07	0.07
12.5 mm plasterboard	0.08	0.08	0.08	0.08	0.08
Internal surface resistance	0.04	0.04	0.04	0.04	0.04
Total thermal resistance, R	3.96	4.33	4.64	4.87	5.02
U-value ($1/\Sigma$ R) (W/m²K)	0.25	0.23	0.21	0.21	0.20

The results show that the use of rubberised concrete and infill rubberised concrete blocks at higher rubber contents achieve the requisite U-value of 0.25 W/m²K for floors.

6 CONCLUSIONS

The addition of rubber aggregate was found to reduce the thermal conductivity of concrete for both rubberised concrete and rubberised concrete blocks. For rubberised concrete, it was observed that the trend of reducing thermal conductivity with increasing rubber content was more marked and more consistent than reported in the previous two studies. Verification analysis was conducted to further justify the results and it was found that the experimental values were in reasonable good agreement with the published models, which are based on a two-phase system in which the dispersed phase of coarse aggregate is enveloped in a solid continuous phase of mortar. It was also found that the experimental results for rubberised concrete blocks for which there are no previous experimental studies were very close agreement with the models. The presence of low thermal conductivity rubber aggregate contributed to the reduced rate of heat transfer along the heat path as compared to natural aggregate.

Potential applications of the improved thermal efficiency achieved by rubberised concrete blocks is in solid masonry wall construction and for rubberised concrete is in beam and block flooring, as it was found to reduce the U-value significantly; hence helping to meet the new and stringent thermal requirements of the UK Building Regulations. This offers potential opportunities in which the productivity and efficiency of the construction could be enhance with the utilisation of the lower density and the large format of rubberised concrete blocks, especially in solid wall construction where a simple construction process is involved.

However, more research and probably demonstration project such as full scale constructions would be a successful way of attracting building related industry and countering the notion that rubberised concrete is an inferior product. These will help alleviate the reservations of concrete producers and assist in the creating of a market for rubberised concrete, which will allow the products to be used to their full potential so that it becomes must more than just a disposal option for a waste material. However, to achieve this, it is essential to develop applications, which have beneficial characteristic, acceptable technical viability and economically viable.

ACKNOWLEDGMENTS

The authors would like to acknowledge the Onyx Environmental Trust for funding this study. The authors are grateful to Charles Lawrence Recycling for supplying the rubber particles used in the study. The authors would also like to thank Dr. M. J. Kenny of University of Strathclyde, Glasgow, UK for his invaluable input in this study.

REFERENCES

ACI 2005. Guide to thermal properties of concrete and masonry systems. In ACI Manual of Concrete Practise, Chapter 122R-02: American Concrete Institute, Detroit.

ASTM D5334 1995. Standard test method for determination of thermal conductivity of soil and soft rock by thermal needle probe procedure. In Annual Book of ASTM Standards, Vol. 04.09. US.

Bougerra, A. 1999. Prediction of effective thermal conductivity of moist wood concrete, Journal of Physic D: Applied Physic, No. 32: 1407–1414.

Brailsford, A.D. & Major, K.G. 1964. The thermal conductivity of aggregates of several phases including porous materials, British Journal of Applied Physics, Vol. 15: 313–319.

British Standard Institution 1986. BS 874: Methods for determining thermal insulating properties with definitions of thermal insulating terms, UK: BSi.

British Standard Institution 2000. BS EN 206-1: Concrete. Specification, performance, production and conformity, UK: BSi.

British Standard Institution 2003. BS EN 771-3: Specification for masonry units: Aggregate concrete masonry units (dense and light-weight aggregates), UK: BSi.

British Standard Institution 1986. BS 1881-125: Testing concrete: Method for mixing and sampling fresh concrete in laboratory, UK: BSi.

Building Regulations 2002. Approved documents Part L: Conservation of fuel and power in dwellings: 2006 Ed. UK: The Stationary Office.

Campbell-Allen, D & Thorne, C.P. 1963. The thermal conductivity of concrete, Magazine of Concrete Research, 15 (43): 39–48.

Carslaw, H.S. & Jaeger, J.C. 1959. Conduction of heat in solids, 2nd Ed, Oxford University Press, 261–262. UK.

Concrete Society 2003. U-values: understanding heat movement. Environmental Working Party of the Concrete Society's Materials Group, 37 (3): 42–43, UK: Concrete.

Design of Normal Concrete Mixes 1975. Department of Environment, Building Research Establishment (BRE), Transport and Road Research Lab. UK: BRE.

Ganjian, E. 1990. The relationship between porosity and thermal conductivity of concrete, PhD Thesis. University of Leeds, UK.

Gibbon, G.J. & Ballim, Y. 1998. Determination of the thermal conductivity of concrete during the early stages of hydration, Magazine of Concrete Research, 50 (3): 229–235.

Hanson, J.L., Neuhaeuser, S. & Yesiller, N. 2004. Development and calibration of a large-scale thermal conductivity probe, Geotechnical Testing Journal, 27 (4): 1–11.

Kew, H.Y. 2007. Investigation into the potential of rubberised concrete products, PhD Thesis, University of Strathclyde, UK.

Kew, H.Y., Cairns, R. & Kenny, M.J. 2004. The use of recycled rubber tyres in concrete. In M.C. Limbachiya and J.J. Roberts (eds), Sustainable Waste Management & Recycling: Used/Post-Consumer Tyres; Proc. intern. conf., Kingston University, 14–15 September 2000. UK: Thomas Telford.

Kim, K.H., Jeon, S.E, Kim, J.K. & Yang, S. 2003. An experimental study on thermal conductivity of concrete, Cement and Concrete Research, No. 33: 363–371.

Kubicar, L. & Bohac, C. 2000. A step-wise method for measuring thermophysical parameters of materials, Measurements Science Techniques: 252–258.

Laidoudi, B., Marmoret, L. & Queneudec, M. 2004. Reuse of rubber waste in cementitious composites: Hygrothermal behaviour, In M.C. Limbachiya and J.J. Roberts (eds), Sustainable Waste Management & Recycling: Used/Post-Consumer Tyres; Proc. intern. conf., Kingston University, 14–15 September 2000. UK: Thomas Telford.

Li, G., Stubblefield, M.A., Garrick, G., Eggers, J., Abadie, C. & Huang, B. 2004. Development of waste tyre modified concrete, *Cement and Concrete Research*, Vol. 34: 2283–2289.

Li, Z., Li, F. & Li, J.S.L. 1998. Properties of concrete incorporating rubber tyre particles, *Magazine of Concrete Research*, 50 (4): 297–304.

Manohar, K., Yarbrough, D.W. & Booth, J.R. 2000. Measurement of apparent thermal conductivity by the thermal probe method, *Journal of Testing and Evaluation*, 28 (5): 345–351.

Maxwell, J.C. 1892. Treatise on electricity and magnetism, 2nd ed. Oxford Clarendon Press, Vol. 1: 440.

Moroney, R.C. 2003. The use of granulated rubber from used tyres in concrete, PhD Thesis, University of Dundee, UK.

Nicolas, J., Andre, P. Rivez, J.F. & Debbout, V. 1993. Thermal conductivity measurements in soil using an instrument based on the cylindrical probe method, *Review of Scientific Instruments*, 64 (3): 774–780.

Rostami, H., Lepore, J., Silverstraim, T. & Zandi, I. 1993. Use of recycled Rubber tyres in concrete. In Dhir R. (ed) Concrete; Proc. intern. conf. University of Dundee, 2000, UK.

Segre, N. & Joekes, I. 2000. Use of tyre rubber particles as addition to cement paste, *Cement and Concrete Research*, Vol. 30: 1421–1425.

Tantala, M.W., Lepore, J.A. & Zandi, I. 1996. Quasi-elastic behaviour of rubber included concrete, in Proceedings of the 12th International Conference on Solid Waste Technology and Management.

Tinker, J.A. 1984. Aspect of mix proportioning and moisture content on the thermal conductivity of lightweight aggregate concrete, PhD Thesis, University of Salford, UK.

Valore, R.C. 1980. Calculation of U-values of hollow concrete masonry, *Concrete International*, 2 (2): 40–63.

Excellence in Concrete Construction through Innovation – Limbachiya & Kew (eds)
© 2009 Taylor & Francis Group, London, ISBN 978-0-415-47592-1

Self-cleaning surfaces as an innovative potential for sustainable concrete

M. Hunger & H.J.H. Brouwers

Faculty of Engineering Technology, University of Twente, Enschede, The Netherlands

ABSTRACT: Concrete technology is subject of continuous development and improvement. One of the very recent contributions to durability and sustainability of concrete is the self-cleaning ability. This effect is achieved by applying photocatalytic materials to the concrete mix. This paper describes the effect of self-cleaning and air purification. Since about 10 years concrete paving stones, provided with this function, are available. With the development of a test setup and using nitric oxide (NO) as model pollutant an approach was found to quantitatively assess the air-purifying ability of those paving stones. This seems to be of interest since a real comparative analysis of air purifying concrete products is not available and the establishment of a measurement standard for concrete products is still in a draft-state. A brief technical description of this test setup will be presented to the reader. Using this innovative setup, the influences on the degradation efficiency are studied and a basic reaction model is derived.

1 INTRODUCTION

Despite intensifying immission control requirements (e.g. EU (1999)) and the increased installation of emission reduction systems, the air pollution and in particular the exhaust gas pollution by nitrogen oxides (NO) will be a serious issue in the near future. The by far largest polluters are traffic and industrial flue gases. In this respect attempts regarding the active reduction of nitrogen oxides can be found in forms of filter devices for industrial stacks (denitrogenization – DENOX plants) or active filter systems for e.g. tunnel exhausts. A further solution according to Matsuda et al. (2001) could be the photochemical conversion of (nitric oxides) NO_x to nitric acid by semiconductor metal oxides due to heterogeneous photocatalytic oxidation (PCO).

In this respect titanium dioxide appears to be the most suitable semiconductor material. Titanium dioxide is one of the oxides of titanium, also called rutile titanium white. It appears in remarkable extent in nature (the ninth most abundant element in the earth's crust). In solid state titanium dioxide can appear in three different crystalline modifications namely rutile (tetragonal), anatase (tetragonal) and the seldom brookite (orthorhombic).

Two electrochemical properties turn the anatase modification to the best suitable catalyst. On the one hand the semiconductor band gap of, E_g, of 3.2 eV is wide and on the other hand the potential for oxidization of the valence band is with 3.1 eV (at pH = 0) relatively high. Both lead to the fact that almost any organic

molecule can be oxidized in the presence of UV-light. This also applies for low oxidizable molecules. Besides the high efficiency anatase, in particular, is suitable for photocatalytic degradation because it is chemically stable, harmless and, compared to other semiconductor metal oxides relatively cheap.

When earlier work mainly dealt with the treatment of waste water, PCO recently has received considerable attention regarding the removal of pollutants in air. Since about 10 years efforts are made, first in Japan, in a large scale application of this photocatalytic reaction for air-purifying purposes. For the degradation of exhaust gases originated from traffic a sheet-like application close to the source would be desirable. Large illuminated surfaces in the road environment are for example road noise barriers and the road or sidewalk surfaces itself. Therefore noise barrier elements and paving stones are interesting substrates to study.

In construction industry products containing titanium dioxide are commercially available since the middle of the 1990s. These are for example window glass and ceramic tiles providing self-cleaning features. The production of the first concrete paving blocks containing titanium dioxide started in 1997 in Japan. The utilization of the self-cleaning abilities of titanium dioxide modified blends of cement was used for the first time in 1998 for the construction of the church "Dives in Misericordia" in Rome. In 2002, investigations to the application of a cement based asphalt slurry seal have been conducted in Italy.

The application of titanium dioxide in paving blocks is patent-protected for the European market by Murata et al. (1997) (Mitsubishi Materials Corporation) as well as Cassar et al. (2004) (Italcementi S.p.A.). The patent owned by Mitsubishi Materials Corporation comprises the application of titanium dioxide in a functional surface layer of a double-layer paving block having enhanced NO_x cleaning capability. The patent held by Italcementi S.P.A. also covers the application of titanium dioxide in double-layer paving blocks capable of abating organic and inorganic pollutants. However, the patent furthermore claims the composition of a dry premix containing a hydraulic binder and a titanium dioxide based photocatalyst capable of oxidizing organic and inorganic pollutants present in the environment.

Besides the application of titanium dioxide for the degradation of organic and inorganic pollutants in paving blocks, noise barriers, and cementitious slurries for asphalt sealing, the material is used also because of its self-cleaning abilities. Here, preserving the original appearance of cementitious stone products is another motivation for the application of titanium dioxide in these products.

Regarding the current market situation, a wide variety of cement based products containing titanium dioxide can be found for horizontal and vertical application. Based on the patents owned by Mitsubishi Materials Corporation and Italcementi S.p.A., products with different properties are available on the European market. These products show varying amounts of titanium dioxide, having preferably the active anatase structure, but also the application of blends of titanium dioxide having anatase and rutile structure is comprised in the patents. These products are promoted regarding their photocatalytic capabilities under laboratory conditions. A comparative assessment in terms of the efficiency of different products as well as application techniques has not been carried out so far. The comparison of different products is rather difficult as different test procedures are used by the manufacturers. These test procedures differ in their execution and therefore a direct comparison of different products is questionable.

In Hüsken et al. (2007) a comparative study on the NO_x degradation of active concrete surfaces is addressed. Thereby a representative profile of the economically available paving block products of the European market is considered. The presented paper is a continuation of this research using the results derived with one of these paving blocks as basis for further analysis. Furthermore the before developed test procedure and setup is applied and therefore briefly described in the following.

The reaction rate is an important parameter to evaluate and model the efficiency of PCO. For fluidized bed reactors and other types of photoreactors the PCO of NO_x in presence of UV-light is well described in literature. However, for large scaled concrete applications there is no sufficient data available.

This paper addresses the development of a test setup capable of analyzing photocatalytic degradation of NO on concrete paving block samples. Using varying volumetric flow scenarios as well as different pollutant concentrations with otherwise constant test conditions a reaction model is derived describing the sample in terms of a reaction rate constant (k) and an adsorption equilibrium constant (K_d). Using these parameters, air-purifying concrete paving blocks can be explicitly described.

2 THE MEASURING PRINCIPLE

The photocatalytic efficiency of a system can be assessed by a number of criteria again being influenced by various circumstances and factors. Therefore a model pollutant was selected and a standard measurement procedure was defined as being explained in the following sections.

From the relevant literature it becomes clear that the degradation of nitric oxide (NO) or more general of nitrogen oxide (NO_x), also referred to as DeNOx-process (denitrogenization), delivers a suitable model to assess the ability of surfaces for air purification. This denitrogenization process can roughly be described as a two-stage reaction on the surface of a photocatalyst, which in most of the cases is titanium dioxide in the anatase modification or variants of it. For this purpose a certain amount of water molecules, supplied by the relative humidity, and electromagnetic radiation are required to start a degradation process. The electromagnetic radiation (E) is expressed by the product of Planck's constant (h) and the frequency (v). Herewith, the two steps can be summarized as follows:

$$NO + OH^\bullet \xrightarrow{\ h\upsilon\ } NO_2 + H^+ \tag{1}$$

$$NO_2 + OH^\bullet \xrightarrow{\ h\upsilon\ } NO_3^- + H^+. \tag{2}$$

The free hydroxyl radicals (OH) originate from the photogenerated water electrolysis on the anatase surface. These two reactions describe the processes on the surface of the sample and therefore define the compounds which have to be measured in order to evaluate degradation ability. With the help of the deployed chemiluminescence NO_x analyzer the amounts of NO_x and NO can be quantified. Subsequently, the amount of NO_2 can be calculated by difference formation. Hence, a quantitative analysis can be conducted.

The detection of nitrogen oxides is based on the reaction of nitric oxide with ozone (O_3). As a result of this reaction an excited nitrogen dioxide is produced. Due to the excited state luminescence, representing the excitation-energy, is emitted in the range from 560–1250 nm. This chemiluminescence is measured by a photo diode with optical filters connected upstream. Ozone required for the oxidation is supplied by an ozone generator using dried ambient air. The percentage of NO_2 contained in NO_x will be reduced to NO in a converter before being conveyed to the measurement chamber. Here, solenoid-controlled valves will feed the measurement chamber in alternating sequence with the sample gas or reduced NO. Evaluation electronics will enable the constant measurement of NO_x and NO this way.

3 TEST APPARATUS

As the pre-standard ISO TC 206/SC N (2004) serves as a sound basis for measurements, its recommendations were largely followed for the practical conduction of this experiment. In fact this standard holds for advanced technical fine ceramics but it satisfies the needs for measurements on concrete specimen as well.

The applied apparatus is composed of a reactor cell housing the sample, a suitable light source, a NO_x analyzer, and an appropriate gas supply.

3.1 Reactor

The core of the experiment is a gas reactor allowing a planar sample of the size 100×200 mm^2 to be embedded. The schematic representation of the gas reactor is given in Figure 1.

Furthermore, the reactor is made from materials which are non-adsorbing to the applied gas and can withstand UV light of high radiation intensity. On top the reactor is tightly closed with a glass plate made from quartz or borosilicate, allowing the UV-radiation to pass through with almost no resistance. Within the reactor the planar surface of the test piece is fixed parallel to the covering glass, leaving an alterable slit height h (most commonly 3 mm) for the gas to pass through. The active sample area used for degradation was, deviating from the pre-standard, enlarged from 49.5 mm $\pm$ 0.5 mm in width and 99.5 mm $\pm$ 0.5 mm in length to $B = 100$ mm and $L = 200$ mm (with similar tolerance), which better complies with standard paving stone dimensions. By means of profiles and seals the sample gas only passes the reactor through the slit between sample surface in longitudinal direction and the glass cover. All structural parts inside the box are designed to enable laminar flow of the gas along the sample surface and to prevent turbulences.

3.2 Light source

The applied light source is composed of three fluorescent tubes of each 25 W, emitting a high-concentrated UV-A radiation in the range of 300 to 400 nm with

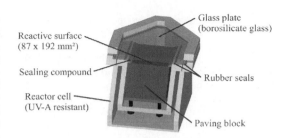

Reactive surface (87 x 192 mm²)

Sealing compound

Reactor cell (UV-A resistant)

Glass plate (borosilicate glass)

Rubber seals

Paving block

Figure 1. Schematic diagram of the gas reactor.

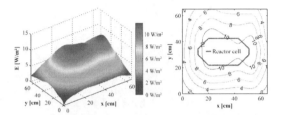

Figure 2. Distribution of UV-A irradiance within the test setup (left) and placement of the reactor box (right).

maximum intensity at about 345 nm. Due to the narrow range in wavelength, an addition of a filter was not necessary. A warming of the reactor by the light source is not expected due to the spatial separation of light source and reactor, and a cooling of the lamp by means of fans. All fluorescent tubes can be adjusted in radiation intensity. With the help of a calibrated UV-A radiometer the radiation intensity was adjusted to 10 W/m^2 at the sample surface. A lead time of about 15 minutes has to be considered for fluorescent tubes till a stable UV-A radiation is approached.

3.3 Testing gas supply and gas types

For the conduction of the experiment two different types of gas, filled in standard gas cylinders, are necessary. First, the model contaminant is discussed. For the pollution of the sample surfaces, nitric oxide (NO) is deployed. The used gas is composed of 50 ppmv NO which is stabilized in nitrogen (N_2). As the concentration of gas, finally applied to the sample, will be adjusted to 1 ppmv, only small quantities of this gas are required. As transport fluid synthetic air, being composed of 20.5 vol.-% of oxygen (O_2) and 79.5 vol.-% of nitrogen, is deployed.

Since the gas cylinders are under high pressure, the gas needs to pass a pressure reducing valve before entering the system. Here pressure is first reduced to 0.3 bar. Before the two gas flows are merged, the model contaminant has to pass a high precision valve in order to adjust a pollution of 1 ppmv NO to the sample. The NO concentration can be monitored with the NO analyzer, connected to the outlet of the reactor box. Furthermore, the synthetic air will be conveyed through a gas-washing bottle, filled with demineralized water, in order to keep the relative humidity of the supplied gas constant at 50%. Using a split gas flow, with one line passing a valve before the gas-washing bottle, one can realize desired humidity. Behind these two stages both gas flows, polluted and transport fluid, are mixed. With the help of a flow controller a volume flow of $Q = 3$ l/min is adjusted. The Reynolds number of the flow reads:

$$\text{Re} = \frac{v_{air} D_h \rho_{air}}{\eta_{air}} = \frac{2 v_{air} h \rho_{air}}{\eta_{air}} = \frac{2Q}{B \upsilon_{air}}. \quad (3)$$

D_h is the hydraulic diameter of the considered channel, defined as four time the cross-sectional area divided by the perimeter, for the slit considered here, $D_h = 2$ h. Substituting $Q = 3$ l/min, $B = 100$ mm and $\upsilon_{air} = 1.54\,10^{-5}$ m^2/s (1 bar, 20°C) yields $Re \approx 65$. Considering $h = 3$ mm, the mean air velocity v_{air} is 0.17 m/s along the sample surface.

This low Reynolds number implies that the flow is laminar. A fully developed parabolic velocity profile will be developed at $L_d = 0.05$ Re 2h, so here $L_d \approx 20$ mm, i.e. only at the first 10% of the slit length there are entrance effects, during the remaining 90% there is a fully developed laminar flow profile.

The gas, mixed and humidified this way, enters the reactor and is conveyed along the illuminated sample surface. At the opposite site of the reactor the gas leaves the chamber and is transported to a flue or outside with the help of an exhaust air duct. The NO analyzer sources the reacted test gas from this exhaust line. An adequate dimensioning of the hose line and, possibly, the installation of non-return valves prevents from suction of leak air from outside via the hose line to the analyzer.

3.4 Analyzer

For the gas analysis a chemiluminescent NO$_x$ analyzer like described in ISO 7996 was deployed. The analyzer is measuring the NO$_x$ and NO concentration in steps of 5 sec while the corresponding NO$_x$ concentration is computed by the difference of the previous two. During the measurement the analyzer is constantly sampling gas with a rate of 0.8 l/min. The detection limit of the deployed analyzer is at about 0.5 ppbv.

4 EXPERIMENTS

4.1 The conduction of measurements

For the development of a reaction model and the herewith related experiments only one sample was used. This way differences in measurement results due

to varying surface roughness or unequal distribution of catalyst can be neglected. The deployed sample is a commercially available concrete paving stone whose photocatalytic properties were tested beforehand (Hüsken et al. (2007)). In preparation of each measurement the sample surface is cleaned in order to remove fouling, contamination and potential reaction products due to a previous NO$_x$ degradation. The cleaning process is always executed following a specified scheme using demineralized water. Subsequently, the sample is dried in a drying oven. For the measurement the sample with the reactive surface upwards is placed in the reaction chamber. With the help of an elastic sealing compound all gaps and joints around the sample are caulked that way, that the fed air could only pass the reactor along the reactive sample surface. In doing so a metal sheet of the dimension 87×192 mm^2 was deployed as a template for the sealing. The active sample surface was kept exactly identical for all measurements this way.

After assembling the sample the reactor is closed and the gas supply is started. The UV-A source is switched on as well in order to start the radiation stabilization, but the reactor stays covered to prevent first degradation. With the help of the controls the flow is now adjusted to e.g. 3 l/m and the relative humidity to 50%. The supplied NO concentration is adjusted to the desired inlet concentration, which is checked by the analyzer. When these conditions appear to be stable the data acquisition is started. Now, for the first 5 minutes the system remained unchanged in order to flush the reactor chamber and to finally eliminate an increase of UV-A radiation. During this time the measured NO outlet concentration of the reactor was first decreasing and then approaching again the original inlet concentration. This phenomenon describes the saturation of surface with NO and was found to be a function of flow velocity, inlet concentration and surface character of the sample. After this period of time the cover sheet was removed to allow the radiation passing through the glass. This was very quickly responded by the analyzer. The degradation for the uncovered reactor lasted for 30 minutes, then the reactor was covered again and the data acquisition was continued for further 5 minutes. Within the last minutes of measurement the NO and NO$_x$ concentrations should ideally return to the original scale. As the reactor can be bridged, i.e. the pollutant can be directly transferred to the analyzer without passing the reaction camber, the NO inlet concentration at the end of a measurement was always compared with the original concentration at the beginning of the measurement. In this way measurement errors due to creeping NO concentrations during the measurement are prevented.

In order to obtain sufficient data for the development of a reaction model varying NO inlet concentrations of 0.1, 0.3, 0.5, and 1.0 ppmv were applied.

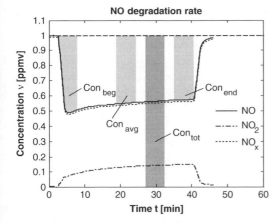

Figure 3. Explanation of the analysis approach for the description of the course of conversion.

The actual conversion for each five-minutes-period used in Equation 4 was determined by integration of the associated, descriptive function in the limits of time by using the trapezoid rule:

$$Con_{beg} = \sum_{i=1}^{n} \frac{t_{i+1} - t_i}{2}(\upsilon_i + \upsilon_{i+1})$$

$$\forall t_i \in [t_{beg}, t_{beg} + 5\ min]\quad. \tag{5}$$

This approach can be assumed to be sufficiently precise, given that the interval $[t_i,\ t_{i+1}]$ only lasts for 5 sec. Here, the first time interval for the beginning conversion (Con_{beg}) starts at zero, when the sample in the reactor was first exposed to UV-A radiation and lasted for five minutes. This way the slope of the starting conversion up to the maximum degradation rate is included and therefore characterizing this first numerical value. In other words, the progress of the degradation development up to maximum conversion is included and evaluated. The second time frame (Con_{avg}) represents the conversion after half of the total time of UV-A exposition, i.e. the chronological middle of measurement ±2.5 minutes. For the last time interval (Con_{end}), the last five minutes before switching off the UV-A source are considered. The delayed decrease of conversion after the obvious inflection point is explained by the inertia of the whole system. As a matter of principle, no conversion can take place after removing the UV-A source. Already formed hydroxyl radicals still can start an oxidation of NO_x but only till they are consumed. For this reason the terminal degradation of NO_x without UV-A exposition is not further assessed. The conversion of the last 5-minutes-range is taken as basis for the further analysis.

Furthermore, for each inlet concentration the flow rate was varied by using flows of 1, 3 and 5 l/min. These different volumetric flows correspond to flow velocities of 0.056, 0.167 and 0.278 m/s at the sample surface inside the reactor box. With these combinations a total number of 12 measurements are executed.

Moreover, to validate the assumption that conversion process is the limiting rate instead of the diffusion step, six more measurements with the above given flow rates but a fixed NO inlet concentration of 0.3 ppmv have been done. In doing so the slit height in the reactor box was varied (2, 3 and 4 mm) for each flow rate. Note that stable and measurable NO inlet concentration was not achieved while applying a flow of 5 l/min and using a slit height of only 2 mm. In that case the resistance of the system turned out to be too high and therefore a reliable measurement was not possible.

4.2 Analysis of the data

The data which is basis for the present analysis was derived applying above explained measurement procedure. Further analysis of the measurement data was conducted following a three stage analysis. For this purpose the course of the three different conversions (NO, NO_x and NO_2) was assessed for a time range of each five minutes equally distributed in the total measurement period. Figure 3 is illustrating this procedure for the formation of NO_2 with an arbitrary sample. The degradation rate [%] in the assessed time range is calculated by means of the ratio between actual conversion and the total conversion of NO (Con_{tot}) as follows:

$$Deg_{beg} = \frac{Con_{beg}}{Con_{tot}} = 1 - \frac{C_{out}}{C_{in}}. \tag{4}$$

5 REACTION MODEL

5.1 Model development

In this section the reaction process in the reactor is modeled. First, one should observe that the NO had to diffuse to the concrete surface, where subsequently the conversion to NO_x takes place. So the process contains two transfer steps, the mass transfer from gas to wall and the conversion at the concrete surface. In line with findings by Zhao and Yang (2003) and Dong et al. (2007), here it will be demonstrated that the conversion is the rate limiting step.

The NO mass flux from gas to surface is governed by:

$$\dot{m} = \frac{ShD}{D_h}(C_g - C_w) = \frac{ShD}{2h}(C_g - C_w), \tag{6}$$

with D_h, as hydraulic diameter, see Section 3.3 and C_g and C_w are the NO concentrations (in mg NO per m^3

549

air) in air (mean mixed or bulk) and on the surface. The relation between C and C_{con} is as follows:

$$C = \frac{M_{NO}\rho_{air}}{M_{air}} C_{con},\qquad(7)$$

Considering that the molecular mass of air (M_{air}) is 29 g/mole (80% M_{N2} and 20% M_{O2}), that of NO (M_{NO}) is 30 g/mole, and that ρ_{air} is about 1 kg/m^3, it follows that C_g (in kg/m^3) is about C_{con} (in mole/mole), or equivalently, C_g (in mg/m^3) is about C_{con} (in ppmv $= 10^{-6}$ mole/mole).

Sh is the Sherwood number, amounting to about 5 for slits with one inert and one exchanging side (Shah and London (1974)), and D the diffusion coefficient of NO in air, which amounts to about $1.5\,10^{-5}$ m^2/s ($\sim v_{air}$).

If it is assumed that the diffusion to the concrete surface is the limiting step (conversion takes place instantaneously and completely), the NO concentration at the surface will be zero. The NO mass balance equation then reads:

$$v_{air}h\frac{dC_g}{dx} = -\dot{m} = -\frac{ShDB}{2h}C_g,\qquad(8)$$

with as boundary condition:

$$C_g = C_{g,in}.\qquad(9)$$

Integrating Equation 8 and application of Equation 9 yields:

$$\frac{C_{g,out}}{C_{g,in}} = e^{-\frac{ShDL}{2v_{air}h^2}},\qquad(10)$$

with $C_{g,out} = C_g(x=L)$, L being the length (200 mm). Substituting all variables, including $v_{air} = 0.17$ m/s and $h = 3$ mm (Section 3.3), into Equation 10 yields $C_{g,out}/C_{g,in} \approx 0.04$. In other words, in the case that the diffusion to the wall would be rate limiting step, the exit concentration would be close 4% of the inlet concentration, i.e. 96% of the NO would be converted. The present measurements and previous research (Hüsken et al. (2007)) learned this is not the case, so that the conversion rate at the surface cannot be ignored (cp. Tables 1 and 2). In the following, it is now assumed *a priori* that the conversion is the rate limiting step, so that C_w now equals C_g, which will be verified *a posteriori*.

For the prevailing photocatalytic gas-solid surface reaction, only adsorbed NO can be oxidized. In the past therefore the Langmuir-Hinshelwood rate model has been widely used, e.g. by Ollis (1993), Dong et al.

Table 1. NO outlet concentrations of the reactor considering varying inlet concentrations and flow rates for the photocatalysis of the paving stone example.

C_{in}	C_{out}			NO$_x$ removal rate [%]		
	Volumetric flow rate Q [l/min]			Volumetric flow rate Q [l/min]		
[ppmv]	1	3	5	1	3	5
0.1	0.011	0.032	0.041	89.0	68.4	59.4
0.3	0.039	0.157	0.197	87.1	47.6	34.3
0.5	0.210	0.309	0.356	58.0	38.3	28.9
1.0	0.334	0.729	0.779	66.6	27.1	22.1

Table 2. NO outlet concentrations and removal rates of the reactor considering constant inlet concentration $C_{in} = 0.3$ ppmv but varying flow rates and slit heights.

Slit height	C_{out}			NO$_x$ removal rate [%]		
	Volumetric flow rate Q [l/min]			Volumetric flow rate Q [l/min]		
[mm]	1	3	5	1	3	5
2	0.093	0.162	–	68.8	46.1	–
3	0.039	0.157	0.197	87.1	47.6	34.3
4	0.065	0.157	0.183	78.3	47.6	39.1

(2007), see also Zhao and Yang (2003), and which will also be applied here. Following this model, the disappearance rate of reactant reads:

$$r = \frac{kK_dC_g}{1+K_dC_g},\qquad(11)$$

with k as reaction rate constant (kg/m^{3}s) and K_d as the adsorption equilibrium constant (m^3/kg). The NO balance equation now reads:

$$v_{air}\frac{dC_g}{dx} = -r = -\frac{kK_dC_gB}{1+K_dC_g}.\qquad(12)$$

Integration and using boundary condition 9 yields:

$$\frac{1}{k}+\frac{1}{kK_d}\frac{Ln\left(\frac{C_{g,in}}{C_{g,out}}\right)}{(C_{g,in}-C_{g,out})} = \frac{L}{v_{air}(C_{g,in}-C_{g,out})} = \frac{V_{reactor}}{Q(C_{g,in}-C_{g,out})}.\qquad(13)$$

550

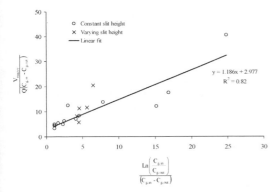

Figure 4. Regression results of data presented in Table 1 and Table 2 for the photocatalysis of the paving stone example.

As $V_{reactor} = LBh$ and $Q = v_{air}Bh$, again, $C_{g,out} = C_g$ ($x = L$).

In Table 1 this $C_{g,out}$ of the experiments with the paving stone is summarized, the inlet concentration $C_{g,in}$ had values of 0.1, 0.3, 0.5 and 1 mg/m^3 (or ppmv), and the flow rate Q was 1, 3 and 5 l/min.

5.2 Validation of the model

In Figure 4, $y = V_{reactor}/Q(C_{g,in} - C_{g,out})$ is set out versus $x = Ln(C_{g,in}/C_{g,out})/(C_{g,in} - C_{g,out})$, and the data fit with the line $y = (1.19\,s)x + 2.98$ m^{3}s/mg. The intersection with the ordinate corresponds to $1/k$, so that $k = 0.37$ mg/m^{3}s, and the slope to $1/kK_d$, so that $K_d = 2.51$ m^3/mg.

Wang et al. (2007) obtained $k = 6.84$ mg/m^{3}s and $K_d = 1.13$ m^3/mg for the NO degradation by woven glass fabrics, their inlet NO concentration being in the range 40 to 80 ppmv. Concerning the photo catalytic ammonia degradation by cotton woven fabrics, Dong et al. (2007) obtained $k = 0.10$ mg/m^{3}s and 0.24 mg/m^{3}s, and $K_d = 0.035$ m^3/mg and 0.112 m^3/mg, whereby the NH$_3$ inlet concentration ranged from 14 to 64 mg/m^3.

With the obtained values of k and K_d the conversion rate and diffusion rate can be compared. Dividing the conversion/transfer rates as governed by Equations 12 and 8 yields kK_d2h^2/ShD, and substituting the prevailing values learns that this ratio is about 0.2, i.e. the diffusion rate is about five times the conversion rate. Note that the employed kK_d is an upper limit for $kK_d/(1 + K_dC_g)$, so that the actual ratio will even be smaller. From this small conversion/transfer rate ratio one can conclude that indeed the degradation rate is much slower than the diffusion rate, and hence it is the limiting rate. This is also confirmed by Matsuda et al. (2001) who found conversion rates proportionally increasing with the specific surface area of the applied titanium dioxide.

Table 3. NO removal and NO$_x$ removal rate for a concrete paving block taken from literature (Mitsubishi (2005)).

NO concentration [ppm]	NO removal [mmol/m^{2}12 h]	NO$_x$ removal rate [%]
0.05	0.2	89.6
0.1	0.4	88.9
0.2	0.8	90.6
0.5	2.0	88.4
1.0	3.7	82.3
2.0	6.1	68.0
5.0	10.0	44.3

In the addressed research this fact was furthermore investigated by executing experiments with slit heights of 2 and 4 mm, $C_{g,in}$ taken as 0.3 mg/m^3 (0.3 ppmv), and the flow rate Q was 1, 3 and 5 l/min. In Table 2, and also in Figure 4, these data are included. With except for the smallest flow rate, the degradation rates match well with the values listed in Table 1, and the computed $V_{reactor}/Q(C_{g,in} - C_{g,out})$ and $Ln(C_{g,in}/C_{g,out})/(C_{g,in} - C_{g,out})$ are also compatible with the previous ones set out in Figure 4 and the fitted trend line.

The above literature review shows that the Langmuir-Hinshelwood model has been frequently applied to characterize photocatalytic gas-solid surface processes using various substrate materials. However, to the authors' knowledge, the kinetics of photocatalytic acting concrete has not been analyzed so far.

Other relevant data on photocatalysis on concrete was found in Mitsubishi (2005). Here, amongst other things, data is presented on the NO$_x$ removal of a paving stone type (NOXER) which is exposed to varying NO concentrations. With the help of background information regarding the setup and conduction of measurement an analysis according to the Langmuir-Hinshelwood model was derived. The information given on the flow properties were sufficient for adequate application of the model.

Using the data on the NO$_x$ removal rate given in Table 3 a linear fit is derived as given in Figure 5. As can be seen the resulting data points fit remarkably well into the proposed model.

With the obtained values again a conversion rate and diffusion rate can derived. For the given data k amounts to 3.54 mg/m^{3}s and K_d to 0.538 m^3/mg, respectively. Compared to the own experiments the conversion rate is notedly higher than the transfer rate in this case. On the conversion side this can be explained with different amounts and types of catalyst (anatase) whereas the transfer could be influenced by different surface morphology of the paving stones. The ratio of conversion/transfer rates therefore amounts to 0.45 for the

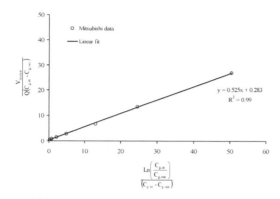

Figure 5. Regression results of data presented in Table 3 for the photocatalysis of a NOXER paving block (data taken from Mitsubishi (2005)).

NOXER case, i.e. the conversion rate is still more than twice the diffusion rate.

6 CONCLUSION

Heterogeneous photocatalytic oxidation is a promising air-purifying technology. Besides this, the self-cleaning aspect coming along with PCO should not be ignored either. Concerning this matter also promising results have been achieved. They are described elsewhere as it is out of the scope of this paper.

The results obtained on the air-purification show that a successful decomposition of NO_x along concrete paving stone surfaces is feasible by using PCO in the presence of UV light. For the assessment of those active paving stones a setup has been developed which is capable of varying and controlling all boundary conditions like for example the volumetric air flow, slit height, NO inlet concentration, UV-irradiance and relative humidity.

Besides the development of the setup a test procedure has been suggested as well. Following this procedure, first time a standard could be obtained which would allow for the real quantitative comparison of concrete paving blocks capable of degrading NO_x from the air.

For this research both experimental and modeling work has been conducted. With the derivation of a reaction model, based on Langmuir-Hinshelwood kinetics, now the treatment performance of certain air-purifying concrete products for decomposition of gaseous pollutants (NO) can be predicted. Furthermore, now the unique description of a photocatalytic material is feasible by the derivation of its conversion and adsorption rate constants. The derived model also confirms that the conversion of NO_x is the rate-determining step in the photocatalytic oxidation of NO_x.

ACKNOWLEDGMENT

The authors gratefully acknowledge the assistance of Dipl.-Ing. G. Hüsken with helping with the conduction and analysis of measurements. Moreover, the authors wish to express their sincere thanks to the European Commission (I-Stone Project, Proposal No. 515762-2) and the following sponsors of the research group: Bouwdienst Rijkswaterstaat, Rokramix, Betoncentrale Twenthe, Betonmortelcentrale Flevoland, Graniet-Import Benelux, Kijlstra Beton, Struyk Verwo Groep, Hülskens, Insulinde, Dusseldorp Groep, Eerland Recycling, Enci, Provincie Overijssel, Rijkswaterstaat Directie Zeeland, A&G Maasvlakte (chronological order of joining).

REFERENCES

Dong, Y., Bai, Z., Liu, R., Zhu, T. 2007. Decomposition of indoor ammonia with TiO2-loaded cotton woven fabrics prepared by different textile finishing methods. *Atmospheric Environment* 41(15): 3182–3192.

EU – The Council of the European Union. 1999. Council Directive 1999/30/EC – *relating to limit values for sulphur dioxide, nitrogen dioxide and oxides of nitrogen, particulate matter and lead in ambient air.*

Hüsken, G., Hunger, M., Brouwers, H.J.H. 2007. Comparative study on cementitious products containing titanium dioxide as photo-catalyst, Proceedings *International RILEM Symposium on Photocatalysis, Environment and Construction Materials-TDP 2007*, 8–9 October 2007, Florence, Italy, 147–154, Eds. P. Baglioni and L. Cassar, RILEM Publications, Bagneux, France (2007).

Matsuda, S., Hatano, H., Tsutsumi, A. 2001. Ultrafine particle fluidization and its application to photocatalytic NOx treatment. *Chemical Engineering Journal* 82(1–3): 183–188.

Mitsubishi Materials Corporation 2005. NOx removing paving block utilizing photocatalytic reaction. In brochure *Noxer – NOx removing paving block.*

Ollis, D.F. 1993. Photoreactors for Purification and Decontamination of air. In D. F. Ollis and H. Al-Ekabi (eds.), *Photocatalytic purification and treatment of water and air.* Elsevier Science: 481–494.

Shah, R.K., & London A.L. 1974. Thermal Boundary-Conditions and Some Solutions for Laminar Duct Flow Forced Convection, *Journal of Heat Transfer* 96(2): 159–165.

Wang, H.Q., Wu, Z.B., Zhao, W.R., Guan, B.H. 2007. Photocatalytic oxidation of nitrogen oxides using TiO2 loading on woven glass fabric. *Chemosphere* 66(1): 185–190.

Zhao, J. & Yang, X.D. 2003. Photocatalytic oxidation for indoor air purification: a literature review. *Building and Environment* 38(5): 645–654.

Excellence in Concrete Construction through Innovation – Limbachiya & Kew (eds)
© 2009 Taylor & Francis Group, London, ISBN 978-0-415-47592-1

Pervious concrete pavement: Meeting environmental challenges

A.K. Jain
Civil Engineering (Selection Grade), S.V. Polytechnic College, Bhopal M.P., India

J.S. Chouhan
Civil Engineering Deptt, SATI (Engineering College) Vidisha, M.P., India

ABSTRACT: The land development process has a documented impact on the quality of watersheds. It affects the whole aquatic ecosystems, in particular, stream hydrology, geomorphology, water quality, and habitat. A simple solution to avoid these problems is to stop installing the impervious surfaces that block natural water infiltration into the soil. Retaining and improving the permeability of the natural land by minimizing the use of impervious materials for site development and paving surfaces is the most pressing need of developmental activities. Recognizing the potential of pervious concrete towards sustainable developments and considering large scale developmental activities going on globally, it will be in the interest of the environmental health of nations to harness the environmental benefits offered by this material by making its extensive use. Global environmental issues have serious adverse effects on sustainable development, and it is beyond the reach of individual nations to resolve them by themselves, therefore the efforts of environmental conservation and sustainable development should positively be promoted through joint efforts by both developed and developing countries.

1 INTRODUCTION

We must have a new ethic, a new attitude towards discharging our responsibility for caring for ourselves and for the earth. We must recognize the earth's fragility and its limited capacity to provide for us. We must no longer allow it to be ravaged. This ethic must be motivated with a great movement, convincing leaders, governments and peoples themselves to effect the needed changes. Obviously it needs the help of the world community of natural, social, economic, and political scientists, world's business and industrial leaders and of the world's peoples to make it happen.

The growing urbanization and large scale housing and infrastructure development activities have influenced the water balance pessimistically, on the earth surface and under it. In a pre-developed setting, much of the rainfall is absorbed by the surrounding vegetation, soil and ground cover but contrary to that, in a developed setting, roofs, sidewalks, parking lots, driveways, streets and other impervious surfaces prevent water from naturally infiltrating soil, filtering pollutants and recharging aquifers, thus changes the water balance and a disproportionate amount of rainfall, get converted in to surface runoff. As a result, large volumes of runoff are required to be handled and directed into waterways. It demands for expensive infrastructure works to control the erosion and flooding that occur when stormwater is concentrated and released into waterways. These volumes of stormwater carry sediment and other contaminants that compromise water quality.

Water is becoming a scare commodity and is being globally considered as the most precious material. The India's National Water Policy (1987) states that water is a prime natural resource, a basic human need and precious national asset. The perennial rivers are becoming dry and most of them getting more polluted. Ground water table is depleting in most of the areas in India. A study undertaken in Tumkur district of Karnataka, India reports that between 1985 to 2001 depth of water table increased from 80 feet to 496 feet, groundwater discharge decreased from 3,500 to 800 gallons/hr, irrigation pump capacity increased from 3 to 7.5 HP. All these figures indicate towards overexploitation and less recharging of groundwater. Situation in other part of India and in many other countries is not much different. This situation warrants concerted efforts to find out means to address these environmental issues.

Increased impermeability of natural permeable land area and degradation of natural area because of developmental activities have caused reduced recharging of ground water table. This coupled with over exploitation of ground water to meet the ever growing demand of water, have caused depletion in ground water table drastically. Paving surfaces with impervious materials leads to increase in quantity of storm water and its pollution level. Disruption of natural water balance, increased flood peaks, more frequent flooding, increased bankfull flows and lower dry weather flows are some of the hydrological effect of paving of more and more land area by using impervious materials. It also put a negative impact on the growth of plants and trees because of absence of supply of sufficient amount of water and air in to its root zone, and caused urban heat island effect. This provides sufficient evidences to be able to conclude that paving of more and more land area with impervious materials is one among the major causes that has triggered many environmental problems.

This pressing environmental demand has made development personals to give a second look at pervious concrete. Pervious Concrete which has the potential to heel the environment has emerged as a sustainable material at the time when increasingly paving of natural surfaces has become an integral part of a construction site development and landscaping.

Pervious paving surfaces provide a desirable combination of stormwater retention and aquifer recharge properties as a substitute for conventional impervious pavements. It retains some of the structural properties of conventional pavements but adds other features such as pollution removal. It's ability of eliminating runoff while providing filtration and ground water recharge has positioned the product as a sustainable construction material. It is a discontinuous mixture of coarse aggregate, Portland cement, admixtures and water with increased porosity due to limited fines. The increased porosity in the mix and 15–20% air voids allows the flow of water through the material. Comparatively low compressive strengths of 2.8 to 28 MPa has restricted its use in areas of light to medium vehicle traffic at low to moderate speeds, such as parking lots, driveways, sidewalks, shoulders for roadways, shoulders for airport taxiways and runways, street and local roads, greenhouse floors, erosion control & slope protection etc.

2 DURABILITY OF THE MATERIAL

The pervious concrete has been used to various degrees and in various capacities around the world. The history of use of this material goes back to 1852. Substantial amount of pervious concrete was also produced in Europe immediately following the Second World War; some are still in service today. Pervious concrete pavement has been used for over 30 years in England and the United States. It is also widely used in Europe and Japan for roadway applications as a surface course to improve skid resistance and reduce traffic noise. It has been specified for pavement in Florida for more than 20 years, earning a fine record of durability and high performance. Hundreds of field performance tests performed and over the long-term, pervious pavements continue to function without signs of structural distress or significant clogging.

3 ENVIRONMENTAL BENEFITS OF PERVIOUS CONCRETE PAVEMENT

The material is receiving renewed interest of the stakeholders associated with construction/developmental activities exclusively in US partly because of Federal Clean Water Legislation in the USA. The US Environmental Protection Agency's (EPA) Phase II Final Rule requires the operators of all municipalities in urban areas to develop, implement, and enforce a program to have a on-site management system for treating all stormwater to reduce pollutants in post-construction runoff from new development and redevelopment projects that result in the land disturbance of greater than or equal to 1 acre, before it leaves for conveyance by the respective local agency. The above is a requirement in order to attain a National Pollutant Discharge Elimination System (NPDES) permit. Among other things, the municipalities are required to develop and implement strategies which include a combination of structural and/or non-structural best management practices (BMPs). Pervious concrete pavement is recognized as a Structural Infiltration BMP by the EPA for providing first flush pollution control and storm water management. In addition to federal regulations there has been a strong move in the USA towards sustainable development. In US, the US Green Building Council (USGBC), through its Leadership in Energy and Environmental Design (LEED) Green Building Rating System fosters sustainable construction of buildings. Projects are awarded Certified, Silver, Gold, or Platinum Certification, depending on the number of credits they achieve. Pervious concrete pavement qualifies for LEED credits and is therefore sought by owners desiring a high LEED certification.

Some of its environmental benefits which have put it in to the list of green construction material are:

I. Besides serving as a parking facility, it helps the process of infiltration of water into the soil by capturing rainwater in a network of voids and allowing it to percolate into the underlying soil, thus directly recharge groundwater aquifer.

II. Current methods of stormwater management are costly on several fronts. To mention a few, Impervious parking lots and roof tops cause more stormwater runoff and pollutant loads than any other type of land use and require stormwater collection and disposal. High volume runoff requires large public drainage facilities and retention/detention tanks to service runoff from developments. Pervious Concrete Pavement serves as a stormwater management system tool because of its ability to eliminate/retain runoff and its filtration action, which significantly improves the quality of water passing through.

III. Stormwater runoff is a major source of the pollutants entering our waterways. About 90 percent of pollutants are carried by the first 38 mm of any rain event generally described as 'first flush'. Use of pervious concrete pavements is a recommended Best Management Practice (BMP) of EPA for first-flush pollution mitigation within the realm of stormwater management. It reduce pollution load of storm water by allowing it to percolate in to soil. During infiltration pollutants get decomposed by biological activity and soil chemistry and comparatively clean water join ground water table. It would have otherwise contaminated watersheds, harm sensitive ecosystems and increase treatment cost of the water to make it potable.

IV. Due to increased dependence of today's society on technology derived from organic chemicals, PAH's (Polycyclic Aromatic Hydrocarbons) pollution from asphalt pavements and sealers have become widespread. Use of Pervious Concrete Pavement eliminates it to enter in to the water ways.

V. Heavy network of traditional asphalt pavements in urban areas, by virtue of its dark color (albedo from 0.05 to 0.10 when new), absorb much of the solar radiation and contribute to the phenomena of "urban heat-island effect" (thermal gradient differences between developed and undeveloped areas). Since Pervious Concrete Pavements are light in color (albedo in the range of 0.35 to 0.45); and have an open-cell structure, does not absorb and store as much heat and raises the surrounding temperature, hence its use helps in reducing the effect. Trees planted in parking lots capture some storm water and offer a cooling effect in the area.

VI. Reduces the need for lighting at night by virtue of its lighter color and reflective characteristics. Therefore it is also beneficial from the energy-savings standpoint.

VII. Improves access of water and air to the root systems of trees. Trees that surround pervious concrete parking lots have been shown to live longer and grow wider than trees in areas with impervious pavements. Also, it channels more water to tree roots and landscaping, so there is less need for irrigation.

VIII. Safer for drivers and pedestrians, because pervious concrete absorbs water rather than allowing it to puddle. It reduces hydroplaning and tyre spray causing drivers to loose control by preventing standing water from pooling on paved surfaces.

IX. Prevents soil erosion and offers excellent pathways to provide disability access for people in wheelchairs.

4 MECHANISM OF FILTERING AND TREATING STORMWATER

Pervious pavement pollutant removal mechanisms include absorption, straining, and microbiological decomposition in the soil. An estimate of pervious pavement pollutant removal efficiency is provided by two long term monitoring studies conducted in Rockville, MD, and Prince William County, VA. These studies indicate removal efficiencies of between 82% and 95% for sediment, 65% for total phosphorus, and between 80% and 85% of total nitrogen. The Rockville, MD, site also indicated high removal rates for zinc, lead, and chemical oxygen demand.

5 SOCIO ECONOMIC IMPACT

Fertilizers and pesticides have entered the water supply through runoff and leaching to the groundwater table and pose a hazard to human, animal and plant populations.

Some of these chemicals include several substances considered extremely hazardous by the World Health Organization (WHO) and which are banned or under strict control in developed countries. Studies on the Ganges River in India indicate the presence of chemicals such as HCH, DDT, endosulfan, methyl malathion, malathion, dimethoate, and ethion in levels greater than recommended by international standards (World Bank 1999). Some of these substances have been known to bioaccumulate in certain organisms, leading to increased risk of contamination where these organisms are used for human consumption and a persistence of the chemicals in the environment over long periods of time.

The implications for human health and security are depressing. Between 0.5 to 1.5 million children under the age of five die yearly from diarrhea in India. (World Bank 1999) Statistics from other South Asian nations also serve as an indicator similar to the situation in India. It is needless to mention that majority of these affected people belongs to socially unprivileged class of the society. Government policies and regulations on water management have so far been unable to stem the growing problems related to water quality and quantity in India. For the most part, this is due to the lack of implementation and enforcement of the existing regulations and improvement in existing regulations as per the demand of the time. Similarly, enforcement of the regulations governing the development and protection of water resources has been poor and serious abuses continue to occur throughout the country. More stringent and enforceable regulations need to be put into place to prevent further degradation and wastage.

Economic incentives and subsidies have not been beneficial in terms of promoting conservation of water resources.

Effective economic and management policies are needed to prevent the crisis that threatens India and many other similar countries in the coming years. Good management of the country's water resources will effectively reduce the amount of pollution and over-exploitation that is currently plaguing the nation's surface, ground and coastal waters. The consequent improvement in water quantity and quality will also have repercussions in terms of ameliorating human and environmental health.

6 DESIGN CONSIDERATIONS

As each project and geographical area has circumstances particular to it, such as soil conditions or available local aggregates, which may require modifications, it is suggested to make sufficient number of trial mixes to achieve the mix proportion that fulfills the specific needs of the project. The American Concrete Institute Committee 522 has published the 522R-06 document, which is a guide to the use of pervious concrete. The committee is also working to complete the 522.1 specification, which will provide guidance for specifiers who may be incorporating the material into their projects. In absence of desired indigenous study, these documents can be referred to take a start to design the mix that fulfills the project requirement. It is important to observe the recommended construction practice to ensure that the objects anticipated and assumptions made during the design stage are fulfilled.

7 MAINTENANCE

Maintenance of pervious concrete pavements is a highly debated issue. The advantages of pervious pavement can only be realized if it is designed and maintained to prevent clogging. To ensure success, design strategies employed should ensure to help prevent clogging. Proper maintenance generally consists simply of vacuum sweeping or power washing. On-going research shows that systems that are not maintained still perform very well over time but obviously not at their original infiltration rates. However, a good cleaning generally will improve the infiltration rate of the system. Similar to nearly all other stormwater treatment tools, proper maintenance is important to keep the system running at higher performance levels.

8 QUALITY ASSURANCE

Since most of the pavement engineers and contractors lack expertise with this technology, it is suggested that the pervious concrete installers be a recognized/certified Pervious Concrete Contractor (or Technician). For achieving quality assurance work should be performed in accordance with ACI 301 and ACI 318 or any other relevant regional code. Recommendations of ACI 306R shall be followed when concreting during

cold weather. Adequate numbers of skilled workmen who are thoroughly trained and experienced in the necessary crafts and who are completely familiar with the specified requirements and the methods needed for proper performance of the work shall be used. Use of proper equipments and machinery required at each stage of the laying and testing of pavement viz. subgrade preparation and its testing, mix design and testing, placing and screeding, compaction, jointing and finishing, curing, flow test etc. is important.

9 INSTALLATION COSTS

In general, initial costs for pervious concrete pavements are higher than those for conventional concrete or asphalt paving, because of the thicker installed size of pervious concrete than regular concrete. But when an overall installation and life-cycle cost is compared, pervious concrete clearly stands as winner. It can not be viewed as just per square meter costs but has to be looked at an overall system costs attaching a certain cost towards its environmental and social benefits.

10 TRAINING AND CERTIFICATION

Lack of widespread transfer of developed and available new technologies has been a major problem in putting them in practice. Research, development and technology transfer to the stakeholders viz. practicing engineers, architects, planners, developers, consultants etc, who contribute in the field of construction and developmental activities at different stages and training and certification of actual field workers is a key to its successful application.

In case of pervious concrete, the tolerance limits of various controlling parameters are quite narrow in comparison to traditional concrete. For example, it is highly sensitive to changes in water content, shape of aggregate etc. Too much water may cause segregation, and too little water will lead to balling in the mixture and slow mixer unloading. It may also hamper adequate curing of the concrete and lead to a premature raveling surface failure. Therefore, for the success of pervious concrete pavement it is of vital importance that the installers/contractors have a well trained and experienced crew with them. It is well known that durable construction is impossible with poor quality of material and/or workmanship, but the reverse can be true. Application of any new material demands for highly trained and motivated taskforce in order to make the material perform in desired way and help other potential user to develop confidence in the material. Taking all these factors into consideration, it is highly recommended to take up activities of Training & Certification of contractors, workers and supervisors before awarding them the work related to pervious concrete.

At present in most of the developing countries, construction industry is highly unorganized and most of the construction workers are seasonal, migrant and itinerant. Their place of work, project and employers, all keep on changing. Employers seldom invest in their training and workers are not willing to sacrifice their earning time for attending courses. Training centers are non-existent and if some training programme is planned its time schedule does not suit to field workers. Professional agencies or NGOs should have to come forward to formulate the scheme of training of the workers of this sector considering practical approach to attract maximum number of workers to undergo such training.

11 LIMITATIONS OF USE

Though the pervious concrete pavement is an excellent option in certain situations, it may not always be a viable choice. Rough-textured honeycombed surface with comparatively less compressive strength, restrict its application on heavily traveled roadways. Any site that is at risk of siltation from adjacent areas, would be a poor choice for pervious concrete unless special measures are taken to protect the pavement. Low strength values and lack of freeze thaw durability test results have limited the use of pervious concrete in hard wet freezing regions.

Special attention must be given to the overall design of the pavement system for pervious concrete to perform as desired. Proper engineering of the substrate beneath the pavement is essential, since together with carrying the pavement having traffic passing over it, it must be able to temporarily store the water while it percolates into the soil. An initial soils site survey and site-specific stormwater calculations should also be done by a stormwater management engineer.

12 CONCLUSION

The seriousness of environmental issues is recognized globally. However many countries and institutions are trying to tackle these problems through various projects but the global trend of the environmental degradation is still continuing. Global environmental issues have serious adverse effects on sustainable development, and it is beyond the reach of individual nations to resolve them by themselves. Therefore the efforts of environmental conservation and sustainable development should positively be promoted through joint efforts by developed, developing and other countries.

We need to reformulate the concept of development by ensuring that there is no net change in the water balance while considering a land area for development

A Global Technology Bank (GTB) should be created which may keep an updated record of the availability of enviro savvy/sustainable technologies and Best Construction Practices around the globe and its suitability/applicability in different regions of the earth addressing the environmental issues. The GTB should also study the environmental problems that need to be addressed in various countries, suggest the best possible available technologies, and coordinate to arrange training workshops, demonstrations, visits in order to disseminate the information/technology to the stake holders to take the full advantage of the technology Just providing grant/loans or subsidies in the name of environmental or social upgradations does not seem to provide any real and sustainable solution to the environmental problems. It has also to be ensured that the time lag between the development of technology in lab and its application in to the field is minimum possible.

Regulations should be framed and enforced to implement Storm Water Managements practices in developmental projects. Looking at the benefits offered by the material and its durability, it will be in favor of the environmental health of the globe to promote the technology for its widespread use.

Different Works Departments of Government at different level, local Municipal and other development authorities of a country should act to include the material in their Specifications and Schedule of Rates.

Projects using this material may also be offered some credit and intensives. This may certainly go in a long way to lead the world towards sustainable development.

REFERENCES

American Concrete Institute, 2006. "Pervious Concrete", ACI Manual of Concrete Practice, 522R-06. Committee. Farmington Hills, MI.

Bob Banka, BASF Admixtures, March 2007, Pervious Concrete Stormwater Management Systems Concrete Solutions, Inc., Northern Virginia Concrete Advisory Council.

Huffman Dan 2005, Understanding Pervious Concrete, The Construction Specifier.

India, the land of holy rivers, is fast becoming a land of highly polluted and even toxic rivers." http://www.devalt.org/water/WaterinIndia/issues.htm, Economic and Policy Issues

Jain A.K et al 2008, 'Pervious Concrete: An Environment Friendly Materials for Sustainable Development', Proceedings of the International Conference on Sustainable Concrete Construction, India.

LEED Credit SS-C7.1, Landscape and Exterior Design to reduce Heat Island Effect, The Ready Mixed Concrete Industry, LEED Reference Guide by the RMC Research Foundation, www.rmc-foundation.org

LEED Reference Guide, The Ready Mixed Concrete Industry, www.rmc-foundation.org

Nagaraj N & Chandrashekar H., Designing Methodologies for Evaluation of Economic and Environmental

Implications of Groundwater Depletion and Quality Degradation Effects: A Study in Karnataka Peninsular India., University of Agricultural Sciences, Bangalore.

Obla Karthik H 2007, Pervious Concrete for Sustainable development, Recent Advances in Concrete Technology, Washington D.C.

Junichiro Koizumi, 2002, Prime Minister of Japan, Speech during The World Summit on Sustainable Development Johannesburg, Republic of South Africa http://www.mofa.go.jp/policy/environment/wssd/2002/kinitiative2.html.

Wanielista Marty et al, Performance of Pervious Pavements, Stormwater Management Academy, University of Florida, Orlando, FL 32816.

World Scientists' Warning to Humanity, 1992 http://www.ucsusa.org/ucs/about/1992-world-scientists-warning-to-humanity.html.

Vernon R. Schaefer, February 2006, Mix Design Development for Pervious Concrete in Cold Weather Climate, Final Report, Iowa State University Ames, IA, USA.

Investigation on the use of burnt colliery spoil as aggregate in low to normal strength concrete

T. Runguphan & P.M. Guthrie
Centre for Sustainable Development, Engineering Department, Cambridge University

ABSTRACT: This paper presents ongoing work currently being conducted on unprocessed Burnt Colliery Spoil, a waste product from the coal mining industry, to investigate its performance as an alternative aggregate material in the production of low to normal strength concrete. This is done as a technical mean of presenting potential value added outlet for the material with economic and environmental benefits. Burnt Colliery Spoil was added as partial substitution to coarse natural aggregate at a range of increments by weight in the conventional production of low and normal strength OPC concrete. Cylinders were produced and tested for unconfined compressive strength, elastic modulus and concrete bulk density. Long term durability was determined on the concrete samples for freezing and thawing resistance, porosity, water absorption and resistance of concrete under aggressive chemical attack. Concrete with acceptable durability performance and with satisfactory strength requirement can be produced with high volume burnt colliery spoil replacing up to 40% of the coarse natural aggregate.

1 INTRODUCTION

1.1 Background

With increasing environmental awareness, the construction industry is constantly seeking ways to reduce the use of virgin aggregates as well as looking to increase the re-use of materials from its own industry (as well as others) rather than disposing of them as waste.

The construction industry worldwide is using the natural resources in large quantities, and statistics show that the UK annual demand for construction aggregates is around 250 million tones[1].

These practices of reducing the demand and dependence on primary aggregates and encouraging the use of alternative construction materials are potentially more sustainable ways to meet future demand. Some industrial by-products and waste materials are available in large quantities and, if technically sound and economically feasible outlets could be found, they might contribute to solving some of the aggregate supply problems.

The usage of Burnt Colliery Spoil (BCS) on low value-added engineering applications has been proven; for example it can be recycled into hydraulically bound mixtures for sub-base and base, unbound mixtures for sub-base capping or embankment fills. This research investigates using this material in higher value-added applications, such as aggregate in the manufacturing of low strength concrete for applications such as blocks, roof tiling, road curb, etc.

2 BURNT COLLIERY SPOIL

Burnt Colliery Spoil is a waste material from coal mining industry, the residue following ignition of coal mine spoil heaps which results in partial to complete combustion of coal particles in the spoil [HD 35/04]. Sherwood (1994) concluded that BCS had high potential as recycled secondary aggregate suitable for use in hydraulically bound mixtures. BCS was also shown to be suited to use as selected granular fill or sub-base or synthesis aggregate in concrete[2].

2.1 Material Supply

Colliery spoils can be founded arising from deep coalmines in the North East, Yorkshire, the Humber, West Midlands, East Midlands and South Wales[3]. Tonnages of the material arising in the UK (excluding Northern Ireland) are:

- 7.52 Mt (England and Wales) and 0.15 Mt in Scotland.

---

[1] Waste and Resources Action Program, Stakeholder update: Aggregate 17/5/04

[2] Use of Coal mining wastes as Aggregate: Aggregate Advisory Service

[3] Aggregain http://www.aggregain.org.uk

- Relevant portion suitable for aggregate use: 7.52 Mt. (100% of arising).
- Aggregate use: 0.81 Mt. (England and Wales) and 0.065 Mt. recycled (including non-aggregate use) in Scotland.

2.2 Characteristics of BCS

2.2.1 Physical properties
Gradation/shape
Spoil can vary in size from 100 mm. (4 in) to 2 mm. (No. 10 sieve) and will consist mainly of flat slate or shale with sandstone or clay intermixed. Coarse colliery spoil (those greater than 4.75 mm or retains in sieve No.4) is reported to be well-graded (Maneval, 1974). Most coarse spoil contains particles that may break down under compaction equipment, resulting in a finer grading following placement.

Particle densities
Several studies have been conducted on determining particle densities of BCS and unburnt spoil. Sherwood (1994) reported the density of BCS to be in the range of $2.65 \, Mg/m^3$ to $2.90 \, Mg/m^3$. Particle densities of $2.0 \, Mg/m^3$ and $2.7 \, Mg/m^3$ are reported for unburnt spoil (Rainbow, 1989)

Moisture-Density Characteristics
The optimum moisture content of coarse colliery spoil ranges from 6 to 15 percent at its maximum dry density ranging from 1300 kg/m^3 to 2000 kg/m^3 (Maneval, 1974).

Durability
Frost susceptibility was investigated for durability characteristic. BCSs are usually highly frost susceptible with the addition of up to 5% cement may reduce the voids content of the material sufficiently to mitigate the problem (Sherwood, 1975).

2.2.2 Chemical properties
pH and water soluble sulphate
The pH values for BCS range from 4.2 to 8.5 and water soluble sulphate contents ranges from 0.6 to $7.0 \, gSO_3/L \, m^3$ (Sherwood, 1994; 1995). Sulphides, as iron pyrites, are common in UCS but less likely to be present in BCS as they are oxidised during combustion.

3 EXPERIMENTAL PROGRAM

The experimental programme sets out to investigate the performance of concrete with a maximum BCS content of 60% by weight substituting coarse natural aggregate.

A direct use of BCS as alternative to primary aggregate in concrete was investigated. The BCS used were obtained locally within the UK. At present BCS used in

Table 1. Typical chemical compositions of burnt colliery spoil.

Chemical component	%
Al_2O_3	45–60
Fe_2O_3	21–31
CaO	4–13
MgO	1–3
Na_2O	0.2–0.6
K_2O	2–3.5
SO_3	0.1–5
Loss in ignition	2%–6%

BCS Sherwood and Ryley, 1970.

the initial test were obtained from Doncaster Colliery, South Yorkshire.

Compressive strength is an important property of concrete as the analysis of many of the mechanical and durability properties of concrete, to a certain extent, are directly related and can be drawn from strength. Unconfined compressive strength tests were investigated at different ages of concrete ranging from 28 days to one year.

Due to small fraction of BCS that was available initially at beginning of the experimental programme, plastic utility pipes with approximate diameter of 50 mm. were adopted as concrete mould. The ratio of diameter to height (d/h) of 2 was selected for the cylindrical concrete specimens in this experiment.

Preliminary mixes were done on standard concrete mix (no replacement of BCS) in order to detect any variation in the unconfined compressive strength obtained from a standard cylinder to that of the prepared pipe-mould. A good correlation with slight strength differences was obtained for concrete made with prepared mould.

3.1 Test materials

3.1.1 Coarse Natural Aggregate
For this experiment 10 mm. crushed gravel were used as coarse natural aggregate for all the concrete mixes. The material is obtained from the Concrete Material Laboratory in the CUED.

3.1.2 BCS
BCS used for the first stage of the laboratory experiment was obtained from Doncaster Colliery in South Yorkshire. Particle density of 2650 kg/m^3 was initially adopted. Surface moisture content of the sample was estimated using moisture reading probe prior to mixing of concrete.

3.1.3 Fine aggregate
In this experiment river-washed sand with fineness modulus of 2.4 was used. Surface moisture content of

Table 2. Trial mix design proportioning C15 concrete.

Constituents	Weight (Kg.) per 0.73 m^3 of concrete*	Weight (Kg.) per 1 m^3 of concrete
Cement	0.20	274
Water	0.17	233
Fine aggregate	0.63	863
Coarse aggregate	0.66	904

*0.73 m^3 of concrete make approximate three 50 mm cylinder samples.

2% measured from electronic moisture probe prior to mixing was adopted for free water correction in the calculation for the mix proportion. The value of particle density for fine aggregate was taken as 2700kg/m^3.

3.1.4 *Cement*

Ordinary Portland cement (ASTM type I) was used in this experiment. The density and specific gravity of cement are taken to be 3150 kg/m^3 and 3.15 respectively.

3.2 *Concrete work*

3.2.1 *Mixing proportion*

The same mixing procedure was used for all mixes. All materials were prepared and weighed in advance. Concrete mix constituents (gravel, sand, cement and BCS) were initially given a dry mix in a pan mixer for 2 minutes when the mixing water was added and further mixed for 3 minutes to ensure a thorough mix of cement paste and aggregate.

Key criteria in mix proportioning

- Standard concrete mix design (as of 100% natural aggregate), 28-day unconfined compressive strength of 15 MPa and 35MPa.
- The slump of 50mm for the concrete mix.
- Maximum aggregate size of 10 mm. was adopted, with the ratio of fine aggregate to total aggregate of 0.5.
- Coarse BCS replaces coarse natural aggregate by weight.
- No additional air content is specified. No chemical additives are used.

3.2.2 *Preparation before mix*

BCS was pre-washed to remove deleterious contaminants, then air-dried and sieved to size range of 9.5 mm.-5 mm. (passed through 10 mm. and retained on the 4.95 mm. sieved) prior to mixing.

Cylindrical moulds to be used in the test are made from utility pipe with diameter of 50 mm. cut at 100 mm in length. The moulds were cut from top to bottom to ease de-moulding. Plastic retentions were

Table 3. Trial mix design proportioning C35 concrete.

Constituents	Weight (Kg.) per 0.73 m^3 of concrete*	Weight (Kg.) per 1 m^3 of concrete
Cement	0.33	452
Water	0.16	221
Fine aggregate	0.66	908
Coarse aggregate	0.53	728

*0.73 m^3 of concrete make approximate three 50 mm cylinder samples.

used to tighten the mould. Plastic cup was placed at the bottom of the mould and acts as base-plate.

3.2.3 *Casting and curing*

Casting of 50 mm diameter cylinder was carried out in 3 equal layers. Compaction was done over the vibrating table. Vibration continued until air stop rising to the surface of the concrete in the mould. Excessive materials were removed with a trowel. Following placement of concrete in the moulds, the exposed surface was covered with cling film to prevent moisture evaporation and stored at room temperature.

After a period of 24 hours the moulds were removed and the specimens placed in water tank for curing at approximately 17°C until testing for strength at the required age (28 days). The specimens were removed from the curing tank after 28 days and further air-cured.

3.3 *Concrete testing*

3.3.1 *Unconfined compressive strength*

The compressive strength of concrete is of significant mainly, since many of the desirable characteristics of concrete are qualitatively relate to strength. Concrete samples were destructively tested for the Standard 28-days unconfined compressive strength and other subsequent concrete ages. The compressive strength of cylinders was determined as per BS 1881: part 110. Tests for compressive strength were carried out at 28, 56, 210 and 365 days. For all tests, each result is the average of the three values.

Control-mix together with concrete mixes with BCS replacement of up to 60% were initially investigated. Four series of two concrete mixes, with BCS content of 0%, 20%, 40%, 60%respectively for concrete UCS of 15MPa and 35MPa were tested. Additional mixes were conducted on concrete C15 with BCS content of 10% and 30%.

3.3.2 *Concrete bulk density*

Experiments for determination of concrete density and porosity were conducted in accordance to BS EN 722-4:1998 Determination real and bulk density and of total and open porosity for natural stone and masonry units.

The specimens were oven-dried at the temperature $70 \pm 5°C$ to constant mass. Wight measurement taken once constant mass is reached and recorded as weight oven dry ($m_{dry,s}$). Concrete sample were then immersed in deionised water at 20°C and left immersed under atmospheric pressure for 24 hour. After 24 hour the samples were weighed in water and recorded as apparent mass (weight immersed, $m_{w,s}$) The surface was then wiped dry and the samples were weighed for weight saturated, $m_{sat,s}$.

The volume of the open pores (in mm^3) is expressed by the equation:

$$V_p = (m_{sat,s} - m_{dry,s})/p_w \qquad (1)$$

The bulk volume (in mm^3) is expressed by:

$$V_b = (m_{sat,s} - m_{w,s})/p_w \qquad (2)$$

Bulk (apparent) density is expressed by the ratio of the mass of the dry specimen and its bulk volume:

$$P_{b,s} = m_{dry,s} \, p_w \times 1000/(m_{sat,s} - m_{w,s}) \qquad (3)$$

Open porosity $P_o(\%) = (V_p/V_b) \times 100 \qquad (4)$

3.3.3 %Water absorption
Absorption of normal concrete is usually expressed by the increase in mass after submersion as a percentage of oven dry mass. Absorption was measured on 6 concrete cylinders for each mix.

3.3.4 Young's Modulus of Elasticity
Tests for determining the modulus of elasticity of concrete were conducted in accordance to BS 1881: part 121:1983 Method for determination of static modulus of elasticity in compression. Modulus of elasticity is taken at 28 days. For all tests, each result is the average of the three values.

The static modulus of elasticity in compression which in Newton per square millimetre, is the measured of the different in stress over differences in strain between a basic loading level of about 0.5 N/mm²) and an upper loading level of about one-third of the compressive strength of concrete. 3 samples were tested from each mix for the average elastic modulus value.

3.3.5 Freeze- thaw resistance
This experiment was conducted in accordance with BS EN 722-18:2000 Methods of tests for masonry units: Determination of freeze-thaw resistance.

The specimens were subjected to fifty freeze and thaw cycles. After the full 50 freeze-thaw cycles the specimens were tested for unconfined compressive strength. The tests were done on concrete sample at the age of 56 days, 6 samples of each mix were tested for one average strength value.

3.3.6 Sulphate attack
Sulphate resistance testing was done by the immersion of concrete specimen after the specified initial curing age in a water tank containing 5% magnesium sulphate solution at standard room temperature ($20 \pm 5°C$) for the period of 16 weeks. Control concretes were kept in lime-saturated solution at standard temperature for the unconfined compressive strength reduction determination.

The degree of sulfate attack was evaluated by visual inspection of concrete specimens to crack and measuring the unconfined compressive strength of the concrete cylinders at the end of sulfate exposure period. XRD analysis is to be conducted to the crushed sample to verify the possible formation of gypsum and ettringite to which attribute to the expected lost in strength of concrete with sulphate attack.

4 RESULTS

4.1 Unconfined Compressive Strength (UCS)
Control-mix together with concrete mixes with BCS replacement of up to 60% were initially investigated and results of the unconfined compressive strength tests at different ages are shown in Table 4.

Figure 1 shows the graphs of unconfined compressive strength of concrete cylinders made with coarse natural aggregate (control-C15REF) and those made with BCS substituting coarse natural aggregate plotting at different ages.

The UCS yield at 28 days for the control concrete was 15.1 MPa which is just achieving the designed strength. This is dropped gradually to 9.8 MPa (approximately two-third of the control strength) with 60% BCS replacing coarse aggregate (C15B60). A gradual reduction trend can be seen with UCS drops from 17.1 Mpa to 14 Mpa (56 days), 23 MPa to 12 MPa (210 days) and 23.5 MPa to 13.5 (365 days) for control concrete and concrete with 60% BCS substituting coarse natural aggregate respectively.

It may be reasonable to postulate that there is a percentage of BCS that would be acceptable, strength wise, for many applications of concrete. The 40% aggregate replacement value when tested at 56 days, is promising as it includes large enough waste material to substitute the use of virgin aggregate to favour the sustainability in aggregate use, without sacrificing the mechanical property requirement(still yielding the required strength). This is valid as long as the specimens are allowed to gain strength for a longer than industrial standard of 28 days which is feasible for concrete applications that does not required high early age strength.

The development with time of UCS of concrete manufactured with increasing BCS addition is given in Figure 2.

Table 4. Unconfined compressive strength test at different ages.

Mix	%BGS	Unconfined Compressive Strength(MPa)			
		28 day	56 day	210 day	365 day
C15REF	0	15.1	17.1	23.0	23.5
C15B10	10	14.0	18.2	21.0	21.9
C15B20	20	11.7	17.1	17.8	18.5
C15B30	30	12.0	17.3	15.8	17.2
C15B40	40	10.8	16.0	15.7	16.5
C15B60	60	9.8	14.0	12.0	13.5

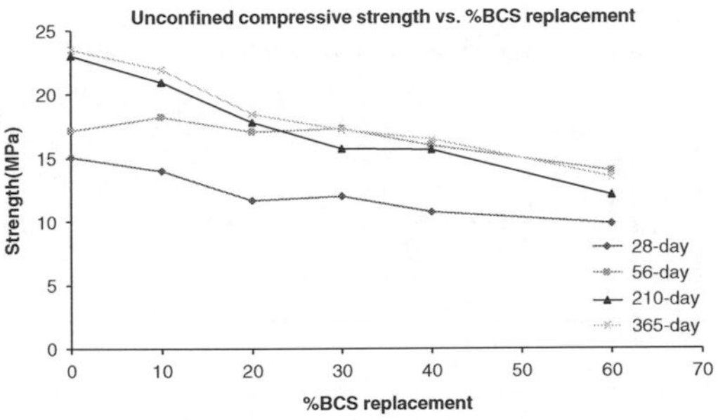

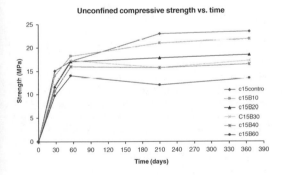

Figure 1. Unconfined Compressive Strength vs. % BCS replacement (C15).

Table 5. Unconfined Compressive Strength for C35 concrete.

Mix	%BGS	Unconfined compressive strength (MPa)		
		28-day	56-day	210-day
C35REF	0	40.8	52.6	53.6
C35B20	20	37.5	45.2	44.8
C35B40	40	34.4	44.0	43.2
C35B60	60	21.8	39.0	38.4

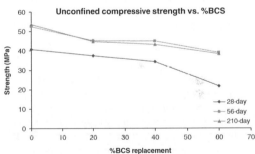

Figure 3. Unconfined Compressive Strength vs. %BCS replacement (C35).

Figure 2. Unconfined Compressive Strength vs. time (C15).

It may be valid to conclude that all BCS addition mixes exhibit similar strength development overtime as that of ordinary concrete (control a 0%BCS) i.e. a very high strength gain(steep gradient) from 0 (demould) to 28 days. This influence of strength increase overtime drops off to a gradual increase as the specimen ages. Concrete with 60%BCS replacement exhibited the least amount of increase of strength overtime as well as the least gradient at 0 to 28days measurement. Apart from control and C15B10, little to no strength increase can be seen as the concrete age more than 56 days. Control and C15B10 gained strength slightly further and this tapered off at about 210 days.

The results of UCS tests on concrete with designed strength of 35 MPa (C35) are given in Table 5 for strength measurement of concrete up to 210 days. Again similar trends can be drawn as with C15 concrete that replacing coarse natural aggregate with BCS resulted in a uniform drop of concrete UCS (Figure 3). The 56-day UCS strength of all specimens surpassed the required strength of 35MPa even at the high per cent BCS inclusion of 60%. The development of strength overtime for BCS concretes form a very uniform trend and similar to that exhibited in control concrete with a steep strength increase from 0 to 56 days and a more gradual increase as the concrete ages further.

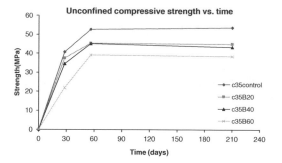

Figure 4. Unconfined Compressive Strength vs. time (C35).

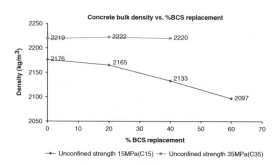

Figure 5. Concrete bulk density vs. % BCS replacement.

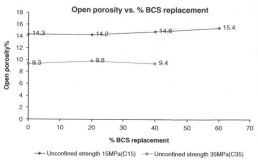

Figure 6. Open porosity vs. % BCS replacement.

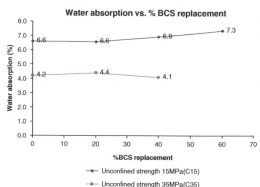

Figure 7. Water absorption vs. % BCS replacement.

4.2 Concrete bulk density and open porosity

The plot of concrete bulk density against the % burnt colliery spoil substitution is shown in Figure 5. C35 concrete exhibited more uniformity with little to no change in bulk density with the increase of % BCS replacement, and has the bulk density range from 2219 kg/m^3 to 2230 kg/m^3 for control (C35REF) and C35B40 respectively.

There is slightly more evidence of the influence of BCS inclusion with C15 concrete where the bulk density of the concrete sample drops from 2176 kg/m^3 (C15REF) to 2097 kg/m^3 (C15B60), although in term of percentage drop, the variation in bulk density of concrete with BCS replacement is considerably essentially unaffected especially at higher designated strength.

Figure 6 shows the influence of BCS replacement to concrete open porosity. The highest % open porosity was exhibited by C15 concrete containing 60% BCS as natural aggregate replacement ($O_p = 15.4\%$). At 40% BCS replacement ratio the % open porosity for C15 and C35 were 9.4 and 14.8, the increase of 1.1% and 3.5% from control concrete respectively.

It can be seen from the results that the concrete open porosity may be viewed as being unaffected by the inclusion of BCS.

4.3 Water absorption

Figure 7 shows the results of water absorption tests at atmospheric pressure on concrete with increasing aggregate volume replacement by BCS addition. Small increase in % water absorption is seen in the C15 concrete as the % replacement BCS to coarse natural aggregate increases. These changes are less as for the concrete with designed strength of 35 MPa where the value of water absorption across the % BCS replacement ranges are considerably constant. At a 40% BCS, the % water absorption are 4.5% and 2.4% respectively.

4.4 Modulus of Elasticity

The Young's modulus of elasticity of the concrete sample was measured at the concrete age of 28 days and is shown in Figure 8. Young's modulus of elasticity is closely related to concrete compressive strength thus similar trend as with UCS is to be for – both control and concrete made with BCS.

Young's modulus of elastic of the control C15 and C35 concrete obtained were 26.1 GPa and 33.7 GPa respectively which are well within the normal rage of elastic modulus of concrete. A gradual decrease in

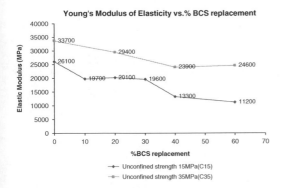

Figure 8. Young's modulus of elasticity vs. % BCS replacement.

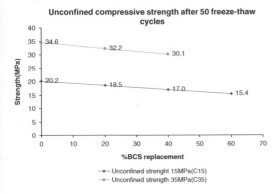

Figure 9. Unconfined compressive strength after 50 freeze-thaw cycles vs. % BCS replacement.

value is seen for C35 concrete as the per cent inclusion of BCS increase, this can be attributed to the weaker granular properties of BCS compared to the natural aggregate. Significant drop in elastic modulus is seen however with the C15 concrete, which may be attributed from the mix having an overall weaker concrete matrix.

4.5 *Freeze- thaw resistance*

All the test specimens still hold to theirs structural integrity even after passing through 50 cycles of freezing and thawing. The UCS tests were conducted and the results are shown in Figure 9.

It is evident from the graph that freezing and thawing resistance of concrete degrades with increasing %BCS content in concrete. This decrease is evident for both C15 and C35 concrete and both exhibit a gradual decreasing trend. At 40% BCS substituting coarse natural aggregate, UCS drops of 16% and 13% from the control concrete were recorded for C15 and C35 respectively after the freeze-thaw testing.

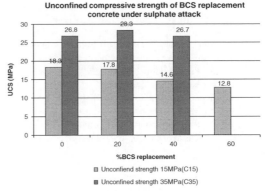

Figure 10. Unconfined compressive strength under sulphate attack vs. % BCS replacement.

4.6 *Sulphate attack*

The results of the UCS tests on concrete after a period of submersion in solution of Magnesium Sulphate are show in Figure 10. A decreasing trend for concrete UCS as the % BCS substitution increases can be drawn for the C15 concrete as is evident in Figure 10. This weaker matrix together weaker granular BCS aggregate could be attributed for the direct influence of replacing natural aggregate with BCS.

The concrete samples designed at C35 yielded higher density, and the concrete matrix could be taken to be more closely packed with less porosity which would result in the sample being less permeable, and the concrete less prone to sulphate ingression. Thus the effect of replacing coarse natural aggregate with BCS is less evident here. The open porosity values were relatively unaffected with the inclusion of BCS even at the high percentage of 60%. At 40% BCS replacement of coarse natural aggregate losses in strength as compared to the control (0%BCS) were −20% and 0.4% for C15 and C35 concrete respectively.

5 CONCLUSION

The replacement of natural aggregate by BCS has been shown to be technically viable. Preliminary results obtained on the mechanical performance of concrete made by replacing coarse natural aggregate by BCS waste from coal industries confirm that acceptable quality concrete can be achieved with the inclusion of burnt colliery spoil as substitute coarse natural aggregate of up to 40%. The resulting BCS concrete has been shown to yield adequate concrete quality, both in term of strength performance and the required durability properties similar to that of concrete made with 100% coarse natural aggregate. This concrete may have certain percentage drop in desirable concrete behavior but useable concrete especially for low strength applications can still be made at this high level of waste inclusion. This fact justifies the efforts to use this waste material, which can contribute to the preservation of the environment and can attain the desired performance at probably lower overall costs than those of ordinary concretes.

This investigation also shows that it is possible to evaluate the influence of BCS addition in mechanical and durability properties of concrete. The results obtained in this work point toward the effective possibility of using burnt colliery spoil in the production of new concrete.

The investigation presents new ways of achieving sustainability of construction materials. As a value contribution, the use of BCS as aggregate replacement would over time significantly help relieve pressure on demand for virgin aggregate resource, and also identify the useful destination for the 7.5 million tons of material that otherwise would have been lost. This practice could provide an adequate solution to environmental issues such as the preservation of virgin aggregate resources and the limited availability of landfill or waste disposal sites. A further advantage of the use of BCS is the lower environmental impact of concrete arising from the partial replacement of virgin aggregate with BCS.

REFERENCES

Aggregates Advisory Service Digest 066: Use of Coal mining wastes as Aggregate, Department of the Environment, Transport and the Regions, Rotherham, UK.

Aggregain: http://www.aggregain.org.uk.

BS 1881: part 121:1983 Method for determination of static modulus of elasticity in compression.

BS 1881–1993. Testing concrete.

BS EN 722-4:1998 Determination real and bulk density and of total and open porosity for natural stone and masonry units.

BS: EN 772-11:2000 Method of test for masonry units – Part 2 Determination of water absorption of aggregate concrete, manufactured stone and natural stone masonry units due to capillary action.

BS EN 722-18:2000 Methods of tests for masonry units: Determination of freeze-thaw resistance.

HD 35/04 Conservation and the use of secondary and recycled materials. Design Manual for Roads and Bridges. London: The Stationery Office.

Maneval, D. R (1974). Utilization of Coal Refuse for Highway Base or Subbase Material. Proceedings of the Fourth Mineral Waste Utilization Symposium. IIT Research Institute, Chicago, Illinois, May, 1974.

Rainbow, A.K.M (1989). Geotechnical properties of United Kingdom minestone. London: British Coal.

Sherwood, P.T and M.D Ryley (1970). The effect of sulphates in colliery shale on its use for roadmaking. RRL Laboratory Report LR 324. Crowthorne: Transport Research Laboratory.

Sherwood, P. T (1975). The use of waste and low-grade materials in road construction: 2. Colliery shale. TRRL Laboratory Report LR 649. Crowthorne: Transport Research Laboratory.

Sherwood, P.T (1994). The use of waste materials in fill and capping layers. TRL Contractor Report CR 353. Crowthorne: Transport Research Laboratory.

Sherwood, P.T (1995). Alternative materials in road construction: a guide to the use of waste, recycled materials and by-products. London: Thomas Telford.

Waste and Resources Action Program, Stakeholder update: Aggregate 17/5/04 (http://www.wrap.org.uk).

Excellence in Concrete Construction through Innovation – Limbachiya & Kew (eds)
© 2009 Taylor & Francis Group, London, ISBN 978-0-415-47592-1

Improvement of characteristics in cement composite sheet with agriculture waste fiber

M. Khorrami
Islamic Azad University (Eslamsshahr branch), Iran

E. Ganjian
Coventry University, UK

M.A. Khalili
Tehran University, Tehran, Iran

ABSTRACT: Waste materials and their applications have been gaining increasing attention in the sustainable development strategies. In this research, to attain these policies, the applications of agricultural-waste fibers (AWF) were considered in producing the cement composite boards (CCB). In this paper, three types of AWF including the bagas fiber, wheat fiber and eucalyptus fiber with 2 and 4% of cement weight were applied to produce CCB. Moreover, the effects of silica fume were studied in flexural and energy absorption characteristics of CCB made of these novel materials. The results showed that there was a good consistency between the natural fibers and cement matrix and also some of them could effectively improve the physical and mechanical properties of CCB. Also it was found that the amount of fibers to be replaced was related to the type, volume weight, length, diameter and the texture of fibers. Low volume-weight or longer fibers were led to inconsistency and bad appearance of the final product. Moreover, it was observed that the silica fume could improve the flexural strength of the CCB; however, it could not be effective in energy absorption or ductility characteristics.

1 INTRODUCTION

In the last two decade, the application of fibers in cement-based composites have been gaining increasing attention and been applied to enhance the properties of theses construction materials. Some of these applications are the fiber concretes or cement-composites like the flat or corrugated boards. In production of cement boards for roofing systems, the mineral fibers are utilizing together with the cement, water and additives. In Iran, the asbestos fibers are frequently used to produce the cement composite boards (CCB).

Asbestos is a mineral stone which its main ingredient is SiO_2 or silica. This stone can be easily converted to needle-shaped and soft fibers. Based on the mineralogy of asbestos, it is classified in two categories: serpentines and amphiboles. Serpentines are commercially named as chrysotile or white asbestos. Amphiboles are divided into three groups: chrosidolites (blue asbestos), amosites (brown asbestos), and antophilites. The mineral fibers, which are used to produce the cement boards, are very harmful for human health. The entrance of the fibers in human bodies would be led to serious illness relative to the type and amount of fiber ingresses. Brown and blue asbestos could be easily converted to the fine particle and pollute the air with fibrous dust, then could be more dangerous than the white asbestos. It should be noted that world health organization (WHO) has been legislated these restrictions for the mineral fibers: Ampzite: 0.5 fiber/cm^3, chrysotile, chrosidolite and other types: 2 fiber/cm^3 (WHO 1989).

In the early 1970s a global effort was initiated to legislate for the removal of asbestos reinforcement from a wide range of products. Fibre cement was a major user of asbestos and as such new reinforcing fibers were being sought as alternatives to asbestos in this class of building material (Coutts 2005). Those countries that recognized the need to legislate against the use of asbestos, on health grounds, have proved to be the ones that have achieved the most advances with respect to asbestos substitution and have thus avoided, in most cases, a downturn in the fiber cement business (Coutts 2005).

On the other hand, as a result of the dangerous effects of asbestos fibers on human health, the majority of developed countries have been prohibited the

Table 1. The annual application of asbestos in Iran.

Year	1960	1970	1980	1990	1995	2000
Average Asbestos Production in tons	1,200	11,000	23,000	72,000	54,000	40,000

use, product and application of them in construction industries. However, in Russia and China there is no regulations respect to this issue yet. In this sense these countries together with Canada, are producing more than 2/3 of the world's total asbestos productions.

Table 1 shows the annual application of asbestos in Iran. Unfortunately, despite the global decrement, we considered a great amount of applications for construction industries in Iran. In this country the superior council for protecting of the living environment, puts some restrictions on the application of the asbestos materials. Based on these limitations, after July 2001, the newly established factories are forbidden to use the asbestos in their products and the factories that were previously using the asbestos as a rough material are ordered to modify their production procedure to replace the asbestos with other allowable material to completely eliminate the applications in the next 7 years.

Over the past few years, lignocellulosic fibers have received considerable attention in the development of asbestos-free fiber reinforced cement composites (MacVicar & Matuana 1999). Much effort has been devoted to wood fibers because they possess many advantages relative to asbestos in terms of availability, lower cost, simple production processes for making cementitious composites of various shapes, renewability and recyclability, non-hazardous nature, and biodegradability (MacVicar & Matuana 1999, Coutts 1983, Cook 1980, Blankenhorn et al. 2001, Bentru & Mindness 1990, Coutts 2005, Bentru & Akers 1989, Pehanich et al. 2004, Aziz et al. 1984, Kaufmann et al. 2004, Mohr et al. 2005, Andonian et al. 1979).

In this research, to assess the applicability of the natural-sourced fibers like wheat fiber (WL), bagas fiber (BL) and eucalyptus fiber (EL), an experimental study was established to investigate the microstructure, flexural strength, and energy absorptions properties of the cement boards produced of these agricultural waste fibers.

2 EXPERIMENTAL PROCEDURE

2.1 Materials

Cement: Type 2 cement supplied by Tehran Cement Factory was used in this study. Standard laboratorial tests (based on Iranian National Standard No.

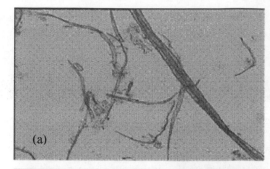
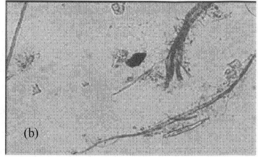

Figure 1. Curing effects on fiber surfaces: a) before and b) after curing.

398) were executed to determine the properties of this cement, which is passed the requirements.

Fibers: The major part of the fibers of this study was prepared from the agricultural wastes. At first the fibers were cured to perform the experiments. To accomplish this, the fibers moistened and mechanically fibrillated with passing them across the moving gear and the other fixed one. It should be noted that the fibrillation has diverse effects on the fibers. As depicted in Fig. 1, if the fibers are fibrillated on their outer surfaces, it could be clearly visible by microscope. In this case, the fibrils which were formed on the surfaces of the fibers would have significant role in developing the mechanical bonds. On the contrary, if the fibrillation have internal effects, it was rarely could be seen by microscopic observations. However, the phenomena may be interpreted similar to the uncoupling the rope with twisting in the inverse direction of texture. Shortening of the fibers could be occurred in the process of curing which should be avoided. During the curing process, dust may be produced because of fibrils dugging out of the external surface of fibers. These substances would have unfavorable effects and should be sieved before use.

From the economical reasons, limited modifications and surface curing are exerted on waste-sourced fibers as recommended below:

Bagas fiber (BL): Bagas fiber (sugerance) with length 0.5–3 and diameter 0.3 were used. These fibers

Table 2. Mix proportion of the specemens.

Mixture no	Cement	Water	Silica fume	Fiber	Proportion
Reference	150	450	–	–	No fiber used
B2	150	450	–	3	2% BL
B2M5	142.5	450	7.5	3	2% BL+5 % SF
B4	150	450	–	6	4% BL
B4M5	142.5	450	7.5	6	4% BL+5 % SF
W2	150	450		3	2% WL
W2M5	142.5	450	7.5	6	2% WL+ %5SF
W4	150	450		6	4%WL
W4M5	142.5	450	7.5	6	4% WL+ %5SF
O2	150	450	–	3	2% EL
O2M5	150	450	–	3	2% EL+ %5SF
O4	150	450	–	3	4%EL
O4M5	150	450	–	3	4% EL+ %5SF
B2	150	450	–	–	–

Table 3. Test results of physical charcteristics of fibers.

Fiber type	Bagas	Eucalyptus	Wheat
Average length (mm)	1.303	1.466	1.238
Average diameter (mm)	0.348	0.480	0.355
Average tensile strength (MPa)	115.947	35.440	8.093
Aspect ratio	3.744	4.054	3.487

were cured in water for 24 hours an then grinded in the laboratory. The griddling process would be fibrilized them. The prepared fibers were refined by 150 micron sieve (Fig. 1).

Eucalyptus fiber (EL): These fibers included the fine particles (dust). After refining with sieve 160 micron, the big fibers with length 0.8–1 and diameter about 0.48 were cured in water for 24 hours.

Wheat fiber (WL): In this case, like the other cases, the fine particles were separated and water cured for 24 hours. The length was 1 1.5 mm and the diameter was 0.3–0.4 mm.

Water: In this study, the used water was drinking water.

2.2 Mix design and sample preparation

The major parameters which considered to study were the type and percentage of fibers to reinforcing the cement boards. The natural fibers of this research were Eucalyptus fiber, Bagas fiber and Wheat fiber which originally sourced of Iranian type. Moreover, the effect of silica fume as an additive was studied. Table 2 shows the samples characteristics plus the characteristic names.

2.3 Test procedure

Water to cement ratio of 0.3 were used to prepare the cement composite boards. A rotary mixer with

horizontal blades was used. The fibers were mixed to separate from each other and better dispassion in cement paste for 5 min. Then the cement, water and fibers (if applicable) were added to mixture and stirred for another 5 min. The well-mixed mortar was poured into molds to form cubes of $8 \times 8 \times 15$ cm for all mixing proportions (Fig. 2).

Excessive water was drawn out from the samples using a vacuum pump (0.9 bar power). During this procedure, a 10 kg load was applied to sample for consolidation purposes. After then, the sample demoded, dried for 1 hour and then cured in steam cabinet (RH = 100%) for 28 days. After curing, the specimens were dried for test in oven for 6 hours, in temperature 75 degree centigrade

3 TESTS

3.1 Fiber tests

3.1.1 Freeness test

One of the main characteristics of the fibers in cement matrix is the Canadian Standard Freeness (CSF) which is using for measuring the drainage properties of wood paste. The results of CSF are depends on many variables such as: the fine particles and small pieces of available wood, fibrillation degree, flexibility of fibers, and the finesse modulus.

The freeness test was carried out according to the Canadian standard AS/NZS 1301.206s:2002 (CSF). CSF was measured for the current study and the value of CSF = 500 were gained for natural fibers in which should be in the range of 400–700.

3.1.2 Physical properties

The physical properties of fibers were tested and summarized in Table 3.

3.2 Cement board tests

3.2.1 Flexural strength

In this research, the strength of samples was tested in flexural loads. The flexural samples were flat

569

Figure 2. Preparation of fibers to cement board composite.

Figure 4. One point flexural strength according to the standard EN12467: 2004.

It could be perceived that the fiber enhances the load bearing capacities of cement boards. The maximum load-bearing capacity in reference sample was 54.42 N while this value for the samples with WL, EL, and BL were enhanced up to 58.8, 76.56 and 96.4 N respectively. To better characterization, stress-strain curves were plotted according to the following relationships.

$$\sigma = M / W \qquad (1)$$

where σ is available stress (MPa), M is flexural moment, W is (—)

$$M = PL / 4 \qquad (2)$$

P is applied concentrated load (N) and L is the length of sample.

$$W = BH^2 / 6 \qquad (3)$$

where B is width (mm) and H is height (mm) of sample.

$$E = \sigma / \varepsilon \qquad (4)$$

where E is modulus of elasticity (MPa) and ε is strain.

Replacing the values of M and W in Eq. (1), the stress will be gained as follows:

$$\sigma = \frac{3PL}{2BH^2} \qquad (5)$$

Deflection can be calculated assuming the linear region as:

$$\delta = \frac{PL^3}{48EI} \qquad (6)$$

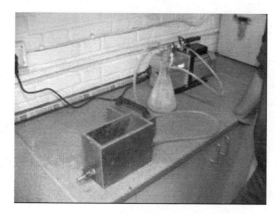

Figure 3. Equipments for preparing the cement boards from cement mortar.

Table 5. Minimum failure load for various classes according to the EN12467:2004

Class	Minimum flexural strength
1	4
2	7
3	10
4	16
5	22

and tested with a point load according to the EN12467:2004 standard test as shown in Figs. 3 and 4. This standard suggests 5 classes for flexural strength as shown in Table 5.

4 TEST RESULTS AND DISCUSSIONS

4.1 Stress-strain curve

Figs. 6 and 7 shows the load-deflection curve of cement boards for reference sample and other samples.

Figure 5. Some samples according to the test EN12467: 2004.

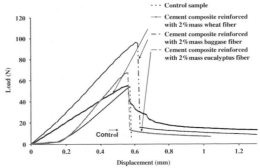

Figure 6. Load-deflection curve for composite cement boards.

where I is moment of inertia and δ is the deflection (mm).

By increasing the load P, the value of δ could be measured in real time. If the linear region was assumed, it would be computed following the:

$$E = \frac{PL^3}{48\delta I} \qquad (7)$$

Where

$$I = \frac{1}{12}BH^3 \qquad (8)$$

Using the Eq. 7, the value of ε can be gained:

$$\varepsilon = \sigma / E \qquad (5)$$

Fig. 6 compares the stress-strain curve of various types of cement composites with reference. As it could be seen, non-reinforced cement boards were brittle under load. In other words, it would be failed suddenly in the yielding stage. In contrast, reinforcing of boards with fibers adds up the flexibility in various stages of stress-strain curve. Unlike the common sense that the fibers will be activated after cracking, here, it was observed that the fibers were active in load bearing phase of the cement board. Moreover, it could be seen that the fiber reinforced cement boards exhibited more flexible behavior than non-reinforced one.

The stress-strain curve demonstrated that adding the natural fiber to the cement paste has been improved its final strength that eventually would be a base to choose the type of fiber, curing process, and the amount of fiber addition. For example as it could be seen, EL had more enhancement properties than WL. Moreover, the post-yielding properties were the main differences between the various types of composites. The non-reinforced sample suddenly failed after its yielding load. However, in fiber reinforced cement composites, this unfavorable sudden failure could not be seen and the samples could withstand more ore less

after yielding. This behavior could be interoperated by fiber-bridging characteristics. Generally, it was considered that all of the specimens which fiber-added has been exhibited ductile performance during the flexural loading.

Fig. 8 demonstrates the flexural strength development in the agricultural-waste fibers in various situations. Adding 2 and 4% of BL to cement composite would be lead to flexural enhancement 37 and 44% respectively in comparison to the non-fibrous cement boards. Also it was observed that up to 2% adding to the cement would offer the most enhancements in the case of BL. In this case more addition of fibers could not be effective. The lightness of BL, originated the poor dispersion and then forming a layered composite with variable fiber distribution amount in each layers. As a result of this problem, other addition amounts were not investigated. This study demonstrated that BL (>2%) would has suitable consistency with cement paste and could help the flexural strength.

Unlike the advantages of silica fume in cement paste properties, the results of this study showed that the application of SF could not improve the flexural properties of BL-composite cement boards. It was observed that substituting 5% of SF in 2 and 4% BL led to a 27 and 6% decrease in flexural strength of samples. In other word, application of silica fume in cement board with BL was not useful. Also, due to the crystals formed in the outer layers of fibers, the thickness of boards containing SF were increase up to 20–30%. The increase of thickness directly decreased the flexural strength according to the Eq. (5).

It could be seen that WL composite boards enhanced the flexural strength up to 2%. The more WL used, the more enhancements occurred in composite boards, where, 4% WL composite would has further strength than non-added composite. In this case the application of silica fume would be effective. The mechanism of enhancement is due to the interconnecting effect of WL that acts as a system of mesh enveloped by cement paste. This effect improved the

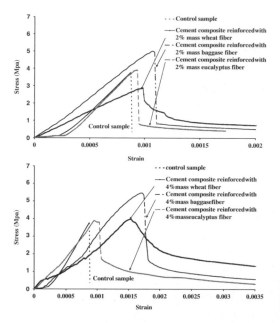

Figure 7. Stress-strain curve for composite cement boards (2 and 4%).

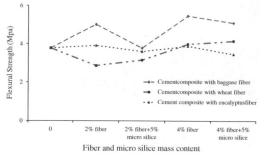

Figure 8. Flexural strength of natural-composite boards.

Figure 9. SEM photograph of BL.

load bearing capacity of samples because of confinement phenomenon. Consequently, in samples with 2% fiber, due to deficit of fibers and mis-dispersion, the bearing capacity dropped and the more percentage of fiber used, the more enhancement took place.

Behavior of EL composite was seen to be different of all other composites. In this case, by increasing the amount of fibers, a decrease was happened in flexural strength. Increasing the amount of EL was resulted in mis-dispersion and then the more WL amount, the more inconsistency occurred leading to a decrees in flexural strength. The assessment of fiber's diameter showed that the EL had greater size in contrast. This properties cause a decrease in available interface zone between cement paste and fibers which consequently led to decrease in bonding and flexural properties.

5   MICROSTRUCTURAL ANALYSIS

Microstructure of developed natural-based cement boards were examined by SEM observations. Figs. 9–11 shows SEM images of BL WL and EL respectively.

As it could be seen in Fig. 8, the used BL has partly rough surfaces. In other word, it has numerous fibrils to interconnect with cement paste matrix. Free of dust fibers, could decrease the slip between the cement-fiber interfacial zones. Hence used BL without dust, 2 mm length and sufficient surface could better develop the boding and flexural performances.

Fig. 10 shows the SEM image of WL. Wheat fibers were impacted and separation or untwisting in single fibers were impossible. This cause to mis-dispersion of fibers in cement paste and led to configuration of non-uniform mesh in various location of composite with different strength development effects. Beside, corresponding to the Fig. 9, the outer surface of fibers were relatively smooth, greasy which led to be water-repellent. Hence, the bonding strength between cement and fiber could not be effectively developed and eventually it was not expected to exhibit good performances.

Despite good fibrilization of EL in some parts and converting the mass fiber to single ones, the outer surface is smoother than BL (Fig. 10). Moreover, the diameter of these fibers were greater and lesser surface were available to bond with cement paste. Hence, EL could not enhance the flexural strength in comparison to BL-made composite.

6   ENERGY ABSORPTION

Energy absorption is characteristics of a material under load to be fracture. The area under the stress-strain

Figure 10. SEM photograph of WL.

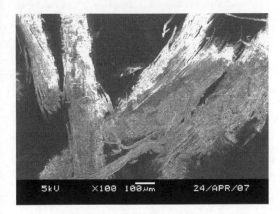

Figure 11. SEM photograph of EL.

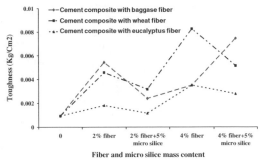

Figure 12. Energy absorption of fiber composites.

curve is generally defied as energy absorption of materials. The values of EA for BL-composite were shown in Fig. 11. In bagas case, the enhancement was seen about 300–400 times in comparison to the references, while the flexural capacity enhancement was gained about 30–40 times. These observations proved the capabilities of fibers in improvement of flexural strength after crack initiation. In other word, this mechanism would enhance the energy absorption properties of composites by delaying the sudden failure. Then, in non-fibrous samples, the failure would be such a brittle while in fibrous samples related to the amount and the type of fibers, the failure mechanism shifted to ductile behaviour.

Figure 12 shows a schematic representation of a cross-section through a fiber reinforced matrix. The diagram shows several possible local failure events occurring before fracture of the composite. At some distance ahead of the crack, which has started to travel through the section, the fibers are intact. In the high-stress region near the crack tip, fibers may debond

from the matrix (e.g. form 1). This rupture of chemical bonds at the interface uses up energy from the stressed system. Sufficient stress may be transferred to a fiber (e.g. form 2) to enable the fiber to be ultimately fractured (as in form 4). When total debonding occurs, the strain energy in the debonded length of the fiber is lost to the material and is dissipated as heat. A totally debonded fiber can then be pulled out from the matrix and considerable energy lost from the system in the form of frictional energy (e.g. form 3). It is also possible for a fiber to be left intact as the crack propagates. This process is called crack bridging.

For the bagas fiber samples, as demonstrated in Fig. 13, the pull-out mechanism was observed during the loading process. As discussed earlier, increasing the BL amount in the range 2–4% would decrease the energy absorption capacity. Non-uniform distribution of BL in 4% case resulted in major decrease in the effectiveness of fibers in cement matrix. Following the full-pulling out of the fiber from the cement paste, the strain energy would be propagated in the length of fibers as heat energy. On the other hand, in pull-out mechanism of fibers, the considerable energy would be relapsed in the form of fractional energy (like sample 3). Also a fiber may be slipped and be intact during the crack propagation. In this case, the intact fiber would be bridged between two fractured surface parts of composite that eventually led to more enhancements. If the bonding was stronger, then the sample beard more loads until tearing of the fiber. Otherwise, the governing fracture mechanism would be of full pull out and consequently lessens the energy absorption characteristics of composite.

The experimental observation (Fig. 10) shows that WL application increases the energy absorption up to 4 times while, the flexural enhancement was lesser. Namely, WL plays as a crack controller in composite and would not enhance the flexural strength.

According to the Fig. 11, 280% and 4% improvement was observed for energy absorption and flexural capacity of composites containing EL respectively. Regarding to these results and SEM micrographs, it

573

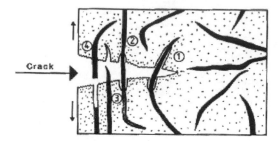

Figure 13. Schematic of crack propagation inside the cement matrix with fibers.

Figure 14. Fracture surface of Bagas-cement composites.

could be say that the big diameter of EL in contrast to the others, would decrease the outer surface leading to decrease in bonding strength and energy absorption characteristics.

In WL and EL samples, the increasing of the replacement from 2 to 4% would be resulted in doubling the energy absorption characteristics. This would be due to the proper dispersion that aid to more fibers contributed in load bearing process. Hence, with increasing the demand of frictional energy, the energy absorption would be increased as a consequence.

Mainly, it was considered that the energy absorption of developed composite was mainly affected by physical characteristics of fibers like aspect ratio and tensile strength. As demonstrated in Table 3, despite the similarity in aspect ratio of WL and BL, the tensile strength of BL was about 14 times of WL. However, energy absorption of BL contained samples only enhanced 16% in contrast to WL samples

As it could be perceived, increasing in aspect ratio, led to fiber-tearing to be the governing failure mechanism. The experimental observations confirmed this mechanism for our cases. Similarly, in samples with WL and EL, the effects of aspect ratio were considerable. In this case, the flexural enhancement was observed to be 2 times while the energy absorption improved up to 2.5 times in contrast. Also, this situation would be deduced for WL and EL samples. The reason here was the pull-out fracture mechanism of fibers that affected by the fibers dimensions in particular their aspect ratios where BL contained sample have

been considered to have the most energy absorption properties.

7 CONCLUSION

Flexural strength of cement boards simply constructed with cement is very low and usually fails under small values of strain. To resolve this deficiency and also improving other demanded characteristics, the fibers are frequently applied.

In Iran, after the beginning of the 21's century, the asbestos fibers were used to produce the cement composites with superior properties. Because of the harmful effects of asbestos on human safety, despite the remarkable properties, the production and applications were forbidden in construction industries.

Nowadays, the natural fibers are used instead of the asbestos which have a proper consistency with cement paste. Also, these fibers have some economical proprieties to be used in contrast to the other types of fibers. In this research, to characterize the properties of the composites made by these natural fibers, the flexural strength and energy absorption properties were considered to study. Three types of agricultural waste including the Bagas Fiber, What Fiber and Eucalyptus Fiber, were used to make the natural-fiber cement composite boards. To study the microstructure, SEM micrographs were analyzed for fibers. Moreover to determine the flexural strength and energy absorption, one-point standard test were used along with the stress-strain curve analyzing.

The results of this study showed that BL would have more beneficial effects on enhancement of the flexural and energy absorption characteristics of cement boards while WL and EL was not exhibited good behavior. Some results of this study are:

- The cellulose fibers (natural-fibers) have a good consistency with cement paste and because of high water adsorption characteristics, could be properly dispersed in cement matrix and bond with cement paste.
- The maximum amount of fibers to be mixed is restricted to the type, length, and diameter of fibers which only 4% is applicable in laboratorial scale.
- Energy abortion of cement boards containing cellules fibers is considerably enhanced where in some cases it was improved up to 4 times.
- The flexural strength enhancement of samples was varied depending on the type and amount of fibers used.
- Bagas fiber could enhance the flexural strength up to 30–40%, however, wheat fiber and eucalyptus fiber did not affect the flexural strength. Nevertheless, expect sample W2, all other samples containing fibers fulfilled the group 2 strength grade of EN12467 standard.

- Based on the results of this experimental research, the order for enhancement priorities were BL, EL and WL.
- The research has potential for future work to study the water absorption, permeability, freeze and thawing cycles resistant and durability accepts.

REFERENCES

Allen H.G., "Tensile properties of seven asbestos cements",

Andonian, R., Mai, Y.W. & Cotterell, B., Strength and fracture properties of cellulose fiber reinforced cement composites. Int. J. Cem. Comp, 1(1979) 151–8.

Aziz, M.A., Paramasivam, P. & Lee, S.L., Concrete reinforced with natural fibers. In concrete Technology and Design, Vol. 2: New Reinforced Concretes, ed. R.N. Swamy. Surrey University Press, 1984, pp.107–40.

Bentur A, Akers SAS. The microstructure and ageing of cellulose fiber reinforced cement composites cured in a normal environment. Int J Cem Comp 1989;11:999109.

Bentur A., Midness S., Fiber Reinforced cementitious composites, Elsevier, 1990.

Cook, D.J., Natural fiber reinforced concrete and cement-recent developments. In Advances in Cement-Matrix Composites, ed. D.M. Roy, A.J. Majumdar, S.P. Shah & J.A. Manson. Proc. Symposium L, Materials Research Society Annual Meeting, Boston, 1980, 251–8.

Coutts R.S.P., "A review of Australian research into natural fiber cement composites", Cem. Comcr. Comp, 27(2005) 518–526.

Coutts RSP. Flax fibres as a reinforcement in cement mortars. Int J Cem Compos 1983;5(4):257–262.

Kaufmann J.; Winnefeld F. and Hesselbarth D., "Effect of the addition of ultrafine cement and short fiber reinforcement on shrinkage, rheological and mechanical properties of Portland cement pastes", Cem. Concr. Comp., 26(2004) 541–549.

Mohr B.J.; Nanko H. and Kurtis K.E., "Aligned kraft pulp sheets for reinforcing mortar", Cem. Concr. Comp., 27(2005) 554–564.

P.R. Blankenhorn, B.D. Blankenhorn, M.R. Silsbee, M. Dicola, Effects of fiber surface treatments on mechanical properties of wood fiber-cement composites, Cem. Concr. Res. 31 (2001) 1049–1055.

Pehanich J.L.; Blankernhorn P.R. and Silsbee M.R., "Wood fiber surface treatment level effects on selected mechanical properties of wood fiber-cement composites", Cem. Conc. Res., 34(2004) 59–65.

R. MacVicar a, L.M. Matuana b9*, J.J. Balatinecz, "Aging mechanisms in cellulose fiber reinforced cement composites", Cement & Concrete Composites 21 (1999) 189–196.

R.S.P. Coutts, "A review of Australian research into natural fibre cement composites", Cement & Concrete Composites 27 (2005) 518–526.

WHO meeting, Oxford, United Kingdom, 10–11 April 1989.

Author Index